GOVERNING DIGITALLY INTEGRATED GENETIC RESOURCES, DATA, AND LITERATURE

The free exchange of microbial genetic information is an established public good, facilitating research on medicines, agriculture, and climate change. However, over the past quarter-century, access to genetic resources has been hindered by intellectual property claims emanating from developed countries under the World Trade Organization's TRIPS Agreement (1994) and by claims of sovereign rights from developing countries under the Convention on Biological Diversity (CBD) (1992). In this volume, the authors examine the scientific community's responses to these obstacles and advise policymakers on how to harness provisions of the Nagoya Protocol (2010) that allow multilateral measures to support research. By pooling microbial materials, data, and literature in a carefully designed transnational e-infrastructure, the scientific community can facilitate access to essential research assets while simultaneously reinforcing the open access movement. The original empirical surveys included here provide a valuable addition to the literature on governing scientific knowledge commons.

Jerome H. Reichman is the Bunyan S. Womble Professor of Law at Duke University School of Law. His research deals with the impact of intellectual property on public health, developing countries, and global science policy. He is the coauthor most recently of *Intellectual Property Rights: Legal and Economic Challenges for Development* (2014).

Paul F. Uhlir, J.D. was Director of the Board on Research Data and Information at the National Academies in Washington, DC, and of the U.S. CODATA until the end of 2014. He is currently a Scholar at the National Academy of Sciences and a consultant on data management.

Tom Dedeurwaerdere is Director of the Biodiversity Governance Unit and professor of philosophy of science at the Universite catholique de Louvain. The editor of two books on the global environmental commons, he was recently awarded a grant from the European Research Council for a project on governing the global genetic resource commons.

Governing Digitally Integrated Genetic Resources, Data, and Literature

GLOBAL INTELLECTUAL PROPERTY STRATEGIES FOR A REDESIGNED MICROBIAL RESEARCH COMMONS

JEROME H. REICHMAN
Duke University School of Law

PAUL F. UHLIR
National Academy of Sciences

TOM DEDEURWAERDERE
Université catholique de Louvain

CAMBRIDGE
UNIVERSITY PRESS

CAMBRIDGE
UNIVERSITY PRESS

University Printing House, Cambridge CB2 8BS, United Kingdom

One Liberty Plaza, 20th Floor, New York, NY 10006, USA

477 Williamstown Road, Port Melbourne, VIC 3207, Australia

4843/24, 2nd Floor, Ansari Road, Daryaganj, Delhi - 110002, India

79 Anson Road, #06-04/06, Singapore 079906

Cambridge University Press is part of the University of Cambridge.

It furthers the University's mission by disseminating knowledge in the pursuit of education, learning and research at the highest international levels of excellence.

www.cambridge.org
Information on this title: www.cambridge.org/9781108433013

First published 2016
First paperback edition 2017

A catalogue record for this publication is available from the British Library

Library of Congress Cataloging in Publication data
Reichman, J. H. (Jerome H.), 1936–, author.
Governing digitally integrated genetic resources, data, and literature : global intellectual property strategies for a redesigned microbial research commons / Jerome H. Reichman, Paul F. Uhlir, Tom Dedeurwaerdere.
p. ; cm.
Includes bibliographical references and index.
ISBN 978-1-107-02174-7 (hardback)
I. Uhlir, P. F. (Paul F.), 1954–, author. II. Dedeurwaerdere, Tom, author. III. Title.
[DNLM: 1. Genetics – legislation & jurisprudence. 2. Access to Information.
3. Intellectual Property. 4. Internationality. 5. Research. QU 33.1]
QH442
572.8–dc23 2015003104

ISBN 978-1-107-02174-7 Hardback
ISBN 978-1-108-43301-3 Paperback

Contents

Preface

This is a book about science policy in a conflicted world, torn between the demands of both the global North and the global South for strengthened protection of their respective intellectual property rights. It presents a strategy and devises new legal and institutional models for making microbiological genetic materials and digital resources readily available from a multilateral regime of facilitated access consistent with the Convention on Biological Diversity (CBD) of 1992.

Tom Dedeurwaerdere, one of the co-authors of this book, is both a science and a law professor who has long been a consultant to leading public microbial culture collections in the European Union. The project began when he consulted Jerome Reichman and Paul Uhlir, the other co-authors of this volume, for two main reasons. He knew that the International Treaty on Plant Genetic Resources for Food and Agriculture (ITPGRFA) of 2001 had adopted a version of Professor Reichman's Compensatory Liability Regime—a "take and pay" automatic royalty scheme initially devised for subpatentable innovations. He wanted to know how this regime might become suitable for exchanges of *ex situ* genetic materials from networks of existing microbial culture collections. He also wanted to know more about data pooling and related digital research issues, about which Reichman and Uhlir had written extensively in the past and in which Paul Uhlir was deeply involved as head of the Board on Research Data and Information at the National Academies.

As the three of us began to engage with these issues, the dimensions of the topic kept expanding in different directions. The holistic New Biology paradigm for the life sciences,[1] as set forth by the National Research Council (NRC) in 2009, made microbiology a central focus in the genomic era. As we note in our book, any shortcomings in the NRC's visionary project are not necessarily to be found in science itself, but rather in tacit assumptions about the enabling nature

1 National Research Council, A New Biology for the 21st Century (Natl' Acads. Press, 2009).

of the external environment in which the desired integration of the life sciences would be rooted. To achieve this unifying goal, researchers working in the relevant scientific subdisciplines must have ready access to essential upstream knowledge assets. Life scientists and microbiologists, in particular, will need to obtain countless biological materials collected and validated from all parts of the world; to make use of vast amounts of data from genomic studies, bioecology, systematics, and from other observational and experimental life-science initiatives; and to access all the knowledge gleaned from an ever-expanding body of scholarly literature.

Although none of us is a microbiologist, we soon found that microbiology has been under stress from numerous sources and for many years. The "soft infrastructure" that currently governs these essential inputs tends to fragment and compartmentalize the building blocks of science in ways that are not conducive to enabling the integrated vision to which the life sciences now aspire. We describe those trends in detail in this volume – from organizational, economic, political, and especially different legal perspectives.

Caught in these cross-currents, the scientific community risks incurring major impediments to public research based on ready access to both *ex situ* and *in situ* microbial genetic materials and related digital resources. A failure to address the threat of privatizing genetic resources previously residing in the public domain for research purposes would have a serious impact on human welfare owing to lost research opportunities. At the same time, these opportunity costs are difficult to quantify or otherwise measure by standard law and economics approaches.

Fortunately, after a lengthy period in which the needs and role of public science were largely ignored by negotiators for both the developed and developing counties, in 2010 the drafters of the Nagoya Protocol to the Convention on Biological Diversity (CBD) of 1992 reopened the door for access to genetic resources and data for public research purposes. The Nagoya Protocol expressly recognizes the importance of scientific research as a provider of both monetary and nonmonetary benefits under the CBD. It expressly validates the multilateral system for facilitated exchanges of plant genetic resources for food and agriculture, for research and breeding purposes, and as a legal alternative to the bilateral access and benefit sharing modalities normally required by the CBD. Above all, the Protocol implicitly invites the microbiological community to follow the path opened by the ITPGRFA and similarly adopt a multilateral regime of facilitated access to microbial genetic resources for public scientific research purposes.

The drafters of the Nagoya Protocol, whose primary task was to tighten the international regime governing misappropriation of genetic resources from biodiversity rich countries under the CBD, thus took a major step to legitimize facilitated access to *ex situ* microbial genetic resources for research and applications under an appropriately designed multilateral regime. The challenges it presented were how to accommodate the existing microbiological infrastructure, built around the World Federation of Culture Collections (WFCC), to the legal pathways provided by the Nagoya Protocol, and how to make that infrastructure more productive in the light of theoretical and empirical knowledge about common pool resources in general that had been emerging from a growing literature.

The point of departure was our realization that science policymakers needed to adapt to the opportunities that the CBD now made available under specified conditions. If public service is to be maintained, it must comply with the Nagoya Protocol. A number of other seminal developments, beyond the legal dictates of the Nagoya Protocol, subsequently informed our investigations and bear emphasizing here.

With regard to microbiological data (also covered by the CBD) and related literature, we analyzed the growing capabilities of digitally networked technologies and their interplay with intellectual property law, as well as institutional models for publishing research results. We undertook an empirical study of more than 300 journals in microbiology to obtain a detailed overview of their open access or subscription approaches. We found a surprisingly large number of open access or partially open publications, which were nonetheless undermined by the legal and institutional hangovers of the print paradigm.

We also examined the policies of both government entities and the academic community with respect to databases compiled for microbial genetic resources and taxonomy, and we looked at some of the costs and benefits of making these data resources more openly available for research purposes. From our analysis of these and other digital publishing developments, we identified a holistic, online approach to complex research endeavors in microbiology and elsewhere that we refer to as Open Knowledge Environments. Efforts to encourage these promising initiatives can be linked to the formation and management of a multilateral knowledge commons for microbial genetic materials.

Finally, we looked at the growing area of infrastructure and knowledge commons theory, as well as at other existing international scientific pooling endeavors, for lessons that they might offer for our project. Of particular interest was a major European demonstration project in transnational microbiology – the Global Biological Research Center Network (GBRCN) – which ended in 2011. The GBRCN endeavored to implement, on a pilot basis, the OECD's earlier proposals to upgrade the WFCC's microbial culture collections – including their digital microbiological resources – in a network of Biological Research Centers. Although laudable in its attempts to implement this major science policy vision, the scheme was flawed – at least initially – by efforts to commercialize upstream microbial genetic resources and related data that the WFCC otherwise provides as a public good. Nevertheless,

GBRCN took important first steps toward organizing a multilateral regime needed to shelter within the ambit of the Nagoya Protocol.

We then combined all these different threads in an effort to propose a redesigned international microbial research commons, building on the WFCC's existing network that would serve the interests of the global public research community, while complying with the Nagoya Protocol to the CBD and supporting downstream commercial users. We conclude this volume with some ideas about how to make such an ambitious international construct sustainable over time.

In addition to presenting our work at numerous conferences in the United States and Europe in the past several years, we organized an international symposium at the National Academies in Washington, DC, which gave us authoritative inputs and led to an initial publication in 2011: viz., *Designing the Microbial Research Commons*.[2] In so doing, we consulted with leading microbiologists, lawyers, economists, and science policymakers about the challenges facing the international research community in this area. We also presented some of our initial findings and proposed solutions and received their sage advice.

How to reconcile the needs of publicly funded microbiological researchers in both the developed and developing world with the new opportunities made available by the Nagoya Protocol is thus the task we undertook in writing this book. We hope that, by explaining the implications of these new and important developments, we can help the public scientific community find a way through a thicket of proprietary claims, in order to implement the visionary goals of the New Biology paradigm that inspired us from the outset.

[2] Designing the Microbial Research Commons: Proceedings of an International Symposium (P.F. Uhlir ed., Nat'l Acad. Press 2011).

Acknowledgments

There are numerous individuals who played an important role in the completion of this book, both in producing the manuscript as well as in researching and reviewing it. Each of the authors had several research assistants who found references, checked footnotes, suggested edits, and generally helped prepare the book, as well as reviewers who helped strengthen our arguments throughout.

We first would like to thank the National Human Genome Research Institute's Center of Excellence for ELSI research, which provided most of the generous funding for this project under NIH Grant No. P50 HG00339 to Duke University's Center for Public Genomics. Jerome Reichman was one of the Principal Investigators under that grant, which was obtained and administered by Duke University in 2005, under the leadership of Robert Cook-Deegan and his colleagues. That grant also paid for the consulting services of Paul Uhlir, and supported the travel expenses of both Paul Uhlir and Jerome Reichman in connection with this project.

We are also grateful for supplementary funding provided by Barry Silverstein and the William James Foundation at a critical point in our work, as well as to David Levi and Kate Bartlett, Dean and former dean of the Duke University School of Law, who consistently supported Professor Reichman's research endeavors.

By far the most assistance for research and references was provided by students and former students at the Duke University School of Law. In particular, the authors are indebted to Heather Ritch, Carla Rydholm, William Ryan, Melissa Turcios, Bill Warren, and Alisha Mehta. Professor Reichman also wishes to thank the dedicated personnel who staff the Reference Desk in the Goodson Library at Duke Law School.

Paul Uhlir engaged the help of several summer Fellows at the National Research Council (NRC), particularly in conjunction with the International Symposium his Board organized on this topic. Lucy Yang and Tania Dutta were instrumental in compiling the database of microbiology journal publishing policies presented in Chapter 7. Other NRC staff who provided some substantive and administrative support for the International Symposium – and by extension, this book – included

Dan Cohen, Subhash Kuvelker, and Cheryl Levey, all of the Board on Research Data and Information, and Fran Sharples of the Board on Life Sciences.

At the Université catholique de Louvain, Louvain-la-Neuve, Belgium, Tom Dedeurwaerdere received assistance from research fellows hired under two EU grants dealing with the governance of global science commons, viz., GENCOMMONS (ERC grant agreement 284) and MICRO B3 (FP7 grant agreement 28758), and under co-funding from a grant by the National Science Foundation, Belgium (F.R.S.-FNRS/MIS Incentive Grant on Governing Global Science Commons). In particular, Christine Frison, Arianna Broggiato, and Arul Scaria provided substantial research inputs to the chapters on the Consultative Group on Int'l Agricultural Research (CGIAR) and the World Federation for Culture Collections (WFCC). These grants also provided for the travel expenses of Tom Dedeurwaerdere in connection with this project. We also thank Heike Rämer and Caroline Van Schendel for their effective assistance with bibliographical research and analysis of survey data.

The authors are also indebted to the many individuals who provided informal reviews, comments, and insights pertaining to different sections and iterations of the evolving manuscript. Microbiologists who were consulted in such capacities included: Joan Bennett, Kevin McCluskey, MA Juncai, Micah Krichevsky, Dagmar Fritze, David Smith, Gerard Verkleij, Stephen McCormack, Philippe Desmeth, Paul De Vos, Peter Dawyndt, Jean Swings, and Lenie Dijkshoorn. We also received guidance on the Crop Commons and CGIAR from Michael Halewood, Emile Frison, and Shakeel Bhatti.

Economic and political insights were provided by Paul David, Anita Eisenstadt, Minna Allarakhia, Daniel Drell, Lita Proctor, and Larry Helfer. Reviews of the legal dimensions were given by Brett Frischman, Jorge Contreras, Katherine Strandburg, Michael Madison, Michael Carroll, James Boyle, and Peter Lee, among many others.

We also would like to express our appreciation to the team at Cambridge University Press, who produced this volume. They included John Berger, Stephen Acerra, Nishanthini Vetrivel, Laura Lawrie, and their colleagues.

Finally, and most important, we are greatly appreciative of Patricia Reichman's unstinting – and voluntary – support of the entire project. Not only did she digitally process the entire manuscript many times over, which consumed enormous amounts of time, but she provided substantive and structural advice throughout the nine years of this endeavor.

For these and many other reasons, we wish to thank her and everyone else who contributed to this project. Needless to say, the views expressed in this volume are those of the authors and not of the institutions with which they are, or were, affiliated at the time of writing.

1

Uncertain Legal Status of Microbial Genetic Resources in a Conflicted Geopolitical Environment

I. INTRODUCTION

Transnational exchanges of plant and microbial genetic resources have played a fundamental role in both agricultural and microbiological research endeavors.[1] Throughout the nineteenth and early twentieth centuries, researchers in either field could freely explore biodiversity-rich environments, often located in colonies or later in developing countries, in order to discover, isolate, and validate new microbial reference strains or new sources of germplasm of interest to their respective scientific disciplines.[2] Particularly important exemplars of these *in situ* genetic resources were then deposited in *ex situ* public repositories, known as culture collections and seed banks.[3] These

[1] *See, e.g.*, Evenson Chege Kamau, *The Multilateral System of the International Treaty on Plant Genetic Resources for Food and Agriculture: Lessons for and Room for Further Development*, *in* COMMON POOLS OF GENETIC RESOURCES: EQUITY AND INNOVATION IN INTERNATIONAL BIODIVERSITY LAW 343, 343 fn. 1 (E. C. Kamau & G. Winter eds. 2013) [hereinafter COMMON POOLS OF GENETIC RESOURCES (2013)] ("No country is self-sufficient: all depend on crops and genetic diversity within these crops from other countries and regions."); Christine Godt, *Networks of* Ex Situ *Collections of Genetic Resources*, *in* COMMON POOLS OF GENETIC RESOURCES (2013), at 246–47 [hereinafter Godt (2013)] (stating that *ex situ* collections of plant, animal, and microbial genetic resources "play an essential role in the preservation and research of biodiversity").

[2] For the low and middle-income countries classified as "developing countries," *see Updated Income Classifications*, WORLD BANK, http://data.world bank.org/news/2015-country-classification/(last visited Jan 14, 2015); for early stages of bioprospecting, *see, e.g.*, Dagmar Fritze [DSMZ, Pres. ECCO], The Proposed Standard MTA of the European Culture Collections' Organization, paper presented to the Microbial Commons Conference, Ghent, Belgium, June 12–13, 2008, at 4 [hereinafter Fritze (2008)]; John H. Barton, *Acquiring Protection for Improved Germplasm and Inbred Lines*, *in* INTELLECTUAL PROPERTY RIGHTS IN AGRICULTURAL BIOTECHNOLOGY 19–20 (F. H. Erbisch & K. M. Maredia eds., CABI 1998); Sélim Louafi & Marie Schloen, *Practices of Exchanging and Utilizing Genetic Resources for Food and Agriculture and the Access and Benefit-Sharing Regime*, *in* COMMON POOLS OF GENETIC RESOURCES (2013), above n. 1, at 193–223.

[3] Godt (2013), above n. 1, at 246–56. Repositories for *ex situ* deposits of horticultural genetic resources possess many of the same characteristics as those dealing with microbiology and agriculture. *See id.*

repositories often added value in the form of catalogues, taxonomic classifications and, more recently, compilations of related genetic data.[4]

Over time, by means of both formal and informal legal arrangements, these plant and microbial genetic resources needed for basic scientific research were painstakingly accumulated, classified, preserved, and made available from *ex situ* public and other repositories around the world.[5] These repositories traditionally supplied their genetic resources to breeders, researchers, and industry at marginal costs of distribution.[6] In so doing, they responded to the risk of market failure that otherwise tends to elicit underinvestment in public goods.[7]

The scientific norms and practices that these collections supported became well-established by the 1950s. They were rooted in the usually tacit assumption that *in situ* plant and microbial genetic resources collected for research purposes belonged to a vast public domain, sometimes characterized as "the common heritage of mankind."[8] Similarly, the publicly funded *ex situ* repositories constituted both scientific infrastructure[9] and a *de facto* "knowledge commons"[10] that enabled the global research

at 251–53 (discussing the International Plant Exchange Network [IPEN] of botanical gardens). This network is beyond the focus of this volume.

4 *See, e.g.*, David Smith, Dagmar Fritze & Erko Stackebrandt, *Public Service Collections and Biological Resource Centers of Microorganisms*, *in* F. ROSENBERG ET AL, EDS. THE PROKARYOTES—PROKARYOTIC AND SYMBOLIC ASSOCIATIONS (4th ed., Springer 2013), Chapter 11; SCOTT STERN, BIOLOGICAL RESOURCE CENTERS: KNOWLEDGE HUBS FOR THE LIFE SCIENCES (Brookings Inst. Press 2004) (discussing resource centers for microbes); Consultative Group on Int'l Agricultural Research (CGIAR), *Research Centers*, CGIAR.ORG, http://www.cgiar.org/cgiar-consortium/research-centers/ (last accessed February 23, 2014) (discussing resource centers for plant genetic resources). For details, see Chapter 2, Sections I.A.–B. The seed banks became particularly important from the beginning of the 1970s on. *See* Barton, above n. 2, at 19–20.

5 David Smith, *Culture Collections*, *in* 79 ADVANCES IN APPLIED MICROBIOLOGY 73–118 (2012); Michael Halewood, Isabel López Noriega & Sélim Louafi, *The Global Crop Commons and Access and Benefit-Sharing Laws: Examining the Limits of International Policy Support for the Collective Pooling and Management of Plant Genetic Resources*, *in* CROP GENETIC RESOURCES AS A GLOBAL COMMONS: CHALLENGES IN INTERNATIONAL LAW AND GOVERNANCE (M. Halewood et al. eds. 2013) [hereinafter CROP COMMONS (2013)].

6 *See, e.g.*, Godt (2013), above n. 1, at 248.

7 *Id.* at 247; Tom Dedeurwaerdere, *Institutionalizing Global Genetic Resource Commons: Towards Alternative Modes for Facilitating Access to the Global Biodiversity Regime* (Working Paper, June 12, 2010), available at http://papers.ssrn.com/sol3/papers.cfm?abstract_id=1611549.

8 *See, e.g.*, Fritze, above n. 2; Stephen B. Brush, *The Demise of "Common Heritage" and Protection for Traditional Agricultural Knowledge*, *in* BIODIVERSITY & THE LAW: INTELLECTUAL PROPERTY, BIOTECHNOLOGY & TRADITIONAL KNOWLEDGE 297–301 (C. McManis ed. 2007).

9 For seminal work on the economics of infrastructure, *see* BRETT M. FRISCHMANN, INFRASTRUCTURE: THE SOCIAL VALUE OF SHARED RESOURCES (Oxford U. Press 2012) [hereinafter INFRASTRUCTURE]. *See generally* YOCHAI BENKLER, THE WEALTH OF NETWORKS: HOW SOCIAL PRODUCTION TRANSFORMS MARKETS AND FREEDOM (2006). *See also* JAMES BOYLE, THE PUBLIC DOMAIN: ENCLOSING THE COMMONS OF THE MIND (Yale U. Press 2008).

10 The term "knowledge commons" is "shorthand for the institutionalized community governance of the sharing and, in some cases, creation, of information, science, knowledge, data and other types

community to access and use the genetic resources to which their investigations naturally led.[11]

As more fully explained in Chapter 9, commons theory is derived from the work of Elinor Ostrom on "common pool resources," which were typically provided by resource holders for use by a specified group of people or by a given community.[12] Empirically, scholarship in this field has evolved from the study of pooled natural resources ("old commons") to the study of "new commons" or knowledge commons, in which nonrivalrous information and other research assets are pooled to avoid the risk of propertization that might otherwise occur.[13] This book draws in part on insights from the study of knowledge commons, and it seeks to further our understanding of their role in basic scientific research.

From a legal perspective, however, the tacit characterization of both plant and microbial genetic resources as freely available research assets was always open to question, particularly after the United Nations Declaration on Permanent Sovereignty Over Natural Resources in 1969.[14] As long as this premise remained unchallenged at

of intellectual and cultural resources." BRETT M.FRISCHMANN, MICHAEL J.MADISON&KATHERINE STRANDBURG, GOVERNING THE KNOWLEDGE COMMONS 1–38 (Oxford U. Press, 2014).

11 *See, e.g.*, DESIGNING THE MICROBIAL RESEARCH COMMONS: PROCEEDINGS OF AN INTERNATIONAL SYMPOSIUM (P.F. Uhlir ed., Nat'l Acads. Press 2011) [hereinafter DESIGNING THE MICROBIAL RESEARCH COMMONS], available at http://www.ncbi.nlm.nih.gov/books/NBK91499/ (last accessed February 23, 2014); CROP COMMONS (2013), above n. 5; COMMON POOLS OF GENETIC RESOURCES (2013), above n. 1. This scientific infrastructure, carefully nurtured by dedicated individuals and academic institutions has played an indispensable, if partly hidden, role in both basic and applied scientific research for the past two centuries at least. *See* Chapter 2, Section I.

12 ELINOR OSTROM, GOVERNING THE COMMONS: THE EVOLUTION OF INSTITUTIONS FOR COLLECTIVE ACTION (Cambridge U. Press, 1990); Gerd Winter, *Common Pools of Genetic Resources and Related Traditional and Modern Knowledge – An Overview, in* COMMON POOLS OF GENETIC RESOURCES (2013), above n. 1. *See* Chapter 10, Section I.

13 *See, e.g.*, Elinor Ostrom & Charlotte Hess, *Framework for Analyzing the Knowledge Commons, in* UNDERSTANDING KNOWLEDGE AS A COMMONS: FROM THEORY TO PRACTICE 41–82 (C. Hess & E. Ostrom eds., MIT Press 2007); Michael J. Madison, Brett M. Frischmann, & Katherine J. Strandburg, *Constructing Commons in the Cultural Environment* 93 *Cornell L. Rev.*, 657 (2010), available at http://www.lawschool.cornell.edu/research/cornell-law-review/upload/Madison-Frischmann-Strandburg-final.pdf (explaining that the term "cultural commons" includes information commons, science commons, knowledge commons, and data commons, among other types of intellectual resource commons). *See also* Winter (2013), above n. 12; Jerome H. Reichman & Paul F. Uhlir, *A Contractually Reconstructed Research Commons for Scientific Data in a Highly Protectionist Intellectual Property Environment*, 66 *Law & Contemp. Probs.* 315 (2003) [hereinafter Reichman & Uhlir (2003)], available at http://scholarship.law.duke.edu/lcp/vol66/iss1/12. In all cases, restrictions on access or use may result in a semicommons rather than a commons open to all. *See, e.g.*, Robert A. Heverly, *The Information Semicommons*, 18 *Berkeley Tech. L.J.* 1127 (2003).

14 *See* Permanent Sovereignty Over Natural Resources, G.A. Res. 1803 (XVII), U.N. Doc. A/RES/1803 (Dec. 14, 1962) [hereinafter 1962 Declaration], available at For a skeptical view of claims to *ex situ* genetic resources, based on misunderstood interpretations of the "common heritage" principle, see JONATHAN CURCI, THE PROTECTION OF BIODIVERSITY AND TRADITIONAL KNOWLEDGE IN INTERNATIONAL LAW OF INTELLECTUAL PROPERTY 9 (Cambridge U. Press 2010) [hereinafter CURCI

the international level,[15] policymakers could indulge in the belief that all countries, including the source or provider countries, benefitted from commercial applications of *in situ* and *ex situ* genetic resources that fostered improvements in agriculture, public health, food security, and human welfare in general.[16]

Beginning in the last quarter of the twentieth century, however, a proliferation of domestic and international intellectual property rights in these same commercial applications rapidly destabilized the preexisting system of transborder exchanges.[17] These new laws threatened the continued availability of genetic resources needed for the emerging paradigm shift in biological sciences.[18]

Already in the 1960s, developed countries had campaigned successfully to protect phenotypical applications of plant genetic resources under a *sui generis* intellectual property regime known as plant variety protection laws.[19] This campaign produced a multilateral treaty under the auspices of the International Union for the Protection of New Varieties of Plants (UPOV) of 1961, which was last amended in 1991.[20] By the mid-1990s, the developed countries had successfully enlarged their demands for globally enforceable intellectual property rights to include patents on applications of both microbial and plant genetic resources, including genes and other products of biotechnology, under what became the World Trade Organization's (WTO) Agreement on Trade-Related Aspects of Intellectual Property Rights of 1994 (TRIPS Agreement).[21]

In response, the developing countries maintained that it was unfair for source genetic materials to be freely taken from their territories without permission,

(2010)]. *See also* GRAHAM DUTFIELD, INTELLECTUAL PROPERTY, BIOGENETIC RESOURCES, AND TRADITIONAL KNOWLEDGE 5–6 (2d ed., 2004) [hereinafter DUTFIELD] (stressing the importance of Resolution 1803).

15 For the demise of the common heritage principle and its implications, particularly for plant genetic resources, see Chapter 2, Sections I.B and III.A.

16 *See, e.g.*, Barton, above n. 2, at 20; CURCI (2010), above n. 14 (noting that this thesis was always a convenient construct of intellectual property systems adopted in industrialized countries).

17 For details, see Chapter 2, Section II and Chapter 3 *passim*.

18 NAT'L RESEARCH COUNCIL (NRC), A NEW BIOLOGY FOR THE 21ST CENTURY (Nat'l Acads. Press 2009) [hereinafter NRC, NEW BIOLOGY]. *See further* Section II.D.

19 *See, e.g.*, JULIANNA SANTILLI, AGROBIODIVERSITY AND THE LAW: REGULATING GENETIC RESOURCES, FOOD SECURITY AND CULTURAL DIVERSITY (Earthscan 2012) [hereinafter SANTILLI (2012)]; DUTFIELD, above n. 14, at 5–6, 11 (stressing importance of Resolution 1803); Barton, above n. 2, at 21–22. Plant Variety Protection systems protect new plant varieties that are distinct, uniform, and stable, for a limited period of time, initially on a copyright-like model, eventually on a patent-like model. *See* Jerome H. Reichman, *Legal Hybrids Between the Patent and Copyright Paradigm*, 94 *Colum. L. Rev.* 2432, 2465–72 (1994).

20 International Convention for the Protection of New Varieties of Plants, Dec. 2, 1961, 33 U.S.T. 2703, 815 U.N.T.S. 89 (as subsequently amended) 1978 and 1991. *See, e.g.*, SANTILLI (2012), above n. 19.

21 Agreement on Trade-Related Aspects of Intellectual Property Rights art. 9.1, April 15, 1994, 108 Stat. 4809, 1869 U.N.T.S. 299 [hereinafter TRIPS Agreement]. For the ambiguity inherent in the provisions, *see, e.g.*, CURCI (2010), above n. 14, at 36–42. *See further* Chapter 3, Section I.B–C.

while commercial applications of these same resources were now to be governed by international intellectual property rights applicable to these same territories.[22] As explained in Chapter 3, these complaints crystallized in the Convention on Biological Diversity of 1992.[23] This Convention asserted territorial sovereignty over all genetic resources, and it challenged the rights of anyone – including scientists – to remove or otherwise use them even for public research purposes without the permission of the relevant government authority.[24]

The professed goal of harmonized intellectual property rights under the TRIPS Agreement was to stimulate higher levels of investment in innovation generally. This initiative responded to opportunities generated by an increasingly integrated global marketplace, in which commercial transfers of technology could occur without territorial governments imposing protectionist trade barriers.[25] The professed aim of the CBD was to support the conservation of genetic resources by provider countries, especially the developing countries, and to reward their indigenous populations whose traditional knowledge may have informed commercial applications of those same genetic resources.[26] Although the relative successes and failures of these endeavors continue to elicit an extensive literature,[27] especially with regard to transfers of

[22] José Esquinas-Alcázar, Angela Hilmi, & Isabel López Noriega, *A Brief History of the Negotiations on the International Treaty on Plant Genetic Resources for Food and Agriculture, in* Crop Commons (2013), above n. 5, at 134, 137. *See also* Barton, above n. 2, at 20; Curci (2010), above n. 14.

[23] United Nations Conference on Environment and Development: Convention on Biological Diversity, *opened for signature* June 5, 1992, 1760 U.N.T.S. 79 [hereinafter CBD].

[24] *See* Godt (2013), above n. 1, at 46–47. *See further* Chapter 3, Sections I.B–C.

[25] *See, e.g.*, Keith Maskus, Private Rights and Public Problems: The Economics of International Intellectual Property in the 21st Century (2d ed., Peterson Inst. For Int'l Econ. 2013); Peter K. Yu, *The International Enclosure Movement*, 82 Ind. L.J. 827 (2007); Jerome H. Reichman, Universal Minimum Standards of Intellectual Property Protection under the TRIPS Component of the WTO Agreement, 29 Int'l Lawyer 345–88 (1998), available at http://scholarship.law.duke.edu/faculty_scholarship/687. *See further* Chapter 2, Section II.

[26] *See further* Chapter 3, Section I.

[27] *See, e.g.*, Susan Sell, Private Power, Public Law: The Globalization of Intellectual Property Rights (Cambridge U. Press 2003) *and* Power and Ideas: North-South Politics of Intellectual Property and Anti-Trust (State U. N.Y. Press 1997); International Public Goods and Transfer of Technology Under a Globalized Intellectual Property Regime (K. E. Maskus & J. H. Reichman eds., Cambridge U. Press 2005) [hereinafter International Public Goods]. *See generally* Graeme B. Dinwoodie & Rochelle C. Dreyfuss, A Neofederalist Vision of TRIPS: The Resilience of the International Intellectual Property Regime (Oxford U. Press 2012); Carolyn Deere, The Implementation Game: The TRIPS Agreement and the Global Politics of Intellectual Property Reform in Developing Countries (Oxford U. Press 2008); Peter Drahos, The Global Governance of Knowledge: Patent Offices and their Clients (Cambridge U. Press 2010); Genetic Resources, Traditional Knowledge & the Law (E. C. Kamau & G. Winter eds., Routledge 2009); Regine Andersen, Governing Agrobiodiversity (2008); Biodiversity & The Law (C. McManis ed., Earthscan 2007); Dutfield, above n. 14.

technology between developed and developing countries,[28] the ancillary negative impact of these same initiatives on the preexisting scientific research infrastructure has elicited much less, if still growing, scholarly attention.[29]

Our work in this volume attempts to address pressing questions about the governance of digitally integrated genetic resources, data, and literature under an international intellectual property regime that now tends to privatize research inputs formerly treated as global public goods.[30] In particular, we document the need for the worldwide microbiological research community to more vigorously address knowledge governance issues that have arisen from the explosion of intellectual property rights since the last quarter of the twentieth century.[31] Drawing on both theoretical and

[28] *See, e.g.*, Crop Commons (2013), above n. 5; Common Pools of Genetic Resources (2013), above n. 1; Designing the Microbial Research Commons, above n. 11. *See also* Tshimanga Kongolo, Unsettled International Intellectual Property Issues 30–61 (Kluwer L. Int'l 2008); Curci (2010), above n. 14; Santilli (2012), above n. 12; Gene Patents and Collaborative Licensing Models: Patent Pools, Clearing houses, Open Source Models and Liability Regimes (G. Van Overwalle ed. Cambridge U. Press 2009); Comparative Issues in the Governance of Research Biobanks: Property Privacy, Intellectual Property and the Role of Technology (G. Pascuzzi et al. eds., Springer 2013). *See generally* David Mowery et al., Ivory Tower and Industrial Innovation: University-Industry Technology Transfer Before and After the Bayh-Dole Act (Stanford U. Press 2004); Keith E. Maskus & Jerome H. Reichman, *The Globalization of Private Knowledge Goods and the Privatization of Global Public Goods, in* International Public Goods, above n. 27, at 1–45.

[29] Common Pools of Genetic Resources (2013), above n. 1; Crop Commons (2013), above n. 5; Designing the Microbial Research Commons, above n. 11. *See also* S. K. Verma, *Plant Genetic Resources, Biological Inventions and Intellectual Property Rights: The Case of India, in* Intellectual Property and Biological Resources 128, 138–41 (B. Ong ed., Cavendish Int'l 2004) (noting the conflicts between TRIPS and the CBD and the negative impacts on research and technology transfer); Bram De Jonge & Niels Louwaars, *The Diversity of Principles Underlying the Concept of Benefit Sharing, in* Genetic Resources, Traditional Resources, Traditional Knowledge & the Law 37, 45–47 (E.C. Kamau & G. Winter eds., Earthscan 2009) (stating that the CBD and related treaties intend to promote benefit sharing and technology transfer, but progress so far has been difficult)

[30] Joseph E. Stiglitz, *Knowledge as a Global Public Good, in* Global Public Goods: International Cooperation in the 21st Century 308, 308–326 (Inge Kaul et al. eds., 1999); Maskus & Reichman (2005), above n. 28.

[31] TRIPS Agreement, above n. 21; CBD, above n. 23; Tenth Meeting of the Conference of the Parties to the Convention on Biological Diversity, Nagoya, Japan, 18–29 Oct. 2010, Nagoya Protocol on Access to Genetic Resources and the Fair and Equitable Sharing of Benefits Arising from their Utilization (ABS) to the Convention on Biological Diversity [hereinafter Nagoya Protocol] (entered into force 2014, after the deposit of the fiftieth instrument of ratification, acceptance, approval, or accession), *available at* http://www.cbd.int/abs/doc/protocol/nagoya-protocol-en.pdf (last accessed 16 Sept. 2014) (favoring the interests of developing countries that maintain vast preserves of *in situ* plant and microbial genetic resources). *See also* WIPO Copyright Treaty, Dec. 20, 1996, 112 Stat. 2860, 2186 U.N.T.S. 152 [hereinafter WCT]; Patrick B. Fazzone, *The Trans-pacific Partnership – Towards a Free Trade Agreement of Asia-Pacific?*, 43 *Geo. J. Int'l L.* 695 (2012) (discussing the proposed Trans-Pacific Strategic Economic Partnership Agreement and predecessor agreements); Rosa Castro, *Intellectual Property Rights in Bilateral Investment Treaties and Access to Medicines: The Case of Latin America*, 9 *J. World Intell. Prop.* 548 (2006) (outlining examples of bilateral investment treaties between the United States and countries in Latin America).

empirical insights from the study of knowledge commons,[32] we will develop far-reaching proposals for redesigning the existing microbial research infrastructure in order to meet the legal and institutional challenges identified above and more fully elaborated in the next three chapters.[33] In so doing, we are confident that the problems and solutions under review with specific regard to research uses of microbial genetic resources will have a wider applicability to other science commons initiatives and to the governance of knowledge commons in general.[34]

II. THE CHANGING NATURE OF MICROBIAL RESEARCH

Technically, microbiology – the study of life and organisms too small to be seen with the naked eye – recognizes six major groupings of unicellular or cell-cluster microscopic organisms, namely, archaea, bacteria, viruses, protozoa, eukaryotes, such as fungi, and prokaryotes.[35] The related disciplines include bacteriology (for the first two groups), virology, protozoology, mycology, and phycology.[36]

Although microbes were used to make beverages and bread for thousands of years, claims about their real world existence remained speculative until the invention of the microscope in the seventeenth century.[37] Until then, microbes were known indirectly by what they did. For example, the ancient Greeks and Romans had already guessed at the role of microbes in disease.[38]

Today, scientists believe that less than one percent of all microbial biodiversity has been identified, and only one percent of those microorganisms can be replicated by growth in cultures.[39] For purposes of systematic research and the development of

32 *See esp.* Chapter 9.

33 For genetic resources see Parts One and Two; for related data and literature, *see* Part Three. For empirical and theoretical and evidence bearing on governance and related proposals, *see* Part Four.

34 *See further* Chapters 7 through 10.

35 Joan W. Bennett, *Microbiology in the 21st Century, in* DESIGNING THE MICROBIAL RESEARCH COMMONS above n. 11, at 3–12 [hereinafter Bennett (2011)]; MICHAEL T. MADIGAN ET AL., BROCK BIOLOGY OF MICROORGANISMS (13th ed., 2010). *See also* George Rice, *Are Viruses Alive?, Microbial Life Educ. Res.* (26 May 2013), *available at* http://sevc.carleton.edu/microblife/yellowstone/viruslive.html.

36 Bennett (2011), above n. 35. Microbiology also typically includes the study of immunology and parasitology. For the role of molecular biology and genomics, *see* Section II.B in this chapter accompanying nn. 54–68.

37 *See* Bennett (2011), above n. 35.

38 *Id.* In 1676, Antoine van Leeuwenhoek used a single lens microscope of his own design to observe bacteria and other microorganisms. Eleven years earlier, Robert Hooke had made the first recorded microbiological observation of molds. *See, e.g.*, MADIGAN ET AL., above n. 35; Howard Gest, *The Remarkable Vision of Robert Hooke (1635–1703): First Observer of the Microbial World*, 48 *Perspectives in Biology & Med.* 266–72 (2005).

39 Bennett (2011), above n. 35; *see also* R.T. Amanni et al., *Phylogenetic Identification and* In Situ *Detection of Individual Microbial Cells Without Cultivation*, 59 MICROBIOLOGY REV. 143–69 (1995); PHAGES: THEIR ROLE IN BACTERIAL PATHOGENESIS AND BIOTECHNOLOGY (M. Waldorf et al. eds. ASM Press, 2005). It should be noted, however, that the emerging field of synthetic biology may ultimately change this paradigm. *See* below n. 122 & accompanying text.

commercial applications, there is also a consensus that microbial biodiversity is now best preserved in *ex situ* culture collections, which presents a formidable challenge for the existing repositories of microbial materials,[40] as discussed throughout this book.[41] Consequently, we can look to the microbial world as a vast, mostly untapped, resource of biotechnological opportunities and challenges.[42]

In a seminal article, published in 2006, Professors Maloy and Schaechter identified critical stages in the evolution of modern microbiology,[43] as briefly summarized below. Their historical review helps to understand the potential role of microbiology in the "New Biology" paradigm,[44] as more recently articulated by the National Research Council. How to implement this paradigm is a primary concern of this book.

A. *The "Wet Lab" Era*

In the nineteenth century, which has been deemed "the first Golden Age of Microbiology,"[45] scientists formulated basic concepts of bacterial physiology (including classifications based on phenotypes), medical microbiology, and immunology. Subsequent applications included the clinical identification of microbes, antimicrobial chemotherapy, vaccines, and industrial fermentations.[46]

40 Organization for Economic Cooperation and Development (OECD), Biological Resource Centers: Underpinning the Future of Life Sciences and Biotechnology 17 (Sci. & Tech. Series, OECD 2001); D. Smith et al. (2013), above n. 4; Rita R. Colwell, *The Future of Microbial Diversity Research*, *in* Biodiversity of Microbial Life 521–34 (2002).

41 *See especially* Chapters 3 and 4 below.

42 *See, e.g.*, *Special Issue on Microbial Research Commons: From Strain Isolation to Practical Use*, 161 Research in Microbiology 407–514 (Dedeurwaerdere et al. eds., 2010).

43 Stanley Maloy [former Pres., Am. Soc'y Microbiology] & Moselio Schaechter, *The Era of Microbiology: A Golden Phoenix*, 9 *Int'l Microbiology* 1 (2006).

44 NRC, New Biology, above n. 18.

45 Maloy & Schaechter, above n. 43, at 1. Although studies conducted during the seventeenth to the nineteenth centuries provided considerable evidence to support and advance early hypotheses, these studies nonetheless remained controversial. Only in the second half of the nineteenth century did microbiology come of age in the sense that, during one twenty-year period alone, "the main bacterial etiological agents of disease in humans and animals were discovered and the field of immunology was developed," which led to many vaccines and serological tests. *Id.* at 2. More generally, it was in this period that the "importance of microbes in the cycles of nature was elucidated," and strain selection was applied for industrial purposes. *Id.* Among the pioneers of the nineteenth century, Ferdinand Cohn, Louis Pasteur, and Robert Koch stand out. Later in the nineteenth century, Martinus Beijerinck and Sergei Winogradsky became the founders of general microbiology, which moved the field beyond its focus on medicine to encompass microbial physiology, biodiversity, and ecology. Gerhart Drews, *Ferdinand Cohn, A Founder of Modern Microbiology*, 65 ASM News 547 (1999); G. Bordenave, *Louis Pasteur (1822–1895)*, 5(6) *Microbes & Infection* 553–60 (2003); Timothy Paustian & Gary Roberts, *Beijerinck and Winogradsky Initiate the Field of Environmental Microbiology*, *in* Through the Microscope: A Look at All Things Small § 1–14 (5th ed., Textbook Consortia, 2014).

46 Maloy & Schaechter, above n. 43.

Of necessity, microbiologists had traditionally focused on the study of single microbial species grown in pure laboratory cultures to the extent possible.[47] Until the second half of the twentieth century, microbiology was thus a "wet lab" science, often dependent upon the observations of naturalists who collected and analyzed locally harvested microbes and, eventually, microbes from all over the world. A major step forward was to devise ways of growing microbes in the laboratory so that scientists could view distinct populations growing together in colonies.[48] It also became clear that a pure wet lab culture did not adequately reflect how microbes lived outside of the laboratory and that the microbial world was "more diverse, more important, and far more interdependent than had previously been imagined."[49]

The organization of the microbiological community mirrored this wet lab foundation. Professional societies, such as the American Society for Microbiology (ASM) in the United States and the Society for General Microbiology in the United Kingdom, were formed at the beginning of the twentieth century. By 1923, the ASM's predecessor organization (the Society of American Bacteriologists) had published a fundamental catalog, known as Bergey's *Manual of Determinative Bacteriology*.[50] These professional societies, in turn, formed the International Union of Microbiological Societies (IUMS) in 1927, which is now one of the 29 scientific unions that constitute the International Council of Science (ICSU). IUMS remains the umbrella organization for the many national microbiology societies.[51]

In 1963, major culture collections holding microbial materials for research and applications in different countries decided to form a cooperative global entity, known as the World Federation for Culture Collections (WFCC).[52] As more fully explained in Chapters 2 and 4, these federated culture collections facilitated cross-border exchanges of microbial genetic resources on which the wet lab era largely depended.

Meanwhile, the next breakthrough period had begun to emerge from the genetic revolution in biology after the Second World War. Even though most of the microbial world still remains invisible in everyday life, the study of the human genome that

47 NRC, New Biology, above n. 18, at 50.

48 Bennett (2011), above n. 35 at 3(observing that many of the early techniques had been developed by the nineteenth century bacteriologists).

49 NRC, New Biology, above n. 18.

50 Bergey's Manual of Determinative Bacteriology (2d ed., Springer 2001). *See also* Int'l Union Microbiological Scis. (IUMS), *Homepage*, http://www.iums.org/ (last accessed 16 Sept. 2014).

51 Bennett (2011), above n. 35.

52 *See* World Federation for Culture Collections (WFCC) (Jan. 20, 2014), http://www.wfcc.info/. The WFCC is a multidisciplinary commission of the IUMS. *See further* Bennett (2011), above n. 35; below Chapter 4, Section I.A.

began in the 1950s led to new analytical techniques that have made microbes more manageable and more valuable for scientific purposes.[53]

B. The Revolution in Genetic Science

Increasingly, the role of microbiology in life science research overlaps with parallel advances in molecular biology and genetics. Until the advent of microbial genetics, many key cellular phenomena remained undecipherable.[54] The discovery of biochemical genetics and of genetic exchange mechanisms in bacteria and viruses ushered in a new period of major advances:

> These discoveries led to modern concepts of the gene and the biochemical basis of genetics, the understanding of how genetic information flows from nucleic acids to proteins, the regulation of gene expression, and how complex structures such as bacteriophages are assembled. These breakthroughs led to a paradigm shift. At that time, anyone who wanted to do modern science, mindful of it or not, had to become a microbiologist. The incipient science of molecular biology was spawned by the use of microbes and, consequently, microbial science was once again recognized as a fundamental scientific discipline.[55]

In this period, which roughly extended from the 1950s to the early 1980s, the primary concepts were bacterial genetics, bacterial physiology, and cellular immunology.[56] Notable applications in microbiology occurred in the fields of genetic engineering, nucleic acid and protein sequencing, microbial classification based on genotypes, and monoclonal antibodies.[57]

An even more transformative phase has been underway since the late 1980s. For example, it was less than two decades ago that the entire genome sequence of the bacterium *Haemophilus influenzae* was completed, and, for the first time, the full set of genetic information about a living organism responsible for a wide range of clinical diseases was discovered.[58] Genome sequencing has accelerated greatly since

[53] Genetics has been defined as "a branch of biology that deals with the heredity and variation of organisms." "Genetics," MERRIAM-WEBSTER.COM, http://www.merriam-webster.com/dictionary/genetics (last accessed 30 Mar. 2014). Genomics has been defined as "a branch of biotechnology concerned with applying the techniques of molecular biology to the genetic mapping and DNA sequencing of sets of genes or the complete genomes of selected organisms, with organizing the results in databases, and with applications of the data (as in medicine or biology ..."). "Genomics," MERRIAM-WEBSTER.COM, http://www.merriam-webster.com/dictionary/genomics (last accessed 30 Mar. 2014).

[54] Maloy & Schaechter, above n. 43, at 2.

[55] *Id.*

[56] *Id.* at 1–2.

[57] *Id.*

[58] Hamilton O. Smith et al., *How Many Genes Does a Cell Need?*, *in* ACCESSING UNCULTIVATED MICROORGANISMS 279–99 (Karsten Zingler ed. ASM Press 2008).

then, while the costs have plummeted, and additional microbial genomes are now rapidly decoded.[59]

This gradual convergence of microbiology and genomics has generated not only enormous amounts of biological materials, from large organisms to miniature genes, but also an explosion of data that are essential for research and development (R&D) in the life sciences.[60] As a result, microbiology has now come to play a unique and fundamental role in essentially every field of the life sciences, and even in other disciplines, ranging from the study of microbes in molecular biology at the cellular level to their role in the broader natural environment.[61]

From the late 1980s on, all these scientific developments were further stimulated and supported by a number of organizational and technological breakthroughs. For example, the advent of digital networks allowed rapid communication and collaboration among scientists and laboratories on a national and global scale. The proliferation of bioinformatics, computational methods, and other automated knowledge discovery tools further hastened the shift from the wet lab approach to *in silico* or virtual modes of knowledge production and dissemination.[62] High-throughput screening, the use of DNA microarrays, and data mining techniques have been particularly fruitful.[63]

These techniques arrived at a time when microbiologists were discovering that the genomic identity of single microbes needed to be understood in relation to their environments, their interdependence with other organisms, and their evolutionary history.[64] For example, microbial ecology has focused particular attention on the

[59] In September 2012, the Gold Online Database listed 3,722 complete bacterial genomes, 218 ongoing genome projects or archaea, and 11,790 ongoing genome projects for bacteria at http://genomesonline.org/cgi-bin/GOLD/index.cgi (last accessed 17 Sept. 2014).

[60] *See, e.g.*, Nikos Kyrpides, *Digital Research: Microbial Genomics, in* DESIGNING THE MICROBIAL RESEARCH COMMONS, above n. 11, Ch. 15; Dagmar Fritze, *The European Initiatives-MINE, CAPRI, EBRCN and ENBI, in* INNOVATIVE ROLE OF BIOLOGICAL RESOURCE CENTERS (M.M. Watanabe et al. eds, WFCC 2004). *See further* Chapter 8 below.

[61] Maloy & Schaechter, above n. 43 (citing authorities). For the role of microbes in other major natural processes, see, e.g., Farooq Azam et al., *Bacteria-Organic Matter Coupling and Its Significance for Oceanic Carbon Cycling*, 28 MICROBIAL ECOLOGY 167–79 (1994); Annette S. Engel et al., *Microbial Contributions to Cave Formation: New Insights into Sulfuric Acid Speleogenesis*, 32 GEOLOGY 369–72 (2004); Andreas P. Teske, *The Deep Subsurface Biosphere is Alive and Well*, 13 TRENDS MICROBIOLOGY 402–04 (2005).

[62] *Cf.* BENKLER, above n. 9, 351–55. *See further* Chapter 6, Section I, below.

[63] *See generally* NRC, NEW BIOLOGY, above n. 18, Ch. 3. *See further* U.S. DEP'T ENERGY, OFFICE OF SCI., SYSTEMS BIOLOGY KNOWLEDGEBASE FOR A NEW ERA IN BIOLOGY, DOE/SC–0113 (2009); Chapter 8, Section III.A.3 below.

[64] *See, e.g.*, NRC, NEW BIOLOGY, above n. 18; Carl R. Woese & George E. Fox, *Phylogenetic Structure of Prokaryotic Domain – Primary Kingdoms*, 74 PROC. NAT'L ACADS. SCI. 5088–90 (1977). *See also* Carl R. Woese & Nigel Goldenfield, *How the Microbial World Saved Evolution from the Scylla of Molecular Biology and the Charybdis of the Modern Synthesis*, 73 *Microbiology & Molecular Biology Rev.* 14–21 (2009). Because some 99 percent of the total microbial population has not been (and cannot be) cultured in the laboratory, the scope of enquiry has thus been vastly expanded to ask "who lives where, who does what, and who is phylo-genetically related to whom." Maloy &

way microbes live in natural communities, some of them with their own kind, others with different types of microbes.[65] Evolutionary genomics is still another field that has begun to pay big scientific dividends.[66] Studies of bacteria-host interactions also provide unique tools for understanding many aspects of cell biology, with corresponding possibilities for practical applications, such as the growing problems of antibiotic resistance, emerging infectious diseases, and bioterrorism threats.[67]

Professors Maloy and Schaechter summarized the important applications emerging from this period in the following list:

- Identification of uncultivated microbes;
- New methods for the rapid identification of microbes;
- New targets for antimicrobial therapies;
- Rational development of probiotics;
- Metabolic engineering;
- Use of microbes as nanomachines;
- Use of microbes for bioremediation.[68]

Box 1.1 juxtaposes in summary form the changing characteristics of the older and newer microbial research approaches, as discussed above.

Schaechter, above n. 43, at 4. Of particular importance here is the role of microbes in modulating host development and interactions among microbes, an emphasis that led to the discovery of an entirely new domain of life, the archaea. *Id.*

65 *See, e.g.*, INST. MEDICINE, THE SOCIAL BIOLOGY OF MICROBIAL COMMUNITIES – WORKSHOP SUMMARY (L.A. Olsen et al., Rapporteurs, Nat'l Acads. Press 2012). Recent studies of microbe-host interactions further show how microbes may influence the hosting behavior in unexpected ways. At the same time, microbial physiology has focused on mechanisms of signal transduction, global regulatory mechanisms, interactions between proteins, and metabolic networks. *See, e.g.*, W. R. Streit & R. A. Schmitz, *Metagenomics – The Key to the Uncultured Microbes*, 7 CURRENT OPINION MICROBIOLOGY 492–98 (2004); Thaddeus S. Stappenbeck et al., *Developmental Regulation of Intestinal Angiogenesis by Indigenous Microbes via Paneth Cells*, 99 PROC. NAT'L ACADS. SCI. 15451–455 (2002); KATHRYN M. CARBONE ET AL., MICROBIAL TRIGGERS OF CHRONIC HUMAN DISEASES (Am. Acads. Microbiology 2005) [hereinafter CARBONE ET AL.]. *See generally* Maloy & Schaechter, above n. 43, at 4–5.

66 Concepts such as lateral gene transfer and "genomic islands" have emerged, whereby sets of genes appear in unrelated microbes and endow them with new functions, such as a virulence trait or a novel metabolic capacity. Maloy & Schaechter, above n. 43. *See, e.g.*, Ulrich Dobrindt et al., *Genomic Islands in Pathogenic and Environmental Microorganisms*, 2 *Nature Revs. Microbiology* 414–24 (2004). Genome sequences have also revealed the importance of single nucleotide polymorphisms (SNPs) in bacterial evolution." One now sees how cellular components are dynamic and "a surprising molecular choreography has been unveiled via fluorescence microscopy and other techniques that allow us to visualize molecules in action." Maloy & Schaechter, above n. 43.

67 *Id.* These applications "require improved ways of rapidly detecting pathogenic agents, preventing their transmission, and effectively treating infected humans, other animals, and plants. Addressing these important challenges presupposes the development of new antimicrobial drugs and vaccines, both of which rely on a deeper understanding of molecular biology, structural biology, and microbial physiology." *Id. See also* INST. MEDICINE, INFECTIOUS DISEASE MOVEMENT IN A BORDERLESS WORLD – WORKSHOP SUMMARY (D.A. Relman et al., Rapporteurs, Nat'l Acads. Press 2010).

68 Maloy & Schaechter, above n. 43, at 5, Table. 3.

Box 1.1 The Changing Characteristics of Contemporary Microbial Research

Pre-1990s	Recent past and future trends
• Phenotype-based inquiry	• Genotype-based inquiry
• Primary focus on single organisms and subsystems	• Increasing focus on inter-dependent and complex systems
• Mostly single-discipline view	• More interdisciplinary
• Atomistic/insular/local	• Integrative/collaborative/global
• *In vitro* research	• *In silico* research
• Print communication	• Networked digital communication
• Data limited	• "Big data," especially genomic
• "Small science" organization	• Increasingly "big science" organization
• Public/basic research largely separated from the private applied	• Distinction between basic research and applications frequently collapsed

C. *Cutting-Edge Applications of Microbiology in Response to Major Global Challenges*

Despite being the smallest living organisms, microbes collectively represent the single largest mass of life on Earth, and they are inextricably intertwined with all other forms of life and the functions they perform.[69] Microbes also manifest the greatest diversity of all living creatures, using biological and chemical processes that exist nowhere else in nature.[70]

[69] Microorganisms live symbiotically with larger organisms that depend on them for nutrients, minerals, and energy recycling, while causing infectious disease when they overlap with susceptible hosts. *See, e.g.*, Maloy & Schaechter, above n. 43. *See also* CARBONE ET AL., above n. 65.

[70] JAMES STALEY & ANNA-LOUISE REYSENBACH, BIODIVERSITY OF MICROBIAL LIFE (Wiley 2001); SUSTAINING LIFE: HOW HUMAN HEALTH DEPENDS ON BIODIVERSITY (E. Chivian & A. Bernstein eds., 2008).

Microbes thus play a major role in the cycle of matter and in the metabolism of this planet, as Professors Maloy and Schaechter have observed:

> With the discovery of the huge microbial biota in subsoil fissures and better estimates of microbial life in the oceans, the microbial biota on Earth is thought to exceed in weight all other living things combined. Bacteria alone account for 50% of the biomass of carbon and over 90% of the biomass of nitrogen and phosphorus ... At least as much photosynthesis is carried out by marine microbes as by terrestrial plants. It has been estimated that if the action of microbes on the nitrogen cycle were to cease, the amount of nitrogen available to plants would become too low to sustain life within about one week.

Maloy & Schaechter, above n. 43, at 3.

The influence of microbiology on daily life is pervasive, and affects us all individually. Consider that human beings have always depended on genetic resources for nitrogen, photosynthesis, food, and medicines.[71]

Although this book primarily addresses legal and institutional obstacles to microbial genetic resources needed for public research purposes, it bears emphasizing at the outset that commercial applications of microbiology have major economic effects. According to one recent report, the global market for microbes used in the healthcare sector alone was valued at $90.5 billion in 2010, with a projected compound annual growth rate of 11 percent.[72] The total global market for microbes and microbial products was worth more than $144 billion in 2010, with a projected total of $259 billion by 2016.[73] Here we highlight a number of the socially and economically most important applications of research and development in the areas of human health, food and agriculture, the environment, and energy.

1. Improving Human Health and Mitigating Pandemics

Microbes are, of course, best known and most feared for the diseases they cause, such as leprosy, bubonic plague, and smallpox, which are no longer the scourges they used to be, or influenza and other bacterial and viral infections for which no cures have yet been found.[74] With the rising burden of antibiotic resistant forms of bacteria and new emerging infectious diseases, the quest for better antimicrobial drugs and vaccines for rapid treatment and prevention of these diseases has become urgent.[75]

Microbes remain the leading cause of infectious diseases that result in periodic public health pandemics. In favorable growth conditions, these organisms can

[71] Winter (2013), above n. 12, at 3–26. *See also* DAN MORGAN, MERCHANTS OF GRAIN: THE POWER AND PROFITS OF THE FIVE GIANT FOOD COMPANIES AT THE CENTER OF THE WORLD'S FOOD SUPPLY (iUniverse 1979); ; Stephen B. Brush, *Bio-Cooperation and the Benefits of Crop Genetic Resources: The Case of Mexican Maize*, 26 WORLD DEV. 755 (1998).

[72] Andrew McWilliams, BCC Research, Microbial Products: Technologies, Applications and Global Markets (Report Code BIO086A) (2011), available at http://www.bccresearch.com/market-research/biotechnology/microbial-products-technologies-market-bio086.html.

[73] *Id.* This reflects a projected annual growth rate of 10.7% between 2010 and 2016.

[74] *See* Bennett (2011), above n. 35; David P. Fidler, *International Law and Equitable Access to Vaccines and Antivirals in the Context of 2009-H1N1 Influenza, in* THE DOMESTIC AND INTERNATIONAL IMPACTS OF THE 2009 H1N1 INFLUENZA PANDEMIC: GLOBAL CHALLENGES, GLOBAL SOLUTIONS, ANNEX A4, 137–54 (D. A. Relman et al., Rapporteurs, Nat'l Acads. Press 2010) [hereinafter IMPACTS OF THE 2009-H1NI PANDEMIC]. Scientists have lately begun to learn that microbes can cause chronic diseases that were previously thought to be due to genetics or the environment – ailments such as ulcers and stomach cancer and many other diseases. *See, e.g.*, Maloy & Schaechter, above n. 43, at 4.

[75] *See* below Chapter 4, Section IV.A (discussing WHO's Pandemic Influenza Preparedness Framework).

be transmitted easily from one susceptible host to another,[76] and from country to country "in a borderless world."[77] Because microbes evolve rapidly, they may escape established drug regimens used for the treatment of disease.[78] Control of infectious diseases and pandemics can have especially important benefits for the populations of developing countries, where their effects are likely to be magnified. Investment in new technologies for providing safe and clean drinking water in developing countries would also mitigate some public health hazards resulting from harmful microbial contamination.[79]

Integrative approaches to biological research have gradually enabled the development of personalized medicine with the possibility of treatments tailored to the specific conditions and genotype of each individual.[80] A major aspect of these personal therapeutics is the growing understanding of the role of the microbiome in dynamic interactions with human health.[81]. The human microbiome refers to "the community of microorganisms, including prokaryotes, viruses, and microbial eukaryotes that populate the human body."[82] It is estimated that there are about ten times as many microbial cells as human cells in each adult person's body, yet relatively little is known about the effects of the human microbiome on health.[83]

76 For instance, more than thirty million people were killed in the influenza epidemic of 1918. C.E. Mills, J.M. Robins & M. Lipsitch, *Transmissibility of 1918 Pandemic Influenza*, 432 NATURE 904–06 (2004).

77 *See, e.g., Global Public Health Governance and the Revised International Health Regulations, in* INST. MEDICINE, INFECTIOUS DISEASE MOVEMENT IN A BORDERLESS WORLD 180–230 (Nat'l Acads. Press 2010) [hereinafter INFECTIOUS DISEASE MOVEMENT].

78 For example, the causative agent in most cases of tuberculosis, *Mycobacterium tuberculosis*, has developed several drug resistant forms that existing treatments cannot overcome. M.D. Iseman, *Evolution of Drug-Resistant Tuberculosis: A Tale of Two Species*, 91 PROC. NAT'L ACADS. SCI. 2428–29 (1994); Rebecca Katz, *Use of Revised International Health Regulations during Influenza A (H1N1) Epidemic*, 15(8) *in* EMERGING INFECTIOUS DISEASES 1165 (2009), *available at* http://wwwnc.cdc.gov/eid/article/15/8/pdfs/09-0665.pdf. *See also* NRC, THE RESISTANCE PHENOMENON IN MICROBES AND INFECTIOUS DISEASE VECTORS – IMPLICATIONS FOR HUMAN HEALTH AND STRATEGIES FOR CONTAINMENT (S.L. Knobler et al., Nat'l Acads. Press 2003) (Workshop Summary).

79 Sameera Al-Tuwaijri et al., *Gender, the MDGs, and Health Research, in* THE 10/90 REPORT ON HEALTH RESEARCH 2003–2004, 127 (Global Forum for Health Research 2004), *available at* http://mercury.ethz.ch/serviceengine/Files/ISN/17141/ipublicationdocument_singledocument/d8b4544b-15a7-4715-8cf3-21a046f91b93/en/1090.04.pdf (last accessed 19 Sept. 2014). *See also* A.L. Polaczyk et al., *Ultrafiltration-Based Techniques for Rapid and Simultaneous Concentration of Multiple Microbe Classes from 100-L Tap Water Samples*, 73(2) J. MICROBIOLOGICAL METHODS 92 (2008).

80 *See, e.g.*, NRC, TOWARD PRECISION MEDICINE: BUILDING A KNOWLEDGE NETWORK FOR BIOMEDICAL RESEARCH AND A NEW TAXONOMY OF DISEASE (Nat'l Acads. Press 2011).

81 NRC, NEW BIOLOGY, above n. 18.

82 Karen E. Nelson et al., *A Catalog of Reference Genomes from the Human Microbiome – The Human Microbiome Jumpstart Reference Strains Consortium*, 328 *Science* 994–99 (2010).

83 Division of Program Coordination, Planning & Strategic Initiatives, Nat'l Inst. Health, "Human Genome Project (HMP): Overview," July 7, 2010.

The assumption is that advances in the understanding of the human microbiome could lead to new therapies.[84]

2. Enhancing Agricultural Production and Food Security

Food security refers to the availability, access, and distribution of safe and healthy food to all households in any given community.[85] Microbes, such as viruses, bacteria, and fungi, as well as other biological and physical factors, interact with and affect the growth of crops in complex ways. For example, although progress has been made over the years in developing high-yield crops that are resistant to insects or diseases, the evolution of pests and microbes in the natural environment results in increasing crop resistance to the improved pesticides and to the susceptibility of crops to new diseases.[86] Microbial communities in the soil provide nutrients and can protect plants from pests and diseases. Understanding these processes to improve predictive capabilities should lead to enhanced plant productivity.[87]

In poor developing countries, food-borne pathogens remain a leading cause of illness and death.[88] To make developing countries more self-sustainable in agricultural production, coordinated research efforts are needed (together with the necessary institutional and economic reforms) for discovering new crop varieties that are resistant to microbial and fungal infections and harsh weather conditions, and for decreasing reliance on petroleum-based pesticides.[89] Even in the developed countries, food-borne pathogens cause millions of cases of infectious gastrointestinal diseases each year, with enormous costs in medical expenses and lost productivity. Pathogenic evolution, changes in agricultural and food production practices, and changes occurring within human hosts are major factors in fostering the emergence

[84] *See* Carl Zimmer, *How Microbes Defend and Define Us*, N.Y. TIMES, 13 July 2010, at D1. For a discussion of the International Human Microbiome Consortium, *see* below Chapter 9, Section II.B.4.

[85] J.V. Braun, *Addressing the Food Crisis: Governance, Market Functioning, and Investment in Public Goods*, 1(1) FOOD SEC. 9–15 (2009).

[86] Timothy Swanson & Timo Goeschl, *On the Economic Limits to Technological Potential: Will Industry Resolve the Resistance Problem?*, *in* THE ECONOMICS OF MANAGING BIOTECHNOLOGIES Ch. 4 (T. Swanson ed., Kluwer Acads. Pubs' 2002); Braun, above n. 85. *See also* Marleni Ramirez et al., *Demonstrating Interdependence on Plant Genetic Resources for Food and Agriculture*, *in* CROP COMMONS (2013), above n. 5, at 39–61.

[87] NRC, NEW BIOLOGY, above n. 18.

[88] FOODBORNE PATHOGENS: MICROBIOLOGY AND MOLECULAR BIOLOGY (P.M. Fratamico et al. eds., Caister Acads. Press 2005) [hereinafter FOODBORNE PATHOGENS].

[89] Biotechnology has discovered diagnostic techniques that enable the detection of toxins, such as aflatoxin, produced by fungi and molds that grow on crops. Pietro Cozzini et al., *Mycotoxin Detection Plays "Cops and Robbers": Cyclodextrin Chemosensors as Specialized Police?*, 9(12) *Int'l J. Molecular Sci.* 2474–94 (2008); S. De Saeger & C. Van Peteghem, *Dipstick Enzyme Immunoassay to Detect Fusarium T-2 Toxin in Wheat*, 62(6) APPLIED & ENVTL. MICROBIOLOGY 1880–84 (1996).

of new food-borne pathogens and diseases everywhere. There are also growing fears that terrorists could use pathogens to contaminate food and water supplies.[90]

Advances in molecular biology have identified some bacteria capable of producing compounds that kill other contaminating bacteria that cause food poisoning and spoilage. Improved diagnostic techniques based on polymerase chain reaction (PCR) have enabled researchers and industry to distinguish between harmful and harmless strains of bacteria.[91] Real-time PCR is also used to detect genetically modified organisms.[92]

Still other uses of microorganisms occur in many areas of food-processing.[93] Genetically modified microorganisms and advances in other scientific fields, such as synthetic microbiology, are promising for the development of enzymes and vitamins that could be used to enhance the nutrient value of some foods.[94]

3. Protecting the Natural Environment and Conserving Biodiversity

Human beings derive most of their food, food supplements, and some medicines from the environment. Conservation of natural resources remains important for diverse processes, such as nutrient storage and recycling, pollution breakdown and absorption, maintenance of the ecosystem, and climate stability.[95] Conservation also preserves biodiversity, which refers to the totality of life forms: the different plants, animals, microorganisms, and the genes they contain.[96]

Microbiology has made important contributions to biodiversity conservation, and it generates many applications for environmental restoration and bioremediation.[97] Microorganisms are used to remedy oil spills, and soil contaminants,[98] and to

[90] Foodborne Pathogens, above n. 88.

[91] Andrea Lauri & Paula O. Mariani, *Potentials and Limitations of Molecular Diagnostic Methods in Food Safety*, 4 *Genes & Nutrition* 1–12 (2009).

[92] Gordon Wiseman, *Real-Time PCR: Application to Food Authenticity and Legislation*, *in* Real-Time PCR: Current Technology and Applications 253 (J. Logan et al. eds., Caister Acads. Press 2009).

[93] K.M. Considine et al., *High-pressure Processing – Effects on Microbial Food Safety and Food Quality*, 281(1) Fed. European Microbiological Soc'ys Microbiology Letters 1–9 (2008).

[94] Sophie Marchand et al., *Selective Determination of the Heat-Resistant Proteolytic Activity of Bacterial Origin in Raw Milk*, 18(5) *Int'l Dairy J.* 514–19 (2007). For synthetic biology, see below n. 122 and accompanying text.

[95] Jordi Bascompte & Daniel B. Stouffer, *The Assembly and Disassembly of Ecological Networks*, 364 Philosophical Transactions of the Royal Soc'y B: Biological Scis. 1781 (2009).

[96] CBD, above n. 23, art. 2. For a detailed discussion of this treaty, see below Chapter 3, sections I.B-C.

[97] *See, e.g.*, *Special Issue: Restoration Ecology*, 325 Science 505–640 (2009). Bioremediation is a process that uses microorganisms, fungi, green plants or their enzymes to return the natural environment altered by contaminants to its original condition. Jim Harris, *Soil Microbial Communities and Restoration Ecology: Facilitators or Followers?* 325 Science 573–74 (2009).}

[98] D.R. Lovley, *Cleaning Up with Genomics: Applying Molecular Biology to Bioremediation*, 1(1) *Nature Revs. Microbiology* 35–44 (2003).

help eliminate environmental degradation from chemicals, such as chlorinated hydrocarbons.[99] The application of genetic engineering to microbial genetic resources, in particular, holds great potential for bioremediation.[100] For example, bacteria can be altered to produce certain enzymes that metabolize industrial waste components, and new pathways can be designed for the biodegradation of various wastes.[101]

Given the growing need for adequate supplies of safe drinking water and the changing epidemiology of waterborne diseases, a major challenge today is the rapid and specific detection of waterborne pathogens. Time-consuming culture methods used to detect such microbes in the past are giving way to new enzymatic, immunological, and genetic approaches that are more efficient.[102]

4. Addressing the Energy Challenge by Producing Biofuels

The warming of the world's climate as a result of the excessive production of greenhouse gases threatens to increase the number of natural catastrophes and to cause widespread economic harm and social dislocation.[103] Hence the pressing need for alternative sources of renewable and non-polluting energy, such as wind and solar energy, and biofuels.

The most commonly used biofuel, ethanol, is produced in many areas using corn or switch grass. With rising food prices, however, using corn for the production of biofuels has become controversial. By combining recent advances in technologies such as high-throughput sequencing, automated gene expression measurement, and metabolic engineering, microbiology could play a major role in developing alternative sources of biofuels. Researchers are now trying to find sources of biomass that microorganisms can

99 E. Cervantes-González et al., *Oil-Removal Enhancement in Media with Keratinous or Chitinous Wastes by Hydrocarbon-Degrading Bacteria Isolated from Oil-Polluted Soils*, 29 *Envtl. Tech.* 171 (2008).

100 For example, the bacterium *Deinococcus radiodurans* has been genetically modified to digest toluene and ionic mercury from highly radioactive nuclear waste. *See* Hassan Brim et al., *Engineering Deinococcus Radiodurans for Metal Remediation in Radioactive Mixed Waste Environments*, 18(1) *Nature Biotechnology* 85 (2000).

101 J.D. Desai & I.M. Banat, *Microbial Production of Surfactants and Their Commercial Potential*, 61(1) Microbiology & Molecular Biology Rev. 47–64 (1997).

102 For example, techniques increasingly used in the water industry include quantitative PCR, protein detection and immunological approaches, loop-mediated isothermal amplification, and microarrays. Environmental Microbiology: Current Technology and Water Applications (K. Sen & N.K. Asbolt eds., Caister Acads. Press 2010).

103 Intergovernmental Panel on Climate Change (IPCC), IPCC Fifth Assessment Report: Climate Change 2014 (AR5) (Cambridge U. Press 2014), *available at* http://www.climatechange2013.org/images/report/WG1AR5_ALL_FINAL.pdf.

convert to ethanol.[104] These include wood residues, paper waste, agricultural residues, and non-edible parts of corn.[105]

If researchers can unravel how to convert this inedible biomass into ethanol, the "food versus fuel debate" would be rendered moot. The challenge for advanced biofuels is thus to find ways to produce fuel more cheaply and with fewer negative externalities than in the past.[106]

D. A New Research Paradigm for the Life Sciences

As the foregoing discussion shows, advances in microbiological and genetic research are generating exciting, new and growing opportunities to integrate disparate sources of scientific knowledge. These opportunities emerged at a time when digital network applications, automated knowledge discovery tools, high throughput screening, full genome sequencing, and other pioneering methods made it possible to produce and process increasingly vast amounts of raw materials, data, and information.[107] According to the 2009 National Research Council report, entitled "A New Biology for the 21st Century," the field has reached "a point of inflection:"

> Years of research have generated detailed information about the components of the complex systems that characterize life – genes, cells, organisms, ecosystems – and this knowledge has begun to fuse into greater understanding of how all these components work together as systems. Powerful tools are allowing biologists to probe complex systems in ever greater detail from molecular events in individual cells to global biogeochemical cycles. Integration within biology and increasingly fruitful collaboration with physical, earth, and computational scientists, mathematicians, and engineers are making it possible to predict and control the activities of biological systems in ever greater detail.[108]

These trends rest on two key premises. First, all organisms are related by evolution, which means that "work on one gene, one cell, one species is directly relevant to

[104] NRC, New Biology, above n. 18, at 30–31.

[105] Charlotte Schubert, *Can Biofuels Finally Take Center Stage?*, 24 Nature Biotechnology 777–84 (2006).

[106] For example, some microorganisms can grow without oxygen to convert biomass into fuel, with minimal energy consumption. Genetic engineering and synthetic biology could also be used to produce photosynthetic bacteria or algae that capture sunlight and convert it into fuels, such as biodiesel. S.K. Lee et al., *Metabolic Engineering of Microorganisms for Biofuels Production: From Bugs to Synthetic Biology to Fuels*, 19 Current Op. Biotechnology 556–63 (2008); *Bacteria to Biofuel: Just Add Sunshine*, Biodesign Inst., http://www.biodesign.asu.edu/research/projects/better-biofuel (last accessed 21 Sept. 2014); *see also* NRC, New Biology, above n. 18, at 30–31.

[107] NRC, New Biology, above n. 18.

[108] *Id.* at 12–13.

understanding all others because processes may be identical or highly similar between different organisms due to their shared descent."[109] Second, the very process of evolution has spawned countless variations on these common themes – a vast array of organisms with myriad adaptations to diverse environments – with all the advantages that inter-disciplinary comparisons can bring.[110]

The NRC report posits that the fundamental unity of biology has now become so clear that scientists may aspire to understand "how all of the parts of living systems operate together in biological organisms and systems."[111] Towards the end of the twentieth century, the sheer volume of emerging knowledge about genetics, cell biology, ecology, microbiology, biochemistry, and molecular biology tended to keep researchers in each subdiscipline on separate, relatively autonomous tracks, with little interaction. As the twenty-first century has proceeded, in contrast, the new tools and concepts arising within single subdisciplines are increasingly applied throughout biology, with a view to disclosing the connections among the different subdivisions of the life sciences.[112] In particular, genomic comparisons "reveal the common descent of organisms and enable researchers to make comparisons of different types of organisms ... while also highlighting ... differences ... in separate evolutionary lineages."[113] The boundaries between fields such as microbiology, botany, and zoology become correspondingly blurred.[114]

This process of integration within the life sciences is further spurred by the parallel integration of techniques and concepts from engineering, robotics, computer science, mathematics, statistics, chemistry, and still other fields.[115] Mathematics has played an especially critical role by providing algorithms and the computational power needed to analyze the massive amounts of data emerging from genomic studies and by enabling the more effective placement of data in digitally accessible collections.[116]

Looking to the future, the NRC report foresees the possibility of much greater integration, with enormous benefits to public health, food security, environmental protection and other urgent socioeconomic needs, owing to expected technological and scientific advances. With regard to the former, information technologies

109 *Id.* at 13.
110 *Id.*
111 *Id.* at 40.
112 *Id.* at 41–42. For example, biochemistry and molecular biology are applied to all subdisciplines; genomic data and techniques shed light on interconnections among fields. *Id.*
113 *Id.* at 42–43, Box 4.2.
114 *Id.* at 42.
115 *Id.*
116 *Id.* at 44–45. Combinatorial algorithms are also essential for understanding genome assembly, sequence alignment, and phylogeny constructions based on molecular data. *Id.* at 62. *See further* Chapter 8 below.

will become more essential than ever in perfecting "the historical transition of the life sciences from a low-throughput descriptive experimental discipline to a high-throughput increasingly quantitative science."[117] Global efforts to collect, archive, and analyze information on living organisms and their myriad components will generate massive amounts of sequence and other data that need to be processed, stored, and analyzed.[118]

Other foundational technologies of particular importance for the future are *in vivo* and real-time imaging of cells, organisms, and ecosystems; high-throughput technologies, including nanotechnology; and engineered biological systems.[119] In a broader perspective, systems biology,[120] computational biology,[121] and synthetic biology[122] are the three foundational sciences to which the NRC's hopes for a new biology are pinned.

[117] *Id.* at 53.

[118] *Id.* "Biological imaging and scanning are producing vast amounts of data about biomolecules, cells, organs, organisms and environments that ... [are] difficult to index and interpret because of ... [their] three-dimensional pictorial or even four-dimensional nature." *Id.*

[119] *Id.* at 52–61. *In vivo* and real-time imaging could provide experimental tools for analyzing "the complicated internal complexes of cells throughout their lifecycles." *Id.* at 54. High-throughput screening techniques have greatly advanced proteomics, i.e., the study of all the proteins in a particular biological sample, with considerable promise for the development of personalized medicines, for accelerated plant breeding, and for monitoring environmental conditions. *Id.* Engineered biological systems techniques promise to provide more "effective experimental systems for analyzing complex human tissue physiology and pathophysiology *in vitro*" and, more generally, "for scaling to the data collection demands of high-throughput screening and systems biology." *Id.* at 60–61. *See also* NRC, Reaping the Benefits of Genomics and Proteomic Research: Intellectual Property Rights, Innovation, and Public Health (Nat'l Acads. Press 2006); Arti K. Rai et al., *Pathways Across the Valley of Death: Novel Intellectual Property Strategies for Accelerated Drug Discovery*, 8 Yale J. Health Pol'y L. & Ethics 53–89 (2008).

[120] Systems biology "seeks a deep quantitative understanding of complex biological processes through dynamic interaction of components that may include multiple molecular, cellular, organismal, population, community, and ecosystem functions." *See* NRC, New Biology, above n. 18, at 61. Systems biology "builds on foundational large-scale cataloguing efforts (e.g., genomics, proteomics, metabolomics, etc.) that specify the 'parts list' needed for constructing models." *Id. See generally* Inst. Medicine, The Science and Applications of Synthetic and Systems Biology (E.R. Choffnes, D.A. Relman & L. Prag, Rapporteurs, Nat'l Acad. Press 2011) [hereinafter Science and Applications of Synthetic and Systems Biology (2011)].

[121] Computational biology has reached the point where it can now provide probabilistic models and Bayesian analysis applicable to gene findings and comparative genomics. NRC, New Biology, above n. 18, at 62. Algorithms are also applied to "genome-wide association studies and to problems of classification, clustering and feature selection arising in the analysis of large-scale gene-expression data." *Id.*

[122] Synthetic biology aims to produce reusable standardized biological molecules ("biobricks") as components with which to engineer new biological systems. This field blurs the distinctions between *in vitro* or *in silico* research, and between the physical and the virtual dimensions, which further increases the potential socioeconomic payoffs. *Id.* at 63. "By standardizing biological parts and the way in which classes of parts can be functionally linked together, this field aims to make large-scale genetic engineering easier and more predictable," which could lead to cells, organisms, or biologically

In this context, microbiology, and particularly microbial genomics, play a major role in laying the foundations for the New Biology that the NRC's 2009 report envisioned.[123] It has become increasingly clear that microbes are far more diverse, interdependent, and important to other life forms than was previously known. Symbiotic relationships are the norm and a key to understanding a broad range of biological functions and processes.[124]

Further important changes are expected to occur from applications of computational science to the integration of genetic, protein, metabolic, and environmental data in microbial ecology. Also promising are the uses of Global Positioning Systems and remote sensing tools in tracking the effects of microbial dispersion. These emerging interdisciplinary methods are likely to offer innovative approaches to microbial surveillance and control, and they can foster research in systemic analysis of microbe-host interactions.[125] Integrating microbiology into other foundational sciences and techniques could, in turn, yield unprecedented benefits for human health, agriculture, ecosystem management, and energy production.[126]

It follows that, as researchers become better connected with each other, particularly through the internet, and as research focuses on issues of global importance, there is a growing need to systematically address problems of access and sharing of materials, data and literature beyond disciplines, and beyond institutional and national boundaries. As we attempt to demonstrate in the rest of this book, coordination and cooperation could create greater value from these basic, upstream knowledge assets. The goals should be to ensure that both researchers and the broader public receive the optimum return on public investments, and to strengthen the value chain of investments in advanced microbiology.[127]

III. LIMITS OF THE EMERGING MOVEMENT TO DIGITALLY INTEGRATE RESEARCH INPUTS INTO THE "NEW BIOLOGY"

The National Research Council clearly recognized that their New Biology paradigm depends on large-scale digital integration of research inputs across

inspired systems "with highly optimized industrial or therapeutic applications." *See, e.g.*, Arti K. Rai. & Sapna Kumar, *Synthetic Biology: The Intellectual Property Puzzle*, 85 TEX. L. REV. 1745–68 (2007). *See generally* THE SCIENCE AND APPLICATIONS OF SYNTHETIC AND SYSTEMS BIOLOGY (2011), above n. 120.

[123] *See* NRC, NEW BIOLOGY, above n. 18, at 50, Box 4.6 ("Microbial Genomics").

[124] *Id.* at 50 (stressing that the many microbes that can only grow in communities were never isolated by classical culturing methods).

[125] *See, e.g.*, NRC, CONTRIBUTIONS OF LAND REMOTE SENSING FOR DECISIONS ABOUT FOOD SECURITY AND HUMAN HEALTH (Nat'l Acads. Press 2007).

[126] NRC, NEW BIOLOGY, above n. 18, at 50. For examples, see above Section II.C.

[127] *Cf.* JOSEPH E. STIGLITZ, PETER R. ORSZAG & JONATHAN M. ORSZAG, THE ROLE OF GOVERNMENT IN A DIGITAL AGE (2000), *available at* http://www.dol.gov/ebsa/pdf/ccia.pdf (study commissioned

previously separate disciplinary boundaries. What the NRC did not address was the legal and institutional framework needed to support the flow and integration of upstream research inputs essential to operationalize its new research paradigm.[128] In reality, any shortcomings in the NRC's visionary project are not necessarily to be found in science itself, but rather in tacit assumptions about the enabling nature of the external environment in which the desired integration of the life sciences would be rooted.

To achieve this goal, researchers working in the various subdisciplines must have ready access to essential upstream knowledge assets. Life scientists and microbiologists, in particular, will need to obtain countless biological materials collected and validated from all parts of the world; to make use of vast amounts of data from genetic sequencing, genomic studies, bioecology, systematics and from other observational and experimental life science initiatives; and to access all the knowledge gleaned from an ever-expanding body of scholarly literature.

The "soft infrastructure" that currently governs these essential inputs, however, tends to fragment and compartmentalize the building blocks of science in ways that are not conducive to enabling the integrated vision to which the life sciences now aspire. In what follows, we briefly identify some of the constraints – i.e., social norms, institutional frameworks, and legal impediments – that endanger the NRC's visionary research enterprise.

A. *Recognizing Institutional and Legal Challenges to the Existing Microbial Research Infrastructure*

The *New Biology* report took for granted the continued stability of a preexisting microbial research infrastructure mainly consisting of public culture collections affiliated with the World Federation of Culture Collections (WFCC), plus a vast array of in-house collections (both academic and industrial) that have always supported basic and applied research in microbiology. The Report also assumed ready access to the massive amounts of scientific data and information that genetic sequencing and automated knowledge discovery tools potentially make available for a unified field approach.

Today, however, the stability of this existing research infrastructure – and the continued availability of the microbial genetic resources it traditionally made available to researchers everywhere – has come under attack from two directions.

by Computer & Commc'ns Indus. Assn.). *See also* Industrial Policy and Development (M. Cimoli, G. Dosi & J. Stiglitz eds., Oxford 2009).

[128] The study does spotlight a number of problems, e.g., stove-piped research and educational approaches, lack of interagency coordination, and insufficient interdisciplinary education. *See* NRC, New Biology, above n. 18. However, the issues raised here are not discussed in that report.

First, public research has been buffeted by an array of privatizing pressures associated with the explosion of globally applicable intellectual property rights since the last quarter of the twentieth century.[129] Second, there has been a host of parallel claims of "biopiracy" by many countries that provide *in situ* and *ex situ* microbial materials in the first place.[130] As will be seen in Chapters 3 and 4, moreover, these propertizing claims now extend well beyond material resources to embrace all the related genetic data that traditional intellectual property laws would have consigned to the public domain.[131]

Beyond these questions about continued access to essential knowledge inputs, there are still other institutional questions about the ability of existing microbial culture collections to upgrade their technical capabilities to the level that would be needed for the "big science" approach envisioned in the New Biology paradigm.[132] Successful integration of the life sciences could also depend on a deliberate elevation of the Mertonian norms of science, especially its precarious sharing ethos, to new and higher levels.[133] As demonstrated in Parts Two and Three, there are many obstacles to the various forms of voluntary collaborations and the sharing of materials and data that digital networks and computational tools could otherwise greatly expand.[134]

[129] *See further* below Chapter 2, Section II.

[130] *See further* below Chapter 3, Section I.A.

[131] *See* below Chapter 3, Section IV.A; Chapter 4, Section II. *See especially* below Chapter 6 ("Legal and Institutional Obstacles Impeding Access to and Use of Scientific Literature and Data").

[132] *See further* PETER LOUIS GALISON, BIG SCIENCE: THE GROWTH OF LARGE SCALE RESEARCH (Bruce Hevig ed., Stanford Univ. Press 1999); ORG. ECON. CO-OPERATION & DEV. (OECD), BIOLOGICAL RESOURCE CENTERS – UNDERPINNING THE FUTURE OF THE LIFE SCIENCES AND BIOTECHNOLOGY 8 (Mar. 2001) [hereinafter OECD REPORT ON BRCS], *available at* http://www.oecd.org/sti/biotech/2487422.pdf; STERN (2004), above n. 4; D. Smith et al (2013), above n. 4.

[133] For the sharing ethos of science, see R.K. Merton, *The Normative Structure of Science*, *in* THE SOCIOLOGY OF SCIENCE 267–78 (R.K. Merton ed., U. Chicago Press 1973). *See also* Paul A. David, From Keeping "Nature's Secrets" to the Institutionalization of "Open Science," 2 (Univ. Oxford, Discussion Paper No. 23, July 2001), *available at* http://www.nuff.ox.ac.uk/economics/history/paper23/23david.pdf (last accessed 23 Sept. 2014); Michael Polanyi, *The Republic of Science: Its Political and Economic Theory*, 1 MINERVA 54, 59–79 (1962); NRC, BITS OF POWER: ISSUES IN GLOBAL ACCESS TO SCIENTIFIC DATA 17–19, 21–22 (Nat'l Acads. Press 2007). For the environmental sciences perspective, see generally NRC, ON THE FULL AND OPEN EXCHANGE OF SCIENTIFIC DATA (Nat'l Acads. Press 1995) and NRC, RESOLVING CONFLICTS ARISING FROM THE PRIVATIZATION OF ENVIRONMENTAL DATA 15–19 (Nat'l Acads. Press 2001) (regarding scientists' views on the need for full and open access to environmental and earth science data).

[134] *See, e.g.*, Stephen Hilgartner, *Access to Data and Intellectual Property: Scientific Exchange in Genome Research*, *in* INTELLECTUAL PROPERTY RIGHTS AND THE DISSEMINATION OF RESEARCH TOOLS IN MOLECULAR BIOLOGY: SUMMARY OF A WORKSHOP HELD AT THE NATIONAL ACADEMY OF SCIENCE, 15–16 Feb. 1996 (Nat'l Acads. Press 1996); Stephen Hilgartner & Sherry I. Brandt-Rauf, *Controlling Data and Resources: Access Strategies in Molecular Genetics*, *in* INFORMATION TECHNOLOGY AND THE PRODUCTIVITY PARADOX (P.A. David & W.E. Steinmueller eds., Hardwood Acad. Press 1998); Reichman & Uhlir (2003), above n. 13; Wesley M. Cohen & John P. Walsh, *Real Impediments to Biomedical Research*, 8 *Innovation, Pol'y & Econ.* 1–30 (2008). *See generally* NRC, THE ROLE AND

Open access to the published results in scholarly literature, including microbiology journals, is still a work in progress.[135]

Moreover, any transition towards the global sharing of upstream data and materials as basic components of an integrated research paradigm is unlikely to succeed without a shift in the way science is organized. Microbiology, for example, would need to move away from uncoordinated small science projects toward big science programs and infrastructures, with a concomitant shift from purely voluntary to mandated sharing policies when necessary. The widespread sharing of *in vitro* and *in silico* genetic resources and data, as well as the resulting literature, is not merely a technical matter, but also a complex social process in which researchers have to balance different pressures and interests.

For example, the competitive advantage that researchers gain from being the first to publish is based on data that have been generated and held in secrecy, which leads to strategic behavior that conflicts with the sharing ethos even after publication. In other cases, a lack of sharing simply results from insufficient investment in the time and resources necessary for releasing and organizing data that underlie research outputs, not to mention comprehensive data standards and ontologies. In all cases, the researcher's willingness to share will depend in large part on the reciprocity benefits to be gained and the costs of holding out over time.[136]

In the late 1980s, research managers around the world began to devise large-scale collaborative research programs in molecular biology and other related disciplines, which increasingly helped to integrate microbiological laboratories that had largely operated autonomously in the past. These big science programs benefited from new digital technologies and networks, and they reinforced the need and drive for greater integration of microbial resources – materials, data and literature – and of its many subdisciplines.[137]

The program that best exemplifies this shift from small to big science in biology was the launch of the Human Genome Project in 1990,[138] which sought to integrate

VALUE OF SCIENTIFIC DATA IN THE PUBLIC DOMAIN: PROCEEDINGS OF A SYMPOSIUM (Nat'l Acads. Press 2003); NRC, THE CASE FOR INTERNATIONAL SHARING OF SCIENTIFIC DATA—A FOCUS ON DEVELOPING COUNTRIES (Nat'l Acads. Press 2012).

135 *See* below Chapter 7.

136 Minna Allarakhia et al., *Modeling the Incentive to Participate in Open Source Biopharmaceutical Innovation*, 40 RESEARCH & DEV. MGMT. 50 (2009); *see also* Minna Allarakhia & Steven Walsh, *Managing Knowledge Assets under Conditions of Radical Change: The Case of the Pharmaceutical Industry*, 31 *Technovation* 105 (2011). *See further* Chapter 8 below.

137 *See, e.g.*, European Bioinformatics Institute (EBI), and European Molecular Biology Laboratory (Embl); for the EU culture collections, Dagmar Fritze, *The European Initiatives*, *in* INNOVATIVE ROLE OF BIOLOGICAL RESOURCE CENTERS (M.M. Watanabe et al. eds., WFCC 2004).

138 Nat'l Human Genome Research Inst., *All About the Human Genome Project* (24 Jan. 2013), http://www.genome.gov/10001772.

the many government and non-profit laboratories and researchers already engaged in molecular biology into a large-scale, multiyear, and international collaborative research effort. Open science, and the rapid sharing of data and materials, were strongly advocated by John Sulston and other leaders of the molecular biology community, and they became core policies of the endeavor from the very beginning. The pledge to share data rapidly was linked to a plea not to patent snippets of the human genome, except when they could foreseeably induce investment in the development of end-products, such as therapeutic proteins.[139]

While the Human Genome Project gave rise to an expanding set of rules that mandated the public disclosure of DNA sequence data, and some important community databases in the life sciences have subsequently emerged on a voluntary basis,[140] the willingness of scientists to more fully engage in data pooling projects remains to be seen, even when it appears to be in their immediate interest.[141] In general, voluntary policies for the broad disclosure and sharing of data and literature – in contrast to the mandated policies discussed in Chapter 8 – have sometimes proven to be relatively ineffective.[142]

In microbiology, moreover, there has not been a similar restructuring of scientific research around big science programs, even if the genomic and taxonomic communities have promoted some sharing of microbiological materials and data within the existing research infrastructure, as discussed in Chapters 4 and 8. The promise of collaborative and integrated *e*-science is not likely to succeed without considering the broader evolution of community norms and the reconfiguration of research processes and infrastructures.[143]

In this context, organizational rules asserting control over research outputs and ownership status often conflict with the individual scientist's freedom to operate.

139 Robert Cook-Deegan & Tom Dedeurwaerdere, *The Science Commons in Life Science Research: Structure, Function and Value of Genetic Diversity*, 188 *Int'l Soc. Sci. J.* 302 (2006). *See further* Chapter 8 below.

140 For examples, see Bryn Nelson, *Data Sharing: Empty Archives*, 461 NATURE 160–63 (2009) (discussing requirements that the underlying sequence data be deposited in respective banks while encouraging researchers to deposit plasmid data in public repositories, such as Addgene (http://www.pnas.org/site/misc/iforc.shtml). Also required are detailed processes of synthesizing a new compound to be made available immediately: authors must provide sufficient information to establish the identity of a new compound and its purity; sufficient experimental details must be included to allow other researchers to reproduce the synthesis; characterization data and experimental details must be included either in the text or the Supporting Information). *Cf. also* D.A.B. Lindberg & B.L. Humphreys, *Rising Expectations: Access to Biomedical Information*, 3 YEAR B MED. INFO. 165–72 (2008).

141 *See, e.g.*, NRC, THE CASE FOR INTERNATIONAL SHARING (2012), above n. 134, at 69–96 ("The Limits and Barriers to Data Sharing"). *See further* below Chapter 8.

142 *See, e.g.*, the PubMedCentral policy and subsequent legislation (which moved from approximately 5% compliance under the voluntary post-publication deposit policy after 12 months of publication to about 70% following the legislative mandate), below in Chapter 8, Section II, A.

143 These topics are addressed at length in Parts Three and Four of this volume.

The communal norms are further confronted with countervailing values, such as the legitimate protection of privacy and national security and, more broadly, with the commercial interests that a proliferation of exclusive intellectual property rights and policies have intensified. As a result, the sharing ethos has been eroding in many fields of publicly funded research in the life sciences due to growing interest in patenting, copyrights, and restrictive licensing by scientific institutions, both in developed and developing countries,[144] as more fully elaborated in Chapters 4 and 6.

B. *Towards a Redesigned Microbial Research Commons*

In the rest of this volume, we analyze the existing microbial research infrastructure and provide detailed proposals for redesigning its disparate components into a more effective, globally integrated contributor to the National Research Council's vision of a "New Biology."[145] In so doing, we take account of theoretical and empirical insights emerging from the study of "knowledge commons" in general,[146] with a view to addressing a broader range of legal and institutional challenges affecting the life sciences and other public research fields.

In Part One, we discuss the "International Regulation of Genetic Resources and the Assault on Scientific Research." We begin by retracing the historical role of both plant and microbial genetic resources as global public goods, with brief snapshots in Chapter 2 of efforts to pool these resources for public research purposes up to the 1990s. The chapter then highlights the proprietary pressures that have subsequently hindered access to genetic resources for both public and private research purposes. The chapter ends with an evaluation of the "bilateral approach" established by the Convention on Biological Diversity (CBD) as a potentially serious threat to public scientific research, even without stricter multilateral regulatory controls that are the subject of the next chapter.

Chapter 3, entitled "Tightening the Regulatory Grip: From the Convention on Biological Diversity in 1992 to the Nagoya Protocol in 2010," begins with a look at the destabilizing effects that the CBD actually imposed on exchanges of pooled genetic resources for research purposes in the decades following its enactment. In this period, the microbiological community began to explore cautious and temporizing measures to defend access to *ex situ* genetic resources. Meanwhile, the critically important Consultative Group on International Agricultural Research (CGIAR) nearly collapsed in the 1990s, only to be rescued by an ambitious and idealistic

[144] *See, e.g.*, Anthony So et al., *Is Bayh-Dole Good for Developing Countries? Lessons from the U.S. Experience*, 6 *PLOS Biology* 2078–84 (2008); Reichman & Uhlir (2003), above n. 13.

[145] *See* above nn. 107–27 & accompanying text.

[146] *See* above nn. 10–13 & accompanying text, below Chapter 9, Section I.

international treaty administered by the United Nations' Food and Agricultural Organization (FAO). The chapter examines the strengths and weaknesses of this treaty – the International Treaty on Plant Genetic Resources for Food and Agriculture (ITPGRFA)[147] – as a reflection of the strengths and weaknesses of the CBD itself, when initially drafted.

Chapter 3 then carries the international regulatory history forward to 2010, when the Nagoya Protocol to the CBD addressed those very weaknesses and sought to strengthen the developing countries' regulatory grip on access to both plant and microbial genetic resources for the future. The chapter finds that the Nagoya Protocol could make access to, and use of, genetic resources for public research purposes far more difficult than before. At the same time, the Nagoya Protocol, created new facilitating possibilities for exchanges of genetic resources in the public research sphere that were not expressly recognized under the CBD as initially drafted.

As a result, the global microbiological community now has the opportunity to redesign its existing research infrastructure at the multilateral level so as to better exploit the favorable opportunities afforded by the Nagoya Protocol, while avoiding the constraints of the bilateral approach as tightened by that same Protocol. How specifically to implement this strategy is the task we undertake in the rest of the book.

Part Two thus focuses on the ways and means of "Preserving the Public Research Functions of Microbial Genetic Resources after the Nagoya Protocol." Chapter 4 first explains how public microbial culture collections have evolved over time from the wet lab era to the genomic revolution, in which they are increasingly asked to become full-fledged Biological Resource Centers. The chapter then presents sobering empirical evidence of the proprietary pressures that threaten to narrow the public good approach that was the hallmark of these collections.

Here we survey contractual restrictions imposed on access to, and use of, upstream microbial genetic resources in both developed and developing countries. The evidence shows that the shifting and relatively uncoordinated efforts by the microbial culture collections to grapple with the implications of the Convention on Biological Diversity are largely insufficient in view of the comprehensive and preemptory enforcement dictates of the Nagoya Protocol. In this context, both the bilateral approach of the European Union's Regulation on Compliance Measures for Users from the Nagoya Protocol (2014)[148] and the multilateral approach of the

[147] International Treaty on Plant Genetic Resources for Food and Agriculture, *opened for signature* 3 Nov. 2001, 2400 U.N.T.S. 303 (entered into force 29 June 2004) [hereinafter ITPGRFA], *available at* http://treaties.un.org/doc/publication/UNTS/Volume%202400/v2400.pdf (last accessed 24 Sept. 2014).

[148] Regulation No. 511/2014 of the European Parliament and of the Council on Compliance Measures for Users from the Nagoya Protocol on Access to Genetic Resources and the Fair and Equitable Sharing of Benefits Arising from their Utilization in the Union, 2014 O.J.L. 150/59.

WHO's Pandemic Influenza Preparedness Framework (2011) are examined and critically evaluated.[149]

The chapter posits that the microbiological community as a whole, together with interested governments, must more aggressively reconcile its upstream research needs with the Access and Benefit Sharing provisions of the CBD by opting into a multilateral approach in order to stimulate more downstream benefits from the bilateral system. This strategy will require devising a legal framework for both formal and informal exchanges of microbial materials, rather than depending on the Conference of the Parties to adopt a Standard Material Transfer Agreement under the Nagoya Protocol, which may not be consistent with the research community's long-term interests.

Chapter 5 then proposes a novel contractual framework for "Facilitating Transnational Exchanges of Genetic Resources within a Redesigned Microbial Research Infrastructure." It envisions the use of a standardized material transfer agreement (MTA) embodying a "take and pay rule" that would enable unfettered public research uses of microbial materials having no known or likely commercial applications at the time of deposit in participating culture collections. While avoiding restrictions based on fuzzy distinctions between commercial and noncommercial research uses, our approach – known as a "Compensatory Liability Regime"[150] – secures equitable compensation from any downstream commercial applications arising from such uses. In so doing, it attempts to address and rectify many of the design flaws in the so-called "Crop Commons" for plant genetic resources that were identified in Chapter 3.

After fleshing out the main components of such a regime, we then develop six detailed scenarios that illustrate how the proposed multilateral system for exchanges of microbial genetic resources would operate on a step-by-step basis. The chapter shows how the proposed regime would significantly improve the prospects for exchanging microbial genetic resources for research purposes over the existing MTAs, based on the bilateral approach, by fully exploiting the new opportunities to promote public research that the Nagoya Protocol affords.

The goal should be to protect all the participants' downstream commercial opportunities without, however, allowing the restrictive practices of the private sector and related intellectual property instruments to seep into and disrupt the

149 World Health Org. (WHO), Pandemic Influenza Preparedness Framework for the Sharing of Influenza Viruses and Access to Vaccines and Other Benefits (2011) [hereinafter PIP Framework Agreement], *available at* http://www.who.int/influenza/resources/pip_framework/en/index.html (last accessed 24 Sept. 2014).

150 *See* Jerome H. Reichman, *Of Green Tulips and Legal Kudzu: Repackaging Rights in Subpatentable Innovation*, 53 *Vand. L. R.* 1743 (2000), *available* at http://scholarship.law.duke.edu/faculty_scholarship/456.

sharing and open-access policies appropriate to the broad public research zone. The guiding principle is to insulate public research from pressures to commercialize research results by establishing standardized contractual templates, procedures, and institutional mechanisms that greatly reduce transaction costs for most upstream research purposes.

In Part Three, we look beyond issues concerning microbial materials to the prospects for "A Digitally Integrated Infrastructure for Microbial Data and Information." For example, just as researchers seeking a cure for specified diseases will often need access to microbial strains from a variety of sources in different countries, so too they will need to become users of databases previously compiled by others, and they will require ready access to increasingly specialized journal articles bearing directly on their specific research projects.

Because microbiology, like other life sciences, has become increasingly data intensive, a redesigned knowledge commons should ensure that the rising tide of precompetitive genomic and other data become widely available to the global research community, and not unduly burdened by intellectual property rights or by assertions of national sovereignty over genomic data under the CBD. By the same token, a properly designed research commons should seek to make the relevant scientific literature openly available to scientists everywhere as a "non-monetary benefit" expressly recognized by the Nagoya Protocol. In short, besides redesigning the existing infrastructure to enable broad and effective access to pooled microbial materials, the proposed research commons should thus also seek to better integrate relevant databases and scientific publications into its digital fabric.

However, that goal will not be easy to obtain, in view of the "Legal and Institutional Obstacles Impeding Access to and Use of Scientific Literature and Data" that are identified in Chapter 6. This chapter opens with a broad summary of the new opportunities for accessing data and information online for purposes of public research, including the use of computational science and automated knowledge discovery tools. But these opportunities are threatened by obsolete and science-hostile copyright and database protection laws at the national and international levels, as well as by highly restrictive licensing practices. Such barriers remain formidable, despite substantial, if fragmented, gains in open access publishing and in the establishment of some open data and literature repositories.

Chapter 7, entitled "Enabling the Microbial Research Community to Control Its Own Scholarly Publications," begins the search for suitable responses to the problems identified in the previous chapter. In Chapter 7, we first present empirical evidence of the extent to which journals that publish microbiological research results have moved towards more open-access options. We then examine a number of more far-reaching proposals for redefining the role of publishing intermediaries in order to avoid the remaining constraints on access to literature under the copyright laws discussed in Chapter 6.

Chapter 8 continues this exploration under the title, "Fully Exploiting Data-Intensive Research Opportunities in the Digitally Networked Environment." We begin by reviewing a number of the open access, public-sector databases germane to microbiological research and their enlightened sharing policies. In our view, the proposed Microbial Research Commons should seek to contractually override existing legal and institutional obstacles in order to facilitate access to a digitally-integrated, ever-expanding pool of materials, data, and literature. To this end, Chapter 8 shows that it has become increasingly advantageous to integrate biological materials, together with relevant data and information, in thematic collections that participating scientists and others can easily access when conducting their investigations.

The objective is an expanding set of federated pools of data, information and microbial genetic materials, open either entirely or partially to the interested research communities, on terms and conditions that these communities have themselves established through a combination of formal and informal governance mechanisms. We call these digitally integrated thematic communities "Open Knowledge Environments," and we identify several existing initiatives in the field of microbiology that have already taken major steps in this direction.

Finally, in Part Four, "Governing Public Knowledge Assets within a Redesigned Microbial Research Commons," we seek to identify and establish the elements of a tailor-made, science friendly governance structure for our proposed research commons at the international level. Chapter 9, entitled "Institutional Models for a Transnational Research Commons," opens with a look at the theoretical research that has enriched our understanding of the economic role and value of common pool resources in general. It focuses especially on "knowledge commons," which have elicited considerable interest in novel organizational structures that combine peer production approaches with networked technologies.

We then embark on an extensive empirical analysis of existing organizational structures that have been used to govern common pooled resources in other fields of scientific endeavor. Although each of these initiatives exhibits some features worth emulating, we ultimately attempt to maximize direct scientific inputs into an innovative governance structure that breaks new ground, as described in Chapter 10.

Chapter 10 is entitled "Governing Digitally Integrated Genetic Resources, Data, and Literature." It begins by analyzing the political economy of our own proposed undertaking in the light of the theoretical insights identified in Chapter 9. In our view, a redesigned knowledge commons for public upstream genetic resources and their digital counterparts should translate the insights and the experience gained from the institutional models empirically examined in Chapter 9 into a more effective transnational governance framework. Such a framework should directly seek to reconcile the needs of public science with the dictates of global intellectual property laws and the development of downstream commercial applications that advance human welfare.

In this context, we emphasize flexibility and self-organization on the part of participating entities, which would benefit from a Governing Body that did not impose novel, ad hoc solutions, but actively stimulated and nurtured them from the bottom up. In other words, we want to move towards a more science-driven organizational model for the digital age. Our proposed governance regime thus envisions a grand bargain, built around an intergovernmental Framework Agreement, that would reconcile the interests of both developed and developing countries under the CBD, while preserving and defending the public research space for the benefit of all relevant stakeholders.

The practical question becomes how to achieve these goals while avoiding possible design flaws that have lately come to light in the FAO's Crop Commons, as examined in Chapter 3. Our implementing proposals, drawn from the preceding theoretical and empirical analyses, as well as from the WHO's PIP Framework examined in Chapter 3, are set out in the section, entitled "Implementing the Multilateral Regime for Facilitated Access to *Ex Situ* Microbial Genetic Resources."[151]

Here we envision an organizational structure that puts science first, as distinct from more politically driven organizational frameworks that tend to alienate the very scientists whose interests they are supposed to advance. Chapter 10 thus places our redesigned Microbial Research Commons in a larger scientific context and looks at the implications for future science policy. The resulting research infrastructure, which seeks to maximize payoffs for the public sector, could also be open to those private-sector players that found it beneficial for their own research needs. However, any private-sector participants must necessarily accept the system on its own terms, and would not be allowed to change the default rules of the research community to conform to their own commercial practices, as they did when negotiating the FAO's Crop Commons, discussed in Chapter 3.

Finally, we also examine the funding strategies necessary to stabilize a redesigned Microbial Research Commons. We argue that this proposed knowledge commons would largely pay for itself by extracting more benefits for all the stakeholders under the CBD than is possible under either the primitive bilateral approach or existing multilateral solutions. We end by stressing the hidden costs of failing to redesign the existing microbial research infrastructure, with the corresponding risk that upstream

[151] *See* Chapter 10, Section III.

and precompetitive microbial genetic resources and data will remain subject to inefficient privatizing tendencies and a poorly organized public institutional framework.

Looking back at this endeavor as a whole, we concede that such a complex and ambitious undertaking along the lines we propose would require carefully nuanced approaches to the management of both microbial genetic resources and digital data and information. Difficulties arise in part because these components are historically governed by different legal regimes and different institutional structures, and also because physical materials pose unique problems of quality control, security, and other limiting factors. Collective action to address these problems and unite the field in an integrated multilateral system could prove extremely challenging in practice.

Nevertheless, we believe that any progress in this direction would constitute a marked improvement over the present situation. It could augment both basic and applied scientific payoffs and provide useful experience for further rationalization and integration of the overall system of microbial exchanges in the future. Above all, it would enable microbiology to better fulfill the critical role envisioned by the drafters of the New Biology paradigm, and it could also provide valuable experience and models for scientific communities in other fields that may seek to move in a similar direction.

PART ONE

International Regulation of Genetic Resources and the Assault on Scientific Research

2

Between Private and Public Goods: Emergence of the Transnational Research Commons for Plant and Microbial Genetic Resources

I. HISTORICAL IMPORTANCE OF GENETIC RESOURCES AS GLOBAL PUBLIC GOODS

In this and the next chapter, we document the conflict between unrestricted cross-border exchanges of genetic resources for purposes of public scientific research and countervailing proprietary claims to such resources supported by recent international conventions dealing with intellectual property and related rights.[1] We begin our enquiry with an overview of the historical role that both plant and microbial genetic resources have played as global public goods.[2] We then show that ongoing transnational efforts to preserve access to both *in situ* and *ex situ* plant genetic resources for public research purposes, despite the privatizing thrust of legal and economic measures identified in this and the next chapters,[3] will bear directly on our proposals to redesign the existing microbial research commons for similar purposes. Both theoretical insights and empirical evidence thus suggest that the future of the microbial research commons largely depends on lessons to be learned

1 *See, e.g.*, Agreement on Trade-Related Aspects of Intellectual Property Rights, April 15, 1994, Marrakesh Agreement Establishing the World Trade Organization, Annex IC, The Legal Texts: The Results of the Uruguay Round of Multilateral Trade Negotiations 320 (1999), 1869 U.N.T.S. 299 33 I.L.M. 1187 (1994) [hereinafter TRIPS Agreement]; Convention on Biological Diversity, *opened for signature* June 5, 1992, 1760 U.N.T.S. 79 [hereinafter CBD]; Tenth Meeting of the Conference of the Parties to the Convention on Biological Diversity, Nagoya, Japan, 18–29 Oct. 2010, Nagoya Protocol on Access to Genetic Resources and the Fair and Equitable Sharing of Benefits Arising from their Utilization (ABS) to the Convention on Biological Diversity [hereinafter Nagoya Protocol] (entered into force in October 2014 after the deposit of the fiftieth instrument of ratification, acceptance, approval, or accession), *available at* http://www.cbd.int/abs/doc/protocol/nagoya-protocol-en.pdf (last accessed 29 Sept. 2014).

2 Joseph E. Stiglitz, *Knowledge as a Global Public Good*, *in* GLOBAL PUBLIC GOODS: INTERNATIONAL COOPERATION IN THE 21ST CENTURY 308, 308–26 (I. Kaul et al. eds. Oxford U. Press 1999).

3 *See* below Chapter 2, Section I.B. and Chapter 3, Sections II–III.

from the existing infrastructure supporting the continued availability of plant genetic resources for food and agriculture.[4]

A. *Dependence of Wet-Lab Microbiology on Cross-Border Exchanges of Validated Reference Strains from Public Culture Collections*

In part because microorganisms replicate frequently and entail the use of special equipment for their study, *in situ* conservation provides an important but unpredictable source of specimens for microbial research.[5] Instead, microorganisms that had been discovered and removed from their natural environments were typically conserved by culture collections and made available for systematic comparative research. In this context it bears reiterating that scientists still know only about one percent of the world's microbial population, and only a relatively small fraction of those known can be grown in cultures and preserved for future use.[6]

To this day, two major categories of culture collections still organize and support the distribution of *ex situ* microbial materials for purposes of research and applications. One category consists of several hundred formally constituted public service collections, typically (but not always) funded by governments. The other consists of the thousands of relatively informal in-house research collections held at universities, hospitals, industrial laboratories, and other research institutes.[7]

[4] The existing research commons for plant genetic resources, as more fully described below and in Chapter 3, is now regulated by an international treaty administered by the United Nations Food and Agricultural Organization (FAO). *See* International Treaty on Plant Genetic Resources for Food and Agriculture, *opened for signature* 3 Nov. 2001, 2400 U.N.T.S. 303 (entered into force 29 June 2004) [hereinafter ITPGRFA], *available at* http://treaties.un.org/doc/publication/UNTS/Volume%20 2400/v2400.pdf (last accessed 29 Sept. 2014). Because this treaty operates in the shadow of the CBD, see n. 1, it casts considerable light on the problems besetting the microbial research commons and the options for responding to this challenge.

[5] *See* Joan W. Bennett, *Microbiology in the 21st Century*, *in* DESIGNING THE MICROBIAL RESEARCH COMMONS: PROCEEDINGS OF AN INTERNATIONAL SYMPOSIUM (P.F. Uhlir ed., Nat'l Acad. Press 2011) [hereinafter DESIGNING THE MICROBIAL RESEARCH COMMONS], *available at* http://www.ncbi.nlm.nih.gov/books/NBK91499/ (last accessed 28 Sept. 2014). *See also* Rudolf I. Amann et al., *Phylogenetic Identification and* In Situ *Detection of Individual Microbial Cells Without Cultivation*, 59 *MICROBIOLOGY REV.* 143, 144–45, 144 table 1 (1995).

[6] Tom Dedeurwaerdere, Arianna Broggiato & Dimitra Manou, *Global Scientific Research Commons under the Nagoya Protocol – Governing Pools of Microbial Genetic Resources*, *in* COMMON POOLS OF GENETIC RESOURCES: EQUITY AND INNOVATION IN INTERNATIONAL BIODIVERSITY LAW 224, 226–227 (E.C. Kamau & G. Winter eds. Routledge 2013) [hereinafter COMMON POOLS OF GENETIC RESOURCES (2013)].

[7] David Smith, *Culture Collections*, 79 ADVANCES IN APPLIED MICROBIOLOGY 73, 75–76 (2012) (stressing that microbial culture collections vary in size, form, and function):

> They can be small and limited in coverage, collected, and maintained by single researchers; they can be based in laboratories within large multifunctional organizations, and they may be

The public culture collections around the world hold a good selection of scientifically validated microbial materials, many of which have been accumulated over a long period of time. The primary purpose of these collections is to ensure the continued availability of microbial materials collected from various sources at different times for both public research and private industrial applications.[8] To this end, a process of lyopholizaton has been developed that enables curators to freeze dry living samples for storage purposes and then to culture them once again for specific research uses.[9]

1. Formation of an International Consortium of Public Service Microbial Culture Collections

In 1963, a large group of dispersed national culture collections agreed to form a cooperative entity, known as the World Federation for Culture Collections (WFCC). The WFCC aims to "promote and support the establishment of culture collections and related services, to provide liaison and set up an information network between the collections and their users, to organize workshops and conferences, publications and newsletters and to work to ensure the long-term perpetuation of important collections."[10] In practice, these collections are formally organized to distribute high-quality microorganisms for research purposes; they maintain public catalogs of their holdings; and they increasingly use standard arrangements for distributing

institutional entities developed with the sole purpose of being public service collections that cover a broad range of organisms from many sources. They can focus on organism type, for example, fungi or bacteria and in some instances specific genera; they may have been established to focus on a specific use, for example, industrial enzymes or antimicrobials or on host crops; they may be linked to a particular sector as the environment health care, education or agriculture.

8 *See, e.g.*, Cletus Kurtzman, *The Agricultural Research Service Culture Collection: Germplasm Accessions and Research Programs*, *in* DESIGNING THE MICROBIAL RESEARCH COMMONS, above n. 5, at 55; Frank Simione, *American Type Culture Collection: A Model for Biological Materials Resource Management*, *in* DESIGNING THE MICROBIAL RESEARCH COMMONS, n. 5, at 63–68.

9 *See, e.g.*, Kurtzman, n. 8, at 57. *See generally* WORLD FED. CULTURE COLLECTIONS (WFCC), GUIDELINES FOR THE ESTABLISHMENT AND OPERATION OF COLLECTIONS OF CULTURES OF MICROORGANISMS, 3d ed., WFCC (Feb. 2010), http://www.wfcc.info/guidelines [hereinafter WFCC GUIDELINES]; ORG. ECON. COOPERATION & DEV. (OECD), BEST PRACTICE GUIDELINES FOR BIOLOGICAL RESOURCE CENTRES (2007) [hereinafter OECD GUIDELINES FOR BRCS], *available at* http://www.oecd.org/sti/biotech/38777417.pdf. Another preservation process relies on the use of liquid nitrogen, whereas some particularly valuable cultures may be preserved by both methods. *See* WFCC *Guidelines*, n. 9. *See also* Kurtzman, n. 8 (emphasizing that specimens deposited for purposes of obtaining product or method patents are preserved by both processes).

10 *See* WORLD FED. CULTURE COLLECTIONS (WFCC), http://www.wfcc.info/about/ (last accessed 1 Oct. 2014). The WFCC is a multidisciplinary commission of the International Union of Biological Sciences (IUBS) and a federation within the International Union of Microbiological Societies (IUMS). *See* Bennett, n. 5, at 8.

over 1.2 million publicly available research samples annually, in both developed and developing countries.[11] Over 200,000 new samples collected from natural environments in all geographical regions of the world are still deposited each year in these collections.[12]

Today, some six hundred of the most qualified public collections operate in 68 countries under the mantle of the WFCC and its agreed quality and security standards.[13] We describe the major holdings and functions of the WFCC and its governance structure in greater detail later in this volume.[14] Of particular importance here is the fact that member collections are obliged to authenticate all their sources and to track all uses in order to avoid flawed research outcomes, unauthorized uses, or misappropriation,[15] and also to comply with the dictates of biological security measures.[16]

In recent years, global collaboration between WFCC culture collections has expanded to include public databases containing information concerning the source of specimens held, scientific publications and patents related to the content of their collections, and linkages to related genomic data.[17] Culture collections offering the

[11] *See* Dedeurwaerdere, Broggiato & Manou (2013), above n. 6, at 227. For quality standards and the governance structure of the WFCC, *see* Chapter 4, Section I.A. and Chapter 9, Section II.B. For licensing practices, see Chapter 4, Section II.

[12] Dedeurwaerdere, Broggiato & Manou (2013), n. 6, at 227.

[13] Scott Stern, Biological Resource Centers: Knowledge Hubs for the Life Sciences 11, 14 (Brookings Inst. Press 2004); David Smith, Dagmar Fritze, Erko Stackebrandt, *Public Service Collections and Biological Resource Centers of Microorganisms, in* The Prokaryotes – Prokaryotic Biology and Symbiotic Associations (E. Rosenberg et al. eds., Springer, 4th ed., 2013), Chapter 11 at 370. *See also About WFCC*, WFCC, http://www.wfcc.info/about/ (last accessed 1 Oct. 2014). *See further* Chapter 4, Section II.A–B.

[14] For general characteristics of the WFCC, see Chapter 4, Section I.A; for the World Data Center for Microbiology (WDCM) and its new digital portal, see Chapter 8, Section II.B, and for its governance framework, see Chapter 9, Section II.B.

[15] *See* Stern, n. 13; WFCC, World Federation For Culture Collections Information Document on Access to *ex-situ* Microbial Genetic Resources within the Framework of the Convention on Biological Diversity (1 Sept. 1996), *available at* http://www.wfcc.info/index.php/wfcc_library/genetic_res/. *See also* Dagmar Fritze, *A Common Basis for Facilitated, Legitimate Exchange of Biological Materials, Proposed by the European Culture Collections' Organization* (ECCO), 4 *Int'l J. Commons* 507, 512 (2010) [hereinafter Fritze (2010)].

[16] *See, e.g.*, WTO Agreement on the Application of Sanitary and Phytosanitary Measures [SPS Agreement], opened for signature 15 Apr. 1994, 1867 U.N.T.S. 493 (entered into force 1 Jan. 1995), http://www.wto.org/english/docs_e/legal_e/15-sps.pdf (setting the framework for trade and rules for managing risks from diseases); Agreement on Technical Barriers to Trade, *opened for signature* 15 Apr. 1994, 1868 U.N.T.S. 120 (entered into force 1 Jan. 1995) (ensuring that such regulations, standards, testing and certification procedures do not create unnecessary obstacles); Convention on International Trade in Endangered Species of Wild Animal and Fauna, *opened for signature* 3 Mar. 1973, *as amended on* 22 June 1979, 993 U.N.T.S. 244 (entered into force 1 July 1975), available at https://treaties.un.org/doc/publication/unts/volume%20993/volume-993-i-14537-english.pdf (last accessed 1 Oct. 2014); Cartagena Protocol on Biosafety, *adopted on* 29 Jan. 2000, 2226 U.N.T.S. 208, 39 I.L.M. 1027 (entered into force on 11 Sept. 2003).

[17] *See* Chapter 8, Section II.B (describing the World Data Center for Microorganisms (WDCM) and the StrainInfo Bioportal). See further D. Smith et al. (2014), n. 13, at 288–89, 295–97.

most advanced technical services are now often designated as Biological Resource Centers, a concept discussed in Chapter 4.[18]

The public culture collections at the national and regional levels have thus played a crucial role in conserving *ex situ* microbial genetic resources, and they have greatly facilitated access to, and distribution of, such materials for purposes of research and development. While serving providers, users, regulatory bodies, and policymakers, they also add value to the deposited biological material (and thus to the corresponding research initiatives) by means of the internal services they provide, as explained in Chapter 4. Due to their limited storage capacity, however, and to the high operating costs of these services, the holdings of the public culture collections represent only a small subset of the total holdings in many other academic research and private working collections. For the public culture collections to function as a cost-effective component of the basic infrastructure for life science research, they must accordingly accept mostly selected materials that have elicited particular interest for present or future research.[19]

As a result, the bulk of *ex situ* microbial materials potentially available for research purposes are not found in public-service cultural collections at all. Rather, vast amounts of publicly undocumented strains are still held in informal or working collections at universities, research institutes, government departments, hospitals, and industrial laboratories all over the world.[20] Most of these microbial materials are of as yet unknown scientific value, although they also include strains that are the object of ongoing research not yet published or that are kept for future follow-on research. These materials are often shared on a confidential basis among academic researchers and working collections, frequently without complying with the same stringent quality management protocols of the formal public culture collections.[21] In what follows, we refer to these practices as a system of informal exchanges, and we discuss the legal and institutional implications of this important phenomenon at length in Chapter 5.[22]

[18] *See* Stern, n. 13; Chapter 4, Section I.B.3.

[19] Stern, n. 13.

[20] *See generally id.*; D. Smith (2012), above n. 7, at 75–76; *see also* Chapter 5, Section I.A.

[21] Informal exchanges tend to be performed on a peer-to-peer basis among researchers and research institutions with small collections. Jeffrey L. Furman & Scott Stern, *Climbing Atop the Shoulders of Giants* 8–9 (Nat'l Bureau Econ. Research Working Paper No. 12523, Sept. 2006), *available at* http://www.nber.org/papers/w12523.pdf (last accessed 1 Oct. 2014). These exchanges typically had low transaction costs, but may have posed greater uncertainties owing to differing quality standards of the research institutes involved, their unharmonized record keeping practices, and above all, to the risk of cross contamination of specimens that has periodically plagued microbiological research. *See, e.g.*, Stern, n. 13, at 1–2, 12–13.

[22] *See* Chapter 5, Section I.A.3.

Since the 1960s, the combined progress of both *in vitro* cell culture technology and then *in silico* molecular biology has led to a tremendous increase both in the quantities of biological resources exchanged through both formal and informal arrangements and in the resulting global research interdependencies.[23] In recent years, more than five hundred thousand microbial samples, collected from various countries, have been exchanged annually throughout the world by public culture collections alone, mostly at marginal costs of distribution.[24] The vast amount of materials exchanged informally among the thousands of non-WFCC collections are not tracked and, therefore, not quantifiable. This aggregated system of both formal and informal exchanges is integrally related to such research advances as the introduction of ever more sophisticated techniques for storing, freezing, and shipping samples; to the genomics revolution; and to the broader impact of globalization on the organization of research in the life sciences.[25]

Until the 1980s, most WFCC collections still routinely exchanged materials among themselves on a relatively informal basis, while carefully tracking the distribution of materials to end users. This practice was rooted in mutual trust, and avoided the need to negotiate ad hoc material transfer agreements (MTAs), which helped to keep transaction costs relatively low.[26] One underlying assumption was that materials exchanged for upstream research purposes were typically of little or no commercial interest as distinct from materials distributed directly to industry. A second, tacit assumption was that *ex situ* genetic resources in general resided in the public domain or were, in effect, the "common heritage of mankind."[27]

In the last quarter of the twentieth century, however, both of these assumptions were severely challenged by adverse legal, political, and institutional developments.

[23] Bronwyn Parry, Trading the Genome 177 (Columbia U. Press 2004); Stern, n. 13.

[24] *See, e.g.*, Tom Dedeurwaerdere et al., *The Use and Exchange of Microbial Genetic Resources for Food and Agriculture* (Comm'n on Genetic Res. Food & Agric., Background Study Paper No. 46, U.N. Doc. UNEP/CBD/WG-ABS/9/INF/13, 9 Mar. 2009), *available at* http://www.cbd.int/doc/meetings/abs/abswg-09/information/abswg-09-inf-13-en.pdf (last accessed 1 Oct. 2014).

[25] *See, e.g.*, Stephen J. McCormack, *Industrial Perspective: Development of an MTA with a Microbial Research Commons, in* Designing the Microbial Research Commons, n. 5, at 25; Daniel Drell, *Research and Applications in Energy and the Environment, in* Designing the Microbial Research Commons, n. 5, at 12. *See further* Chapter 4, Section I. *passim* (contrasting and comparing these early exchanges between qualified public culture collections with the system of informal exchanges among researchers, discussed at length in Chapter 5).

[26] Tom Dedeurwaerdere, *Institutionalizing Global Genetic Resource Commons: Towards Alternative Models for Facilitating Access in the Global Biodiversity Regime* (2010) (unpublished manuscript), *available at* http://ssrn.com/abstract=1611549 (last visited 1 Oct. 2014). *See also* Tom Dedeurwaerdere, *Microbial Commons: Overview of the Governance Considerations – A Framework for Discussion, in* Designing the Microbial Research Commons, n. 5. *See* further Chapter 4.

[27] *See, e.g.*, Fritze (2010), n. 15, at 516–18 (describing the "common heritage" principle and how it may no longer apply to microbial cultures). *Cf.* below Section B.2 (FAO's "common heritage" treatment of plant genetic resources).

The norms and practices pertaining to exchanges of microbial materials – at both the national and transnational levels – have been changing rapidly as a result. For example, the assertion of globally enforceable intellectual property rights on microbial-related innovation emanating mainly, but not exclusively, from developed countries has discouraged holders of *ex situ* microbial genetic resources from making them available for unrestricted public scientific research.[28] In this environment, biological materials exchanged by the collections for any purpose are increasingly perceived as potentially valuable commodities in their own right,[29] and ever more cumbersome legal restrictions on use and reuse have consequently been introduced.[30]

Meanwhile, the developing countries have aggressively challenged the legality of cross-border exchanges of *ex situ* genetic resources taken from their territories under the public-domain rationale. These countries collectively argued that unauthorized use of genetic resources originating in their territories without permission and without the sharing of benefits from commercial applications was a form of "biopiracy" that directly violated their sovereign rights to all natural resources located within their territorial boundaries.[31]

In 1992, the Convention on Biological Diversity (CBD) officially enshrined this view in public international law.[32] Since then, the developing countries have demanded that national patent offices everywhere adopt examination procedures and other regulatory measures to deter the misappropriation of their genetic resources.[33] As will be seen in the next chapter, however, in seeking to implement these demands the drafters of the CBD ignored the need to preserve the preexisting research infrastructure based on formal and informal exchanges of microbial genetic

[28] *See* Section II.A. *Cf.* Subha Ghosh, *How to Build a Commons: Is Intellectual Property Constrictive, Facilitating, or Irrelevant?*, *in* UNDERSTANDING KNOWLEDGE AS A COMMONS 209 (C. Hess & E. Ostrom eds., MIT Press 2007); James Boyle, *The Second Enclosure Movement and the Construction of the Public Domain*, 66 LAW & CONTEMP. PROBS. 33 (2003).

[29] *See, e.g.*, GRAHAM DUTFIELD, INTELLECTUAL PROPERTY, BIOGENETIC RESOURCES AND TRADITIONAL KNOWLEDGE 3–5 (Routledge 2004) [hereinafter DUTFIELD (2004)]; Robin J. R. Blatt, *Banking Biological Collections: Data Warehousing, Data Mining, and Data Dilemmas in Genomics and Global Health Policy*, 3(4) CMTY. GENETICS 204 (2000); KENNETH I. BERNS, RESOURCE SHARING IN BIOMEDICAL RESEARCH (Nat'l Acads. Press 1996).

[30] For details, see generally Chapter 4, Sections I and II.

[31] DUTFIELD (2004), n. 29, at 3–11, 52; *see further* in Section III; Chapter 3, Section I.

[32] CBD, n. 1.

[33] *See, e.g.*, Catherine Saez, *Developing Countries Urged to Beat Biopiracy with Patent Examination, Regulatory Frameworks*, IP WATCH (Feb. 7, 2014), http://www.ip-watch.org/2014/02/07/developing-countries-urged-to-beat-biopiracy-with-patent-examination-regulatory-frameworks/; Catherine Saez, *New WIPO Text on Genetic Resources Misappropriation: Disclosure Still Uncertain*, IP WATCH (Feb. 6, 2014), http://www.ip-watch.org/2014/02/06/new-wipo-text-on-genetic-resources-focuses-on-misappropriation-disclosure-still-uncertain/.

resources having no known or likely commercial value other than as subjects of public scientific research.[34]

2. An Ancillary Research Commons for Influenza Viruses

Meanwhile, another major component of the basic institutional infrastructure governing exchanges of microbial genetic resources was the Global Influenza Surveillance and Response System (GISRS), established by the World Health Organization (WHO) in 1957.[35] Global influenza pandemics killed between fifty and one hundred million people in the period 1918–1920, and millions more in the 1950s and 1960s.[36] The development of vaccines and other medical treatments "critically depends on access to the original virus, not only for research but more importantly, as direct input for vaccines in the form of dead organisms."[37] The WHO's GISRS was accordingly devised to enable countries to coordinate their surveillance efforts for seasonal influenza epidemics, and it operated successfully for more than sixty years.[38]

Under the GISRS, national influenza centers (NICs) would submit local virus samples collected from hospitals, clinics, and other laboratories to WHO collaborating centers[39] for monitoring and research purposes. The collaborating centers would use these samples to develop diagnostic kits and to identify candidate viruses that could be suitable for the development of vaccines. The collaborating centers would further provide the WHO with relevant epidemiological information, which all other participating laboratories (NICs and WHO laboratories) could access.[40] The Global Influenza Surveillance Network (GISN) was thus

> a network of national and WHO laboratories that cooperated to monitor the spread of seasonal influenza and to develop appropriate responses. Where relevant the GISN system would also be used for monitoring influenza viruses with pandemic potential, though the primary objective was the coordination of responses to seasonal influenzas.[41]

[34] *See further* Chapter 3, Sections I.C & IV.

[35] *See WHO Global Influenza Surveillance and Response System (GISRS)*, World Health Org., http://www.who.int./gho/epidemic_diseases/influenza/virological_surveillance/en/index.html (last accessed 2 Oct. 2014).

[36] *See, e.g.*, Marie Wilke, *The World Health Organization's Pandemic Influenza Preparedness Framework as a Public Health Resource Pool, in* Common Pools of Genetic Resources (2013), n. 6, at 315–43.

[37] *Id.* at 315.

[38] *Id.* at 316.

[39] *Id.* at 316–17. These centers were situated in Australia, China, Japan, the United Kingdom, and the United States.

[40] *Id.* at 317. At the time of writing, Marie Wilke headed the international trade law program at the International Center for Trade and Sustainable Development (ICTSD).

[41] *Id.*

However, as Professor Peter Yu has recently explained, the collaborative efforts of the WHO's GISN were first challenged by the appearance of the "highly contagious and panic inducing" coronavirus responsible for SARS in the spring of 2003.[42] Although the WHO promptly established a Multi-Center Collaborative Network on SARS Aetiology and Diagnosis, the appearance of the SARS coronavirus sparked a race to discover its genome, and a second race to patent the isolated gene sequences associated with the virus.[43]

According to Professor Yu, in April 2003, the United States Centers for Disease Control and Prevention (CDC), the BCCA in Canada, and Versitec, Ltd. (the commercial arm of Hong Kong University) were all competing to patent technology pertaining to the isolated sequences of the SARS coronavirus.[44] A private firm also filed patent applications claiming ownership of the key isolated sequences of two SARS genes thought to control reproduction of the virus in its infectious state.[45]

This race to patent the SARS virus raised public health concerns about possible blocking effects on future research, and about the hoarding of information needed to protect the public in an eventual SARS crisis. Both the CDC and BCCA, publicly funded agencies in the United States and Canada respectively, justified their patent applications as a defensive ploy to ensure that the scientific and medical communities retained open access to the virus for research purposes.[46] However, there were no guarantees that the winner of the race – apparently the CDC – would share either their findings or their royalties with institutions in other countries. Questions also arose about attribution and reputational benefits from the CDC's claims to have identified a new coronavirus as the likely source of SARS.[47]

The SARS outbreak eventually subsided, and the relevant patent applicants agreed to collaborate with each other through the establishment of a patent pool. As Professor Yu observes, patent pools have become a useful tool to combat patent thickets, especially in the public health sphere.[48] Nevertheless, the practice

[42] Peter K. Yu, *Virotech Patents, Viropiracy, and Viral Sovereignty*, 45 Ariz. St. L.J. 1563, 1589 (2014). Over 8,000 people from 29 countries became ill from this virus and about 774 died. *Id.* at 1590 (citing authorities). *See also* Dana Beldiman, *Patent Chokepoints in the Influenza-Related Medicines Industry: Can Patent Pools Provide Balanced Access?*, 15 *Tul. J. Tech. & Intell. Prop.* 31 (2012), *available at* http://papers.ssrn.com/s013/papers.cfm?abstract_id=2049035.

[43] Yu (2014), n. 42, at 1591.

[44] *Id.* at 159–192.

[45] *Id.* at 592 (discussing the firm Combimatrix, a subsidiary of Acacia Research Corp., and citing Paul Elias, *Race to Patent SARS Virus Renews Debate*, Assoc. Press (May 5, 2003).

[46] Yu (2014), n. 42.

[47] *Id.* at 1596; *see also* E. Richard Gold, *SARS Genome Patent: Symptom or Disease*, 361 Lancet 2002 (2003) (observing the need for defensive patenting in that case to preserve the public domain).

[48] Yu (2014), n. 42, at 1602 (discussing the Medicines Patent Pool initiative to promote affordable treatment for HIV infections in poor countries); *see also* Frederick Abbott & Graham Dukes, Global Pharmaceutical Policy: Ensuring Medicines for Tomorrow's World 29 (2011). *But see* Geertrui Van Overwalle, *Of Thickets, Blocks and Gaps: Designing Tools to Resolve Obstacles in the Gene*

of patenting viruses essential to research on influenza pandemics once established, fueled both the hoarding instincts of scientists (and their commercial sponsors) in the developed world and growing claims of sovereignty from holders of microbial genetic resources in developing countries.[49]

When the next influenza crisis struck, with the arrival of the H5N1 strain of avian influenza ("bird flu") in 2007, these privatizing tendencies overwhelmed the WHO's existing research commons, the GISN, and threatened its very existence. As explained below in Section III.A, the WHO ultimately responded to the crisis by forming a new multilateral research commons, to be known as the Pandemic Influenza Preparedness (PIP) Framework which deliberately aimed to comply with the dictates of the Convention on Biological Diversity.[50]

However, to fully appreciate the political and institutional nuances of the WHO's newest pandemic influenza initiative, one must first take account of the contemporaneous controversy surrounding plant genetic resources in this same period and of the international legal infrastructure that controversy ultimately generated. We turn now to this topic, which is of central importance to the overall thesis of this book.

B. *Early Efforts to Form an Agricultural Research Commons for Plant Genetic Resources*

Crop domestication began about 12,000 years ago, and it moved rapidly across and between continents. Colonialism and imperial trade in the 1500s and 1600s accelerated the diffusion of cultivars, to the point where people were eating much the same staples all over the world.[51] This relative interdependence of common food sources has, however, conditioned food security on the availability of both basic plant cultivars and microbial cultures from diverse parts of the world, which are needed to treat new diseases afflicting both plants and animals.[52]

Patents Landscape, *in* GENE PATENTS AND COLLABORATIVE LICENSING MODELS: PATENT POOLS, CLEARING HOUSES, OPEN SOURCE MODELS AND LIABILITY REGIMES (G. Van Overwalle ed. 2009) (noting disadvantages of patent pools); Diane Nicol & Jane Nielsen, *Opening the Dam: Patent Pools, Innovation and Access to Essential Medicines*, *in* INCENTIVES FOR GLOBAL PUBLIC HEALTH: PATENT LAW AND ACCESS TO ESSENTIAL MEDICINES 237, 253–57 (T. Pogge et al. eds. 2010) (expanding on these disadvantages).

49 *See, e.g.*, Yu (2013), n. 42, at 27–37; Beldiman, n. 42, at 38–43.

50 WORLD HEALTH ORG., PANDEMIC INFLUENZA PREPAREDNESS FRAMEWORK FOR THE SHARING OF THE INFLUENZA VIRUSES AND ACCESS TO VACCINES AND OTHER BENEFITS, World Health Assembly Res. WHA645 (May 24, 2011) [hereinafter PIP FRAMEWORK], *available at* http://www.who.int/influenza/resources/pip_framework/en/index.html (last accessed 2 Oct. 2014). For details, see Chapter 4, Section IV.A.

51 1 Fernand Braudel, *The Structure of Everyday Life*, *in* CIVILIZATION AND CAPITALIZATION, 15TH–18TH CENTURY *passim* (2d prtg. 1992).

52 Maria Jose Amstalden Sampaio, Microbial Genetic Resources for Food and Agriculture – Interdependence, paper presented at the Biodiversity International Side Event, Conference of

Plant genetic resources for food and agriculture (PGRFA)[53] thus traditionally depended on both natural and artificial selection carried out by farming communities, who domesticated plant species and adapted them to the changing needs of farmers and consumers. Migration, trade, and colonization further spread cross-border exchanges and also stimulated the processes of selection.[54]

Since the mid-nineteenth century, professional seed suppliers, followed by specialized plant breeders and biotechnologists, have developed even more advanced methods for selecting plant genetic resources based on phenotypes, genotypes, and molecular biology. Although large corporations increasingly dominate the commercial seed market for some major and high value crops, such as maize and certain vegetables, small and medium-sized firms continue to operate in niche markets. At the same time, basic research remains largely centered at universities, which also interact with plant breeders and seed producers, both large and small.[55]

1. Emergence of an International Consortium for the Preservation and Improvement of Cultivars Essential for Food Security

As breeders improve plant varieties to adapt to local conditions and population growth, hybrids emerge that become subject to new diseases, climate fluctuations, and other stresses over time. Plants may become weaker in some respects as they become stronger in others (a process known as convergence). This challenge requires constant resort by scientists to basic food stocks and basic genetic materials that are held in public repositories all over the world.[56] Increasingly, these research efforts combine newer genetic materials with old varieties, land uses, and wild crop relatives that have been preserved and made available by these repositories.[57] Exchanges of germplasm from seed banks, gene banks, and other public repositories thus "amount to several tens of thousands of transfers annually and play … an

the Parties, CBD Access and Benefit Sharing (22 Mar. 2010), *available at* http://www.cbd.int/abs/side-events/abs-9/id1690-embrapa.pdf (last accessed 2 Oct. 2014).

53 For reasons of space, time, and relevance, this study does not focus on other important types of genetic resources for food and agriculture, such as animal (AnGR), aquatic (AgGR), and forest (FGR). Genetic resources for biological control (BC) in pest management are also not directly covered, except insofar as they overlap with PGRFA and microbial genetic resources. *See, e.g.*, Sélim Louafi & Marie Schloen, *Practices of Exchanging and Utilizing Genetic Resources for Food and Agriculture and the Access and Benefit Sharing Regime, in* Common Pools of Genetic Resources (2013), n. 6, at 193–212. Looking ahead, nevertheless, we note that all these genetic resources – plant, animal, and microbes – fall within the ambit of the Convention on Biological Diversity of 1992. *See* Christine Godt, *Networks of* Ex Situ *Collections of Genetic Resources, in* Common Pools of Genetic Resources (2013), n. 6, at 248 (noting that CBD, n. 1, art. 2, excludes only human tissue and blood).

54 Louafi & Schloen, n. 53, at 198–99.

55 *Id.* at 199. *See also* Dutfield (2004), n. 29, at 11–24.

56 Amstalden Sampaio, n. 52.

57 Louafi & Schloen (2013), n. 53, at 199.

important role in conservation and research and development, both in developing and developed countries."[58]

Beginning in the early 1970s, efforts were made to link major research holdings of basic plant cultivars in different countries within a federated research infrastructure to be known as the Consultative Group on International Agricultural Research (CGIAR).[59] This entity, founded in 1971 by the Rockefeller Foundation, holds large collections of *ex situ* essential plant genetic resources available to the public, and it also conducts research on the breeding of new plant varieties.[60]

When first established, the CGIAR was a loosely organized network of autonomous research institutes that depended on multiple donors from the public sector and nonprofit organizations, especially the Rockefeller Foundation, the World Bank, and FAO, among others. The member centers were managed informally, on a mostly consensus basis until 2010, and they received independent scientific advice from a Technical Advisory Committee (TAC). In 1974, the CGIAR established an oversight body – the International Board of Plant Genetic Resources (IBPGR) – which is now known as Bioversity International. This entity has its own budget and Secretariat and closely coordinates its activities with the United Nations Food and Agriculture Organization's Plant Genetic Resource Unit.[61]

In its third decade (1991–2000), the CGIAR's mission statement was expanded as follows:

> Through international research and related activities, and in partnership with national research systems, ... [it will] contribute to sustainable improvements in the productivity of agriculture, forestry, fisheries in developing countries in ways that enhance nutrition and well-being, especially of low-income people.[62]

[58] *Id.* at 200. However, breeding pools within a single region account for most uses of genetic materials for breeding and variety development, while new "exotic" materials are less frequently accessed. *Id.*

[59] Derek Byerlee & Harvey J. Dubin, *Crop Improvement in the CGIAR as a Global Success Story of Open Access and International Collaboration*, 4 *Int'l J. Commons* 452, 456–57 (U. of Mich. Press 2010). *See generally* Elinor Ostrom et al., Rules, Games, and Common-Pool Resources (1994); Michael Halewood, *Governing the Management and Use of Pooled Microbial Genetic Resources: Lessons from the Global Crop Commons*, 4 *Int'l J. Commons* 404–36 (2010) [hereinafter Halewood (2010)].

[60] The collections held by the CGIAR gene banks "are among the largest in the world and arguably the most important for the livelihoods of the poor and global food security." The CGIAR invests $6 million annually to maintain these resources as global public goods. CGIAR, *Crop Genebank Knowledge Base*, http://cropgenebank.sgrp.cgiar.org [hereinafter CGIAR, *Crop Genebank*] (last accessed 23 Dec. 2014).

[61] *History*, Bioversity International, http://www.bioversityinternational.org/about-us/who-we-are/history/ (last accessed 6 Oct. 2014). *See further* Bioversity Int'l, *Constitution*, *available at* http://www.bioversityinternational.org/fileadmin/user_upload/about_us/Governance/IPGRIConstitution.pdf (last accessed 3 Oct. 2014).

[62] Technical Advisory Comm'n (TAC) to the Consultative Group on Int'l Agric. Research (CGIAR), Review of CGIAR Priorities and Strategies Ch. 2 (1992), *available at*

A later amendment added the duty "to contribute through its research, to promoting sustainable agriculture for food security in developing countries."[63] In pursuit of this mission, the CGIAR formulated its own policies through a coordinating body known as the Genetic Resources Policy Committee (established in 1994), and it recently transformed the TAC into a full-fledged Science Council.[64]

Over time, the CGIAR enlarged the range of cultivars gathered from all parts of the world and organized them as a secure common pool resource that greatly facilitated cross-border exchanges for research bearing on food and agriculture.[65] To this end, the CGIAR supports a network of some sixteen International Agricultural Research Centers (IARCs), including the International Plant Genetic Resources Institute (IPGRI based in Rome), the International Rice Research Institute (IRRI, based in the Philippines), the International Maize and Wheat Improvement Center (CIMMYT, based in Mexico), and the International Center for Tropical Agriculture (CIAT, Mexico).[66]

The international collections conserved and hosted by the CGIAR centers were built up over decades and, as of 2011, they held about 750,000 accessions of crops and forages originating from over 100 countries.[67] The CGIAR's plant genetic resources are made freely available to researchers anywhere in the world. Disregarding transfers between the centers themselves, more than 80 percent of the materials distributed went to developing countries and countries in transition. The bulk of these transfers – benefit public research organizations, universities, regional organizations, germplasm networks, and gene banks.[68]

http://www.fao.org/wairdocs/tac/x5756e/x5756e05.htm. Over time the CGIAR affiliated twenty-five research institutes in the Southern Hemisphere and twenty-seven in the Northern Hemisphere.

63 CGIAR, Nourishing the Future Through Scientific Excellence: Annual Report 1997 1 (1997).

64 *See* Science Council Secretariat, *Key Achievements in 2003*, CGIAR, http://www.cgiar.org/web-archives/www-cgiar-org-soar-2003-2003 sc-html (last accessed 6 Oct. 2014)

65 *See generally* Dan Morgan Merchants of Grain: The Power and Profits of the Five Giant Companies at the Center of the World's Food Supply (1 Universe 1979); Stephen B. Brush, *Bio-Cooperation and the Benefits of Crop Genetic Resources: The Case of Mexican Maize*, 26 *World Dev.* 755 (1998). *See also* Dutfield (2004), n. 29, at 7–8 (stating that CGIAR's mission is to contribute to food security and poverty eradication in developing countries through research, partnerships, capacity building, and policy support, promoting sustainable agriculture development "based on the environmentally sound management of natural resources").

66 Dutfield (2004), n. 29, at 8. For more detailed information about the CGIAR, see Section II and Chapter 9, Section II.A.2 (Implementation of the Multilateral Regime).

67 Michael Halewood et al., *Changing Rates of Acquisition of Plant Genetic Resources by International Gene Banks*, *in* Crop Genetic Resources as a Global Commons – Challenges in International Law and Governance 99 (M. Halewood et al., Routledge 2013) [hereinafter Crop Genetic Resources as a Global Commons (2013)].

68 *Id.* at 99. Because "[m]ost countries do not have the resources to assemble and maintain collections of similar size and diversity ... they need to rely on access to these international collections." *Id.*

The genetic resources held in CGIAR's affiliated collections supply an important source of biodiversity for farmers, plant breeders, and researchers seeking to develop crops and forages that are capable of resisting pests and diseases, of withstanding climate stresses, and of growing in degraded soils. Access to such biological diversity is expected to become even more important as countries strive to meet the challenges of climate change.[69]

Despite its outstanding record of research and applications, the CGIAR entered a period of crisis toward the end of the twentieth century. One prong of the crisis was financial in nature. It obliged the group to move beyond informal management "by consensus" and to adopt "legally binding funding and performance agreements" that would produce a formally constituted Consortium and Fund Council.[70] The second and more lasting prong of the crisis was the mounting impact of proprietary claims to plant genetic resources emanating from both developed and developing countries in the last quarter of the twentieth century, as described later in this chapter.[71]

2. Short-Lived Recognition of Plant Genetic Resources as the Common Heritage of Mankind

From a legal perspective, the operating assumption of the CGIAR's Agricultural Research Centers was that all their *ex situ* genetic resources belonged to the public domain.[72] With the advent of international plant breeders' rights in the 1960s, however, followed by patents on living matter,[73] and then on biological products of genetic research,[74] questions were raised about who actually owned samples of germplasm taken from tropical and subtropical regions in developing countries that ended up in *ex situ* collections often in the developed countries.[75] Spurred by Bioversity International, and by growing concerns among developing country members of FAO about the assertion of intellectual property rights in plant genetic resources, intense negotiations were held in the early 1980s, with a view to clarifying

69 *Id.*

70 CGIAR, *CGIAR Reform*, http://www.cgiar.org/who-we-are/history-of-cgiar/cgiar-reform/ (last accessed 9 Oct. 2014); CGIAR, *Who We Are*, http://www.cgiar.org/who-we-are/cgiar-fund/ (last accessed 1 July 2014).

71 *See further* Chapter 3, Section II ("Destabilizing the Exchange of Plant Genetic Resources as Global Public Goods").

72 *See, e.g.*, José Esquinas-Alcázar et al., *A Brief History of the Negotiations for the International Treaty on Plant Genetic Resources for Food and Agriculture*, *in* CROP GENETIC RESOURCES AS A GLOBAL COMMONS (2013), n. 67, at 136–39 [hereinafter Esquinas-Alcázar et al. (2013)].

73 *See. e.g.*, Diamond v. Chakrabarty, 447 U.S. 303 (1980).

74 *See further* Sections II and III. *See generally* Sabrina Safrin, *Hyperownership in a Time of Biotechnology Promises: The International Conflict to Control the Building Blocks of Life*, 98 Am. J. Int'l L. 641 (2004).

75 Esquinas-Alcázar et al. (2013), n. 72, at 137.

the legal status of gene banks in general, particularly the CGIAR's own *ex situ* holdings.[76]

In a tense atmosphere, with the CGIAR beginning to encounter serious financial and political difficulties, the FAO took steps to develop the first multilateral legal instrument to support its efforts to maintain basic stocks of plant genetic resources for food and agriculture as a global public good. Known as the International Undertaking on Plant Genetic Resources of 1983, this idealistic and nonbinding arrangement[77] sought to establish the CGIAR on a sound legal footing as an internationally coordinated network of national, regional, and international centers "under the auspices or the jurisdiction of the FAO."[78] Governments, seed banks, and other institutions having plant genetic resources under their control were "to allow access to samples of such resources, and to permit their export, where the resources have been requested for the purposes of scientific research, plant breeding or genetic resource conservation."[79] The CGIAR, in turn, was "to hold, for benefit

[76] *See id.* at 136–39. The International Bureau for Genetic Resources (IBGR) became part of the International Program of the CGIARs and has its headquarters at the FAO in Rome. Its creation followed a recommendation of the UN Conference on the Human Environment, held in Stockholm in 1972, and of the CGIAR's own Technical Advisory Committee. *Id.* Biodiversity International has played an important role in the negotiations leading to the multilateral actions discussed in this Chapter and Chapter 3.

[77] International Undertaking on Plant Genetic Resources, FAO Res. 8/83, 22d Sess. (5–23 Nov. 1983) [hereinafter FAO International Undertaking]. The Undertaking defined "plant genetic resources" for the first time as "the reproductive or vegetative propagating material" used in plant breeding for food and agriculture. *Id.* at 2.1. Specifically, it covered the following:

i. Cultivated varieties (cultivars) in current use and newly developed varieties;
ii. Obsolete cultivars;
iii. Primitive cultivars (land races);
iv. Wild and weed species, near relatives of cultivated varieties;
v. Special genetic stocks (including elite and current breeders' line and mutants).

Article 2.2 states that the Understanding "relates to the plant genetic resources described in par. 2.1(a), of all species of economic and/or social interest, particularly for agriculture at present or in the future, and has particular reference to food crops." *Id.* art. 2.2. The core focus was on "reproductive or vegetative propagating material," which is basically a reference to seeds used as breeding material for conventional breeding methods. Because the International Undertaking did not refer to "functional units of hereditary ... [of] potential or actual value," a term later be used in the CBD, above n. 1, it arguably applied a narrow concept of plant genetic resources pertaining to food security. Morten Walløe Tvedt & Oliver Rukundo, *Functionality of the ABS Protocol*, UNEP/CBD/WG-ABS/9/INF/20 (26 Aug. 2010), *available at* https://www.cbd.int/doc/meetings/abs/abswg-09-3rd/information/abswg-09-3rd-inf-20-en.pdf (last accessed 14 June 2014).

[78] FAO International Undertaking, above n. 77, art. 7.

[79] *Id.* art. 5 (adding that samples were to "be made available free of charge on the basis of mutual exchange or mutually agreed terms"). In this respect, the International Undertaking was largely meant to become a global management scheme for *ex situ* collections under FAO auspices. *See, e.g.*, Esquinas-Alcázar et al. (2013), above n. 72, at 138, Box 6.1.

of the international community," its plant genetic resources "on the principle of unrestricted exchange," that is to say, as the "common heritage of mankind."[80]

The International Undertaking thus stressed the virtues of a global crop commons precisely at the moment when the developed countries were about to push for a strengthened multilateral regime of intellectual property rights at the World Intellectual Property Organization (WIPO) (and ultimately at the World Trade Organization (WTO)). Meanwhile, the developing countries, alarmed by growing applications of intellectual property rights to plant genetic resources taken without permission from their own countries,[81] were already asserting strong claims of national sovereignty over these same resources in negotiations that led to the Convention on Biological Diversity of 1992.[82] Not surprisingly, the grand aspirations embodied in the International Understanding quickly succumbed to the Access and Benefit Sharing demands of provider countries embodied in the CBD, as explained below in Chapter 3.[83]

II. IMPINGING INTELLECTUAL PROPERTY RIGHTS PROMOTED BY THE DEVELOPED COUNTRIES

By 1994, the annual market for products derived from all genetic resources was estimated to be worth between $500 and $800 billion.[84] In the rest of this chapter, we explain how the prospects of financial gain from commercial applications of plant and microbial materials has led to a proliferation of intellectual property rights at both the national and international levels, with mounting restrictions on the availability and use of genetic resources for upstream research purposes.[85] These restrictions are further intensified by growing concerns about biosafety and biosecurity, primarily in the developed world,[86] and by increasingly successful efforts of developing country

[80] FAO International Undertaking, above n. 77, arts. 1, 7. *See also* Safrin, above n. 74. For other examples of this concept, *cf.* Antarctic Treaty, 1 Dec. 1959, 402 U.N.T.S. 71; Treaty on Principles Governing the Activities of States in the Exploration and Use of Outer Space, including the Moon and Other Celestial Bodies, 27 Jan. 1967, 610 U.N.T.S. 205; United Nations Convention on the Law of the Sea, 10 Dec. 1982, 1833 U.N.T.S. 396. Although the United States signed all three treaties, it formally ratified only the first two.

[81] For the role of patents and plant breeders' rights, see below Section II.A.

[82] CBD, above n. 1. *See further* Sections II–III. and Chapter 3, Section I.

[83] *See* Chapter 3, Section I.B & IV (Nagoya Protocol).

[84] DUTFIELD (2004), above n. 29, at 18–19 (citing Kate and Laird [1999]). These returns derived from pharmaceuticals, botanical medicines, agricultural produce (including seeds), ornamental horticultural products, crop protection products, and biotechnologies in various fields. *Id.*

[85] *See, e.g.*, Sikina Jinnah & Stefan Jungcurt, *Could Access Requirements Stifle Your Research?*, 323 SCIENCE 464 (2009).

[86] *See, e.g.*, DAVID FIDLER & LAWRENCE GOSTIN, BIOSECURITY IN THE GLOBAL AGE: BIOLOGICAL WEAPONS, PUBLIC HEALTH, AND THE RULE OF LAW (2007); INST. MEDICINE, FORUM ON

governments to assert sovereignty over genetic resources originating from their respective territories.[87]

A. Sui Generis *Plant Breeders' Rights and Related Biotechnology Patents*

In the industrialized countries, plant patents were available only from a few national systems, notably that of the United States, but the standards of eligibility – novelty and especially nonobviousness – were traditionally set too high for most commercial breeders to qualify.[88] Resistance to plant patents was also rooted in the opposition to patents on living matter generally, especially in Europe.[89] In response to these obstacles, some countries introduced a new, *sui generis* form of protection for commercial breeders of plant varieties in the 1950s, which operated on a lower, more flexible standard of eligibility. This intellectual property regime was designed to stimulate and protect investments by professional plant breeders who employ scientific methods to develop high-yielding, genetically homogeneous, and stable varieties that are well-adapted to an industrialized agricultural model.[90]

1. Strengthened International Protection for Commercial Plant Breeders

Plant breeders' rights (PBRs) were embodied in the International Convention for the Protection of New Varieties of Plants in 1961,[91] known as "UPOV" because of

MICROBIAL THREAT SERIES (National Academies Press). However, as noted earlier, we do not address the biosafety and biosecurity aspects in this volume.

87 *See generally* Chapter 2 below.

88 *See, e.g., Ex parte* Hibberd, 227 U.S.P.Q. (BNA) 443 (1985); Bowman v. Monsanto Co., 133. S.Ct. 1761 (2013). *See generally* Jerome H. Reichman, *Legal Hybrids Between the Patent and Copyright Paradigms*, 94 *Colum. L. Rev.* 2432, 2467–721 (1994) [hereinafter Reichman, *Legal Hybrids* (1994)], *available at* http://scholarship.law.duke.edu/faculty_scholarship/97 (last accessed 12 June 2014). *See also* Jerome H. Reichman, *Of Green Tulips and Legal Kudzu: Repackaging Rights in Subpatentable Innovation*, 53 *Vand. L. Rev.* 1743 (2000) [hereinafter Reichman, *Green Tulips*], *available at* http://scholarship.law.duke.edu/faculty_scholarship/456 (last accessed 14 Oct.2014). Nevertheless, by the end of 2001, there were more than 1,800 U.S. patents with claims to plants, seeds, or plant parts or tissues. DUTFIELD (2004), above n. 29, at 23.

89 DUTFIELD (2004), above n. 29, at 22.

90 *See, e.g.,* JULIANA SANTILLI, AGROBIODIVERSITY AND THE LAW: REGULATING GENETIC RESOURCES, FOOD SECURITY AND CULTURAL DIVERSITY (Earthscan 2012) [hereinafter SANTILLI (2012)] (stressing that plant breeders ignore farmers' traditional breeding methods and assume that professional breeders are the primary innovators in agriculture). For the exclusion of classical plant breeding methods from patentability under art. 53(b) of the European Patent Convention, see European Patent Office, Enlarged Board of Appeals Cases G2/07 and G1/08 (9 Dec. 2010).

91 Technically, the International Convention for the Protection of New Varieties of Plants (the UPOV Convention) was adopted in November 1961. International Convention for the Protection of New

the French acronym for the International Union created by this Convention.[92] The UPOV Convention was subsequently amended in 1972, 1978, and 1991,[93] with the latter two versions of particular importance for present purposes. Professor Reichman has elsewhere characterized the differences between these versions in terms of a "copyright-like approach" confirmed in 1978 and a "patent-like approach" adopted in 1991.[94]

Under UPOV 1978, eligible propagating material had to be novel, stable, homogeneous, and distinct in at least one characteristic.[95] The protection then afforded by PBRs is phenotypical, not genotypical, in the sense that "the object of protection is the complete living organism, with its entire set of characteristics, some new, some old."[96] However, the protected variety, which is not "described" as in patent law, but is instead deposited for evaluation, may be freely accessed and used as a source of discovery for still other new and protectable varieties that

Varieties of Plants, *adopted on* 2 Dec. 1961, 815 U.N.T.S 89 (entered into force 10 Aug. 1968) [hereinafter UPOV 1961], *available at* http://www.upov.int/en/publications/conventions/1961/act1961.htm (last accessed 27 June 2014). It was revised in 1972, and, importantly, in 1978, which Act entered into force in 1981. International Convention for the Protection of New Varieties of Plants, *as revised on* 10 Nov. 1972 (entered into force 11 Feb. 1977) [hereinafter UPOV 1972], *available at* http://www.upov.int/en/publications/conventions/1961/act1972.htm (last accessed 19 Oct. 2014); International Convention for the Protection of New Varieties of Plants, *as revised on* 23 Oct. 1978, 33 U.S.T. 2703, 1861, U.N.T.S 281 (entered into force 8 Nov. 1981) [hereinafter UPOV 1978], *available at* http://www.upov.int/en/publications/conventions/1978/pdf/act1978.pdf (last accessed 19 Oct. 2014). For the 1991 revision, see International Convention for the Protection of New Varieties of Plants, *as revised on* 19 Mar. 1991 (entered into force 24 April 1998) [hereinafter UPOV 1991], *available at* http://www.upov.int/upovlex/en/conventions/1991/act1991.html (last accessed 19 Oct. 2014).

92 Union Internationale pour la Protection des Obtentions Végétades (International Union for the Protection of New Varieties of Plants).

93 *See* above n. 91 & accompanying text.

94 *See generally* Reichman, *Green Tulips* (2014), above n. 88.

95 *See, e.g.*, Santilli (2012), above n. 90, at 80. In effect, under UPOV, eligible varieties must be novel, distinct, stable, and "homogeneous" (1978) or uniform (the term under the 1991 Act that is now widely accepted), but there is no requirement of nonobviousness. *Cf.* Dutfield (2004), above n. 29, at 34 (listing UPOV's requirements). A plant variety was initially defined in 1978 as including "any variety, clone, line, stock or hybrid which is capable of cultivation," but Article 1 of the 1991 version has a more refined technical definition, viz.

"a plant grouping within a single botanical taxon of the latest known rank, which grouping … can be:

- defined by the expression of the characteristics resulting from a given genotype or combination of genotypes;
- distinguished from any other plant grouping by the expression of at least one of the said characteristics, and
- considered as a unit with regard to its suitability for being propagated unchanged."

UPOV 1991, above n. 91, art. 1(4)(iv).

96 Santilli (2012), above n. 90, at 79.

vary from prior varieties in at least one major trait.[97] In this respect, the relatively weak, copyright-like approach of the 1978 Act prohibited wholesale duplication but not derivative varieties; it did not prevent farmers' use of seeds; and it allowed even protected varieties to be used in the research and breeding of independently protectable subsequent varieties.[98]

This lightweight, anticopying regime was replaced in 1991 by a much tougher, patent-like model.[99] The 1991 Act increased the term of protection from fifteen to twenty years (25 years for grapevines and trees), and it extended the scope of protection to harvested material from the protected variety (including entire plant and parts of plants).[100] Signatory countries could also extend protection to "products made directly from harvested material of the protected variety, e.g., soya oil, soya flour."[101]

Under the 1991 Act, the scope of protection covers not merely production or reproduction of the propagating material for commercial purposes, but also the "conditioning, offering for sale, selling, exporting, importing or stocking" of such material.[102] This Act then simultaneously cuts back on both the research exemption and farmers' rights under the 1978 text.

For example, instead of allowing free access to protected materials for research and discovery purposes, the 1991 Act's definition of infringement now covers the initial breeder's right to a derived variety by prohibiting reproduction and sale of any variety developed from, and expressing the essential characteristics of, a protected variety.[103] While the mode of implementing this provision remains controversial, the net effect was to narrow the preexisting breeders' exemption and to expand the rights of first-generation breeders[104] along the lines that bring UPOV "closer to patent laws."[105]

97 *Id.* (discussing the "breeders' exemption" or privilege).

98 *See id.* at 91–94; *see also* Reichman, *Legal Hybrids* (1994), above n. 88, at 246–69 (describing the United States Plant Variety Protection Act of 1970, as amended to conform to UPOV 1978 in 1981).

99 *Accord.* SANTILLI (2012), above n. 90, at 91.

100 *Id.*

101 *Id.* at 93.

102 *Id.* at 91.

103 *See, e.g.*, UPOV 1991, above n. 91, art. 14.5(c) (stating that "essentially derived varieties" may be obtained, for example, by the selection of a natural or induced mutant, or of a semiclonal variant individual from plants of the initial variety, back crossing, or transformation by genetic engineering). *See also* Reichman, *Legal Hybrids*, above n. 88, at 247. This provision addressed the predicament of plant breeders who had little protection when second comers inserted bioengineered genes into a protected variety and were denied multiple royalties from a chain of creations "predominantly derived" from the originally protected variety.

104 Laurence R. Helfer, *Intellectual Property Rights in Plant Varieties: International Legal Regimes and Policy Options for National Governments* 38 (FAO Legislative Study No. 85, 2004).

105 SANTILLI (2012), above n. 90, at 95.

Another practical effect of the 1991 Act was to repeal a de facto exemption for farmers that allowed them freely to exchange seeds, and to replace it with a provision that allows governments, at their discretion, to deny farmers the right to reuse saved seeds in future harvests or to require them to pay royalties to breeders for this purpose.[106] National laws can also limit the size of the lands, the quantity of seed, and the plant species to which the farmers' right to save seeds, if any, are applicable.[107] These provisions affected small farmers in developing countries, such as India, where farmers reportedly produced two-thirds of the country's annual seed requirements.[108]

Finally, the 1991 Act expunged the preexisting prohibition against overlapping protection for patents and plant breeders' rights. Instead, it explicitly allowed cumulative protection under both regimes for a single variety.[109] UPOV 1991 thus ceased to operate as an alternative to the patent system, in conformity with its original purpose, and is "actually becoming increasingly, similar to the patent system," which it reinforces.[110] After 1998, moreover, new adherents to UPOV must ratify the 1991 version, as the 1978 text no longer remains open for signature.[111]

Meanwhile, since the genomic revolution, breeders have also been marketing genetically modified plants, which can attract gene patents, plant patents, and plant variety protection, with tensions arising among all these regimes.[112] In 1994, the TRIPS Agreement further legitimized these practices and obliged WTO members to at least allow patents on transgenic plants and genetic engineering processes.[113] The TRIPS Agreement also obliged all WTO members that did not

[106] *See* UPOV 1991, above n. 91, arts. 14–15; Agrow Seed Co. v. Winterboer et al. , 513 U.S. 179 (1995). *See generally* DUTFIELD (2004), above n. 29, at 35–36; JONATHAN CURCI, THE PROTECTION OF BIODIVERSITY AND TRADITIONAL KNOWLEDGE IN INTERNATIONAL LAW OF INTELLECTUAL PROPERTY 62 (Cambridge U. Press 2010) [hereinafter CURCI (2010)].

[107] SANTILLI (2012), above n. 90, at 93. *See, e.g.*, European Council Regulation 2100/94, On Community Plant Variety Rights, 1994 O.J. L 227, art. 14. *But see* Commission Directive 98/44, On the Legal Protection for Biotechnological Invention, 1998 O.J. L. 213/219, art. 11.1 (preserving farmers' rights to save and reuse seeds of plant varieties that contain patented components).

[108] *See, e.g.*, S.K. Verma, *TRIPS and Plant Variety Protection in Developing Countries*, 17 E.I.P.R. 281–89 (1995). *See also* KEITH AOKI, SEED WARS: CASES AND MATERIALS ON INTELLECTUAL PROPERTY AND PLANT GENETIC RESOURCES (Carolina Acad. Press 2007). However, some critics have questioned the economic importance of UPOV. *See, e.g.*, M.D. Janis & J. P. Kesan, *U.S. Plant Variety Protection: Sound and Fury . . .*, 39 HOUS. L. REV. 727–78 (2002); Dwijen Rangnekar, Kingston Univ., Faculty of Human Sciences, Kingston-on-Thames, U.K., *Plant Breeding Biodiversity Loss and Intellectual Property Rights*, Economics Discussion Paper 00/5, Oct. 10, 2000.

[109] *Compare* UPOV 1978, above n. 91, art. 2.1 *with* UPOV 1991, above n. 91, art. 2. *See, e.g.*, J.E.M. Ag. Supply v. Pioneer Hi-Bred Int'l, 534 U.S. 124 (2001).

[110] SANTILLI (2012), above n. 90, at 94.

[111] *Id.*

[112] *See, e.g.*, DUTFIELD (2004), above n. 29, at 33–37.

[113] *See* TRIPS Agreement, above n. 1, art. 27.3(b).

grant plant patents to provide some *sui generis* protection for unpatentable plant varieties, although adoption of the UPOV regime was not expressly mandated.[114] As a result, plant varieties protected under a *sui generis* regime can also incorporate a patented, genetically engineered component. Conflict between the two regimes may thus occur, especially if the patent blocks access to plant genetic resources needed to develop new varieties or to regulate cases of interdependence of plant biotechnologies.[115]

To address these and other blocking effects, the European Communities' Directive on the Legal Protection for Biotechnological Inventions of 1998 provided for the possibility of a compulsory license to enable either the plant breeder to avoid infringing the gene patent or, when necessary, to enable the patent holder to avoid infringing an otherwise blocking PBR.[116] The avowed object of the compulsory license was to ensure a "remunerated open access" to the underlying genetic resource in either case.[117]

2. The Developing Countries Assert Countervailing Proprietary Rights of Their Own

Armed with biotechnology patents and universal protection of plant breeders' rights, the big seed companies in the developed countries soon dominated the global market for technologically advanced agricultural products. As the late Professor Keith Aoki succinctly phrased it,

> The commodification of germplasm is a story of the transformation and privatization of agriculture from an economic sector where the state primarily supplied seeds and subsidized agricultural plant-and-seed research to benefit farmers into a global economic enterprise effectively controlled by a handful of multinational corporations. Notably, those "Promethean" value-bringers rewarded with expansive intellectual property rights tend to be employees of large corporations that benefit from the conferral of ... patents.[118]

[114] *See id.*; Jerome H. Reichman, *Intellectual Property in the Twenty-First Century: Will the Developing Countries Lead or Follow?*, 46 *Hous. L. Rev.* 1115–85 (2009), *available at* http://scholarship.law.duke.edu/cgi/viewcontent.cgi?article=2748&context=faculty_scholarship.

[115] SANTILLI (2012), above n. 90, at 98–99.

[116] *See* Directive 98/44/EC of the European Parliament and of the Council of 6 July 1998 on the Legal Protection of Biotechnological Inventions, 1998 O.J. L 213/217 [hereinafter EU Biotech Directive].

[117] SANTILLI (2012), above n. 90, at 99 (citing A. Van Wijk & N. Louwaars, *Framework for the Introduction of Plant Breeders' Rights in Countries with an Emerging Plant Variety Protection System, Center for Genetic Resources, Naktumbouw,* The Netherlands (2002)). Cross-licensing of both parties to the dispute is the envisioned outcome.

[118] AOKI, above n. 108, at 8.

In response, opponents of expanded intellectual property protection began to argue that industrial crop development and agricultural policies were destroying the genetic base essential to plant breeding.[119] Concentrated ownership of gene libraries combined with intellectual property rights were also said to inhibit both farmers' abilities to reproduce seeds used in their own crops and the transfer of new crop development technology to farmers and breeders in developing countries.[120]

Developing country governments, in turn, began to formulate a three-pronged defensive strategy of their own. First, they declined to issue patents on living matter, and narrowly defined relevant patent obligations under the TRIPS Agreement of 1994.[121] Second, many countries enacted *sui generis* plant breeders' rights laws tailored to their own needs, without adopting the UPOV 1991 model, as allowed under Article 27.3(b) of the TRIPS Agreement.[122] Still other countries that had previously adhered to UPOV in 1978 refused to ratify UPOV 1991, although pending Free Trade Agreements may change their minds.[123]

Above all, both non-governmental organizations (NGOs) and governments in developing countries began to attack the legal premise that plant genetic resources were a public good that commercial firms in the industrialized countries could privatize at will. Of particular importance here was the fact that both plant breeders' rights and gene patents obtained in developed countries were often based on, or derived from, plant genetic resources that had originated from developing countries and that may or may not have been made available by public seed banks. Questions were accordingly raised about "the uncompensated appropriation of plant genetic diversity from the developing countries of the global South to the industrialized countries of the global North."[124] Who had authorized the commercial use of such resources, and, in the absence of any such authorization in the source countries, who actually owned the underlying genetic resources under public international law?[125]

[119] *Id.* at 38; ROBIN PISTORIUS & JEROEN VAN WIJK, THE EXPLOITATION OF PLANT GENETIC INFORMATION: POLITICAL STRATEGIES IN CROP DEVELOPMENT 8 (Oxford U. Press 1999).

[120] *See* PISTORIUS & VAN WIJK, above n. 119, at 8–10.

[121] *See, e.g.*, JEROME H. REICHMAN & CHRISTOPH SPENNEMANN, U.N. CONFERENCE ON TRADE & DEV. (UNCTAD), USING INTELLECTUAL PROPERTY RIGHTS TO STIMULATE PHARMACEUTICAL PRODUCTION IN DEVELOPING COUNTRIES: A REFERENCE GUIDE (2011), *available at* http://unctad.org/en/Docs/diaepcb2009d19_en.pdf; UNCTAD, THE TRIPS AGREEMENT AND DEVELOPING COUNTRIES (1996), *available at* http://unctad.org/en/docs/ite1_en.pdf.

[122] *See* SANTILLI (2012), above n. 90, at 94–98 (citing cases of India, Bangladesh, Pakistan, Sri Lanka, Nepal, Namibia, and Uganda). *See also id.* at 229–39 (discussing farmers' rights in the African Model Law and in Ethiopia).

[123] *See id.* at 98 (citing Brazil, Argentina, Paraguay, Uruguay, Chile, Columbia, Ecuador, and Mexico).

[124] AOKI, above n. 108, at 38.

[125] *See, e.g.*, CURCI (2013), above n. 106, at 57; Safrin (2004), above n. 74; DUTFIELD (2004), above n. 29, at 52–59. *See further* below Chapter 3, Sections I.A & I.C.2.a (selected cases of alleged biopiracy).

The resulting tensions soon engulfed the United Nations Food and Agricultural Organization. Because the FAO's International Undertaking of 1983 was largely geared to promoting collaboration for research and public breeding efforts to improve food security, it had not really addressed problems emanating from direct commercial use of the plant genetic resources to be managed and conserved by designated host members of the CGIAR repositories.[126] For this and other reasons, the International Undertaking could not adequately deal with the potential impact of intellectual property rights on these same resources. Initially, this gap obliged the FAO, in 1989, expressly to clarify that plant breeders' rights, as regulated by UPOV, were not inconsistent with the International Undertaking because "free access" for research purposes did not necessarily mean "free of charge" in the case of commercial applications.[127]

Meanwhile, the patenting of genetically modified plants, accompanied by controversial cases of alleged "biopiracy" in which the patents in question were based on genetic materials taken from developing countries without permission, were revealing an even bigger gap in the FAO's International Undertaking.[128] Thus, by 1991, another FAO Resolution had conceded that the "common heritage of mankind" concept embodied in the International Undertaking of 1983 *did not displace the sovereignty of states over genetic resources existing in their territories*.[129] These circumstances logically led provider states – especially the developing countries – to ask why they should continue to contribute plant cultivars to the CGIAR's crop commons at all, if the end result was to be patented agricultural products in developed countries that were sold back to the same developing countries at exorbitant prices.[130]

Caught between these rising tensions, the FAO's nonbinding International Undertaking was about to crumble as the foundations were laid for two major new international treaties – the Convention on Biological Diversity of 1992 (CBD)[131] and the World Trade Organization's (WTO) Agreement on Trade-Related Aspects of Intellectual Property Rights (TRIPS Agreement) of 1994.[132] As more fully explained in Chapter 3, the CBD would endorse the developing countries' claims to sovereignty over genetic resources emanating from their territories and establish a rudimentary legal regime of misappropriation to regulate access to and benefit sharing from the use of such resources.[133] The TRIPS Agreement, in contrast, would

[126] *See* FAO International Undertaking, above n. 77.
[127] Agreed Interpretation of the International Undertaking, FAO Res. 4/89, 25th Sess. (29 Nov. 1989).
[128] *See* Chapter 3, Section I.A.
[129] Report of the Conference of the FAO, FAO Res. 3/91, 26th Sess. (9–27 Nov. 1991).
[130] Halewood (2010), above n. 59.
[131] *See* CBD, above n. 1.
[132] *See* TRIPS Agreement, above n. 1.
[133] *See* Chapter 3, Sections I.B.2 & I.C.

require all WTO Members, including the developing countries (but not the group of Least Developed Countries) to provide, and strictly enforce, either plant patents or some form of *sui generis* protection for plant breeders' rights (not necessarily UPOV), in conformity with Article 27.3.[134] It would also require these same WTO countries (including most developing countries) to strengthen their domestic patent laws in conformity with relatively high minimum standards of patent protection or risk being sued for damages before WTO dispute-resolution panels.[135]

The stage was thus set for a confrontation between the different and conflicting interests promoted respectively by the CBD and the TRIPS Agreement.[136] Not surprisingly, the notion of a global research commons for plant genetic resources was – temporarily at least – eclipsed by the shadow of these two treaties, as was the notion that public scientific research required careful attention and the preservation of such resources in the ensuing struggle to commoditize everything.

B. Mandatory Protection of Some Microbial-Related Inventions Under the TRIPS Agreement of 1994

These same trends inevitably affected the field of microbiology, and elicited growing opportunities for commercially profitable applications of genetic resources

[134] TRIPS Agreement, above n. 1, art. 27.3; above nn. 108–09 & accompanying text. *See, e.g.*, Protection of Plant Varieties and Framers' Rights Act 2001, No. 53, Acts of Parliament, 2001 (India), *available at* http://agricoop.nic.in/PPV&FR%20Act,%202001.pdf.

As to Least Developed Countries (LDCs) , they were exempted from the substantive provisions of TRIPS until 2013 (for patents) and 2016 (for TRIPS obligations generally). *See* TRIPS Agreement, above n. 1, arts. 65–66; World Trade Org., Council for Trade-Related Aspects of Intellectual Property Rights, Extension of the Transition Period Under Article 66.1 for Least Developed Country Members, June 11, 2013, IP/C/64, *available at* http://www.ipi.gov.mz/IMG/pdf/Decisao_do_Conselho_do_TRIPS_da_OMC_sobre_a_extensao_do_prazo_de_transicao_para_os_PMA.pdf. The 2013 deadline has now been extended to July 1, 2021. *See* Technical Note, Frederick M. Abbott, *The LDC TRIPS Transition Extension and the Question of Rollback*, INT'L CTR. FOR TRADE & SUSTAINABLE DEV. (ICTSD) PROGRAMME ON INNOVATION, TECH. & INTELLECTUAL PROP. (ICTSD Pol'y Brief No. 15, May 2013), *available at* http://ictsd.org/i/publications/164907/?view=document.

[135] *See* TRIPS Agreement, above n. 1, arts. 27–35; DSU, Dispute Settlement Rules: Understanding on Rules and Procedures Governing the Settlement of Disputes, Marrakesh Agreement Establishing the World Trade Organization, Annex 2, THE LEGAL TEXTS: THE RESULTS OF THE URUGUAY ROUND OF MULTILATERAL TRADE NEGOTIATIONS 354 (1999), 1869 U.N.T.S. 401, 33 I.L.M. 1226 (1994); Jerome H. Reichman, *Universal Minimum Standards of Intellectual Property Protection under the TRIPS Component of the WTO Agreement*, 29 INT'L LAWYER 345–88 (1998), *available at* http://scholarship.law.duke.edu/faculty_scholarship/687; Jerome H. Reichman & Rochelle C. Dreyfuss, *Critical Reflections on the Proposed Substantive Patent Law Treaty: Harmonization Without Consensus*, 57 DUKE L.J. 85–130 (2007), *available at* http://scholarship.law.duke.edu/faculty_scholarship/2228.

[136] *See, e.g.*, CURCI (2013), above n. 106, at 51–62 ("The private property regime on biological diversity established by TRIPS may undermine the implementation of benefit-sharing provisions of the CBD ...," *id.* at 56).

in areas such as biomedical, agricultural, and renewable energy innovation.[137] Once scientific investigators discover that any given microbial material or class of materials has known or likely commercial applications, it becomes logical for the researchers involved – and any potential commercial partners – to seek to exploit the incentives for risky investment that intellectual property rights and contractually imposed licensing agreements make possible. Resort to proprietary rights in such cases, even with regard to federally funded research results, is consistent with both the spirit and the provisions of the Bayh-Dole Act of 1980 in the United States and its progeny elsewhere.[138]

In this connection, one should recall that, in 1980, the first patent on a living organism to be upheld by the U.S. Supreme Court covered a genetically altered microbe used to remediate oil spills.[139] Since then, patents on both isolated DNA (genomic DNA or gDNA) and purified DNA (complimentary or cDNA) sequences from microbes and other living organisms were allowed in the United States, followed by the European Union and Japan, on the theory that locating, isolating, and describing molecular biological matter requires ingenuity and costly investment.[140] Claims pertaining to synthetic or complementary DNA sequences were also justified on analogy to synthetic chemicals, although state practice varies considerably with regard to these issues,[141] and the United States Supreme Court has recently begun to take a more skeptical view of some patents on genetic sequences.[142] The *Chakrabarty* decision in 1980 and the ensuing genomic revolution, stimulated interest in patentable applications of microbial genetic resources in general, especially for high payoff medical uses.[143] Technical problems bearing on the need for public disclosure of microbe-related inventions were resolved by the Budapest Treaty on the International Recognition of the Deposit of Microorganisms for Purposes of Patent Procedure of 1977.[144]

137 *See* above Chapter 1, Section II.C.

138 Patent and Trademark Law Amendments Act (Bayh-Dole Act), 35 U.S.C. § 200 (1980). *See* DAVID MOWERY ET AL., IVORY TOWER AND INDUSTRIAL INNOVATION: UNIVERSITY-INDUSTRY TECHNOLOGY TRANSFER BEFORE AND AFTER THE BAYH-DOLE ACT (Stanford Bus. Books, 2004); Anthony So et al., *Is Bayh-Dole Good for Developing Countries? Lessons from the U.S. Experience*, 6 *PLoS Biology* 2078–84 (2008).

139 Diamond v. Chakrabarty, 447 U.S. 303 (1980).

140 *See, e.g.*, GRAHAM DUTFIELD, INTELLECTUAL PROPERTY RIGHTS AND THE LIFE SCIENCE INDUSTRIES: PAST, PRESENT, AND FUTURE 194–211 (2d ed., World Scientific Pub. Co. 2009) [hereinafter DUTFIELD (2009)].

141 *See id. See generally* Linda L. Demaine & Aaron Xavier Fellmeth, *Reinventing the Double Helix: A Novel and Nonobvious Reconception of the Biotechnology Patent*, 55 *Stan. L. Rev.* 303 (2002); Safrin, above n. 74.

142 Assn. for Molecular Pathology v. Myriad Genetics, Inc., 133 S.Ct. 2107 (2013); Arti K. Rai & Robert Cook-Deegan, *Moving Beyond 'Isolated' Gene Patents*, 341 *Science* 137–38 (2013).

143 *See, e.g.*, below n. 164.

144 Budapest Treaty on the International Recognition of the Deposit of Microorganisms for Purposes of Patent Procedure of 1977, 19 Aug. 1980, *as amended on* 26 Sept. 1980, 32 U.S.T. 1241, 1861 U.N.T.S.

Needless to say, gene patents in general have become extremely controversial,[145] and the problems that they may – or may not – pose for scientific research have generated a vast literature of their own.[146] Much of this debate concerns the extent to which such patents have generated overly broad claims, thickets of rights, and anticommons effects that inhibit future research, especially with regard to both research tools and diagnostics.[147] Recent commentary has also emphasized the extent to which the drive to patent research results in molecular biology makes access to biological materials for further research increasingly problematic,[148] a topic of primary importance throughout this book.[149]

In its recent decisions on this subject, the United States Supreme Court held that genes and the information they contain are not patentable subject matter "simply because they have been isolated."[150] In so doing, the Court invoked the "product of nature" doctrine to invalidate patents covering genomic DNA (gDNA), while retaining eligibility for patents on complimentary DNA (cDNA).[151] From a policy perspective, the Court's rejection of claims on isolated DNA sequences seemed to rely on a distinction between "patent claims that create incentives for innovation

361 [hereinafter Budapest Treaty], *available at* http://www.wipo.int/treaties/en/registration/budapest/trtdocs_wo002.html (last accessed 3 Nov. 2014).

145 *See, e.g.*, DUTFIELD (2009), above n. 140, at 209–17 ("Are patents appropriate for biotechnological inventions?").

146 *See, e.g.*, Rebecca Eisenberg, *Why the Gene Patenting Controversy Persists*, 77 ACAD. MED. 1381–7 (2002); *Re-Examining the Role of Patents in Appropriating the Value of DNA Sequences*, 49 *Emory L. J.* 783–800 (2000); *but see* Christopher M. Holman, *Debunking the Myth that Whole Genome Sequencing Infringes Thousands of Gene Patents*, 30 NATURE BIOTECH 240 (2012). See generally INTELLECTUAL PROPERTY AND BIOTECHNOLOGY (Arti K. Rai, ed. Elgar, 2011), 111–162.

147 *See, e.g.*, Demaine & Fellmeth, above n. 141; Michael A. Heller & Rebecca S. Eisenberg, *Can Patents Deter Innovation? The Anticommons in Biomedical Research* 280 *Science* 698 (1998). *See also* above n. 48.

148 *See, e.g.*, Note, Lisa Larrimore Ouellette, *Access to Bio-Knowledge: From Gene Patents to Biomedical Materials*, 2010 *Stanford Tech L. Rev.* N1, *available at* http://papers.ssrn.com/sol3/Delivery.cfm/SSRN_ID1575705_code1231602.pdf?abstractid=1431580&mirid=1; Zhen Lei, Rakhi Juneja, & Brian D. Wright, *Patent Versus Patenting: Impediments of Intellectual Property Protection for Biological Research*, 27 *Nature Biotechnology* 36 (2009); John P. Walsh, Wesley M. Cohen & Charlene Cho, *When Exclusivity Matters: Material versus Intellectual Property in Academic Biomedical Research*, 36 RES. POL'Y 1184 (2007). *See generally* Rebecca S. Eisenberg, *Noncompliance, Nonenforcement, Non-problem? Rethinking the Anticommons in Biomedical Research*, 45 *Hous. L. Rev.* 1059, 1063–75 (2008).

149 *See especially* Chapter 4, Sections II, III.

150 Assn. for Molecular Pathology v. Myriad Genetics, Inc., 133 S.Ct. 2107 (2013); Mayo Collaborative Servs. v. Prometheus, 132 S.Ct. 1289 (2012). *See generally* Rai & Cook-Deegan, above n. 142, at 137–38; Robert Cook-Deegan & Tom Dedeurwaerdere, *The Science Commons in Life Science Research: Structure, Function and Value of Genetic Diversity*, 188 *Int'l Soc. Sci J.* 302 (2006).

151 Rai & Cook-Deegan, above n. 142, at 137.

and claims that block further innovation" because of the quantum of information they contain.[152] Although the full implications of these decisions remain to be seen, synthesized DNA and other genetically engineered DNA molecules would seem to remain patentable,[153] provided they meet the now universal standards of novelty, nonobviousness, and utility.[154]

Meanwhile, the patentability of microbial-related inventions at both the national and multilateral levels has become – perhaps paradoxically – more firmly established over time, especially for profitable medical uses.[155] Since the landmark *Chakrabarty* decision in 1980, laws allowing patents on microbes were adopted or confirmed in numerous countries, including Japan and members of the European Patent Convention.[156] As noted, technical problems bearing on the need for public disclosure of microbe-related inventions were ultimately resolved by the Budapest Treaty on the International Recognition of the Deposit of Microorganisms for Purposes of Patent Procedure of 1977.[157] In 1994, moreover, the drafters of Article 27.3(b) of the WTO's TRIPS Agreement expressly mandated patent protection for microorganisms, and for non-biological and microbiological processes, while otherwise ducking the question of the patentability of life forms in general. All members of the World Trade Organization (except LDCs) must, accordingly, ensure that their domestic patent laws recognize some microorganisms as patentable products, assuming they satisfy other eligibility requirements.

However, neither genes – nor the information contained in genes – necessarily fall within generally accepted definitions of a "microorganism," within the meaning of that provision,[158] and the ambiguous language used in Article 27.3(b) remains open to very different interpretations. As a result, state practices regarding how microorganisms should reasonably be defined for purposes of Article 27.3(b) vary widely even in countries that favor patents on living organisms, notably the United

[152] *Id. See also* Arti K. Rai, *Biomedical Patents at the Supreme Court: A Path Forward*, 66 STAN. L. REV. ONLINE 114–16 (Oct. 14, 2013).

[153] *See, e.g., id.* at 114–15; Rai & Cook-Deegan, above n. 142, at 137–38.

[154] *See* TRIPS Agreement, above n. 1, art. 27.1.

[155] *See* Douglas Robinson & Nina Medlock, *Diamond v. Chakrabarty: A Retrospective on 15 Years of Biotech Patents*, 17(10) *Intell. Prop. & Tech. L.J.* 12 (2005), *available at* http://bannerwitcoff.com/_docs/library/articles/Chakrabarty.pdf. *See also* DUTFIELD (2009), above n. 140, at 201–11.

[156] *See* Robinson & Medlock, above n. 155.

[157] Budapest Treaty, above n. 144.

[158] Frederick M. Abbott, *The Definition of Pharmaceutical Substance and Exclusion of Micro-organisms under the WTO TRIPS Agreement* 13 (Study for Indian Pharmaceutical Assoc., 25 Apr. 2005) (citing Technical Board of Appeal, European Patent Office and other authorities). There is no commonly accepted definition of patentable subject matter in international law. *See, e.g.*, Reichman & Dreyfuss, above n. 135.

States, the European Union (as members of the European Patent Convention[159]), and Japan.[160]

In contrast, most developing countries still oppose patents on living organisms and invoke various doctrinal justifications to avoid them. For example, a case can be made that substances occurring in nature are not "inventions," at least in their isolated form,[161] or that patenting them is otherwise contrary to the public policy of the state.[162] Developing countries also tend to view the proliferation of patents on living organisms, and on applications of genetic resources generally, as an open invitation to "biopiracy," especially in view of the fact that most developed countries still refuse to require would-be patentees to disclose the origins of genetic resources underlying their claimed inventions.[163] At least one distinguished authority contends that states may invoke the *ordre public* doctrine to exclude patents based on biological materials that were not obtained in conformity with the Convention on Biological Diversity.[164]

Nevertheless, the proliferation of patents on microbial-related inventions remains a fact of life in developed countries, as demonstrated below with particular regard to

[159] The unitary patent is a recent development for the European Union, although it does not replace existing EU patent options. The European Patent Office (EPO) is a creature of the Convention on the Grant of European Patents (European Patent Convention), Oct. 5, 1973, 1065 U.N.T.S. 199 *as revised* on Nov. 29, 2000, *available at* http://www.epo.org/law-practice/legal-texts/html/epc/2013/e/ma1.html, to which most EU countries also belong.

[160] *See* CURCI (2013), above n. 106, at 40–42; DUTFIELD (2004), above n. 29, at 28–30. For example, patent offices in all these countries have been known to allow patents on animal and plant cells as microorganisms. *See, e.g., id.* at 29; CURCI (2013), above n. 106, at 41. *See also* M. Adcock & M. Llewellyn, *Microorganisms – Definitions and Options under TRIPS* (Quaker United Nations, Geneva, Nov. 23, 2000) (suggesting a fixed legal definition is untenable because scientific classifications evolve continuously). The European Patent Office allows patents on microbiological processes "if there is a technical invention by man in the process" and if it "plays a significant part in determining or controlling the result it is desired to achieve." But conventional plant and animal breeding methods are excluded. *See* DUTFIELD (2004), above n. 29, at 29–36.

[161] *See, e.g.*, UNITED NATIONS CONFERENCE ON TRADE & DEVELOPMENT (UNCTAD), USING INTELLECTUAL PROPERTY RIGHTS TO STIMULATE PHARMACEUTICAL PRODUCTION IN DEVELOPING COUNTRIES 48–49 (2011) (discussing laws of Argentina and Brazil, plus a Decision of the Andean Community). However the process used for isolating biological substances remains patentable subject matter, if to the eligibility requirements are met. *Id.* at 49.

[162] *See* TRIPS Agreement, above n. 1, art. 27.2 (allowing exclusions on grounds of *ordre public* or morality, "including to protect human, animal or plant life or to avoid serious prejudice to the environment"). However, an exclusion on grounds of *ordre public* may require the state to prohibit commercial sales of the product in question. For an exhaustive analysis of the *ordre public* exception, see CURCI (2013), above n. 106, at 233–74.

[163] *See* below Chapter 3, Sections I, II.B. and IV.

[164] Marco Ricolfi, *Biotechnology, Patents and Epistemic Approaches*, J. BIOLAW & BUS., SPEC. SUPP. 77 (2002). *See also* Paolo Spada, *Liceità dell'invenzione brevettabile ed esorcismo dell'innovazione*, 5(1) *Rivista di Diritto Privato* 5, 18 (2000), *available at* http://www.biblio.liuc.it/scripts/essper/ricerca.asp?tipo=scheda&codice=11032237.

the United States. As will be seen, a growing array of problems for public scientific research has resulted from these practices.

1. Increasing Reliance on Patents and Trade Secrecy Laws to Protect Commercial Applications of Microbial Genetic Resources

From a purely commercial perspective, the tendency to patent microbe-related inventions is indicative of the economic impact that basic microbiology is having, as previously evidenced in Chapter 1.[165] With specific regard to the United States, for example, a search of patents related to microorganisms that were granted by the United States Patent and Trademark Office (USPTO) between 1980 and 2010, revealed the following data:

Box 2.1. USPTO Data on Microbe-Related Patents by Subject Matter (1980–2010)

Subject Matter	Number of Patents
• Microbe + genetics	1,200
• Bacterial + genetics	4,800
• Fungal + genetics	1,330
• Viral + genetics	4,700

Delphion database 2010, compiled by Carla Rydholm. Some patents in this survey may fall into overlapping categories.

The types of microbial-related patents covered in these statistics may be summarized as follows: genes; compounds; organisms; methods of use; methods of discovery/isolation kits for diagnostic tests; nanotechnology and microbiology manufacturing; disease detection and monitoring; and biological materials for information processing.[166]

With specific regard to third-generation biofuel development, some 341 U.S. patents were granted in 2003, which grew to some 1,878 patents issued in 2008.[167] Lab-on-a-chip nanotechnology – including microarrays – was also a field

[165] *See* Chapter 1, Section II.C.

[166] *See also* Robinson & Medlock, above n. 155.

[167] *See* Thomas Reuters, *Mining Data for Tomorrow's Breakthroughs*, Innovation Hot Spots: IP Market Report (2009). Of these, however, only three biofuel patents based on the use of algae were granted in 2003, while 63 algae biofuel patents were issued in 2008. Moreover, the technology at issue in these patents sometimes used green algae, which are plants, rather than "blue-green algae," which are

with the potential to become critical to global public health initiatives. This type of innovation facilitates comprehensive diagnostic analysis using minimum equipment and a technique of specimen sampling that facilitates the identification of infectious diseases.[168] These technologies include both microbial genetic bioengineering methods and products, which lead to utility and composition of matter patents.[169]

Biofuels and microarrays are technologies attracting venture capital that could yield still more real-world products (and attendant litigation). Recently, Craig Venter(who had helped to decode the human genome in the 1990s) continued to push the boundaries of life sciences and related technology with his 2011 patent application on a minimal bacterial genome, which applied research that had led to the creation of the first self-replicating synthetic bacterial cell.[170] Related patent applications were pending in the EU and Japan (and under the Patent Cooperation Treaty at WIPO). Venter's patent application included broad claims, plus a variety of claims on genes, organisms, and methods. Even if such claims were subsequently narrowed, there is reason to believe that this patent could rival the Cohen-Boyer patents for recombinant DNA in terms of its foundational nature for future research.[171]

Perhaps the most famous patent dispute involving microbes after the Supreme Court's *Chakrabarty* decision in 1980 was *the Taq polymerase* case in the 1990s.[172] This case was notable for the public source of the microbe at issue and also because it revealed the possibilities of using exclusive rights to disrupt upstream academic research dependent on PCR technology. It is summarized in Box 2.2 below.

The *Taq polymerase* case illustrates some of the hurdles that upstream patents can erect in the path of academic research. It also suggests the need for preventive measures that funders of basic research can take to avoid such problems later on, as

actually cyanobacteria. *Id.* No significant litigation concerning this microorganism-based type biofuel was found at the time of writing.

168 For major cases challenging the validity and scope of these patents, *see* Assn. for Molecular Pathology v. United States PTO, 2012 U.S. App. LEXIS 17679 (Fed. Cir. Aug. 16, 2012); Assn. for Molecular Pathology v. Myriad Genetics, Inc., 133 S.Ct. 2107 (2013); and Mayo Collaborative Servs. v. Prometheus, 132 S.Ct. 1289 (2012).

169 Reportedly, some 766 patents covering lab-on-a-chip technology were granted in 2003, while some 1,682 patents on this technology issued in the period January 2008 to April 2009.

170 *See* Daniel G. Gibson et al., *Creation of a Bacterial Cell Controlled by a Chemically Synthesized Genome*, 329 *Science* 52 (2010). The J. Craig Venter Institute has filed for 13 patent family applications related to this breakthrough, *First Self-Replicating Bacterial Cell – Frequently Asked Questions*, *J. Craig Venter Inst.*, http://www.jcvi.org/cms/research/projects/first-self-replicating-synthetic-bacterial-cell/faq#q11 (last accessed 11 Nov. 2014), and the patent (U.S. Patent Application No. 20,070,122,826 (filed May 31, 2007) regarding the Mycoplasma laboratorium genome (Mycoplasma mycoides JCVI-syn1.0)).

171 *Cf.* Jonathan Kahn, *Synthetic Hype: A Skeptical View of the Promise of Synthetic Biology*, 45 *Val. U. L. Rev.* 29 (2011), *available at* http://scholar.valpo.edu/vulr/vol45/iss4/2.

172 Hoffman La Roche, Inc. v. Promega Corp., 319 F. Supp. 2d 1011 (2004).

Box 2.2. Taq Polymerase Case

The isolate of *thermos aquaticus*, used to purify a Taq polymerase for PCR, was obtained from the American Type Culture Collection (ATCC). That isolate, however, had initially been obtained from a Yellowstone National Park hot spring. Patents for the PCR technique and purified *Taq polymerase* resulted in PCR becoming an enormous commercial success, with the market for *Taq polymerase* estimated to be in the range of $80-$85 million[s] annually.[a] However, Yellowstone received no payments for the *Taq polymerase* patents.[b]

Diversa (now Verenium) was the bio-prospecting company that contracted with Yellowstone in the 1990s to collect samples in the park.[c] Non-profits challenged this agreement in court, but lost.[d] The courts held such bio-prospecting licensing in national parks was not illegal, and there was no provision for royalties to be paid to the government as owner of the park. This case may thus be viewed as an early form of legalized "biopiracy" in a domestic context, which has then recurred with serious political consequences at the international level.[e]

The patents on PCR also threatened to block upstream research for a period of time, owing to the high prices that the patent holders initially charged would-be academic users. This crisis was narrowly averted when subsequent purchasers of the patents, under pressure from the NIH, made them available for upstream research on more reasonable terms and conditions.[f]

a. Marcia Baringa, *Promega Wins Round in Fight Over Taq*, 273 SCIENCE 1039 (1996).
b. Source: A. Ian Cohen, *A Strange Brew*, 4(7) IP LAW & BUSINESS 33 (2004).
c. Edmonds Institute v. Babbit, 93 F.Supp.2d 63 (2000); *see* ALAN T. BULL, MICROBIAL DIVERSITY AND BIOPROSPECTING 451–53 (2003).
d. Robert Cook-Deegan & Tom Dedeurwaerdere, *The Science Commons in Life Science Research: Structure, Function and Value of Genetic Diversity*, 188 *Int'l Soc. Sci J*. 302 (2006); Jerome H. Reichman & Jennifer Giordano-Coltart, "A Holistic Approach to Patents Affecting Frontier Sciences: Lessons from the Seminal Genomic Discovery Studies," paper presented at the CEER Retreat, Duke University Center for Genetics, Ethics & Law (April 2008).
e. *See later*, Chapter 3, Section I.A.
f. *See earlier*, n. d.

well as the need for greater government use of safeguards already set out in both the Bayh-Dole Act and the Patent Act when holders of patented research tools threaten to disrupt basic research.[173]

The extent to which microorganism-related patents will spawn future litigation that complicates or hinders upstream research remains to be seen. Existing

[173] *See, e.g.*, Arti Rai & Rebecca Eisenberg, *Bayh-Dole Reform and the Progress of Biomedicine*, 66 *Law & Contemp. Probs*. 289–315 (2003); *see also* Rebecca Eisenberg (2008) above n. 148.

litigation indicates growing friction between specific private interests,[174] but not necessarily signs of pervasive patent thicket formation. One hypothesis in this regard is that the Budapest Treaty, requiring deposits of patented microbes for purposes of disclosure,[175] makes it easier to avoid possible research barriers in ways that a greater degree of secrecy might impede.

Disregarding the unknown extent to which scientists ignore potentially blocking patents when conducting basic research,[176] another variable is that treatments and

[174] *See, e.g.*, Genentech, Inc. v. Amgen, Inc., 289 F.3d 761 (Fed. Cir. 2002) (Alleged infringement of three patents for methods and cloning vehicles of expressing "genetic information," including DNA, into unicellular organisms that would not naturally contain that DNA; essentially, Genentech patented the process of expressing mammalian DNA in microbes, such as *E. coli*, an important technique in the biotech industry); Carnegie Mellon Univ. v. Hoffman-La Roche, Inc., 2004 U.S. Dist. LEXIS 4975 (N.D. Cal., 2004) (in a previous proceeding two of Carnegie Mellon's patents on methods and organisms for bacterial DNA polymerases were held invalid; CMU sued on a third patent and lost again; any potential thicket was resolved); Microbes Inc. v. Espoma Co., 2011 U.S. Dist. LEXIS 39705 (E.D. Tex 2011) (involving a patent fight over fertilizers containing live probiotic fungal spores); Glaxo v. Genentech, 2010 U.S. Dist. LEXIS 46440 (N.D. Cal., 2010) (involving upstream research tools to produce antibodies in host cells as a cancer therapy; this case bears watching since the research tool does affect upstream research). For a series of cases dealing with probiotics and animal feeds, *see* Nutrition Physiology Corp. v. Enviros., Ltd., 87 F. Supp. 2d 648 (N.D. Tex 2000); Ajinomoto v. ITC, 597 F.3d 1267 (Fed. Cir. 2010); Nestec v. Wysong, 2010 U.S. Dist. LEXIS 130773 (E.D. Mo. 2010).

[175] Budapest Treaty, above n. 78. *See, e.g.*, EPO Policy, in "Notice from the European Patent Office dated 7 July 2010 concerning inventions which involve the use of or concern biological material," Eur. Patent Office (7 July 2010), http://xepc.eu/node/oj2010-498; USPTO policy, in "Office of Policy and External Affairs: Budapest Treaty," USPTO (2 Oct. 2013), http://www.uspto.gov/ip/global/patents/ir_pat_budapest.jsp. For microbial related patents in Europe, a search of the EPO database revealed few apparently troublesome patents in this area at the time of writing. *See, e.g.*, Method for Generating a Genetically Modified Microbe, Eur. Patent Office (EPO) Patent No. 2,438,156, https://register.epo.org/espacenet/application?number=EP10721095; Process for the Hydrolysis of Cellulose Mediated by Ternery Complexes of Cellulose, Clostridium Thermocellum Cells, and Cellulase Expressed by these Cells, Eur. Patent Office Patent No. 2,013,355, http://worldwide.espacenet.com/publicationDetails/biblio?CC=EP&NR=2013355&KC=&locale=en_EP&FT=E (Dartmouth is the assignee); Novel Brevibacillus Choshinesis and Process for Producing Protein with Use of the Microbe as a Host, Eur. Patent Office (EPO) Patent No. 1,686,170, https://data.epo.org/publication-server/pdf-document?pn=1686170&ki=B1&cc=EP (a research tool alternative to *E. coli* genetic transformation using a modified fungus; could be quite useful if successful); Antibiotic Producing Microbe, Eur. Patent Office (EPO) Patent No. 0,906,336, https://data.epo.org/publication-server/pdf-document?pn=0906336&ki=B1&cc=EP (some overlapping patents but little litigation in the U.S.); Microbially Derived Rennin Having Enhanced Milk Clotting Activity and Method of Producing Same, Eur. Patent Office (EPO) Patent No. 0,805,866, https://data.epo.org/publication-server/pdf-document?pn=0805866&ki=B1&cc=EP (enhanced cheese making method and microbial composition; there was US litigation in this area in the 1980s).

[176] *Cf.* Madey v. Duke Univ., 307 F.3d 1351 (2002). For survey evidence regarding the impact of blocking patents on conducting new research, see Walsh, Cohen & Cho, above n. 148; John P. Walsh, Ashish Arora & Wesley M. Cohen, *Science and the Law: Working through the Patent Problem*, 299 *Science* 1021 (2003); John P. Walsh, Charlene Cho & Wesley M. Cohen, *The View from the Bench: Patents, Material Transfers and Biomedical Research*, 309 *Science* 2002 (2005) [hereinafter Walsh et al., *View from the Bench*].

therapies may not yet have fully evolved to a point where directly protecting microbes, fungi, and bacteria as such are often worth the cost. They might be protected as components of larger systems or products, such as bacterially-transformed cell culture lines, but the bacteria themselves seem to be rarely protected, or patented, and not yet in a manner that throttles downstream product development.[177] Policies that facilitate access to genetic materials may also be helping both research and applications.[178]

Conversely, there is some concern that firms may be wary of the difficulties of proving infringement of microbe-related patents owing to the lack of foolproof tracking mechanisms at this time.[179] Firms increasingly rely on trade secrecy laws rather than patents to protect such innovations, thereby sidestepping the public disclosure function of the Budapest Treaty. Research prospects may then become contingent on any given researcher's ability to reverse engineer any particular invention, which is costly and time consuming even when feasible.[180] Scientists who publish articles based on their knowledge of alleged trade secrets may be liable for claims of misappropriation and even subject to prosecution under the Economic Espionage Act in the United States if the theft benefits a foreign government.[181]

Another reason that firms increasingly rely on trade-secret protection of microbial-related research applications is to avoid mounting pressures to publicly disclose the source of relevant genetic materials when filing patent applications at home or abroad.[182] At the same time, firms that rely on trade secrets, rather than

177 *See, e.g.*, Clemens Kerle, *International IP Protection for GMO – A Biotech Odyssey*, 8 COLUM. SCI. & TECH. L. REV. 147, 155–57 (2007).

178 *See* Robin Feldman & Kris Nelson, *Open Source, Open Access, and Open Transfer: Market Approaches to Research Bottlenecks*, 7 NW. J. Tech. & Intell. Prop. 14 (2008), *available at* http://scholarlycommons.law.northwestern.edu/njtip/vol7/iss1/2. In fact, there are far more patents and ensuing litigation for anti-microbial compositions than microbials. *See* above n. 146 & accompanying text; *see also* Rebecca S. Eisenberg, *Technology Transfer and the Genome Project: Problems with Patenting Research Tools*, 5 *Risk: Health, Safety & Env't* 163–75 (1994); *Wisdom of the Ages or Dead-Hand Control? Patentable Subject Matter for Diagnostic Methods After* In Re Bilski, 3 CASE W. RES. J. L. TECH. & INTERNET 1 (2012).

179 Defense Advanced Research Projects Agency (DARPA) Meeting in Arlington, VA on March 28, 2011 (attended by Paul Uhlir). For a fuller discussion of tracking methods and their implications, *see* below Chapter 5, Section II.C.3.

180 *See* Uniform Trade Secrets Act § 1, cmt. to § 1.1 (*amended* 1985) (allowing reverse engineering of trade secrets by honest means). *See also* Jerome H. Reichman, *How Trade Secrecy Law Generates a Natural Semicommons of Innovative Know-How, in* LAW AND THEORY OF TRADE SECRECY: A HANDBOOK OF CONTEMPORARY RESEARCH 185 (R. Dreyfuss et al. eds. 2011) [hereinafter Reichman (2011)].

181 *See* Chris O'Malley, *Ex-Dow Sciences employee accused of stealing trade secrets*, INDIANAPOLIS BUS. J., http://www.ibj.com/exdow-agrosciences-employee-accused-of-stealing-trade-secrets/PARAMS/article/21421 (last accessed 14 June 2014). The alleged employee co-authored a 2008 article on "Recent Advances in the Biochemistry of Spinosyns," published by Hunan Normal University in China. *Id.*

182 WORLD INTELLECTUAL PROP. ORG. (WIPO), TECHNICAL STUDY ON PATENT DISCLOSURE REQUIREMENTS RELATED TO GENETIC RESOURCES AND TRADITIONAL KNOWLEDGE (2004),

microbial-related patents, run the risk that competitors may reverse engineer the unpatented know-how either by honest or dishonest means. There is accordingly a serious and growing risk of misappropriation, especially when companies extend their supply chains overseas.[183]

Although misappropriation of trade secrets became an international business tort under Article 39 of the TRIPS Agreement in 1994,[184] governments have so far made little use of this provision,[185] and private parties cannot invoke it before WTO tribunals. The extent to which reliance on trade secrets ultimately challenges basic microbial research may also depend on the development of better tracking mechanisms, which in turn could stimulate more patenting of microbes with disclosure and deposit under the Budapest Treaty.

2. Possible Patent Thickets

At the time of writing, the literature on litigation with respect to microbial-related patents did not yet seem to reflect the level of controversy found in certain other areas of molecular biology and biochemistry, particularly with regard to patent thickets, anticommons effects, and blockage of research tools.[186] However, there was at least one active patent thicket forming around *Agrobacterium tumefaciens*, a genus of bacteria that was discovered to cause tumors in plants by transferring DNA from itself to the plant. Labs at the University of Washington in Seattle and

available at http://www.wipo.int/export/sites/www/freepublications/en/tk/786/wipo_pub_786.pdf?; *See* Dan L. Burk & Mark A. Lemley, *Policy Levers in Patent Law*, 89 VA. L. REV. 1575 (2003); *but see* Lisa Larrimore Ouellette, *Do Patents Disclose Useful Information?*, 118 HARV. J. L. & TECH. 2007–28 (2005), *available at* http://jolt.law.harvard.edu/articles/pdf/v25/25HarvJLTech531.pdf; Jeanne C. Fromer, *Patent Disclosure*, 94 IOWA L. REV. 539, 549 (2009) (arguing that disclosure stimulates "inventing around, improving upon, and inspiring both during and after the patent term").

[183] Firms that rely on trade secrets may, to some extent, hope to reduce the risk of free-riding by avoiding disclosure in patent applications. CREATE.org, Center for Responsible Enterprise and Trade, Trade Secret Theft – Managing the Growing Threat in Supply Chains (CREATE White Paper, 2012), *available at* https://create.org/resource/trade-secret-theft-managing-the-growing-threat-in-supply-chains/. Alternatively, their innovation may not meet the eligibility requirements of patent law, which leaves them no other way to obtain protection. *See* Reichman (2011), above n. 181.

[184] TRIPS Agreement, above n. 1, art. 39.

[185] Obviously, measures to preserve actual secrecy have high costs and serious limitations. Trade secret protection, instead, has been greatly strengthened by new international norms set out in Article 39 of the TRIPS Agreement, above n. 1. In principle, however, such protection would not prevent third parties from reverse engineering unpatented innovations by honest means. Moreover, the efficacy of trade secret protection may be further diminished by the recent patent reform legislation in the U.S., which adopts the first-to-file patent system already used in the rest of the world. *See* Robert Maier, *The Big Secrets of the America Invents Act*, INTELLECTUAL PROPERTY TODAY (2013), http://www.iptoday.com/issues/2011/12/the-big-secret-america-invents-act.asp.

[186] *See, e.g.*, above 175–179 & accompanying text; INTELLECTUAL PROPERTY AND BIOTECHNOLOGY, above n. 146, 111–162.

at Washington University, St. Louis, performed pivotal work in the early 1980s to harness this natural phenomenon for genetic engineering purposes. A. *tumefaciens* was found to be an extremely useful and efficient way to transfer genes from itself into higher plants; and it was one of the only such inter-kingdom transfers to prove successful.[187]

By 2000, more than 125 patents had sprung up around this technology, and most observers now agree that a genuine thicket exists. Monsanto, Syngenta, Japan Tobacco, and Pioneer HiBred each hold six or more patents in the field. Monsanto and Syngenta engaged in a series of battles at the USPTO and in the U.S. District Court for the District of Delaware, but in 2004 agreed to settle all disputes and cross-license their patents to one another. Most scientists agree that progress in the area now requires a license from one or both of these companies.[188] While some public interest organizations, including non-profit Cambia, have attempted to fund and forward alternatives, few believe that any substitutes will be found that are as efficient as A. *tumefaciens*, with growing concerns that this thicket may harm research in plant biotechnology.[189]

From a more general perspective, gene patents that are sought early in the R & D process may often be either overly broad in scope[190] or overly narrow and fragmented. In the first case, a single invention may monopolize an entire field, including subsequent improvements, and thus hinder both upstream research possibilities as well as the pace of downstream commercial applications.[191] In the second case, other firms seeking commercial applications of the same genetic resource may find themselves obliged to license and combine multiple patents in order to achieve

[187] Jeffrey Schell & Marc Van Montagu, *The Ti-plasmid of Agrobacterium tumefaciens, a natural vector for the introduction of nif genes in plants?*, 9 *Basic Life Sci.* 159 (1977). We are grateful to Ryan O'Quinn (Ph.D (Biology) Univ. North Carolina, J.D. Duke Law School), for his work on this and other topics.

[188] Japan Tobacco Inc. , through U.S. Patent No. 5,591,616 (issued Jan. 7, 1997), has claimed methods using the bacterium to transfer all monocots, and Washington University—St. Louis claimed methods for transforming all dicots through U.S. Patent No. 6,051,757 (issued Apr. 18, 2000), http://www.syngentabiotech.com/bio/images/US6051757.pdf. Syngenta later licensed this technology from Washington University, and patented other such dicot technology, e.g., U.S. Patent No. 6,162,965 (issued 19 Dec. 2000). *But see Japan Tobacco gets US patent*, AGROW (17 Jan. 1997), https://www.agra-net.net/agra/agrow/japan-tobacco-gets-us-patent-50136.htm.

[189] Amy Yancey & C. Neal Stewart, *Are University Researchers at Risk for Patent Infringement?*, 25 *Nature Biotechnology* 1225, 1226–28 (2007), *available at* http://www.nature.com/nbt/journal/v25/n11/full/nbt1107-1225.html (last accessed 22 Nov. 2014).

[190] *See, e.g.*, Michael Heller & Rebecca Eisenberg, *Can Patents Deter Innovation? The Anticommons in Biomedical Research*, 280 SCIENCE 698 (1998); *see also* Eisenberg (2008), above n. 174.

[191] *See, e.g.*, Yancey & Stewart, above n. 189. *But see* Randal Scott. Testimony on gene patents and other genomic inventions, before the House Judiciary Subcommittee on Courts and Intellectual Property (July 13, 2000), available at http://commdocs.house.gov/committees/judiciary/hsu66043000/hs466043Of.htm, suggesting that broad patents may turn out to be weaker under judicial scrutiny; *accord.* Nautilus, Inc. v. Biosig Instruments, Inc., 134 S. Ct. 2120 (2014).

the desired end product.[192] Synthetic biology, which combines the problems arising from multiple software patents on the same product with those of gene patents, is another area that bears watching in this regard.[193]

To the extent that more microbial genetic resources become patentable under the TRIPS Agreement[194] and actually patented under domestic laws (which vary considerably in their approaches and receptivity),[195] the twin problems of patent thickets and anticommons effects[196] are likely to grow in the future. If so, they could block the development of commercial applications, especially when multiple licenses must be obtained from competitors in the same market, some of whom may hold out or otherwise refuse to deal.[197] These conditions can thus tend to discourage investment in risky R&D, especially in the pharmaceutical sector, where promising early stage products may ultimately fail costly clinical trials, with the risk that few, if any, potentially interested firms undertake the relevant R&D process at all.[198]

III. MOUNTING IMPEDIMENTS TO RESEARCH USES OF GENETIC RESOURCES

Growing empirical evidence suggests that the proliferation of gene patents and related intellectual property rights has intensified concerns that holders of genetic materials may lose out on profitable gains from downstream commercial applications

[192] *See, e.g.*, Ed Levy et al., *Alternative Intellectual Property for Genomics and the Activity of Technology Transfer Offices: Emerging Directions in Research*, 16 *B.U. J. Sci. & Tech. L.* 194 (2010); Carl Shapiro, *Navigating the Patent Thicket: Cross Licenses, Patent Pools, and Standard Setting, in* 1 *Innovation Pol'y & Econ.* 118 (A.B. Jaffe et al. eds. 2001).

[193] *See, e.g.*, Arti Rai & James Boyle, *Synthetic Biology: Caught Between Property Rights, the Public Domain, and the Commons*, 5(3) *PLoS Biology* e58 (2007). For the view that traditional patent rights are generally inappropriate for information products and that a different paradigm sounding in liability rules is needed to accommodate them, *see* Pamela Samuelson et al., *A Manifesto Concerning the Legal Protection of Computer Programs*, 94 *Colum. L. Rev.* 2308 (1994); Reichman, *Legal Hybrids (1994)*, above n. 88.

[194] *See* above nn 155–160 and accompanying text.

[195] *See, e.g.*, UNCTAD, above n. 161; Abbott, above n. 158.

[196] In the U.S. much also depends on recent cases in the federal courts and elsewhere – not involving microbial patents as such – that may have narrowed the eligibility of method patents in biotechnology and which may also reshape DNA based patents generally (both composition of matter and method patents). *See, e.g.*, Bilski v. Kappos, 561 U.S. 593, 130 S.Ct. 3218 (2010); Assn. for Molecular Pathology v. Myriad Genetics, Inc., 133 S.Ct. 2107 (2013); Mayo Collaborative Servs. v. Prometheus, 132 S.Ct. 1289 (2012).

[197] *See, e.g.*, Hillary Greene, *Patent Pooling Behind the Veil of Uncertainty: Antitrust, Competition Policy*, 90 *B.U. L. Rev.* 1397 (2010); Eisenberg (2008), above n. 173, at 1073; Daniel Burk & Mark Lemley, *Biotechnology's Uncertainty Principle*, 54 *Case W. Reg. L. Rev.* 691 (2004); *see also* Mark Lemley & Carl Shapiro, *Frontiers of Intellectual Property: Patent Holdups and Royalty Stacking*, 85 *Tex. L. Rev.* 1991 (2007).

[198] *See, e.g.*, Beldiman, above n. 42 (citing authorities).

if they do not restrict access to these resources even for public research purposes.[199] Academic researchers in developed countries also fear they may lose reputational benefits as well as follow-on research opportunities if they release biological materials to other researchers on an informal basis. The technology transfer offices at their respective universities will likely insist that ever more restrictive material transfer agreements should accompany such transactions even when researchers are otherwise willing to exchange genetic materials.[200]

For example, one survey of academic biomedical researchers in 2004 found that over a two-year period, 18 percent of requests by academics for materials, such as cell lines, genes, or organisms, had been denied, while 33 percent of similar requests by industry had also been denied.[201] The authors of this survey concluded that non-compliance with requests for materials had grown over time, and that commercial incentives, the efforts involved, and scientific competition were all factors affecting non-compliance.[202]

Another survey of some ninety-three American academics in the field of agricultural biology found that most of those surveyed believed that intellectual property rights negatively impacted their research, even though patent infringement as such was not usually a concern.[203] Most problems reportedly arose from the MTAs that universities imposed to protect their intellectual property. Negotiating these restrictive MTAs caused serious delays and complicated research endeavors, and could even result in contractual restraints on publication and academic freedom.[204]

Still another recent study, partly carried out at Duke University, on microbicides used to prevent HIV infections revealed serious challenges to the sharing of materials, especially patented materials, for research purposes.[205] Researchers in this field need access to a diverse set of patented materials, such as active pharmaceutical ingredients, delivery systems, chemical reagents, polymers, and animal models from different sources. Those seeking to develop combination microbicides had trouble initially in obtaining sufficient information about the properties of needed materials, in part because providers did not make that information available in journals or on the internet.

199 *See, e.g.*, Ouellette (2010), above n. 148, ¶¶ 71–77 (citing authorities).

200 *See generally* Walsh, Cohen, & Cho, above n. 148. *See further* below Chapter 4, Section III.

201 Walsh, Cohen & Cho, above n. 148, at 1191.

202 *Id. See generally* Ouellette (2010), above n. 148, ¶¶ 71–77.

203 Lei et al. (2009), above n. 148, at 38–39.

204 Thirty-four faculty members (42%) experienced a total of ninety-seven delays in research with an average delay of 8.7 months. *Id.* at 38.

205 *See* Tejen Shah, Intellectual Property Challenges for Microbicides: R&D Approaches to Increasing Innovation and Access to Microbicides Used to Prevent Sexually Transmitted Infections (2013), unpublished paper submitted to Seminar on Access to Medicines: Intellectual Property & Global Public Health, Duke Univ. Sch. Law, Spring 2013 (on file with the authors).

A second concern was that legal negotiations for the transfer of materials could be time consuming and unpredictable. Some evidence suggests that, faced with these obstacles, scientists in this field may simply ignore patents or arrange for informal sharing of materials that circumvent institutional management of intellectual property rights, which exposes the relevant universities to legal risks.[206]

The same study also showed that non-standard Material Transfer Agreements (MTAs) often included restrictions on publications that act as barriers to the sharing of both knowledge and materials. Case studies indicated that progress in developing new microbicides had been slowed by the practice of pharmaceutical companies to make available only materials of inferior efficacy and to withhold internal safety evaluations.[207] Whether proposals to resolve these problems through patent pools and tiered pricing strategies in the global market will prove successful remains to be seen.[208]

Meanwhile, in the developing countries, suspicions that gene patents and related intellectual property rights obtained in developed countries were often based on plant and microbial genetic resources that had originated in developing countries led the latter to impose serious restrictions on access to such resources in their domestic laws.[209] Because these fears of "biopiracy" were often fuelled by the past behavior of academics, as we demonstrate in Chapter 3, requests for access to plant and microbial genetic resources from academic institutions were increasingly treated with suspicion and subject to difficult negotiations.[210]

At the same time, governments in the developing countries began formally to assert sovereign rights to their genetic resources at the multilateral level. Negotiations to establish an international regulatory regime to prohibit the misappropriation

[206] Shah (2013), above n. 205, at 8–10 (citing sources). *See generally* Walsh et al., *View from the Bench*, above n. 176; John P. Walsh, Charlene Cho & Wesley M. Cohen, *Patents, Material Transfers, and Access to Research Inputs in Biomedical Research*, Final Report to the *Nat'l Acads. Scis. Comm.* on Intellectual Property Rights in Genome & Protein-Related Inventions (2005).

[207] Shah (2013), above n. 205, at 12–16 (case studies of Gilead licensing of Tenovir and Tenopovir Disoproxil Fumarate and Janssen Pharmaceutical's Rilpivirine (NNRII)).

[208] *See, e.g.*, William Fisher & Talha Syed, *Differential Pricing*, *in* The Health Crisis in the Developing World and What We Should Do About It (forthcoming Stanford Univ. Press), *available at* http://cyber.law.harvard.edu/people/tfisher/Infection.htm; Anthony So & Cecilia Oh, *Approaches to Intellectual Property and Innovation that Meet the Public Health Challenge of AIDS* (Technical Advisory Group of the Global Comm'n on HIV & the Law, Working Paper 7–9 July 2011), *available at* http://hivlawcommission.org/index.php/working-papers?task=document.viewdoc&id=87. *See also* Josh Lerner & Jean Tirole, Efficient Patent Pools, No. 9175, Nat'l Bureau Econ. Research (NBER, 2002), http://www.nber.org/papers/w9175.pdf.

[209] *See, e.g.*, Gurdev S. Khush, *Biotechnology: Public-Private Partnerships and Intellectual Property Rights in the Context of Developing Countries*, *in* Biodiversity and the Law 174–79 (2007); Juliana Santilli, *Genetic Resources Common Pools in Brazil*, *in* Common Pools of Genetic Resources (2013), above n. 6, at 103–08; *see generally* below Chapter 3, Section I.B. (citing authorities).

[210] *See* below Chapter 3 Sections I.A & B.

of genetic resources from source countries led to the Convention on Biological Diversity of 1992 and the Nagoya Protocol on enforcement in 2010,[211] a topic dealt with at length in Chapter 3.[212] These negotiations called into question the very concept of treating plant and microbial genetic resources as global public goods, given that developing countries that supplied such resources received no benefits when firms in developed countries privatize research results based on these same resources via patents and related intellectual property rights.

Already by the mid-1980s, these tensions had begun to undermine the ability of existing plant and microbial research commons to continue to deliver their traditional support for public scientific research. As documented in the next two chapters, public service repositories holding such resources began to impose ever more restrictive conditions on access and use even for the most basic upstream research purposes. They were also forced to re-examine the legal and institutional foundations of the public services they had traditionally performed on behalf of both the academic and industrial research communities.[213]

In extreme cases, fundamental questions about who had the power to determine the public good characteristics of genetic resources needed for research purposes led to a revolt against the existing institutional infrastructure that supported transnational exchanges of plant and microbial genetic resources. Perhaps the most telling and clamorous of these cases arose in the context of hoarded pandemic influenza viruses.[214]

A. *The Revolt Against the WHO's First Pandemic Influenza Research Commons*

As explained earlier in this chapter, patent applications on genetic sequences of the SARS virus had already raised questions about the viability of the WHO's cooperative efforts under the aegis of the Global Influenza Surveillance Network (GISN) in 2003.[215] When, in 2005, cases of the highly pathogenic avian influenza pertaining to sub-type H5N1 began to emerge,[216] these concerns were quickly transformed into a bitter North-South conflict.

[211] CBD, above n. 1; Nagoya Protocol, above n. 65.

[212] *See generally* Chapter 3, Sections I & IV.

[213] *See generally* below Chapter 3, Section II and Chapter 4, Section II.

[214] For a similar crisis affecting the stability of the CGIAR's seed banks, see below Chapter 3, Section II.B.

[215] *See* above Section I.A.2.

[216] Wilke (2013), above n. 36, at 316. The reference is to the Asian lineage, subclate 2 of HPAI-A (H5N1). *Id.* at 337, fn. 2. *See generally* INST. MEDICINE, THE DOMESTIC AND INTERNATIONAL IMPACTS OF THE 2009 – H1N1 INFLUENZA PANDEMIC: GLOBAL CHALLENGES, GLOBAL SOLUTIONS (Nat'l Acads. Press, 2010).

The most important clade or category known to have caused the infections since 2005 (the so-called Asian lineage subclade 2) had first been discovered in Indonesian poultry in 2003. The human infections in 2005 were also first reported in Indonesia, from which they spread to China, Turkey, Egypt, and elsewhere. By the end of 2007, more than 100 cases of such infections had occurred in Indonesia alone, with a fatality rate then exceeding 80 percent. By February 2012, nearly 600 known cases had occurred in 15 countries, with about 350 casualties.[217]

In conformity with the agreed practices of the WHO's GISN, member states had initially shared samples of potentially pandemic H5N1 strains. As the crisis mounted, Indonesian samples became indispensable for the identification, assessment, and monitoring of the virus, as well as for the development of diagnostic kits. But the Indonesian authorities refused to continue sharing samples, reportedly after learning that an Australian company had applied for a patent on a vaccine developed from Indonesian specimens.[218] Instead, Indonesia agreed to allow a private company, Baxter Healthcare SA, to develop human vaccines with the Indonesian strain, under an exclusive license.[219] In so doing, Indonesia expressly invoked territorial sovereignty over its genetic resources, a principle which by then had become embodied in the Convention on Biological Diversity of 1992.[220]

According to Professor Peter Yu, Indonesia's claim to "viral sovereignty" shocked the international public health, diplomatic, and academic communities because the samples accumulated in Indonesia were thought to be vital for ongoing research efforts and to avoid a global pandemic.[221] Nevertheless, Indonesia had its defenders[222] and, according to Professor Yu, it had at least four legal and economic justifications for its actions.

First, Indonesia feared that the WHO would share its viruses with major pharmaceutical companies in developed countries. Among other concerns, vaccines patented by these companies might then be sold to developing countries, including Indonesia, at prices few could afford.[223]

[217] Wilke (2013), above n. 36, at 316 (citing WHO figures).

[218] *Id.* at 317.

[219] Yu (2013), above n. 42, at 1606.

[220] *See* Wilke (2013), above n. 36, at 317; Yu (2013), above n. 42, at 32–37. For the CBD, see below Chapter 3, Section I.B. Indonesia also claimed that there had been technical violations of GISN's internal procedures, which may or may not have been specious. For details see Wilke (2013), above n. 36, at 317.

[221] Yu (2014), above n. 42, at 1611–15. Over a four-year period most outbreaks in humans and poultry had occurred in that country, and at least 53 types of known H5N1 bird flu viruses had been detected there. *Id.* (citing authorities).

[222] *See, e.g.*, Editorial, *Global Solidarity Needed in Preparing for Pandemic Influenza*, 369 LANCET 532 (2007). *See also* David Fidler, *Influenza Virus Samples, International Law and Global Health Diplomacy*, *in* EMERGING INFECTIOUS DISEASES, Jan. 2008.

[223] Yu (2014), above n. 42, at 1606–08.

Second, Indonesia feared that it and other developing countries might not even succeed in obtaining vaccines they could afford owing to limitations on supply. These limitations were magnified by the developing countries' lack of capacity to manufacture vaccines and antivirals on their own.[224] Third, once Indonesia had released its virus samples, "there was no guarantee that the pharmaceutical companies would develop drugs that respond to the needs of the Indonesian population," as distinct from the more wealthy populations in the developed world.[225] Fourth, international law now gave Indonesia the authority to determine access to genetic resources under its own domestic laws.[226]

Indonesia had thus confronted the developed world with a well-founded fear of "viropiracy" just when the developed world felt most threatened by a possible pandemic for which it had no remedy. As Indonesia's claims of "viral sovereignty" spread to other countries,[227] the WHO began to grapple with this confrontation in a series of measures that, while conceding the principle of national sovereignty, culminated in a Resolution of the World Health Assembly, that established the foundations of a second research commons, to be known as the Pandemic Influenza Preparedness (PIP) Framework.[228]

Under the resulting intergovernmental framework agreement, the National Influenza Centers of WHO member states pledged to make biological materials, such as H5N1 viral samples, available to other WHO Centers. The Centers would, in turn, transfer these genetic resources to qualified third parties for purposes of developing influenza-related medicines.[229] Because recipients are contractually obliged to meet certain benefit-sharing obligations in favor of the National Centers that make virus samples available for these purposes, the PIP Framework strives expressly to remove obstacles to the exchange of microbial genetic resources that would otherwise arise under the Convention on Biological Diversity.[230]

224 *Id.* at 1609.

225 *Id.*

226 *Id.* at 1611 (citing both art. 15.1 of the CBD and art. 10.1 of the International Treaty on Plant Genetic Resources for Food and Agriculture (2001)). For an explanation of these treaties, see Chapter 3, Sections I.B. and III. For evidence supporting Indonesia's arguments, see Wilke (2013), above n. 36, at 318.

A fifth argument attributed to Indonesia was that some of its earlier sequences submitted to WHO had been shared with the Los Alamos National Laboratory in the U.S., which raised fears of eventual use in research for biological warfare. *See* Yu (2014), above n. 42, at 1610.

227 *Id.*, at 1613, 1642.

228 PIP FRAMEWORK, above n. 50.

229 *See* PIP FRAMEWORK, above n. 50, art. 5.1.2 (collaborating centers and WHO H5N1 reference laboratories); *id.* arts. 5.4 & 6.3; Beldiman, above n. 42, at 36. Qualified third parties may include vaccine manufacturers, laboratories of the originating and other member states, or other laboratories meeting the biosafety standards. Beldiman, above n. 42, at 36 (discussing PIP FRAMEWORK, above n. 50, arts. 5.4 & 6.3).

230 "In exchange, recipients of the virus sample material are required to comply ... with certain benefit sharing obligations in the form of monetary support, medicine donations, or technology transfer

The PIP Framework thus constitutes the newest component of the existing Microbial Research Commons, one that has many lessons for our efforts to redesign that Commons as a whole, as elaborated in Part Four. We will accordingly take a closer look at the PIP Framework in Chapter 4, in conjunction with our discussion of the public culture collections' response to the challenges of the CBD in general.[231] For present purposes, it bears noting that the PIP Framework Agreement leaves qualified recipients of the sample viruses free to pursue their own intellectual property strategies with respect to the commercialization of end-product medicines resulting from research on the viruses in question.[232] As a result, Professor Beldiman predicts that the kind of patent thickets and anticommons effects that gene patents have elsewhere generated could, in practice, impede investment in influenza vaccines even under the PIP Framework Agreement, unless appropriate governance strategies to regulate the resulting intellectual property rights were ultimately incorporated into that same agreement.[233]

B. *Implications for the Present Study*

The developing countries' challenge to the WHO's initial efforts to pool pandemic influenza viruses was an extreme manifestation of a much larger trend that threatens to disrupt basic research in the life sciences generally. As we will document in the next two chapters, the tensions between developed and developing countries fueled by unauthorized uses of plant and microbial genetic resources for research and applications has destabilized the existing scientific infrastructure that traditionally supports both basic and applied research.

The tightened regulatory apparatus at the multilateral level that now restricts access to genetic resources originating from developing countries even for public research purposes is described and evaluated in the next chapter. As further documented in Chapter 4, both the public culture collections and the far more numerous university and research laboratories that also manage microbial genetic resources increasingly rely on material transfer agreements that are often costly to negotiate and that

or licenses ..." Beldiman, above n. 42, at 36(discussing PIP FRAMEWORK, above n. 50, art. 78, and Standard Material Transfer Agreement 2, arts. 4.1.1.A5 & A6). *See also* Wilke (2013), above n. 36, at 323–25.

[231] *See* Chapter 4, Section IV.A.

[232] *See* PIP FRAMEWORK, above n. 50, art. 78, pmbl. The Agreement does recognize, however, that private industry incentives under intellectual property rights may not adequately address the needs of "small and uncertain markets." PIP FRAMEWORK, above n. 50, arts. 78 & 2.1.1; Beldiman, above n. 42, at 40–44.

[233] *See id.*, at 44–52 (discussing open-source and compulsory license approaches, liability rules, and patent pools); *see generally* FREDERICK M. ABBOTT, AN INTERNATIONAL LEGAL FRAMEWORK FOR THE SHARING OF PATHOGENS: ISSUES AND CHALLENGES (Int'l Ctr. Trade & Sustainable Dev. 2000).

burden the upstream research process with onerous conditions.[234] In their eagerness to obtain a piece of the action from future commercial applications, providers of genetic resources in both developed and developing countries thus perversely tend to hoard the very inputs that support basic research on which downstream applications ultimately depend. The sharing ethos of scientists working with genetic resources has also suffered because of this proprietary ethos.[235]

When analyzing these restrictions on access and use, however, one must carefully distinguish between the relatively few microbial genetic resources that have some known or likely commercial value, such as the H5N1 influenza viruses covered by the WHO's PIP Framework Agreement,[236] and the bulk of such resources that had no known or likely commercial value at the time they were deposited in public collections, or were otherwise made available for research purposes. In the first case, the providers of given genetic resources have a definable interest in a specific enterprise based on preexisting scientific knowledge. For example, Indonesia – one of the countries most affected so far by the H5N1 influenza virus – nonetheless initially refused to release virus samples located on its territory for external research purposes without guarantees that it would receive access to the resulting technologies and medicines under reasonable terms and conditions.[237] Such concerns must be reconciled with the incentives that exclusive intellectual property rights otherwise provide commercial investors who defray the costs of risky research and development efforts.

In the second case, however, where the genetic resources in question possess no known or likely commercial value, scientists cannot even make basic research discoveries at all if they cannot access and use essential research inputs, given the providers' fears of lost opportunity costs with respect to end-use outcomes.[238] From a science policy perspective, these two sets of problems will normally require quite different approaches.

With regard to genetic resources having some known or likely commercial value, the WHO's PIP Framework Agreement was itself a response to overcome the unwillingness of member countries to make H5N1 viruses available for research on influenza treatments without specific guarantees concerning access to the resulting medicines and technology.[239] More generally, the proliferation of patents,

[234] *See* below Chapter 4, Section II, for details.

[235] Walsh, Cohen & Cho, above n. 148.

[236] *See* PIP FRAMEWORK, above n. 50.

[237] *See, e.g.*, Beldiman, above n. 42, at 35; Tom Dedeurwaerdere, *Global Microbial Commons: Institutional Challenges for the Global Exchange and Distribution of Microorganisms in the Life Sciences*, 161 *Research in Microbiology* 407–13 (2010). *See* above nn. 217–226 & accompanying text.

[238] *See further* Chapters 3 and 4.

[239] Beldiman, above n. 42, at 36–38. *See also* Adam Kamradt-Scott & Kelly Lee, *The 2011 Pandemic Influenza Preparedness Framework: Global Health Secured or a Missed Opportunity*, 59 POLITICAL STUDIES 831 (2011).

and especially gene patents, with all their ensuing real or perceived obstacles to basic research, have focused considerable scholarly attention on measures to avoid or overcome patent thickets and anticommons effects, if and when they arise, both as barriers to research and to the efficient development of downstream or end products.[240]

For example, research funders or other relevant knowledge governance entities may contractually impose nonexclusive licenses and built-in experimental use clauses to address specific R&D obstacles.[241] They may also build in contractual provisions that facilitate the formation of patent pools, when needed, to clear the blocking effects of too many patents covering the same knowledge assets.[242] Similarly, domestic patent and competition laws may themselves be structured to provide governments with compulsory licenses and other measures that can achieve the same objectives.[243] This important topic lies beyond the primary scope of this study.

The second set of problems concerning microbial genetic resources possessing no known or likely commercial value has not been sufficiently studied in the past, and it does constitute the core concern of this volume. Because such genetic resources could conceivably trigger one of tomorrow's medical blockbusters, measures to incentivize providers to make them available for basic research purposes would have to be devised, even if there were no binding international treaties, such as the Convention on Biological Diversity, to defend the interests of provider countries. More to the point, the bulk of these same genetic resources remain inherently susceptible to unauthorized appropriation by bioprospectors, especially in developing countries that control the world's reserves of unexplored *in situ* biodiversity,[244] or to the hoarding of both *ex situ* and *in situ* specimens because governments fear lost opportunity costs.

[240] *See* above Section II.

[241] *See, e.g.*, Peter Lee, *Contracting to Preserve Open Science: Consideration-Based Regulation in Patent Law*, 58 EMORY *L.J.* 889, 895 (2009); So et al., above n. 75. *See also* Engelberg Ctr. Innovation L. & Pol'y, New York Univ., DEFENSIVE PATENT LICENSE PROJECT, http://www.law.nyu.edu/centers/engelbergcenter/conferences/thedefensivepatentlicenseproject; *Homepage*, THE DEFENSIVE PATENT LICENSE (DPL), http://www.defensivepatentlicense.com/ (last accessed 23 Nov.2014).

[242] *See, e.g.*, Beldiman, above n. 42; Halewood (2010), above n. 59; Patrick Gaulé, *Towards Patent Pools in Biotechnology?* CDM Working Papers, CEMI Report No. 2006–010. (2006), *available at* http://papers.ssrn.com/sol3/papers.cfm?abstract_id=1427751 (last accessed 23 Nov.2014); Josh Lerner & Jean Tirole, *Public Policy Towards Patent Pools, in* 8 *Innovation Pol'y & Econ.* 157 (A.B. Jaffe et al. eds., Univ. Chicago Press 2008).

[243] *See, e.g.*, Greene, above n. 117; Jerome H. Reichman, *Intellectual Property in the 21st Century*, above n. 114; Jerome H. Reichman, *Compulsory Licensing of Patented Pharmaceutical Inventions: Evaluating the Options*, 37 *J.L. Med. & Ethics* 247 (2009).

[244] *See* below Chapter 3, Section I.A. *See generally* NICHOLAS BRAHY, THE PROPERTY REGIME OF BIODIVERSITY AND TRADITIONAL KNOWLEDGE: INSTITUTIONS FOR CONSERVATION AND INNOVATION 195–214 (2008).

As shown in Chapter 3, the resulting fears of what has been termed "biopiracy," if left unchecked, threatened to destabilize the preexisting systems of formal and informal exchanges of both *ex situ* and *in situ* genetic resources on which both microbiological and agricultural research and applications have traditionally depended. By the same token, overzealous regulatory measures to defend sovereign rights to these genetic resources could perversely shut down that same system of exchanges, with potentially serious consequences for global scientific research.[245]

[245] *See further* Chapter 3, passim.

3

Tightening the Regulatory Grip: From the Convention on Biological Diversity in 1992 to the Nagoya Protocol in 2010

I. REGULATORY MEASURES CONTROLLING ACCESS TO GENETIC RESOURCES PROMOTED BY THE DEVELOPING COUNTRIES

Users and providers of genetic resources have overlapping but increasingly divergent interests.[1] Broadly speaking, the literature identifies three major subcategories of users, namely, traditional communities, commercial enterprises, and academic researchers.[2] Traditional users include local and indigenous communities "proximate to, and dependent on, the biological resources in their natural environment." Their reliance on such resources "may constitute an integral part of their day-to-day existence."[3]

Commercial users obviously include pharmaceutical companies, agro-technology firms, and chemical and petrochemical conglomerates or startups,

[1] *See, e.g.*, Marco Ricolfi, *Intellectual Property and Biodiversity: A Review of Legal and Conceptual Issues and of Policy Options, Atti del Seminario*, Instituto Agronomico per l'Oltremare Firenze 3040 (2009), http://brasile/109.florence.it/documenti/ricolfi.pdf. *See generally* JONATHAN CURCI, THE PROTECTION OF BIODIVERSITY AND TRADITIONAL KNOWLEDGE IN INTERNATIONAL LAW OF INTELLECTUAL PROPERTY 95–102 (Cambridge U. Press 2010) [hereinafter CURCI (2010)] 95–102.

[2] Burton Ong, *Harnessing the Biological Bounty of Nature: Mapping the Wilderness of Legal, Socio-Cultural, Geo-Political and Environmental Issues*, *in* INTELLECTUAL PROPERTY AND BIOLOGICAL RESOURCES 1, 3–4 (B. Ong ed., Cavendish Square Pub. 2004). *See also* BIODIVERSITY AND TRADITIONAL KNOWLEDGE: EQUITABLE PARTNERSHIPS IN PRACTICE (S.A. Laird ed., Routledge 2002).

[3] Ong, n. 2, at 4. *See generally* Graham Dutfield, *Legal and Economic Aspects of Traditional Knowledge*, *in* INTERNATIONAL PUBLIC GOODS AND TRANSFER OF TECHNOLOGY UNDER A GLOBALIZED INTELLECTUAL PROPERTY REGIME 495–505 (K.E. Maskus & J.H. Reichman eds., Cambridge U. Press 2005) [hereinafter INTERNATIONAL PUBLIC GOODS]; Anthony Taubman, *Saving the Village: Conserving Jurisprudential Diversity in the International Protection of Traditional Knowledge*, *in* INTERNATIONAL PUBLIC GOODS, above, at 521–35; Thomas Cottier & Marion Panizzon, *Legal Perspectives on Traditional Knowledge: The Case for Intellectual Property Protection*, *in* INTERNATIONAL PUBLIC GOODS, at 565, 576–81.

among others, which require access to physical samples of biological materials for research testing and product development purposes. These firms may sometimes draw on the knowledge of traditional users when conducting their own R&D, including high-throughput screening and other more advanced techniques.[4]

The distinction between "commercial" and "academic" users, however, is far more tenuous than science policy and practice have commonly acknowledged, at least for legal purposes, a topic we will address in Chapter 4. In principle, commercial users include the researchers and scientists working on products for various branches of industry, whereas academic users are understood to focus primarily on scientific discoveries and the acquisition of new knowledge, such as taxonomic studies. Yet, researchers and scientists working for industry can make scientific discoveries and add to the store of knowledge, whereas academic scientists now often direct their attention to applied research. Academics may also serve as a bridge between industry and traditional users, as well as collaborators on end-use products. Both academic and commercial scientists may likewise become involved in the pursuit of intellectual property rights bearing on genetic resources and derived applications.[5]

As regards the providers of genetic resources, legal analysis differentiates stakeholders that control *in situ* resources from those that distribute *ex situ* specimens held in repositories of one kind or another. The typical providers of today's *in situ* resources are nation states whose agencies and indigenous communities are responsible for preserving the natural habitats in which plants, animals, and microbes survive.[6]

As Professor Burton Ong has explained, providers of *ex situ* genetic resources typically "control repositories of biological specimens ... removed from their original environment and housed in an artificially assembled collection," for example, microbial culture collections, seed banks, botanical gardens, and gene banks, among others.[7] Such entities may be public or private; they may or may not engage in commercial activities; and they may have "correspondingly different objectives and policies regarding the provision of genetic resources to the users who request ... access [to them]."[8]

4 Ong, n. 2 at 3.

5 *Id.* at 5. *Cf.* Arti K. Rai et al., *Pathways Across the Valley of Death: Novel Intellectual Property Strategies for Accelerated Drug Discovery*, 8 *Yale J. Health L. Pol'y & Ethics* 1 (2008).

6 However, "the place of geographical origin does not always coincide with the country of initial origin." Curci (2013), n. 1, at 96, 96–99 (noting importance of further distinctions among the genetic resource, traditional knowledge referring to it, and technology applied to it).

7 Ong, n. 2, at 5; *see* Chapter 2, Section I.

8 Ong, n. 2, at 5.

When plant or microbial genetic resources identified for purely scientific objectives are later used in commercial applications not foreseen at the time of discovery, these applications may lead to questions about the origins of the underlying specimens, the legitimacy of the means by which they were acquired, either *in situ* or *ex situ* as the case may be, and the knowledge of potential commercial uses that may have been gained, directly or indirectly, from indigenous populations in the countries of origin. As users in developed countries began to acquire and enforce more patents and related intellectual property rights (especially plant breeders' rights) based on derivatives of plant and microbial genetic resources,[9] governments in developing countries observed that a growing number of claims bore on genetic materials that had originally been taken from their territories, along with observations by collectors on how indigenous populations used them.[10]

A. *Bioprospecting or Biopiracy?*

Both patents and plant breeders' rights, often of considerable economic value, were challenged by developing-country governments as a form of so-called "biopiracy" that violated their sovereign rights over natural resources.[11] Technically, the term "biopiracy" refers to the unauthorized extraction of biological resources, as well as associated traditional knowledge from developing countries, or to the patenting of inventions and the acquisition of other intellectual property rights on such knowledge or resources.[12]

Most of the best known cases of alleged biopiracy pertained to plant genetic resources, with some notable claims also involving microbes,[13] especially from Africa.[14] For example, plant genetic resources taken from India figured in more

9 For example, in 1998, the Canadian NGO, ETC Group (then called RAFI), denounced some 147 cases in which mostly public institutions had claimed plant breeders' rights in varieties acquired from the network of seed banks managed by the Consultative Group on International Agricultural Research (CGIAR). *See* Graham Dutfield, Intellectual Property, Biogenetic Resources and Traditional Knowledge 55 (Routledge 2004) [hereinafter Dutfield (2004)].

10 *See generally* Carlos M. Correa, Intellectual Property Rights: The WTO and Developing Countries (Zed Books, 2001); Dutfield (2004), n. 9, at 57–59.

11 *See* Permanent Sovereignty over Natural Resources, G.A. Res. 1803 (XVII), U.N. GAOR, 17th Sess., Supp. No. 17, U.N. Doc. A/5217, at 15 (1962) [hereinafter 1962 Declaration].

12 *See* Dutfield (2004), n. 9, at 52. *See further* Charles R. McManis, *Fitting Traditional Knowledge Protection and Biopiracy Claims into the Existing Intellectual Property and Unfair Competition Framework*, *in* Intellectual Property and Biological Resources 425–510 (2014).

13 *See, e.g.*, Tomme Young et al., *Analysis of Claims of Unauthorized Access and Misappropriation of Genetic Resources and Associated Traditional Knowledge*, *in* Governing ABS: Addressing the Need for Sectoral Geographical, Legal, and International Integration in the ABS Regime 117 (T. Young ed., Int'l Union for the Conservation of Nature 2009), *available at* http://data.iucn.org (last accessed 14 June 2014).

14 *See, e.g.*, Jay McGown & Beth Ellen of Burroughs, Out of Africa: Mysteries of Access and Benefit Sharing (Edmonds Inst., Washington & African Center for Biosafety, Richmond, S. Africa 2006), *available at* http://www.newscastmedia.com/4investors_africa.pdf (last accessed 14 June 2014).

than eleven major cases, with mounting push back from the Indian government.[15] Complaints by the Philippines, Thailand, Indonesia, and China have all led to public disclosure of unauthorized and possibly illicit uses of genetic resources,[16] as have complaints from Latin American[17] and African governments.[18]

However, subsequent research shows the extent to which evaluating claims of biopiracy requires a certain degree of discrimination in view of factual and legal details that are often overlooked.[19] One must always distinguish between arguments rooted in domestic intellectual property laws and practices, especially those pertaining to the patents or plant breeders' rights in question, and countervailing claims of illicit appropriation and use of genetic resources or traditional knowledge rooted in public international law, especially after 1992, when the Convention on Biological Diversity was adopted.

As we saw in Chapter 2, the tendency in developed countries was to view *in situ* genetic resources as part of some vast, unexplored public domain – or "nobody's land" (*terra nullius*), as it was sometimes designated – and to treat traditional knowledge itself as "prior art" or "know-how," also freely available from that same public domain.[20] However, even disregarding the impact of the CBD for a moment, such a thesis ignored the fact that Article 39 of the TRIPS Agreement had made transnational violations of trade secrecy law an international tort from 1995 on, and that the use of some genetic resources and some traditional knowledge appropriated from developing countries could fall under this provision.[21] More generally, invoking the concept of a "public domain" often begs the question of the scope of the relevant

15 Young et al., above n. 13, at 102 table 1.

16 *Id.* at 102, table 1. Philippines: Snail Conus magnus, Thailand: Bitter Melon; Horn mali (Jasmine Rice); Kaw Kew (compound Pueraria Mivica); Plao-Noi; Indonesia, Kemuku Popes cuebeba and Sambiloto Angrographia panicurata. China, Snake Gourd).

17 *See id.*; Michael Blakeney, *Bioprospecting and Biopiracy in Intellectual Property and Biological Resources, in* INTELLECTUAL PROPERTY AND BIOLOGICAL RESOURCES, n. 2.

18 *See* MCGOWN & BURROUGHS, n. 14.

19 *See, e.g.*, DUTFIELD (2004), n. 9, at 52.

20 *See, e.g.*, STEPHEN LADAS, PATENTS, TRADEMARKS, AND RELATED RIGHTS: NATIONAL AND INTERNATIONAL PROTECTION Ch. 44 (Harvard Univ. Press 1975); Jerome H. Reichman & Tracy Lewis, *Using Liability Rules to Stimulate Local Innovation in Developing Countries: Application to Traditional Knowledge, in* INTERNATIONAL PUBLIC GOODS, n. 3, at 337–67; *see also* Jerome H. Reichman, *How Trade Secrecy Law Can Generate a Natural Semicommons of Know-How Applied to Industry, in* THE LAW AND THE THEORY OF TRADE SECRECY: A HANDBOOK OF CONTEMPORARY RESEARCH (R. Dreyfuss et al. eds. 2011).

21 *See* Agreement on Trade-Related Aspects of Intellectual Property Rights, art. 39 April 15, 1994, 108 Stat. 4809, 1869 U.N.T.S. 299 [hereinafter TRIPS Agreement]; McManis, n. 12, at 434, 438, 445–47, 447–450; *see also* Jerome H. Reichman, *Rethinking the Role of Clinical Trial Data in International Intellectual Property Law: The Case for a Public Goods Approach*, 13 *Marq. Intell. Prop. L. Rev.* 1 (2009), *available at* http://scholarship.law.marquette.edu/iplr/vol13/iss1/1 (last accessed 14 June 2014). Mandatory protection of geographical indications of origin under the TRIPS Agreement also became relevant for some genetic resources. *See, e.g.*, McManis, n. 12.

public domain, the legal foundations on which it rests, and how or by whom those foundations were established,[22] including the role of colonization in constructing a particular view of any such domain.[23]

One may accordingly ask why domestic assertions of sovereignty or even tribal norms, customs, and laws that ostensibly "protected" the genetic resources or traditional knowledge at issue were less worthy of respect than intellectual property norms emanating from other countries and cultures, as Professor Dutfield has observed.[24] With regard to international law, moreover, claims by developing countries sounding in both sovereignty and human rights had more substance than was usually acknowledged, as will be seen in the next Section, even if no established transnational legal regime to enforce such claims existed before 1992.

In a helpful comment on this topic, Professor Burton Ong pointed out that, in most cases, where controversies arose from products that were developed from biological resources, the providers were actually local or indigenous communities:

> [They] facilitated the process by which the genetic resource ... was identified and extracted, usually because these communities have had a long history of using the resources themselves for particular purposes or in particular ways that may not have been commonly known to those outside of these communities.[25]

The question logically asked is whether such providers should be entitled to share in the resulting economic benefits from downstream commercial applications, based on their preservation and accumulated know-how over time, including the traditional knowledge exploited in the product development process.[26]

Where, instead, the provider of the genetic resources at issue was an *ex situ* repository, Professor Ong cautions that much depends on the legal status of these same resources; on the mission and goals of the repository itself; and on the relationship between that repository and different users at different times.[27] These uncertainties blur the line between legitimate "bioprospecting" for, say, scientific purposes, and so-called "biopiracy" (a term that has "no specific legal meaning").[28]

As Professor Ong aptly put it, the rhetorical function of the term "biopiracy" was to "challenge the legitimacy of bioprospecting activities which involve the participation of locals who share ... traditional knowledge and practices, but do not

[22] James Boyle, The Public Domain: Enclosing the Commons of the Mind (Yale Univ. Press, 2008). As will be seen, claims of free use based on "public domain" theory must now contend with the CBD and the Nagoya Protocol. *See* below, Sections III.B.-IV.C. *See generally* Young et al., n. 13.

[23] Dutfield (2004), n. 9, at 58–59.

[24] *See, e.g., id., See also* Taubman, n. 3.

[25] Ong, n. 2, at 6.

[26] *Id.*

[27] *Id.*

[28] *Id.* at 7.

receive equitable compensation for their contributions."[29] Allegations of biopiracy thus require careful "scrutiny into the details of the process by which the genetic resources were discovered, identified, extracted and developed."[30]

B. *Foundations of an International Regime of Misappropriation to Govern Genetic Resources*

When assessing the types of loss or harm suffered by countries that reported cases of unauthorized takings of plant and microbial genetic resources, the International Union for Conservation of Nature (IUCN) identified the following categories of claims or complaints:

- Direct harm to commercial/livelihood interests;
- Potential harm to commercial/livelihood interests and expectations;
- Inequitable actions, e.g., gaining a benefit from genetic resources obtained without permission from national authorities and/or holders of traditional knowledge, and without any sharing of benefits;
- Unauthorized publishing or transfer of genetic or biochemical information; and
- Damages or lack of rights in specimen collection.[31]

The question that needs to be addressed, however, was the nature of the legal foundations underlying or supporting such complaints in a decentralized international system that relies primarily on territorial laws. In the next two sections, we shall see how the international community finally responded to this dilemma by laying the foundations for a global regime of misappropriation to regulate access to, and use of, both *in situ* and *ex situ* genetic resources.

1. Indigenous Communities (and Their State Sponsors) as Emerging Stakeholders

The process of decolonization under way since the Second World War had led the developing countries to assert strong claims of sovereignty over all the natural resources located within their territories,[32] which the colonial powers had formerly exploited and exported at will. Pressures to embody these claims in a binding

[29] *Id.* at 8.

[30] *Id.*

[31] *See* Young et al., n. 13, at 98–116 (invoking authority of the Convention on Biological Diversity). Most claims filed by plaintiff governments usually include more than one of these alleged torts. *Id.* at 111. Most claims also assert patent invalidity, *id.* at 115, although some are not patent related at all. *Id.* at 111.

[32] *See, e.g.*, NICO J. SCHRIJVER, SOVEREIGNTY OVER NATURAL RESOURCES: BALANCING RIGHTS AND DUTIES (Cambridge U. Press 1997).

international legal instrument culminated in the United Nations General Assembly's Declaration on Permanent Sovereignty over Natural Resources of 1962.[33]

Besides proclaiming the "inalienable right of states ... to dispose of ... natural resources ... in accordance with their national interests," the 1962 Declaration affirmed that "profits derived must be shared in the proportions freely agreed upon ... between investors and the recipient state."[34] Although no mention of genetic resources was made at that early date, they would arguably have been included in the claim of sovereign rights had their economic potential been recognized at the time.[35] In any event, the principles of mutually agreed terms and benefit sharing established in Paragraph 3 of the 1962 Declaration were destined to be expressly applied to genetic resources once the Convention on Biological Diversity was finally adopted in 1992.[36]

Another set of principles of eventual importance for regulating access to genetic resources was first established in the Declaration of the United Nations Conference on the Human Environment, held at Stockholm in 1972.[37] Here the right to exploit natural resources was expressly linked with a duty of sovereign nations to conserve and use resources rationally for the benefit of future generations.[38] This Declaration also forged a link between the goal of securing an adequate return to developing countries from economic exploitation of certain commodities and the duty to promote sound environmental management.[39] Above all, at least for present purposes, the Stockholm Declaration expressly recognized the importance of scientific research in the context of environmental problems "both national and

33 *See* 1962 Declaration, n. 11 (adopted by 87 votes in favor to 2 against, with 12 abstentions). For earlier resolutions, *see, e.g.*, G.A. Res. 523 (VI), U.N. GAOR, 6th Sess., Supp. No. 26, U.N. Doc. A/2052, at 20 (1952) (linking the right to use national resources with economic development policy); G.A. Res. 626 (VII), U.N. GAOR, 7th Sess., Supp. No. 25, U.N. Doc. A/2332, at 18, pmbl. (1952) (stressing rights of states to use and exploit natural resources as inherent in sovereignty). The notion of permanent sovereignty over natural resources is derived from the broader international legal principles of the equality of states, nonintervention, and self-determination of peoples, as embodied in the UN Charter itself. *See* Nico J. Schrijver, *Permanent Sovereignty over Natural Resources*, *in* Max Planck Encyclopedia of Public International Law (Oxford U. Press 2008), online ed. (last accessed 14 June 2014).

34 1962 Declaration, n. 11, pmbl. ¶ 3.

35 Despite intensive work on natural resources in recent decades, no general definition exists of the term "natural resources" in international law.

36 *See* Convention on Biological Diversity, *opened for signature* June 5, 1992, 1760 U.N.T.S. 79 [hereinafter CBD] art. 15.

37 Stockholm Declaration, G.A. Res. 2998 (XXI), U.N. Doc. A/CONF/48/14 (Dec. 15, 1972), *reprinted in* 11 I.L.M 1416 [hereinafter 1972 Stockholm Declaration].

38 *See id.* Principle 21 (adding the "responsibility to ensure that activities within their jurisdiction or control do not cause damage to the environment of other States or areas beyond the limits of national jurisdiction").

39 *See id.* Principle 10.

multinational," and it advocated "the free flow of up-to-date scientific information and transfer of experience ... to facilitate the solution of environmental problems."[40]

Against this background, the developing countries in the 1980s began to organize collective resistance to the unauthorized appropriation of genetic resources from their territories, which they denounced as "biopiracy," in opposition to complaints about "piracy" in the developing countries of products and processes covered by intellectual property rights emanating from the developed countries. As the developed countries campaigned to strengthen international patent protection under the Paris Convention for the Protection of Industrial Property from 1979 to 1986,[41] and then under the Uruguay Round of Multilateral Trade Negotiations that began in 1986,[42] the developing countries became more determined to repress unauthorized uses of *ex situ* and *in situ* genetic resources that had been taken without permission from their territories.

As previously explained, developing country governments feared that, without such a formal legal regime, their genetic resources would end up in patented products, protected under the pending WTO TRIPS Agreement, to be signed in 1994, that would be sold back to them at high prices, even when based on the traditional knowledge of their own indigenous populations.[43] The OECD countries were also pressing the developing countries to preserve *in situ* genetic resources – at the expense of urban expansion and economic development projects generally – because humanity depended on the maintenance of biodiversity. That burden was said to be the responsibility of those developing countries in which such resources were predominantly located.[44]

Meanwhile, the developing countries had begun to implement sovereign control over their genetic resources in at least three types of responses. First, their governments began to make it harder for foreign scientists and other bioprospectors

[40] *Id.* Principle 20 (adding that "environmental technologies should be made available to developing countries on terms which would encourage their wide dissemination without constituting an economic burden on developing countries." *See also id.* Principle 18 (stressing the role of science and technology).

[41] Paris Convention for the Protection of Industrial Property, Mar. 20, 1883, *as last amended on* Sept. 28, 1979, 21 U.S.T. 1583. The Diplomatic Conference to revise the Convention broke down in 1986, which led OECD countries to shift the issues to the GATT negotiating forum.

[42] The Uruguay Round of Multilateral Trade Negotiations, which produced the TRIPS Agreement as part of the Agreement Establishing the WTO, *see* TRIPS Agreement, above n. 21, began with the Ministerial Declaration on the Uruguay Round, Punta del Este, Uruguay, Sept. 20, 1986, GATT B.I.S.P. (33rd Supp) (1987).

[43] *See, e.g.*, DUTFIELD (2004), n. 9, at 18–20.

[44] *See* COMM'N ON INTELLECTUAL PROPERTY RIGHTS (IPRC), INTEGRATING INTELLECTUAL PROPERTY RIGHTS AND DEVELOPMENT POLICY (2002) [hereinafter IPRC (2002)], *available at* http://www.iprcommission.org/papers/pdfs/final_report/ciprfullfinal.pdf; AGRICULTURAL VALUES OF PLANT GENETIC RESOURCES (R. E. Evenson et al., eds., CABI 1998).

to obtain plant, animal, and microbial materials from their territories.[45] Second, developing country governments began to challenge the legitimacy of continued use of *ex situ* specimens of plant and microbial genetic resources held in the two major existing research commons, i.e., the CGIAR's holdings of plant materials and the WFCC's holdings of microbial materials. Increasingly, developing country governments were demanding the return of these *ex situ* plant and microbial resources on the grounds that they might become the object of patents or related intellectual property rights in the developed countries.[46]

Third, developing countries pressed for formal international regulation of access to genetic resources under a binding international treaty, in opposition to the pending TRIPS Agreement emerging from the Uruguay Round of Multilateral Trade Negotiations. These countervailing negotiations ultimately produced the Convention on Biological Diversity of 1992 (CBD).[47] In so doing, the developing countries successfully argued that sovereignty over natural resources entitled them not only to control access to, but also to demand a share of all the benefits deriving from commercial use and applications of, genetic resources by foreign investors and enterprises.[48] They further established the basic principle that developing countries were entitled to compensation for efforts to preserve the world's biodiversity resources.[49]

[45] *See, e.g.*, Flora Katz, *Proposal for a Microbial Semi-Commons: Perspectives from the International Cooperative Biodiversity Groups*, *in* DESIGNING THE MICROBIAL RESEARCH COMMONS: PROCEEDINGS OF AN INTERNATIONAL SYMPOSIUM 129–35 (P.F. Uhlir ed., Nat'l Acad. Press 2011) [hereinafter DESIGNING THE MICROBIAL RESEARCH COMMONS].

[46] Michael Halewood, *Governing the Management and Use of Pooled Microbial Genetic Resources: Lessons from the Global Crop Commons*, 4 *Int'l J. Commons* 404–36 (2010) [hereinafter Halewood (2010)]; Tom Dedeurwaerdere et al., *The Use and Exchange of Microbial Genetic Resources for Food and Agriculture* (Comm'n on Genetic Res. Food & Agric., Background Study Paper No. 46, U.N. Doc. UNEP/CBD/WG-ABS/9/INF/13, 7 Mar. 2009), *available at* http://www.cbd.int/doc/meetings/abs/abswg-09/information/abswg-09-inf-13-en.pdf (last accessed 1 Oct. 2014); IPRC (2002), n. 44.

[47] United Nations Conference on Environment and Development, Rio de Janeiro, Brazil, 3–14 June 1993, Rio Declaration on Environment and Development, U.N. Doc. A/CONF.151/26/Rev. 1 (Vol. I), Annex 1 (12 Aug. 1992). The Rio Earth Summit, *convened in* June 1992, promulgated the Convention on Biological Diversity, and the Rio Declaration on Environment and Development. Art. 15 of the CBD expressly recognizes "the sovereign rights of States over their natural resources." CBD, n. 36.

[48] *See, e.g.*, Juliana Santilli, *Genetic Resources Common Pools in Brazil*, *in* COMMON POOLS OF GENETIC RESOURCES: EQUITY AND INNOVATION IN INTERNATIONAL BIODIVERSITY LAW 112–13 (E.C. Kamau & G. Winter eds., Routledge 2013) [hereinafter COMMON POOLS OF GENETIC RESOURCES (2013) (citing authorities)].

[49] *See e.g.*, CBD, n. 36, arts. 8, 15, 20; REGINE ANDERSON, GOVERNING AGROBIODIVERSITY – PLANT GENETICS AND DEVELOPING COUNTRIES 117–35 (Ashgate, 2008). *See also* Julia Fraser, *New CBD Access And Benefit Sharing Clearing-House Website Presented at WIPO*, IP WATCH (7 Feb. 2014), *available at* http://www.ip-watch.org/2014/02/07/new-access-and-benefit-sharing-clearing-house-website-presented-at-wipo/ (last accessed 23 Dec. 2014).

Some 193 countries have adhered to the CBD, with the notable exception of the United States, which signed but has not ratified the treaty.[50]

2. Access and Benefit Sharing Under the Convention on Biological Diversity

The Convention on Biological Diversity, which entered into force on 29 December 1993, has three main objectives: to promote the conservation of biological diversity, the use of biological diversity in a sustainable fashion, and the fair and equitable sharing of benefits from the use of genetic resources.[51] It also seeks to protect indigenous peoples against unauthorized uses of traditional knowledge pertaining to genetic resources and to secure compensation for commercial uses of such knowledge.[52]

To these ends, the drafters of the CBD explicitly rejected the "common heritage of mankind" concept and replaced it with the principle that "conservation of biological diversity" is a "common concern of mankind."[53] This latter concept, first used in the UN climate change negotiations of 1988,[54] underscored the responsibility of states to cooperate in solving issues that adversely affected them all.

At the same time, the CBD expressly recognized the sovereign rights of states to their natural resources, and it invested national governments with the authority to determine the conditions of access to genetic resources under their domestic laws.[55] These precepts were then further elaborated in a series of articles that, among other things, require each state to:

- Develop national strategies for the conservation and sustainability of biological diversity, in keeping with the express intent of the Convention;[56]
- Identify and monitor important components of biodiversity, especially those subject to adverse impacts;[57]
- Regulate and establish facilities for *ex situ* conservation of plants, animals, and microorganisms, preferably in country;[58] and

[50] *Status – Convention on Biodiversity*, U.N. TREATY COLLECTION (June 27, 2014), https://treaties.un.org/pages/ViewDetails.aspx?src=TREATY&mtdsg_no=XXVII-8&chapter=27&lang=en.
[51] CBD, n. 36, art. 1.
[52] *See id.* art. 8, obliging states to promote wider application of traditional knowledge, practices/know-how, subject to prior informed consent and equitable sharing of benefits. *See, e.g.*, Halewood (2010), n. 46, at 17; Blakeney, n. 17, at 406.
[53] CBD, n. 36, pmbl.
[54] *See* Protection of Global Climate for Present and Future Generations of Mankind, G.A. Res. 43/53, U.N. Doc. A/RES/43/53 (5 Dec. 1988), *available at* http://www.un.org/documents/ga/res/43/a43r053.htm (last accessed 14 June 2014).
[55] CBD, n. 36(l), art. 15.
[56] *Id.* art. 6.
[57] *Id.* art. 7.
[58] *Id.* art. 9.

- Adopt incentive measures for the conservation and sustainable use of biodiversity.[59]

As recompense for undertaking these measures, the CBD expressly entitled countries rich in biodiversity to share in all of the benefits from commercial applications of their genetic resources and related traditional knowledge, as an incentive for bearing the burdens of conserving and managing these same resources in their *in situ* state.[60]

Article 1 of the CBD thus envisions "appropriate access to genetic resources" and the "fair and equitable sharing of benefits arising out of the utilization of genetic resources."[61] Article 2 defines "genetic resources" as "genetic material of actual or potential value,"[62] which includes "any material of plant, animal, microbial, or other origin containing functional units of heredity."[63] The Convention does not further distinguish between subsets of genetic resources, and it establishes the same legal framework for plants, animals, and microorganisms. According to Professor Blakeney, the Convention would thus apply "to seeds and cuttings and DNA extracted from a plant, such as a chromosome, gene, plasmid or any part of these, such as the promoter part of a gene," but not necessarily to "biochemical extracts which do not contain DNA or RNA."[64]

Given this broad subject matter coverage, the "actual or potential value" in question implicitly extends beyond a purely economic calculus to encompass unspecified "nonmonetary benefits" as well, the nature of which will be spelled out in later legal instruments.[65] At the same time, the Convention deliberately regards

59 *Id.* art. 11. As previously noted, "'Biological diversity' means "the variability among living organisms from all sources including, *inter alia*, terrestrial, marine and other aquatic ecosystems and the ecological complexes of which they are part; this includes diversity within species, between species and of ecosystems." *Id.* art. 2.

60 *See* CBD, n. 36, arts. 1, 8, 15–16.

61 CBD, n. 36, art. 1; Blakeney, n. 17, at 405.

62 CBD, n. 36, art. 2.

63 *Id.*; *see, e.g.*, Peter Johan Schei & Morten Walløe Tvedt, *Genetic Resources in the CBD – The Wording, the Past, the Present, and the Future*, *FNI* Report 4/2010 (Fridtjof Nansen Inst. 2010), *available at* http://www.fni.no/doc&pdf/FNI-R0410.pdf (last accessed 14 June 2014). Animal genetic resources are beyond the scope of this volume.

64 Blakeney, n. 17, at 405.

65 *See, e.g.*, Sixth Meeting of the Conference of the Parties to the Convention on Biological Diversity, The Hague, Neth., 17–19 April 2002, Bonn Guidelines on Access to Genetic Resources and Equitable Sharing of the Benefits Arising out of their Utilization, U.N. Doc. UNEP/CBD/COP/6/20, Annex 2 (27 May 2002) [hereinafter Bonn Guidelines]; Tenth Meeting of the Conference of the Parties to the Convention on Biological Diversity, Nagoya, Japan, 18–29 Oct. 2010, Nagoya Protocol on Access to Genetic Resources and the Fair and Equitable Sharing of Benefits Arising from their Utilization (ABS) to the Convention on Biological Diversity [hereinafter Nagoya Protocol] (entered into force on October 2, 2014, after the deposit of the fiftieth instrument of ratification, acceptance, approval, or accession), *available at* http://www.cbd.int/abs/doc/protocol/nagoya-protocol-en.pdf (last accessed 14 Feb. 2014); *see* further Section IV.

the potential economic value of genetic resources as reaching beyond the time of access to encompass downstream applications that advance both knowledge and technology.[66]

Article 8(j) thus imposes three primary obligations on the Contracting Parties to the CBD:

- To respect, preserve and maintain knowledge, innovations and practices of indigenous and local communities embodying traditional lifestyles relevant for the conservation and sustainable use of biological diversity;
- To promote the wider application (with the approval and involvement of the holders) of such knowledge, innovations, and practices; and
- To encourage the equitable sharing of the benefits arising from the utilization of such knowledge, innovation and practices.[67]

As regards *ex situ* genetic resources, Article 9 deals with the "conservation of components of biological diversity outside their natural habitats," for example in seed banks, botanical gardens, museums, laboratories, and agricultural or microbial research institutions.[68] This article calls for national legislation to provide for the acquisition, conservation, storage, and management of *ex situ* collections.[69] Whether sovereignty claims under the CBD apply to genetic resources of provider countries that were collected prior to the entry of the CBD into force in those countries remains controversial in practice, though plausible in theory.[70]

Core provisions to implement all the foregoing obligations are established in Articles 15 and 16.[71] In order to discourage the unregulated appropriation of genetic resources by unauthorized commercial interests, both foreign and domestic, Article 15 of the CBD conditions both access and use on prior informed consent (PIC) of the provider country and mutually agreed terms (MAT), including obligations covering the benefits ultimately to be shared and the transfer of relevant technology.[72] The

66 Schei & Tvedt, n. 63. The extent to which derivatives should be treated as "potential value" of genetic resources within the scope of the CBD, implicit in article 2, was an issue that will be clarified under the Nagoya Protocol. *See* nn. 427–39 and accompanying text.

67 CBD, n. 36, art. 8(j); DUTFIELD (2004), n. 9, at 37–38.

68 *Id*. art. 9.

69 *Id*.; Blakeney, n. 17, at 405.

70 For the view that ABS provisions do not retroactively apply under art. 15(3), *see* Blakeney, n. 17, at 405. *Accord*. Matthias Buck & Clare Hamilton, *The Nagoya Protocol on Access to Genetic Resources and the Fair and Equitable Sharing of Benefits Arising from their Utilization to the Convention on Biological Diversity*, 20 *Rev. Eur. Cmty. Int'l Envtl.* 47, 57 (2011). For the view that the CBD does apply to prior *ex situ* genetic resources, see Frein & Meyer, *Wer kriegle was? Das Nagoya-Protokoll gegen Biopiraterie – Eine politische Analyse*, Evangelischa Entwick Lungsdienst eV. (EED), Bonn, at 57. For the view that the matter remains uncertain, see Godt (2013), below n. 85, at 246–47.

71 CBD, n. 36, art. 15(1), 16.

72 *Id*. art. 15(2).

literature characterizes these provisions as establishing a bilateral system of access and benefit sharing.[73] Research is specifically mentioned in Article 15.7, which further obliges states to take legislative and administrative policy measures, "as appropriate with the aim of sharing in a fair and equitable way the results of research and development and the benefits arising from the commercial and other utilization of genetic resources with the Contracting Party providing such resources … upon mutually agreed terms."[74]

Article 16 then emphasizes the importance of technology transfer as a form of benefit that provider countries may receive.[75] It thus explicitly makes the connection with intellectual property rights, the only such reference in the Convention.[76] Under Article 16, parties to the Convention must undertake to provide and facilitate access to, and transfer of, relevant technologies to other parties, under fair and equitable terms.[77] Where patents or other intellectual property rights apply to the technologies in question, art. 16.2 concedes that access must be "on terms which recognize and are consistent with the adequate and effective promotion of intellection property rights."[78] At the same time, Article 16.5 insists that parties cooperate to ensure that patents and other intellectual property rights "are supportive of and do not run counter to" the objectives of the CBD.[79]

In response to the CBD, more than 60 countries have established Access and Benefit Sharing (ABS) regimes in their domestic laws.[80] Intense negotiations were

73 *See, e.g.*, Santilli (2013), n. 48, at 114; *see further* Chapter 4, Section IV ("From the Bilateral to the Multilateral Approach").

74 CBD, n. 36, art. 15.7.

75 *Id.*, art. 16.1.

76 Dutfield (2004), n. 9, at 38.

77 Only biotechnology is explicitly referenced, but art. 16 speaks of any technologies "that are relevant to the conservation and sustainable use of biological diversity *or make use of genetic resources* and do not cause significant damage to the environment." CBD, n. 36, art. 16 (emphasis supplied).

78 *Id.* art. 16.2.

79 *Id.* art. 16.5. *See* Dutfield (2004), n. 9, at 38 (stating that this "reflects the profound disagreement during the negotiations between those who believed that IPRs conflict with the CBD's objectives and others that saw no contradiction."). For potential conflicts between the CBD and the TRIPS Agreement, see, e.g., Curci (2013), n. 1, at 50–62.

80 For a discussion of pioneering laws in the Philippines, Costa Rica, the Andean Community, plus a proposed African model legislation, *see* Graham Dutfield, Intellectual Property Rights and the Life Science Industries: Past, Present, and Future 138–64 (2d ed., World Scientific Pub. Co. 2009) [hereinafter Dutfield (2009)]. For a discussion of laws in Kenya, Brazil, South Africa, China, and Australia, *see* Genetic Resources, Traditional Knowledge and the Law: Solutions for Access and Benefit Sharing 173–142, 271–210 (E.C. Kamau & G. Winter eds., Routledge 2009) [hereinafter Genetic Resources, Traditional Knowledge and the Law (2009)] (articles by A.N. Anjweni, Juliana Santilli, B. Wynberg & A. Taylor, T. Qin & G. Barton). *See also* M.S. Suneetha & Balakrishna Pisupati, *Benefit Sharing in ABS-Options and Elaborations*, *in* United Nations University-Institute of Advanced Studies Report 28 (United Nations U. 2009), *available at* http://www.ias.unu.edu/resource_centre/UNU_ABS_Report_Final_lowres.pdf

also underway at the World Intellectual Property Organization (WIPO) and other forums with a view to bolstering enforcement of these provisions at the international level.[81] One major proposal would obligate state patent offices to require disclosure of the country of origin whenever patents on genetic resources were filed.[82] Another

(last accessed 14 June 2014). According to Convention on Biological Diversity, *Access and Benefit-Sharing Measures*, CBD, http://www.cbd.int/abs/measures/groups.shtml (last accessed 5 July 2014), there are 57 countries with Nagoya-consistent regimes and 7 regions. Although the majority are developing countries, there are several developed countries, including Australia, Denmark, France, Germany, Japan, Belgium, Canada, Italy, Norway, and Sweden. WIPO is currently compiling a database of all of the biodiversity-related access and benefit-sharing agreements, *Biodiversity-related Access and Benefit-sharing Agreements*, WIPO. http://www.wipo.int/tk/en/databases/contracts/ (last accessed 23 Dec. 2014). The CBD website also has a comprehensive compilation, including legislation, regulation, statements of policy, etc. on the regional, national, and subnational level, at http://www.cbd.int/abs/measures/default.shtml.

For a complete overview and description of international, regional and national ABS measures (as well as of their main gaps and difficulties for their implementation) *see* CENTER FOR INTERNATIONAL SUSTAINABLE DEVELOPMENT LAW (2005); Executive Secretary of the Convention on Biological Diversity (CBD), *Analysis of Gaps in Existing National, Regional and International Legal and Other Instruments Relating to Access and Benefit-Sharing*, at 12–13, U.N. Doc. No. UNEP/CBD/WG-ABS/5/3 (8–12 Oct. 2007) [hereinafter *Analysis of ABS Gaps*]; *Overview of Recent Developments at National and Regional Levels Relating to Access and Benefit-Sharing*, U.N. Doc. No. UNEP/CBD/WG-ABS/5/4 (2007); *Overview of Recent Developments at the International Level Relating to Access and Benefit-Sharing*, U.N. Doc. No. UNEP/CBD/WG-ABS/5/4/ADD1 (2007); *Compilation of Submissions Provided by Parties and Other Relevant Organizations on Issues of Relevance to the International Regime on Access and Benefit-sharing*, U.N. Doc. No. UNEP/CBD/WG-ABS/5/INF/1 (2007); *Compilation of Submissions by Parties on Experiences in Developing and Implementing Article 15 of the Convention at the National Level and Measures Taken to Support Compliance with Prior Informed Consent and Mutually Agreed Terms*, U.N. Doc. No. UNEP/CBD/WG-ABS/5/INF/2 (2007); *Compilation of Submissions by Parties on Experiences in Developing and Implementing Article 15 of the Convention at the National Level and Measures Taken to Support Compliance with Prior Informed Consent and Mutually Agreed Terms (Addendum)*, U.N. Doc. No. UNEP/CBD/WG-ABS/5/INF/2/ADD1 (2007); *Compilation of Submissions by Parties on Experiences in Developing and Implementing Article 15 of the Convention at the National Level and Measures Taken to Support Compliance with Prior Informed Consent And Mutually Agreed Terms (Addendum)*, U.N. Doc. No. UNEP/CBD/WG-ABS/5/INF/2/ADD2 (2007). Most national measures may be accessed in the CBD ABS database, *available at*: http://www.cbd.int/abs/measures.shtml.

[81] *See, e.g.*, Julia Fraser, n. 49; Catherine Saez, *Protection of Folklore Joins TK, GR On Way To WIPO General Assembly*, IP WATCH (7 April 2014), http://www.ip-watch.org/2014/04/07/protection-of-folklore-joins-tk-gr-on-way-to-wipo-general-assembly/. *See also* World Intellectual Property Organization, General Assembly, Matters Concerning Intellectual Property and Genetic Resources, Traditional Knowledge and Folklore, 25 Aug. 2000, WIPO Doc. WO/GA/26/6, *available at* http://www.wipo.int/meetings/en/doc_details.jsp?doc_id=1460 (last accessed 14 June 2014). The Intergovernmental Committee (IGC) on Intellectual Property and Genetic Resources, Traditional Knowledge and Folklore seeks to ensure the effective protection of traditional knowledge (TK), traditional cultural expressions (TCEs)/folklore, and genetic resources, and it maintains a database of its meeting sessions and draft articles at http://www.wipo.int/tk/en/igc/index.html.

[82] *See, e.g.*, Evanson C. Kamau, *Disclosure Requirements – A Critical Appraisal*, *in* GENETIC RESOURCES, TRADITIONAL KNOWLEDGE AND THE LAW (2009), n. 80, at 399–418 [hereinafter Kamau (2009)]; U.N. Conference on Trade & Dev (UNCTAD), Analysis of Options for Implementing Disclosure of Origin

proposal envisioned at least a soft law declaration recognizing that the taking of genetic resources for any purpose without prior informed consent amounts to tortious conduct under an emerging international regime of misappropriation.[83]

Meanwhile, in 2002, the Conference of the Parties adopted the Bonn Guidelines on Access to Genetic Resources and Fair and Equitable Sharing of the Benefits Arising out of their Utilization.[84] These Guidelines were intended to assist the Contracting Parties in establishing administrative, legislative, or policy measures on Access and Benefit Sharing (ABS) and in negotiating contractual arrangements for access to genetic resources and benefit sharing. However, these guidelines were viewed as largely provisional, pending further negotiation and adoption of a protocol on these and other issues, which eventually resulted in the Nagoya Protocol of 2010, discussed in Section IV.

C. *Critical Evaluation of the CBD*

Under the CBD's "bilateral approach" to transactions between providers and users of genetic resources, every transaction must, in principle, trigger a negotiated outcome that generates prior informed consent plus an agreement concerning the sharing of benefits from eventual commercial applications. In retrospect, this simple-minded approach traded a market-like methodology for the preexisting public goods regime,[85] without any serious evaluation of the likely social costs and benefits. It also ignored the fact that "stakeholders can act both as providers and users of genetic resources" with "no clear-cut line … between them,"[86] and that the value added to both *in situ* and *ex situ* genetic resources by public research scientists was typically a *sine qua non* in the production of any benefits to be shared, as will be seen.[87] That the bilateral approach might constitute a disincentive, or even

Requirements in Intellectual Property Applications, UNCTAD/DITC/TED/2004/14 (2006), *available at* http://unctad.org/en/Docs/ditcted200514_en.pdf. Some countries, such as Switzerland and Norway, among others have implemented this request.

83 Intergov't Committee on Intellectual Prop. & Genetic Resources, Traditional Knowledge, & Folklore, Protection of Traditional Knowledge, Summary of Draft Policy Objectives and Core Principles, WIPO/GRTKF/IC/7/5 (2004), Annex 1, at 6.

84 Bonn Guidelines, n. 65. These Guidelines listed examples of benefits and distinguished between monetary and non-monetary benefits. See ANDERSON (2008), n. 49, at 137.

85 *See, e.g.*, Sélim Louafi & Marie Schloen, *Practices of Exchanging and Utilizing Genetic Resources for Food and Agriculture and the Access and Benefit Sharing Regime, in* COMMON POOLS OF GENETIC RESOURCES (2013), above n. 48, at 205–07; Christine Godt, *Networks of* Ex Situ *Collections of Genetic Resources, in* COMMON POOLS OF GENETIC RESOURCES (2013), above n. 48, at 249 (contrasting contractually constructed or tailor-made regimes with the bilateral approach). *See further* Section I.C.2.b.

86 *See, e.g.*, Louafi & Schloen (2013), n. 85, at pt. 2, "Potential impact of ABS measures on the exchange of genetic resources for food and agriculture."

87 *See, e.g.*, Godt (2013), n. 85, at 246. For the value-adding practices of *ex situ* microbial collections, see Chapter 4, Section I.A–B.

a major obstacle, to scientific research seems to have been overlooked in the course of negotiations triggered largely by adverse reactions to ongoing negotiations concerning the TRIPS Agreement.

1. The CBD as an Incomplete International Regime of Misappropriation

Leaving aside concerns about the social costs of this bilateral approach for a moment, a more immediate question posed by the signing of the CBD in 1992 was how exactly its new regime of PIC and MAT was to be enforced at the international level. More precisely, if the CBD attempts to establish the foundations of an international regime of misappropriation to punish unauthorized uses of genetic resources emanating from the developing countries, how were the aggrieved providers to discipline violators in a decentralized universe of national states operating under their respective territorial laws?

As countries that possessed a wealth of biodiversity took steps to implement the CBD in their domestic laws and administrative regulations, permission to access and use genetic resources under these laws was increasingly required. Obtaining such permission from the relevant authorities became ever more difficult in practice.[88] Even scientists within the countries concerned began to experience difficulties in obtaining microbial specimens for research from national culture collections, and especially in arranging cross-border exchanges of such materials with other scientists.[89]

At the same time, national laws implementing the CBD had to be reconciled with national laws protecting intellectual property rights under the TRIPS Agreement of 1994, especially patents and plant breeders' rights, as discussed in the previous chapter. Unlike violations of the CBD, alleged violations of the TRIPS Agreement subjected the offending party to actions before WTO dispute settlement panels, with the risk that cross-sectoral damages would be awarded to the aggrieved state.[90] How to achieve the objectives of the CBD without undermining the objectives of

88 *See, e.g.*, Evanson C. Kamau & Gerd Winter, *Streamlining Access Procedures and Standards*, *in* Genetic Resources, Traditional Knowledge and the Law (2009), n. 80, at 365–79 [hereinafter Kamau & Winter (2009)].

89 Ninth Meeting of the Ad Hoc Open-Ended Working Group on Access and Benefit-Sharing in the Convention on Biological Diversity, Cali, Colombia, 22–28 March 2010, Side Conference Presentations [hereinafter Cali Presentations], *available at* http://www.cbd.int/wgabs9/events/se-abs9.shtml#tab=0; Ad Hoc Open-Ended Working Group on Access and Benefit-Sharing, Report of the First Part of the Ninth Meeting of the Ad hoc Open-Ended Working Group on Access and Benefit-Sharing UNEP/CBD/WG-ABS/9/3 (26 Apr. 2010), Annex I. *See also, e.g.*, Santilli (2013), n. 48, at 118 (stressing effect of discouraging research).

90 TRIPS Agreement, n. 21, art. 64; WTO Panel Report, *Canada – Patent Protection of Pharmaceutical Products* (Complaint by the EU), WT/DS114/R (17 Mar. 2000), *available at* http://www.wto.org/english/tratop_e/dispu_e/7428d.pdf, and Report of the Arbitrator, *Arbitration under Article 21.3(c) of the DSU*, WT/DS114/13, 18 Aug. 2000.

the TRIPS Agreement thus remained an unresolved problem for both policymakers and scholars.[91]

Meanwhile, the battery of domestic laws enacted in the wake of the CBD had only territorial effects, without any means of enforcement at the international level. Taken together, these national laws did reinforce claims of misappropriation against unauthorized use of genetic resources generally, in keeping with the principles set out in the CBD.[92] However, efforts to put the enforcement of a full-fledged international regime of misappropriation on a more solid legal foundation bogged down in interminable theoretical discussions at WIPO and in informal negotiations under the auspices of the Conference of the Parties to the CBD (COP).[93] As a result, actual enforcement typically depended on the laws and regulations adopted in user states, on their willingness to recognize the objectives of the CBD or to cooperate with provider states, and on their own interpretations of intellectual property laws sanctioned by the TRIPS Agreement.[94] As the selected cases of alleged biopiracy discussed in the next section reveal, this was often a very uncertain process for aggrieved parties seeking redress in provider states.

As a practical matter, governments in provider countries asserting complaints about "biopiracy" in user countries had to rely heavily on protests through diplomatic representation rather than legal proceedings in domestic courts. Bolstered by reference to international law as codified in the CBD, these channels sometimes produced negotiated settlements that were presumably

[91] *See* Curci (2013), n. 1, at 50–86; Dutfield (2009), n. 80, at 165–203 (case of India), 204–18 (case of Kenya), 219–21 (lessons from the case studies). *See also* S.A.S. Kishi, *PIC in Access to TK in Brazil, in* Genetic Resources, Traditional Knowledge and the Law (2009), n. 80.

[92] *See, e.g.*, Blakeney, n. 17, at 404–05; McManis, n. 12.

[93] These negotiations came to a dramatic conclusion at Nagoya in 2010, where the COP unexpectedly succeeded in codifying a powerful version of that regime, to be discussed in Chapter 4, Section IV.

[94] For example, the patent granted to W.R. Grace and Co. in 1992 for a pesticide whose active ingredient was derived from the neem tree (*Azadirachter indicia*) did not actually cover the neem seed itself nor its use as a natural insect repellent in keeping with customary uses in India. Rather, according to Professor McManis, the patent covered a method of production and a resulting storage stable solution that was actually an improvement on the natural product. *See* U.S. Patent No. 5,124,349 (filed 21 Oct. 1990), for "a storage stable *azadirachtin* formulation," the active ingredient of which is derived from the neem tree. *See also* McManis, n. 12, at 425–510, 454. The patent did not prevent use of the neem extract in the traditional manner or by other nonpatented methods. The fact that use of the neem seed as a natural insect repellent was widely known in India did not necessarily negate novelty for such an improvement under the then "relative novelty" standard of U.S. patent law, although some neem-related European patents were subsequently revoked under different eligibility criteria. *See* McManis, n. 12, at 455–56. The U.S. novelty standard became absolute, not relative, after TRIPS; *see* Christoph Spennemann, UNCTAD Div. on Inv. & Enter., TRIPS Pre-Grant Flexibilities: Patentability Criteria, *available at* http://ictsd.org/downloads/2010/01/patentability-criteria-rev.pdf (last accessed 14 June 2014).

satisfactory to the parties. For example, widespread bioprospecting for species or compounds in Brazil during the period 1980–1996 had reportedly resulted in pharmaceutical uses by major companies in the United States and Europe. The government of Brazil, together with several public interest groups, lodged vigorous protests in 1996, and the cases were "apparently resolved," although few details are available.[95]

When disputes were amicably resolved in this way, the CBD provided at least moral support for the complainants. More often, however, complaints about illicit use of genetic resources – whether lodged through diplomatic channels or in domestic courts – bogged down in the intricacies of domestic intellectual property laws, as well as contracts and conflicts laws, which raised issues that were left largely unresolved by the Convention itself. Further complications arose because the United States never ratified the CBD and because, many of the unauthorized uses complained of were, initially at least, of a purely scientific character, which the CBD did not expressly address.[96]

Because, the CBD embodied no specific enforcement measures in international law, the likelihood of success in any given protest thus depended on the vagaries of domestic laws, a defect later to be addressed by the Nagoya Protocol of 2010.[97] As will be seen in Section IV, that Protocol would, if ratified, impose a strong regime of enforcement at both the multilateral and domestic levels, from which few escapes would be possible in the future.

[95] Young et al., n. 13, at 133 (Annex 1). Also in Brazil an indigenous community objected to patents filed between 1994–2000 on Cuani and Tipir (product name "Cuanio)." Samples had been collected in the 1990s. This case is reportedly closed (but on what basis is not known to us yet). *Id.* at 131 (Annex 1). Public disclosure was also reported with regard to the following items in the Amazon region:

- *Acai Enterpe precatoria*;
- Cat's Claw, Sangre de Drago, Quebra Pedras, and wormseed;
- *Cupuaca Theobroma grandisflorum*

Claims were also recorded with regard to the following items in the Andean Region:

- Nuna Bean
- Quinoa.

Young et al., n. 13, at 101–02, tbl. 1.

[96] *See, e.g.*, Christine Godt, *Enforcement of Benefit-Sharing Duties in User Countries*, *in* GENETIC RESOURCES, TRADITIONAL KNOWLEDGE AND THE LAW (2009), n. 80, at 419–38 [hereinafter Godt (2009)]; Hiroji Isozaki, *Enforcement of ABS Agreements in User States*, *in* GENETIC RESOURCES, TRADITIONAL KNOWLEDGE AND THE LAW (2009), n. 80, at 439–54.

Although research and science are recognized, see CBD n. 36, arts. 12 and 18, no provisions addressed the specific needs of scientists for facilitated access to genetic resources for basic research.

[97] *See* Section IV.C. For hopeful references to jurisdictional scope and cooperation, see CBD, n. 35, arts. 4 & 5.

2. The Threat to Public Scientific Research on Plant and Microbial Genetic Resources

Although most of the better known cases of alleged biopiracy after 1992 concerned plant genetic resources, a number have also dealt with microbes, but less is usually known about the relevant details. The cases summarized here dramatically illustrate the ways in which research scientists have been trapped between overlapping proprietary claims emanating from both the CBD and the TRIPS Agreement since the early 1990s. Similar claims – and the attendant risks for both scientists and research institutes – will become far more formidable now that the Nagoya Protocol has taken effect. That prospect, and the need to obviate the attending risks, is a primary reason why we have written this book.

A. SELECTED CASES OF ALLEGED BIOPIRACY INVOLVING ACADEMIC RESEARCHERS AFTER 1992. One of the earliest cases of alleged biopiracy in Asia arose under Australia's Plant Breeders Rights Act of 1994, when two national agricultural research institutes sought to protect certain species of chickpeas that had been bred from material originally provided by the International Crop Research Institute for the Semi-Arid Tropics (ICRISAT), a CGIAR member.[98] In 1998, a leading scientific journal criticized the relevant government agencies for seeking "property rights in chickpeas grown by subsistence farmers in India and Iran,"[99] and there were protests from an Asian NGO about "privatizing seeds that belong to our farmers and selling them back to us."[100] The applications were eventually withdrawn, and regulations later adopted required applicants for plant breeders' rights in varieties derived from germplasm obtained from CGIAR centers to demonstrate that permission had been obtained from the relevant center.[101]

Similarly, a patent on blight resistant rice, filed by the University of California in 1995 and granted in 1999,[102] was traced back to a strain of rice from Mali, *Oryza longistaminata*, which a researcher in India had identified as resistant to bacterial blight in the 1970s. In 1978, the resistant sample was taken to the International Rice Research Institute (IRRI) in the Philippines for further investigation, where researchers spent fifteen years developing a high-yield, blight-resistant strain of rice by conventional breeding methods. After IRRI identified a single focus, called Xa21, as the locus of resistance, a postdoctoral fellow from the University of California at

[98] *See* Blakeney, n. 17, at 393, 395.
[99] *See id.* (citing NEW SCIENTIST, Feb. 14, 1998).
[100] *Id.* at 398.
[101] *Id.*
[102] *See, e.g.*, Blakeney, n. 17, at 404–05.

Davis, who was working at IRRI, arranged to map, sequence, and clone the Xa21 gene.[103]

The patent obtained by the University of California, which listed Dr. Pamela C. Ronald and her co-workers as inventors, was viewed as compromising CGIAR's research efforts in rice-producing regions of Asia.[104] There were also questions about compensation for the traditional farmers of Mali who had conserved *O. longistaminata*. These concerns prompted U.C. Davis to establish a benefit-sharing fund for commercial use of its patent; to allow noncommercial researchers access to the gene (subject to a noncompetition clause); and to allow IRRI full rights "to develop new rice varieties incorporating cloned Xa21 and [to] distribute this material as well as the clone to developing countries."[105] The university acknowledged that, absent such measures, it would have been "more difficult for ... [them] in the future to obtain research access to developing countries' national genetic materials."[106]

Particularly controversial cases concerning plant genetic resources originating from Latin American provider countries further reveal the kind of pressures that the CGIAR research centers operated under, once the CBD had taken effect.[107] For example, one case concerned five traditional varieties of yacon held at the CGIAR's International Potato Center (CIP) in Peru, which the Ministry of Agriculture had sent to researchers in Japan. Yacon (*Smallantus sonchifolias*) is an ancient Andean crop, eaten as a fruit, with a high fructose content and a high percentage of insulin.[108] Cultivation of yacon in Japan for use as a vegetable, pickles, and juices was subsequently reported.

However, once plant breeders' rights were obtained on the first commercial variety named Sarada-Otome in 2000, Japanese researchers refused to send the relevant germplasm to Peru for testing in local farmers' fields.[109] Although an inquiry ultimately held CIP blameless for having sent the germplasm to Japan in the first place,[110] this case – like many others – reveals tensions between the public good role of genetic resource repositories and claims by provider countries against

[103] *Id.* at 396.

[104] U.S. Patent No. 5,859,339 (issued Jan. 12, 1999),

[105] Blakeney, n. 17, at 396–97.

[106] *See id.* at 397–99. The application covered "nucleic acids from *Oryza sativa*, which encode leucine-rich repeat polypeptides and enhance *Xanthomanas* resistance in plants." *Id.* at 398 (citing U.S. Patent No. 5,859,339 (issued Jan. 12, 1999)).

[107] For the political ramifications of such pressures, see the discussion of the Crop Commonsin Section III.

[108] Blakeney, n. 17, at 400. Its leaves reportedly have anti-diabetic properties. *Id. See* Nat'l Research Council, Lost Crops of the Incas: Little Known Plants of the Andes with Promise for Worldwide Cultivations (Nat'l Acads. Press 1989).

[109] Blakeney, n. 17, at 400–01.

[110] *Id.* at 401.

downstream intellectual property holders' commercial derivatives from cultivars deposited in these repositories.[111]

In much the same vein, a U.S. patent was granted to two agronomists at Colorado State University in 1994 for "Cytoplasmic Male Sterile Quinoa," and it was later assigned to a commercial technology company.[112] Although quinoa (*Chenopodium quinoa*) had long been grown as a drought resistant food crop in elevated regions of the Andes, and male sterile quinoa lines have been reported in technical literature, the patent claimed that a reliable system of cytoplasmic male sterile plants had not previously been "available for commercial production of quinoa hybrids."[113]

According to Professor Dutfield, one of the inventors, Professor Sarah Ward, stated that she had found the cytoplasm in question in quinoa plants of the Bolivian Apelena variety growing in a field in Colorado. She argued that the cytoplasm did not exist in quinoa plants growing in South America, but had been transferred naturally from a related weed species growing nearby in Colorado. Because this claim was never made clear in the patent, "the failure to indicate the non-Bolivian provenance of the cytoplasm inducing male sterility or to refer to the discovery made it possible to interpret the patent very broadly in ways that the inventors may not have intended."[114]

A certain NGO – the Rural Advancement Foundation International (RAFI)[115] – campaigned against this patent, partly on the grounds that if a larger market should develop for high-yielding hybrids of quinoa derived from a traditional Bolivian variety, it would displace Bolivia's existing export market.[116] The patent ultimately became worthless, however, and was abandoned, because commercial production proved unfeasible.[117] Nevertheless, the case raised troubling questions about uses that may or may not be made of genetic resources ostensibly in the public domain,[118] and about how the public domain was actually to be defined – or reconfigured – in an emerging world order.

Another case that elicited public outcry, particularly in Peru, arose from patents issued in the United States on two compounds derived from *Maca Lepidium meynii*

[111] *See further* Section III (discussing the International Treaty on Plant Genetic Resources for Food and Agriculture).

[112] U.S. Patent No. 5,304,718 (issued April 19, 1994); McManis, n. 12, at 460.

[113] *Id.* (citing the patent).

[114] Dutfield (2004), n. 9, at 54.

[115] Now known as the Action Group on Erosion, Technology and Concentration (ETC). *See* McManis, n. 12, at 460.

[116] *Id.* at 461.

[117] *Id.*

[118] *See, e.g.*, McManis, n. 12, at 161 (arguing that RAFI's claim conflicted with "the oft-stated view of public research organizations, as well as many proponents of farmers' rights, that plant genetic resources should be freely available as 'the common heritage of mankind.'").

in 2000 and 2001 for Viagra-like therapeutic effects.[119] This plant, of the *cruciferae* mustard family, had been grown for centuries by indigenous peoples in the Puna highlands of Peru, both as a staple food crop and for medicinal purposes.[120] Despite many legal ambiguities,[121] the case raised troubling questions about uses of maca seed deposited in the repository managed by the International Potato Center mentioned above;[122] about the incentives for future deposits that may or may not be made to similar repositories under applicable international agreements;[123] and about obligations to share the benefits of commercial applications from uses of traditional knowledge under these agreements.

Some thirty-three Andean varieties of beans from Peru, Bolivia, Ecuador and Colombia were reportedly the source from which a "bean-nut popping bean" – the subject of a U.S. patent in 2000 – had been derived.[124] Indigenous communities condemned these patents as contrary to their own commercial and environmental interests and as a violation of their previous efforts to keep the varieties public to "ensure continued maintenance of the world's seed biodiversity."[125] Nine of the varieties in question had, in fact, been held at the International Center for Tropical Agriculture (CIAT), a CGIAR Center in Cali, Columbia, and that Center was urged to keep farmer-bred bean varieties in the public domain.[126] Such statements, however, beg the questions raised earlier about who configures the public domain under what authority, and about the duties of public repositories that may actually be imposed under emerging international laws.[127]

119 Young et al., n. 13, at 132 (Annex 1); Blakeney, n. 17, at 402; McManis, n. 12, at 464 (citing Extract of *Lepidium Megenii* Roots for Pharmaceutical Applications, U.S. Patent No. 6,267,995 (issued July 31, 2001)); Maca & Antler for Augmenting Testosterone Levels, U.S. Patent No. 6,093,421 (issued July 25, 2000); Compositions & Methods for Their Preparation from *Lepidium*, U.S. Patent No. 6,552,206 (issued April 22, 2003); Treatment of Sexual Dysfunction with an Extract of *Lepidium meyenii* Roots, U.S. Patent No. 6,428,824 (issued Aug. 6, 2002). For details, see McManis, n. 12, at 464.

120 Blakeney, n. 17, at 402. Complaints from the Peruvian government and NGOs representing indigenous tribes and farmers had reportedly not produced results. Young et al., n. 13, at 132 (Annex 1).

121 According to Professor McManis, the patents in question are technically not "patents on Maca," as claimed by RAFI/ETC, but entail process patents for producing specified derivatives for specified purposes. On this view, the U.S. patents did not foreclose the market for imports of Peruvian farmers' maca, which cannot be grown in extremely high altitudes, but would only bar imports of the specified extracts or mixtures and may have helped to "create a market for a plant that was in danger of going extinct." McManis, n. 12, at 464–65 (citing authorities).

122 McManis, n. 12, at 464.

123 *See* CBD, n. 36, and see also International Treaty on Plant Genetic Resources for Food and Agriculture, opened for signature 3 Nov. 2001, 2400 U.N.T.S. 303 (entered into force 29 June 2004) [hereinafter ITPGRFA], discussed in Section III; Halewood (2010), n. 46.

124 Blakeney, n. 17, at 401 (citing U.S. Patent No. 6,040,503 (issued July 16. 2002)); Patent Cooperation Treaty Patent No. WD99/11115.

125 News Release, RAFI, "Bracing for 'El Nuna,' Andean Groups Hopping Mad About Popping-Bean Patent," Mar. 20, 2001.

126 Blakeney, n. 17.

127 *See* n. 21–25 & accompanying text.

The government of Mexico also challenged patents on Enola bean, including the varieties *Azufrato* and *Mayocoba*, granted to Larry Proctor, president of a Colorado-based seed company, in 1999.[128] The patented bean was allegedly developed from a bag of dry beans purchased and brought over from Mexico (although the data about importation remain uncertain and disputed).[129] Professor Dutfield argues that the patent claims were excessively broad in view of the prior art and that Proctor "had employed conventional crossing and selection breeding methods that were not novel. Yet the patent prevented others from using the bean and other beans with similar characteristics in their own breeding programs."[130] Proctor's company reportedly demanded royalties from importers of Mexican beans to the United States (which threatened a serious drop in sales), and at least two infringement actions against sellers of similar beans in the U.S. were apparently filed.[131]

CIAT, the gene bank affiliated with CGIAR, also filed a formal request for re-examination of this patent, alleging that the Enola bean was the same as the Mexican yellow bean, and that widely available prior art from the literature should have defeated the patentee's novelty and nonobviousness claims.[132] CIAT declared that it maintained some 260 bean samples with yellow seeds, six of which were substantially identical to claims set out in the patent.[133] CIAT also argued that the patented genetic resources had been "misappropriated" from Mexico in violation of the Convention on Biological Diversity,[134] which, however, the United States had never ratified.[135] Apparently, the patent was later invalidated for obviousness.[136]

With specific regard to microbial genetic resources, some of the relevant case studies deal with medicines and cosmetics that were allegedly based on materials originating from Africa.[137] Among them are at least five U.S. patents bearing on a micro-bacterium collected from Uganda in the 1970s and used to fight chronic viral infections, including HIV. The material in question – *Mycobacterium vaccae*

[128] *See* Young et al., n. 13, at 132 (Annex 1); Blakeney, n. 27, at 399–400 (citing U.S. Patent No. 5,894,079 (issued Apr. 13, 1999)). Proctor also obtained a U.S. plant variety protection certificate on the same bean variety. *Id.* at 399.

[129] McManis, n. 12, at 465.

[130] DUTFIELD (2004), n. 9, at 54–55.

[131] Blakeney, n. 17, at 399–400 (citing authorities).

[132] *Id.*

[133] *Id.* (citing News Release, RAFI, "Enola Bean Patent Challenged," Jan. 5, 2001, *available at* http://www.rafi.org). However, CIAT did not have evidence that the patent owner had obtained yellow beans from CIAT's gene bank. Blakeney, n. 17, at 400.

[134] Blakeney, n. 17, at 400.

[135] For implications of the subsequently adopted, n. 123, *see* Section III. A & B.

[136] *See* Young et al., n. 13, at 132 (Annex 1).

[137] An early study provides details concerning thirty-six case studies of medicines, cosmetics, and agricultural products that allegedly originated from biodiversity (including plants, marine life, and microbes) situated in African countries. *See* MCGOWN & BURROUGHS, n. 14.

R877.R – was originally isolated from mud samples in central Uganda without permission. The owner of the patent is a British company that reportedly had more R877R-related patents in the pipeline.[138]

In another case concerning extremophile microbe[s] found in Kenya, the conflict arose when a U.K.-based company, Genecor, listed derived products – known as "IndiAge Neutra" and "Puradax" – in its sales catalogs. Samples of this microbe, collected by the University of Leicester in 1992, were subsequently transferred to the company, which first described its unpatented products in its annual corporate report. The Government of Kenya, acting through its Wildlife Services, claimed that the collector who had obtained the samples in a protected area did not have the government's permission to take the material from this area, and that he could not provide any evidence to the contrary.[139] Thus, the core of the claim lies in the illegal collection practices, which could arguably invalidate the rights of the user, even though no issues of traditional knowledge were at stake. As of 2009, the parties were reportedly still in negotiation.[140]

Still another case of African microbes arose in Zimbabwe, where the University of Lausanne (Switzerland) had obtained research access to *Swartzia madagascariensis* in 1995. A U.S. patent was granted in 1999 on isolated compounds, known as "*antimicrobial diterpenes*." Two local NGOs and one Swiss NGO protested against the unauthorized use of traditional knowledge, among other complaints; but these protests were apparently unavailing, and the patent reportedly expired at the end of its term.[141]

A similar case concerned a Mexican microbe, *Bacillus Subtilis*, used in Pozol, a Mayan drink derived from fermented corn, which generations of traditional knowledge had associated with nutritional and medical benefits. In 1999, a Dutch corporation, Quest International, and the University of Minnesota jointly obtained a patent, which they claimed only covered an isolated microorganism and not the Pozol itself. The patentees also denied using any of the traditional knowledge that they had access to in the 1990s.[142] This case has never been resolved, and is no longer active, as the patent expired.

Many other instances of alleged biopiracy after 1992 might be cited, some with happier endings than others.[143] The selected cases show that the CBD strengthened

[138] Young et al., n. 13, at 115,132 (Annex 1).

[139] *Id.*

[140] *Id.* at 133 (Annex 1).

[141] *See* Chakravarthi Raghavan, *Biopiracy in Zimbabwe, Patenting by Swiss University Denounced* (Jan. 2002), *available at* http://www.twnside.org.sg/title/denounced.

[142] Marcia Ellen DeGeer, *Biopiracy: The Appropriation of Indigenous Peoples' Cultural Knowledge*, 9 *New Eng. J. Int'l & Comp. L.* 201 (2002).

[143] *See, e.g.*, DUTFIELD (2004), n. 9, at 52–53 (the Hoodia case in South Africa); McManis, n. 12, at 457–58 ("wine of the soul" case in Ecuador); *id.* at 462–63 (the turmeric and basmati rice patents challenged by the Indian Council of Scientific and Industrial Research).

the developing countries' legal position when protesting against unauthorized use of plant and microbial genetic resources in developed countries after 1992. It did not, however, give aggrieved parties standing to sue under the Convention in the domestic courts of the Contracting Parties; nor did it provide any guidance or forum for resolving conflicts between the laws of user and provider countries, particularly with regard to intellectual property rights in conflict with access and benefit sharing claims.

Meanwhile, under the influence of the CBD, cases of alleged biopiracy had multiplied so fast that, as one commentator put it, "it seems there are no legitimate cases of valid access."[144] This tendency left the scientific community and its basic research infrastructure at the mercy of the political conflicts and cross-currents generated by tension between the CBD and the TRIPS Agreement.

B. MAJOR WEAKNESSES OF THE "BILATERAL APPROACH". To its drafters' credit, the CBD took steps towards a more rational process of conserving *in situ* biological resources that are of potential benefit to all signatory countries. The CBD also established the first set of basic international legal principles to govern access to genetic resources.[145] However, in attempting to implement these principles, the *demandeur* governments have tended to adopt a Coasean model,[146] according to which genetic resources, once invested with de facto property rights, would be traded in response to market forces.

From a theoretical perspective, the PIC, MAT, and ABS regime provides both incentives to access and use genetic resources and disincentives in the form of potentially burdensome legal, financial and administrative burdens. High transaction costs for both users and providers then lead logically to cost-benefit analyses with uncertain outcomes. Although more information about available resources eventually might be produced and shared, where costs appear to exceed measurable benefits fewer legitimate exchanges may be made and more backdoor or hidden exchanges may be attempted.[147]

In reality, genetic resources rarely possess market value in themselves, but rather typically constitute precompetitive inputs into both basic and applied research. Moreover, their ultimate research value, under modern methods, often depends

144 M. Ribadeneira Sarmiento, *Biopiracy or Fallacy? Identifying Genuine Biopiracy Cases in Ecuador*, *in* GENETIC RESOURCES, TRADITIONAL KNOWLEDGE AND THE LAW (2009), n. 80, at 141, 141–48 (stating that the "existing reports give the … impression that biopiracy is a political denomination … but not a national or international legal entity that could be presented to courts in order to set reparation or compensation for the country of origin of the G[enetic] R[esources].").

145 *See esp.* CBD, n. 36, art. 15.8 (i) (establishing PIC and ABS principles).

146 Ronald Coase, *The Problem of Social Cost*, 3 *J. Law & Econ.* 1–44 (1960).

147 *See* Louafi & Schloen (2013), n. 85, at 209–12.

on the researchers' ability to aggregate large quantities of genetic materials that all bear on the subject of enquiry, especially in early stages of research when still uncharacterized organisms may have to be exchanged for screening.[148] As a result, the Coasean model issuing from the ongoing negotiations of the Parties directly conflicts and interferes with the needs of basic scientific research in both developed and developing countries, with potentially serious impediments to eventual commercial applications that the system is designed to produce.

In this context, research activities dependent on public funding will suffer more from costly and lengthy exchange transactions than the private sector, which can often fall back on its own collections and genetic derivatives. Even the public *ex situ* repositories face mounting transaction costs (as documented in Chapter 4 for microbial culture collections), as well as challenges to the legitimacy of their operations.[149] At the same time, these collections lack opportunities to share in the ensuing benefits owing to their status as intermediaries,[150] unless they shift to a more proprietarial business model. Evidence marshaled in this book shows that considerable movement towards a more proprietary model does not bode well for the public good mission these intermediaries have traditionally advanced.[151]

Because the CBD's approach invests single cultivars or microbes with a property interest under sovereign control, countries that conserve biodiversity tend to treat both their *in situ* and *ex situ* holdings as if they were all of great potential value. That approach is understandable in view of the unforeseen blockbuster applications that sometimes actually managed to reach the market place, only to elicit the kind of *ex post* recriminations visible in the "biopiracy" allegations reviewed in the previous section.

In practice, however, under the emerging legal regime, scientists and governments must bargain case-by-case for access to ever more restricted resources, with mounting transaction costs, lengthy delays and growing reports of refusals to deal even for local scientists working within their home countries.[152] For example, a collaborative project with entomologists in India to study insects of the Western Ghats was aborted by the

[148] *Id.* at 213. Breeding of plant varieties may also rely on recurrent exchanges of germplasm. *Id.*

[149] *See* Godt (2013), n. 85, at 15–16. *See also* James S. Miller, *Impact of the Convention on Biological Diversity: The Lessons of Ten Years of Experience with Models for Equitable Sharing of Benefits, in* BIODIVERSITY AND THE LAW: INTELLECTUAL PROPERTY, BIOTECHNOLOGY AND TRADITIONAL KNOWLEDGE 58, 64 (C. McManis ed. 2007) (stating that "[u]p-front payments, expensive permit fees, and/or significant commitments to training or capacity building may be reasonable expectations of research efforts conducted by large corporate entities, but they may be prohibitive impediments for individual non-commercial research programs or small commercial programs.").

[150] *See, e.g.*, Godt (2013), n. 85.

[151] *See* Chapter 4, Section II.

[152] *See, e.g.*, Santilli (2013), n. 48, at 118; Katz (2011), n. 48.

Indian National Biodiversity authority because of biopiracy concerns.[153] Similarly, Juliana Santilli reports that the CBD failed to take into consideration the complexity of social and cultural processes that enrich agrobiodiversity:

> It tends to undermine the free circulation of plant genetic material, encourage monopolies, and restrict [the] public domain, and potentially can have a negative impact on local agricultural systems.[154]

Besides breeding formidable transaction costs, the bilateral proprietary approach potentially invests numerous stakeholders with claims and possible veto power over the use of genetic resources for virtually all purposes.[155] For example, one major project for bioprospecting in Mexico, sponsored by the International Cooperative Biodiversity Group (ICBG) and several pharmaceutical companies, failed because it could not obtain the unanimous consent of the relevant villages (and of the confederation of local healers' organizations), who all claimed "ownership" interests.[156] In this respect, both provider governments and local entities are treating their *in situ* genetic resources more or less on a par with the "special collections" that microbial culture collections keep for private industry or applied research teams,[157] often without having added any extra value to the resources in question and increasingly with so many restrictions on research that no licensing transaction ever takes place.[158]

Altogether missing in this initial approach was any recognition of, or provision for, the role and needs of public science and the culture collections as indispensable intermediaries between the providers of raw materials and the commercializers of end products.[159] During the negotiations, on the contrary, scientific research – when not ignored, despite vigorous protests from the CGIAR, Bioversity, and other research-oriented NGOs – was generally treated as a potential cash cow, whose use of genetic resources would likely generate "actual or potential benefits."[160] Also

[153] K.S. Jayaraman, *Entomologists Stifled by Indian bureaucracy*, 452 NATURE 7 (2008), *available at* http://www.nature.com/news/2008/080305/full/452007a.html (last accessed 14 June 2014); *see generally* Sikina Jinnah & Stefan Jungcurt, *Could Access Requirements Stifle Your Research?*, SCIENCE 464 (2009).

[154] Santilli (2013), n. 48, at 118.

[155] *See id.* at 117.

[156] Sabina Safrin, *Hyperownership in a Time of Biotechnology Promises: The International Conflict to Control the Building Blocks of Life*, 98 *Am. J. Int'l L.* 641 (2004). *See also* the Philippines and Brazil access-restricting regimes (plant genetic resources), discussed in *id.*

[157] *See* Chapter 4, Section I.A.

[158] *See* Safrin, n. 156.

[159] *See* Section II.A, B.

[160] Complementary provisions in Articles 16(1) and 16(2) of the CBD emphasize the importance of access to resulting biotechnologies and, indeed, to all relevant technologies on "fair and equitable terms," if not "concessional and preferential terms." CBD, n. 36, arts. 16(1) & 16 (2). Article 19(1) requires parties

missing was any realistic appraisal of the fact that developing countries "are not net providers but net receivers of PGRFA."[161]

The primary legal instrument chosen for this exercise was the Prior Informed Consent rule embodied in the Convention. On this approach, bioprospecting scientists would presumably divulge their intentions to the local regulatory authorities or their intermediaries, and providers of genetic resources would impose the controls needed to secure benefit sharing in suitably negotiated contracts governing access. This approach, however, suffered from at least two critical errors.

First, it seems to have assumed that all research was potentially commercial, hence all research goals could and should be disclosed at the time of access. It thus ignored the fact that most public scientific research is speculative in nature, and that most publicly available genetic resources were valuable only insofar as they served as research tools, and not as generators of financial gain. For example, a report on two publicly funded bioprospecting programs by the National Cancer Institute (NCI) and the International Cooperative Biodiversity Group (ICBG), with the support of major United States funding agencies, demonstrated "an obvious trend in bioprospecting, namely that marketable discoveries are rare and, despite screening more than 50,000 plant samples, none have yet yielded a new drug."[162]

The second mistake was to ignore the way upstream scientific research actually operates today, with its insatiable need for unfettered investigation of multiple knowledge inputs long before any "benefits" are known or even suspected.[163] In other words, the drafters of the CBD ignored the fact that public scientific research depends on the freedom to operate on the broadest possible spectrum of upstream knowledge assets, and it accordingly made no provisions to facilitate or support such endeavors.

On the contrary, by squeezing public scientific research between the vice of negotiated access and negotiated uses, it implicitly delegitimized any use of genetic

to take appropriate measures to "provide for the effective participation in biological research activities by ... Contracting Parties especially developing countries, which provide the genetic resources for such research." Article 19(2) drives the point home by insisting that parties "take all practicable measures to promote and advance priority access on a fair and equitable basis ... especially developing countries, to the results and benefits arising from biotechnologies based upon genetic resources provided by those Contracting Parties" on mutually agreed terms.

[161] Regine Anderson (2008), n. 49, at 138 (arguing for "smoothest possible access to PGRFA" of benefit to all countries).

[162] Miller, n. 149, at 63 (adding that the "experience of these two programs is consistent with other discovery efforts, all of which suggest that the realization of marketable products requires many years.").

[163] *Cf.* Rai et al., n. 5.

resources that was not expressly licensed and authorized *ex ante*. In so doing, it burdened upstream researchers with a duty to negotiate when nothing foreseeable was necessarily on the table. It thus threatened to make upstream scientific research on genetic resources having no known or likely commercial value increasingly so burdensome as to not be worth undertaking.

Besides potentially high transaction costs, the current approach is thus inconsistent with innovation processes that rely on research uses of large quantities of genetic materials, as for example, in high throughput screening of microbial populations.[164] It also obstructs computational science, because automated knowledge generation and integration cannot proceed if access barriers must be removed for each bit of raw material and if restrictions on use rights encumber even the resulting data.[165] Given the highly protectionist intellectual property regimes in OECD countries,[166] this confluence of factors could lead to the erection of a protected digital fortress in developed countries overlooking a vast expanse of restricted raw materials in developing countries, with greatly diminished scientific collaboration and productivity between them.[167]

The CBD as initially drafted in 1992 thus failed to distinguish between the different uses to which genetic resources might be put, and in particular, it provided no support for public scientific research uses.[168] To their credit, the CBD's own Working Group on ABS subsequently recognized this omission as a possible gap, which regional and national regimes had failed to address,[169] and scientists began to express concerns about the harmful effects that restrictive access regulations might have on future research.[170]

Meanwhile, all the controversies engendered by the CBD were destabilizing the public repositories that traditionally hold the *ex situ* plant cultivars and microbial materials on which public science has long depended. How these intermediaries responded to preserve their public interest services in the face of this challenge is explained in the rest of this chapter.

164 *Cf. id.*

165 *See* Chapter 8.

166 *See further* Chapters 6 & 7.

167 *See, e.g.*, Santilli (2013), n. 48, at 118 (stating that the "CBD did not provide a solution to the negative impacts of IP rights on biodiversity, and, at the same time, legitimized them (indirectly). Access to genetic resources and associated knowledge became more restricted …").

168 However, the CBD did recognize the possibility of "non-monetary benefits," which will become of crucial importance under the Nagoya Protocol, n. 4. *See* Section IV.B.

169 *See Analysis of ABS Gaps*, n. 80, at 34.

170 *See, e.g.*, Jinnah & Jungcurt, n. 153; Carolina Roa-Rodriguez & Thom Van Dooren, *Shifting Common Spaces of Plant Genetic Resources in the International Regulation of Property*, 11 *J. World Intellectual Prop.* 176 (2008).

II. DESTABILIZING THE EXCHANGE OF PLANT AND MICROBIAL GENETIC RESOURCES AS GLOBAL PUBLIC GOODS

Under the CBD, obtaining permission to conduct the kind of wide-ranging, *in situ* bioprospecting that had been customary in the past now entailed cumbersome and burdensome negotiations, even when provider countries had established the necessary administrative procedures, which was often not the case.[171] As a result, the continuity of agricultural and microbiological research became more dependent than ever on the *ex situ* genetic resources heretofore made available by public seed banks and microbial culture collections.[172] In the minds of their administrators, these basic research inputs – accumulated by generations of dedicated researchers – had been held in trust for "the common heritage of mankind."[173]

Apart from patented or other microbial specimens deposited by industry and research institutes under special access conditions,[174] this view meant that both the microbial culture collections and the agricultural seed banks managed subsets of a disjointed, but nonetheless vast public domain in which "ownership" of specified genetic resources had not been an issue. These resources functioned as inputs into public and private research of both a basic and applied character. That normative position, however, was now increasingly contested by developing country governments on the grounds that *ex situ* genetic resources held either in repositories located within their territorial boundaries or previously exported from their territories without express permission were not legitimately the property of the research institutes in which they resided.[175]

As a result, continued use of such resources for public research purposes was now subject to claims of misappropriation and eventually "biopiracy" emanating from the countries where the resources in question had originally been discovered, as demonstrated earlier.[176] The Convention also raised questions about the possibility of multiple claims of ownership to existing and future microbial cultures, in which proprietary rights could be asserted by or against bio-prospectors, depositors,

[171] *See, e.g.*, Santilli (2013), n. 48, at 118; Jinnah & Jungcurt, n. 153.

[172] *See* Chapter 2, Section I. A–B.

[173] For plants, see, e.g., International Undertaking on Plant Genetic Resources, FAO Res. 8/83, 22nd Sess. (Nov. 5–23) [hereinafter International Undertaking]. For microbes, see, e.g., Dagmar Fritze, *A Common Basis for Facilitated, Legitimate Exchange of Biological Materials Proposed by the European Culture Collections (ECCO)*, 4 *Int'l J. Commons* 518, 519 (2010) [hereinafter Fritze (2010)].

[174] *See* Chapter 4, Section I.A.

[175] Godt (2013), n. 85; *see also* Safrin, n. 156.

[176] For examples, *see* Section I.C.2.a. *See also* Halewood (2010), n. 46.

collection managers, or users who had not acquired such material in a manner consistent with its provisions. No country remained entirely exempt from these and other concerns generated by the regulatory sweep of the CBD, even if its government had not formally adhered to that treaty.[177]

In this charged legal atmosphere, we shall see that the impact of the CBD on the existing microbial research infrastructure was somewhat more attenuated and less immediately threatening than it was for the CGIAR's agricultural research infrastructure. In both cases, past modes of operation were nonetheless called into question, and new conceptual approaches had to be devised.

A. *The Public Microbial Culture Collections Consider Defensive Options*

On the whole, and with some notable exceptions,[178] the public microbial culture collections traditionally pursued open-exchange networking policies that are more fully explained in the next chapter.[179] The advent of the CBD in 1992 challenged these open-exchange policies in at least two ways. First, the CBD posed still unresolved questions about the ownership of *ex situ* microbial genetic resources acquired before 1992.[180] Second, the Convention clearly required that all acquisitions and exchanges of *ex situ* materials after 1992 should conform to its PIC and ABS mandates.[181]

Faced with these challenges, the microbial culture collections located in both developed and developing countries began to reexamine their operational

177 For example, even though the United States is not a party to the Convention on Biological Diversity, its research institutes and repositories that hold plant and microbial genetic resources are indirectly affected by the tenets of the Convention. This follows because they collect and receive materials from other countries, all of which are Contracting Parties, under collaborative research initiatives. The Smithsonian Institute has reportedly recognized these facts of life and advised U.S. culture collections to act as if the CBD were binding. Telephone interview with Prof. Kevin McClusky, Curator Fungal Genetic Stock Center, http://www.w.f.g.c.ll, April 24, 2012. Moreover, with 193 countries on board, there exist grounds for claiming that the CBD expresses customary international legal norms rooted in the Declaration on Sovereignty over Natural Resources. *See* 1962 Declaration, n. 11.

178 The American Type Culture Collection (ATCC), one of the world's most important sources of high-quality microbial materials, does not conform to the public, noncommercial model under which most WFCC collections operate. For details, see Chapter 4 Section II.A.

179 *See* Chapter 4 Section I. We use the term "open exchange" rather than "open access" because public culture collections could only release microbial materials to qualified scientific recipients or other collections that met preestablished quality and security standards even before the advent of the CBD. *See further* Chapter 4, Section I.A. *See also* Godt (2013), n.85, at 249.

180 *See* n. 70 & accompanying text. The Nagoya Protocol to the CBD, discussed in Section IV, will give new life to this controversy. *See, e.g.*, Godt (2013), n. 85, at 246–47 (citing authorities) (2009). *But see* Buck & Hamilton (2011), n. 70 (claiming the issue is settled).

181 *See* ANDERSON (2008), n. 49, at 135–44.

premises and to evaluate different kinds of responses.[182] In particular, both single collections and regional networks have been trying to formulate common rules and principles to govern their services "from accessioning of biological material to ... authentication ... preservation and maintenance, through to ... ultimate release to the scientific community."[183]

One important milestone was the 1996 WFCC Information Document, which specified the special characteristics of microorganisms that distinguish them from plants and animals and the consequences of such characteristics for inventorying, tracking, and benefit sharing.[184] This document also recommended that access to *ex-situ* microbial genetic resources should remain unimpeded for the purposes of scientific research, industrial application, education and health care.[185] Some regional entities, such as the EU Culture Collections' Organization (ECCO), also made efforts to devise harmonizing guidelines that would help to standardize procedures and to provide a framework for compliance with the CBD and with growing legislation concerning biosafety and security.[186]

Also important was a voluntary code of conduct, first published in 1999, to introduce access and benefit sharing procedures for microbial resources within the framework of the CBD. This code, known as Micro-Organism Sustainable Use and Access Regulation International Code of Conduct (MOSAICC), focused attention on the need for model Material Acquisition Agreements and model Material Transfer Agreements (MTAs) for culture collections as well as for research scientists.[187] This code, updated in 2009, was produced under the leadership of the Belgian Coordinated Collections of Micro-Organisms (BCCM) with eleven other partners from both developed and developing countries, as well as representatives from the nonprofit and commercial sectors.[188] These and other important initiatives concerning the management of microbial materials are discussed in greater detail in Chapter 4.

For present purposes, the common thread underlying them all is a set of operational standards that required the public culture collections to adopt a defensive strategy

[182] *See, e.g.*, Godt (2013), n. 85, at 248–56. *See also* Dedeurwaerdere et al. (2009), n. 46.

[183] *See, e.g.*, Fritze (2010), n. 173, at 518–19.

[184] *See* WFCC, Access to *ex-situ* Microbial Genetic Resources within the Framework to the Convention on Biological Diversity, Sept. 1, 1996, *available at* http://www.wfcc.info/index.php/wfcc_library/genetic_res/ (last accessed 14 June 2014).

[185] *Id. See further* Fritze (2010), n. 173.

[186] These initiatives are described in Chapter 4, Section III.A.2.

[187] For the text of MOSAICC, see http://bccm.belspo.be/projects/mosaicc/docs/code2011.pdf (last accessed 14 June 2014). The project was funded by the European Commission.

[188] MOSAICC, Micro-Organisms Sustainable use and Access regulation International Code of Conduct Mosaic Website, http://bccm.belspo.be/projects/mosaicc/ (last accessed 14 June 2014). For details, *see* Chapter 4, Section III.A.2.

based on four major premises. First, most collections will deny "ownership" of *ex situ* microbial materials by self-characterizing themselves as "custodians" serving only an intermediary role between providers and users.[189] Second, with regard to new acquisitions, the public collections tend to require disclosure of geographical origin and information about PIC and mutually agreed terms (MAT). Third, exchanges are made under ever more complicated Material Transfer Agreements (MTAs) that tend to restrict both the recipients' own uses and further transfers of the materials in question. Finally, these MTAs tend to inform users of possible liability for violations of ABS provisions under the CBD.[190]

In effect, the public culture collections thus attempted to adapt by shifting responsibility for compliance with the CBD's benefit-sharing mechanisms to users of their *ex situ* materials while denying responsibility for themselves as intermediaries or brokers.[191] Whether this defensive approach would have satisfied the much stiffer requirements of the Nagoya Protocol, adopted in 2010,[192] will become clearer once we proceed to analyze that instrument in Section IV.[193] That, indeed, is another major factor behind our proposals for a redesigned Microbial Research Commons at the multilateral level.[194]

Meanwhile, in tandem with these defensive efforts to accommodate the CBD, there has also been a notable trend among public culture collections everywhere, including the WFCC member collections and those affiliated with that umbrella organization, to devise Material Transfer Agreements that increasingly restrict access to microbial genetic resources even for public research purposes.[195] As will be seen, these restrictions on access are motivated both by concerns to comply with the provisions of the CBD and by parallel concerns in developed countries to preserve their users' own intellectual property claims (as reinforced by the TRIPS Agreement of 1994) to downstream commercial applications of microbial materials made available from the collections. We explore these and related trends more fully in Chapter 4, entitled "The Existing Microbial Research Commons Confronts Proprietary Obstacles."

[189] Fritze (2010), n. 173. Increasingly, the collections now reject the notion of "ownership" and see themselves as custodians of strains with a right or license to reproduce copies for distribution. *Id. But see* the contrary view taken by ATCC, as discussed in Chapter 4.

[190] *See, e.g.*, Godt (2013), n. 85, at 254–55. For empirical evidence of specific MTAs, see Chapter 4, Section III.A.

[191] *See* Godt (2013), n. 85, at 254–58.

[192] Nagoya Protocol, n. 65.

[193] For the view that the public collections' defensive measures may not satisfy the Nagoya Protocol, see Godt (2013), n. 85, at 256–61.

[194] *See* Chapter 4. For governance of the Microbial Research Commons, see generally Part Four.

[195] *See* Chapter 4, Section II.

B. The CGIAR's Agricultural Research Infrastructure on the Verge of Collapse

The impact of the CBD on the preexisting agricultural research infrastructure was far more immediately disruptive of its mission than was the case with the microbial culture collections.[196] On the one hand, the International Agricultural Research Centers (IARCs) that are affiliated with the Consultative Group on International Agricultural Research managed strategic inputs for crop breeding through farmer selection, conventional plant breeding, and modern biotechnological techniques.[197] They played a critical role in a world where all countries' depended on genetic resources that were domesticated and subsequently developed in other countries or regions for their own food and sustainable agricultural development.[198] Foreseeably, plant breeders and farmers would need an ever greater supply of genetic diversity in the future, in order to adapt to new conditions that climate change was expected to impose.[199]

On the other hand, the proliferation of patents and plant breeders' rights covering products derived from genetic resources found in developing countries' increasingly made their governments willing to pursue well publicized claims of alleged biopiracy that involved academics, IARC seed banks, or both, as we saw earlier in this Chapter. However shortsighted it may seem in retrospect, these governments began to view the CGIAR's own holdings as potential lottery tickets that should be returned to their rightful owners.[200]

[196] *See, e.g.*, José Esquinas-Alcázar et al., *A Brief History of the Negotiations for the International Treaty on Plant Genetic Resources for Food and Agriculture, in* Crop Genetic Resources as a Global Commons: Challenges in International Law and Governance 135, 142 (M. Halewood et al., Routledge 2013) [hereinafter Esquinas-Alcázar et al. (2013)] at 135, 142. For the CGIAR's mission, see Chapter 2, Section I.B.

[197] *See* Gerald Moore & Emile Frison, *International Research Centers: The Consultative Group on International Agricultural Research and the International Treaty on Plant Genetic Resources for Food and Agriculture, in* Plant Genetic Resources for Food and Security: Stakeholder Perspectives on the International Treaty on Plant Genetic Resources for Food and Agriculture 149–54 (C. Frison, F. Lòpez & J.T. Esquinas-Alcázar eds. 2011) [hereinafter Moore & Frison (2011)]. The highest rates of germplasm acquisition and distribution occurred in the period 1983–1985, under the International Undertaking. *Id.* at 153.

[198] *See* Marleni Ramirez et al., *Demonstrating Interdependence on Plant Genetic Resources for Food and Agriculture, in* Crop Genetic Resources as a Global Commons (2013), n. 196, at 39 (citing authorities). "Even the world centres of crop diversity ... which coincide with the centres of domestication mostly rely on non-indigenous crop genetic resources to meet their food needs ... " *Id.* (citing authorities).

[199] *Id.* (citing Sam Fujisaka et al., *The Impact of Climate Changes on Countries' Interdependence on Genetic Resources for Food and Agriculture*, Background Study Paper No. 48 (2009), *available at* ftp://ftp.fao.org/docrep/fao/meeting/017/ak532e.pdf (last accessed 14 June 2014)); *see also* Julian Ramirez-Villegas et al., *Crop and Forage Genetic Resources: International Interdependence in the Face of Climate Change, in* Crop Genetic Resources as a Global Commons (2013), n. 196, at 78–98.

[200] *See e.g.*, Anderson (2008), n. 49, at 139. *Cf.* Santilli (2013), n. 48, at 118 (stressing that "CBD raised unrealistic expectations in many biodiversity-rich countries," which did not materialize in most cases).

The CGIAR's research centers, which had operated as providers of global public goods under the FAO's nonbinding International Undertaking on Plant Genetic Resources for Food and Agriculture (International Undertaking) since 1983,[201] thus found themselves caught in a propertizing tug of war between developed and developing countries once the CBD and the TRIPS Agreement had both taken effect, in 1993 and 1995, respectively.[202] Seed bank managers could no longer ignore questions about who should be deemed the rightful owners of genetic resources that had been obtained from developing countries, and why the rights of those who applied technology to these genetic resources were to be recognized under international law, but not the rights of the providers of those same genetic resources.[203]

As Michael Halewood, General Counsel for Bioversity, describes it, the CGIAR were now operating in a tumultuous legal atmosphere in which rumors of conspiracies to take advantage of the situation spread quickly, amidst fears that the Centers' plant genetic resources would face demands to return them to the countries from which they were originally acquired, and the World Bank, which funded the Centers, was accused of attempting to take over these same resources.[204] In 1993, new acquisitions dropped to under 10,000, about one quarter of the total in 1984, when the FAO International Undertaking had set the tone.[205]

Faced with these challenges, the CGIAR sought a refuge within the protective embrace of the FAO's Commission on Plant Genetic Resources (renamed in 1995 as the Commission on Plant Genetic Resources for Food and Agriculture (CPGRFA)). Discussions in this and related fora during the 1990s had led to resolutions that recognized both claims of national sovereignty over genetic resources and the rights of farmers and intellectual property owners to share in the benefits of commercial applications of those same resources.[206]

Twelve of the CGIAR's International Agricultural Research Centers signed formal agreements with FAO on 26 October 1994, which placed their *ex situ* seed banks in an international network operating under FAO auspices.[207] Under these

[201] *See* Chapter 2, Section I.B.2.

[202] Esquinas-Alcázar et al. (2013), n. 196, at 142; Moore & Frison (2011), n. 197, at 154.

[203] *Id.* at 135–39. *See also* Safrin, n. 156. Actually these questions had been on the table at FAO deliberations since 1979. *See* Esquinas-Alcázar et al. (2013), n. 196 at 137.

[204] Halewood (2010), n. 59. According to Halewood, "clearly something was necessary to preserve the goodwill of the global community and to ensure that the [CGIAR] centers were able to continue providing facilitated access to the collections they were hosting." *Id.* Michael Halewood is legal counsel for Bioversity International, one of the institutions that constitute the CGIAR.

[205] Moore & Frison (2011), n.197, at 154.

[206] *Id.* at 139.

[207] *Id.*

"In Trust" agreements, the IARCs would continue to provide the international community with germplasm "for the benefit of developing ... countries," and the CGIAR retained its *ex situ* collections. The FAO, in turn, recognized the IARCs as "trustees" of the international community who would not claim ownership of the designated germplasm and would not seek intellectual property rights in that germplasm or related information.[208]

After adopting a Global Plan of Action on Plant Genetic Resources for Food and Agriculture in 1996,[209] the FAO took steps to solidify the status of the international network of seed banks now operating under its own auspices, technically under authority of the nonbinding International Undertaking that was still in effect. The CPGRFA accordingly opened formal negotiations to clarify the boundaries between public and proprietary genetic resources, while conforming the now obsolete International Undertaking to the Access and Benefit Sharing mandate of the CBD.[210] Spurred on by unrelenting pressures from technology-exporting countries for higher intellectual property protection for plant breeders and the seed industry, the negotiators ultimately found it possible to unite all the stakeholders around a common objective, namely, to reconcile the need to conserve important crop genetic resources for future generations with "fair and equitable access to them for research and genetic improvement."[211]

After seven years of arduous negotiations, the FAO Council arrived at a consensus on November 2, 2001, which led to the adoption of a binding International Treaty on Plant Genetic Resources for Food and Agriculture (ITPGRFA) by the General Assembly "in a climate of euphoria" on November 3, 2001.[212] This historic treaty, which entered into force on June 29, 2004, provided the IARCs with a binding international legal framework that aimed to support plant breeding, conservation,

[208] Ramirez-Villegas et al., n. 199, at 89; Moore & Frison (2011), n. 197, at 154–55 (stressing the importance of recognizing the FAO's intergovernmental policy authority). These agreements were renewed at four-year intervals.

[209] Global Plan of Action for Conservation and Sustainable Use of Plant Genetic Resources for Food and Agriculture (GPA), Fourth International Technical Conference on Plant Genetic Resources, Leipzig, Germany, June 23, 1976 [hereinafter First Global Plan].

[210] *See, e.g.*, Laurence E. Helfer, *Comment II: Using Intellectual Property Rights to Preserve the Global Genetic Commons: The International Treaty on Plant Genetic Resources for Food and Agriculture*, *in* INTERNATIONAL PUBLIC GOODS, n. 3, at 217, 219. *See also* Santilli (2013), n. 48, at 114 (stressing difficulties augmented by different approaches of agricultural specialists affiliated with FAO and environmentalists affiliated with UNEP responsible for the CBD).

[211] Esquinas-Alcázar et al. (2013), n. 196, at 142.

[212] *Id. See* ITPGRFA, n. 123, ratified by 126 countries plus the European Union (not including the U.S.) and entered into force, June 29, 2004. *See* Danielle Manzella, *The Design and Mechanics of the Multilateral System of Access and Benefit Sharing*, *in* CROP GENETIC RESOURCES AS A GLOBAL COMMONS (2013), n. 196, at 151; Kal Raustiala & David G. Victor, *The Regime Complex for Plant Genetic Resources*, 58 INT'L ORG. 277 (2004).

and research based on facilitated access to the genetic resources held in their collections.[213]

III. AN INTERNATIONAL TREATY TO RESCUE AND EXPAND "THE GLOBAL CROP COMMONS"

Although the FAO's binding International Treaty generally promotes lofty goals concerning the "Conservation, Exploration, Collection, Characterization, Evaluation and Documentation" of plant genetic resources for food and agriculture (PGRFA)[214] and the "Sustainable Use" of them,[215] its formal objective emphasizes the "fair and equitable sharing of benefits arising out of their use, in harmony with the Convention on Biological Diversity, for sustainable agriculture and food security."[216] To this end, the treaty seeks to create a multilaterally governed gene pool, with facilitated access to selected plant genetic resources for purposes of research, breeding, and training, that are publicly available from both virtually networked *ex situ* collections and *in situ* locations around the world.[217] In exchange for this facilitated access to PGFRA under the Multilateral System, the Contracting Parties "agree that benefits accruing there from shall be shared fairly and equitably" in accordance with the provisions of the Treaty.[218] However, "benefit sharing" need not only consist of monetary and other benefits of commercialization, but may also include the exchange of information, including the characterization, evaluation and utilization of PGRFA; access to the transfer of relevant technology; and capacity building, especially in developing countries.[219]

Under the bilateral approach of the CBD, every transaction concerning access to plant genetic resources and eventual benefit-sharing could entail case-by-case negotiations, with prohibitive transaction costs. Under the International Treaty and the logic of its multilateral system, a country that agrees "to include the PGRFA that are under its management and control and in the public domain in a common pool... will gain access to the PGRFA that are under the management and control of all

[213] *See, e.g.*, Helfer, n. 210, at 139; Esquinas-Alcázar et al. (2013), n. 196, at 145; Moore & Frison (2011), n. 197, at 156–59.

[214] ITPGRFA, n.123, art. 5.

[215] *Id.*

[216] *Id.* art. 1.1.

[217] *Id.* arts. 12, 12.3; *see* Shakeel Bhatti, *The International Treaty on Plant Genetic Resources*, *in* DESIGNING THE MICROBIAL RESEARCH COMMONS, n. 45, at 137–43; Michael Halewood, *International Efforts to Pool and Conserve Crop Genetic Resources in Times of Radical Change*, *in* INTELLECTUAL PROPERTY RIGHTS: LEGAL AND ECONOMIC CHALLENGES FOR DEVELOPMENT (Mario Cimoli et al., Oxford U. Press 2014), at 289, 307 [hereinafter Halewood (2014)].

[218] ITPGRFA, n. 123, art. 13.1.

[219] *Id.* art. 13.2 (a), (b), (c), (d).

other Contracting Parties" (127 to date).[220] In principle, moreover, this facilitated access – implemented by means of a Standard Material Transfer Agreement – also applies to *ex situ* germplasm collections of the CGIAR's research institutions,[221] which were placed under the treaty by agreements signed on October 16, 2006.[222] Apart from some mostly European gene banks, however, member states on the whole have been reluctant to provide new germplasm or even to share information about what materials are nominally available from the public domain under their territorial jurisdiction.[223]

In retrospect, the International Treaty on Plant Genetic Resources for Food and Agriculture was thus an ambitious and idealistic undertaking, whose implementation depends on a full-fledged intergovernmental organization embedded within the larger administrative framework of the United Nations Food and Agricultural Organization.[224] Each of these features – its oversized ambitions, its idealistic aspirations, and its top-down institutional structure – are sources of both strengths and weaknesses, as explained in Section C.

A. *Basic Concepts of the International Treaty on Plant Genetic Resources for Food and Agriculture (ITPGRFA)*

As noted above, the ITPGRFA, which is administered by the FAO officially promotes the conservation of plant genetic resources for food and agriculture and the equitable sharing of benefits from the use thereof for sustainable agriculture and food security.[225] To this end, Articles 5 and 6 commit the Contracting Parties to a program of conservation and sustainable use of all crops, and not just those subject to the multilateral regime described in Annex I.[226] This program is based on principles and guidelines covering both *in situ* and *ex situ* resources that were first articulated in the Global Plan of Action negotiated in 1996 and subsequently updated in the

[220] *Id.* art. 11.2; Michael Halewood, "What Kind of Goods are Plant Genetic Resources for Food and Agriculture? Towards the Identification and Development of New Global Commons," paper presented at the Conference on Governing Pooled Knowledge Resources: Building Institutions for Sustainable Scientific, Cultural and Genetic Resources Commons for the International Association for the Study of the Commons (IASC), Louvain-le-Neuve, Belgium, Sept. 12–14, 2012 [hereinafter Halewood (Louvain 2012)], at 9. The International Treaty covers only the PGRFA of 64 crops and forages that were agreed during the negotiations. *See* ITPGRFA, n. 123, Annex 1.

[221] ITPGRFA, n. 123, art. 11.5 (mandating this coverage).

[222] *See id.* art. 15; CGIAR, Crop *Genebank Knowledge Base*, http://www.cropgenebank.sgrp.cgiar.org [hereinafter Crop Genebank] (last accessed Dec. 23, 2014).

[223] Halewood (Louvain 2012), n. 220, at 9.

[224] Helfer, n. 210, at 217–24. *See further* Chapter 9, Section II.A (describing the evolution of the governance structure for the International Treaty and relations with the CGIAR).

[225] *See* ITPGRFA, n. 123, pmbl. ¶8.

[226] *Id.* arts. 5–6. *See further* Section B.

revised Global Plan of Action that the FAO's Commission on Genetic Resources for Food and Agriculture adopted in July 2011.[227]

For example, Article 5.1(c) commits the contracting parties to promote and support farmers' and local communities' efforts to manage and conserve on-farm PGRFA. *In situ* conservation is more generally addressed in Article 5.1(d), which obliges contracting parties to promote the conservation of wild crop relatives and wild plants for food production in both protected and unprotected areas.[228]

Article 5.1(e), instead, deals with *ex situ* conservation. It requires collective action to promote the development of an efficient and sustainable system of *ex situ* conservation, with specific regard to the need for adequate documentation, characterization, regeneration, and evaluation of genetic resources. An important development here was the creation of the Global Crop Diversity Trust in 2004, to support *ex situ* conservation of plant genetic resources.[229]

More generally still, Article 6 of the treaty obliges the contracting parties to develop and maintain requisite policies and legal measures to promote the sustainable use of PGRFA. Such measures would encourage diverse farming systems, research, and plant breeding efforts (especially when directed at farmers), broadening the genetic base of crops, and supporting the wider use of diverse varieties and species in on-farm activities.[230] Taken together, these provisions recognize that "agrobiodiversity is the result of complex and dynamic management of agricultural crops by farmers, and that public policies and legal instruments must promote an integrated approach to agrobiodiversity, which takes both biological and cultural diversity into consideration."[231]

[227] *See* Global Plan of Action (GPA), n. 209, a voluntary instrument adopted by 150 countries during the Fourth International Technical Conference on Plant Genetic Resources, Leipzig, Germany, 17–23 June 1996. For details, see JULIANA SANTILLI, AGROBIODIVERSITY AND THE LAW: REGULATING GENETIC RESOURCES, FOOD SECURITY AND CULTURAL DIVERSITY 126–30 (Earthscan 2012) [hereinafter SANTILLI (2012)], at 126–30. For an updated version of the GPA, approved by the Commission on Genetic Resources for Food and Agriculture in July 2011, See FAO Council, Second Global Plan of Action for Plant Genetic Resources for Food and Agriculture (Second GPA), Nov. 29, 2011. *See also* ITPGRFA, n. 123, art. 4 (committing Member states to conform national legislation to these obligations) and art. 7 (committing member states to implement arts. 5–6 in local laws).

[228] ITPGRFA, n. 123, art. 5.1(c), (d). *In situ* conservation enables "cultivated plants [to] maintain their capacity to evolve and adapt," while maintaining entire ecosystems as an integrated whole. SANTILLI (2012), n. 227, at 127.

[229] ITPGRFA, n. 123, arts. 5.1(e), 6. For details, see SANTILLI (2012), n. 227, at 127–29 (citing authorities and also stressing importance of the Global Seed Vault established in 2008). For details, *see id.*, Chapter 3.

[230] ITPGRFA, n.123, art. 6. *See also id* arts. 4 (obliging member countries to conform their laws, regulations and procedures with treaty obligations), and art. 7 (obliging contracting parties to incorporate principles of arts. 5 and 6 into their agricultural development programs).

[231] SANTILLI (2012), n. 227, at 129.

B. Establishing the Multilateral System for Access and Benefit-Sharing

Beyond these efforts to promote conservation and improve the status of farmers, the primary achievement of the ITPGRFA was to establish a multilateral system of access and benefit-sharing that deliberately aims to deviate from, and improve on, the bilateral system for exchanging plant genetic resources that constitutes the default regime under the CBD. To lay a foundation, the Preamble declares that PGRFA "are a common concern of all countries, in that all countries depend very largely on plant genetic resources for food and agriculture that originated elsewhere."[232] At the same time, the Preamble takes pains to distinguish PGRFA from other genetic resources subject to the CBD, and thereby seeks to justify the establishment of a *sui generis* multilateral regime for managing these resources that would operate in the shadow of the CBD. Here the Preamble emphasizes:

- The critical role of farmers in domesticating wild plants for food and agriculture;
- The extent to which key varieties tend to be composites of genetic materials with different geographical origins;
- The extent to which even commercially bred PGRFA are likewise composites of many varieties; and
- The extent to which all countries depend on PGRFA that originate from sources beyond their territorial jurisdiction.[233]

Given these premises, the Preamble to the International Treaty – unlike the CBD – expressly emphasizes the importance of scientific research. In so doing, it highlights the needs of all countries to access and use PGRFA from other countries for both scientific research and breeding purposes, as well as for direct use in their agricultural and food systems. Maintaining open access to, and the exchange of, such resources thus becomes essential for breeders, farmers, and consumers.[234]

To this end, the treaty as a whole has been conceived as a sector specific regime under the umbrella of the CBD that covers both the Contracting Parties' *in situ* resources as well as *ex situ* resources held by national collections.[235] Although expressly recognizing the sovereign rights of states in plant genetic resources originating from their territories, the Contracting Parties agree to establish a multilateral system of ABS so as to facilitate access to PGRFA and to share the benefits from the use of these resources.[236] As a result, all the specific seeds and materials pertaining to

[232] ITPGRFA, n. 123, pmbl. ¶3.
[233] ITPGRFA, n. 123, pmbl. ¶¶7–9; *see also* Santilli (2012), n. 227, at 119–120.
[234] *Id.*, pmbl. ¶3; Santilli (2012), n. 227, at 120.
[235] Godt (2013), n. 85, at 250 (citing examples and sources).
[236] ITPGRFA, n. 123, arts. 10.1, 10.2.

the crops listed in Annex 1 that are "under the management and control of the Contracting Parties and in the public domain"[237] become, by virtue of the treaty, the "limited common property" of the Multilateral System that it creates, which natural and legal persons in contracting parties may freely access.[238]

Annex I provides a detailed list of 35 food crops and 29 forage species (legumes, grapes, and others) that resulted from contentious negotiations.[239] According to Juliana Santilli, the broader list of desired crops originally presented was substantially reduced, with many important crops excluded, and some crops of lesser importance for food security included.[240] Nevertheless, the Annex I sources thus committed to the open-access policies of the Multilateral System cover "the 64 most important staple crops, which account for 80 percent of all human consumption (including wheat and rice in various collections)."[241]

Also included in the Multilateral System were the *ex situ* plant genetic resources listed in Annex 1 that are held by the eleven International Agricultural Research Centers (IARCs) affiliated with the CGIAR.[242] Under Article 15, the Contracting Parties recognize the importance of the collections held in trust by the CGIAR centers and "call … on the centres to sign agreements with the Governing Body placing these collections within the purview of the Treaty."[243]

Article 15 sets out the main terms and conditions that such agreements will apply to these centers. In principle, all the Annex I materials (PGRFA) held by the CGIAR Centers will be made available under the same Standard MTA (SMTA) to be negotiated between the Contracting Parties themselves.[244] Non-Annex I materials were to be made available on terms consistent with those mutually agreed between

[237] This deliberately excludes genetic materials that are the subject of intellectual property rights. Manzella, n. 212, at 150, 153.

[238] ITPGRFA, n. 123, arts. 10.2, 11.2; Halewood (2014), n. 217, at 307–08 (noting that minimum administrative fees may be charged). *Cf.* Carol Rose, *The Several Futures of Property: Of Cyberspace and Folktales, Emission Trades and Ecosystems*, 83 *Minn. L. Rev.* 129, 132 (1998) (defining "limited common property" as "property held as commons amongst the members of a group, but exclusively vis-a-vis the outside world").

[239] ITPGRFA, n. 123, Annex I.

[240] For details, *see* SANTILLI (2012), n. 227, at 123 (citing authorities).

[241] Godt (2013), n. 85, at 250 (citing authorities). Inclusion of a new crop in Annex I requires a consensus of the parties.

[242] ITPGRFA, n. 123, arts. 11.5, 15; Halewood (2014) n. 217, at 308 (stating that 650,000 ex situ accessions were thus made available from the multilateral system).

[243] Moore & Frison (2011), n. 197, at 157. The centers are not States, and they "possess their own independent international legal personality." Hence, this provision was necessary because the centers cannot be bound directly by the treaty nor can they become parties to it in their own right. *Id.*

[244] ITPGRFA, n. 123, art. 15; Moore & Frison (2011), n. 197, at 157. In principle, there was to be a distinction between materials collected before and after the International Treaty entered into force, *see id.*, but this was muted by the practice of the centers subsequently to apply the Standard MTA to all their resources. *See* nn. 271–282 & accompanying text.

the centers and provider countries (or countries of origin) in keeping with the CBD.[245] In practice, since the SMTA for Annex I PGRFA was negotiated, the CGIAR centers have made all these resources – Annex I or not – available under that SMTA.[246]

1. The "Facilitated Access" Regime

The cardinal principle of the system is that all member states, their nationals, and other Contracting Parties obtain "facilitated access" to all the seeds and germplasm thus contributed to the multilateral system for purposes of breeding, research, conservation, and training, but not for "chemical, pharmaceutical and/or other nonfood/feed industrial uses."[247] Uses not permitted still require case-by-case bilateral negotiations under the CBD. Article 15 then stipulates that the Contracting Parties agree to provide the CGIAR centers that have signed agreements with the Governing Body with "facilitated access" to Annex I PGRFA.[248] The Contracting Parties further agreed, however, that recipients who thus obtain facilitated access to the system must share, in a fair and equitable way, the benefits arising from their utilization of these same plant genetic resources.[249]

The multilateral system thus constitutes what amounts to a distributed, global gene pool of some sixty-four food and feed crops that account for the bulk of human nutrition, although some important crops remain excluded for political reasons.[250] In principle, the treaty mandates that national governments and the institutions they control must provide *in situ* and *ex situ* samples of the listed crops that are in the public domain, along with similar contributions from natural and legal persons, such as the national collections and universities, that operate within the jurisdiction of the Contracting Parties.[251] Voluntary contributions of genetic resources not in the public domain or otherwise mandated by the treaty are also encouraged, and some governments have made substantial deposits in this regard.[252]

[245] *See* Moore & Frison (2011), n. 197 (citing art. 15 of the Treaty).

[246] For the SMTA, *see* Section III.B.2.

[247] ITPGRFA, n. 123, art. 12.3(a). Uses other than those permitted would require case-by-case negotiations.

[248] Moore & Frison (2011), n. 197, at 157 (citing art. 15).

[249] ITPGRFA, n. 123, art. 10.2.

[250] *See* Santilli (2012), n. 227, at 122–23; *see also* Manzella, n. 212, at 150–51.

[251] ITPGRFA, n. 123, arts. 11.3, 11.4. *See* Carlos M. Correa, *Plant Genetic Resources Under the Management and Control of the Contracting Parties and in the Public Domain: How Rich the ITPGRA Multilateral System?*, *in* Crop Genetic Resources as a Global Commons (2013), n. 196, at 177–86.

[252] ITPGRFA, n. 123, art. 11.2; Michael Halewood et al., *Changing Rates of Acquisition of Plant Genetic Resources by International Gene Banks*, *in* Crop Genetic Resources as a Global Commons (2013), n. 196, at 99–131. However, anticipated voluntary deposits by individuals and corporate entities have seldom materialized, Halewood (2014), n. 217, at 308 (citing ITPGRFA, n. 123, art. 11.4).

In practice, however, the bulk of the resources actually deposited and exchanged under the treaty so far consist of the *ex situ* holdings managed by the CGIAR's International Agricultural Research Centers. Under an agreement between the IARCs and the FAO (acting on behalf of the Governing Body of the Treaty), these collections were formally placed within the multilateral system in 2006.[253] Although reaffirming the status of IARC resources as global public goods, this agreement recognized the authority of the Governing Body under the treaty to provide policy guidelines to the IARCs pertaining to these collections, which were now subject to the provisions of the treaty.[254] The Governing Body has also authorized the IARCs to apply the Standard MTA governing exchanges of plant genetic resources under the multilateral system to non-Annex 1 materials that were collected before the International Treaty entered into force in 2004.[255] Other important agricultural research centers not affiliated with CGIAR have also joined the gene pool – now increasingly referred to as the Crop Commons[256] – under similar agreements, notably the Tropical Agricultural Research and Higher Education Center (CATIE).[257]

Once plant genetic resources are deposited in the pool, known as the "multilateral system," they become subjected to the "facilitated access" regime of Article 12.[258] Under this article, the provider must grant access expeditiously either free of charge or at the marginal cost of distribution,[259] solely for the purpose of conservation for research, breeding, and training for food and agriculture.[260] Provider states have no obligation to track single accessions, however, although related nonconfidential descriptive information should be provided, if available.[261]

In exchange for free access, the recipient may not commercialize, claim, or establish intellectual property rights in PGRFA or their genetic parts or components *in the form received from the multilateral system*.[262] The recipient may, however,

[253] *See* ITPGRFA, n. 123, art. 15; Esquinas-Alcázar et al. (2013), n. 196, at 140; Manzella, n. 153, at 153 (stressing the importance of IARC collections to the system as a whole).

[254] Moore & Frison (2011), n. 197, at 158–99.

[255] Santilli (2012), n. 227, at 138. However, non-Annex I material collected after 2004 "fall outside the scope of the multilateral system in the IARCs and are subject to the CBD and to the Nagoya Protocol." *Id. See* ITPGRFA, n. 123, art. 15.3.

[256] *See generally* Crop Genetic Resources as a Global Commons (2013), n. 67.

[257] *See* n. 313 and accompanying text.

[258] ITPGRFA, n. 123, at 12.

[259] *Id.* art. 12.3(b); Evanson Chege Kamau, *The Multilateral System of the International Treaty on Plant Genetic Resources for Good and Agriculture: Lessons and Room for Development*, *in* Common Pools of Genetic Resources (2013), n. 48, at 342, 347–48 [hereinafter Kamau (2013)].

[260] *Id.* at 348; ITPGRFA, n. 123, art. 12.3(a). Access is excluded for chemical, pharmaceutical and/or other nonfood or feed industrial uses. *Id.*

[261] Kamau (2013), n. 259, at 347 (citing ITPGRFA, n. 4, art. 12.3(b), (c)).

[262] *Id.* art. 12.3(d); Kamau (2013), n. 259, at 348. Holders of PGRFA protected by intellectual property rights must respect those rights and relevant international agreements and national laws. *Id.* at 348 (citing ITPGRFA, n. 123, art. 12.3).

seek its own intellectual property protection for newly developed products, based on plant genetic materials obtained from the multilateral system, in return for specified obligations to share both noncommercial and commercial benefits.[263] If the recipient uses the material for noncommercial purposes, the expected nonmonetary benefits include the sharing of relevant information and access to and transfer of relevant technology, as well as capacity building.[264]

If, instead, the recipient puts the plant genetic material to commercial uses, he or she must expect to share monetary and related benefits under Article 13.2(d) of the Treaty. This duty is triggered when a commercial product either incorporates material from the system or the recipient has otherwise used the material to make a product and then claims intellectual property rights in that product.[265] These benefit-sharing obligations are then further spelled out in the SMTA, described in the next section.[266]

2. Notification, Benefit Sharing, and the Standard Material Transfer Agreement

To render the multilateral system of facilitated exchanges operational, two ancillary measures become necessary. One is the project to build an information database to facilitate notification of and access to the materials available from the Crop Commons. The second is a negotiated SMTA to implement the benefit-sharing options that the treaty stipulates.

In order to build a global information database enabling users to know what materials are available from where, the Contracting Parties must notify the Secretariat of the Treaty about the genetic diversity under their management and control (or jurisdiction) and in the public domain.[267] Of critical importance here is the "provision of information or passport data related to the crops stored in the gene banks of contracting parties through adequate and public documentation," which effectuates de facto inclusion of material in the multilateral system.[268] In practice, however, this notification process imposes a burdensome obligation on the parties, apart from the difficulties of even a good faith attempt to distinguish resources under the control of governments or in the public domain from those that are not.[269] Not

[263] *See, e.g.*, Kamau (2013), n. 259, at 349.

[264] *See* ITPGRFA, n. 123, arts. 13, 19.3(f) (including improved varieties and genetic material developed from PGRFA under the multilateral system, subject to applicable intellectual property rights); Kamau (2013), n. 259, at 349, 349 nn. 33–35.

[265] *See id.* at 349 (citing ITPGRFA, n. 123, art. 13.2(d)).

[266] *See* nn. 275–298 & accompanying text.

[267] ITPGRFA, n. 123, art. 11,2; Halewood (2014), n. 217, at 311–12; Kamau (2013), n. 259, at 345.

[268] *Id.*

[269] *See, e.g.*, Claudio Chiarolla & Stefan Jungcurt, *Outstanding Issues on Access and Benefit Sharing under the Multilateral System of the ITPGRFA*, Background Study Paper, Berne Declaration &

surprisingly, the multilateral system to date primarily consists of the plant genetic resources held in the *ex situ* collections of the CGIAR's IARCs, which currently amount to about 7,000,000 specimens, plus one-third more contributed by parties and institutions within the European region.[270]

As for implementing the benefit-sharing obligations under the International Treaty, the Governing Body adopted the SMTA in 2006,[271] which establishes the terms and conditions for the transfer of genetic materials between providers and recipients thereof.[272] The SMTA is a bilateral, legally enforceable contract between providers and recipient institutions, and not the participating countries themselves.[273] As a result, single researchers and research institutions in countries that have not ratified the treaty can nonetheless access plant genetic resources from the multilateral system.[274] Once a country ratifies the treaty, "the adoption of the SMTA becomes mandatory for crops listed in Annex I."[275]

The SMTA operates as a "viral license" that obliges recipients of plant genetic resources to impose its conditions on all subsequent transferees of those same resources under new MTAs.[276] Although initial recipients must notify the Governing Body of further transfers, they are not responsible for the actions of subsequent recipients.[277]

In principle, the SMTA obliges recipients to disclose all nonconfidential information resulting from R&D performed on the material received from the multilateral system to other users of that system.[278] In practice, implementing this provision depends on what is or is not deemed confidential, and the SMTA itself establishes no binding criteria for this determination.[279] More realistically,

Development Fund (Mar. 2011), http://www.eub.ch/con_data/ ITPGR_ABS_Study_1.pdf [hereinafter Chiarolla & Jungcurt (2011)]. For recent developments and a pilot test project, see Halewood (2014), n. 217, at 311–12.

[270] *Id.*; Kamau (2013), n. 259, at 346.

[271] Governing Body of the ITPGRFA, FAO, Standard Material Transfer, Resolution GB/ITPGRFA) (2006) Appendix G; *adopted* Oct. 16, 2006; see also GB-1-07 Resolution; Santilli (2012), n. 227, at 137.

[272] FAO Conference, Comm'n on Genetic Resources for Food and Agriculture, Standard Material Transfer Agreement, ¶6.8 [hereinafter SMTA], available at http://www.planttreaty.org/content/drafting-standard-material-transfer-agreement.

[273] Santilli (2012), n. 227, at 133 (noting that the participating countries are contracting parties to the treaty, not to the SMTA).

[274] *Id.* (noting that a literal reading of ITPGRFA, n. 123, art. 12.2 might suggest a contrary result).

[275] Santilli (2013), n. 48, at 137, 157 n. 40 (noting that other MTAs can be used only for plant genetic resources not included in the treaty).

[276] SMTA, n. 272, art. 6.4 (implementing ITPGRFA, n. 123, art. 12.4); Kamau (2013), n. 259, at 349.

[277] Santilli (2013), n. 48, at 137, 157 n. 40 (noting that providers must periodically inform the Governing Body about subsequent MTAs, which information is made available to the FAO as third-party beneficiary in charge of monitoring compliance with the SMTA).

[278] SMTA, n. 272, art. 5; *see* Santilli (2013), n. 48, at 137.

[279] *Id.*

Contracting Parties pledge to share nonconfidential information among themselves, such as "catalogues and inventories, information technologies, and the results of technical, scientific and socioeconomic research, including characterizations, evaluation and utilization ... regarding those PGRFA under the multilateral system."[280]

Beyond these lofty premises, the primary function of the viral SMTA is to implement the obligations to share monetary and other benefits of commercialization otherwise established by Article 13.1 of the treaty.[281] These duties arise from specific acts of access or transfer, as a result of which a recipient commercializes a product that incorporates material from the multilateral system, and thereby triggers a potential duty to pay royalties into the Benefit Sharing Fund.[282]

As drafted and approved in 2006, the SMTA somewhat confusingly conditions this duty to pay, and the duration of that duty, on the extent to which the commercializing entity's licensing terms restrict access to, and use of, the end product for purposes of research, breeding, and conservation.[283] If the downstream commercial user agrees not to impose any such restrictions on the use of the product for further research and breeding, there is no corresponding duty to pay any tithe at all.[284] Conceivably, this exemption might suit university and other public research institutes that license plant genetic resources to industrial users, assuming that the latter would accept such a condition, even if one could be certain of the precise meaning and outer limits of the provision as drafted.[285]

Conversely, if recipients who accessed resources from the multilateral system commercialize an end product,[286] but are unwilling to allow unrestricted research and breeding uses of that product, they must choose between two mandatory payment options or modalities. Under the first option, and presumably the more typical case, the SMTA would impose a version of what Professor Reichman, in

[280] ITPGRFA, n. 123, art. 13.1. This information would presumably be shared via the Global Information System on Plant Genetic Resources for Food and Agriculture envisioned in *id.*, art. 17. Article 13.1 also prescribes cooperation for capacity building among Contracting Parties.

[281] ITPGRFA, n. 123, art. 13.1.

[282] *See* ITPGRFA, n. 123, art. 16(d)(ii); SMTA, n. 272, art. 6.7; Halewood (2014), n. 217, at 308–11.

[283] *See* SMTA, n. 272, art. 2, which defines a "product" as PGRFA that incorporates (for example, by pedigree or gene insertion) the material or any of its genetic parts or components that are ready for commercialization, excluding commodities and other products used for food, feed, and processing.

[284] *See* SMTA, n. 272, art. 6.8; ITPGRFA, n. 123, art. 13.28(d)(ii)(3); Manzella, n. 212, at 156; Kamau (2013), n. 259, at 35.

[285] SMTA, n. 272, art. 6.8. "In the case that the Recipient commercializes a Product that is a Plant Genetic Resource for Food and Agriculture and that incorporates Material as referred to in Article 3 of this Agreement and where that Product is available without restriction to others for further research and breeding, the purpose in accordance with Annex 2 to this Agreement." Voluntary payments are nonetheless encouraged. *See, e.g.*, Santilli (2012), n. 227, at 143.

[286] The end product must also be a plant genetic resource as defined in art. 2 of the SMTA.

other writings, has described as a "Compensatory Liability Regime."[287] Under this "take and pay" option, would-be users remain free to commercialize downstream applications of specimens taken from the multilateral system under the SMTA on condition that they pay a net royalty of 0.77% of gross sales into the Benefit Sharing Funds for as long as the product remains unavailable for research and breeding purposes.[288]

Alternatively, the downstream user could opt to pay a discounted rate of 0.5 percent on the aggregate sales of all its products from the same crop.[289] This second payment option is valid for ten years, and is renewable for additional periods of five years unless the user notifies the Governing Body of an intention to opt out, which entitles the user to restrict use of the product for breeding or research purposes. It also entitles that user to access all of the genetic material of that same crop to be found in the multilateral system, without additional payment, even though this additional material from the same crop would otherwise be subject to separate MTAs. By the same token, however, this payment obligation applies "not only to the sales of the producer that incorporated material received from the multilateral system, but to any products belonging to the same crop to which material received from the multilateral system belongs."[290] According to Professor Correa, this option

[287] Jerome H. Reichman, *Of Green Tulips and Legal Kudzu: Repackaging Rights in Subpatentable Innovation*, 53 *Vand. L. Rev.* 1743 (2000) [hereinafter Reichman, *Green Tulips*], available at http://scholarship.law.duke.edu/faculty_scholarship/456. This proposal antedated the drafting of the SMTA and there is reason to believe it influenced the adoption of a liability rule. *See* JONATHAN CURCI, THE PROTECTION OF BIODIVERSITY AND TRADITIONAL KNOWLEDGE IN THE INTERNATIONAL LAW OF INTELLECTUAL PROPERTY 290–91, 293–96 (Cambridge U.P. 2010) *See also* Reichman & Lewis (2005), n. 20; Rai et al., n. 5. For proposed applications to microbial genetic resources with major modifications, see generally Chapter 5.

[288] *See* SMTA, n. 272, art. 6.11, Annexes 2 & 3; Manzella, n. 212, at 156. Technically, the royalty is assessed at 1.1 percent of gross sales minus 30 percent for costs. Presumably, if plant breeders' rights or patents applied, the duration would last for 20 years, *id.*, but this hypothesis ignores the duration potentially imposed by contracts, which could be longer. The extent to which the plant breeder's privilege to breed under UPOV 1991, International Convention for the Protection of New Varieties of Plants, adopted on 2 Dec. 1961, 815 U.N.T.S. 89 (entered into force 10 Aug. 1968, as revised on 19 Mar. 1991, S. Treaty Doc. 104–17, 815 U.N.T.S. 89 (entered into force 24 April 1997, could excuse the breeder from even this duty to pay royalties remains controversial. *See, e.g.*, SANTILLI (2012), n. 227, at 143.

[289] Carlos M. Correa, *An Innovative Option for Benefit-Sharing under the International Treaty on Plant Genetic Resources for Food and Agriculture: Implementing Article 6.11 Crop-Related Modality of the Standard Material Transfer Agreement*, *in* PLANT GENETIC RESOURCES AND FOOD SECURITY 249, 253 (C. Frison, F. López & J. Esquinas-Alcázar et al. eds. 2011), *available at* http://www.planttreaty.org/sites/default/files/PGR_FS_FINAL21dec2012.pdf; SMTA, n. 272, art. 6.11, Annexes 2 and 3. In this case, payment is made per crop, not per accession or per product, and would be due "even if the products were made available for breeding and regardless of whether they incorporated the material received." Halewood (2014) n. 217, at 309.

[290] SANTILLI (2012), n. 227, at 143.

would be easier for all sides to administer because only a "straight and simple annual calculation of the royalty payment to be made," and there is no further need for the recipient to track the incorporation of the material received from the multilateral system.[291]

Under either royalty payment option, the providers of relevant materials remain under the obligation to notify every use of materials covered by the SMTAs, and the terms of exchange, to the Governing Body of the Treaty, which keeps that information secure and confidential.[292] The SMTA imposes a similar duty to report uses and sales on all recipients of materials taken from the multilateral system who develop commercial applications.

However, the International Treaty expressly obviates the need for providers to track the uses of specimens they distribute,[293] while obliging them not to charge more than the marginal costs of distribution for their services, as previously noted.[294] Because royalties, when mandatory, are paid directly to the Trust Fund, providers receive no direct financial benefits from downstream applications and thus arguably lack incentives to monitor and enforce compliance with the SMTAs.[295] Technically, the SMTAs are also private contracts, not directly subject to international law, which regulates relations between nation states.[296]

For these and other reasons, developing country stakeholders pressed for a specific enforcement vehicle, rooted directly in the treaty, which could monitor compliance and trigger dispute resolution mechanisms in case of noncompliance. In response, the SMTA recognized a "Third Party Beneficiary," which it entrusted with the duty to enforce these contractual obligations on behalf of the multilateral system.[297] The FAO, on behalf of the Governing Body, has in turn agreed to act as the Third Party Beneficiary under the ITPGRFA for enforcement purposes.[298]

Funds eventually accruing from the SMTAs, including both mandatory and voluntary contributions to the Benefit Sharing Fund, are placed under the direct

[291] Correa (2011), n. 289, at 255.

[292] SMTA, n. 272, art. 5(e); (allowing notice by transmission of a copy of the SMTA to the Governing body or by notifying it directly to the Third Party Beneficiary that has been appointed to enforce the schemes), *See* nn. 368–71 and accompanying text.

[293] ITPGRFA, n. 123, art. 12.3(b); SMTA, n. 272, art. 5(a).

[294] ITPGRFA, n.123, arts. 12.3(b) and (c). Providers must also supply all available passport data and related nonconfidential descriptive information. *Id.*

[295] Manzella, n. 212, at 157.

[296] *See, e.g.*, Gerald Moore, *Protecting the Interests of the Multilateral System Under the Standard Material Transfer Agreement – The Third Party Beneficiary*, *in* Crop Genetic Resources as a Global Commons (2013), n. 196, at 164, 168 [hereinafter Moore (2013)].

[297] *See id.*; SMTA, n. 272.

[298] *See, e.g.*, Moore (2013), n. 296, at 164–176; Manzella, n. 212, at 158.

control of the Governing Body. That Body has subsequently determined that these funds should be allocated according to the following priorities:

(1) Information exchange, technology transfer, and capacity building;
(2) Managing and conserving PGRFA on farms;
(3) The sustainable use of PGRFA.[299]

In principle, funds are supposed to be used to benefit farmers in all countries, especially in developing countries, who conserve plant genetic resources for food and agriculture.[300] In any event, eligible governmental and nongovernmental organizations, including the gene banks and research institutions, can apply for grants from this fund. Plant genetic resources developed under such grants should, in theory, go back into the multilateral system, thus "completing a loop" in which genetic resources "are accessed, improved, conserved, shared, used and made available for facilitated access."[301]

Finally, the ITPGRFA entrusts responsibility for implementing the multilateral system as a whole to a Governing Body, a Secretariat, and a relatively complex administrative apparatus. It has thus created a full-fledged intergovernmental governance architecture that became an integral part of the FAO's own rather intricate operational framework. The nature and design of this governance architecture are more fully explained and critiqued in Chapters 9 and 10.

C. *Strengths and Weaknesses of the International Treaty on Plant Genetic Resources for Food and Agriculture*

The creation of the multilateral system for PGRFA out of the ruins of the FAO's preexisting International Understanding[302] represented a major achievement of both public international law and global scientific policymaking. In return for allowing facilitated access to plant genetic resources placed under the authority of that system, Contracting Parties – especially developing countries rich in biodiversity – become eligible to receive different kinds of benefits guaranteed by the treaty. The primary benefit to all concerned was, of course, the fact that the International Treaty made "it possible for farmers, plant breeders and researchers, in both the public and private sectors, to have access to the widest possible range of PGRFA."[303]

[299] FAO, Report of the Second Session of the Governing Body of the International Treaty on Plant Genetic Resources for Food and Agriculture, Doc. I/GB-2/07/Report, Appendix D.1(2007), http://www.fao.org/Ag/cgrfa/cgrfa11.htm?. See also Manzanella, n. 212, at 156.

[300] Kamau (2013), n. 259, at 350 (citing ITPGRFA, n. 123, art. 13.3).

[301] Manzella, n. 12, at 156–57 (citing FAO (2009), Appendix A.3).

[302] *See* Chapter 2, Section I.B.2.

[303] Manzella, n. 212, at 155 (citing ITPGRFA, n. 123, art. 13.1).

At the same time, the International Treaty, constructed in the shadow of a relatively weak CBD that lacked the teeth later to be addressed by the Nagoya Protocol in 2010,[304] suffers from serious legal and technical defects that undermine its usefulness as a model for the redesigned Microbial Research Commons we envision in Part Four of this volume. In the next sections, we describe both the strengths and weaknesses of this treaty.

1. Demonstrable Achievements

On the positive side, the multilateral system created by the 127 signatories to the International Treaty on Plant Genetic Resources for Food and Agriculture[305] addressed, and partly resolved, the very real risk that tensions between the *demandeurs* of the TRIPS Agreement of 1994 and those of the Convention on Biological Diversity of 1992 could destroy the global agricultural research infrastructure. Avoiding this risk was all the more essential at a time when climate change had already threatened to destabilize food security on an unprecedented scale.[306] The 2006 agreements between the Governing Body of the ITPGRFA Treaty and the CGIAR's International Agricultural Research Centers (IARCs)[307] "reaffirmed the status of *ex situ* collections held by the centres as global public goods," and placed their collections under the auspices of the Treaty.[308] Thus sheltered politically from conflicting proprietary claimants, the IARCs continued to supply researchers and breeders with plant genetic resources under the Standard Material Transfer Agreement.[309]

[304] *See* Section IV.

[305] Manzella, n. 212, at 151.

[306] Esquinas-Alcázar et al. (2013), n. 196, at 136–37 (stressing questions about ownership and benefits of countries that provide plant genetic resources to public research centers); Isabel López Noriega et al., *Assessment of Progress to Make the Multilateral System Functional: Incentives and Challenges at the Country Level*, *in* Crop Genetic Resources as a Global Commons (2013), n. 196, at 199, 211–12 (stressing incentives for countries to pool plant genetic resources from external suppliers to avoid devastating pests and climate change conditions).

[307] The following CGIAR affiliated IARCs are covered, with a total of 693,000 *ex situ* accessions: Africa Rice Center; Bioversity International; International Center for Tropical Agriculture (CIAT); International Maize and Wheat Improvement Center for Agricultural Research in Dry Areas (ICARDA; International Crops Research Institute for the Semi-Arid Tropics (ICRISAT); International Institute for Tropical Agriculture (IITA); International Livestock Research Institute (ILRI); International Rice Research Institute (IRRI), World Agroforestry Centers. López Noriega et al., n. 306, at 199, 205 (Fig. 11.1).

[308] Esquinas-Alcázar et al. (2013), n. 196, at 140, Box 6.2. In so doing, the IARCs recognized the authority of the Governing Body under the Treaty to provide policy guidelines concerning *ex situ* resources that are subject to the provisions of the Treaty. *Id.*

[309] *Id.* In the first nine months of the multilateral systems' operation, more than 100,000 transfers were reported by the CGIAR system alone. Sélim Louafi & Shakeel Bhatti, *Efforts to Get the Multilateral System Up and Running – A Review of Activities Coordinated by the Treaty's Secretariat*, *in* Crop Genetic Resources as a Global Commons (2013), n. 196, at 164, 190.

By the year 2011, the CGIAR's Centers, built up over decades, had cumulatively managed more than 746,000 accessions of crops and forages collected from over 100 countries.[310] In the first nineteen months of operation after the 2006 Agreement with FAO took effect in 2007, the centers distributed about 550,000 samples of PGRFA under the SMTA. Of these, nearly three-quarters were materials that the Centers had helped to improve, and most of them went to developing countries.[311]

Significantly, after more than a decade of decline, there has also been a palpable increase in materials sent to the centers from developing countries since 2010.[312] A number of important agricultural research institutes not affiliated with the CGIAR have also joined the multilateral system, which collectively made more than 30,000 additional *ex situ* accessions available for facilitated exchanges.[313] Substantial voluntary contributions of plant genetic resources from both contracting and noncontracting governments have further enriched the multilateral system.[314] Altogether, some 460,000 accessions of germplasm held by national entities were officially available from the multilateral system in 2013, and even "much more material appears to be available" as a practical matter.[315]

The availability of these plant genetic resources from the system, under the oversight of the Governing Body, resulted in free and simplified access for research purposes, without state interference and without case-by-case negotiations with gene banks. Use of the SMTA thus avoided lengthy, bureaucratic, and uncertain or nontransparent procedures "that drastically increase transaction costs."[316]

[310] Halewood et al. (2013), n. 252, at 99.

[311] Moore & Frison (2011), n. 197, at 159.

[312] Halewood et al. (2013), n. 252, at 101. However, the discrepancy between the total CGIAR holdings (about 746,000) and the much smaller, officially available number available from the Multilateral System confirms reports that provider countries often restrict access to their contributions.

[313] These included the following centers, with the number of accessions shown in parentheses:

- Centre for Pacific Crops and Trees (CePACT)-SPC Community (15,000);
- Tropical Agricultural Research and Higher Education Center (CATIE) (11,000);
- Mutant Germplasm Repository of the FAO/IAEA Joint Division 92,500);
- International Cocoa Gene Bank (2,000);
- International Coconut Gene Banks (158).

López Noriega et al., n. 306, at 205 (Fig. 11.1).

[314] *See, e.g.*, López Noriega et al., n. 306, at 205–06; *see also* Michael Halewood, Isabel López Noriega & Sélim Louafi, *The Global Crop Commons and Access and Benefit-Sharing Laws: Examining the Limits of International Policy Support for the Collective Pooling and Management of Plant Genetic Resources*, *in* Crop Genetic Resources as a Global Commons (2013), n. 196, at 1–36 (recognizing importance of voluntary contributions from nonsignatory countries).

[315] López Noriega et al., n. 306, at 205 (citing authorities).

[316] Kamau (2013), n. 259, at 353. *See also* Louafi & Schloen (2013), n. 85, at 206 (stressing avoidance of multiplication of transaction costs as exchanges occur across the value chain, with multiple exchanges and a range of providers and recipients).

Another major achievement was the Contracting Parties' decision to codify a set of "take and pay" rules for the use of precompetitive research inputs at the international level[317] in place of case-by-case bilateral negotiations or exclusive property rights intrinsically ill-suited for such purposes.[318] Under the SMTA adopted to implement this regime, upstream knowledge assets continue to flow from the existing agricultural research infrastructure to the global research community, without costly and wasteful negotiations for genetic resources having no known or likely commercial value at the time of accession.[319] For these and other reasons, we contend that "take and pay" rules could greatly enhance a redesigned Microbial Research Commons, as envisioned in Chapter 5 of this book, on condition that serious design flaws in the ITPGRFA, identified below were avoided or rectified in the process.

At least three additional benefits of the International Treaty deserve to be highlighted here. First, the treaty obliges participating governments to make substantial, upfront financial commitments to a Benefit-Sharing Trust Fund, for the purposes of capacity building, conservation, and technical assistance in developing countries. In this and other ways, the treaty expressly seeks to reward farmers in provider countries for their breeding and conservation efforts over time.[320] To drive this point (and related obligations) home, Article 18.4(b) expressly conditions the effective implementation of the treaty by developing countries on the extent to which the developed countries party to the treaty effectively allocate the resources referred to in this article.[321] The ITPGRFA also recognized a range of nonmonetary benefits that can and should be provided to contracting parties, especially the developing countries, as previously mentioned.[322]

Another expected benefit for the Contracting Parties as a whole is that no recipient of any plant genetic materials exchanged through the multilateral system can claim "any intellectual property or other rights that limit the facilitated access to … (PGRs) or their genetic parts or components in the form received from the multilaterals system."[323] In other words, genetic resources once deposited into the

[317] *See* text & accompanying nn. 281–91.

[318] *See, e.g.*, Safrin, n. 156. *See especially* Reichman, *Green Tulips*, n. 287. *See also* Reichman & Lewis, n. 20; Rai et al., n. 5. *But see* Cottier & Panizzon, n. 3.

[319] For the importance of this principle in regard to microbial culture collections, see Chapter 5.

[320] *See* ITPGRFA, n. 123, arts. 18.1–18.5. These and other benefits to farmers are seen as a form of "farmers' rights."

[321] ITPGRFA, n. 123, art. 18.4(b). By the same token, developing countries pledge to make capacity building a planning priority. *Id.*

[322] *Id.* art. 13.1(a)–(c); *see* nn. 232–34 & accompanying text.

[323] ITPGRFA, n. 123, art. 12.3(d). This same obligation was included in standardized Material Transfer Agreements that all private parties seeking to access the multilateral system must execute. *Id.* art. 12.4. *See* Halewood (2014) n. 217, at 308 (citing GB/ITPGRFA 2006, art. 6.2).

gene pool must remain in that pool and cannot in principle become privatized in their original form.[324]

Third, the logic of the monetary form of benefits clearly remains central to the multilateral system.[325] In principle, the ITPGRFA thus envisions that, once recipients working on materials provided for research and breeding purposes reach the product stage where, for example, "a new plant variety is generated and this variety contains genetic material accessed under the multilateral system, ... and if such a product is commercialized, part of the generated revenue is [to be] shared."[326] These commercial applications do become legitimate subjects of intellectual property protection, and any resulting financial benefits to be shared with the multilateral system should flow directly or indirectly to farmers in all participating countries, especially in developing countries "who conserve and sustainably utilize plant genetic resources for food and agriculture."[327] In practice, however, these returns are paid into a multilateral fund – the Benefit Sharing Fund[328]– maintained by the Governing Body, which should in principle redistribute them in a manner consistent with the aims and objectives of the Treaty.[329]

Finally, the decision to encase the multilateral system within a formal international treaty administered by a full-fledged international organization has allowed the facilitated exchange process to operate at both the macro and micro levels under consensually developed policy guidelines, without the kind of interference from disgruntled governments that could undermine or challenge the workings of the system as a whole.[330] In this context, various entities have made some contributions, mostly voluntary, to the Benefit-Sharing Fund, and the Governing Body has allocated some of these funds to support capacity building projects in developing countries.[331]

Although disputes can and do arise concerning uses of specific resources provided by the multilateral system,[332] the intergovernmental legal architecture supporting that system intrinsically channels complaints from Contracting Parties to their duly constituted Governing Body, which at least in principle,

[324] However, implementation in practice raises some tricky questions not clearly resolved in the Treaty. *See, e.g.*, id., at 308–09 (noting vague language used to paper over disagreements that await arbitration of disputes to resolve). See further Section IV.

[325] *See* Manzella, n. 212, at 155.

[326] *Id.* at 156 (noting exception for products made freely available for research and breeding); *see* ITPGRFA, n. 123, art. 13.2.

[327] ITPGRFA, n. 123, art. 13.3.

[328] *See id.* art. 13.2(d)(ii); Manzella, n. 212, at 156, 156 fn. 30.

[329] ITPGRFA, n. 123, art. 19.3(f). For the role of the Governing Body, see Chapter 9, Section II.A.

[330] Manzella, n. 212, at 159–60.

[331] *Id.* at 159.

[332] *See* the cases on alleged biopiracy, discussed in Section I.C.2.a.

indirectly spurs governments outside the system to join, so as to have a voice in the Governing Body.[333] By the same token, the integration of the multilateral system into the FAO's own complex legal framework helps to stabilize the implementation process and ensures a modicum of funding for that purpose, at least in principle.[334]

2. Major Weaknesses

As previously intimated, however, the ambitious scope of the treaty, the idealistic aspirations that informed its design, and the complex legal architecture erected to sustain it are all sources of weakness that have made themselves increasingly visible. For example, expectations that the multilateral system would continue to expand as governments added both their national *ex situ* collections and their vast *in situ* resources have largely been disappointed.[335] Even efforts to expand the relatively limited list of mandatory crops covered in Annex I have proved fruitless, although some observers had taken such a move for granted at the time of negotiating the Treaty.[336] According to Dr. Kamau, "the number of Contracting Parties and natural or legal persons that have been notified is very small," and fewer than one-quarter of all the Parties had made any notifications at all.[337] Yet, without notifications to trigger the operations of the International Treaty, potential users lack the knowledge about availability to make their use feasible.[338]

As a result, most of the materials exchanged through the Multilateral System are sourced from the *ex situ* collections of the CGIAR gene banks.[339] More than one-half of all notified collections (about 1.3 million specimens) are lodged in the CGIAR's IARCs. Otherwise, there has reportedly been "meager" notification of other collections held by the Contracting Parties, and "[m]ost Annex I PGRFA are still not available for use or facilitated access through the Multilateral System."[340]

[333] *Cf.* Halewood et al. (2013), n. 252; *but see id.* at 26 (countervailing incentives to free ride on the system).

[334] *See, e.g.*, Manzella, at 157–58; Halewood (2014), n. 277, at 310 (noting agreed funding target, but querying mode of implementation).

[335] *See, e.g.*, Godfrey Mwila, *From Negotiations to Implementation: World Review of Achievements, Bottlenecks and Opportunities for the Treaty in General and for the Multilateral System in Particular*, *in* Crop Genetic Resources as a Global Commons (2013), n. 67, at 226, 236–37; *see also* Halewood et al. (2013), n. 252.

[336] Mwila (2013), n. 335, at 227–29, 236–37.

[337] Kamau (2013), n. 259, at 353. *See also* Chiarolla & Jungcurt (2011), n. 212 (noting that no private plant breeding companies had voluntarily placed their collections of Annex I materials directly in the multilateral system as of 2011).

[338] Kamau (2013), n. 259, at 354.

[339] Manzella, n. 212, at 236.

[340] Kamau (2013), n. 259, at 354, 357 (stating that "success in including material held by natural and legal persons in the Multilateral System is negligible").

To some extent, this reluctance to notify plant genetic resources potentially available to the multilateral system stems from objective technical difficulties of compiling the relevant data and information, especially for materials held by natural and legal persons (including universities and entities engaged with local farmers).[341] Capacity to properly characterize these holdings may be in short supply in developing countries, and there may be uncertainty about which materials at any given institute are actually covered by the Treaty, especially in the case of public-private initiatives.[342]

Such excuses, however, fail to explain why the Contracting Parties themselves – i.e., their relevant government agencies – have not performed the duties of notification any better than "natural and legal persons,"[343] which suggests that technical and administrative burdens are only a part of the problem. A more persuasive explanation is that developing countries witnessing the expanding use of plant breeders' rights and gene patents (especially on isolated genetic resources) in developed countries see mounting restrictions on access to their own PGRFA, without corresponding benefits accruing to them under the International Treaty.[344] In effect, the corresponding obligations to contribute proceeds from the sales of end products to the Trust Fund have so far proved to be largely theoretical (as explained in Chapter 9).

Meanwhile, nothing in the Treaty prevents either governments or natural and legal persons that fail to meet their obligations to contribute to the system from using the resources otherwise available from the system for their own purposes.[345] There are no requirements or conditions of reciprocity to stop what many observers regard as free-riding practices.[346] Nor does the Treaty prevent transfers of materials from the Multilateral System to nonparties on terms and conditions other than those contained in the SMTA.[347]

In practice, the CGIAR's Centers continue to allow facilitated access to all their *ex situ* collections on an open-access basis, as part of their institutional mandate, regardless of whether recipients are bound under the International Treaty or not. Yet, nearly one-half of all the contributions held by the Multilateral System are PGRFA controlled by the CGIAR Centers. In Dr. Kamau's view, the fact that users can access so much validated material from the system without undertaking legal obligations under the Treaty operates as a disincentive either to contribute to the system or to adhere to the Treaty.[348]

[341] *Id.* at 352.
[342] *Id.* at 352–57; *see* ITPGRFA, n. 123, art. 13.2(c).
[343] Kamau (2013), n. 259, at 358.
[344] *See, e.g.*, SANTILLI (2012), n. 227, at 123 *et seq* ; *see also id.* at 137–48; Kamau (2013), n. 259, at 355.
[345] *Id.* at 356–58.
[346] *See, e.g.*, *id.*; Halewood et al. (2013), n. 252.
[347] Kamau (2013), n. 259, at 358.
[348] *Id.* at 359–61 (addressing "cracks in the Multilateral System").

Whether it was ever likely that the Contracting Parties would diligently survey, list, and contribute vast reserves of *in situ* plant genetic resources to the multilateral system is a matter for conjecture. States sometimes adhere to international agreements full of noble aspirations with little intent to fulfill specific obligations, and the realm of science is no exception in this regard. In the case of the ITPGRFA, knowledgeable observers attribute the lack of follow through on the part of the developing countries to persistent skepticism about the ability of the treaty to serve their interests.[349]

Apart from the problems identified, this skepticism is partly fueled by the indirect nature of the benefits likely to flow to provider countries from their participation in the multilateral system under the best of circumstances. Precisely because the global regime of misappropriation that the CBD envisioned in 1992 was objectively weak,[350] the Contracting Parties under the International Treaty deemed it necessary to establish their own internal compliance machinery, dependent on the good offices of the FAO acting as a Third Party Beneficiary.[351] In so doing, the parties agreed that all benefit-sharing funds from both voluntary contributions and commercial use of the pooled genetic resources would flow back to the multilateral system itself, for consensual redistribution to support farmers and local conservation efforts.[352] Accordingly, they deprived themselves of any direct financial incentives as national or local providers of the plant-genetic materials in question,[353] and they renounced any rights of control over the resources distributed via confidential SMTAs, including measures to track specific uses of those same resources.[354]

Given the logical apprehension this scheme already generates, the method of distributing even the mostly voluntary contributions so far made to the Benefit-Sharing Trust Fund has done nothing to dissipate the developing countries' anxieties in this regard. On the contrary, the Fund's administrators seem to have appointed themselves as *de facto* funders of research, with the funds going to those who draft the best grant proposals pertaining to issues of food-security, climate change, and agro-biodiversity conservation.[355] Reportedly, the grantees are usually those who write the best proposals (a developed country writing skill) and not necessarily the best projects, so that those who actually invest in preparing materials of use to others for purposes of breeding and research do not necessarily benefit from

[349] Mwila (2013), n. 335, at 227.
[350] *See* Section I.C.2.b. *See also* López Noriega et al., n. 306, at 201–02, 213.
[351] *See generally* Moore, n. 296.
[352] *See, e.g.*, López Noriega et al., n. 306, at 215 (contrasting immediate benefits to users of gene banks with long-term and speculative benefits to farmers); Kamau (2013), n. 259, at 363 (citing ITPGRFA, n. 123, art. 13.3).
[353] *See, e.g.*, López Noriega et al., n. 306, at 211–18.
[354] ITPGRFA, n.123, art. 13(3). General notification of use is sent to the Governing Body.
[355] Kamau (2013), n. 259, at 313.

their contributions.[356] Meanwhile, hold-out governments and national collections that continue to access and use materials from the multilateral system without contributing resources of their own can also continue to benefit directly from negotiated access agreements to their own materials under the bilateral approach of the CBD.

Predictably, all these factors have engendered a "wait and see" attitude on the part of the developing countries, which, in theory, should have been attenuated by substantial voluntary and mandatory contributions to the Benefit-Sharing Fund from developed countries for purposes of capacity building, conservation, food security, and farmers' benefits in general.[357] In practice, and despite the Governing Body's strenuous efforts to develop apposite funding strategies, their implementation has fallen far short of the expected goals,[358] although some new funding is expected from a Global Crop Diversity Trust with support from CGIAR and the FAO.[359]

Given that the multilateral system possesses no central repositories of its own, and depends entirely on the willingness of Contracting Parties to allow their *ex situ* and *in situ* resources to be plugged into a distributed and virtual global gene bank, this perceived failure of developed-country funders to keep their side of the bargain bears directly on the reluctance of developing-country governments to further implement the treaty at the national level.[360] One may surmise that this foot-dragging has been reinforced by continuing complaints about biopiracy pertaining to specimens taken at one time or another from collections affiliated with the CGIAR, whether or not such complaints withstand sound legal analysis on their facts.[361]

Meanwhile, no benefit-sharing funds have so far accrued from royalties on commercial applications of plant genetic resources drawn from the multilateral system, and few, if any of the SMTAs accompanying these facilitated exchanges have triggered even a duty to pay.[362] The conventional wisdom holds that not enough time has elapsed for the kind of commercial applications likely to trigger such obligations under the Compensatory Liability Regime and that this situation will resolve itself in due course.[363] We believe, however, that design flaws embodied

[356] *Id.* at 364.

[357] *See, e.g.*, Mwila (2013), n. 335, at 229 (stressing developing countries' insistence on prioritizing measures to ensure that the Benefit Sharing Fund becomes quickly operational). *Id.* at 235.

[358] *Id.* at 235–36. As of March 2012, only some voluntary contributions from Norway, Italy, Spain, and Switzerland had been reported. *See* Kamau (2013), n. 259, at 352 n. 46.

[359] *See* Halewood (2014) n. 217, at 311 (stating that GCDT is expected to raise hundreds of millions of dollars to support long-term conservation of 18 *ex situ* crop collections).

[360] Mwila (2013), n. 335, at 236–37. *See also* López Noriega et al., n. 306, at 215. The fact that such funds would be funneled through FAO is also viewed by some as a further disincentive.

[361] *See, e.g.*, Section I.C.2.a. for selected cases of alleged biopiracy by academics. *Cf.* López Noriega et al., n. 306, at 214. But see Chapter 9, Section II.A.1.c ("Long Term Funding Arrangements").

[362] *See, e.g.*, Mwila (2013), n. 335, at 235; López Noriega et al., n. 306, at 215; Halewood et al. (2013), n. 314, at 25.

[363] Mwila (2013), n. 335, at 235.

in the SMTA may be partly responsible for the lack of royalty payments under the facilitated exchange mechanism.

First of all, the SMTA as drafted purports to waive the downstream user's duty to pay royalties if he or she allows third parties freely to access the end products for research and breeding purposes.[364] As a practical matter, determining when an end user triggers this exemption by making a product "available to others for further research and breeding" is inherently ambiguous and "raises many controversies."[365]

More to the point, a research exemption for biological products derived from materials in the Crop Commons should never have been traded away for a waiver of royalties in the first place. On the contrary, a qualified research exemption should have been built into the legal Framework as a quid pro quo for facilitated access to the genetic resources managed by the multilateral system, as we explain in Chapter 5.[366] The duty to pay benefit-sharing royalties into the Trust Fund for commercial applications would then have remained absolute, and not conditional.[367] As matters stand, even if the duty to pay royalties were triggered, those payments would not go to the provider country, as shown , whereas commercial users can avoid the duty to pay royalties altogether by invoking the research exemption among a number of other possible out cards.

More generally, there are mounting concerns about the distance that separates downstream commercial users from upstream providers of germplasm under the SMTAs,[368] and about the corresponding difficulties of enforcing the benefit-sharing obligations within the confines of the Treaty as drafted. These concerns persist despite the FAO's commitment to act as a Third Party Beneficiary for enforcement purposes.[369] For example, given that the Treaty expressly absolves the multilateral system from any duty to track the uses of the PGRFA it makes available,[370] would-be commercial users may be tempted to breed off of alternative supplies of plant genetic resources outside the multilateral system the moment their applied research looks promising.[371] We discuss these and other possible flaws in the design of the

[364] *See* SMTA, n. 272, art. ¶6.8; *see also* Kamau (2013), n. 239, at 362.

[365] Santilli (2012), n. 227, at 144. *See also* Chiarolla & Jungcurt (2011), n. 269.

[366] *See* Chapter 5, Sections II & III.

[367] *See, e.g.*, Jerome H. Reichman, A *Compensatory Liability Regime to Promote the Exchange of Microbial Genetic Resources for Research and Benefit Sharing*, *in* Designing the Microbial Research Commons, n. 45, at 43. *Accord.* Santilli (2012), n. 272, at 147–48 ("After all, it would be fair that all users/recipients of the multilateral system channel part of their profits obtained from sales of their products to the conservation of plant genetic resources, and ... contribute to a more solid and sustainable funding strategy in the long term").

[368] *See, e.g.*, Godt (2013), n. 85, at 248–49 (noting a problem of "'genetic proximity' of the *ex situ* collections to the originally accessed materials").

[369] *See esp.* Mwila (2013), n. 335, at 235; *see also* Moore, n. 296, at 164–76.

[370] *See* n. 354 & accompanying text.

[371] *Cf.* López Noriega et al., n. 306, at 217.

Crop Commons when we describe our own proposals for a redesigned Microbial Research Commons in Chapters 5 and 10. For present purposes, it suffices to observe that the very success of the Crop Commons' SMTA in fostering exchanges of plant genetic resources under the ITPGRFA only serves to deepen developing country misgivings about the extent to which that treaty adequately protects their interests.

On the one hand, the CGIAR and other affiliated research institutes have relied heavily on the SMTAs when distributing plant genetic materials from the multilateral system to recipients everywhere, regardless of whether those recipients operate in countries that have adhered to the International Treaty.[372] The importance of the multilateral system for global agricultural research is thus amply demonstrated, whereas the global public goods approach that was a hallmark of the CGIAR has been preserved and expanded.[373]

On the other hand, the developing countries tend to view this very success as a continuation of the colonial exploitation heritage, because it relies on uncompensated uses of natural resources that they have – willingly or unwillingly – been induced to provide.[374] Moreover, the CGIAR's open access policy may conflict with one express term of the Treaty, which actually envisions the possibility of invoking the principle of reciprocity to limit access to the multilateral system to countries or entities that have signed the Treaty and complied with both its funding and resource-providing commitments.[375] Even representatives of the CGIAR recognize that their continued willingness to supply the world at large with plant genetic resources actually encourages nonmember countries to free ride on the multilateral system, which in turn reinforces the reluctance of developing countries to add substantially to the resources available from that system.[376] These contradictions have led the CGIAR leadership to call for sanctions against non-members as well as against member countries that fail to meet their obligations under the Treaty.[377]

From a broader perspective, the difficulties in implementing the Crop Commons that have recently come to light suggest that its rigid international legal architecture – devised to withstand assault from proprietary interests in both developed and developing countries – constitutes an obstacle in itself to a more cohesive and

[372] *See, e.g.*, Mwila (2013), n. 335, at 235; López Noriega et al., n. 306, at 211.

[373] *See id.* at 211 (noting that 75% of materials distributed by the IARCs are materials that the centres had improved, which confirms that "improved materials and breeding lines are exchanged more frequently than materials held in gene banks." (citing authorities)).

[374] *See, e.g.*, Mwila (2013), n. 335, at 229, 232 (stating that most developing countries remain skeptical of the agreements between the Governing Body and the IARCs, "viewing them as mechanisms designed to enhance facilitated access from the multilateral system while doing little for benefit sharing").

[375] ITPGRFA, n. 123, art. 11(4); Mwila (2013), n. 335, at 233; Halewood et al. (2013), n. 314, at 11–15. *See also* Santilli (2012), n. 227, at 114–15.

[376] *See esp.* Halewood et al. (2013), n. 314, at 15–23.

[377] *Id.* at 19–27.

effective agricultural research mission. With no transnational enforcement machinery under the CBD as it stood in 1992, and with relatively few national focal points in operation at the time the ITPGRFA was drafted,[378] the Contracting Parties understandably dedicated both their publicly controlled plant genetic resources and the relevant benefit-sharing funds to "the multilateral system." That system was a substitute for "the Common Heritage of Mankind" principle underlying the FAO's nonbinding International Undertaking of 1983.[379]

As a result, neither the research institutes that maintain, improve, and provide *ex situ* genetic resources nor the governments that maintain and provide *in situ* resources become direct beneficiaries of the system, except insofar as they subsequently qualify for grants consensually approved by the Governing Body.[380] Apart from the lengthy, expensive, and mind-numbing negotiations and tradeoffs necessary to obtain any consensus for further action at the political level, would-be beneficiaries must also navigate four different bureaucratic strands that are woven into the cumbersome legal governance structure created by the Treaty.[381]

Viewed with hindsight, the elephantine legal infrastructure created by the ITPGRFA reflects the practices of a top-down United Nations Specialized Agency far more than those of an entity devised by scientists to manage a research infrastructure on behalf of the global scientific community. Particularly worrisome in this regard are recent complaints that the governance structure established under the Treaty has shunted scientists and agricultural research in general to an outlier and largely irrelevant status.[382] We shall return to this complaint in particular when we consider the choice of governance models for the proposed Microbial Research Commons in Part Four, with a view to avoiding most of the same problems that confronted those who devised the Crop Commons in 2001.

Before doing so, however, we must take account of another major legislative development that dramatically changed the international legal landscape in 2010, namely, the adoption of the Nagoya Protocol to the Convention on Biological Diversity.[383] As will be seen, the Nagoya Protocol has attempted to provide the kind of enforceable international regime of misappropriation for genetic resources that was lacking at the time that the ITPGRFA was adopted. This new state of affairs, in turn, immediately compels us to ask how the relevant research commons – whether they deal with plants, microbes, or other genetic resources – might be better designed if

378 *See, e.g.*, Manzella, n. 212; *see also* Moore, n. 296, at 168–72.
379 *See generally* Esquinas-Alcázar et al. (2013), n. 196; Chapter 2, Section I.B.2.
380 *See, e.g.*, Halewood, n. 314, at 23.
381 *See further* Chapter 9, Section II.A.
382 *See esp.* Halewood (Louvain 2012), n. 220.
383 Nagoya Protocol, n. 65 (entered into force, October 12, 2014). *See generally* next section.

they were to build on the strengths of the Nagoya Protocol rather than attempting to avoid the weaknesses of the CBD.[384]

IV. NEW CONSTRAINTS AND OPPORTUNITIES FOR SCIENTIFIC RESEARCH UNDER THE NAGOYA PROTOCOL

The International Treaty on Plant Genetic Resources for Food and Agriculture of 2001 had responded to the failure of the CBD's drafters in 1992 to understand the pivotal role of pooled genetic resources in the life sciences generally. They had neglected to provide appropriate legal support for the kind of global research infrastructure that would benefit both developed and developing countries by stimulating research and innovation.[385] As we have seen, however, the design and implementation of the FAO's Crop Commons was itself hampered by another design flaw in the CBD, namely, its lack of any agreed legal machinery for cross-border enforcement of the international regime of misappropriation that it had sought to establish.[386]

The CBD firmly established state sovereignty over both genetic resources and the traditional knowledge of indigenous local communities associated with the preservation and sustainable use of these same resources.[387] However, implementing and enforcing the Access and Benefit-Sharing (ABS) regime, which was a pillar of the Convention,[388] remained uncertain in actual state practice. Serious questions arose concerning the implications of the bilateral approach to ABS for scientific research, economic development, and industrial uses of genetic resources. Equally uncertain were the modalities that should govern the obligations of both providers and would-be users to secure Prior Informed Consent (PIC) and Mutually Agreed Terms (MAT) from indigenous local communities in appropriate cases.[389]

National legislation implementing ABS principles had only been enacted in a relatively small number of countries with wide discrepancies between the different

[384] *Cf.* Clive Stannard, *The Multilateral System of Access and Benefit Sharing: Could It Have Been Constructed Another Way?, in* CROP GENETIC RESOURCES AS A GLOBAL COMMONS (2013), n. 196, at 243–64.

[385] *See, e.g.*, Godt (2013), n. 85; Gerd Winter, *Knowledge Commons, Intellectual Property, and the ABS Regime, in* COMMON POOLS OF GENETIC RESOURCES (2013), n. 48 at 296 [hereinafter Winter, *Knowledge Commons* (2013)]; *see generally* COMMON POOLS OF GENETIC RESOURCES (2013), n. 48.

[386] *See* Section I.C.1.

[387] *See, e.g.*, CBD, n. 36, arts. 8(j) and 15.

[388] *See id.* art. 1, establishing three pillars of the Convention, viz. 1) conservation of biological diversity; 2) sustainable use of its components; 3) fair and equitable sharing of the benefits arising from the utilization of genetic resources.

[389] *See, e.g.*, THOMAS GREIBER ET AL., AN EXPLANATORY GUIDE TO THE NAGOYA PROTOCOL ON ACCESS AND BENEFIT SHARING 3, 8–10, 13 (Int'l Union for Conservation of Nature & Natural Resources (IUCN)), Envtl Pol'y & L. Paper No. 13, 2012) [hereinafter IUCN, GUIDE TO THE NAGOYA PROTOCOL (2012)].

laws and gaps in the global legislative landscape.[390] The common denominator in the existing legislative and administrative framework, moreover, was seriously to restrict access to genetic resources for both commercial and noncommercial purposes. Yet, these very restrictive practices contradicted the CBD's express undertaking to provide "facilitated access" to genetic resources for environmentally sound uses that depended on both research and applications.[391]

As a result, the Conference of the Parties had become increasingly aware that there was no clear line between "providers" and "users" of genetic resources[392] and that the bilateral approach to ABS, left to itself, could compromise both research and applications in the end, with fewer monetary benefits to divide.[393] Implicit in this reconsideration of ABS modalities was the possibility that, where scientific research was concerned, non-monetary benefits might outweigh the value of monetary benefits in many, if not most, situations.[394]

Meanwhile, the FAO's International Treaty on Plant Genetic Resources for Food and Agriculture, as adopted in 2001, had established a de facto multilateral exception to the bilateral approach to ABS enshrined in the CBD. Because this exception was a *fait accompli*, the parties to both the CBD and the ITPGRFA needed to reconcile the former with the multilateral approach embodied in the latter.[395] The very existence of the ITPGRFA prodded the Conference of the Parties to consider the need and scope for more multilateral action with respect to ABS within the confines of the CBD itself.[396]

Above all, a major defect of the ABS regime as enacted in 1992 remained the inability of provider countries to seek redress for misappropriation of genetic resources, or misuse of related contractual agreements, directly from the legal authorities in user countries.[397] This weakness stemmed from the territorial approach to sovereign control of genetic resources inherent in the CBD itself. Under this regime, governments in user countries had assumed no formal legal obligations to assist complainants in provider countries who sought remedies for noncompliance with the ABS provisions of the CBD.[398]

390 *Id.* at 3, 14–15.

391 *Id.* at 6, 8. *See* CBD, n. 36, art. 15(2) ("Each Contracting Party shall endeavor to create conditions to facilitate access to genetic resources for environmentally sound uses by other Contracting Parties and not to impose restrictions that ran counter to the objectives of this Convention.").

392 IUCN, Guide to the Nagoya Protocol (2012), n. 389, at 6.

393 *Id.* at 6, 12–13.

394 *Id.* at 17.

395 *See id.* at 33–35.

396 *See id.* at 12–18.

397 *Id.*

398 *See id.* at 29–33.

To their credit, the Conference of the Parties (COP) soon turned its attention to these and related issues. In 2000, the CBD's Ad Hoc Open-Ended Working Group on ABS (AHWG) started to elaborate and negotiate an international regime that would enforce the CBD's Access and Benefit-Sharing principles.[399] In 2001, the AHWG submitted the draft Bonn Guidelines,[400] which were adopted with some changes in 2002.[401] However, the Bonn Guidelines turned out to be only a voluntary and relatively contentious starting point that did not satisfy the concerns of provider countries.[402] Acting on decisions taken at the United Nations World Summit on Sustainable Development in 2002,[403] the COP in 2004 charged the AHWG with elaborating and negotiating an international regime that would effectively implement Articles 8(5) and 15 of the CBD.[404] Several international meetings were held to discuss the process, nature, scope, elements, and modalities of such a regime,[405] although the academic research community was little involved in these discussions.[406] In 2007, the 9th Conference of the Parties to the CBD (COP 9) decided to extend the Working Group's ABS's mandate for an additional three sessions, with the expectation that it would present an agreed text to COP 10, in 2010, for adoption.

COP 9 also decided that, to assist the negotiation process, there should be a Meeting of the Group of Legal and Technical Experts on Concepts, Terms, Working Definitions and Sectoral Approaches. At a meeting in December 2008,[407] that Group recognized significant differences between uses of genetic resources in the agricultural and food sectors, on the one hand, and the pharmaceutical sector on the other. The experts also noted the special nature of microbial genetic resources within the area of food and agriculture,[408] but they lacked time to take the next

399 *See* CBD COP5 decision 2/26, Access to genetic resources; Draft Bonn Guidelines on Access to Genetic Resources and Fair and Equitable Sharing of Benefits Arising out of Their Utilization (Draft Bonn Guidelines).

400 *See id.*

401 Bonn Guidelines, SCBD 2002, CPD COP6 decision VI/24, Access and benefit=sharing as related to genetic resources.

402 IUCN, Guide to the Nagoya Protocol (2012), n. 389, at 19.

403 *See* CBD COP5 decision VII/19, Access and benefit-sharing as related to genetic resources (art. 15). *See generally* IUCN, Guide to the Nagoya Protocol (2012), n. 389, at 19–20.

404 *Id.* at 19.

405 *See id.* at 20–21: Working Group on Access and Benefit Sharing, Convention on Biological Diversity (CBD), http://www.cbd.int/abs/wgabs/ (last accessed 14 June 2014).

406 *Cf.* Jinnah & Jungcurt, n. 153, at 464.

407 *See Ad Hoc Open-Ended Working Group on Access & Benefit-Sharing, Report of the Meeting of the Group of Legal and Technical Experts on Concepts, Terms, Working Definitions and Sectoral Approaches*, 7th mtg., UNEP/CBD/WG-ABS/7/2 (April 2009).

408 From a general perspective, the creation of a practical and workable international regime for the different sectors had some support from industry. *See, e.g.*, Int'l Chamber of Commerce, 2008. *Objective, scope, fair and equitable benefit sharing, access and compliance*, Submission

planned-for step, which was to examine whether the differences noted between the sectors had any implications for how access and benefit-sharing norms should be developed or adapted.[409]

The Group of Legal and Technical Experts also considered the possibility that noncommercial research might constitute a discreet regulatory component in the future. Documents prepared by the CBD's Secretariat emphasized the need to simplify procedures for access to genetic resources to be used for basic research. The Working Group on Access and Benefit Sharing (WG-ABS) also recognized that administrative procedures pertaining to Prior Informed Consent and Mutually Agreed Terms rarely differentiated between access for scientific or commercial purposes.[410] This body concluded that failure to make this distinction may have created a disincentive to scientific research and reduced the potential sharing of non-monetary benefits through non-commercial scientific cooperation, such as the exchange of researchers and joint research projects.[411]

However, the Working Group also feared that adoption of a simplified procedure for access to genetic resources for basic research could potentially create a loophole through which resources initially accessed for research purposes could subsequently be used for commercial purposes, without having obtained the consent of the providers.[412] A possible solution to this problem could entail the drafting of a standard MTA that did allow for *ex post* compensation in case of commercial applications, a solution that has since been widely employed.[413]

to the Secretariat of the Convention on Biological Diversity for the 7th Ad Hoc Open Ended Working Group on Access and Benefit Sharing, Paris, France, 2–8 April 2009. In addition to the representatives of culture collections, some commentators have given their support for a global system for microbial genetic resources. *See, e.g.*, MOSAICC, MICRO-ORGANISMS SUSTAINABLE USE AND ACCESS REGULATION INTERNATIONAL CODE OF CONDUCT. ELABORATION AND DIFFUSION OF A CODE OF CONDUCT FOR THE ACCESS TO AND SUSTAINABLE USE OF MICROBIAL RESOURCES WITHIN THE FRAMEWORK OF THE CONVENTION ON BIOLOGICAL DIVERSITY (2009). *See* J. Schmidt, C. Hsin & M. Esteban, Workshop Report on Sectoral Linkages and Lessons Learned on Access and Benefit Sharing (ABS): Moving the ABS Agenda Forward, 28 Nov. 2008, Tokyo, at 4. In a more concrete proposition, Dr. Sumida, from the Japan Bioindustry Association, had even suggested that governments could delegate their authority to determine access to genetic resources to *ex situ* collections. While government agencies could observe and control the process, *ex situ* collections, in his view, could enhance compliance with the CBD and facilitate exchange. *Id.* ¶4. Note, however, that this proposition seemed to refer to all kinds of *ex situ* collections and not to those only dealing with microbial genetic resources.

409 *See further* Chapter 10, "Governing Digitally Integrated Genetic Resources, Data, and Literature."

410 ABS WG9, "Report of the First Part of the Ninth Meeting of the Ad hoc Open-Ended Working Group on Access and Benefit-Sharing" (UNEP/CBD/WG-ABS/9/3, 26 April 2010).

411 *Cf. Analysis of ABS Gaps*, n. 80, ¶ 7; *see also* IUCN, GUIDE TO THE NAGOYA PROTOCOL (2012), n. 389, at 17.

412 *Analysis of ABS Gaps*, n. 80 ¶ 31.

413 *See, e.g.*, Chapter 4, Section III.

After six years of tortuous negotiations, the Nagoya Protocol on Access to Genetic Resources and the Fair and Equitable Sharing of Benefits Arising from Their Utilization to the Convention on Biological Diversity (the "Nagoya Protocol") was adopted at the tenth meeting of the Conference of the Parties on 29 October 2010, in Nagoya, Japan.[414] The Protocol was opened for signature on February 2, 2011, to February 1, 2012, following which states could adhere by means of accession.[415] Fifty instruments of ratification were required for the Protocol to enter into force, as occurred on October 12, 2014.[416]

The Nagoya Protocol aims to provide greater legal certainty and transparency for both providers and users of genetic resources.[417] Specifically, the Protocol reaffirms and clarifies the broad economic scope of the CBD, while at the same time expressly addressing the needs of scientific research, which had largely been neglected in the CBD itself. As explained below, these provisions could make it easier for the scientific community to operate within the boundaries of the Convention, with the proviso that these boundaries must be carefully taken into account when designing any given research project involving cross-border exchanges of microbial genetic resources.[418] Major new enforcement provisions for the CBD have also been formulated, which both provider and user countries will have to implement in their national legislation and practice.[419] These provisions are briefly summarized here, and they are subsequently re-examined when we discuss governance of the proposed Microbial Research Commons in Part Four of this volume.

A. *Clarifying the Broad Economic Scope of the CBD*

The language originally adopted in article 15 of the CBD left considerable room for arguments about the extent to which its ABS provisions applied to all downstream applications of raw genetic resources and related information. Technically, these provisions required the fair and equitable sharing of benefits from "the utilization of genetic resources."[420] But the downstream reach of this term was left undefined,

414 Nagoya Protocol, n. 65.

415 *Id.* art. 35(1).

416 *Id.* art. 35(1). *See* Parties to the Nagoya Protocol, Convention on Biological Diversity, https://www.cbd.int/abs/Nagoya-protocol/signatories.

417 Nagoya Protocol, n. 65, pmbl.

418 *See, e.g.*, Margo A. Bagley & Arti K. Rai, *The Nagoya Protocol and Synthetic Biology Research: A Look at the Potential Impacts*, SYNBIO6/November 2013.

419 *See, e.g.*, Morton W. Tvedt & Oke K. Fauchald, *Implementing the Nagoya Protocol on ABS: A Hypothetical Case Study on Enforcing Benefit Sharing in Norway*, 14 *J. World I.P.* 383 (2011); Tianbao Quin, *Common Pools of Traditional Chinese Medical Knowledge in China*, *in* COMMON POOLS OF GENETIC RESOURCES (2013), n. 48, at 150, 164; *see also* SANTILLI (2012), n. 227.

420 *See, e.g.*, Nagoya Protocol, n. 1, art. 65; CBD, n. 36, art. 15(1). (7).

which in turn rendered the obligations to obtain prior informed consent and mutually agreed terms undetermined.[421]

For example, the extent to which the CBD's ABS obligations applied to biological products based on or derived from isolated gene sequences was unclear. Equally unclear was the status of "naturally occurring chemical compounds resulting from the metabolism of cells (such as aromas used in the cosmetic industry or biochemicals used in pharmaceutical research).[422] Because the CBD "provides scant detail on how transactions of genetic resources are to take place [so as] to be consistent with the Convention" while ostensibly applying its ABS obligations "to a broad range of research and development activities," it posed complicated problems of interpretation that reportedly help to explain the relatively "low level of implementation by Contracting Parties."[423]

Article 2 of the Nagoya Protocol provides new definitions that resolve some of these ambiguities and aggressively attempt to expand the economic scope of the CBD. For example, the term "utilization of genetic resources" now expressly means to conduct research and development on the genetic and/or biochemical composition of genetic resources, including through the application of biotechnology.[424] Biotechnology is defined as "any technological application that uses biological systems, living organisms, or derivatives thereof, to make or modify products or processes for specific use."[425] In this context, derivatives are specifically included within the definition of "biotechnology" and further defined as "a naturally occurring biochemical compound resulting from the genetic expression or metabolism of biological or genetic resources, even if it does not contain functional units of heredity."[426]

According to Professor Winter, the chemical compounds encoded by the genes are thereby brought within the scope of the ABS regime, at least by virtue of the provider state's ability to determine the conditions of utilizing genetic resources contractually.[427] The developing countries had rightly perceived that the "potential value," as covered by Article 2 of the CBD, would lie primarily in the naturally occurring compounds that result from the activity of genes.[428] The explicit inclusion of biochemical compounds resulting from the expression of genetic resources in

[421] *See, e.g.*, IUCN, Guide to the Nagoya Protocol (2012), n. 389, at 58 (noting that the "controversy over the exact content of these provisions and the complexity of implementing them" motivated the negotiation of the Protocol).

[422] Buck & Hamilton (2011), n. 70, at 50; Godt (2013), n. 85.

[423] Buck & Hamilton (2011), n. 70, at 48.

[424] Nagoya Protocol, n. 65, art. 2(c).

[425] *Id.* art. 2(d).

[426] *Id.* art. 2(e).

[427] Winter, *Knowledge Commons* (2013), n. 385, at 296. *See also* IUCN, Guide to the Nagoya Protocol (2012), n. 389, at 66.

[428] CBD, n. 36, art. 2; Buck & Hamilton (2011), n. 70.

the definition of "utilization of genetic resources" under Article 2 of the Nagoya Protocol means that parties needing prior informed consent to use genetic resources

> "will be expected to regulate – in that context – research and development on both the genes and on any naturally occurring biochemical compounds contained in or derived from material acquired under their domestic access and benefit-sharing framework."[429]

The Nagoya Protocol thus brings within the scope of the CBD not only research and development on gene sequences, but also, for example, R&D "on naturally occurring biochemicals in the pharmaceutical industry or the development of perfumes on the basis of naturally occurring aromas that were collected from the wild."[430] The Protocol also applies to "traditional knowledge associated with genetic resources ... and to the benefits arising from the utilization of such knowledge."[431]

Arguably, the Protocol even extends to collections of data pertaining to genetic resources or derivatives, as defined therein. The Group of Legal and Technical Experts in 2008 proposed that the term "derivative" could legitimately include "information or knowledge derived from genetic materials in general or a specific gene sequence in particular."[432] Although the term "genetic material" can be interpreted narrowly, the definition of "utilization of genetic resources" implies a "broad and dynamic understanding of the concept ... that would encompass digital information."[433]

Given the growing success of, and potential for, *in silico* research in marine microbiology, for example, there is a growing consensus that the definition of "utilization of genetic resources" in the Nagoya Protocol encompasses not only *in situ, ex situ*, and *in vitro* research uses, but also *in silico* access to genetic resources by means of data exchanges or databases.[434] This conclusion follows from the fact

429 Buck & Hamilton (2011), n. 70; Nagoya Protocol, n. 65, arts. 2, 5(1). Notably, the Protocol does not support additional prior informed consent requirements for access to biochemicals that are not contained in genetic material, but, for instance, are kept in a laboratory. Buck & Hamilton (2011), n. 70. However, information about biochemicals derived from genetic resources in a database remains subject to benefit-sharing obligations under art. 5 of the Nagoya Protocol. *See* text & accompanying nn. 433–36.

430 Maria Julia Oliva, Nagoya Protocol on Access and Benefit Sharing –Technical Brief, Union of Ethical Biotrade (2010), *available at* http://ro.unctad.org/biotrade/congress/BackgroundDocs/UEBT%20ABS% 20Nagoya%20ProtocolFINAL.pdf (last accessed 14 June 2014).

431 Nagoya Protocol, n. 68, art. 3.

432 Report of the Group of Legal and Technical Experts on Concepts, Terms, Working Definitions and Sectoral Approaches, UNEP/CBD/WG-ABS/712, 12 Dec. 2008, *reproduced in* IUCN, Guide to the Nagoya Protocol (2012), n. 389, Box 9 at 67.

433 Report of the Group on Legal and Technical Experts, n. 432. *But see* Bagley & Rai, n. 418 (noting that the term "genetic material" can be read so as to exclude intangibles and citing authority).

434 *See, e.g.*, Arianna Broggiato et al., *Fair and Equitable Sharing of Benefits from the Utilization of Marine Genetic Resources in Areas Beyond National Jurisdiction: Bridging the Gaps between Science and Policy*, 49 *Marine Pol'y* 176–185, 178 (2014) (noting that the *in silico* pathway "corresponds to the use of knowledge of a nucleic acid sequence for any purpose other than ... *in vitro* synthesis").

that, like the other pathways, *in silico* research "required the use of functional units of heredity at some step of the biotechnological development and/or production process," which brings it within the broad definition of genetic resources set out in the Nagoya Protocol.[435]

Professor Gerd Winter accordingly reports that the Conference of the Parties will strongly push for inclusion of genetic sequence data within the ABS provisions of the Nagoya Protocol by means of both technical legal interpretation and implementing MTAs at both the domestic and regional levels.[436] This view is further confirmed by implementing legislation in both the Andean Community and Brazil, which broadly cover genetic information of value or of real or potential use.[437] That said, the enforcement of such claims – if authorized by national legislation or contract – could prove technically difficult in practice.[438]

Article 2 of the Nagoya Protocol thus fully expands the focus of the CBD's ABS obligations from product marketing to research and development.[439] It could accordingly hinder or at least complicate virtually all scientific research endeavors that depend on use of genetic resources,[440] including synthetic biology.[441] However, the Nagoya Protocol also deliberately establishes offsetting principles and procedures to facilitate scientific research within the ambit of the CBD for the first time.

B. *Facilitating Scientific Research*

Having extended the reach of the CBD to virtually all research and applications in which use of genetic resources becomes a factor, the Nagoya Protocol expressly identifies legal pathways for the conduct of research that can immunize public scientists and their institutions from liability for violations of the CBD. At the same time, the Protocol introduces tough new enforcement measures to overcome prior weaknesses in the CBD's global regime of misappropriation.[442] These measures, in turn, greatly augment the risks of legal liability and sanctions for those researchers and research institutes that fail to comply with the CBD's ABS provisions as thus

[435] *Id.*
[436] Remarks of Prof. Gerd Winter, 2d. Gen. Assembly Meeting, Micro B3 Project, Max Planck Inst. Marine Microbiology, Bremen, Germany, 23–25 April 2014.
[437] *See, e.g.*, Bagley & Rai, n. 418, at 20 (citing authorities).
[438] *See, e.g.*, Winter, *Knowledge Commons* (2013), n. 385, at 296–98; Remarks of Matthias Buck, at International Workshop on Access to Genetic Resources: Improving Effectiveness, Justice, and Public Research in ABS." Bremen, Germany, 15–16 Sept. 2011.
[439] *See* Godt (2013), n. 85, at 258.
[440] *See* Jinnah & Jungcurt, n. 153.
[441] Bagley & Rai, n. 418, at 20–24.
[442] *See* Sections I.C.1 & IV.C.

strengthened. Let us first survey the immunizing opportunities provided to righteous researchers before examining the greater risks of noncompliance in Section C.

1. Recognizing the Link Between Public Science and Commercial Benefits

As we saw earlier, the CBD's failure to acknowledge the importance of public scientific research implicitly reinforced the tendencies of developing country governments to restrict access to their microbial genetic resources even for such purposes.[443] These restrictions – grounded in fears of potential lost benefits from unknown commercial applications – perversely magnified the prospects for actual lost benefits by inhibiting the kinds of upstream research likely to lead to these same applications.[444] The scientific community was thus arguably one of the stakeholder groups most adversely affected by the CBD's ABS provisions as initially drafted.[445]

After protracted and difficult negotiations, with persistent inputs from various scientific entities, notably the CGIAR, the drafters of the Nagoya Protocol were persuaded that facilitating scientific research – within the context of tightened enforcement provisions – would likely augment the economic benefits that developing-country providers hoped to share in the long run. To this end, Article 8(a) of the Protocol requires that each Party, when devising and implementing its ABS legislation or regulatory requirements, shall

> create conditions to promote and encourage research which contributes to the conservation and sustainable use of biological diversity, particularly in developing countries, including *through simplified measures on access for non-commercial research purposes*, taking into account the need to address a change of intent for such research.[446]

This provision aims to free noncommercial research from the tentacles of overly zealous access restrictions, while recognizing that the line between commercial and noncommercial research has become tenuous, especially in the life sciences.

[443] The CBD does mention research, *see* CBD, n. 36, art. 12, but no provisions were made to safeguard the needs of researchers for access to genetic resources.

[444] *See, e.g.*, Bagley & Rai, n. 418, at 21 ("making the NP applicable to derivative products could have the negative effect of disincentivizing the use of genetic resources in research and commercialization endeavors.").

[445] Buck & Hamilton (2011), n. 70; *see also* Jinnah & Jungcurt, n. 153.

[446] Nagoya Protocol, n. 65, art. 8(a) (emphasis supplied). Article 8(b) further urges Parties to "pay due regard to cases of present or imminent emergencies that threaten or damage human, animal or public health," particularly the need for "expeditious access to genetic resources" and also for "access to affordable treatments by those in need, especially in developing countries." Finally, art. 8(c) urges consideration of "the importance of genetic resources for food and agriculture and their special role for food security."

To close this possible loophole, the provision admonishes both providers and users to take into account the possibility that any noncommercial public research project could subsequently generate the kind of downstream commercial activities that should trigger the normal access and benefit sharing provisions of article 15 of the CBD.[447] The "change of intent" language in Article 8(a) of the Protocol puts a burden on researchers to clarify their intentions as they proceed, in order to respect the ABS obligations of the CBD if and when their research outputs elicit prospects of financial gain.

This burden can be addressed in different ways that bear directly on the relevant transaction costs of both the research community and providers of genetic resources. As before, every contract dealing with the use of *ex situ* and *in situ* genetic resources for noncommercial research purposes could entail bilateral negotiations bearing on unknown prospects for future financial gain, with endless amounts of speculation resulting in ever higher transaction costs for all the parties.[448] At the opposite extreme, vast amounts of *in situ* and *ex situ* genetic resources could be deemed to reside in a contractually constructed public pool, in the way that the FAO's Crop Commons has consigned certain plant genetic resources for food and agriculture to "the multilateral system."[449] The Nagoya Protocol now explicitly authorizes this approach.[450]

Between these two extremes, research communities that depend heavily on *ex situ* genetic resources made available through public repositories have already envisioned systems of cross-border exchanges that facilitate access for public research purposes without case-by-case negotiations, while simultaneously preserving and defending the provider countries' ABS rights. For example, networks of microbial culture collections, in consultation with key providers in the developing countries, have formulated a standard form MTA, with built-in, mutually acceptable provisions covering both access and use for public research purposes as well as the eventuality that some financial gains might accrue from the research outputs in the end.[451] Absent a multilateral agreement, however, the extent to which such intermediate deviations from the standard bilateral transactions between contracting parties are consistent with the Protocol remains to be seen.[452]

Of capital importance for this volume is the fact that Article 4 of the Nagoya Protocol, prompted by the creation of the FAO's multilateral system for exchanges of plant genetic resources, has directly addressed the legal status of multilateral

[447] Nagoya Protocol, n. 65, art. 8(a).
[448] *See, e.g.*, Katz, n. 45.
[449] *See* Section III.B.
[450] Nagoya Protocol, n. 65, art. 4(1)-(4).
[451] *See* Chapter 4, Section III.A.2.
[452] For doubts, *see* Godt (2013), n. 85, at 249.

approaches to common pools of genetic resources (potentially including microbial genetic resources) for the first time. For starters, Articles 4(1) and 4(4) expressly legitimize the FAO's multilateral system by suspending the CBD's bilateral approach "[w]here a specialized international access and benefit sharing instrument applies that is consistent with, and does not run counter to the objectives of the Convention and this Protocol."[453] Article 4(3) further drives the point home by requiring the Protocol to be implemented "in a mutually supportive manner with other international instruments" and with due regard to "useful and relevant ongoing work or practice under such international instruments and relevant international organizations" that are "supportive of" the objectives of the CBD and the Protocol.[454]

Article 4(2) then expressly allows the Contracting Parties to formulate and implement still "other relevant international agreements, including other specialized access and benefit sharing agreements," provided that they, too, are "supportive and do not run counter to the objectives of the Convention and this Protocol."[455] In other words, the Protocol expressly allows other common pool initiatives, operating under other specialized international instruments, to deviate from the rigors of the CBD's general ABS rules, if they are supportive of the CBD's overall objectives.[456] According to Professor Christine Godt, there is a consensus that these "specialized instruments" refer to *ex situ* collections operating in various sectors, under different governance models, each responding to the specific needs of a given community, and with a corresponding opportunity to develop tailor-made regimes "that retain public openness while respecting the CBD rationale."[457]

The Protocol thus envisions common pool resources as a means of decoupling access from benefit sharing; i.e., the duty to share benefits may be decoupled from the provider state and redefined as a benefit to everybody, including the provider state.[458] However, the *sine qua non* condition of any such tailor-made policy arrangement is that its provisions facilitating access and use of genetic resources are

453 Nagoya Protocol, n. 65, art. 4(4) (stating that, in such a case, "this Protocol does not apply for the Party or Parties to the specialized instrument in respect of the *specific genetic resources covered by and for the purposes of the specialized instrument*") (emphasis supplied). *See also id.* art. 4(1) (nonderogation from other international agreements consistent with the protection of biological diversity).

454 *Id.* art. 4(3). *See generally* IUCN, Guide to the Nagoya Protocol (2012), n. 389, at 75–81 (noting relevance of the United Nations Convention on the Law of the Sea, the Cartagena Protocol on Biosafety, and the International Treaty on Plant Genetic Resources for Food and Agriculture (at 77)).

455 Nagoya Protocol, n. 65, art. 4(2).

456 *See* Godt (2013), n. 85, at 247, 256. *See also* Winter, *Knowledge Commons* (2013), n. 385 (distinguishing "knowledge commons" from the "public domain" for this purpose).

457 Godt (2013), n. 85, at 247, 249 (noting possible extension of this principle to public culture collections held by universities that have been challenged by the CBD's ABS rationale).

458 Winter, *Knowledge Commons* (2013), n. 385, at 296–99; *see also* Godt (2013), n. 85, at 259–60.

in fact supportive of and consistent with the objectives of the CBD and the Protocol. Whether the defensive measures so far undertaken by the public microbial culture collections satisfy this standard is questioned further here, and with empirical evidence, in Chapter 4.[459] Meanwhile, regardless of whether one focuses on single investigators or *ex situ* repositories of genetic resources, the Nagoya Protocol has further eased the ABS obligations for scientific research by opening up the very notion of "benefits" to include a wide array of "non-monetary benefits" of particular interest to science.

2. Recognizing the Importance of Non-Monetary Benefits

As noted, the master principle of Article 5(1) provides that "benefits arising from the utilization of genetic resources as well as subsequent applications and commercialization shall be shared in an equitable way with the Party providing such resources, that is the country of origin … or a Party that has acquired the genetic resources in accordance with the Convention."[460] When formulating this provision, the drafters of the Nagoya Protocol expressly added Article 5(4) to the effect that such "[b]enefits may include monetary and non-monetary benefits,"[461] examples of which are set out in the Annex to the Protocol.[462] These provisions could further serve to accommodate the needs of scientific researchers with the otherwise tough compliance measures that the Protocol elsewhere puts in place, as explained in the next section.

A preliminary observation is that, for purposes of ABS, the master principle of the Protocol carefully distinguishes between "the utilization of genetic resources" and "subsequent applications and commercialization."[463] This can be interpreted to mean that even noncommercial uses of genetic resources for research purposes, sanctioned by Article 8(a), require some fair and equitable *quid pro quo*.[464] The Annex suggests that access fees or upfront payments, of the kind that culture collections in OECD countries charge for *ex situ* microbial specimens, could be applied for this purpose.[465]

[459] *See especially* Godt (2013), n. 85, at 254–61. *See further* Chapter 4, Section III.A.

[460] Nagoya Protocol, n. 65, art. 5(1) (adding that "[s]uch sharing shall be upon mutually agreed terms.").

[461] *Id.* art. 5(4).

[462] *Id.*, Annex.

[463] Nagoya Protocol, n. 65, art. 5(1).

[464] *See, e.g.*, Buck & Hamilton (2011), n. 70, at 52.

[465] Nagoya Protocol, n. 65, Annex, ¶ 1(a), (b), (c), (d), (e). Fees in excess of the marginal cost of distribution may become controversial, however. *See, e.g.*, Chapter 4, Section II (dealing with proprietary models adopted by some microbial culture collections).

However, the Annex also recognizes the possibility of microbial exchanges based on non-monetary benefits, such as:

- The sharing of research and development results; and
- Collaboration, cooperation and contribution in scientific research and development programs, particularly biotechnological research activities where possible in the Party providing genetic resources.[466]

This provision cuts two ways. It enables individual scientists to trade such nonmonetary benefits for noncommercial research uses of microbial genetic resources in lieu of more onerous monetary payments. It also enables tailor-made pooling arrangements for *ex situ* microbial genetic resources to reward providers in developing countries that participate in a research commons with nonmonetary benefits, apart from any monetary benefits otherwise made available through the commons.[467] At the same time, it enables the providers of genetic resources to seek nonmonetary benefits as a condition of negotiating mutually agreed terms.

From a broader perspective, the illustrative list of both monetary and nonmonetary benefits set out in the Annex to the Nagoya Protocol makes it possible to design long-term scientific cooperation programs between research organizations in both developed and developing countries on a more stable and predictable basis than was possible under the CBD as initially drafted. Even the provisions on monetary benefits, such as royalties and licensing fees, also recognize that "research funding," "joint ventures," and "joint ownership of relevant intellectual property rights," can qualify as monetary benefits given in exchange for access and use of microbial genetic resources.[468]

More to the point, the Annex further recognizes an array of nonmonetary benefits that could strengthen the scientific research infrastructure of the providing countries as a whole.[469] Besides the sharing of research results and collaborative research opportunities just mentioned, this list includes:

- Collaboration, cooperation and contribution, in education and training;
- Admittance to *ex situ* facilities and to databases;
- Transfer to providers of genetic resources of knowledge and technology under preferential terms, particularly with regard to the conservation and sustainable utilization of biological diversity;
- Institutional capacity building;
- Training related to genetic resources;

[466] Nagoya Protocol, n. 65, Annex, ¶ 2(a), (b).
[467] *See* further Chapters 5 & 10.
[468] Nagoya Protocol, n. 65, Annex, ¶ 1(h), (i), (j).
[469] *Id.*, ¶ 2.

- Access to scientific information relevant to conservation and sustainable use of biological inventories and taxonomic studies;
- Research directed towards priority needs, such as health and food security;
- Institutional and professional relationships that can arise from an access and benefit sharing agreement and subsequent collaboration activities;
- Joint ownership of relevant intellectual property rights.[470]

It seems clear from the care with which this list has been compiled that provider countries engaged in the Nagoya negotiations envisioned the possibility of long-term collaboration with user countries in establishing transborder scientific infrastructures as potentially another major product of the ABS regime installed by the CBD. They have accordingly lifted their sights beyond the prospects of mere financial returns under a rigid bilateral regime.[471]

In our view, this opening to the possibilities of collaborative scientific research under the Nagoya Protocol could become a key, if not an indispensable, factor in redesigning a global Microbial Research Commons along the lines envisioned in this book. By the same token, the tough and stringent compliance measures established for the first time in other provisions of the Protocol make construction of such a research commons all the more necessary, in view of the penalties likely to be inflicted on future users of microbial genetic resources that fail to comply with the Protocol's strengthened ABS provisions, as explained in the next section.

C. *Prescriptions for Strict Enforcement of the Newly Codified Global Regime of Misappropriation*

At long last, the much discussed and heretofore inchoate legal regime prohibiting the misappropriation of *in situ* and *ex situ* genetic resources has acquired a solid legal foundation in Articles 5 to 18 of the Nagoya Protocol,[472] now that it has been ratified by at least fifty signatory countries.[473] For example, Article 5(3) obliges each Contracting Party to "take legislative, administrative or policy measures, as appropriate" to implement the sharing of benefits "arising from the utilization of genetic resources as well as subsequent applications and commercialization,"[474] in

470 *Id.*, ¶ 2(d), (e), (f), (h), (j), (k), (m), (n), (q).

471 This follows from reading articles 5(4) and 8 of the Nagoya Protocol in conjunction with both the Annex and Article 4.

472 For endless discussion of a supplementary implementing treaty concerning traditional knowledge associated with genetic resources under the auspices of WIPO's Intergovernmental Committee on Intellectual Property, Genetic Resources, Traditional Knowledge, and Folklore, *see most recently*.

473 *See* Nagoya Protocol, n. 65, art. 33(2). WIPO, http://www.wipo.int/tk/en/igc/snapshot.html (last accessed 1 Jan. 2015).

474 Nagoya Protocol, n. 65, art. 5(3), 5(1).

the form of either monetary or nonmonetary compensation.[475] Such sharing "shall be upon mutually agreed terms."[476] Under Article 6.2, user states must also ensure that access to genetic resources complies with the requirement of prior informed consent, especially that of indigenous and local communities, where needed, while provider countries must establish clear and transparent regulatory measures for such requests.[477]

The drafters then subdivide the newly codified enforcement measures into two categories: namely, measures dealing with genetic resources "that occur in trans-boundary situations or for which it is not possible to obtain prior informed consent"[478] and measures applicable to exchanges of genetic resources directly subject to regulation under the Protocol now that it has taken effect.[479] Article 10 thus reopens the controversy surrounding the massive amounts of *ex situ* microbial and other genetic resources accumulated prior to 1992 by culture collections, research institutes, universities, hospitals, and private-sector laboratories without prior informed consent ever having been obtained from the provider countries.[480] The Protocol does not renounce proprietarial rights in such resources as a *fait accompli*. On the contrary, it posits a legal obligation to consider a "global multilateral benefit-sharing mechanism" for continued or new uses of such genetic resources.[481] In effect, this provision seeks to establish a duty of Parties to the Protocol to provide retroactive compensation in the form of monetary or nonmonetary benefits for such unauthorized acquisitions and use thereof.[482]

At the same time, the drafters of Article 10 attempt to attenuate the controversy by proposing that any monetary compensation resulting from retroactivity should be paid into a general fund charged with supporting biodiversity conservation in provider countries as a group.[483] Nevertheless, there is in principle no free ride

[475] *Id.* art. 5(4).
[476] *Id.* art. 5(1).
[477] *Id.* art. 6(2), (3).
[478] *Id.* art. 10 ("Global Multilateral Benefit-Sharing Mechanism").
[479] *See* n. 65. *Id.* art. 33. The term "transboundary situations" refers to situations where genetic resources are known to have migrated from certain countries where there are no longer *in situ* specimens to be made available. *See* IUCN, GUIDE TO THE NAGOYA PROTOCOL (2012), n. 389, at 128–29.
[480] For conflicting views on retroactivity, see n. 70 & accompanying text.
[481] Nagoya Protocol, n. 65, art. 10.
[482] *See, e.g.*, Tvedt & Fauchald (2011), n. 419, at 387–88 (warning that art. 10 could lead to "obligations to ensure that users of genetic resources contribute to the mechanism where the origin of the genetic resource remains unknown"). Nagoya Protocol, n. 65, art. 10. Putting such microbes in a public culture collection could qualify as a nonmonetary benefit under this provision. *See id.*, Annex ¶ 2.
[483] *Id.* (stating that the benefits "shared by users of genetic resources and traditional knowledge associated with genetic resources through this mechanism shall be used to support the conservation of biological diversity and the sustainable use of its components globally").

for acquisitions prior to the CBD at least in the minds of the drafters, but only a more flexible modality for the sharing of benefits still accruing from use of those previously acquired genetic resources in a manner to be worked out in greater detail by future negotiations.[484]

With respect to transborder exchanges of microbial genetic resources occurring after the Nagoya Protocol took effect, noncompliance with ABS, MAT and PIC requirements have become an international delict capable of triggering the responsibility of any Contracting Party in whose territory the violatory acts transpire.[485] To this end, stringent enforcement measures are required at both ends of any given transaction; that is, in both the provider and recipient countries.[486]

As regards both genetic resources and related traditional knowledge, under Articles 6 and 7,[487] provider countries (i.e., either the country of origin or "a Party that has acquired the genetic resource in accordance with the Convention"[488]) must, *inter alia*:

- Provide for a clear and transparent written decision [allowing or denying the requested use] by a competent national authority, in a cost-effective manner and within a reasonable period of time;[489]
- Provide for *the issuance at the time of access of a permit* or its equivalent as evidence of the decision to grant prior informed consent and mutually agreed terms.[490]

Moreover, the supplying Party's duty to provide "clear rules and procedures for requiring and establishing mutually agreed terms," should encompass:

(i) A dispute resolution clause;
(ii) Terms on benefit-sharing, *including in relation to intellectual property rights*;
(iii) Terms on subsequent third party use, if any; and
(iv) Terms on changes of intent, where applicable, for example, from a noncommercial to a commercial use.[491]

484 Nagoya Protocol, n. 65, art. 10 ("Parties shall consider the need for and modalities of a global multilateral benefit sharing mechanism . . ."). *See also id.* art. 11 (requiring cooperation between states "where the same genetic materials are found *in situ* within the territory of more than one Party").
485 *Id.* arts. 5–7, 12–18.
486 Nagoya Protocol, n. 65, arts. 6–7, 15(1)-15(3).
487 *See id.* arts. 6(1)-(2) & 7.
488 Nagoya Protocol, n. 65, arts. 6(1) & 6 (2).
489 *Id.* art. 6(3)(d).
490 *Id.* art. 6(3)(e) (emphasis supplied).
491 *Id.* art. 6(3)(g) (emphasis supplied). For "Changes of intent," *see also id.* art. 6(3)(g)(iv). *See generally* IUCN, Guide to the Nagoya Protocol (2012), n. 389, at 93–124.

It follows that Article 6(3) of the Nagoya Protocol has thus stipulated the basic premises for an evolving future global MTA potentially applicable to all genetic resources covered by the CBD. This goal is further supported by Article 19(2), which proposes to streamline the ABS procedures by means of "sectoral and cross-sectoral model contractual clauses,"[492] which could help user countries enforce contracts between providers and users in domestic courts under Article 18(2).[493]

In the interest of both access and enforcement, all Parties to the Protocol must designate "a national focal point" to provide applicants seeking access to genetic resources or related traditional knowledge with all necessary information concerning procedures for PIC and MAT.[494] They must also identify at least one or more "designated national authorities" to be responsible for granting access, obtaining prior informed consent, and entering into mutually agreed terms.[495] Ideally, all the relevant information, including access permits or their equivalents, as well as certificates of compliance,[496] could be stored and made available – with due regard for the need to protect confidential information – through a Clearing House envisioned under Article 14.[497]

With specific regard to receiving countries, Article 15 of the Nagoya Protocol obliges each Party to "take appropriate, effective and proportionate legislative, administrative or policy measures *to provide that genetic resources utilized within its jurisdiction have been accessed in accordance with prior informed consent and that mutually agreed terms have been established as required by the domestic access and benefit-sharing legislation* or regulatory requirements of the other [i.e. providing] Party."[498] Parties are further obliged "*to take effective measures to address situations of non-compliance*"[499] and to *cooperate in cases of alleged violation of domestic access and benefit-sharing legislation or regulatory requirements*.[500] Article 16 then imposes similar and parallel measures to ensure compliance with ABS and PIC provisions concerning traditional knowledge associated with genetic resources.[501]

Article 15, in conjunction with Article 2(c), further mandates that genetic resources used for research and development within the jurisdiction of a Contracting Party

[492] Nagoya Protocol, n. 65, art. 19(1), (2).
[493] *Id.* arts. 15, 16(3), 18(2); Tvedt & Fauchald (2011), n. 419, at 398.
[494] Nagoya Protocol, n. 65, art. 13.
[495] *Id.*, art. 13(1), (2). A single entity may fulfill the functions of both "focal point" and "competent national authority." *Id.* art. 13(3).
[496] *See id.* art. 17(3).
[497] *Id.* art. 14 ("The Access and Benefit-Sharing Clearing House and Information Sharing").
[498] *Id.* art. 15(1) (Compliance with Domestic Legislation or Regulatory Requirements on ABS) (emphasis supplied).
[499] *Id.* art. 15(2) (referencing measures adopted under art. 15(1)) (emphasis supplied).
[500] *Id.* art. 15(3) (emphasis supplied) (referencing measures adopted under art. 15(1)).
[501] *Id.* art. 16.

must comply with the rules of the provider party with respect to PIC and MAT.[502] These provisions expressly apply to "applications of biotechnology," including the use of living organisms or derivatives to make or modify products or processes for specific use.[503] According to Professor Godt, these provisions apply retroactively "if continuous possession and new forms of utilization of pre-CBD material fall under the scope of the CBD," although this viewpoint remains controversial.[504]

Taken together, Articles 15 and 16 make states Parties to the Protocol directly responsible for violatory acts committed within their territories. The Nagoya Protocol then puts still more teeth into these compliance measures by requiring each Party to actively monitor the utilization of genetic resources within its jurisdiction.[505] To this end, each party must designate "one or more checkpoints" that would require users of genetic resources to provide information about their compliance with ABS and PIC requirements[506] and to "*take appropriate, effective, and proportionate measures to address situations of noncompliance.*"[507] Such checkpoints could demand "internationally recognized certificates of compliance" where available,[508] and would be empowered to monitor "*the utilization of genetic resources or ... the collection of relevant information at ... any stage of research, development, innovation, pre-commercialization or commercialization.*"[509]

The drafters further foresaw that the clearest and most efficient means for users of genetic resources to prove compliance with these enforcement provisions is for them to obtain "a permit or equivalent" document issued in accordance with Article 6 of the Protocol and made available to the Clearing House to be established under Article 14.[510] Such a permit (or its equivalent) "shall constitute an internationally recognized certificate of compliance"[511] to serve as evidence that the genetic resource it covers has been accessed in accordance with prior informed consent and that mutually agreed terms have been established, as per domestic laws or regulations.[512]

To provide such a warranty, however, the internationally recognized certificate of compliance must identify the issuing authority, the date of issuance, and the provider. It would also need to specify the unique identifier covering any given genetic resource, where available; the person who granted prior informed consent;

[502] *Id.* arts. 2(c) (definition of "utilization of genetic resources"), 15; *see* Godt (2013), n. 85, at 247.
[503] Nagoya Protocol, n. 65, art. 2(d).
[504] Godt (2013), n. 85, at 247. *But see* Bagley & Rai, n. 418, at 17–18 (citing authorities).
[505] Nagoya Protocol, n. 65, art. 17.
[506] *Id.* art. 17(a)(i).
[507] *Id.* art. 17(a)(ii) (emphasis supplied).
[508] *Id.* art. 17(a)(iii) (excluding "confidential information" where appropriate).
[509] *Id.* art. 17(a)(iv) (emphasis supplied).
[510] *See id.* arts. 6(3)(c), 14(2)(c).
[511] *Id.* art. 17(2).
[512] *Id.* art. 17(3).

and confirmation that mutually agreed terms were established, along with prior informed consent.[513] The certificate must also specify whether the use is commercial or noncommercial.[514]

The Nagoya Protocol further encourages users and providers to specify a forum for dispute resolution, a choice of law clause, and even options for alternative modes of dispute resolution, such as mediation or arbitration.[515] More to the point, Article 18(2) obliges all Parties to "ensure that an opportunity to seek recourse is available under their legal systems, consistent with applicable jurisdictional requirements, in cases of disputes arising from mutually agreed terms."[516] In short, and assuming that the Contracting Parties fully implement these provisions in their domestic laws,[517] aggrieved provider governments or their agents should in principle be allowed access to the courts of user countries for purposes of enforcing contracts regulating access to and use of *in situ* and *ex situ* genetic resources originating from their territories.[518] A multilateral regime to supervise the compliance of the Parties – as distinct from individual users of genetic resources – is also envisioned in Article 30.[519]

V. CHALLENGING PROSPECTS FOR THE EXISTING MICROBIAL RESEARCH COMMONS

Assuming that the Contracting Parties to the Nagoya Protocol fully implement its mandate in their domestic laws, access to, and use of, both *in situ* and *ex situ* genetic resources and related traditional knowledge for research and applications will become subject to detailed international regulatory measures for which few, if any, escape hatches remain available. Further efforts to harmonize and standardize these measures are also underway at the World Intellectual Property Organization, where the responsible Committee had reportedly reached a sufficient consensus in 2014 as to warrant convoking a full-fledged Diplomatic Conference in the near

513 *Id.* art. 17(4). Microbial genetic resources held in public repositories are normally assigned unique identifiers. *See* Chapter 4, Section I.A. However, plant genetic resources under the FAO's Crop Commons are not assigned similar identifiers. *See* Section III.C.2.

514 Nagoya Protocol, n. 65, art. 17(4).

515 *Id.* art. 18(1) (Compliance with Mutually Agreed Terms).

516 *Id.* art. 18(2). Enforcement of judgments and access to justice are recommended by art. 18(3).

517 For efforts by the Norwegian government to implement the Nagoya Protocol, and corresponding difficulties, *see generally* Tvedt & Fauchald (2011), n. 419.

518 *See, e.g.*, IUCN, Guide to the Nagoya Protocol (2012), n. 389, at 186–87 (citing Claudio Chiarolla, *The Role of Private International Law under the Nagoya Protocol, in* The 2010 Nagoya Protocol on Access and Benefit-sharing in Perspective: Implications for International Law and Implementation Challenges (E. Margerer, M. Buck & E. Tsioumani eds. 2012)).

519 Nagoya Protocol, n. 65, art. 30. *See also id.* art. 10 (Global Multilateral Benefit).

future to consider a supplementary international treaty dealing with implementation issues.[520]

Microbiologists who depend on access to *ex situ* and *in situ* genetic resources are particularly exposed to the pitfalls that this transnational regulatory regime will have created. Those contemplating projects likely to fall within the scope of the CBD, as amplified by the Nagoya Protocol, should accordingly understand that "the sole apprehension of being accused of misappropriation or misuse of genetic resources has already become a serious impediment to research and bioprospecting activities."[521] As the authors of the IUCN's *Explanatory Guide to the Nagoya Protocol* admonish,

> [r]esearchers as well as private industries fear image problems in case of public outcries. Allegations of "biopiracy" would make it difficult for them to negotiate legitimate ABS agreements with other parties and gain access to potential funding sources, likely causing significant loss of commercial opportunities that may be available to a competitor. Potential users are also concerned about possible administrative appeals or formal lawsuits that might render their activities unprofitable or at least unpredictable.[522]

Nor should researchers seeking transborder access to microbial genetic resources further indulge the assumption that countries not adhering to the CBD and the Nagoya Protocol – such as the United States – remain legally immune from their regulatory prescriptions. On the contrary, even in the short and medium terms, non-adhering countries that flout the international ABS/PIC regime for genetic resources may expect to be treated as exporters of contraband goods, much as those who export counterfeit trademarked goods are currently treated under the TRIPS Agreement and posterior enforcement measures.[523] In that event, products such as pharmaceuticals and cosmetics derived from microbes not obtained under ABS/PIC provisions risk being seized in foreign ports under private international law provisions that have recently been invoked to reinforce respect for global intellectual property rights and public international law generally.[524]

[520] *See* WIPO Intergovernmental Committee on Intellectual Property, Genetic Resources, Traditional Knowledge, and Folklore, Twenty-Eighth Session, WIPO/GRTDF/TC/28 (7–9 July 2014).

[521] IUCN, Guide to the Nagoya Protocol (2012), n. 389, at 12.

[522] *Id.* at 12–13.

[523] *See* TRIPS Agreement, n. 21, arts. 51–60. *Cf. also* Anti-Counterfeiting Trade Agreement, Dec. 3, 2010, *opened for signature* Mar. 1, 2011, *available at* http://www.ustr.gov/webfm_send/2417 [hereinafter ACTA]; Margot Kaminski, An Overview and the Evolution of the Anti-Counterfeiting Trade Agreement (CPIIIP Research Paper No. 171 American University Washington College of Law, Washington, D.C. 2011).

[524] *See, e.g.*, Directive 2004/48 of the European Parliament and of the Council on the Enforcement of Intellectual Property Rights, 2004 O.J. (L 157) (EC), *available at* http://eur-lex.europa.eu/LexUriServ/LexUriServ.do?uri=CELEX:320048R%2801%29:EN:NOT; Jerome H. Reichman, *Securing Compliance*

By the same token, Professors Bagley and Rai warn United States researchers who enter provider countries in search of new genetic resources or who even seek "to obtain intellectual property rights in any country … that has PIC/ABS legislation over inventions developed with genetic resources accessed and/or used in violation of a provider country's domestic legislation may be subject to the range of legal action specified in such legislation, including, in some cases, imprisonment."[525] They accordingly advise researchers to "inquire as to the origin of genetic resources used in research and seek to comply with the domestic legislation of the identified provider country regarding PIC/ABS/MAT."[526]

In the medium and long-term, moreover, noncomplying countries must also be wary of claimed violations of customary international law. Such claims could be rooted in the 1962 Declaration of Sovereign Rights over Natural Resources[527] – signed by virtually all United Nations members at the time – on the theory that the later Convention on Biological Diversity and the Nagoya Protocol further specify preexisting legal obligations.[528] In this connection, one should remember that even though the United States never ratified the Vienna Convention on the Law of Treaties, it later recognized the provisions of that treaty as binding customary international law.[529]

From a global perspective, the Nagoya Protocol presents the microbial research community with both a challenge and an opportunity that should not be ignored. While directly subjecting individual scientific researchers, public and private culture collections, and research institutes to its fortified ABS/PIC regime, the Protocol simultaneously invites the scientific community to devise appropriate legal regimes that respect both the regulatory goals of the Protocol and the needs of public scientific research. If, however, science policymakers ignore the challenges and opportunities posed by the Nagoya Protocol, the regime that eventually governs transborder exchanges of microbial genetic materials will be established by the Conference of the Parties to the CBD itself, rather than by the microbiological research community.[530] In this connection, Appendix I to the Nagoya Protocol already sets out a lengthy blueprint of proposed elements for standardized material

with the TRIPS Agreement After U.S. v. India, 4 *J. Int'l Econ. L.* 585 (1998); Frederick M. Abbott, The Definition of Pharmaceutical Substance and Exclusion of Micro-organisms under the WTO TRIPS Agreement 13 (Study for the Indian Pharmaceutical Assoc., 25 April 2005).

525 Bagley & Rai, n. 489, at 22.

526 *Id.*

527 1962 Declaration, n. 11.

528 *See* Catherine Tinker, *Responsibility for Biological Diversity Conservtion under International Law*, 28 *Vand. J. Transnat'l L.* 777, 780 (1995). *But see* Curtis Bradley & Mitu Gulati, *Customary International Law and Withdrawal Rights in an Age of Treaties*, 21 *Duke J. Comp. & Int'l L.* 1–30 (2010) *and* Curtis Bradley & Mitu Gulati, *Withdrawing from International Custom*, 120 *Yale L. J.* 202–75 (2010).

529 *See* Evan Cridale, *The Vienna Convention on the Law of Treaties in U.S. Treaty Interpretation*, 44 *Va. J. Int'l L.* 431, 432 (2004).

530 *See* Nagoya Protocol, n. 65, art. 14(3).

transfer agreements,[531] presumably without inputs from either the public microbial research collections or the relevant research communities.[532]

Of primary importance in this regard are the responses to the Nagoya Protocol that will be made by managers of the public culture collections on which microbial science traditionally depends for access to *ex situ* genetic resources, especially those collections affiliated with the WFCC that constitute what we call "the existing microbial research commons."[533] The empirical evidence gathered in Chapter 4 will show that these collections find themselves caught between privatizing pressures emanating from both the globalization of intellectual property rights under the TRIPS Agreement of 1994 and the global regulation of biodiversity under the CBD and the Nagoya Protocol.[534] Not surprisingly, a growing number of the most technically advanced culture collections have been tempted to adopt market-like distribution schemes they hope will at least generate new funding opportunities for themselves in a world increasingly inclined to treat genetic resources as profit-maximizing commodities.[535]

Meanwhile, from a legal perspective, the evidence marshaled in the next chapter will also show that most of the WFCC collections – whether embracing market-like methods or attempting to preserve their traditional public good missions – have largely tried to opt out of the regulatory pitfalls of the Nagoya Protocol by defining themselves as intermediaries between provider and user countries with regard to cross-border exchanges of microbial genetic resources. On this approach, both single collections and regional groups of collections have developed standard MTAs that seek to shift the burden of meeting ABS obligations onto users, while attempting to immunize themselves from direct responsibility under the CBD.[536] As we will show, these delaying tactics are unlikely to satisfy the provisions in the Nagoya Protocol that allow providers and users of genetic resources to avoid the rigors of the bilateral approach embodied in the CBD.[537] In our view such an approach seems likely to generate more – not less – legal uncertainty surrounding cross-border exchanges of genetic resources for public research purposes.

[531] *See id.*, Appendix I, reproduced in IUCN, GUIDE TO THE NAGOYA PROTOCOL (2012), n. 389, at 335–36.

[532] The drafters of the IUCN, GUIDE TO THE NAGOYA PROTOCOL, n. 389, at 335–36, "Possible Ways Forward," expressly declare that voluntary norms could be valuable tools to support implementation of the Nagoya Protocol, including those of academic researchers.

[533] *See* Chapter 2, Section I.A; below Chapter 4.

[534] *See* Chapter 4, Sections II, III, & IV. *Cf.* Keith Maskus & Jerome H. Reichman, *The Globalization of Private Knowledge Goods and the Privatization of Global Public Goods, in* INTERNATIONAL PUBLIC GOODS n. 3.

[535] *See* Chapter 4, Section II.A; *see also* SCOTT STERN, BIOLOGICAL RESOURCE CENTERS: KNOWLEDGE HUBS FOR THE LIFE SCIENCES 11, 14 (Brookings Inst. Press, 2004).

[536] *See* Chapter 4, Sections II & III.

[537] *See* Chapter 4, Section IV.C; *see also* Godt (2013), n. 85.

In contrast, the express language of the Nagoya Protocol can be read as potentially allowing the public microbiological research community to opt out of the bilateral approach under the CBD if, but only if, it opts into a tailor-made, transnational arrangement consistent with the spirit and goals of Articles 4 and 8(a).[538] Article 4 was specifically devised to accommodate the International Treaty on Plant Genetic Resources for Food and Agriculture. There is a growing consensus that it could legitimate other intergovernmental agreements that deviate from the bilateral approach by establishing common pool resources for public research purposes if they respect the needs of both provider and user countries under the research-friendly provisions of the Protocol.[539]

Such arrangements could also enable participating culture collections to bring related data and technical services within the ambit of both the monetary and nonmonetary benefits now expressly recognized by the Protocol.[540] This strategy would furnish both provider and user countries with streamlined possibilities for coordination and cooperation, including networks of national contact points, clearing houses for information sharing, and mechanisms to ensure capacity building.[541] Once established, these multilateral arrangements could ultimately help to broker a more standardized and simplified approach to accessing even *in situ* microbial genetic resources than any that seems likely to emerge from a rigid bilateral approach.[542]

For present purposes, what matters most is that the Nagoya Protocol has finally and clearly recognized the pivotal role of public scientific research for all stakeholders. It has also provided a unique opportunity for the scientific community to fashion its own response to the challenge of a globally enforceable regime of misappropriation for unauthorized uses of genetic resources and traditional knowledge. In our view, the microbiological research community should not simply stand its ground and attempt to hide from that Protocol. It should, instead, accept the invitation to develop a workable, cross-border regime for exchanges of genetic resources and related data within the space that the Protocol itself delineates for this purpose. How to redesign the existing Microbial Research Commons for this purpose is the task undertaken in the rest of this book.[543]

[538] *See* Nagoya Protocol, n. 65, arts. 4 and 8(a).

[539] *See, e.g.*, Godt (2013), n. 65; Winter, *Knowledge Commons* (2013), n. 385. For details, see Chapter 4, Section IV (From the Bilateral to the Multilateral Approach). For an example, *see id.*, Section A. (discussing WHO's Pandemic Influenza Preparedness Framework).

[540] Gorch Detlef Bevis Fedder, *Biological Databases for Marine Organisms: What They Contain and How They Can Be Used in ABS Contexts, in* Common Pools of Genetic Resources (2013), n. 48, at 268–84.

[541] *See, e.g.*, Tvedt & Fauchald (2011), n. 419, at 388, 398. *See further* Part Four.

[542] *Cf.* Fedder (2013), n. 540.

[543] *See especially* Chapters 4, 5, 7, and 10.

PART TWO

Preserving the Public Research Functions of Microbial Genetic Resources after the Nagoya Protocol

4

The Existing Microbial Research Commons Confronts Proprietary Obstacles

I. EVOLUTION OF MICROBIAL CULTURE COLLECTIONS AS BASIC SCIENTIFIC INFRASTRUCTURE

Historically, microbes made available for research purposes were dependent on either *ex situ* or *in situ* conservation. *Ex situ* conservation typically occurs when microbes that can be grown are preserved for scientific purposes in culture collections having the requisite storage capacity, technical infrastructure, and expertise. Microbes of known interest that could not be grown were available only from their natural habitats.[1] In that case, if the occurrence of such microbes depends on the functioning of certain ecosystem features or species, active *in situ* conservation of these related ecosystems will be needed.

Microbiologists around the globe obtain large numbers of *ex situ* organisms by means of both formal and informal exchange arrangements. Formal exchanges of microbial materials are typically managed by public culture collections, which may be either governmental or nongovernmental entities. Informal exchanges result from less structured arrangements among trusted individual research groups.[2] In general, research institutes and universities are collectively the major users of *ex situ* genetic resources. They typically conduct both academic and applied research and often act as intermediaries for industry by collecting materials.[3]

[1] *See* David Smith, *Culture Collections*, *in* 79 ADVANCES IN APPLIED MICROBIOLOGY, Ch. 4 (2012) [hereinafter Smith, *Culture Collections*]. For an historical perspective, *see* Kate Davis, Eliana Fontes & Luciane Marinoni, Ex Situ *Collections and the Nagoya Protocol: A Briefing on the Exchange of Specimens Between European and Brazilian* Ex Situ *Collections, and the State of the Art of Relevant ABS Practices*, *in* THE ROLE TO BE PLAYED BY BIOLOGICAL COLLECTIONS UNDER THE NAGOYA PROTOCOL AS PART OF THE 6TH EU/BRAZIL SECTORAL DIALOGUE SUPPORT FACILITY 30–38 (2013) [hereinafter Davis, Fontes & Marinoni (2013)], *available at* http://sectordialogues.org/sites/default/files/acoes/documentos/background_paper.pdf. The genomic revolution has enlarged the availability of microbes initially found only *in situ*. *See, e.g.*, *id.* at 47–49; below Chapter 8, Sections II & III.

[2] *See, e.g.*, Smith, *Culture Collections*, above n. 1; *see also* Chapter 5, Section I.A.3. below.

[3] Davis, Fontes & Marinoni (2013), above n. 1, at 48 (citing U.K. DEPT. ENV'T FOOD & RURAL AFF. (DEFRA), U.K. IMPLEMENTATION OF THE NP: ASSESSMENT OF THE AFFECTED SECTORS, FINAL

The role of public culture collections in recent decades has remained fairly constant, even though the uses of genetic materials have changed significantly as microbiology moved from phenotypes to genotypes. As summarized by a leading authority, these culture collections:

- Provide a mechanism for *ex situ* conservation of organisms;
- Act as custodians of both local and foreign genetic resources thereby "providing the living materials to underpin the science base;"
- Maintain repositories of strains that are subjects of published research;
- Carry out safe, confidential, and patent-related deposit services for both researchers and industry;
- Generally supply microbial materials for discovery, study and innovation.[4]

What has changed are the methodologies to maintain and add value to such resources, a task that is complicated by an ever evolving legal and legislative operational environment and by the demands of users that increase both in terms of quantity and technical quality requirements.

Exchanges of microbial materials for research purposes have become an essential resource for the life sciences in general and biotechnology in particular.[5] Authentic strains with reproducible properties are needed for many reasons, but especially "as voucher specimens, that is, type species, in taxonomy, reference strains for standards and representative research strains for confirmation of findings and further work discoveries published in the scientific literature."[6]

By making available biological materials and information of guaranteed identity and quality, the culture collections, now sometimes referred to as Biological Resource Centers (BRCs) when technically qualified, serve an essential infrastructural function for both scientific investigation and commercial R&D.[7] The availability of materials in public, certified repositories, instead of minimally curated, in-house

Report to Defra from ICF GHK, *available at* http://randd.defra.gov.uk/Default.aspx?Menu=Menu&Module=More&Location=None&Completed=0&Project ID=17827); *see also* Christine Frison & Tom Dedeurwaerdere, Public Infrastructure and Regulations on Access to Genetic Resources and the Sharing of Benefits Arising Out of Their Utilization for Innovation in Life Sciences Research 92 (2006), Belgian federal survey, *available at* http://www.academia.edu/attachments/8944658/download_file?st=MTM5Njk5NTIxMiwxNTIuMy40My4xODI%3D&ct=MTM5Njk5NTIxMw%3D%3D).

4 Smith, *Culture Collections*, above n. 1, at 24–25.

5 Org. Econ. Cooperation & Dev. (OECD), Biological Resource Centers – Underpinning the Future of the Life Sciences and Biotechnology 8 (Mar. 2001) [hereinafter OECD Report on BRCs], *available at* http://www.oecd.org/sti/biotech/2487422.pdf.

6 *Id.* For the role of the World Federation of Culture Collections, see above Chapter 2, Section I.A. and below Section I.A.

7 *See* Scott Stern, Biological Resource Centers: Knowledge Hubs for the Life Sciences 42 (Brookings Inst. Press 2004).

research and private collections, has become *a sine qua non* for building upon previously validated knowledge. Using certified materials from established culture collections reduces the costs of mistakes in cumulative research as well as the search costs of finding appropriate materials.[8]

Today, most microbial genetic resources held by the public culture collections originate from *in situ* sources, as the bulk of biodiversity otherwise remains unknown. Approximately half of these resources were acquired directly by the public culture collections from *in situ* locations.[9] Other resources are provided by researchers who deposit their materials upon publication of research results, or from informal research or working collections that transfer some valuable parts of their holdings to the public culture collection when such materials are considered of high scientific value and the collections in question have space and means to accommodate them. About 20 percent of their aggregate holdings result from exchanges among different public collections.[10] Some of these entities will also consolidate preexisting microbial collections from university institutes, and in this capacity, they may rescue valuable materials that might otherwise have been abandoned by academic researchers who retire.[11]

8 *Id.* at 42. *See also* R. E. Evenson & Y. Kislev, *A Stochastic Model of Applied Research*, 84 *J.* POLITICAL ECON. 265 (1976); *cf.* B. Visser et al., *Transaction Costs of Germplasm Exchange under Bilateral Agreements*, FAO/Global Forum on Agric. Research (Doc. No.: GFAR/00/17-04-04) (2000)).

9 Tom Dedeurwaerdere, Arianna Broggiato, Sélim Louafi, Eric Welch & Fulya Batur, *Global Scientific Research Commons under the Nagoya Protocol: Governing Pools of Microbial Genetic Resources*, *in* THE NAGOYA PROTOCOL IN PERSPECTIVE: IMPLICATIONS FOR INTERNATIONAL LAW AND IMPLEMENTATION CHALLENGES 389–422 (E. Morgera et al. eds., 2013) [hereinafter Dedeurwaerdere, Broggiato, Louafi, Welch & Batur]; Tom Dedeurwaerdere et al., *The Use and Exchange of Microbial Genetic Resources for Food and Agriculture*, Background Study Paper of the Comm'n on Genetic Res. for Food & Agric. No. 45, U.N. Doc. UNEP/CBD/WG-ABS/9/INF/13 (9 Mar. 2009), at 22, *available at* http://www.cbd.int/doc/meetings/abs/abswg-09/information/abswg-90-inf-13-en.pdf (last accessed 3 July 2014).

10 According to a survey of 117 public culture collections, members of the World Federation of Culture Collections (WFCC). Per M. Stromberg, Tom Dedeurwaerdere & Unai Pascual, *The Heterogeneity of Public* Ex Situ *Collections of Microorganisms: Empirical Evidence about Conservation Practices, Industry Spillovers, and Public Goods*, 33 ENVTL. SCI. & POL'Y 19–27 (Nov. 2013); Tom Dedeurwaerdere, *Global Microbial Commons: Institutional Challenges for Global Exchange and Distribution of Microorganisms in the Life Sciences*, *in* 161(6) RESEARCH IN MICROBIOLOGY 407–413 (2010).

Dedeurwaerdere, Broggiato, Louafi, Welch & Batur, above n. 9, at 389–422 report the following breakdown:

- Origin of the material in the public culture collections: own collecting effort (45%); from research and working collections (30%); from other public culture collections (20%); other (5%).
- Recipients of the materials in the public culture collections: research and working collections (58%); private sector (23%); to other public culture collections (10%); other (9%).

11 *See, e.g.*, Cletus P. Kurtzman, *The Agricultural Research Service Culture Collection: Germplasm Accessions and Research Programs*, *in* DESIGNING THE MICROBIAL RESEARCH COMMONS: PROCEEDINGS OF AN INTERNATIONAL SYMPOSIUM 55–63 (P.F. Uhlir ed., Nat'l Acads. Press, 2011)

A. The Pivotal Role of the World Federation for Culture Collections

No single collection could possess all the microbial cultures needed for research and applications today, nor could it perform all the functions that collections scattered around the world collectively undertake. As mentioned in Chapter 2, the World Federation for Culture Collections acts as an umbrella organization that aims to rationalize and coordinate the activities of some 678 member collections in 71 countries. In particular, the WFCC aims to:

- Encourage the study of procedures for the isolation, culture, characterization, conservation, and distribution of microorganisms and to publicize these procedures;
- Promote the training of personnel in the operation of culture collections;
- Promote the establishment of a world network of data services pertaining to the location of, and information about, microorganisms in culture collections;
- Promote the establishment of special reference collections and identification services;
- Find solutions for problems of distribution of microbial cultures arising from postal regulations, quarantine rules, patent laws, public health problems and other factors of international importance.[12]

As a global organization of individual scientists and affiliated culture collections, the WFCC thus seeks "with limited means to help the community of culture collections."[13] Through its Newsletter and triennial conferences, it provides a forum for discussion and development of the culture collection community. The WFCC also provides training schemes and courses, many associated with its International Conferences for Culture Collections (ICCC), as well as other ad hoc training courses.[14] It also has work programs on capacity building, teaching and education, patents, implementation of applicable legislation, endangered collections, and standards. These programs are devised to assist both new and established collections.[15]

[hereinafter DESIGNING THE MICROBIAL RESEARCH COMMONS] (noting that the Northern Regional Research Laboratory tries to take "some of the more prominent abandoned collections" that were built with considerable grantee funds over whole careers).

12 World Fed. Culture Collections (WFCC), *Statutes*, art. VI [hereinafter WFCC, *Statutes*], *available at* http://www.wfcc.info/index.php/about/statutes/.

13 Philippe Desmeth, *News from the Secretary*, WFCC Newsletter No. 47, Jan. 2010) at 3.

14 World Fed. Culture Collections (WFCC), *Guidelines for the Establishment and Operation of Collections of Cultures of Microorganisms* 11, ¶ 14.3 (3d. ed., Feb. 2010), *available at* http://www.wfcc.info/guidelines/ [hereinafter WFCC, *Guidelines*].

15 *Id.* at 11, ¶14.3, and at 12, ¶ 16,4.

The operations of the WFCC are central to the themes we explore in this book. We accordingly provide a more detailed perspective on that organization and its members in the next subsections.

1. Aggregate Holdings and Capacity

Taken together, all the registered collections in the WFCC held over 1.8 million strains in 2012.[16] These aggregate strains included more than 770,000 bacteria, over half a million fungi, over 19,000 viruses, and more than 7,000 cell lines.[17] According to David Smith, former president of the Federation, these WFCC collections can be distinguished from thousands of other non-members "in their commitment to provide high quality resources for research and development. They provide a public service and agree to operate to WFCC guidelines."[18] However, only about 240 member collections produce catalogs of their holdings.[19] Table 4.1 subdivides the aggregate WFCC holdings by numbers of species and subspecies.

Most of the centers or collections affiliated with the WFCC are designated as not-for-profit institutions, many at universities, although some are located in for-profit institutions, notably hospitals.[20] The WFCC member collections are staffed by some 5,400 employees.[21] The role of the private sector also deserves mention in part because of the large proprietary collections it maintains and also because its operations intersect with those of the WFCC members in different ways.

All WFCC members maintain general public collections whose deposits of strains are available to all qualified users, and most of their holdings reside in these open collections for research purposes. In addition, many members also hold strains in non-public collections as safe-keeping deposits for private owners' use only.[22] About 40 major culture collections in 22 countries also serve as International Depository Authorities (IDAs) that accept microorganisms cited in patents under the Budapest Treaty on the International Recognition of the Deposit of Microorganisms for the Purposes of Patent Procedure.[23] IDAs normally comply with secrecy requirements,

[16] Smith, *Culture Collections*, above n. 1.

[17] *Id.*

[18] *Id.*

[19] *Id.*

[20] WFCC, *Culture Collections Information Worldwide*, http://www.wfcc.info/ccinfo/home/ (last accessed January 29, 2015) [hereinafter WFCC Website].

[21] *Id.* (This list includes ATCC in the semigovernmental category.).

[22] Smith, *Culture Collections*, above n. 1, at 104. The proprietary collections may be deemed "specialist," "institutional," or "research collections," available only by permission of depositors. *See id.* at 79.

[23] Budapest Treaty on the International Recognition of the Deposit of Microorganisms for Purposes of Patent Procedure of 1977, 19 Aug. 1980, *as amended on* 26 Sept. 1980, 32 U.S.T. 1241, 1861 U.N.T.S. 361 [hereinafter Budapest Treaty], *available at* http://www.wipo.int/treaties/en/registration/budapest/trtdocs_wo002.html (last accessed 3 July 2014).

TABLE 4.1. *Aggregate WFCC Holdings by Numbers of Species and Subspecies*

Strain	Species/Subspecies
Algae	3,060
Archaea	460
Bacteria	16,495
cDNA	15
Cell lines, animal	401
Cell lines, plant	0
Fungi	25,611
Hybridomas, animal	0
Hybridomas, plant	0
Lichens	0
Plasmids	1,783
Protozoa	60
Vectors	1,783
Viruses, animal	66
Viruses, bacteria	976
Yeasts	1,216
Viruses, plant	84

WFCC, *Strains*, WORLD FED. CULTURE COLLECTIONS, http://www.wfcc.info/ccinfo/search/strain_search/ (last accessed 3 July 2014). For a breakdown of the number of culture collections per country and their aggregate holdings, *see id.*

and they will supply samples of deposited materials only to those persons entitled to receive them.[24]

It is worth emphasizing that many important WFCC collections are also affiliated with research institutes and universities that maintain their own independent culture collections for research purposes. These affiliated research collections are not subject to WFCC quality standards and procedures, but they often feed the public collections with deposits related to published research results. By the same token, affiliated but non-public collections have traditionally exchanged microbial materials among themselves, informally, and without the tracking, validation, and other safeguards that WFCC members provide. Whether these practices remain

[24] Smith, *Culture Collections*, above n. 1, at 104. As of 2010, more than 73,000 cultures were deposited in IDAs under Budapest Treaty rules, and more than 15,000 samples of patented materials had been supplied. *Id. See also id.* at 105–17, tbl. 4.4 (showing details of patent strains in these holdings).

feasible under the Convention on Biological Diversity seems doubtful, as discussed in Chapters 3 and 5.[25] Finally, we reiterate in passing that there are literally tens of thousands of other culture collections in hospitals, industrial laboratories, and university departments throughout the world, about which little information is publicly available.

The WFCC members vary considerably in their capacity to receive deposits. One study of 47 WFCC member collections that receive deposits of between 1,000 and 3,000 strains each year showed that the selected group as a whole could absorb over 20,000 strains per year, but with many reaching their capacity in two to five years, assuming costs were covered.[26] Potential limits on the capacity of highly qualified culture collections will thus pose a recurring problem that the proposals set out later in this book attempt to address.[27]

Conversely, recent studies show that relatively few of the strains used in obtaining published research results are actually made available for confirmation of those results and for further study.[28] Key strains that can demonstrate newly described properties or that figure in new scientific hypotheses or other research results may never be deposited in any public service collection, or even supplied on request by other researchers or preserved over time.[29] Funders of microbial research may accordingly need to devise mandatory rules requiring deposits of key strains in public collections, comparable to the growing obligations to deposit research publications and data in open access repositories.[30]

2. Servicing the Broad Microbiological Research Community

Generally speaking, the WFCC culture collections endeavor to perform some or all of the following services:

- isolate, identify and possibly study microbial genetic resources to be deposited with them;
- maintain and conserve such deposits;
- allocate unique strain identifiers (identity codes) that remain constant even in cases of taxonomic changes;

25 Smith, *Culture Collections*, above n. 1, at 87; *see also* above Chapter 3, Section I; Chapter 5, Section I.A.3.

26 Smith, *Culture Collections*, above n. 1, tbl. 4.2 (with details on these collections).

27 These limits on capacity are sometimes known as "the Big Refrigerator Problem." *See* Fiona Murray, "Institutional Foundations of Scientific Progress: Implications for Collaboration and Participation," paper presented at Global Science and the Economics of Knowledge-Sharing Institutions, 2d. Communia Int'l Conference, Turin, Italy, 29–30 June 2009.

28 Smith, *Culture Collections*, above n. 1, at 85; E. Stakebrandt, *Diversification and Focusing: Strategies of Microbial Culture Collections*, 18 TRENDS MICROBIOLOGY 283–87 (2010).

29 Smith, *Culture Collections*, above n. 1, at 85.

30 *See* STERN, above n. 7; *see further* Chapters 7 and 8 below.

- keep records of depositors, taxonomic information and other properties;
- keep records of recipients of biological material;[31]
- since 1993 (when the CBD entered into force), keep records of information about the persons involved in isolating and identifying any given genetic resource, the date of isolation, and the country of origin;
- sponsor research on systematics and ecology, often conducted by multinational collaborators.[32]

The culture collections' most essential responsibility is the long-term storage of microorganisms for research and applications. The availability of strains maintained in a genetic and physiologically unchanged state must be guaranteed over time. In particular, major production strains, reference strains for research and testing, as well as other strains with valuable properties should be available for comparative analyses even after decades. Similarly when new uses of old strains are discovered, both research and applications may depend on ready availability of reference stains preserved in their original state.[33] It is important to reiterate that type strains deposited in the collections may not be patented.

Considerable attention has lately focused on elevating and harmonizing the quality standards that affiliated collections should maintain, especially in the light of the OECD's Best Practices Guidelines for Biological Resource Centers, to be discussed in the next section. These guidelines also cover biosecurity, capacity building, preservation of biological resources and data management.[34]

The WFCC's own revised Guidelines, issued in 2010, were meant to provide "a first step towards implementation of the OECD Best Practices." In requiring all its member collections to meet these WFCC standards "in a reasonable time frame," the Guidelines emphasize "that high standards of scientific service can be achieved in laboratories with modest resources and that sophisticated equipment is not a prerequisite for good microbiological practice."[35] As a baseline reality, however, as of

[31] WFCC, *Guidelines*, above n. 14, ¶ 9.5. These records of service as supply collections are usually more complete than those kept by in-house research collections. Dagmar Fritze, President of ECCO. The Proposed Standard MTA of the European Culture Collections' Organization, paper presented to the Microbial Commons Conference, Ghent, Belgium, June 12–13, 2008, at 4 [hereinafter Fritze (2008)].

[32] Fritze (2008), above n. 31, at 4.

[33] David Smith, Dagmar Fritze, Erko Stackebrandt, *Public Service Collections and Biological Resource Centers of Microorganisms*, *in* The Prokaryotes – Prokaryotic Biology and Symbiotic Associations (E. Rosenberg et al. eds., Springer, 4th ed., 2013) [hereinafter D. Smith et al. (2013)]; *see also* Fritze (2008), n. 31.

[34] *Id.* at 2. *See* OECD, OECD Best Practice Guidelines for Biological Resource Centers (2007) [hereinafter OECD Best Practices], *available at* http://www.oecd.org/sti/biotech/38777417.pdf.

[35] WFCC, *Guidelines*, above n. 14.

January 2010, only about 120 of its 647 members were designated as "full affiliated" member collections, which means that they had fully met the quality standards set out in the WFCC bylaws.[36]

The WFCC Guidelines strictly require careful documentation for each strain held by member collections.[37] The registration of collections with the World Data Center for Microorganisms (WDCM), a subsidiary of the WFCC, further facilitates access to and traceability of microorganisms and associated data in the member collections. The registration process "provides for an efficient coding of the strains by defining a collection acronym and WFCC number," which allows each culture collection to give "Globally Unique Identifiers (GUID) to each strain of its holdings, combining their acronym with their own internal numbering."[38]

As a result, when the unique acronym attached to the strain in any given collection appears in published scientific literature, it allows instant recognition of the source collection and associated data, including the country of origin of that strain. When the strains are accessioned into other collections, citation of the WDCM acronym enables users to link the information generated on different lines and avoids duplication of materials in scientific studies.[39]

As noted, qualified WFCC collections must retain their materials in an unchanged condition for the long term to ensure reproducible results and repeated use.[40] Above all, accepted microorganisms must be validated as to authenticity and properly preserved, "and any associated information must be valid and sufficient to enable ... confirmation of its identity and to facilitate its use."[41] Authenticity requires the public collections to control for human error as well as genetic changes that

[36] *See further* Chapter 9, Section II.B.1.a (Governance aspects).

[37] WFCC, *Guidelines*, above n. 14, ¶ 11.1. The following information should be kept:

- Place
- Substrate or host
- Date of isolation
- Name of person isolating the strain
- Depositor (or other source of the strain, such as from another collection)
- Name of the person identifying the strain
- Preservation procedures used
- Optimal growth media and temperatures
- Data on biochemical or other characteristics
- Applicable regulatory conditions.

[38] *Id.*, ¶ 11.1.

[39] Smith, *Culture Collections*, above n. 1, at 74.

[40] *Id.* at 76. "Taxonomic reference strains must never be lost, and thus, their maintenance must be coordinated and in the hands of organizations that have solid foundations and a sound project for a long-term future." *Id.*

[41] *Id.*

evolve over time. Organisms must also be collected and distributed in compliance with international and domestic laws, and they must be shipped to users in a safe and secure manner, with due regard to health, safety, and "dual use" biosecurity requirements.[42]

To this end, the recent WFCC Guidelines stress the importance of having "a clearly defined accessions policy in which new strains are to be taken into the collection," in order not to overburden storage capacities, personnel, and financial resources. The Guidelines strongly recommend collaboration and networking among collections with a view to implementing a coordinated accession policy.[43] Toxic or pathogenic holdings must be clearly labeled and securely kept in conformity with all applicable safety regulations.[44]

In stressing the importance of curation and management,[45] preservation,[46] and culture authentication,[47] the WFCC Guidelines state "that each culture should, whenever practical, be maintained by at least two different procedures, one of which should preferably be by freeze-drying or storage at ultra-low temperatures," to minimize the risk of genetic change.[48] Absent proper authentication, users may employ "the wrong organism in their investigations, which could prove time-wasting, expensive and lead to invalid published results," not to mention biosecurity concerns.[49]

Member collections that list cultures as available in their service catalogues are expected to provide those resources on a nondiscriminatory basis, subject to import, quarantine containment, and other regulations.[50] The WFCC requires all member collections, and recommends to all others, that type strains must be made available without restriction to the scientific community.[51] Charges are determined by the supporting entities, but may differentiate according to purpose, for example, teaching versus industrial applications.[52]

42 *Id.* at 76–77. *See also* D. Smith et al (2013), n. 33, at 293.

43 WFCC, *Guidelines*, above n. 14, ¶¶ 5.2, 5.5.

44 *Id.*, ¶ 5.3.

45 *Id.*, ¶¶ 6.1–6.4.

46 *Id.*, ¶¶ 7, 1–7.3.

47 *Id.*, ¶¶ 8.1–8.6. *See also* Smith, *Culture Collections*, above n. 1, at 92–94 (authentication and characterization), 94–102 (preservation), and 102–09 (culture supply and services).

48 WFCC, *Guidelines*, above n. 14, ¶ 7.1.

49 *Id.*, ¶ 8.1 ("WFCC member collections have a responsibility to provide resources with accurate identities as reference materials if they offer a public service and must make every effort to ensure that organisms they supply are authentic."). For biosecurity issues, see D. Smith et al. (2013) n. 33, at 291–93, Box 11.7.

50 WFCC, *Guidelines*, above n. 14, ¶ 9.2.

51 *Id.*, ¶ 9.7. *But see* the case of ATCC, discussed below in Section II.A.

52 WFCC, *Guidelines*, above n. 14, ¶ 9.2.

Pathogenic or toxic microbes can only be distributed if quarantine, biosafety, and biosecurity regulations have been satisfied.[53] Patent laws and other intellectual property rights must be respected,[54] and considerable efforts are made to ensure that WFCC collections operate in conformity with the Convention on Biological Diversity, as will be seen below.[55]

Beyond providing these traditional services, some WFCC and non-WFCC culture collections may sponsor or conduct their own research on systematics and ecology, while fulfilling ever more onerous responsibilities under regulations governing biosafety and biosecurity.[56] They may also promote new products, such as DNA, enzymes, metabolites and other derivatives from authenticated strains.[57]

Finally, considerable efforts in developed countries have focused on measures to upgrade the public collections in order to boost their potential contribution to ongoing and future research initiatives rooted in molecular biology and genomic sequencing methods, often with a focus on authenticity, while conforming to the CBD.[58] Public service collections have traditionally provided information through their catalogues for decades, and these are now increasingly accessible on the internet. More recently, major efforts are underway to draw these and other data together, under the auspices of the WDCM, in order "to gain the benefits of a larger data landscape."[59] These initiatives are described in Chapter 8.

3. The Perennial Problem of Funding

Both the technical capacity and the longevity of microbial culture collections depend on their having adequate financial resources. Besides governmental or semigovernmental support, universities often play a major role in funding public culture collections, although some are also supported by the private sector or foundations, and many benefit from the voluntary services of dedicated microbiologists.

The WFCC's own estimates are shown in Table 4.2. In practice, the WFCC collections seldom rely on a single source of funding. For example, the government supported collections may also receive some income from various services and

53 *Id.*, ¶ 9.4. *See also id.*, ¶¶ 15.1–15.4 (safety and security).

54 *Id.*, ¶¶ 9.4, 10.3.

55 Desmeth, above n. 13, at 2–3; *see also* Smith, *Culture Collections*, above n. 1. *See further* Chapter 5, Section I. below.

56 Fritze (2008), above n. 31, at 4–6.

57 Davis, Fontes & Marinoni (2013), above n. 1, at 47–49; Smith, *Culture Collections*, above n. 1, at 79.

58 *See* Davis, Fontes & Marinoni (2013), above n. 1, at 47–51; *see also* discussion of Biological Resource Centers in Section I.B.

59 Smith, *Culture Collections*, above n. 1, at 112–13.

TABLE 4.2. *Funding Sources of the WFCC Collections*

Supported by	No. of Collections
Governmental	253
University	253
Semi-governmental	56
Private	40
Industry	22

WFCC website, above n. 20, last visited January 29, 2015.

products they provide to either the research community or industry. Even then, the level of government support may sometimes be reduced by the amount of service income any given collection accrues, which may then leave little room for investment in expanded coverage or for new technological capabilities.[60]

David Smith, one of the founders of the demonstration project for a Global Biological Resource Centres Network (GBRCN), emphasizes the need for "sound and innovative business plans to allow them to keep pace with ever increasing demands of science and their users."[61] One proposed solution is for research funders to provide payments for deposits of research results in public collections, as part of their granting process.[62] Another approach is to develop and charge for providing research tools, such as DNA, enzymes, metabolites, and other derivatives from authenticated strains.[63] Collections can also sometimes develop commercial products through the provision of biotechnological solutions that result from the discovery of active compounds and from the funding of such activity by public-private initiatives.[64]

However, David Smith – who generally endorses these more commercial approaches to funding – recognizes that this approach can "divert the collections[s] too far away from their responsibilities in delivering their public services."[65] In his view, the collections will increasingly require a combination of governmental support and commercial income from products and services, as well as direct support by research funders. Efforts to conform the practices of the culture collections with

60 Smith, *Culture Collections*, above n. 1, at 79.
61 *Id.*
62 *Id.* at 80.
63 *Id.* at 79.
64 *Id.* at 79–80.
65 *Id.* at 80–81. He notes that the cost of acquiring greater bioinformatics capacity may be shared or covered by cooperating universities or other institutions as well as the private sector. *Id.*

regulations implementing the Nagoya Protocol have also reportedly fostered a risk of unfunded government mandates.[66]

When government support of the public culture collections shrinks, they may die – despite their research value – unless they fall back upon large-scale commercialization efforts. In fact, the aggregate number of collections registered with the WFCC has reportedly declined by some 400 in number since the WFCC began issuing identification numbers to each member collection.[67] By the same token, when a major collection is driven by government neglect to commercialize its services, that collection may simply abandon the public good approach altogether, as is seen in the case of ATCC below.[68] A recent Directive of the Office of Science and Technology Policy (OSTP) reminded departments of the United States government of their responsibility to defray the costs of any biological resource centers under their management.[69]

One reason for improving the microbial research infrastructure along the lines proposed in this study is to devise ways of rationalizing the financial burdens of coverage by means of collective action at the multilateral level. If qualified collections can be federated via digitally integrated networks under the authority of a single, transnational entity, it becomes possible to envision "cost efficient sharing of facilities, technologies, and expertise."[70] Such a framework should elicit adequate and sustainable funding directly by both participating governments and users, as part of an overall settlement to support microbiological research in the shadow of the Convention on Biological Diversity and the Nagoya Protocol. Some of these proposals are set out in the next chapter, while the difficult question of funding is reexamined in Chapter 10, in connection with governance of a redesigned Microbial Research Commons.

B. From Culture Collections to Biological Resource Centers

Beginning in 1998, leading microbiologists and senior science policy officials from different countries belonging to the Organization for Economic Cooperation and Development (OECD) began to press for a reevaluation of the role and functions of

66 *See, e.g.*, Technical Report on the Workshop in Brazil, SIXTH EU BRAZIL SECTORAL DIALOGUE (2013) above n. 1, at 12 (Recommendation No. 6); *see further* Section III.A.3 (EC Regulation).

67 Smith, *Culture Collections*, above n. 1, at 79.

68 *See* below Section II.A.

69 Memorandum from John P. Holdren on Increasing Access to the Results of Federally Funded Scientific Research to the Heads of Exec. Depts. & Agencies, EXEC. OFFICE OF THE PRESIDENT, OFFICE OF SCI. & TECH. POL'Y, Feb. 22, 2013, *available at* http://www.whitehouse.gov/sites/default/files/microsites/ostp/ostp_public_access_memo_2013.pdf

70 Smith, *Culture Collections*, above n. 1, at 112. *See further* Chapter 10, Section IV.

culture collections as traditionally constituted in a universe of discourse increasingly configured by advances in molecular biology, on the one hand, and geopolitical pressures, on the other.[71] After a preparatory meeting in 1999, a Task Force consisting of distinguished scientists and administrators from different countries was appointed to examine this topic under the aegis of the OECD's Working Party on Biotechnology. After two years of work, the Task Force and the Working Party published their findings and proposals in a 2001 report, entitled "Biological Resource Centers – Underpinning the Future of the Life Sciences and Biotechnology."[72]

This report began with the premise that the "revolution in molecular biology has given us greatly increased ability to obtain and to modify ... biological resources" – namely, living organisms, cells, genes, and related information – "and to use them for the benefit of all humankind."[73] Given the sequencing and associated analysis of gene functions for a growing number of genomes, including the human genome, both governments and industry were making large investments in recovering biological resources from nature and in exploring and engineering these resources. It was accordingly of crucial importance that "[t]hese investments must not be lost and their results must remain accessible so as to reap scientific, economic and medical benefits."[74]

In this context, the Task Force's main finding was that the traditional culture collections on which microbiology had long depended were not connected to each other and were often inadequate to meet the challenges of a big science approach.[75] They would, accordingly, have to transform themselves into better equipped, more scientifically ambitious repositories that would become responsible for preserving and distributing biological materials and related information within a complex transnational regulatory framework.[76] To this end, the OECD would undertake a major project to devise new and higher quality operational standards for existing and future microbial resource repositories.[77] In principle, these proposals would change the underlying paradigmatic concept from that of "Culture Collection" to one of "Biological Resource Center" ("BRC").[78]

71 The first request for such action came from Japan, in 1998, and this effort led to the "Workshop on Scientific and Technological Infrastructure for BRCs, OECD," 17–18 Feb. 1999, Tokyo, Japan [hereinafter Tokyo Workshop]. *See* OECD Report on BRCs, above n. 5. For the task force members and other signatories, *see id.*

72 OECD Report on BRCs, above n. 5. *See also* Chapter 1 above, Section II.B. ("The Revolution in Genetic Science").

73 OECD Report on BRCs, above n. 5, at 11.

74 *Id.* at 11. *See also* OECD, The Bioeconomy to 2030: Designing a Policy Agenda (2011).

75 *See* above Chapter 1, Section III (discussing shift to "Big Science").

76 OECD Report on BRCs, above n. 5, at 11–12.

77 *Id.*

78 *Id.*; *see generally* Stern, above n. 7. The term Biological Resource Center was already used at the Tokyo Workshop, above n. 71.

In the OECD's vision, Biological Resource Centers would become a key element of the infrastructure that supports the life sciences and biotechnology.[79] As defined in 1999, BRCs

> … consist of service providers and repositories of the living cells, genomes of organisms, and information relating to heredity and the functions of biological systems. BRCs contain collections of culturable organisms (e.g., micro-organisms, plant, animal and human cells), replicable parts of these (e.g., genomes, plasmids, viruses, cDNAs), viable but not yet culturable organisms, cells, and tissues, as well as data bases containing molecular, physiological and structural information relevant to these collections and related bioinformatics.[80]

To this baseline definition, formulated at the 1999 Tokyo Workshop on Biological Resource Centers, the Task Force and Working Party added the following mission statement in 2001:

> BRCs must meet the high standards of quality and expertise demanded by the international community of scientists and industry for the delivery of biological information and materials. They should provide access to biological resources on which R&D in the life sciences and the advancement of biotechnology depends.[81]

More generally, BRCs would focus on adding value to their biological materials and linking more closely with the life sciences and bio-industry.[82] To this end, the drafters of the OECD report emphasized that the proposed BRCs would have to provide greater quality assurances than most of the existing culture collections and databases. This could be accomplished by raising the level of quality to an international standard that was later to be defined in the OECD Best Practice Guidelines of 2007.[83] In this connection, long-term preservation of living resources was to be made a crucial function of BRCs.[84] For example:

> It is necessary to improve the infrastructure and to develop techniques for storing DNA samples from diverse ecosystems in which 'molecular signatures' are found but the organisms themselves have yet to be cultured. Ensuring the accuracy of the genetic data associated with these living resources is a further crucial function of BRCs.[85]

79 OECD Report on BRCs, above n. 5, at 11.
80 *Id.* at 7.
81 *Id.*
82 D. Smith et al. (2013), above n. 33, at 276; *see also* OECD, The Bioeconomy to 2030 (2011), above n. 74.
83 OECD Best Practices, above n. 34.
84 OECD Report on BRCs, above n. 5, at 14–20.
85 *Id.* at 29. *See further* D. Smith et al. (2013), n. 33, at 277–79.

Besides access to living materials, genes, and genetic elements, the BRCs were tasked with providing accurate information about such biological resources[86] and integrating data more fully into the research mission of these centers, both in-house and with regard to the services they would provide to the scientific community.[87] *In silico* research advances, especially recent developments in synthetic biology, thus made it necessary to rationalize the relationship between *in vitro* and *in silico* approaches.[88]

Genomic sequencing projects (even of one single human gene or a single bacteria) can generate tens of thousands of new biological entities[89] that may have to be preserved, identified, and duplicated for further research in other laboratories and for creating cumulative research in genomics on well recognized models.[90] These so-called "derived" biological entities include the replicable parts of organisms, such as plasmids, rDNA, or viruses. High-throughput screening and sequencing then dramatically increase the amount of materials that might be preserved by the culture collections and potentially made available for follow-on research.

These integrating tasks, however, would become more difficult as the amount and diversity of living materials to be incorporated into BRCs expands.[91] Even before the advent of genomic science, the culture collections operated with severe limitations on physical capacity. These limitations always meant that careful selection had to be made of the type strains and reference strains that would be preserved. In other words, the collections were never like the Library of Congress, which takes deposits of all books; rather, they faced difficult choices about how and where to expend their limited resources, a constraint sometimes known as "The Big Refrigerator Problem."[92]

The drafters of the OECD report envisioned the formation of international linkages that could enhance global accessibility to information and biological materials as a necessary step to addressing this problem of capacity. As a group of European culture collection managers have rephrased it more recently, the transformational change from national though networked repositories for biological materials "toward a multilateral facility being part of a global infrastructure for the emerging knowledge-based bioeconomy requires not only an enlargement of managerial requirements but also a new mutual standard in quality management."[93]

86 OECD Report on BRCs, above n. 5, at 29.
87 *See id.* at 14–20.
88 *See* Chapter 1 above, Section II.D. *See also* Inst. Medicine, The Science and Applications of Synthetic and Systems Biology (Nat'l Acads. Press, 2011).
89 OECD Report on BRCs, above n. 5, at 19.
90 Stern, above n. 7, at 46.
91 OECD Report on BRCs, above n. 5, at 29.
92 Fiona Murray has called this "the Big Refrigerator Problem." *See* Murray, above n. 28.
93 D. Smith et al. (2013), n. 33 at 278; *see generally* OECD Report on BRCs, above n. 5, at 43.

For such networking to succeed, however, the means to coordinate and combine catalogues and databases that could support the requirements of science in the post-genomics era would need to be implemented. The coordination of curation efforts, together with the development of networked informatics tools for data analysis and visualization, would also become important.[94]

The networking possibilities that digital technologies make possible thus suggests a logical and feasible path towards linking existing culture collections in a federated, distributed system beyond that already pioneered by the WFCC members.[95] Ideally, such a system would reach across national and regional borders, with a view to making the world's aggregate *in vitro* microbial resources accessible and available for public research purposes anywhere. A digitally integrated system – a major goal of this book – could further help ameliorate "the Big Refrigerator Problem" by linking existing physical resources in what Elinor Ostrom has called "a common pool resource."[96]

At the same time, the drafters of the OECD report clearly understood that existing culture collections were not adequately coping with either the proliferation of intellectual property rights or the dictates of biodiversity regulation under the CBD, with the risk of diminishing worldwide access to biological resources for both scientific research and industry.[97] As we pointed out in Chapter 3, both developed and developing countries have experienced difficulties in integrating and implementing the CBD and the Nagoya Protocol.[98] Communication among research facilities concerning the specification of the genetic identity of organisms will become essential in order to ensure transparency of lineage for cross-border exchanges of genetic resources under the Nagoya Protocol.[99]

To address these challenges, the OECD's Task Force and Working Party envisioned a multi-pronged strategy for BRCs. Besides upgrading the quality of their services and extending their scope to include relevant genomic data, the BRCs would have to respect intellectual property rights and comply with the CBD, all without compromising their public-good mission, as some collections had already

94 For recent efforts to implement these proposals, *see further* D. Smith et al. (2013), n. 33, 275–83.

95 OECD REPORT ON BRCs, above n. 5, at 41–48; *see also* Anita Eisenstadt, *International Developments: A Context for the Creation of a Microbiology Commons*, *in* DESIGNING THE MICROBIAL RESEARCH COMMONS, above n. 11, at 188. *See further* Chapters 9 and 10 in this volume.

96 *See generally* ELINOR OSTROM ET AL., RULES, GAMES, AND COMMON-POOL RESOURCES (Univ. Mich. Press 1994); Charlotte Hess & Elinor Ostrom, "Artifacts, Facilities, and Content: Information as a Common-Pool Resource," paper presented at the Conference on the Public Domain, Duke Law School, Durham, North Carolina, Nov. 9–11, 2001. *See further* Chapter 9, Section I. (discussing theory of knowledge commons).

97 *See generally* OECD REPORT ON BRCs, above n. 5, at 14–20 *passim* and 38–39.

98 *See further* below Sections II & III.

99 *See* above Chapter 3, Section IV.C, below Chapter 10, Section III.B & C.

begun to do. In this connection, carefully drawn material transfer agreements would be needed. If properly done, the drafters believed that "[c]ountries may find BRCs a unique mechanism for coping with the demands of the CBD, especially if they are joined in a coordinated system of BRCs but still preserve the national sovereignty of their biological resources."[100]

To make such a project work, however, bridges would have to be built between collections with insufficient funds to attain the highest standards of a BRC and full-fledged BRCs themselves.[101] Over time, a successful, distributed network of cooperating BRCs would also seek to integrate important collections now held at universities, again assuming that minimum quality and verification standards could be met and sustained. These specialized collections are frequently the work product of a single dedicated research scientist. At the death or retirement of given researchers, there is a question of what will become of their valuable collections. Sometimes a well-known collection may be transferred to a national culture collection.[102] Often, however, the collection may suffer curation and quality lapses and even some established collections have been lost for lack of funds or because their incorporation into another collection was stymied by onerous regulatory requirements.[103]

The drafters of the OECD report stressed that the long-term stability of BRCs would require adequate and reliable sources of funding, a need that will become more critical as more biodiversity resources are deposited in BRCs.[104] Failure to rationalize and stabilize such funding would inevitably result in more biological resources being transferred to entities likely to charge excessive prices and to restrict access even for research purposes.[105] Indeed, the drafters argued that, if BRCs with little government funding were forced to transfer most of their costs to users, it could "create obstacles to the exchange of cultures and harmonization and give an advantage to users who can afford to pay for expensive strains, penalizing those who

100 OECD Report on BRCs, above n. 5, at 18. *See also id.* at 39.

101 Recent proposals in this regard envision the formulation of three different sets of best practice guidelines, namely, one for BRCs, one for lesser qualified repositories of microorganisms in general, and one for biosecurity. *See* D. Smith et al. (2013), above n. 33 at 277–78.

102 See Smith, *Culture Collections*, n. 1, 75–76.

103 *See* D. Smith et al (2013), above n. 33; *see also* Kurtzman, above, n. 11; Frank Simione, *American Type Culture Collection: A Model for Biological Materials Resource Management*, *in* Designing the Microbial Research Commons, above n. 11 (ATCC special collections). The WFCC has a committee on endangered collections, but it lacks funds to address the problem adequately. *See* above Section I.A.3 (discussing WFCC funding). At the national level, funds may or may not be available to rescue deserving collections.

104 OECD Report on BRCs, above n. 5, at 23. For the latest proposals concerning the funding of BRCs, *see* D. Smith et al. (2013), above n. 33, at 280–83; *see also* Chapter 10, Section IV.

105 *Cf.* OECD Report on BRCs, above n. 5 at 24–25 (focusing only on high-value resources).

cannot, particularly those in developing countries." Hence, "most BRCs will require a significant government funding component, and some guarantees of continued funding to ensure that their essential functions remain reliable for R&D and the support of biotechnology."[106] This statement can be read as implicitly critical of the American Type Culture Collection in the United States and the strategies it adopted to ensure its survival without adequate government support, as explained later in this chapter.[107]

Given these premises, the drafters of the OECD Report concluded with the view that the essential infrastructure they envisioned – with BRCs at the core – would require national governments to undertake the following actions, in concert with the international scientific community:

- Selectively seek to strengthen existing *ex situ* collections of biological data and materials, create collections of new resources, including in non-OECD countries, and elevate those collections to the quality required for accreditation as national BRCs.
- Support the development of an accreditation system for BRCs, based upon scientifically acceptable, objective international criteria for quality, expertise and financial stability.
- Facilitate international coordination among national BRCs by creating an agreed system of linkage. This should be based on modern informatics systems that link biological data to biological materials across national BRCs and upon common technological frameworks.
- Take into account the objectives and functioning of BRCs when establishing and harmonizing national or international rules and regulations.
- Develop policies to harmonize the operational parameters under which BRCs function, including those governing access to biological resources as well as their exchange and distribution, taking into account relevant national and international laws and agreements
- Support the establishment of a global BRC network that would enhance access to BRCs and foster international cooperation and economic development.[108]

The question this ambitious set of proposals leaves under theorized, however, is how to achieve a globalized BRC infrastructure under real world conditions without generating social costs that outweigh the perceived benefits in the end. We return to these and related themes generally in Part Four and with detailed governance proposals in Chapter 10.

[106] *Id.*

[107] *Id.* at 28. *See* below Section II.A.

[108] *Id.* at 9 (Executive Summary). *See further* D. Smith et al. (2013), above n. 33, at 276–83.

C. Beyond the WFCC: Regional and Global Networks of BRCs

The OECD's vision of cross-border collaboration between qualified BRCs fits neatly into the National Research Council's (NRC) vision of the "New Biology" paradigm discussed in Chapter 1. That paradigm envisioned digitally integrated research approaches that freely traversed preexisting disciplinary borders. In that project, microbiology was assigned an important role, and the coordinated linkage of highly qualified BRCs across national and regional borders, each supplementing the assets of the others, could provide the infrastructure needed to fulfill that assignment.[109]

While recognizing the accuracy and prescience of the OECD's 2001 vision, however, we caution that its implementation in practice is likely to prove far more difficult and complex than its drafters foresaw. The OECD's proposed quality standards for BRCs are stringent, costly to implement, and beyond the reach of many, if not most, collections. In response to this challenge, networks of culture collections have been formed to enable multiple entities to support each others' projects, as discussed below. The WFCC has also developed and promulgated operational guidelines of their own that move in the same direction without requiring a large financial investment as a precondition.[110] These guidelines were examined in the previous section.

1. Disparities Among the WFCC Member Collections

BRCs preserve their holdings using long-term storage techniques, such as cryopreservation and lyophilization, depending on the type of organism. Often these techniques require optimization to enable both survival of the cultures and retention of their properties. However, single collections are seldom able to invest in preservation research. The improvement and testing of new techniques is also limited by disparities in technical capacity between culture collections in single countries, let alone the differences in this regard between countries at different levels of economic development. These differences are magnified by funding and limited awareness of the role that culture collections play in contemporary life sciences research. These constraints further limit the number of skilled personnel available to provide the services attributed to BRCs on a sustained and acceptable level.[111]

[109] *See* above Chapter 1, Section II.D. *See further* D. Smith et al. (2013), above n. 33, at 290–91.

[110] *See* above Section I.A.2.

[111] *See, e.g.*, D. Smith et al. (2013), above n. 33, at 290; Smith, *Culture Collections*, above n. 1, at 77 (stating that staff "must be appropriately trained in authentication, preservation, and supply of strains with research expertise related to the aims of the collection" plus a research knowledge base regarding taxonomic studies, international regulations, shipping, etc.).

Even in advanced OECD member countries, for example, only a relatively few of the most robust collections will employ a team consisting of scientists, information technicians, and administrative staff. Most of the others would lack both the staff and resources to qualify for BRC certification, while some are maintained by a single concerned scientist.

These disparities become far more acute if one views the existing microbial culture collection from a global perspective. The legacy collections in Europe, the United States, Japan and other developed countries have high quality standards, deep technical capacity, and holdings drawn from all over the world (including the developing countries) that were accumulated over a long period of time. Until recently, these *ex situ* collections often benefited from funding commitments by government agencies and other sources. They have traditionally viewed their duties as providing a public good, in the pursuit of which the sharing of materials with other collections, research institutes, and individual microbiologists was a logical corollary. Adequate funding has become an issue, however, and, as we have already noted, some collections have been irretrievably lost.[112]

A. LEGACY COLLECTIONS IN THE EUROPEAN UNION AND THE UNITED STATES. The European Union, in particular, has a strong and evolving public infrastructure for the preservation and exchange of microbial materials, with a set of agreed standards that traditionally have promoted open access for use and reuse, with tracking, on a formal basis.[113] This infrastructure has two major types of institutional components, namely, the government-funded public collections, and the university held, closed research collections, with the latter also often sponsored and supported by government. The long-term preservation efforts of public culture collections are thus supplemented by a few highly specialized private collections, as well as by a network of informal collections that operate outside the WFCC.

Table 4.3 shows that Belgium, Denmark, France, Germany, the Netherlands, Russia, Sweden and the United Kingdom are the major European holders of microbial materials affiliated with the WFCC. Although the European microbial collections predominantly hold biological resources from Europe, they also collectively manage major legacy collections of materials originating from the developing countries, and they are actively involved in obtaining new materials from these same countries.[114] As a result, they operate in the shadow of the CBD – the EU are members singly and

[112] Smith, *Culture Collections*, above n. 1, at 72.

[113] Altogether, there were some 226 collections operating in Europe as a whole, in 2015, including Russia, Ukraine, Turkey, and other non-EU members. Their total holdings amounted to 841,276 specimens. WFCC Website, above n. 20. Russia alone holds over 60,000 specimens. *Id.*

[114] Dedeurwaerdere (2010), above n. 9, at 407–13.

TABLE 4.3. *European Culture Collections Registered with the World Data Center for Microorganisms*

Countries and Regions	*Culture Collections*	*Cultures*
Armenia	1	11,520
Austria	2	6,070
Belarus	1	1,175
Belgium	7	76,496
Bulgaria	4	12,979
Czech	13	11,053
Denmark	3	102,066
Estonia	5	14,304
Finland	2	10,588
France	38	86,350
Germany	13	94,882
Greece (Hellenic Rep.)	6	6,377
Hungary	8	13,962
Ireland	1	380
Italy	13	26,054
Kazakhstan	2	398
Latvia	1	1,361
Netherlands	6	90,775
Norway	2	3,028
Poland	9	8,545
Portugal	6	10,135
Romania	2	760
Russian Federation	22	60,168
Slovak	3	4,616
Slovenia	3	15,992
Spain	4	10,321
Sweden	3	52,700
Switzerland	4	3,965
Turkey	10	5,607
U.K.	19	84,109
Ukraine	8	11,569
Uzbekistan	3	2,074
Yugoslavia (former)	2	897
Total	226	841,276

WFCC Website, above n. 20.

collectively of that treaty – and they are concerned about the need to stabilize the legitimacy of their holdings.[115] The European culture collections have also formed collaborative networks of major importance, as discussed below.

In contrast to the EU, whose major WFCC collections largely adhere to a public-good approach supporting noncommercial research, the U.S. culture collection scene is dominated by the American Type Culture Collection (ATCC). This nominally nonprofit, public-private entity operates on a proprietary basis, and it uses restrictive licensing terms similar to those of the private sector, which can impede academic research, despite its formal not-for-profit status.[116]

The ATCC is one of the most technically and scientifically advanced culture collections in the world, and to that extent it is perceived as a model for the BRC concept. However, because the ATCC holds some 75,000 specimens amounting to more than a third of the total held by all U.S. members of the WFCC,[117] its licensing policies have a major effect on researchers everywhere and – as discussed below – its proprietary licensing model has also begun to influence major collections in other countries.[118] At the moment, there is no institution in the EU that plays a role analogous to that of ATCC in the United States, although without some international framework of the kind we envision in Part Four, other countries seem likely to emulate that model.[119]

Several agricultural research collections located at the U.S. Department of Agriculture and one collection at the Centers for Disease Control (CDC)[120] are members of the WFCC. However, government departments in the United States often maintain their own microbial culture collections for a variety of purposes,[121] and these

[115] *See* below Section I.C.2 ("The Emerging BRC Networks") and Section III.A.2 (discussing EU Culture Collections' Organization (ECCO)). *See also* The ECCO Core Material Transfer Agreement for the Supply of Samples of Biological Material from the Public Collection, Feb. 2009 [hereinafter ECCO MTA], *available at* http://www.eccosite.org/wp-content/uploads/2014/07/ECCO_core-MTA_V1_Feb09.pdf.

[116] *See* below Section II.A (discussing ATCC licensing conditions).

[117] WFCC Website, above n. 20.

[118] *See* Section II.B below.

[119] *See esp.* the discussion of the proposed Global Biological Resource Center Network (GBRCN), Chapter 9, Section II.C.1 below. *See also* the CABI Materials Transfer Agreement, *available at* https://www.pdffiller.com/en/project/10335834.htm (last accessed 3 July 2014).

[120] Kurtzman, above n. 11 (Four USDA collections belong to the WFCC, viz.: ARS Collection of Entomopathogenic Fungi; ARS Rhizobium Germplasm Resource Collection; Agricultural Research Service Culture Collection (NRRC); and Oregon Collection of Methanogen Southern Regional Research Centers (SSRC)). The Center for Disease Control's Division of Vector-Borne Infectious Diseases Collection is also a WFCC member.

[121] Agencies known to maintain important collections include the Environmental Protection Agency (EPA); Department of Energy (DOE); Department of Defense (DOD); Homeland Security (HDS); and the National Institutes of Health (NIH). These collections have not affiliated with the WFCC, according to its website. WFCC Website, above n. 20.

collections have not sought affiliation with the WFCC. Nor, until recently, has there been any effort systematically to catalog their holdings or licensing practices.[122]

In the United States, there are only 21 microbial collections registered with the WFCC, including ATCC and the government collections mentioned above. Together they hold over 210,000 specimens, most of which are located at universities.[123] These university-held collections sometimes make their materials available at relatively low cost, or even free of cost, to academic researchers, but their access and use policies vary considerably.

On the surface, the microbial genetic resource infrastructure in the United States thus appears quite different from that of the European Union for at least three reasons. First, major collections in the United States are government owned rather than not-for-profit or university based. Second, the dominant player in the United States – the ATCC – operates on a self-sustaining proprietary model. Third, the United States is the only major country that has not ratified the Convention on Biological Diversity (1992), which to some extent shielded its major collections from some of the pressures that developing countries brought to bear on the European collections, at least until the advent of the Nagoya Protocol.[124]

On closer analysis, however, many culture collections owned and managed by diverse U.S. government departments and agencies traditionally operated rather like the public culture collections funded by governments in the European Union. In other words, they have made available microbial materials to the public for research and applications under varying, but generally permissive legal conditions.[125] Lately, however, the privatizing pressures discussed throughout this book have also affected the collections owned by the U.S. government. For example, representatives of the relevant departments have stated that they find it increasingly difficult to obtain materials from both foreign and domestic collections for research purposes, despite the practice of customary exchanges in the past.[126]

The U.S. collections have also become increasingly concerned about lending their resources to others without carefully drafted Material Transfer Agreements that consider a range of possible restrictions on the use and reuse of their materials. These

[122] Some important collections held by the government were not aware of the WFCC as late as 2011.

[123] WFCC Website, above n. 20.

[124] *See further* above Chapter 3, Section IV. On the whole, we have not found publicly reported concerns about possible violations of CBD principles with respect to the microbial culture collections in the U.S. as such, although there have been numerous and clamorous complaints and lawsuits about U.S. patents issued on plant genetic resources extracted from developing countries without prior informed consent. *See* Chapter 3, Section I.A. U.S. officials have observer status at the CBD.

[125] *See, e.g.*, Kurtzman, above n. 11; Statements made at a Smithsonian Meeting of Government-Owned Collections in February, 2010 (attended by Prof. Reichman).

[126] Smithsonian Meeting above n. 125.

concerns are exacerbated by heightened awareness of risks to biosecurity, possible criminal or terrorist activity, as well as by the need to foster industrial competitiveness and also to seek some financial return on government funded research. These and related concerns have recently elicited a White House mandate regarding microbial resources,[127] which has triggered a major inventory of the contents of all government held microbial culture collections. The drafting of a standard MTA is also envisioned within this project.

In the United States, as in Europe, there is also a large unregulated network of culture collections at universities, hospitals, and industrial laboratories with no common standards. These collections tend to follow licensing strategies of their own, subject to pressure on universities from government funders to preserve the sharing ethos and access to genetic resources for public research purposes. However, no recent comprehensive survey of their contents has been undertaken.[128] As will be seen below, academics at such institutions (both in the U.S. and the EU) often participate in an informal, club-based exchange system that operates in parallel with (and in opposition to) the highly regulated formal system of material transfer practices run by the technology transfer offices of their respective universities.

B. WIDE DISPARITIES AMONG COLLECTIONS IN OTHER REGIONS. Looking at the rest of the world, the most likely candidates for BRC status are selected culture collections located in Australia, Brazil, Canada, China, India, Japan, Korea, Taiwan and Thailand. The WFCC collections in both Australia and Canada hold respectively more than 80,000 aggregate specimens,[129] which ranks them thirteenth and fourteenth in the world.[130]

In Asia, Japan stands out for the magnitude of its holdings and for its longstanding leadership in attempting to deal with issues pertaining to the CBD. Japan has twenty-four culture collections affiliated with the WFCC, and their aggregated holdings amount to some 252,000 microbial cultures.[131] Japan's WFCC holdings are thus the largest in Asia by a wide margin and also the largest of such national assets in the world (disregarding the EU as a whole).

[127] National Security Council, Exec. Office of the President, National Strategy for Countering Biological Threats (2009), *available at* https://www.fas.org/irp/offdocs/ppd/bio-strategy.pdf (last visited 7 Jan. 2015).

[128] There is one out-of-date publication that attempted to survey these collections in the U.S. *See* L. R. Hill & Micah Krichevsky, *International Strain Data Networks*, 2 World J. Microbiology & Biotech. 341 (1986).

[129] *Statistics*, WFCC, http://www.wfcc.info/ccinfo/statistics/ (last accessed 3 July 2014) [hereinafter *Statistics*, WFCC]. Australia has 34 WFCC member collections, Canada has 18, and Korea has 21.

[130] *Id.*

[131] *Statistics*, WFCC, above n. 129.

TABLE 4.4. *Top 20 Strain Holding Countries According to WDCM (2014)*

Rank	*Countries and Regions*	*Total hold*
1	U.S.A.	257,060
2	Japan	252,339
3	China	170,346
4	India	160916
5	Korea (Rep. of)	158,528
6	Brazil	109,560
7	Denmark	102,066
8	Thailand	97,401
9	Germany	94,882
10	Netherlands	90,775
11	France	86,350
12	U.K.	84,109
13	Australia	82,946
14	Canada	82,315
15	Belgium	76,496
16	Taiwan	67,227
17	Russian Federation	60,168
18	Sweden	52,700
19	Italy	26,054
20	New Zealand	25,045

WFCC Website, above n. 20.

In 1999, Japan initiated and funded the OECD task force on Biological Resource Centers (BRCs), discussed above, and its government played a leading role in the formation and implementation of the Global Biodiversity Information Facility (GBIF).[132] Japan's approach to the constraints of the CBD led it to elevate its need for foreign microbial resources to a high scientific priority. In this regard, the Japanese national culture collection (NITE) has forged a bilateral agreement with Thailand's Biotec Collection, with a view to sharing and exploiting that country's microorganisms *in situ* and *ex situ*,[133] partly in exchange for capacity building of culture collections in

[132] The movement to convert culture collections into better-equipped BRCs is discussed above in Section I.B. For a discussion of GBIF, see Chapter 9, Section II.B.2 below.

[133] This Agreement is between NITE in Japan and Biotec in Thailand, which are respectively the main culture collections in those countries. *See* Nat'l Inst. Tech. & Evaluation (NBRC), *Biological Resource*

Thailand.[134] Japan has thus begun to forge the kind of formal bilateral and plurilateral ties that will be necessary to ensure that the research needs of the global microbiological community can be rendered consistent with the goals of the CBD.

As the Japanese culture collections forge collaborative agreements with collections in other countries, however, it remains to be seen whether they will follow the WFCC's public good model or a more proprietary approach. For example, NITE is already participating in a regional initiative to devise a standard MTA, which could bridge relations between at least one OECD country and a number of developing countries within the ambit of the CBD.[135] But the extent to which the resulting MTAs will resemble the ATCC's proprietary model remains to be seen.[136]

India ranks fourth in the WFCC's list of top twenty national strain holders (disregarding the EU as a whole),[137] with more than 160,000 microbial specimens. The two most prominent collections in India are the Indian Type Culture Collection (ITCC) at IARI, New Delhi, and the Microbial Type Culture Collections (MTCC) at IMTCCH, Chandigarh.[138] Rapid development in microbial research took place after independence, when a chain of national laboratories and regional research laboratories were established by the Council of Scientific and Industrial Research (CSIR).

Since then, there has been a steady increase in the laboratories under the CSIR, the Indian Council of Agricultural Research (ICAR), the Indian Council of Medical Research (ICMR), and certain defense agencies, "whose basic and molecular microbiological research is carried out to find solutions to … problems in industry, agriculture, medicine, food, environment, etc."[139] Research in industrial microbiology is reportedly being carried out at almost every Indian university,

Center, World Data Centre for MicroOrganisms, http://www.wfcc.info/ccinfo/collection/by_id/825 (last accessed 3 July 2014) and *BIOTEC Culture Collection*, World Data Centre for Microrganisms, http://www.wfcc.info/ccinfo/collection/by_id/783 (last accessed 3 July 2014).

134 *Cf.* Oliver E. Williamson, *Markets and Hierarchies*, 63(2) *Am. Econ. Rev.* 316–25 (1973) (who says economists must go beyond Coasean bargaining where the option value of resources is highly uncertain).

135 *See About Us*, Asian Biological Resource Network (ABRCN), http://www.abrcn.net/aboutus.html; *see further* Section II.B below. Until recently, Japan hosted the World Data Center for Microorganisms (WDCM), which serves as the official information node for the WFCC and also serves as the Japanese hub of the International Nucleotide Sequence Database Collaboration, DDBJ/EMBL/GenBank. *See History*, WDCM, http://www.wdcm.org/history.html (last accessed 3 July 2014).

136 *See* Nat'l Inst.of Tech. & Evaluation, Materials Transfer Agreement (Form A), *available at* http://www.nbrc.nite.go.jp/pdf/mta_type_a.pdf; *also see* ATCC MTA, below n. 198.

137 *Statistics*, WFCC, above n. 129.

138 *See* Geeta Sumbali & R.S. Mehrotra, Principles of Microbiology 116 (McGraw Hill 2009). This source lists an additional 18 Microbial Culture Collection Centers in India. *Id.*

139 *Id.* at 35.

various national research institutes, and in those industries that use microorganisms for the production of high value products.[140]

China plays an increasingly important role, especially through its Institute of Microbiology of the Chinese Academy of Sciences (IMCAS).[141] IMCAS maintains the largest microbiological culture collections in China, with more than 17,000 strains.[142] Since 2008, IMCAS has reorganized its R&D activities into a "value chain" that consists of a Biological Resource Center with a Biosafety Level 3 laboratory, a scientific research system, and a technology transfer center for both basic and applied research.[143] With its 487 staff members, of whom nearly 300 are researchers, IMCAS clearly fits the OECD's BRC model. Altogether, China has 25 culture collections affiliated with the WFCC, with more than 170,000 strains registered with the WDCM.[144]

However, China has lagged behind in converting its own traditional medicinal knowledge into proprietary and protectable assets. Thus, despite China's attention to, and investment in, the preservation of genetic resources, its traditional medicinal knowledge has reportedly been siphoned off, especially from publicly available databases, by both foreign and national interests, without benefit sharing under the CBD.[145] New laws under review at the time of writing may address this problem, although the national authorities appear not to have yet devised a strategy to fully implement the Nagoya Protocol.[146]

Other major players in Asia include Thailand, the Republic of Korea, and Taiwan. Of these, Thailand has some sixty WFCC members holding about 97,000 specimens, while Korea has 21 collections with about 158,000 specimens.[147] According to the WFCC, Korea thus ranks fifth in the list of top twenty strain holders worldwide, just below India. Korea is also a leader in designing digital infrastructure to link its collections.[148]

Taiwan, with just two WFCC collections, nonetheless holds about 67,000 specimens. Indonesia with eighteen WFCC collections holds over 11,000 specimens.

[140] *Id.* (listing a number of major research entities).

[141] *About IMCAS*, Instit. Microbiology, Chinese Acad, Scis., IMCAS, http://www.im.ac.cn/english1/about.htm (last accessed 3 July 2014) [hereinafter *About IMCAS*].

[142] *Id.* It also holds the largest fungal herbarium in Asia, with some 400,000 specimens. *Id.*

[143] *See About IMCAS*, above n. 141.

[144] *Statistics*, WFCC, above n. 129.

[145] Tianbao Qin, *Common Pools of Traditional Chinese Medical Knowledge in China*, *in* COMMON POOLS OF GENETIC RESOURCES: EQUITY AND INNOVATION IN INTERNATIONAL BIODIVERSITY LAW 150–67 (E.C. Kamau & G. Winter eds., Routledge 2013) [hereinafter COMMON POOLS OF GENETIC RESOURCES (2013)].

[146] *Id.* at 164.

[147] *Statistics*, WFCC, above n. 129.

[148] *Id. See also* below Section I.C.2.

The remaining seven countries – Iran, Malaysia, Pakistan, the Philippines, Singapore, Sri Lanka, and Vietnam – collectively manage thirty-six WFCC collections holding 26,000 specimens.[149]

Only one country in Latin America seems to possess the institutional, human, and technical capabilities to enter the BRC sweepstakes on a par with the leaders in Asia, namely, Brazil. Brazil has some sixty-five public culture collections affiliated with the WFCC, with a total of almost 110,000 strains at the time of writing, which makes them the sixth largest holder of *ex situ* cultures in the world.[150] A 1969 decree established the National Council for Scientific and Technological Development (CNP), as the responsible body for authorizing foreigners to collect biological specimens for research, and it continues in that capacity under a more comprehensive decree adopted in 1999.[151]

Since then, major efforts have been made to ensure that taxonomic material of national importance should be deposited in public collections for use by Brazilian scientific institutions. In the last quarter of the twentieth century, Brazil's Ministry of Science and Technology (MIT), in collaboration with the WFCC, sponsored a program to train experts on collection management, preservation techniques, and microbial taxonomy. In the twenty-first century, Brazil has made concerted efforts to implement the OECD's Guidelines for BRCs. The Brazilian strategy is currently focused on the consolidation of a distributed network of specialized resource centers to meet the growing demands of the user community.[152] However, recent legislation has reportedly clouded the legal status of genetic resources held in common pool arrangements, and it has also generated unresolved tensions between public and private interests that may potentially complicate public research endeavors.[153]

Other important countries in this hemisphere include, in alphabetical order, Argentina, Chile, Columbia, Cuba, Ecuador, Mexico, and Venezuela. These countries collectively host fifty WFCC member culture collections with aggregate holdings of about 35,000 specimens.[154] In comparison, the microbial culture collections in South Africa, with only three registered collections and about 10,860 strains, is the largest holder of approved cultures in all of Africa.[155] Apart from South Africa, there are six other African countries that have WFCC members, namely,

[149] *Statistics*, WFCC, above n. 129.

[150] *Id.*

[151] *See* Davis, Fontes & Marinoni (2013), above n. 1, at 36 (discussing Decree 98.830 (1990) and Decree 65.057 (1969)).

[152] *Id.* at 36. See generally D. Smith et al. (2013), above n. 33, at 287–88.

[153] *See* Juliana Santilli, *Genetic Resources Common Pools in Brazil, in* COMMON POOLS OF GENETIC RESOURCES (2013), above n. 145, at 103, 105–07 (discussing Provisional Act 2186–16 of 2001).

[154] *Statistics*, WFCC, above n. 129. Chile and Mexico are recent OECD member states, but are included here because of their borderline economic status and their geographical and cultural location.

[155] *Statistics*, WFCC, above n. 129.

Egypt, Morocco, Nigeria, Senegal, Uganda, and Zimbabwe. Taken together, these collections hold fewer than 5,000 specimens.[156]

Nevertheless, African countries on the whole have become acutely conscious of the implications of the CBD. A growing number of African governments have taken steps to consolidate and improve their holdings of plant genetic resources and related traditional knowledge, with a view to participating in commercial applications of basic and applied research.[157] It remains to be seen if and when these developments lead to greater investment in upgrading their collections of microbial genetic resources for public research purposes.

Meanwhile, when evaluating the lesser technical capacities of developing countries in Latin America, Asia, and Africa, one must distinguish between the *ex situ* microbial resources discussed above and the vast reserves of *in situ* resources that remain subject to their territorial sovereignty.[158] It is largely these *in situ* resources that fuel the drive for greater protection of genetic resources generally, in such forums as the CBD, the World Intellectual Property Organization, and other UN specialized agencies.

Today, most developing countries strictly limit access to both the genetic resources and related indigenous know-how extant within their territorial borders even for public scientific research. As primary supporters of the CBD, the developing countries in particular are eager to share the benefits of commercial exploitation, once prior informed consent and mutually agreed terms have been obtained, and to receive technology transfers from those private or public entities whose permissible explorations lead to downstream applications. These expectations, coupled with real or perceived losses suffered from instances of alleged biopiracy in the past,[159]

[156] *Id.*

[157] *See, e.g.*, Gino Cocchiaro & Britta Rutert, *Common Pools of Traditional Knowledge: The Story of the Kukula Traditional Health Practitioners of Bashbackridge, Kruger to Canyons (K2C) Biosphere Reserve*, *in* COMMON POOLS OF GENETIC RESOURCES (2013), above n. 145, at 29; Evanson Chege Kamau, *Common Pools of Traditional Knowledge and Related Genetic Resources: A Case Study of San-Hoodia*, *in* COMMON POOLS OF GENETIC RESOURCES (2013), above n. 145, at 373–408.

[158] With respect to *ex situ* microbial culture collections in developing countries, it is worth noting that many of them in the past benefited from concerted capacity building efforts under the aegis of UNESCO, but much more would be needed to build a truly functional scientific research commons. *See, e.g.*, *Microbial Resources Centres (MIRCEN)*, U.N. EDUC., SCI. & CULTURAL ORG. (UNESCO), April 15, 2008, http://portal.unesco.org/science/en/ev.php-URL_ID=2491&URL_DO=DO_TOPIC&URL_SECTION=201.html; *see also* Jerome H. Reichman et al., *Access to Scientific and Technological Knowledge: UNESCO's Past, Present and Future Roles*, *in* STANDARD SETTING IN UNESCO, 1 NORMATIVE ACTION IN EDUCATION, SCIENCE AND CULTURE 323–50 (A. A. Yusuf ed., Martinus Nijhoff Pub., 2007); *see also* Flora Katz, *Proposal for a Microbial Semi-Commons: Perspectives from the International Cooperative Biodiversity Groups*, *in* DESIGNING THE MICROBIAL RESEARCH COMMONS, above n. 11. Nevertheless, there are relatively few bottom-up science initiatives in the poorer countries, their overall research capacities remain weak, and such funding as is available may not generally be distributed on the basis of merit.

[159] *See* above Chapter 3, Section I.A.

make them prone to adopt highly restrictive MTAs and to impose strict controls covering access and use of all microbial resources, whether these materials are commercially promising or not. Such restrictions are often inconsistent with public research needs, and they may impede the very commercial payoffs they seek.[160] Moreover, their approach to information disclosure, and their policies regulating government-generated and government-funded scientific data and information, also tend to be restrictive or still embedded in a culture of secrecy.

One may question the viability of this fortress mentality. Today, the fruits of microbial prospecting in the open areas surrounding many developing countries are already being exploited without need for permission, and microbes do not respect territorial boundaries.[161] Enterprises from some OECD countries, and especially from some BRICs countries, are purchasing land in poor countries, which could enable them to circumvent some of the formal restrictions on access to biodiversity.[162] Most developing countries – other than the BRICs countries – also lack the technical skills and funds even to bring their culture collections up to minimum international quality standards, let alone to establish the kind of advanced scientific research infrastructure envisioned in the OECD's concept of Biological Resource Centers. Meanwhile, digital networking and genomic science in OECD countries can increasingly derive new applications from existing microbial resources, without necessarily tapping into the developing countries' *in situ* resources.

In sum, hoarding microbial genetic resources, rather than exchanging them for bigger and mutually advantageous scientific payoffs, could turn out to be a dead end that leaves microbiological research and applications in these countries further behind.[163] This assessment, in turn, provides a basis for envisioning a more innovative and cooperative approach with respect to the pooling of microbial genetic resources for public research purposes than has occurred in the context of plant genetic resources. We discuss these prospects fully in Chapters 5 and 10.

[160] *See, e.g.*, Katz, above n. 158; below Section II.

[161] *See* J. Craig Venter Instit., Research Voyage of the Sorcerer II Expedition, J. Robert Beyster and Life Technologies Foundation 2009–2010. *See* Chapter 8 below, Section III.A.2; *see also Homepage*, THE MICRO B3 PROJECT ON MARINE BIOLOGY AND BIODIVERSITY, http://www.microb3.eu/ (last accessed 2 July 2014).

[162] *Cf.* Paris Convention for the Protection of Industrial Property, arts. 2–3, Mar. 20, 1883, *as amended on* Sept. 28, 1979, 21 U.S.T. 1583, 828 U.N.T.S. 305, which give national treatment to enterprises established anywhere in the union. It should also be noted that these provisions of the Paris Convention are incorporated into the TRIPS Agreement, art. 2.1, and that the WTO Appellate Body has elevated national treatment to a cardinal principle of the post-TRIPS global IP system. Agreement on Trade-Related Aspects of Intellectual Property Rights, Apr. 15, 1994, Marrakesh Agreement Establishing the World Trade Organization, Annex 1C, THE LEGAL TEXTS: THE RESULTS OF THE URUGUAY ROUND OF MULTILATERAL TRADE NEGOTIATIONS 320 (1999), 1869 U.N.T.S. 299, 33 I.L.M. 1197 (1994) [hereinafter TRIPS Agreement].

[163] *Cf.* The poor state of agriculture in those African countries' that steered clear of the "green revolution."

2. The Emerging BRC Networks

Against this background, the OECD's call for a "global BRC network" has induced leaders in the field to develop ambitious plans for the formation of regional and global networks that would federate and coordinate the resources and functions of a selected number of the world's most technically advanced microbial culture collections. These proposals build on the fact that networking at the national level is already a well-established practice in Europe, Asia, and to some extent, in Latin America.[164] They also build on preexisting efforts to foster regional collaboration among culture collections in Europe, notably the European Union Culture Collections' Organization (ECCO),[165] and on efforts by the Asian Biological Resource Center Network (ABRCN) to develop a common database covering some 25,000 strains held at seven major collections.[166] Also relevant is the Asian Network of Research Resource Centers, which some 13 countries have joined and the Asian Consortium for Conservation and Sustainable Utilization of Microbial Resources (ACM), founded in 2004, with thirteen member countries.[167]

At the multilateral level, in particular, a major effort to establish a Global Biological Resource Center Network (GBRCN) obtained funding from the German Science Ministry in 2008 for a Demonstration Project that is examined in detail in Chapter 9.[168] The Demonstration Project, with selected partners from North and South America, Africa, Asia, as well as a strong base in Europe, aimed to build "a structured long-lasting global network" designed to pave the way for these technically advanced collections to better meet user needs."[169] While this

[164] *See* Smith, *Culture Collections*, above n. 1, at 109–11, and *id.*, tbl. 4.5, at 110 (listing 14 national networks, including Cuba and Brazil).

[165] *See* ECCO, http://www.eccosite.org/ (last accessed 3 July 2014); Smith, *Culture Collections*, above n. 1, at 109 (characterizing ECCO as "an incubator for pan-European initiatives"). *See also* ECCO MTA, below Section III.A.2.

[166] *E.g.*, General Microbiological Culture Collection Center (CGMCC), China; NITE Biological Resource Center (NBRC), Japan; Korean Collection for Type Cultures (KCTC), Republic of Korea; National Center for Engineering and Biotechnology Culture Collection (BCC), Thailand; Philippine National Collection of Microorganisms (PNCM), Philippines; and Microbial Culture Collection – Museum of Natural History (MCC-MNH), Philippines. *See* Smith, *Culture Collections*, above n. 1, at 113.

[167] *See* Asian Network of Research Res. Ctrs., Member Countries, ANRRC, http://www.anrrc.org/rrc/rrcCountry.jsp (last accessed 3 July 2014); Yeowhee Lee, Asian Network of Research Resource Centers: Future Directions, presentation at WDCM Second Symposium, Beijing, China, June 7–8, 2012. For ACM, see D. Smith et al. (2013), above n. 33, at 286–87, Box 11.5.

[168] *See Demonstration Project for a Global Biological Resource Centre Network*, Global Biological Resource Ctr. Network (GBRCN), http://www.gbrcn.org (last accessed 3 July 2014); below Chapter 9, Section II.C.1.

[169] Smith, *Culture Collections*, above n. 1, at 111.

Demonstration Project ended in 2011,[170] the European Commission has recently funded a full-fledged regional version of the GBRCN project to be known as the European Culture Collections Resource Research Infrastructures (MIRRI), which was just getting underway at the time of writing.[171]

Also at the formation stage are: an emerging Brazilian network consisting of eleven culture collections;[172] the Asian Biological Resource Center Network (ABRCN),[173] mentioned above; and a United States Culture Collection Network (USCCN),[174] which was initially funded by the National Science Foundation. The latter entity aspires to coordinate the efforts of some 26 non-governmental collections in the United States.[175]

These projects to unite selected BRCs in regional and global networks are important for this study. They could, at least conceptually, constitute a foundation on which to erect the redesigned Microbial Research Commons that we elaborate in Part Four. However, these same proposals, which in effect envision a global hierarchy of elite BRCs operating beyond the governance apparatus of the WFCC,[176] raise troubling questions about the nature and goals of the microbial infrastructure that might result from such initiatives and about the relative costs and benefits they would entail, especially if they were to drift away from the traditional public-good service model of most WFCC collections. These issues are explored more fully in Part Four of this book.

II. CONTRACTUAL RESTRICTIONS ON ACCESS TO AND USE OF UPSTREAM MICROBIAL GENETIC RESOURCES IN BOTH DEVELOPED AND DEVELOPING COUNTRIES

When, in 2001, the OECD Task Force recommended a shift from culture collections as traditionally constituted to Biological Resource Centers that would "underpin … the future of life sciences and biotechnology," they partly anticipated the National

[170] *See* GBRCN Demonstration Project Secretariat, Final Report on the GBRCN Demonstration Project (November 2009–November 2011), *available at* http://www.gbrcn.org/fileadmin/gbrcn/media/downloads/GBRCN_Final_Report/GBRCN-FinalReport2012.pdf

[171] European Culture Collections Resource Research Infrastructures (MIRRI), *About MIRRI*, MIRRI, http://www.mirri.org/about-mirri/implementation-steps.html (last accessed 3 July 2014); Smith, *Culture Collections*, above n. 1, at 113. For more on MIRRI, *see* below Chapter 9, Section II.C.3.

[172] D. Smith et al. (2013) n. 33, at 287–88.

[173] *See* above Section I.C.1.b; Smith, *Culture Collections*, above n. 1, at 113.

[174] *See About the USCCN*, U.S. Culture Collection Network (USCCN), http://www.usccn.org/about/Pages/default.aspx (last accessed 3 July 2014).

[175] U.S. Culture Collection Network, RCN Proposal Summary: A Community of Ex Situ Microbial Germplasm Collections (2011), *available at http://www.usccn.org/about/Pages/RCN-Proposal-Summary.aspx* (last visited 7 Jan. 2015).

[176] For the WFCC's governance structure, *see* below Chapter 9, Section II.B.1.

Research Council's later vision of the enhanced role of microbiology within the framework of its "New Biology" paradigm.[177] At the same time, the OECD Task Force kept its feet on the ground by insisting that the proposed network of Biological Resource Centers should play a major role in reconciling the needs of the microbiological research community for global access to genetic resources with the regulatory demands of provider countries under the Convention on Biological Diversity of 1992. To this end, the Task Force recommended that the public culture collections begin immediately to draft Material Transfer Agreements covering exchanges of their *ex situ* holdings, with a view to meeting this challenge on a regional, and eventually, global basis.[178]

Use of MTAs can greatly facilitate the tracking and control of microbial genetic resources and helps to enforce biosafety and security laws.[179] They can also enable the public culture collections to record the specific rights and obligations of providers as well as users for purposes of compliance with the CBD. Such records should be particularly important for industries that will eventually seek to patent microbial-related inventions derived from upstream research at universities and other scientific entities.[180]

What the Task Force failed to anticipate, however, was that the proprietary pressures described in Chapters 2 and 3 would induce a growing number of public culture collections around the world to draft Material Transfer Agreements that were inspired more by the prospects of commercial payoffs than by the needs of the scientific research community. In keeping with this trend, both university technology transfer offices and culture collection managers have drafted MTAs tending to impose ever more restrictive conditions that can hinder research, spawn high transaction costs, and even defeat the research goals of the public collections as originally constituted.

While there are a number of variations on this theme, and most public culture collections still seek to preserve some space for not-for-profit research,[181] the standard-form MTA of the American Type Culture Collection (ATCC) for exchanges in connection with not-for-profit research uses is particularly instructive

[177] *See* OECD Report on BRCs, above n. 5; NAT'L RESEARCH COUNCIL, A NEW BIOLOGY FOR THE 21ST CENTURY (Nat'l Acad. Press 2009), discussed above in Chapter 1, Section II.D.

[178] *See* OECD Report on BRCS, above n. 5.

[179] *See, e.g.*, Dagmar Fritze, *A Common Basis for Facilitated Legitimate Exchange of Biological Materials Proposed by the European Culture Collections' Organization* (ECCO), 4 INT'L J. COMMONS 507 (2010), *available at* http://www.thecommonsjournal.org.

[180] *See* Chapter 5, Sections II.C.3 & III below.

[181] *See, e.g.*, Belgian Coordinated Collections of Microorganisms (BCCM), *General Conditions of Material Transfer*, Jan. 2007 [hereinafter BCCM MTA] *available at* http://bccm.belspo.be/services/bccm_mta.pdf.

in this regard. Because ATCC is one of the largest and most technically advanced culture collections in the world,[182] its controversial practices merit careful attention.

A. *The Advent of a Proprietary Model in Response to Government Neglect in the United States*

ATCC was originally founded in 1925, when it took responsibility for the Winslow Collection at the Museum of Natural History. That collection had constituted the first major response to an earlier call from the Society of American Bacteriologists for the formation of a public culture collection to serve the needs of microbiology.[183] Less than a century later, ATCC had become one of the world's most important suppliers of high-quality, standardized reference materials, including microbes, cell lines, and derived materials, such as DNA and related products.[184] It has also radically changed its business model over time from the public-good approach, which characterizes most of the publicly funded culture collections elsewhere, to that of a highly proprietarial business enterprise that recoups all its operating costs plus a sizeable profit, which it claims to reinvest in improvements to its own infrastructure.[185]

ATCC has in fact developed new technologies to improve delivery of a full range of services that includes acquisition, authentication, preservation, production, development, and distribution of relevant microbial genetic resources.[186] Unlike most of the public culture collections in the rest of the world, which remain largely government-funded initiatives, ATCC's government subsidies were terminated at a critical point in its history. Today, ATCC no longer receives any dedicated government support for any of its collection activities,[187] although some fifteen per

[182] *See* below Section II.A. The American Type Culture Collection is one of the largest culture collection in the world, and it operates on a nominally not-for-profit basis. *See* Am. Type Culture Collection (ATCC), *About ATCC*, ATCC, http://www.atcc.org/en/About/About_ATCC.aspx (last accessed 3 July 2014).

[183] Simione, above n. 103.

[184] *Id.* at 64. ATCC reportedly holds some 70,000 specimens of the approximately 1,754,290 specimens held collectively by all Members of the WFCC. *See Who We Are*, ATCC, http://www.atcc.org/en/About/About_ATCC/Who_We_Are.aspx (last accessed 3 July 2014).

[185] Simione, above n. 103, at 64–65. Technically, ATCC is one of a few "private non-profit collections." Brian J. Tindall, Paul De Vos, & Hans G. Trüper, Judicial Commission of the International Committee on Systematics of Prokaryotes, XIth International (IUMS) Congress of Bacteriology and Applied Microbiology: Minutes of the meetings, 23, 24 and 27 July 2005, San Francisco, CA, USA.

[186] Simione, above n. 103, at 65.

[187] *Id.* at 64 (stating that "[w]e receive no government subsidy for our collections, and all of our financial resources are generated either through the distribution of the cultures and related products, or from other activities in biological materials management.") However, ATCC also has a major division that "manages government contracts and commercial contracts," which generate additional revenue for support of the collections." *Id.*

cent of its core funding flows from direct government grants.[188] ATCC thus competes with the large culture collections held internally by various government agencies in the United States, on the one hand, and with the externally funded WFCC member collections mentioned above, on the other.[189]

Deprived of government funding, ATCC nearly went bankrupt in 1973, when it received temporary support from the government followed by a number of time-limited grants. By the early 1990s, ATCC found itself unable to maintain its facilities or to upgrade its data management capacity to meet the needs of molecular biology.[190] New management then opted out of the public culture collection model and became a self-supporting "private nonprofit collection."[191]

In practice, ATCC embraced a market-driven model rooted in the assertion of ownership rights to the contents of its repositories; in contractually asserted restrictions on access and use of these materials; and in intellectual property and contractual claims to future commercial applications derived from the resources it makes available.[192] ATCC also developed the pay-per-use model for its services that has enabled it to reverse its financial situation, while reinvesting substantial amounts in upgrading the quality and variety of services it offers.[193]

The end result is that ATCC, while technically remaining a "non-profit" entity for tax purposes, has become "a knowledge and technology transfer broker between research and commercial entities."[194] In so doing, it claims to maintain an "uncompromising commitment to quality standards, biosafety, biosecurity, and regulatory compliance."[195] ATCC thus prides itself on having endeavored to ensure its own continuity and on the preservation of the biological materials under its care, unlike many other nonprofit collections.[196]

However, that characterization harbors a self-serving ambiguity. Most endangered culture collection are not WFCC members, but are instead research collections held at universities that are put at risk when the responsible academics retire. The WFCC actually seeks to rescue and preserve those endangered collections, at least when their scientific value justifies the effort.[197] ATCC does not perform this function.

[188] STERN, above n. 7.
[189] *See* above Section I.C.1.
[190] Tindall, De Vos & Trüper, above n. 185. (ATCC claims it ploughs its profits back into services and infrastructure); Simione, above n. 103, at 64.
[191] *Who We Are*, ATCC, above n. 184.
[192] Simione, above n. 103, at 64–65.
[193] *Id.* at 65.
[194] *Id.*
[195] *Id.* at 64.
[196] *Id.*
[197] *See* Chapter 9, Section II.B.1 below (noting WFCC's permanent endangered collections committee).

Moreover, the tradeoff implicit in ATCC's successful proprietarial model is that it has adopted some of the most restrictive access and use conditions in its standard MTA to be found anywhere in the world of culture collections. These practices are widely perceived to be inconsistent with, or hostile to, the needs of public scientific research. They are also spreading to other culture collections in different countries at the expense of the public-goods approach that had been the norm until recently, as will be seen below.

For example, the ATCC's standard MTA available to not-for-profit affiliates included the following prescriptions at the time of writing:

- use in a single laboratory only;
- no redistribution to other persons or entities, including other culture collections, without permission;
- limitations on derivatives, with a built-in reach through claim on any unmodified derivatives and progeny, including material used in modifications; and
- a duty to negotiate and obtain permission for each and every posterior transaction pending in connection with any given material, whether commercial or not, with an implicit reach through claim on all commercial applications.[198]

In the case of industry-sponsored academic research, the authorized use extends only to research carried out at the designated university and by its employees. Any use of the biological materials by the industry sponsor requires a separate ATCC license.

The ATCC's MTA thus restricts redistribution of materials by both single scientific users and by other culture collections. In other words, the general practice of sharing materials among qualified culture collections has been aborted, and all would-be users are forced to deal with ATCC.[199] One consequence of this restriction is that ATCC's charges are roughly twice those of most other collections.[200] Another is that the would-be researcher must negotiate permission for the specific research he or she wishes to undertake, and runs the palpable risk that such permission will

198 ATCC, Material Transfer Agreement, Nov. 15, 2011 [hereinafter ATCC MTA] *available at* http://www.atcc.org/~/media/PDFs/MTA_2.ashx.

199 ATCC claims that this restraint is necessary to minimize the number of times a culture is sub-cultured and to prevent nefarious uses of their specimens. *See* Simione, above n. 103, at 65. But other WFCC members maintain high quality and guard against unauthorized uses without disrupting the global system of exchanges that the WFCC supports.

200 This is an approximate number based on an analysis of the prices on the ATCC website. However, in many cases the price difference is much higher. General bacterial strains from the public BCCM collection, for example are charged between 60 and 70 USD, while a lot of identical strains are charged around 200/250 USD at ATCC. For an identical freeze-dried ampoule of "Sphingomonas trueperi Kämpfewr" (a type strain from a microbe isolated from U.S. soil), different prices are charged – in the ATCC 12417 (427 euros when ordered in the EU), DSM 7225 (65 euro), LMG 2142 (54 euro), and NCIBM 9391 (95 pounds).

be conditioned on negotiating a reach-through agreement with ATCC that makes them a de facto partner in all future commercial exploitations.[201]

These restrictions apply to all "biological materials," which besides ATCC materials (in the form in which they were deposited), include "progeny, unmodified derivatives and any unmodified derivatives within modifications, either individually or jointly." For this purpose, modifications mean "substances created by Purchaser which contain and/or incorporate a significant or substantial portion of ATCC material." Progeny means "an unmodified descendant from the ATCC materials, such as virus from virus, cell from cell, or organism from organism." Unmodified derivatives are "substances created by Purchaser that constitute an unmodified, functional sub-unit or product not changed in form or character and expressed by the ATCC material provided by ATCC."[202]

ATCC's MTA categorically affirms its "ownership" of the materials deposited in and distributed from its collection,[203] unlike most other public collections, which prefer to define themselves as "custodians" of biological resources in the public interest[204] Any recipient that violates any of the terms of its MTA, for example, by distributing type strains received from ATCC to a third party or using them in a different laboratory from the one stipulated in the contract, is thus acting illegally and risking a lawsuit. Moreover, the ATCC's ownership claims attach to any material used in the making of a modification, with a view to obtaining joint ownership of the end result.[205]

ATCC's ownership policy begs a number of important questions. If the deposits were made by collectors who initially took the samples from CBD countries without permission after 1992, they are arguably subject to access and benefit-sharing claims from members of the CBD even though the United States has not ratified that Convention. By the same token, future distribution of these same materials across borders, without conforming to the CBD's access and benefit sharing provisions, will raise serious questions of violating the international regime of misappropriation

201 In so doing, ATTC's practices thus approximate those of the developing countries under the bilateral approach of the CBD. *See, e.g.*, above Chapter 2, Section II.A.2; Chapter 3, Section I.B; below Section II.B. How flexible these conditions are in practice was not known at the time of writing.

202 ATCC MTA, above n. 198, at 1. "Unmodified Derivatives include, but are not limited to, subclones of unmodified cell lines, purified or fractional subsets of materials provided by ATCC, proteins expressed by DNA/RNA supplied by ATCC, or monoclonal antibodies secreted by a hybridome cell line."

203 With the exception of samples deposited by the National Park Service in a so-called "special collection," *Special Collections*, ATCC, http://www.atcc.org/Products/Collections/Special%20Collections.aspx (last accessed 3 July 2014), ATCC's MTA states that the collection and/or its contributors retain ownership of all rights to the original material included in modifications, and it in effect asserts a joint ownership claim to such modifications. ATCC MTA, above n. 198.

204 *See, e.g.*, Fritze (2008), above n. 31; Fritze (2010), above n. 179.

205 *See* above nn. 201–202.

codified in the Nagoya Protocol of 2010, whether or not the U.S. adheres to that Protocol.[206]

In fairness, the ATCC's proprietary approach has enabled it to maintain very high quality standards, including a large storage capacity and related services. It has also enabled ATCC to remain largely self-sufficient and financially sustainable over time, which was not the case when it was totally dependent on government funds.[207] Nevertheless, this model, which applies both to not-for-profit and for-profit users, encumbers the formal system of material exchanges with high transaction costs and burdensome restrictions on research that break with the traditional sharing practices of the public culture collections. By operating outside of the CBD, given its location in the United States (the only major country not to ratify that convention), the ATCC also indirectly incites developing-country suppliers of microbial genetic resources to adopt similar practices while generating mistrust in those same countries that undermines the international system of cooperative exchanges as a whole.

B. *Diffusion of a More Proprietary Approach to Other Public Culture Collections*

The ATCC model reflects the growing awareness that upstream research on microbial genetic resources may lead to commercially valuable downstream applications, and the now universal temptation of upstream providers to seek a share of the proceeds from such applications. Faced with the need to self-finance preservation activities that the government had previously funded, ATCC logically looked to the possibility of sharing in future commercial exploitation of its genetic resources as a medium- and long-term basis of sustainability. ATCC's success then made other public and private culture collections more aware of possible market-driven means of bolstering their financial resources over and above the cost-recovery fees charged for distributing specimens to industry and research institutions.

In principle, the WFCC culture collections have generally operated under a public-good model that emphasized service to the research community rather than generating revenues. Aspiring to share their microbial resources with both domestic and foreign researchers as part of the common heritage of mankind, their primary goal was to serve the needs of scientists in preparing and validating their publications.[208] The public collections also sought to support research conducted at hospitals and in the private sector.

206 *See* above Chapter 3, Section IV.C.

207 *See* Simione, above n. 103, at 63.

208 *See, e.g.*, Fritze (2008), above n. 31. The collections did want attribution for their services, and this still remains a concern.

How fully this sharing ethos was actually implemented in practice, however, varied considerably from collection to collection,[209] even before the 1980s, when they began to employ formal MTAs more extensively.[210] One should also recall that the WFCC members hold only a small fraction of the total number of microbial genetic resources governed by non-WFCC collections around the world, including universities, hospitals, and industry, and that it is difficult to make generalized statements about the sharing practices of these other institutions before the 1980s.

Given their public-good aspirations, the WFCC collections largely depended on government funding and not on generating revenues of their own. Their parastatal legal status often limits what they can charge for their services, although some collections not so limited do charge relatively high fees.[211] Nevertheless, even the provision of pure public goods necessarily remains contingent on the willingness and capacity of governments to provide adequate funds, a condition that can easily vary over time. As constraints on government resources mount, some of the public culture collections would have been tempted to consider and – to varying degrees imitate – the propertizing practices of one of the world's largest and most qualified suppliers of microbial genetic resources, even if there had been no comparable pressures emanating from the provider countries that promoted the Convention on Biological Diversity.

As we have seen, the developing countries under the aegis of the CBD were themselves translating its assertion of sovereignty over national genetic resources into restrictions on access and use that – consciously or not – paralleled those of the ATCC model, for largely the same reasons. Although the culture collections in the developing world are almost entirely funded as public goods by governments, these same governments logically began to view both their *ex situ* collections and *in situ* genetic resources as a potential means of rebalancing outward trade flows to intellectual property rights holders in OECD countries under the TRIPS Agreement of 1994. By controlling access to and use of local genetic resources under the ABS provisions of the CBD, and thereby securing a share of the returns from downstream applications, provider governments could at least hope to secure

209 Even a cursory glance at the roster of WFCC members suggests that an inherent tension likely exists between open access norms and proprietary interests among the participating members.

210 *See, e.g.*, Rebecca Eisenberg, *Bargaining Over the Transfer of Proprietary Research Tools, in* EXPANDING THE BOUNDARIES OF INTELLECTUAL PROPERTY (R. Dreyfuss, D. Zimmerman & H. First eds., Oxford Univ. P., 2001); Victor Rodriguez, *Governance of Material Transfer Agreements*, 30(2) *Tech. in Soc'y* 122–128 (2008). This tension has been growing over time in response to the commoditizing pressures described herein.

211 *See, e.g., Microbial Services*, Ctr. for Agric. Biosci. Int'l (CABI), Jan. 1, 2012, http://www.cabi.org/default.aspx?site=170&page=4456; *see also e.g.*, ATCC, *Deposit Services*, ATCC'http://www.atcc.org/en/Services/Deposit_Services.aspx (last accessed 3 July 2014).

funds needed to support the costs of maintaining their own *ex situ* and *in situ* biological resources.[212]

Given these proprietary trends on both sides of the development divide, it is hardly surprising that microbial materials provided by public culture collections were subjected to a growing number of restrictions on use that have potentially complicated public research endeavors. Many of the governmental and not-for-profit WFCC members have thus abandoned previous policies facilitating exchanges of materials with minimum conditions on use and reuse and moved to more formal relationships consistent with both the applicable statutory regimes and the evolving MTAs governing these matters at home and abroad.[213]

As a result, even in the pre-Nagoya period, most MTAs covering the exchange of microbial materials specified that only non-commercial research was allowed,[214] with ever more detailed and diverse provisions that tried to pin down the elusive distinction between what is and is not potentially commercial.[215] Efforts to regulate modifications and derivatives are growing in number, with a tendency to ensure that a provider collection can claim at least some benefits from, if not a form of joint ownership in, the resulting organism.[216] Every potential commercial use – however broadly defined – must then typically be negotiated with the collection in advance, which conjures up visions of mounting transaction costs, and legal fees, as well as all the problems familiar from nondisclosure agreements for the presentation of unpatented ideas to would be financiers.

This duty to negotiate upfront permission for potential commercial uses with the provider collection intrinsically begs the question of that collection's right to make the resources available for commercial uses in the first place. While a number of collections assert ownership over their materials in the manner of ATCC (operating in one of the last major non-CBD countries),[217] many other MTAs simply remain

[212] Supporting these maintenance costs in developing countries is an express foundational premise of the CBD. *See* Convention on Biological Diversity, *opened for signature* June 5, 1992, 1760 U.N.T.S. 79 [hereinafter CBD], Preamble, art. 1.

[213] Imitation of the ATCC model is also motivated by other collections seeking to address their potential relations with purely commercial clients as well as by mounting concerns to abide by the CBD. *Cf.* Comm. Resource Sharing in Biomedical Research, Instit. of Med., Resource Sharing in Biomedical Research (Nat'l Acad. Press 1996).

[214] *See e.g.*, Tom Dedeurwaerdere et al., *Global Scientific Research Commons Under the Nagoya Protocol – Governing Pools of Microbial Genetic Resources*, *in* Common Pools of Genetic Resources *(2013)*, above n. 145, at 224–245 (noting that the ECCO MTA, which was developed before the Nagoya Protocol, allows only for non-commercial use).

[215] In addition to the ATCC (U.S.), *see* MAS of BCCM (Belgium); CABI (U.K.); CCMM (Morocco); BTT Culture Collection (Finland); and Coleccion Española de Cultivos Tipo (CECT) (Spain).

[216] *See, e.g.*, the MTAs for the ATCC (U.S.); CCMM (Morocco); CSIRO (Australia); University of Koln (Germany); University of Cape Town (South Africa).

[217] *See, e.g.*, ATCC (U.S.); CSIRO (Australia); Czech Collection of Microorganisms (Czech Republic); INRA, Centre international des resources microbiennes (France); Kiln University (Germany); SAG

silent about who the ultimate beneficiary of negotiations with the collection for this purpose may be. Perhaps some MTAs that beg this question merely reflect the fact that the relevant institutions have not yet fully addressed this issue.

Still other collections expressly declare that ownership of the specimens to be made available remains with the depositor,[218] a provision that reflects the influence of the CBD. Pressures from developing countries under the CBD have obliged a growing number of public culture collections to reconsider the precise legal nature of their holdings and to better define their rights and duties with respect to depositors of the specimens in question. To this end, in 2009, the European Biological Resource Centers Network (EBRCN) recommended that its member collections more fully elaborate efforts to comply with the CBD in their standard MTAs.[219]

Even in the United States, which has not yet ratified the CBD, pressures on culture collections to comply with its terms have been mounting. For example, the Smithsonian Institution has reportedly recommended that the government-owned culture collections behave as if U.S. law had implemented the CBD.[220]

Provisions that directly or indirectly invoke obligations under the CBD and under the various national laws implementing its mandate are thus universally encountered.[221] Some MTAs, for example, invoke "the sovereign rights over genetic resources [that] remain with their country of origin" or even specify that "Prior Informed Consent" may be needed.[222] Many more require compliance with

(Germany); KACC/Genebank (Korea); Faculty of Medicine, Chiang Mai University (Thailand); and HPACC, Health Protection Agency Culture Collections (U.K.).

218 *See, e.g.*, Landcare Research Manaaki Whenua (New Zealand) (stipulating that depositor must be the rightful owner to claim ownership). Other collections declare that ownership is the property of the "provider" (but this is ambiguous at the moment).

219 European Biological Resource Center Network (EBRCN) Information Resource, Convention on Biological Diversity, Draft Sept. 2009, WFCC, at 2 [hereinafter EBRCN, *CBD*] ("Culture Collections must put in place mechanisms to comply with the provisions of the Convention … covering Collection, Use, Distribution").

220 Telephone interview with Kevin McCluskey, June 12, 2012.

221 For example, EBRCN suggests that culture collections can include a request for relevant information under the Acceptance Criteria set out in their Accession Forms, in the following manner:

> Under the CBD, a Culture Collection is obliged to ask senders if any of the material being sent in for deposit was collected after December 1993. If that is the case you need to inform us if prior informed consent was received to collect it … The sender also needs to establish whether they have the authority to deposit their collections in a Culture Collection and if there are conditions regarding third party supply described in the PIC.

EBRCN, CBD, above n. 219. EBRCN also recommended that its Members follow the Bonn Guidelines' voluntary code of practice for access and benefit sharing. *Id. See* Chapter 3, Section IV.A.

For other direct invocations of the CBD, *see, e.g.*, Culture Collection Micoteca URM (Brazil); UTT Culture Collection (Finland); Landcare Research Manaaki Whenua (New Zealand); CECT (Spain); CCAP (U.K.).

222 *See, e.g.*, CABI (U.K.); CCAP, National Botanical Garden of Belgium (U.K.).

applicable national and international laws, regulations or guidelines,[223] which implicitly invokes the Access and Benefit Sharing provisions, as well as the Prior Informed Consent principles, of the CBD.[224] Other MTAs, however, refer only to "compliance with national laws," which avoids the question of whether the national laws conform to the CBD and the Nagoya Protocol or not.[225] Still other collections used MTAs that made no reference whatsoever to any duties arising under the CBD.[226]

With specific regard to redistribution, a growing temptation to impose some form of reach-through claims on potential commercial uses of microbial specimens seems to have induced many MTA drafters to limit every foreseeable possibility of unauthorized redistribution of the material in question.[227] Sometimes, the clauses restricting redistribution go so far as to virtually indicate the room in which research must be carried out.[228] Still more disheartening was a tendency to restrict public culture collections from redistributing their microbial materials to other similar collections without express permission of the providers.[229] Such clauses directly conflict with the basic public-good service functions that characterized public culture collections, especially those affiliated with the WFCC.

The foregoing survey shows that, over time, commoditizing pressures, in combination with a shortage of public funds and assertions of sovereignty under the CBD, have weakened the commitment of many public culture collections to the sharing ethos that was their original *raison d'être*. This trend culminated with the project to unite some of the highest-quality collections in the Global Biological

[223] *See, e.g.*, BCCM (Belgium); CABI (U.K.); CCM (Czech Republic); INRA (France); CFBP (France); CCCRYO (Germany); Agricultural University of Athens (Greece); DBUPG (Italy); NITE (Japan); Riken BRC (Japan); Hut Culture Collection (Japan).

[224] Some MTAs actually refer to prior informed consent as such. *See, e.g.*, BCCM MTA, above n. 181.

[225] *See, e.g.*, KCLB, Korean Cell Line Bank; KCTC, Korean Collection for Type Cultures; KBPV, Korea Bank for Pathogenic Viruses; HPKTCC, *Heliobacter pylori* Korean Type Culture Collection; BIOTECH (Thailand); Faculty of Medicine Ching Mai University (Thailand); HPACC, Health Protection Agency Culture Collections (U.K.).

[226] *See* T. Dedeurwaerdere, A. Broggiato, & D. Manou, *Global Scientific Research Commons Under the Nagoya Protocol: Governing Pools of Microbial Genetic Resources in* COMMON POOLS OF GENETIC RESOURCES n. 145 at 224–245.

[227] *See id. See also* Sean O'Connor, *The Use of MTAs to Control Commercialization of Stem Cell Diagnostics and Therapeutics*, 21 *Berkeley Tech. L.J.* 1017 (2006) (case of WARF and WiCell); Note, Lisa Larrimore Ouellette, *Access to Bio-Knowledge: From Gene Patents to Biomedical Materials*, 2010 *Stanford Tech L. Rev.* N1, *available at* http://papers.ssrn.com/sol3/Delivery.cfm/SSRN_ID1575705_code1231602.pdf?abstractid=1431580&mirid=1.

[228] *See, e.g.*, University of Cape Town MTA (South Africa), http://www.rcips.uct.ac.za/usr/rcips/ip/Materials_Disclosure_Form.DOCX, which is used to draft an individual MTA.

[229] *See* O'Connor, above n. 227, at 1026. *See, e.g.*, the standard BCCM MTA, above n. 181, adopted in 2007. However, this MTA was recently replaced by a new pan-European model (ECCO), which reverses this trend, as discussed in Section III.A.2 below; *see also* ECCO MTA, above n. 115.

Resource Center Network, which – for a time at least – overtly planned to imitate the ATCC model, in conflict with the principles of the WFCC, from which it was a spinoff organization. This proposed network is discussed at length in Chapter 9.

III. THE RESEARCH COMMUNITY PUSHES BACK

The contradictions between the proprietarial MTAs discussed in the preceding section and the traditional service goals of the WFCC's affiliated culture collections did not go unobserved by leading members of the microbiological research community.[230] Most public culture collections view themselves as research institutions, and they are often integrated into the operational framework of specific universities. In this capacity, they were accustomed to a more open and flexible approach to the handling of microbial cultures,[231] one that was now challenged by both the universities' own concerns to benefit from downstream commercial applications and by the growing need to protect themselves from legal attacks rooted in the CBD and the Nagoya Protocol.[232]

At the same time, hundreds, and perhaps thousands of other culture collections around the world also operate within university departments, under the aegis of particular scientific investigators, without attempting to meet the standards of the public culture collections affiliated with the WFCC. Traditionally, scientists holding such research collections under conditions of actual or legal secrecy were accustomed to exchanging microbial materials among themselves, on an informal basis, in the expectation that other, similar laboratories would reciprocate in the future. Such exchanges were typically undertaken without MTAs, much as many public culture collections themselves used to exchange microbial materials without formal MTAs before the 1980s.[233]

Beginning in that same period, however, the commoditizing pressures that drove many public culture collections to adopt MTAs with ever more restrictive conditions on access to and use of biological materials struck unaffiliated research collections with perhaps even greater force. On the one hand, university administrators, like those of the public culture collections, had increasingly to take account of pressures from governments concerning the anti-biopiracy movement that culminated in the CBD of 1992. On the other hand, enactment of the Bayh Dole Act of 1980 in the U.S.,[234]

[230] *See, e.g.*, O'Connor, above n. 227; Ouellette, above n. 227.

[231] Fritze (2010), n. 179, at 521.

[232] *Id.* at 513–14.

[233] *See, e.g.*, Fritze (2010), above n. 179.

[234] Patent and Trademark Law Amendments Act (Bayh-Dole Act), 35 U.S.C. § 200 (1980). *See generally* R. Nelson et al., Ivory Tower and Industrial Innovation: University-Industry Technology Transfer Before and After the Bayh-Dole Act (Stanford Bus. Books 2004).

and exaggerated reports of its success,[235] induced universities everywhere to erect legal fences around their research assets in the hopes of sharing in the benefits of future downstream commercial applications.[236] Technology transfer offices were soon established for this purpose at universities around the world, with the result that private or semi-private research collections were gradually covered by ever more restrictive, ad hoc MTAs that made exchanges among microbiologists time consuming, burdensome, and subject to suffocating bureaucratic legal constraints.[237]

Despite these pressures – or indeed, because of them – both research scientists and their informal culture collections sometimes continued to ignore formal MTAs and have reportedly developed an alternative method of exchange that bypasses their restrictions.[238] To this end, single laboratories or research units informally have exchanged biological resources among themselves for public research purposes, on the basis of mutual trust and reciprocally recognized quality controls, without entering into any formal legal undertakings whatsoever. In effect, this informal network, which reportedly accounted for approximately 60 percent of all microbial materials exchanged in 2008,[239] converts the private goods of the single participants into a type of "club goods" available to trusted members. The informal system has also tried to maintain the original sharing norms of the WFCC Federation.[240]

However, the bulk of the research collections participating in this informal system of material exchanges were not subject to WFCC standards at all.[241] On the contrary, these collections, usually the results of particular research projects, tend to be relatively disorganized. They operate without the quality controls, tracking mechanisms, and validation/authentication practices of the formally organized public collections, and they rarely, if ever, publish catalogs of their holdings.

[235] *See, e.g.*, Anthony So et al., *Is Bayh-Dole Good for Developing Countries? Lessons from the U.S. Experience*, 6 *PLoS Biology* 2078–84 (2008) (questioning the extent of successes claimed for Bayh-Dole).

[236] For progeny of Bayh-Dole even in India and South Africa, see, e.g., India, Utilization of Public Funded Intellectual Property Bill, Bill No. LXVI (2008), *available at* http://www.prsindia.org/uploads/media/1229425658/1229425658_The_Protection_and_Utilisation_of_Public_Funded_Intellectual_Property_Bill__2008.pdf (India); Republic of South Africa, Intellectual Property Rights from Publicly Financed Research and Development Bill, Bill No. B46B-2008 (2008), *available at* http://www.pmg.org.za/bill/20080815-intellectual-property-rights-publicly-financed-research-and-developmen-0.

[237] *See, e.g.*, Peter Lee, *Contracting to Preserve Open Science: Lessons for a Microbial Research Commons*, *in* Designing the Microbial Research Commons, above n. 11, at 69–76.

[238] Dedeurwaerdere, Broggiato, Loufi, Welch & Batur, above n. 9.

[239] *See* Per M. Stromberg et al., Ex situ *collections for microbial research: The contribution of public networks for search tools* (2008) (survey report) (on file with authors).

[240] *Id.*

[241] Stromberg, Dedeurwaerdere & Pascual, above n. 10.

A. *Efforts to Negotiate More Research Friendly Material Transfer Agreements*

Needless to say, this entrenched system of informal exchanges among research collections at universities in different parts of the world has ignored the growing network of international laws and regulations governing exchanges of microbial materials for research purposes. By the same token, the dependence of continued scientific progress on such informal methods of exchange has itself spawned vigorous efforts to negotiate more research friendly MTAs than those surveyed in the preceding pages of this chapter. These efforts and their implications are briefly surveyed below.

1. The Uniform Biological Material Transfer Agreement in the United States and Its Progeny

Over time, American universities became more aware of the problems that restrictive MTAs had been creating for the life sciences. A major attempt to alleviate some of these obstacles in the United States was the Uniform Biological Material Transfer Agreement (UBMTA) of 1995,[242] which garnered some 250 signatory institutions.[243]

The UBMTA leaned heavily on a sharp distinction between commercial and non-commercial research, plus some elevated language invoking the public-good mission of research universities. To make it more workable in practice, the Science Commons component of the Creative Commons movement devised a complementary project known as the Biological Materials Transfer Project.[244] This initiative, later abandoned, aimed to facilitate use of the UBMTA by providing a standardized toolkit for "listing, searching, contracting and tracking [the] downstream impact" of exchanged materials.[245] Funding agencies, such as the U.S. National Institutes of Health (NIH) and comparable institutions in Europe, have also exerted considerable pressure on grantees to avoid allowing patents to obstruct research uses of biological materials in certain cases.[246]

[242] Nat'l Instit. Health (NIH), *Uniform Biological Material Transfer Agreement*, March 8, 1995, *available at* http://www.autm.net/AM/Template.cfm?Section=Technology_Transfer_Resources&Template=/CM/ContentDisplay.cfmContentID=1405 [hereinafter UMBTA].

[243] *See* Assn. Univ. Tech. Managers (AUTM), In the Public Interest: Nine Points to Consider in Licensing University Technology, AUTM, March 6, 2007, *available at* http://www.autm.net/Nine_Points_to_Consider1.htm/ [hereinafter Nine Points to Consider].

[244] *See, e.g.*, Robert Cook-Deegan et al., *The Dangers of Diagnostic Monopolies*, 458 *Nature* 405, 405 (2009); O'Connor, above n. 226 (case of WARF and WiCell).

[245] Thinh Nguyen, *Science Commons: Material Transfer Agreement Project*, 2(3) *Innovations* 137–143 (Summer). 2007, at 141. Science Commons also provided a set of MTAs for transfers from universities to industry. *Id.*

[246] Nguyen, above n. 245.

In practice, however, the UBMTA was widely deemed a failure, as numerous university technology transfer offices ignored it in favor of their own, tailor-made, and highly proprietary licensing agreements.[247] Its deterrent effect on so-called "black market" or informal exchanges among cooperating research collections was never demonstrated and often questioned.

As a result, some of the leading research universities in the United States, whose past practices had largely undermined the Uniform Biological Material Transfer Agreement, adopted a new set of licensing guidelines in 2007, entitled In the Public Interest: Nine Points to Consider in Licensing University Technology.[248] This document states that, when licensing their technologies, universities should generally "reserve the right to practice licensed inventions and to allow other non-profit and governmental organizations to do so."[249] Under this principle, universities would specifically reserve the right to transfer both tangible biological and other research materials and intangible outputs (such as computer software, databases and know-how) to others in the non-profit and governmental sectors on preferential terms.[250]

Other recommendations in the Nine Points document would have university technology transfer agreements avoid reach-through clauses that restricted future improvements[251] and generally employ non-exclusive licenses to enable broad access to research tools.[252] Preferential access to technologies that addressed unmet needs of the developing countries was also stressed in Point 9.[253]

If implemented, these proposals could reduce some restrictions on research uses of biological materials generally and help to reinforce the sharing ethos. Whether these principles will be effectively implemented or not remains to be seen. For example, the National Research Council found it necessary, in 2010, to remind university technology transfer offices to avoid impediments to uses of biological materials for research purposes in their licensing practices.[254]

Moreover, none of these initiatives refer to, or otherwise deal directly with, the implications of the Convention on Biological Diversity for microbial research

[247] NINE POINTS TO CONSIDER, above n. 243. However, many universities in Europe have reportedly used the UMBTA.

[248] *Id.* (explicitly acknowledging the failure of the UBMTA and seeking "to reduce some restrictions on use of biological materials and help to reinforce the sharing ethos").

[249] *Id.* at 2 (point 1).

[250] *Id. See also* A.B. Bennett, W.D. Streitz, & R.A. Gacel, *Specific Issues with Material Transfer Agreements, in* INTELLECTUAL PROPERTY MANAGEMENT IN HEALTH AND AGRICULTURAL INNOVATION (A. Krattiger et al. eds. 2007).

[251] NINE POINTS TO CONSIDER, above n. 243, at 4 (point 3).

[252] *Id.* at 5 (point 5).

[253] *Id.* at 8 (point 9).

[254] *See* NAT'L RES. COUNCIL (NRC), MANAGING UNIVERSITY INTELLECTUAL PROPERTY IN THE PUBLIC INTEREST (2010), http://www.nap.edu/catalog.php?record_id=13001.

collections held at universities in the United States, despite the growing complaints about biopiracy by university professors discussed earlier in this volume.[255] Ostensibly, this omission arises from the failure of the United States to ratify the CBD and, accordingly, the lack of federal laws or regulations to implement it. As pointed out earlier however, the wisdom of continuing to operate under such a parochial outlook in the face of the Nagoya Protocol seems questionable.[256]

2. The Core MTA of the European Union Culture Collections' Organization

In contrast, the implications of the CBD as a challenge to their public-good mission became a major concern of many of the WFCC affiliates in Europe, especially the 61 collections from 22 countries that belonged to the European Union Culture Collections Organization (ECCO).[257] While acknowledging that public microbial culture collections had to "face the problem of illegitimate or dishonest users,"[258] ECCO's representatives decided that it was necessary to devise a core MTA that would harmonize the different regulations used by their members, with a view to the historical goal of promoting the exchange of biological materials between collections for research purposes.[259]

ECCO's drafting committee, formed in 2005, understood that the CBD "had altered the way in which individuals, scientific institutions and private companies were entitled to access genetic resources," and that unless the collections devised "new and flexible approaches to adapt to recent developments," there were serious risks that the established process of exchanging microbial genetic resources would be impeded or disrupted.[260] Building on earlier WFCC initiatives[261] as well as on the

[255] *See* above Chapter 3, Section I.A.

[256] *See* above Chapter 3, Section IV.

[257] *See* ECCO, *Homepage*, www.eccosite.org (last accessed 5 July 2014). Founded in 1982 to promote regional collaboration among participating microbial culture collections, ECCO's members provide professional public service on demand and without discrimination, accept cultures for deposit, provide catalogues, and are housed in countries with microbiological societies affiliated with the Federation of European Microbiological Societies and registered with the WFCC. Davis, Fontes & Marinoni (2013), above n. 1, at 43, 43–50.

[258] Fritze (2010), above n. 179, at 513.

[259] *Id.* at 508.

[260] *Id.* at 511. However, there was initially no unanimous agreement that such a separate document was needed, partly because, as late as 2005, even the term "MTA" was still relatively new, although many of the records that collections had previously kept would have met the definition of an MTA. *Id.* at 515.

[261] *See, e.g.*, WFCC Info. Doc., Access to ex-situ Microbial Genetic Resources within the Framework of the Convention on Biological Diversity, background document to the UNEP/CBD/COP/3/Int.19 (1996), http://wdcm.nig.ac.jp/wfcc/Info.Doc.html. *See also* the BCCM MTA, above n. 181, and in use until December 2006, which attempted to institute a public research exemption, even if it stopped short of fully implementing it. In this standard contract, further distribution of material by

MOSAICC Code of Conduct in 2001,[262] the ECCO working group produced a core MTA in 2009 to cover key issues in need of greater uniformity,[263] while leaving the single collections room to add their own specific requirements.[264]

A broad segment of this standard MTA is devoted to biosafety and biosecurity aspects, with various clauses that emphasize the responsibility of recipients in this regard.[265] Quality management is also emphasized in Article 10, which requires the collections to certify "that the material shall be viable and pure upon shipment from the [relevant] Collection."[266]

Beyond these operational standards, the ECCO Core MTA focused mainly on ways to comply with both the spirit and norms of the CBD, while respecting the needs of scientific research. It first defines "recipients" as both end users and "intermediaries," who place an order on behalf of the "end user" and to which the collection addresses the material.[267] Both end users and intermediaries, such as wholesalers, importers, or other agents unrelated to the end-user institution, as well

organizations that acquire microbial organisms from a culture collection is not allowed, except under some very specific exceptions, including distribution of modified material (that is having acquired new properties as a result of value adding research) for research and education purposes only).

The new standard MTA of the BCCM, adopted in January 2007, adopted a different approach to these exceptions (*cf.* document available on line at http://bccm.belspo.be/services/bccm_mta.php (last accessed 3 July 2014)). On this approach, recipient organizations are not allowed to sell, lease, license, lend, supply, distribute or otherwise transfer the microbial material to any others, save those involved in so-called legitimate exchanges. The contract distinguishes between two categories of legitimate exchange. The first category designates the transfer of material within the Research Group, which are entitled scientists working in the same laboratory, or contractually bound to work on the same research topic, for non-commercial purposes. The second category designates the transfer of material between named culture collections/biological resources centers for accession purposes, provided that further distribution by the receiving culture collections/biological resources center is under MTA provisions compatible with and equivalent to those in place at the supplying collection. This approach has now also been adopted in the model MTA of the European Union Culture Collection Organization (ECCO), agreed upon at their board meeting on June 11, 2008 in Ghent, Belgium.

262 Belgian Fed. Sci. Pol'y Office Microorganism Sustainable Use and Access Regulation International Code of Conduct (MOSAICC) (2011), www.belspo.be/bccm/mosaicc/index.htm.

263 ECCO MTA, above n. 115.

264 *Id.* at 10. This document was approved by 61 member collections of ECCO in February 2009. *Introduction: The ECCO core Material Transfer Agreement for the supply of samples of biological material from the public collection*, http://www.eccosite.org/MTA_core.html (last accessed 3 July 2014).

265 *See, e.g.*, ECCO MTA, above n. 115, art. 3 (requiring recipients to agree that the handling of material "will be conducted under their responsibility and in compliance with all applicable laws and regulations); art. 4 (specifying particular safety precautions).

266 *Id.* art. 10; Fritze (2010), above n. 179, at 524.

267 ECCO MTA, above n. 115, definitions of end user, intermediary, recipient. The definition of "material" includes "original material, progeny, and unmodified derivatives, but not modifications, i.e., substances produced by the recipient by using the material (which are not the original material supplied by the depositor, progeny (i.e., subcultures or replicates), or unmodified derivatives) and that have new properties. Modifications include, but are not limited to, recombinant DNA clones. *Id.*

as recipient culture collections themselves are deemed to "accept . . . the terms and conditions of this material transfer agreement by placing an order with the collection."[268]

The traceability of samples of biological materials is then addressed by an obligation on intermediaries to forward to end users "the present MTA and the material in unchanged form and quantity as received from the collection" and to use for this shipping purpose "the proper packaging, a trained shipper, and an authorized carrier" under applicable laws and regulations.[269] The principle is that transparency of movement between source country and end user is central to the requirements of the CBD, for purposes of "safeguarding individual collections as well as for securing countries' access and benefit sharing (ABS) rights."[270] The unique identifiers that WFCC members apply to all microbial specimens in their collections further secure implementation of this principle.[271]

ECCO's Core MTA then specifies that recipients "may use the material in any lawful manner for non-commercial purposes."[272] Uses of the material or modifications thereof for commercial purposes obliges the recipient to negotiate in advance and in good faith "the terms of any benefit sharing with the appropriate authority in the country of origin of the material, as indicated by the collection's documentation."[273]

These paragraphs added little to prior practice, partly in view of an acknowledged possible future need to adopt further provisions in light of the COP 10 (Japan, 2010),[274] which would produce the Nagoya Protocol, discussed above in Chapter 3.[275] What the ECCO Core MTA does add is a clear statement to the effect that nothing in its terms "grants recipient[s] any rights under any patents, proprietary intellectual property, or other rights with respect to the Material" conveyed.[276]

The point here is to inform recipients of biological material "that they have no immediate rights to the received material other than working with it."[277] This term

[268] *Id.* Legend at the end of the definitions.

[269] ECCO MTA, above n. 115, Definition C, Recipient.

[270] Fritze (2010), above n. 179, at 522.

[271] *See* above Section I.A.2.

[272] ECCO MTA, above n. 115, art. 6.

[273] *Id.* art. 7. "Commercial Purpose" is defined as "The use of the material for the purpose of profit." *Id.*, Definition l. In principle, the ECCO MTA does not require the collection to be involved in the benefit-sharing negotiation. *See* ECCO MTA, above n. 115. *Cf.* EBRCN, CBD, above n. 219, which recommended that the relevant culture collection should refer would-be commercial users to the proper authority in the country of origin. Alternatively, the EBRCN apparently authorized the collection to negotiate a specific agreement based on then applicable Bonn Guidelines. *Id. See* Sixth Meeting of the Conference of the Parties to the Convention on Biological Diversity, The Hague, Neth., 17–19 April 2002, Bonn Guidelines on Access to Genetic Resources and Equitable Sharing of the Benefits Arising out of their Utilization, U.N. Doc. UNEP/CBD/COP/6/20, Annex 2 (27 May 2002) [hereinafter Bonn Guidelines].

[274] Fritze (2010), above n. 179, at 522.

[275] *See* above Chapter 3, Section IV.

[276] ECCO MTA, above n. 115, art. 6.

[277] Fritze (2010), above n. 179, at 523.

is thus consistent with the view that any collection using ECCO's Core MTA operates as a "custodian" of the biological resources deposited in that collection, within the limits of national and international law, and not as an "owner" of such material.[278] Hence, would-be commercial users must negotiate with "the appropriate authority in the country of origin," and not with the collections as proprietors of those same materials.[279] In other words, ECCO's MTA envisions the role of the public culture collections as that of intermediaries between providers and users of microbial genetic resources who shift the burden of compliance with ABS obligations under the CBD to end users.[280]

Here, moreover, ECCO's Core MTA endeavors to ensure that biological materials will be widely exchanged for research purposes, and not remain subject to the tight restrictions on redistribution of many MTAs discussed in the preceding section.[281] To this end, Article 5 restricts further redistribution of the material except to intermediaries (i.e., wholesalers, importers and other agents[282]) and to other recipients "involved in legitimate exchanges as defined above."[283] The novel concept of "legitimate exchange" is then further defined as follows:

> The transfer of the material between scientists working in the same laboratory, or between partners in different institutions collaborating on a defined joint project, for non-commercial purposes; this also includes the transfer of material between public culture collections/BRCs for accession purposes, provided that the further distribution by the receiving collection/BRC is under MTA conditions equivalent and compatible to those in place at the supplying collection.[284]

The Core MTA thus envisions a network of qualified culture collections operating under a common viral license, which may freely exchange microbial materials among themselves for research purposes, as well as complementary networks of qualified scientific collaborators also governed by similar viral contractual conditions.[285]

[278] *Cf.* Fritze (2010), above n. 179, at 518.

[279] ECCO MTA, above n. 115, art. 6.

[280] *See* Christine Godt, *Networks of* Ex Situ *Collections of Genetic Resources*, *in* COMMON POOLS OF GENETIC RESOURCES (2013), above n. 145, at 258 [hereinafter Godt (2013)].

[281] *See also* O'Connor, above n. 227; Ouellette, above n. 227.

[282] ECCO MTA, above n. 115, art. 5. EBRCN, *CBD*, above n. 219.

> It is agreed that ownership will not be claimed over the genetic resources received from the Culture Collection, nor to seek intellectual property rights over them or related information. In case of commercial utilization or exploitation of these resources, suitable and adequate recompense as required by the Convention on Biological Diversity will first be discussed with the Culture Collection and/or country of origin.

[283] *See* ECCO MTA, above n. 115, art. 5; *id.*, Definition E.

[284] *Id.*, Definition M.

[285] *Cf.* EBRCN, CBD, above n. 219 (recommending that the collections adopt a viral license approach by requiring recipients to agree "that the genetic resources concerned are not distributed outside their own organization without written consent of the Culture Collection. In that case the recipient will [be] bound by the same provision."

This approach particularly facilitates exchanges and research collaboration across national and institutional boundaries within Europe.

The ECCO Core MTA expressly strives to strike a balance between concerns to avoid illegitimate or unlicensed duplication of microbial material by third parties[286] and its "main goal," namely, "the legitimate exchange of biological material for research and application."[287] ECCO thus views its Core MTA as part of the larger efforts of the WFCC to formulate a "balanced system" for the exchange of materials and information, one that could play "a major role in developing a microorganism approach to implementing an ABS regime [under the CBD]."[288]

ECCO's innovative MTA has influenced a number of major culture collections in Europe, which have begun to implement the concept of "legitimate exchange."[289] In particular, key provisions of ECCO's standard MTA were incorporated into the 2009 revision of the MOSAICC Code of Conduct.[290] It has also influenced at least one major culture collection in Thailand, namely, the National Center for Genetic Engineering and Biotechnology (BIOTECH), whose standard MTAs now enable material to be released to colleagues in other institutions under the recipients' direct supervision via a viral license and also allow further distribution of material by public collections that receive material from BIOTECH under a similar viral license.[291]

Some other culture collections in countries outside Europe have also reportedly adopted licenses that permit other public collections to receive, use, and further distribute materials on a non-exclusive basis.[292] These commendable efforts to facilitate the exchange and distribution of strains among the scientific community will, however, need to be re-examined in light of the Nagoya Protocol, as discussed below.

286 Fritze (2010), above n. 179, at 525 (citing ECCO MTA, above n. 115, arts. 5–7).

287 *Id.*

288 *Id.*; *see also* David Smith & Philippe Desmeth, *Access and Benefit Sharing: A Main Preoccupation of the World Federation of Culture Collections* 68–70, http://r4d.dfid.gov.uk/PDF/Outputs/CABI/CBD-2007-Smith-Desmeth.pdf (citing Canada: UNEP/CBD in UNEP/CBD/W6-ABS/6/Inf/3, 13 Dec. 2007, Compilation of Submission Provided by Parties, Governments, Indigenous and Local Communities and Stakeholders on Concrete Options on Substantive Items on the Agenda of the Fifth and Sixth Meetings of the ad hoc Open Ended Working Group on Access and Benefit Sharing).

289 Dedeurwaerdere, Broggiato, Louafi, Welch & Batur, above n. 9.

290 *Id.*

291 *See* Nat'l Ctr. for Genetic Eng'g & Biotechnology (BIOTECH), *Material Transfer Agreement*, BIOTECH, http://www.biotec.or.th/tncc/mta_tncc.pdf.

292 *See, e.g.*, Dedeurwaerdere et al., above n. 9. Nevertheless, important culture collections in various countries still continue to employ ad hoc MTAs that tend to impose ever more restrictive conditions on uses of microbial materials for research purposes. *See, e.g.*, O'Connor, above n. 227; Ouellette, above n. 227. For details, see above Section II.A & B.

3. The European Commission's Regulation on Access to and Use of Genetic Resources

The Commission of the European Union quickly recognized the seriousness of the Nagoya Protocol and the legal implications of noncompliance once it took effect, in 2014.[293] In particular, the Commission understood that all parties to the Protocol must take measures to ensure "that only legally acquired genetic resources and associated traditional knowledge are utilized within their jurisdiction."[294] In this regard, it noted that Parties must:

- Monitor the compliance of users within their jurisdiction and designate one or more checkpoints for this task;
- "Take appropriate, effective, and proportionate measures in cases where users within their jurisdiction do not comply with their ABS-related obligations:"
- Ensure that disputes arising from specific benefit-sharing contracts can be taken to court;
- Establish a National Focal Point on ABS "to liaise with the international Secretariat [of the CBD] and to respond to information requests by shareholders;"
- Designate one or more competent National Authorities responsible for granting access and advising on applicable procedures for requiring prior informed consent and entering into mutually agreed terms.[295]

The Commission further observed that Parties to the Protocol would need to give special consideration to non-commercial research, to the exchange of genetic resources having pathogenic properties, and to genetic resources for food and agriculture.[296]

At the same time, the Commission perceived that unless Parties moved expeditiously to adopt implementing measures of their own, as implicitly allowed by the Nagoya Protocol, such measures would likely be adopted in other fora.[297] After intensive consultations with stakeholders, the Commission accordingly launched its

[293] European Commission, Proposal for a Regulation of the European Parliament and of the Council on Access to Genetic Resources and the Fair and Equitable Sharing of Benefits Arising from Their Utilization in the Union, COM (2012)576 final, Oct. 4, 2012 [hereinafter Draft EC Regulation (2012)], at 2–3 (adopted by the European Parliament and Council as Regulation No. 511/2014 on Compliance Measures for Users From the Nagoya Protocol on Access to Genetic Resources and the Fair and Equitable Sharing of Benefits Arising from their Utilization in the Union, 2014 O.J. L 150/59 [hereinafter Regulation No. 511/2014 on Compliance Measures]).

[294] Draft EC Regulation (2012), above n. 293, at 3.

[295] *Id.* at 3–4.

[296] *Id.* at 4.

[297] *See id.* at 3.

draft Proposal for a Regulation of the European Parliament and of the Council on Access to Genetic Resources and the Fair and Equitable Sharing of Benefits Arising from Their Utilization in the Union, on October 4, 2012.[298] The ultimate goal was to implement the Nagoya Protocol in the Union and to enable Union ratification of this treaty.[299]

The draft proposal was discussed by EU member states and the Council of Europe, with a view to obtaining an agreed version in time for the next meeting of the CBD's Council of the Parties in 2014, and it was also discussed and criticized at a meeting of EU and Brazilian experts in Brazil, in June 2013.[300] The final version, entitled Regulation No. 511/2014 of the European Parliament and of the Council on Compliance Measures for Users from the Nagoya Protocol on Access to Genetic Resources and the Fair Equitable Sharing of Benefits Arising from Their Utilization in the Union, was adopted on April 16, 2014.[301]

A. UNDERLYING PREMISES. Two preliminary observations are necessary to clarify the precise thrust of the Regulation as finally adopted. First, the Regulation does not seek to establish a multilateral regime among the European Union's member states within the purview of Article 4 of the Nagoya Protocol.[302] On the contrary, the Regulation leaves the existing microbial research infrastructure intact and subject to both national and international laws already governing the culture collections and the MTAs they adopt. What the Regulation proposes to achieve, instead, is a harmonized approach to supplying the research community with diverse types of genetic resources that would lower the transaction costs of single collections while ensuring overall compliance with the Nagoya Protocol within the EU.[303]

Second, the Regulation builds on language in Article 2 of the Nagoya Protocol which, in defining "utilization of genetic resources,"[304] seems to shift the focus of

[298] *See* above n. 293 and accompanying text.

[299] Draft EC Regulation (2012), above n. 293, at 2.

[300] *See* Davis, Fontes & Marinoni (2013), above n. 1, at 9–11, 55.

[301] Regulation No. 511/2014, above n. 293.

[302] *See* Tenth Meeting of the Conference of the Parties to the Convention on Biological Diversity, Nagoya, Japan, 18–29 Oct. 2010, *Nagoya Protocol on Access to Genetic Resources and the Fair and Equitable Sharing of Benefits Arising from their Utilization (ABS) to the Convention on Biological Diversity*, art. 4 [hereinafter Nagoya Protocol], entered into force October, 2014, *available at* http://www.cbd.int/decision/cop/?id=12267 (last accessed 3 July 2014).

[303] *See* EU Regulation 511/2014 on Compliance Measures, above n. 293 Preamble ¶¶9, 21, 24, 28; *id.*, arts. 4(7), 5.

[304] *See* Nagoya Protocol, above n. 302, art. 2(c) ("Utilization of genetic resources" means to conduct research and development on the genetic and/or biochemical composition of genetic resources, including through the application of biotechnology as defined in article 2 of the Convention); *id.* art. 2(d) ("Biotechnology ... means any technological application that uses biological/systems, living organisms, or *derivatives thereof*, to make or modify products or processes for specific use") (emphasis supplied).

the CBD's ABS regime away from product marketing as such to "research and development."[305] One consequence, of course, is that all institutions engaged in R&D – not just industry – "bear a responsibility to ensuring that the ABS mechanisms . . . function,"[306] which is precisely what triggers the need for a regulation governing culture collections in the first instance.

At the same time, Article 2 of the Nagoya Protocol can be read so as to uncouple "access" from "benefit sharing,"[307] with emphasis on the latter in a research context. On the basis of this still controversial interpretation, the Commission reads the Protocol as leaving "Parties discretion whether they wish to regulate access and required prior informed consent and benefit-sharing for the use of their genetic resources or not."[308] As a result, the Commission's Draft Proposal left the microbial culture collections, among others, free to pursue their own policies and practices with regard to accessing microbial genetic resources. The proposed Regulation would instead establish "an EU platform for discussing access to genetic resources and sharing best practices as the preferable option for access."[309]

In drafting its final Regulation, the Commission accepted the culture collections' own self-characterization as "intermediaries" between providers and users of genetic resources. In this capacity, the collections may determine *their own access procedures* as well as the extent to which they will allow users to undertake commercial applications of the specimens they supply, provided that overall compliance with the Nagoya Protocol is maintained.[310] With these enabling premises on the table, the Regulation as adopted focused its full attention on user compliance with the strictures of the Protocol.

B. BASIC CONCEPTS AND METHODS. Operationally, Regulation 511/2014 hinges on two complementary approaches. On the one hand, all users of genetic resources and related traditional knowledge to be acquired after the Nagoya Protocol entered into force are subjected to a relatively onerous set of due diligence obligations. On the other hand, users may attenuate the costs and risk of these same due diligence obligations by obtaining the genetic resources they need from collections admitted to a system of trusted intermediaries that the Regulation endeavors to establish.[311]

305 Godt (2013), above n. 280, at 258.
306 *Id.*
307 *See, e.g.*, Davis, Fontes & Marinoni (2013), above n. 1, at 52.
308 Draft EC Regulation (2012), above n. 293, at 3.
309 *Id.* at 3; for details, *see id.* art. 13.
310 Draft EC Regulation (2012), above n. 293, pmbl., §18; EU Regulation 511/2014, above n. 293, Preamble Paragraphs 18–24; Davis, Fontes & Marinoni (2013), above n. 1, at 55.
311 EU Regulation 511/2014, above n. 293, arts 4 (user compliance), 5(Register of Collections); see esp. id. art. 4(7) ("users obtaining a genetic resource from a collection included in the register . . . shall be considered to have exercised due diligence as regards the seeking of information listed in paragraph 3

To these ends, the Regulation obliges all "users" to exercise due diligence to ascertain that the genetic resources and associated traditional knowledge to be "used" were accessed in accordance with applicable legal requirements and to ensure that the resulting benefits are shared, where applicable.[312] For this purpose, a "user" is any "natural or legal person that utilizes genetic resources or traditional knowledge associated with genetic resources," while "utilization of genetic resources" means "to conduct research and development on the genetic and/or biochemical composition of genetic resources."[313] This due diligence obligation then applies irrespective of the size of the user, "including micro, small and medium-sized enterprises" because excluding these actors from the system would entirely undermine its effectiveness.[314]

Having thus shifted the burden of due diligence with respect to both access and benefit-sharing under applicable legislation or regulatory requirements onto users,[315] the Regulation specifies the information that due diligence requires users to seek, keep, and transfer to subsequent users as follows:

- The date and place of access of genetic resources and associated technical knowledge;
- The description of such resources or traditional knowledge;
- The source from which the resources or knowledge were directly obtained as well as information about subsequent users;
- The presence or absence of rights and obligations pertaining to access and benefit sharing; and
- Access permits and mutually agreed terms, where applicable, or an internationally recognized certificate of compliance.[316]

Due diligence also requires users to obtain additional information or evidence where uncertainties about the legality of access and use persist,[317] and to obtain a proper access permit, establish mutually agreed terms, or discontinue the use where it appears that access was not in accordance with applicable legislation or regulatory requirements.[318] Records of the relevant information are to be kept for 20 years after the end of the period of use.[319]

of this article."). These measures should be "complemented by awareness and training activities, work on contractual model clauses, work on technical tools for monitoring and tracking genetic resources flow, and where appropriate through bilateral cooperation with other countries or regions." *See id.*, arts. 6–9.

312 *Id.*, pmbl., §21; *see id.* art. 4(1).

313 *Id.* arts. 3(5), 3(6).

314 *Id.*, pmbl., §23. As will be seen, however, the trusted culture collections, operating as intermediaries between providers and users, will play a different and moderating role. *See* below text at nn. 319–322.

315 *Id.* art. 4(1).

316 *Id.* art. 4(3)(a), (b).

317 *Id.* art. 4(5).

318 *Id.* arts. 4(1), 4(5).

319 *Id.* art. 4(6).

Clearing these due diligence requirements could become both onerous and costly for would-be users operating on their own, as would presumably occur if they negotiated access to *in situ* genetic resources directly with foreign governments or with their designated authorities.[320] However, the drafters of the Regulation assume that most users will prefer to obtain *ex situ* genetic resources from intermediaries, especially "collections, or agents that acquire genetic resources in third countries."[321] It accordingly invites users seeking to fulfill their due diligence obligations at the lowest cost to acquire genetic resources from one of the trusted intermediary collections listed in the EU Register to be established for this purpose.[322] The payoff for users who accept this invitation is that, in so doing, they will be "considered to have exercised due diligence" as regards the seeking of information otherwise required of all users of genetic resources.[323]

Once the Commission has established an internet-based, easily accessible EU register of trusted collections,[324] member states are given the authority to vet collections under their jurisdiction and to inscribe those that meet the specified criteria onto that Register.[325] Qualifying collections must demonstrate their capacity to:

- Apply standardized procedures for the exchange of samples and related information with other collections or third persons for use;
- Only supply genetic materials and related information to third persons with documents providing evidence that they were accessed legally, in accordance with prior informed consent and mutually agreed terms or other regulatory requirements;
- Keep records of all samples and information supplied to third parties; and
- Use "unique identifiers, where possible, for samples ... supplied" and "appropriate tracking and monitoring tools" for exchanging samples of genetic resources and related information with other collections.[326]

The EU's member states must verify that each collection they submit for inclusion in the Register meets these conditions, and they must remove those that fail to comply.[327] The member states must also designate competent authorities

[320] *See id.*, pmbl., §27 ("The collections of genetic resources in the wild is mostly undertaken for noncommercial purposes by academic, university, and noncommercial researchers or collectors"). In that case, the Commission expressly aims to repress biopiracy. *See id.* at pmbl., §§3, 6, 9, 10.

[321] *Id.*, pmbl., §27.

[322] *Id.* arts. 4(7), 5(1).

[323] *Id.* art. 4(7).

[324] *Id.* art. 5(1).

[325] *Id.* art. 5(2).

[326] *Id.* art. 5(3). *See, e.g.*, Davis, Fontes & Marinoni (2013), above n. 1, at 55.

[327] Regulation 511/2014 above n. 293, arts. 5(3)(d),€, 5(4).

responsible for the application of the proposed regulation.[328] The Commission itself will designate a focal point on access and benefit sharing in order to provide information to applicants seeking access to genetic resources in the European Union.[329]

Articles 7 through 11 of the Regulation set out detailed monitoring and compliance provisions. For example, all publicly funded research grantees that use genetic resources and related information must agree to meet the due diligence requirements of Article 4.[330] Users must certify compliance with the due diligence obligations when they request marketing approval for the products developed from genetic resources and related traditional knowledge, or when they otherwise commercialize their use of such resources even if market approval is not required.[331] The competent authorities, to whom these declarations are made, will in turn transmit the information received concerning due diligence to the Access and Benefit Clearing House, established under Article 14(1) of the Nagoya Protocol, to the Commission, and "where appropriate," to the national authorities referred to in Article 13(2) of the Nagoya Protocol.[332]

Member states are to check and verify that users comply with the due diligence obligations, and compliance with a recognized set of best practices[333] "may reduce that user's risk of 'non-compliance.'"[334] The competent authorities should also accept an internationally recognized certificate of compliance as evidence that genetic resources were legitimately accessed and that mutually agreed terms had been established according to both domestic and foreign laws, and they should intervene when there is evidence of noncompliance, especially when provider countries raise such concerns.[335] Penalties for noncompliance may include fines, suspension of use activities, and confiscation of illegally acquired genetic resources,[336] while other interim measures – including seizure of illegally acquired genetic resources – may be available.[337]

Viewed as a whole, the Regulation envisions that an EU-wide system of trusted intermediary collections would substantially lower both the costs of the due diligence

[328] *Id.* art. 6(1).
[329] *Id.* art. 6(3).
[330] *Id.* art. 7(1).
[331] *Id.* art. 7(2).
[332] *Id.* art. 7(3).
[333] EU Regulation No. 511/2014, above n. 293, art. 9(1) See Nagoya Protocol, above n. 302, arts. 13–14. The Commission allows associations of users to develop best practices for fulfilling these obligations, and they may obtain official recognition from the Commission for this purpose. *Id.* art. 8.
[334] *Id.* art. 9(1).
[335] *Id.* arts. 4(3)(a), 7(2)(a), 9(1), 9(2)(b).
[336] *Id.* art. 11.
[337] *See, id.* art. 9(6).

obligations imposed on users and the risk that illegally acquired genetic resources will be used in member states.[338] The system of trusted collections to be established under Article 5 would be "particularly beneficial for academic, university, and noncommercial researchers as well as small and medium-sized enterprises."[339] How, and to what extent, the microbial culture collections and their clients will respond to this challenge remains to be seen.[340]

B. *Opting Out or Opting In? Limits of the Trusted Intermediary Approach*

The response of the public microbial culture collections in Europe to the challenges posed by the Nagoya Protocol's global regime of misappropriation seems both ingenious and logical from an historical perspective. The *ex situ* collections dedicated to a public good mission want to continue to support scientific research – both public and private – with the fewest possible disruptions to their established methods of operation. They have, accordingly, redefined themselves as "intermediaries" between providers and users of genetic resources, as defined in the CBD and the Nagoya Protocol, who perform services to the research community as a whole.[341] The intermediaries' service function is further rationalized as the "creation of a separate market between two other distinct markets," with benefits to both providers and users of genetic resources.[342]

In their capacity as intermediaries, the culture collections undertake to create a record – a chain of title – applicable to all exchanges of *ex situ* microbial genetic resources that would enable users and providers to negotiate and enforce benefit-sharing agreements for relevant commercial applications. In effect, this self-defined intermediary status thus allows some room for shifting the risk of liability for violations of the CBD onto users,[343] while supplying providers with the records needed to enforce their rights to prior informed consent (PIC) and the sharing of end-use benefits.

As a practical matter, the microbial culture collections are better positioned to perform this role than many *ex situ* collections in other fields, such as those that

[338] *Id.* Preamble §§21, 28.

[339] *Id.* §28.

[340] *See, e.g.*, Davis, Fontes & Marinoni (2013), above n. 1, at 55 (stating that "*ex situ* collections in all sectors are in the process of determining whether, and how, they will need to change their practices to account for a possible increase in demand from commercially-orientated users, and whether the costs involved in being a 'trusted collection' outweigh the benefits."). *See further* below Section IV.

[341] Godt (2013), above n. 280, at 259–60; *see also id.* at 255.

[342] *See id.* at 259–60; *see also* Gerd Winter, *Knowledge Commons, Intellectual Property, and the ABS Regime*, *in* COMMON POOLS OF GENETIC RESOURCES (2013), above n. 145, at 285–303.

[343] Godt (2013), above n. 280, at 255–56.

supply plant or animal genetic resources. For purely scientific reasons, as well as considerations of public health and security, all WFCC affiliates routinely attach unique identifiers to the microbial specimens in their collections, and current scientific practice requires them to validate and track the taxonomic characteristics and quality of specimens provided throughout the research process.[344] In effect, traditional research methods in microbiology thus anticipated the tracking demands of the Nagoya Protocol and the EU Regulation, unlike the situation under the International Treaty for Plant Genetic Resources for Food and Agriculture (ITPGRFA), which expressly abjured obligations to track exchanges of plant genetic resources.[345]

On closer inspection, however, serious legal infirmities may still undermine this attempt to engraft a "tailor-made regime of openness" onto the bilateral contractual approach of the CBD and the Nagoya Protocol.[346] With respect to access, for example, the microbial culture collections normally do not accept materials of uncertain origin, in conformity with the Nagoya Protocol. But they do not demand typically certificates of prior informed consent or declarations of mutually agreed terms that would seem mandatory under the bilateral approach. Whether more stringent requirements in this regard will emerge from implementation of Article 5 of the Regulation remains to be seen.[347]

How the culture collections will modify their existing interpretations of the CBD's access provisions in light of E.U. Regulation 511/2014 also remains uncertain. For starters, the Commission appears to read the Nagoya Protocol so as to uncouple "access" from "utilization," in accordance with a scholarly view in European circles[348] that is not shared in leading developing countries.[349] On this view, the duty to share benefits from *ex situ* resources is decoupled from access as such and redefined as a benefit to everybody, including the provider state.[350] In practice, this interpretation relies on declarations of geographic origin for *ex situ* resources, in conformity with current practices. It does not necessarily require proof of legal access, despite the fact that the Nagoya Protocol states that host collections of *ex situ* genetic resources are responsible for "access and benefit sharing" under the CBD.[351] As Professor Christine Godt has observed, any failure to clear accessions for ABS

344 For details, see above Section I.A.2.
345 *See* above Chapter 3, Section III.B.
346 Godt (2013), above n. 280, at 249.
347 *Compare* Godt (2013), above n. 280, at 255 *with* Regulation No. 511/2014, above n. 293, art. 5(3)(b).
348 Winter, above n. 342.
349 *See, e.g.*, Davis, Fontes & Marinoni (2013), above n. 1, at 9 (noting that "the Brazilian and European definitions of 'access' are fundamentally different").
350 Godt (2013), above n. 280, at 259 (citing Winter, above n. 342).
351 Godt (2013), above n. 280, at 256.

conformity may inherently impair the possibilities for the sharing of benefits in the future.[352]

The premises underlying the European culture collections' relatively hands-off approach to the sharing of benefits in the past then raises other troubling legal issues. By defining themselves as "intermediaries" rather than "users,"[353] the public collections deliberately remove themselves from the stringent ABS obligations that the Nagoya Protocol – and now the EU Regulation – impose on "users." Yet, the Nagoya Protocol can be read so as to allow concessions to *ex situ* collections as "users" only if they opt into a suitable multilateral regime whose open-access approach remains consistent with Article 4.[354] Professor Godt has accordingly questioned the legality of the culture collections' decision to opt out of the bilateral approach by tailor-made deviations for "intermediaries" without opting into a full-fledged multilateral regime within the ambit of Article 4 of the Nagoya Protocol.[355]

Any infirmities in this legal maneuver are then magnified by the culture collections' reliance on the distinction between commercial and noncommercial research in their MTAs. All the viral licenses, like those modeled on ECCO's core MTA,[356] claim to be conserving and sharing microbial genetic resources and related biological data for noncommercial research purposes only. Yet, the distinction between commercial and noncommercial research has become unworkably blurred in today's research environment.[357] The intermediaries' reliance on that same distinction, coupled with their self-proclaimed exemption from the status of "users," conjures up the possibility of complicated disputes in specific cases. It also generates mistrust in provider countries about the long-term credibility of this approach.[358]

The public collections' MTAs do limit their recipients' ability to transfer microbial materials to third parties for both scientific and safety reasons.[359] However, their approach to the eventual sharing of benefits from downstream commercial applications remains relatively passive. The collections merely notify recipients that the latter may need to obtain PIC and MAT from parties in the country of origin before proceeding to use the genetic resources in question.[360] The collections have

[352] *Id.* at 259.
[353] *Id.* at 258.
[354] *See* Nagoya Protocol, above n. 302, art. 4; Godt (2013), above n. 280, at 259–61.
[355] Godt (2013), above n. 280, at 249, 259–61.
[356] *See* above Section III.A.2.
[357] *See, e.g.*, Rebecca S. Eisenberg, *Proprietary Rights and the Norms of Science in Biotechnology Research*, 97(2) *Yale L.J.* 177 (1987); Arti Rai & James Boyle, *Synthetic Biology: Caught Between Property Rights, the Public Domain, and the Commons*, 5 *PLoS Biology* 58 (2007).
[358] *See, e.g.*, Godt (2013), above n. 280, at 259.
[359] *See* above Section III.A.2 (ECCO's core MTA).
[360] *See* Godt (2013), above n. 280, at 254–56.

thus reduced their own transaction costs by shifting the risk of liability under the CBD to recipients *qua* users, while shielding themselves from liability as self-defined "intermediaries."[361]

As Professor Godt points out, the net result was to leave industry responsible for sharing the gains from commercial use with provider countries, without directly addressing the risks of biopiracy, the extent to which provider countries' own access regulations have been satisfied, or the consequences of the downstream user's eventual resort to exclusive intellectual property rights.[362] Yet, under the CBD's bilateral approach, the specimens in the *ex situ* collections remain subject to the national ABS authorities in provider states, who should be kept informed of the users and uses to which they are put.[363] The net result is that, while the collections do nothing to undermine the Nagoya Protocol's ABS regime, the role of intermediaries embodied in their MTAs risks losing the trust of the provider states over time.

To some extent, the European Commission's Regulation 511/2014 indirectly recognizes this possible defect in the degree of support given to the ABS regime. For example, the Regulation expressly obliges users to obtain a proper access permit, establish mutually agreed terms, or discontinue the use if access does not conform to the CBD. Moreover, culture collections qualifying for the Register established by Article 5 must have demonstrated capacity to "supply genetic resources and related information to third persons for their utilization only with documentation providing evidence that the genetic resources and the related information were accessed in accordance with applicable access and benefit sharing legislation or regulatory requirements and, where relevant, with mutually agreed terms."[364]

Nevertheless, the culture collections are not treated as "users" under the Regulation, even if as trusted intermediaries they must provide documentary evidence that the resources and information in question were legally accessed and that, where relevant, mutually agreed terms for the fair and equitable sharing of benefits had occurred.[365] That is why users acquiring genetic resources from a trusted intermediary collection under the Regulation are presumed to have fulfilled all the due diligence requirements that the Regulation otherwise imposes.[366]

The ambiguity here resides in the unknown extent to which registered culture collections must actually police their acquisitions and distributions of genetic materials for conformity to the CBD, and how capable they are of undertaking such a mission, given their limited financial resources and their traditional dependence on

[361] *Id.* at 256 ("overall, the collections aim to avoid benefit sharing").
[362] *Id.* at 257.
[363] *Id.* at 261.
[364] EU Regulation No. 511(2014), above n. 293, art. 5(3)(b); *see also id.*, art 4(5).
[365] *Id.* art. 5(b).
[366] *Id.* art. 4(7).

voluntary as well as paid staff. How burdensome this provision will actually become for collections seeking to qualify as trusted intermediaries under the Regulation thus remains to be seen.[367]

There is some evidence that authorities in developing countries will expect the Commission to require trusted collections to become more deeply involved in the enforcement of ABS benefits than in the past, with the risk of correspondingly elevated transaction costs. For example, the managers of some Brazilian collections wanted the trusted collections operating under the European Commission's Regulation to monitor and implement so-called "change of intent" decisions by users of genetic resources who subsequently decide to shift from non-commercial to commercial research endeavors.[368] The Brazilian experts also interpret the concept of "traceability" as an undertaking that "starts at the provider end, and necessitates a system that allows information to flow back to providers over use and user chains."[369]

To satisfy the enforcement expectations of provider countries, in other words, the trusted collections qualifying under the Regulation might have to become de facto agents of the bilateral system at their own expense, with none of the benefits that a multilateral regime established under Article 4 of the Nagoya Protocol might otherwise confer. Not surprisingly, questions about funding were of paramount importance at the follow up consultation on the draft Regulation that were held in Brazil in 2013.[370] By the same token, many European culture collections were reportedly uncertain about whether the costs of attaining trusted intermediary status under the draft Regulation would be worth the benefits.[371]

For these and other reasons, one unintended consequence of Regulation No. 511/2014 could actually be to drive more of the most technically advanced microbial culture collections – especially those aspiring to become Biological Resource Centers – towards ATCC's market-like model in the United States, and away from their traditional public-good model. Logically, these technically advanced

[367] *See, e.g.*, Davis, Fontes & Marinoni (2013), above n. 1, at 55 (noting that the draft Regulation provided no prescription as to exactly how collections should implement ABS, as long as those that are registered as trusted "can fulfill the legal and tracking requirements ... Hence ABS measures will likely continue to be developed and implemented on a voluntary sector-specific basis").

[368] *See* Davis, Fontes & Marinoni (2013), above n. 1, at 10.

[369] *Id.* at 9.

[370] *See, e.g.*, *id.*, at 12, Recommendation 6 (stating that "unfunded mandates should be avoided" and that "the government that requires traceability should provide the required infrastructure (clearinghouse, regulatory body) and funding ..."). *See also id.*, Recommendation 7 ("The degree of effort and resource expended on tracing should be proportional to the risk of misuse").

[371] *See, e.g.*, Davis, Fontes & Marinoni (2013), above n. 1, at 55 (noting possible increase in demand from commercially oriented users with the result that *ex situ* collections in all sectors were considering "whether the costs of implementing comprehensive monitoring mechanisms, and of negotiating with providers terms that might need to extend to later commercialization involved in being a 'trusted collection' outweigh the benefits.").

collections may decide that they, too, want a piece of the action from commercial applications, like the authorities in provider countries and industrial patentees in user countries. Indeed, this possibility was already an integral component of the Demonstration Project for the Global Biological Resource Center Network, and it may well surface again in current negotiations concerning the Microbial Resource Research Infrastructure (MIRRI) project that constitutes its latest embodiment.[372]

The possible fallacy, of course, is that the more that stakeholders drive up the costs of sharing genetic resources for scientific purposes in a quest for speculative financial gains, the harder it becomes for public science to conduct the basic research that actually leads to valuable commercial applications. Rather than a bonanza of benefits to divide, a plausible result is a no-win situation with diminishing returns for all stakeholders over time, for the reasons set out more fully in Chapter 5.

Meanwhile, the European responses in practice – whether leaning more towards ECCO's Core MTA or the Commission's Regulation – will likely trigger an array of *sui generis*, overlapping, and often conflicting agreements hammered out by different networks of microbial culture collections in different parts of the world, with a corresponding rise of transaction costs attendant upon case-by-case implementation of the bilateral approach. Every BRC, every national group of culture collections, every regional network will be tempted to negotiate their own MTAs, with a resulting patchwork quilt of regulatory practices that could complicate rather than simplify the existing modalities supporting cross-border exchanges of microbial genetic resources.[373] Without any overarching governance framework and common dispute resolution mechanism, the end result could be a flood of litigation to resolve ABS disputes before domestic courts that are now obliged to hear such cases under the Nagoya Protocol and the EU Regulation, with no common policy guidelines or commonality of interests to assist them.

The deeper problem is that the "opt out" MTAs of the public culture collections in Europe are fundamentally legal instruments that tend to preserve the bilateral approach embodied in the CBD, as initially drafted in 1992. Yet, formidable obstacles to public scientific research are rooted in that same bilateral approach, as spokesmen for the CGIAR's seed banks have demonstrated over and over again with respect to plant genetic resources for food and agriculture.[374] Like their counterparts in agricultural research, microbiologists need facilitated access to multiple genetic resources for diverse upstream research projects, only some of which will occasionally

[372] *See* below Chapter 9, Section II.C (discussing both GBRCN and MIRRI). For MIRRI, a pan-European distributed research infrastructure, see http://www.miri.org/. For ATCC, see above Section II.A.

[373] *But see* Davis, Fontes & Marinoni (2013), above n. 1, at 31 (hoping that the EU Regulation will elicit sectoral codes of conduct, model contractual clauses, guidelines, and best practices to facilitate exchanges of materials).

[374] *See* above Chapter 3, Section II.B.

become of commercial importance. The bilateral approach, with its prescription for case-by-case negotiations at the pre-competitive stage when the prospect for future applications remain purely speculative, risks becoming a research inhibitor that discourages the scientific inputs needed to produce end products subject to the benefit-sharing norms of the CBD.

That was precisely the lesson that the drafters of the Nagoya Protocol learned from the conflict over plant genetic resources that produced the International Treaty on Plant Genetic Resources for Food and Agriculture in 2001.[375] Article 4 of the Nagoya Protocol accordingly facilitates public research on plant genetic resources in a multilateral framework precisely because it pays off in non-monetary benefits that all countries need, over and above any commercial benefits that may or may not also accrue.

By the same token, Article 4 invites the microbiological research community to consider emulating tailor-made deviations from the bilateral approach comparable to those devised for the Crop Commons.[376] From this perspective, if the public culture collections opted into a well-designed transnational legal framework that improved upon the shortcomings of the model adopted for the Crop Commons, they could produce a much sounder legal and institutional foundation for science than any variant of the trusted-intermediary models that primarily seek to opt out of the bilateral approach embodied in the CBD.[377]

For the first time since the debate about ownership of genetic resource began a half century ago, there is an emerging consensus in both developed and developing countries that a robust public domain in biological science promotes human welfare generally. Every effort should accordingly be made to ensure that the opportunities that the Nagoya Protocol miraculously provided for constructing a multilateral regime of facilitated access to microbial genetic resources for research purposes are not missed because of inertia or for lack of imagination.

IV. FROM THE BILATERAL TO THE MULTILATERAL APPROACH

Not long after the Nagoya Protocol was adopted in 2010,[378] the World Health Organizations' efforts to negotiate a sounder legal platform for dealing with the

[375] *See* International Treaty on Plant Genetic Resources for Food and Agriculture, *opened for signature* 3 Nov. 2001, 2400 U.N.T.S. 303 (entered into force 29 June 2004) [hereinafter ITPGRFA], *available at* http://treaties.un.org/doc/publication/UNTS/Volume%202400/v2400.pdf (last accessed 3 July 2014); *see further* above Chapter 3, Section III.A.

[376] *Accord*. Godt (2013), above n. 280, at 261.

[377] *Cf. id.*, at 247–49 (questioning the extent to which tailor made regimes of facilitated access may shelter "as a deviation from the narrow bilateral contractual CBD approach").

[378] *See* Nagoya Protocol, above n. 302.

avian flu crisis than in the past reached fruition. As we indicated in Chapter 2, the WHO's cooperative response to pandemic influenzas had relied on its Global Influenza Surveillance Network (GISN) for more than 60 years.[379] Under this system, a network of National Influenza Centers and WHO laboratories had cooperated to monitor the spread of seasonal influenza and to develop suitable responses, especially vaccines.[380] The GISN network had also been used for monitoring influenza viruses with pandemic potential until it nearly collapsed during the H5N1 crises in the period 2005–2008. At that time, Indonesia refused to share essential samples of the virus with WHO repositories because onward transfers of previously donated samples had led to patented vaccines produced in the private sector, which were then not made available to Indonesia in quantities it needed or at prices it could afford.[381]

In 2011, the WHO's efforts to rescue the GISN network and restore confidence in a global system of exchanging influenza viruses finally resulted in the Pandemic Influenza Preparedness Framework for the Sharing of Influenza Viruses and Access to Vaccines and Other Benefits (PIP Framework Agreement).[382] This instrument, adopted by the WHO World Health Assembly (WHA) in May 2011, thus became the first multilateral initiative governing exchanges of microbial genetic resources to be adopted after the Nagoya Protocol was signed in 2010. Although the WHO negotiations were conducted separately from those of the Nagoya Protocol,[383] the end result – developed in the shadow of the CBD[384] – helps to clarify the larger issues underlying post-Nagoya exchanges of microbial materials in general.

In what follows, we briefly sketch the main concepts and methods of the PIP Framework Agreement. We then discuss its implications for a redesigned Microbial Research Commons, bearing in mind the critically different purposes that underlie the PIP initiative from those that underlie the existing microbial research commons as a whole.

379 *See* above Chapter 2, Section III.A.

380 *See* Marie Wilke, *The World Health Organization's Pandemic Influenza Preparedness Framework as a Public Health Resources Pool, in* COMMON POOLS OF GENETIC RESOURCES (2013), above n. 145, at 315, 315–16 [hereinafter Wilke (2013)].

381 *See id.* at 316–18; Dana Beldiman, *Patent Chokepoints in the Influenza-Related Medicines Industry: Can Patent Pools Provide Balanced Access?*, 15 *Tul. J. Tech. & Intell. Prop.* 31 (2012), *available at* http://papers.ssrn.com/s013/papers.cfm?abstract_id=2049035.

382 World Health Org. (WHO), *Pandemic Influenza Preparedness Framework for the Sharing of the Influenza Viruses and Access to Vaccines and Other Benefits*, World Health Assembly Res. WHA645 (May 24, 2011) [hereinafter PIP FRAMEWORK], *available at* http://www.who.int/influenza/resources/pip_framework/en/index.html (last accessed 23 Feb. 2014).

383 *See generally* Frederick M. Abbott, *An International Legal Framework for the Sharing of Pathogens*, 14–17, *Int'l Ctr. for Trade & Sustainable Dev.* (ICTDSD), Issue Paper No. 30 (2010).

384 *See, e.g.*, Wilke (2013), above n. 380.

A. *Basic Concepts of the WHO's Pandemic Influenza Preparedness Framework Agreement (2011)*

The primary purpose of the PIP Framework Agreement was to establish a standing mechanism for exchanging viral samples and related epidemiological information, and also for sharing the resulting benefits, especially vaccines.[385] From a legal standpoint, the Framework Agreement is a nonbinding instrument based on the premise that WHO member states are committed to sharing influenza viruses and the resulting benefits as part of their collective action for global public health.[386] Emphasizing the principle that benefits should be shared on the basis of "public health risk and need,"[387] the Agreement commits member states to the "rapid, systematic and timely provision of biological materials as feasible" while reaffirming the sovereign rights of provider countries in their biological resources.[388]

From a practical standpoint, the Agreement builds on the preexisting GISN network, now renamed the Global Influenza Surveillance and Response System (GISRS), while "clarifying the rights and obligations of the different public and private actors involved."[389] It also adds new benefit sharing obligations and specifies modalities of distribution.[390]

In this context, the substantive reach of the Agreement is limited to "H5N1 and other influenza viruses with human pathogenic potential."[391] It thus excludes seasonal influenza and non-influenza pathogens.[392] According to Marie Wilke, these and still other definitions ensure that countries "have absolute certainty about the type of information they share when submitting a specimen." By limiting the scope to influenza viruses with pathogenic potential, any other valuable information contained in a blood sample, for instance, even if "accidentally transmitted, will not be covered by the agreement."[393]

[385] *PIP Framework*, above n. 382, arts. 1.3 (sharing as part of collective action for global public health), 1.55 (global vaccine production insufficient to meet anticipated needs in pandemics).

[386] *See PIP Framework*, above n. 382, arts. 1.3.

[387] *Id.* art. 1.8.

[388] *Id.* art. 1.10. It also reaffirms WHO parties' obligations under the International Health Regulation of 2005 World Health Org. (WHO) International Health Regulation (2005), *available at* http://apps.who.int/iris/bitstream/10665/43883/1/9789241580410_eng.pdf (last accessed 24 Nov. 2015). *See PIP Framework*, above n. 382, art. 1.6.

[389] Wilke (2013), above n. 380, at 318–19.

[390] *Id.*

[391] *PIP Framework*, above n. 382, art. 3.1. For definitions of "influenza viruses with pathogenic potential," *see id.* art. 4.2 ("wild-type influenza viruses ... found to infect humans" that are distinct from seasonal viruses), 4.1 (defining PIP biological material to include human clinical specimens, virus isolates, and "unmodified viruses with human pandemic potential developed by WHO GISR laboratories"). *See also id.* art. 4.1 (including extracted RNA and complementary CDNA "that encompasses the entire coding region of one or more viral genes"). *See generally* Wilke (2013), above n. 380, at 324–25.

[392] *Id.* at 325. *See also* Abbott, above n. 383.

[393] Wilke (2013), above n. 380, at 325 (citing *PIP Framework*, above n. 382, arts. 4.1, 4.2).

In practice, these materials are to be shared through the GISRS and the National Influenza Centers (NICs) that provide original samples to WHO Collaborating Centers (CCs) and to other WHO laboratories more or less as before.[394] Technically, Article 5.1.1 declares that member states through their NICs and other authorized laboratories "should in a rapid, systematic and timely manner provide PIP biological materials from all cases of H5N1 and other influenza viruses with human pandemic potential, as feasible," to WHO Collaborating Centers or designated WHO reference laboratories.[395] Member states also pledge to share genetic sequence and other related data pertaining to the covered viruses,[396] and to operate under "a transparent traceability mechanism … to track in real time the movement of PIP biological materials into, within, and out of the WTO CISRS."[397]

Article 5.1.2 then stipulates that providers of PIP biological materials under the above provisions "consent … [to] … the onward transfer and use" of covered materials and data, "to institutions, organizations, and entities," subject to provisions in the Standard Material Transfer Agreements."[398] Article 5.1.3 further requires providers to ensure that their materials are viable and that they are accompanied by sufficient information to satisfy traceability requirements as well as by "other clinical and epidemiological information needed for risk assessment."[399] As Marie Wilke (who participated in the negotiations) explains it, the WHO Collaborating Centers thus share "diagnostic reagents, candidate vaccine viruses and test kits" with both the National Influenza Centers, and "other interested laboratories and … institutions, whether commercial or for non-commercial research, that are not formally part of the GISRs."[400]

Here, however, the PIP Framework Agreement imposes benefit-sharing obligations on both the GISRS, as public service providers, and especially on those other entities that are not affiliated with the GISRS.[401] Two standard MTAs – $SMTA_1$ and $SMTA_2$ – have been devised for this purpose,[402] and both SMTAs are thought to be legally enforceable despite the fact that the PIP Framework Agreement itself is a nonbinding intergovernmental undertaking.[403]

394 *See PIP Framework*, above n. 382, arts. 5.1.1–5.1.3.
395 *Id.* art. 5.1.1.
396 *Id.* art. 5.1.2.
397 *Id.* art. 5.3.1.
398 *Id.* art. 5.1.2.
399 *Id.* art. 5.1.3.
400 Wilke (2013), above n. 380, at 319.
401 *Id. See generally PIP Framework*, above n. 382, art. 6.
402 *See id.* art. 5.4.1 (covering all transfers of PIP biological materials within the WHO GISRS system) and *id.* art. 5.4.2 (covering agreements between WHO Director General "with entities outside the WHO GISRS").
403 *See* Wilke (2013), above n. 380, at 320–322. For details, see PIP *Framework*, n. 382, Annexes 1 and 2.

The benefit-sharing obligations assumed by WHO member states under Article 6 take the form of "best effort" clauses that constitute goals rather than binding legal commitments. These provisions:

- Commit the WHO to coordinate pandemic influenza preparedness and responses,[404] and the GISRS to undertake pandemic risk assessment and risk responses;[405]
- Ensure that WHO collaborating centers provide the candidate vaccine virus upon request to influenza vaccine manufacturers on a no-preference basis, and to any other laboratories that meet quality and safety standards;[406]
- Ensure that the WHO CCs continue to make available noncommercial diagnostic reagents and test kits to NICs and other authorized labs without charge;[407]
- Ensure that regulatory laboratories provide reference reagents to determine the potency of vaccines;[408]
- Urge developed Member States to help build capacity for influenza surveillance, especially in developing countries;[409]
- Encourage developed Member States to assist in building regulatory capacity for rapid approval of safe and effective vaccines, diagnostic, and pharmaceutical products developed from the use of PIP biological materials;[410]
- Commit the WHO to seek contributions from public and private sources in order to develop stockpiles of antiviral medicines and equipment[411] as well as vaccines,[412] for outbreaks of potential pandemic influenzas;
- Oblige Member States to press influenza vaccine manufacturers to set aside a portion of each production of relevant vaccines for stockpiling and/or use by developing countries, and to ensure that these countries are supplied on a par with developed countries and "on the basis of public health risk and needs and at tiered prices;"[413]

[404] *PIP Framework*, above n. 382, art. 6.1.
[405] *Id.* art. 6.2.
[406] *Id.* art. 6.3.
[407] *Id.* art. 6.4. *See also id.* art. 6.4.2 (urging influenza diagnostic manufacturers who receive PIP materials to make available diagnostic reagents and test kits at no or low charge).
[408] *Id.* art. 6.5.
[409] *Id.* art. 6.6.
[410] *Id.* art. 6.7.
[411] *Id.* art. 6.8.
[412] *Id.* art. 6.9.
[413] *Id.* art. 6.10; *see also id.* art. 6.12 (tiered pricing).

- Oblige Member States to press vaccine manufacturers to set aside a portion of each production of relevant vaccines for developing countries;[414]
- Commit the WHO to work closely with Member States and influenza vaccine manufacturers to increase vaccine supply, including strategies to build new production facilities in developing and/or industrialized countries and "through the transfer of technology skills and know-how."[415]

The PIP's $SMTA_1$ then binds the influenza laboratories operating within the WHO's Global Influenza Surveillance and Response System to comply with agreed WHO terms of reference.[416] It further binds both providers and suppliers of biological materials to a viral license covering WHO guidelines and national biosafety standards.[417] Neither providers nor recipients are allowed to obtain any intellectual property rights on the materials covered by $SMTA_1$.[418]

Providers expressly agree to the onward transfer of materials to both WHO GISRS *and outside entities* on the same terms and conditions as those set out in $SMTA_1$,[419] with a duty to subject all such shipments to the WHO's Influenza Virus Tracking Mechanism.[420] Recipients likewise agree to subject further shipments of the same materials to entities outside the WHO GISRS to the same Tracking Mechanism,[421] and to subject further transfers within the WHO GISRS to the terms of $SMTA_1$.[422] Recipients are also obliged to co-involve scientists from the provider countries in their resulting research and publications, with due regard for attribution of all concerned, especially the collaborators and laboratories that provide clinical specimens or influenza viruses with pandemic potential.[423]

If $SMTA_1$ thus carries forward and enlarges upon the public service benefits that the WHO's pandemic influenza preparedness machinery had been providing for more than 60 years,[424] the more novel provisions of the PIP Framework Agreement bear on the benefits that non-GISRS entities are now expected to provide. First and foremost is the obligation imposed on influenza vaccine, diagnostic, and pharmaceutical manufacturers that make use of the WHO's GISRS to provide annual financial contributions covering 50% of the running costs of these same

414 *Id.* art. 6.11.
415 *Id.* art. 6.13.
416 PIP *Framework*, above n. 382, Annex 1, Standard Material Transfer Agreement 1 [hereinafter SMTA 1], art. 1.1.
417 SMTA 1, above n. 416, arts. 4.1.1–4.1.2.
418 *Id.* art. 6.1. *See also id.* arts. 5.1.1–5.1.2.
419 *Id.* arts. 4.2, 4.3.
420 *Id.* art. 4.4.
421 *Id.* art. 5.1.3.
422 *Id.* art. 5.1.4.
423 *Id.* arts. 5.2, 5.3.
424 *See, e.g.*, Wilke (2013), above n. 380, at 323.

GISRS to begin in 2012.[425] As of 2010, the running costs of the GISRS amounted to about $56.5 million annually.[426]

All WHO Members, including provider countries that contribute valuable biological materials to the PIP Framework, thus benefit concretely from private-sector contributions to the operational costs of avoiding global influenza pandemics under a multilateral preparedness regime. $SMTA_2$, which governs relations between the WHO and relevant non-GISRS entities,[427] then spells out additional benefits that all WHO members may expect from downstream applications of the genetic resources to be pooled under the PIP Framework Agreement.[428]

Under $SMTA_2$, the non-GISRS users of genetic resources from the pool pledge to share the positive benefits of their commercial research – vaccines, medical treatments including pharmaceuticals for the first time, relevant production licenses, and private capacity building efforts – with the WHO, which in turn pledges to share them with the membership as a whole.[429] These multilateral sharing arrangements under $SMTA_2$ vary with the nature of the external entity that receives PIP biological materials,[430] "such as influenza vaccine, diagnostic and pharmaceutical manufacturers, as well as biotechnology firms, research institutions and academic institutions."[431]

For example, manufacturers of vaccines or antiviral treatments who have received materials from the system must commit to at least two of the following options:

- Donate at least 10% of real-time pandemic vaccine production to WHO, or at least a specified percentage of treatment courses of antiviral medicines needed in the pandemic;[432]
- Reserve at least 10% of real-time pandemic vaccine production to the WTO at affordable prices or donate a specified percentage of treatment courses of antiviral medicines needed for the pandemic;[433]

[425] *PIP Framework*, above n. 382, art. 6.14.3 (stating that "specific amounts to be contributed by each company as well as the mechanism for implementing" are to be determined by the Director General and the Advisory Group). However, the relevant companies have reportedly failed to meet these commitments. I.P. Watch, 2015.

[426] *PIP Framework*, above n. 382, art. 6.14.3 n. 1.

[427] *Id*. Annex 2, Standard Material Transfer Agreement Outside the WHO Global Influenza Surveillance and Response System (GISRS) [hereinafter SMTA 2].

[428] *See* Wilke (2013), above n. 380, at 325–27 (characterizing genetic resources under the WHO PIP Framework as "common pool resources").

[429] SMTA 2, above n. 427, art. 4 (obligations of the Recipient); Wilke (2013), above n. 380, at 319, 323 (stressing the multilateral dimension of this sharing, which displaces the bilateral approach of the CBD).

[430] SMTA 2, above n. 427, art. 4.

[431] *Id*. art. 1 n. 1. The selection must be made in consultation with the WHO and the Advisory Group with a view to "optimal pandemic preparedness and response considerations." *Id*. art. 1.1.

[432] *Id*. art. 4.1.1 A1, A2.

[433] *Id*. art. 4.1.1 A3, A4.

- Grant fair and reasonable licenses on mutually agreed terms to manufacturers in developing countries – taking into account development levels in the country of end use – in regard to technology, know-how, products and processing covered by intellectual property rights for the production of influenza vaccines, adjuvants, antivirals, or diagnostics;[434]
- Grant royalty-free licenses to manufacturers in developing countries, or grant to WHO, royalty free, non-exclusive licenses on intellectual property rights, which can be sublicensed to manufacturers in developing countries on appropriate terms, for the production of pandemic influenza vaccines, adjutants, antivirals, products, and diagnostics needed in a pandemic.[435]

Manufacturers of other relevant products may choose any one of the following options:

- Donate to WHO at least a specified percentage of diagnostic kits needed for pandemics;[436]
- Reserve for WHO at least a specified percentage of relevant diagnostic kits at affordable prices;[437]
- Support, in coordination with WHO, the strengthening of influenza-specific laboratory and surveillance capacity in developing countries;[438]
- Support, in coordination with WHO, transfer of relevant technology, know-how and processes to developing countries.[439]

$SMTA_2$ then prohibits the recipient from any further transfer of PIP biological materials unless the prospective recipient has also concluded an SMTA with the WHO.[440] However, recipients may freely exchange such materials with any other holder of an SMTA concluded with the WHO.[441]

B. Governance and Related Issues

Notwithstanding the ambitious nature of the PIP undertaking as a whole and its complexity, the Framework Agreement establishes a relatively lightweight governance

[434] *Id.* art. 4.1.1 A5.
[435] *Id.* art. 4.1.1 A6 (stating that under options A5 and A6, recipients must provide information on licenses granted and their implementation to the WHO).
[436] *Id.* art. 4.1.1 B1, 4.1.1. B1 n. 1 (specifying a range of 5–20% as appropriate).
[437] *Id.* art. 4.1.1 B2, 4.1.1. B2 n. 2 (stressing need for flexible negotiations).
[438] *Id.* art. 4.1.1 B3.
[439] *Id.* art. 4.1.1 B4. Recipients are also urged, but not required, to consider additional contributions, such as donations of vaccines, pre-pandemic vaccines, antivirals, medical devices, and diagnostic kits, among other facilitations listed in *id.* art. 4 on a purely voluntary basis.
[440] *Id.* art. 4.4. Credits for and attribution of the contribution of WHO laboratories are also mandated. *Id.* art. 4.3.
[441] *Id.* art. 4.5.

structure. Implementation of the PIP Framework Agreement as a whole is entrusted to the oversight of the World Health Assembly, with advice from the Director General.[442] This oversight machinery also includes an independent Advisory Group to be composed of international experts who are supposed to serve WHO exclusively and provide "evidence-based reporting, assessment and recommendations regarding the functioning of the Framework," but will not engage in administrative functions.[443]

The Advisory Group will consist of 18 Member States drawn from three Member States in each WTO Region, "with a skill mix of internationally recognized policymakers, public health experts, and technical experts in the field of influenza."[444] Each member will serve a three-year term, with a renewal of one-third of the members each year.[445] The Group will appoint a Chairperson and Vice-Chairperson from its own ranks, who each serve for a two-year term.[446]

The basic task of the Advisory Group is to "monitor access, and report on how the different functions of the Framework are implemented by the components."[447] They are also to evaluate "the system for sharing H5N1 influenza viruses and other influenza viruses with human pandemic potential as well as access to vaccines and other benefits of the Framework."[448] In carrying out this task, the Advisory Group will obtain relevant information from the WHO's own Secretariat as well as from other sources, if necessary,[449] and it will present an annual report evaluating implementation of the PIP project.[450] The Advisory Group may also make recommendations concerning the use of financial and non-financial contributions.[451]

Given the oversight of the WHO itself (through the Director General and the World Health Assembly) and the monitoring of the institutional components by technical experts on the semi-autonomous Advisory Group, the PIP Framework largely depends on the decentralized operations of its constituent institutional components. The scientific missions, and corresponding sharing and capacity building obligations of each institutional component are then spelled out in

[442] *PIP Framework*, above n. 382, art. 7.1.2(i).
[443] *Id.* art. 7.1.2 (iii), 7.2.1.
[444] *Id.* art. 7.2.3.
[445] *Id.*, Annex 3, Advisory Group–Terms of Reference, art. 3.2 ("replacements must maintain the equitable representation of the six WHO regions and affected countries").
[446] *Id.*
[447] *PIP Framework*, above n. 382, Annex 3, art. 2. These institutional components include the National Influenza Centers, other authorized laboratories, WHO Collaborating Centers, H5 Reference Laboratories, and Essential Regulatory Laboratories. *Id.* art. 21.2; *see also* PIP *Framework*, above n. 382, art. 4.
[448] *Id.*, Annex 3, art. 1.2 ("[T]he pharmaceutical industry, although not included, can be consulted by the Advisory Group."). *Id.*
[449] *Id.*, Annex 3, art. 7.2.
[450] *Id.*, Art. 7.2.5. For details concerning the Advisory Group, *see also* id., Annex 3, Advisory Group–Terms of Reference.
[451] *Id.*, Annex 3, art. 2.4.

detailed terms of references applicable to each component.[452] In principle, each term of reference contractually binds the relevant institutional components to the methods and objectives set out in the Framework Agreement.

For example, Collaborating Centers, designated as such by the WHO, must *inter alia*:

- Rely on governmental or non-governmental sources of support;
- Use the WHO Influenza Virus Traceability Mechanism;
- Comply with the $SMTA_1$;
- Maintain the capacity to exchange materials and information on a regular and timely basis;
- Meet specified quality standards;
- Actively collaborate with National Influenza Centers and WHO;
- Provide expertise, training and laboratory support, especially to NICs in developing countries;
- Conduct detailed scientific research on relevant influenza viruses, with sharing of results and of genetic sequence data;
- Help to develop candidate influenza vaccine viruses;
- Select, maintain and update a group of reference influenza viruses and develop related diagnostic reagents.[453]

Perhaps above all, Collaborating Centers must share clinical specimens and relevant viruses with other institutional components; select and distribute candidate viruses to appropriate recipients (including influenza vaccine manufacturers and research institutes); and distribute both reference viruses and corresponding antisera on request, for noncommercial research and surveillance activities.[454]

This decentralized system of research repositories, held together by interlocking chains of contractual obligations, depends obviously on the external support of the WHO. It nonetheless appears to retain a relatively high degree of operational autonomy, subject to the direct supervision of technical experts – representing different regional interests – who staff the Advisory Group and are subject to oversight by the Director General and the Assembly.[455]

[452] *PIP Framework*, above n. 382, Annex 4, Guiding Principles for the Development of Terms of Reference for Current and Potential WHO Global Influenza Surveillance and Response (GISRS), Laboratories for H5N1 and Other Human Pandemic Viruses. *See also id.* Annex 5, WHO Centers for Influenza–Terms of Reference Related to Work with Pandemic Influenza Preparedness Biological Materials (covering GISRS, National Influenza Centers, WHO H5 Reference Laboratories, and Essential Regulatory Laboratories).

[453] *See id.*, Annex 5, at 44–48 (Core Terms of Reference for WHO Collaborating Centers for Influenza).

[454] *Id.*, Annex 5, Core Terms of Reference for WHO Collaborating Centers for Influenza, items A12–14, at p. 45. For other terms of reference applicable to National Influenza Centers, *see id.* at 49–52; to WHO H5 Reference Laboratories, *see id. at* 53–56; to Essential Regulatory Laboratories, *see id.* at 57–60.

[455] *See PIP Framework*, above n. 382, arts. 7.1, 7.2.

In keeping with the relatively lightweight governance structure adopted for the PIP Framework as a whole, the only dispute resolution provisions in the agreement apply to the SMTAs. Disputes arising under SMTA1 are subject to mediation by the Director General and the Advisory Group.[456] Disputes arising under SMTA2, which envisions exchanges of genetic resources to parties outside the government-supported Framework, are subject to binding arbitration.[457]

The absence of any more formal dispute resolution machinery in the overall governance structure may seem anomalous. However, such disputes would almost certainly co-involve governments, especially provider governments, who had not undertaken binding obligations in the first place[458] and whose continued cooperation would be essential in the face of a real influenza pandemic.[459] Given the nonbinding status of the Framework Agreement, reliance on the mediating skills of the Director-General and the Advisory Group are thus a concession to legal and political realities. All the same, the ability of Member States to complain to the Director-General about alleged acts of non-compliance by participating institutions and laboratories partly offsets the lack of a more formal dispute resolution process.[460]

Finally, the funding of the PIP Framework's operations, as previously reported, depends heavily on mandatory contributions from influenza vaccine, diagnostic, and pharmaceutical manufacturers who make use of the WHO GISRS and are correspondingly obligated to cover 50 percent of the costs.[461] Specific amounts per company will be determined by the Director-General and the Advisory Group. Voluntary contributions are also formally requested – and presumably expected – from both Member States and other stakeholders.[462]

C. *Lessons for a Redesigned Microbial Research Commons*

As the most ambitious multilateral agreement to facilitate exchanges of genetic resources to be adopted after the Nagoya Protocol in 2010,[463] the PIP's Framework

[456] *See PIP Framework*, above n. 382, Annex 1, art. 7.

[457] *Id.* Annex 2, art. 5. Disputes arising under SMTA 1, above n. 416, are subject to mediation by the Director General and the Advisory Group. *Id.*, Annex 1, art. 7. For disputes arising under SMTA 2, above n. 427, *see id.*, Annex 2, art. 5.

[458] *See* above n. 386 and accompanying text.

[459] *See* the case of Indonesia, above nn. 381 and accompanying text.

[460] *See PIP Framework*, above n. 382, art. 7.3.3.

[461] *Id.* art. 6.14.3. *See* above nn. 425–426 and accompanying text. Whether they will in fact defray these costs remains to be seen.

[462] *See id.* arts. 6.14.3.1, 6.14.2–6.14.9.

[463] However, regional undertakings to facilitate the exchange of microbial genetic resources after the Nagoya Protocol are underway in Europe, Asia, and elsewhere, as reported earlier in Section I.C.2 of this chapter. *See, e.g.*, above nn. 171–176 and accompanying text (dealing with MIRRI and the Asian Biological Resource Center Network, among others).

sheds considerable light on all the topics to be covered in this book. Equally important is the fact that the WHO's initiative has been crafted at a time when operations of the FAO's multilateral system to facilitate exchanges of plant genetic resources have engendered growing complaints from both seed bank administrators and developing country providers.[464]

On the surface, the PIP Framework and the redesigned Microbial Research Commons as envisioned in this book might appear to be comparable organizational endeavors, from both a geopolitical and a science policy perspective. To begin with, both initiatives deal directly with exchanges of microbial genetic resources, rather than other subcategories of genetic resources having their own legal and economic subcultures,[465] and both build on relatively successful institutional foundations.[466] More importantly, both initiatives seek to implement technically advanced iterations of "common pool resources" – now often designated as "knowledge commons" – in the sense that we use these terms elsewhere in this study.[467] In so doing, both initiatives respond to market failures that otherwise impede the sharing of microbial genetic resources for purposes of research and applications under existing legal and economic conditions.[468]

On closer inspection, however, the contrasting differences between the PIP Framework and the Microbial Research Commons envisioned in this volume far outweigh these similarities. Most critical in this regard are the different economic values of the genetic resources at stake and the different focus of research inputs and outputs subject to market failure in each of these undertakings. In particular, the PIP Framework promotes *downstream pooling of high-value genetic resources* for the equitable sharing of otherwise unattainable public goods, namely, the eradication of influenza viruses with pandemic potential.[469] In contrast, the proposed Microbial Research Commons promotes upstream pooling of genetic resources having no

464 *See* above Chapter 3, Section III.C.2.

465 *See, e.g.*, Godt (2013), above n. 280 (differentiating museums, botanical gardens, seed banks, and microbial culture collections among others, in this respect).

466 For the existing Microbial Research Commons loosely organized by the WFCC, see above Section I.A ; for the preexisting network of WHO influenza preparedness components, see above Chapter 2, Section III. A; *see also* Wilke (2013), above n. 380, at 327.

467 *See* below Chapter 9, Section I; *see generally* COMMON POOLS OF GENETIC RESOURCES (2013), n. 145; BRETT M. FRISCHMANN, MICHAEL J. MADISON & KATHERINE J. STRANDBURG, GOVERNING THE KNOWLEDGE COMMONS (Oxford U. Press, 2014) [hereinafter FRISCHMANN et al. (2014)]. For the view that the PIP Framework constitutes a common pool resource, see Wilke (2013), above n. 380, at 325–27.

468 *See* above Chapter 2, Section II (discussing impact of TRIPS Agreement and CBD on microbial culture collections, and *id.* at Section III.A (discussing collapse of WHO's influenza system during H_5N_1 scare in 2005–2007).

469 *See, e.g.*, Wilke (2013), above n. 380.

known or likely commercial value at the time of deposit, for purposes of both basic research and downstream commercial applications.[470]

1. Trading Downstream Benefits from the Bilateral System for Essential Public Goods

In an insightful article about the PIP Framework, Marie Wilke rightly distinguishes that initiative from other common pool systems operating in the public health sector.[471] For example, unlike the Pool for Open Innovation against Neglected Tropical Diseases[472] and the Medicine Patent Pool,[473] she observes that exchanges under the WHO Framework do not concern a homogeneous group of resources. Rather, the PIP pool encourages exchanges of original biological samples, diagnostic antigens, epidemiological information, candidate vaccine viruses, vaccine production know-how, and actual medical treatments and vaccines.[474]

Under the PIP Framework, moreover, multiple stakeholders, with differing and complex power asymmetries, are co-involved in very different relationships, not just as between developed and developing countries, but also as between old and new pharmaceutical companies, private and public laboratories, and the WHO's own entourage of component institutions.[475] Despite these and other factors differentiating the PIP initiative from many other knowledge commons, Dr. Wilke emphasizes that the "common pool" methodology provides the "potential to regulate ... identified exchanges in an effective and equitable manner," two indicia of sustainability.[476]

[470] The importance of this distinction was first pointed out to meetings of WFCC culture collection administrators and first published in Jerome H. Reichman, *A Compensatory Liability Regime to Promote the Exchange of Microbial Genetic Resources for Research and Benefit Sharing*, *in* Designing the Microbial Research Commons, above n. 11, at 43–54. For insightful applications of this distinction to the PIP Framework, see Wilke (2013), above n. 380.

[471] Wilke (2013), above n. 380, at 325–26.

[472] *See* Bio Ventures for Global Health, *Pool for Open Innovation against Neglected Tropical Diseases – Core Principles*, *available at* http://www.bvgh.org/LinkClick.aspx?fileticket-Bob.mgnC-QGM=.

[473] *See Homepage*, Medicines Patent Pool, http://www.medicinespatentpool.org (last accessed 5 July 2014).

[474] Wilke (2013), above n. 379, at 326.

[475] *Id.* at 326.

[476] *Id.* at 321 (viewing a common pool resource as a "system regulating pooled resources that are available for those inside the system but are restricted for outsiders"). For a discussion of common pool resources and knowledge commons, see below Chapter 9, Section I.A. *See generally* Michael J. Madison, Brett M. Frischmann, & Katherine J. Strandburg, *Constructing Commons in the Cultural Environment*, 93 *Cornell L. Rev.* 657 (2010), *available at* http://www.lawschool.cornell.edu/research/cornell-law-review/upload/Madison-Frischmann-Strandburg-final.pdf; Frischmann, et al. (2014), above n. 467.

As for the market failure that the PIP Framework addresses, Dr. Wilke points first to the default legal obligation on every country that possesses influenza viruses of interest in any given pandemic to negotiate MTAs under the bilateral system of the CBD with would-be users of those same viruses in other countries. Given this state of affairs, would-be users might logically try to obtain similar viruses outside the reach of the CBD, especially in developed countries, where they might be free of benefit-sharing obligations, such as a commitment to help build sufficient capacity for vaccine production in the country that supplies specimens under the CBD.[477] She notes, moreover, that "the benefits of such bilateral agreements only materialize if the company actually ends up being the one to develop vaccines and treatment," which is hard to predict. Absent the PIP multilateral framework, moreover, "only resource providing countries can benefit, while other countries which might be more valuable to [addressing] a pandemic would be left out."[478]

Still another indicator of market failure under the preexisting WHO system arises from the inability of vaccine producers to meet global demand for vaccines during a pandemic, coupled with a corresponding failure to make those vaccines that are produced available at prices poor countries can afford. Hence the need for stockpiling and tiered pricing, as addressed under the PIP Framework, whose efficacy remains to be seen.[479] In other words, although the CBD's bilateral system was designed to ensure that countries providing genetic resources to fight pandemic influenzas would share in the ultimate benefits, there were very few such benefits actually obtained under the preexisting WHO system. After the troubles with Indonesia in the period 2005–2007, indeed, the public health benefits expected from that system were likely to evaporate.[480]

In adhering to the WHO's PIP Framework Agreement, the Member States have thus opted out of a dysfunctional bilateral system in order to improve the prospects for one critical sector of global public health at the expense of a handful of countries that would otherwise provide needed specimen viruses only in exchange for a share of the resulting benefits.[481] The multilateral system is predicated on global pandemic response objectives rather than on the prospects for benefit-sharing from downstream commercial activities under the bilateral approach. As Dr. Wilke explains, it is precisely the most affected countries, and those with limited access to needed vaccines in times of emergency, that would benefit from global research and production efforts, regardless of whether they had provided the relevant virus

[477] Wilke (2013), above n. 380, at 320–21.

[478] *Id.*

[479] *Id.* at 327–31.

[480] *See* above Chapter 2, Section III; *see also* Beldiman, above n. 381; Peter K. Yu, *Virotech Patents, Viropiracy, and Viral Sovereignty*, 45 Ariz. St. L. J. 1563, 1606 (2014).

[481] *See* Wilke (2013), above n. 380, at 332.

specimens.[482] The stockpiling of vaccines for use of H5N1 pandemics and other related initiatives could thus presumably benefit both developed and developing countries. Still other benefits would accrue from tiered pricing obligations, from the transfer of relevant technology, and from capacity building, all of which are express goals of the multilateral approach.[483]

Whether the WHO's multilateral approach under the PIP Framework will actually attain its goals or not remains to be demonstrated. For example, viruses deposited in the system could escape its control, despite the standard MTAs, unless adequate measures to notify the WHO of multiple exchanges were put in place. How to ensure that private companies will fulfill obligations concerning tiered pricing,[484] technology transfer, capacity building, and funding, is another open question. The extent to which global vaccine production can be ramped up to meet global demand depends on the elimination of formidable structural obstacles,[485] irrespective of facilitating exchanges of candidate viruses under the multilateral system, although considerable efforts to expand production capacity in developing countries are under way.[486]

Professor Dana Beldiman fears that the drafters of the WHO's PIP Framework may have underestimated the potential blocking effects of intellectual property rights, as countries with both advanced and emerging technological capacities implemented the TRIPS Agreement of 1994. Precisely because candidate viruses for vaccine production are high-value research inputs and outputs, she worries that multiple overlapping patents, especially gene patents and related patents on diagnostic processes, could generate anticommons effects that would complicate and frustrate the WHO's pooling efforts.[487] Without an intellectual property policy that imposed rigorous and mandatory obligations to pool even patented materials for global pandemic responses, Professor Beldiman predicts that the PIP Framework may succumb to the same excesses of intellectual property protection that have hampered progress in other areas, notably software patents.[488]

482 *Id.*
483 *Id.* at 332–34.
484 *See id.* at 333–34.
485 *See id.* at 332–33; *see also* Abbott, above n. 383.
486 *See, e.g.*, United Nations Conference on Trade and Development (UNCTAD), Local Production of Pharmaceuticals and Related Technology Transfer in Developing Countries – A Series of Case Studies, UNCTAD/DIAE/PCB/2011 (2011). *See also* Christophe Spenneman & Jerome H. Reichman, Using Intellectual Property Rights To Stimulate Pharmaceutical Production In Developing Countries – A Reference Guide (U.N. Conference on Trade & Dev., 2011).
487 Beldiman, above n. 381.
488 *Id.*

For these and other reasons, one cannot yet estimate the aggregate value of the global public health benefits likely to ensue from the PIP multilateral approach, however noble and well-intentioned the drafting process may have been. What cannot be denied is that, by opting out of a dysfunctional bilateral system in the pandemic influenza sector and into a multilateral response to a global public health threat, the WHO has taken major steps to address the market failure that impeded downstream production of vaccines, diagnostic tools, and medicines under the preexisting regime.[489] At the very least, this multilateral approach avoids the need for endless bilateral negotiations among multiple stakeholders, and it optimizes the potential contributions of diverse stakeholders in ways that could not be coordinated or harmonized before.[490] The ultimate goal is equitable access to downstream resources vital for global public health.

In the end, as Marie Wilke observes, the global need for vaccines and medical treatments in times of pandemic influenzas outweighs any bilateral interests because of the risk to all populations from inaction at the global level. In times of emergency, "collaboration and a pool approach are needed instead of bilateral bargaining to generate a meaningful response."[491]

2. Opting into a Multilateral Approach in Order to Stimulate More Downstream Benefits from the Bilateral System

A redesigned Microbial Research Commons as envisioned in the rest of this book also depends on a multilateral approach to address market failures that impede the benefit-sharing aspirations of the CBD's bilateral system. However, its very different conceptual premises differentiate this broader effort to facilitate the exchange of microbial genetic resources from that of the WHO's PIP Framework.

Like the federated collections that exchanged pandemic influenza viruses under WHO auspices prior to the PIP Framework of 2011, the *ex situ* microbial collections loosely affiliated with the World Federation of Culture Collections have established a common pool resource to promote exchanges of genetic resources for research and applications.[492] Unlike either the WHO's preexisting initiative, or the PIP Framework that replaced it, however, which regulate access to and use of high-value genetic resources, the WFCC member collections at issue in this study facilitate access to and use of microbial genetic resources that *have no known*

489 *See* Wilke (2013), above n. 380, at 345.
490 *See id.* at 336.
491 *Id.*
492 *See* Chapter 2, Section I.A.1. For the political economy of the existing microbial research commons as a common pool resource, see below Chapter 9, Section I.C and Chapter 10, Section I.A.

or likely commercial value at the time they were made available to the commons.[493] The WFCC members thus provide both public research institutes and private laboratories with raw materials for pre-competitive research.[494]

As a result, the market failure addressed by the WFCC differs from that addressed by the WHO pools of influenza viruses with pandemic potential. As we have seen, the inability of the bilateral approach under the CBD to ensure exchanges of high-value influenza viruses to generate essential public health benefits leads to a multilateral commons dedicated entirely to that goal. In contrast, the market failure afflicting the WFCC members arises from the provider countries' resistance to sharing microbial genetic resources of unknown value for upstream research purposes without *ex ante*, bilaterally negotiated guarantees of benefit-sharing in the ultimate downstream commercial products.[495] As will be demonstrated in Chapter 5 these efforts to negotiate *ex ante* benefit-sharing agreements for microbial genetic resources having no known or likely commercial value on a bilateral basis actually impede the very upstream research needed to identify and develop those genetic resources that are truly the best candidates for downstream commercial research and development.[496]

The goal of a redesigned Microbial Research Commons is, accordingly, quite different from that of the WHO's PIP Framework, even though both initiatives adopt common pool resource methodologies to address market failure. Providers of high-value influenza viruses to the PIP Framework opt out of the bilateral system because the benefits accruing from the exploitation of private goods pale in comparison with the resulting loss of collective action to provide essential public goods. In contrast, providers of low-value genetic resources to the Microbial Research Commons outlined in this volume expect initially to obtain non-monetary payoffs from the public research enterprise. Ultimately, they expect to share in much larger payoffs from downstream commercial applications than could otherwise have

[493] WFCC collections also act as repositories for high-value microbes deposited by both academic institutes and private companies, as well as repositories for patented microbes under international law. *See* above Section I.A; *see also* David Smith, *Culture Collections*, *in* 79 Advances in Applied Microbiology, Ch. 4 (2012): David Smith, "Networking Collections to Provide Facilitated and Legislation Compliant Access to Microbial Resources," paper presented at the Conference on Governing Pooled Knowledge Resources: Building Institutions for Sustainable Scientific, Cultural and Genetic Resource Commons, International Association for the Study of the Commons (IASC), Louvain-le-Neuve, Belgium, Sept. 12–14, 2012 [hereinafter David Smith (2012)]. However, exchanges of these high-value resources are directly controlled by the depositors and are not generally available for public research purposes.

[494] *Cf.* Arti K. Rai et al., *Pathways Across the Valley of Death: Novel Intellectual Property Strategies for Accelerated Drug Discovery*, 8 *Yale J. Health L. Pol'y & Ethics* 1 (2008) (identifying the importance of this distinction).

[495] *See* above Chapter 3, Section I.C; below Chapter 5, Section I.

[496] For details, *see* below Chapter 5, Section I.A.

been obtained without facilitating the pre-competitive research process through a multilateral regulatory framework.

In other words, the Microbial Research Commons envisioned here is not a multilateral regime that trades the lesser benefits from high-value private goods for greater benefits expected from public goods, as occurs under the PIP Framework. On the contrary, it enables providers of low-value genetic resources to opt into a multilateral regime precisely because that will generate more high-value downstream private goods subject to bilaterally negotiated benefit-sharing agreements in the end.

In effect, under the multilateral approach to exchanges of low-value microbial genetic resources elaborated in this study, developing countries that provide such resources may expect to obtain far greater monetary and non-monetary benefits than would otherwise accrue from their maintaining a pure bilateral approach under the Nagoya Protocol. These greater benefits follow precisely from a strategy that liberates upstream research from proprietary constraints while channeling the positive role of intellectual property rights and the sharing of benefits they provide to bilateral negotiations concerning the development and marketing of downstream end products. How to achieve these results is explained in the next chapter, which also provides detailed scenarios illustrating how the proposed regime would operate.[497]

Still another valuable lesson to be learned from the PIP Framework bears on the WHO's efforts to integrate relevant genomic data into the architecture of its common pool resource, along with candidate vaccine viruses and other institutional components.[498] As will be seen, we devote several chapters in Part Three to this same topic, with corresponding proposals to integrate both data and literature into the multilateral system we envision.[499]

The WHO's PIP Framework also teaches valuable lessons about the governance of a knowledge commons that is specifically directed to attaining science policy goals.[500] When combined with insights from our empirical review of other relevant commons initiative in Chapter 9,[501] these lessons can help avoid many design pitfalls that might otherwise occur. For example, the WHO's adoption of a nonbinding Framework Agreement, rather than an international treaty, appears fully consonant with our own independent analysis of the options to be considered in this regard.[502]

497 *See* below Chapter 5, Sections II & III. *See* further Chapter 10 (Governance and Implementation).

498 *See* above nn. 453 and accompanying text.

499 For details, see especially below Chapter 7 ("Enabling the Microbial Research Community to Control Its Own Scholarly Publications") and Chapter 8 ("Fully Exploiting Data Intensive Research Opportunities in the Digitally Networked Environment"). For implementation, *see* Chapter 10, Section III.C.5 & D.

500 For more on the theory of the knowledge commons, see below Chapter 9, Section I.A.

501 *See* below Chapter 9, Sections II & III.

502 *See especially* Chapter 10 below.

Moreover, the relatively light and flexible governance structure adopted for the PIP initiative, reinforced by a science-driven Advisory Group, complement our own independently generated recommendations in this regard.[503]

In our view, the time has come to accept the challenge that the Nagoya Protocol poses by redesigning the existing Microbial Research Commons in ways that can reconcile the goals of the CBD with the needs and goals of public scientific research, to the benefit of human welfare everywhere. The opportunity to do so has been created by the Nagoya Protocol itself, which is now in force and which, for the first time, recognizes both the role and importance of scientific research in ways that the CBD, as initially drafted in 1992, had overlooked.[504]

To the extent that the "New Biology" paradigm discussed earlier in this study[505] also depends on relatively unrestricted access to pooled genetic resources for public research purposes, the scientific community can no longer afford to rely on the good offices of the WFCC as previously constituted, nor on the system of informal exchanges that was widely used in the past. By the same token, if the microbial research community fails to accept this challenge and to take the initiative now, especially in view of the WHO's groundbreaking multilateral initiative, the likely result is that the CBD's Conference of the Parties will write their own rules for these same purposes,[506] with or without the participation of the scientific community. We trust that the remaining chapters in this volume will help to facilitate that undertaking.

503 *See* below Chapter 10, Section II.
504 *See* above Chapter 3, Section IV.B.
505 *See* above Chapter 1, Section II.D.
506 *See, e.g.*, Nagoya Protocol, above n. 302, art. 19 ("Model Contractual Clauses").

5

Facilitating Transnational Exchanges of Genetic Resources within a Redesigned Microbial Research Infrastructure

I. RECONCILING UPSTREAM RESEARCH NEEDS WITH BENEFIT-SHARING UNDER THE NAGOYA PROTOCOL

Disregarding the WHO's recourse to a multilateral regime for its Pandemic Influenza Preparedness Framework,[1] the empirical evidence reviewed in the preceding chapters shows that, as matters stand, transnational exchanges of *ex situ* microbial genetic resources under the evolving international legal framework are largely confined to two procedural options. One option was to negotiate research uses on a case-by-case approach, within the ambit of relatively standardized clauses that tolerate so-called noncommercial uses to varying degrees, under different responses to the CBD's bilateral approach.[2] A second option was to avoid *ex ante* negotiations altogether by arranging informal and relatively undocumented exchanges of materials among "clusters" of trusted researchers on the basis of reciprocity.[3] Let us further probe the disadvantages of both these options before proceeding to explore the premises for a third approach that could operate far more efficiently under a different legal and economic calculus.

A. *How the Existing Modalities of Exchange Fail the Needs of Scientific Research*

Under the status quo, the single public culture collections make specimens available to users either under their own MTAs or, in the case of some collections, under regionally negotiated or network agreements, such as those discussed in Chapter 4. This means that the culture collections themselves are initially responsible for compliance with the CBD, with the risk that they could incur some international legal responsibility

[1] *See* Chapter 4, Section IV. A & B.

[2] *See id.*, Section III.A.

[3] *See further* below Section I.B.

under the Nagoya Protocol for allowing unauthorized possession and use of microbes imported from other CBD member countries. This Protocol envisions common rules and procedures in order to qualify for international certificates of compliance applicable to all bilateral transfers between signatory countries.[4]

1. Social Costs of the Case-by-Case Transactional Approach

Given these risks, the collections must protect themselves by means of conditions imposed on users; researchers and others must ensure that the specimens they use have been legally obtained; and both users and arguably collections must, in principle, notify the Designated Authorities in provider countries when benefits are obtained in user countries.[5] If the microbial materials in question are available from major collections in more developed countries, includes BRICS countries, standard MTAs may lessen some of these burdens while attempting to limit the potential risks of the relevant provider collections.[6] Even then, how big an administrative burden will thus have been shifted to single researchers could depend on their own (and their respective universities') capacities for risk aversion, especially with respect to post-1992 acquisitions.[7]

If, instead, the needed *ex situ* materials are to be obtained from culture collections in developing countries (or that were acquired from developing countries after 1992), matters could quickly become much more complicated. For example, to comply with these obligations, a researcher seeking access to *ex situ* microbial genetic resources in a provider state may be asked to enter a research cooperation agreement with an institution in that state, especially in a developing country. Such agreements are viewed as "a core means of capacity building" under the Nagoya Protocol.[8]

4 Evanson Chege Kamau et al., *The Nagoya Protocol on Access to Genetic Resources and Benefit Sharing: What Is New and What Are the Implications for Provider and User Countries and the Scientific Community*, 613 *L. Env't & Dev. J.* 246, 258 (2010), *available at* http://www.lead-jounal.org/content/10246.pdf.

5 *See* Chapter 3, Section IV.A & C. Developing country member are also pressing for an international agreement to require patentees to disclose the origin of genetic resources at the time of filing. A group of countries has introduced this proposal formally at WIPO. *See* World Intellectual Prop. Org. (WIPO), Technical Study on Patent Disclosure Requirements Related to Genetic Resources and Traditional Knowledge (2004), *available at* http://www.wipo.int/export/sites/www/freepublications/en/tk/786/wipo_pub_786.pdf.

6 *See* Chapter 4, Section III.A.

7 Provider countries may clearly claim improper possession and use after 1992 (when the CBD entered into force) and may conceivably claim for improper possession of specimens obtained without consent prior to the CBD under international legal theories not rooted in the formal treaty. *See* Permanent Sovereignty over Natural Resources, G.A. Res. 1803 (XVII), U.N. GAOR, 17th Sess., Supp. No.17, U.N. Doc. A/5217, at 15 (1962) [hereinafter 1962 Declaration].

8 Kamau et al., above n. 4, at 258 (citing Tenth Meeting of the Conference of the Parties to the Convention on Biological Diversity, Nagoya, Japan, 18–29 Oct. 2010), Nagoya Protocol on Access

Irrespective of any such cooperative research agreement, the prospective researcher seeking *ex situ* (or *in situ*) materials must obtain administrative authorization for access from the competent state body, a process that will vary with the different national laws. At the very least, notice to a Designated Authority must be given for noncommercial research. Even with such notice, an express research authorization may be required from many states, with a "change of intent" clause to address potential industrial applications.[9] If the microbial genetic resource in question belongs to a local community, or if its know-how is directly or indirectly to be used, that community's express authorization is also required.[10]

In principle, the researcher must also negotiate an ABS contract with the provider state specifying the conditions of access; the permitted uses; and the so-called "mutually agreed terms" concerning the duty to share a percentage of any monetary returns from royalties and sales of end products.[11] According to one source, the "stricter the obligation to share benefits," in such an agreement, "the broader the list of allowed R&D activities a provider state will be willing to grant."[12]

Once given approval under such an agreement, the researcher must eventually arrange a transfer of the microbial material in question to third parties, such as a culture collection in his or her own country. This complicated issue must be covered by the ABS agreement, and in effect it requires a negotiated MTA that subjects the receiving culture collection to the same contractual terms and obligations as the researcher.[13]

Finally, the researcher will have to agree to terms regarding the duty to make research results available to researchers in the provider country, and he or she must also give assurances that use of the microbes will be properly monitored during the research process. These "checkpoints" must satisfy both the provider and the user states, which means that the researcher "will also have to enquire about the relevant rules of the user state where his or her research is conducted."[14]

to Genetic Resources and the Fair and Equitable Sharing of Benefits Arising from their Utilization (ABS) to the Convention on Biological Diversity, arts. 22(4)(d) & 23 [hereinafter Nagoya Protocol] (entered into force after the deposit of the fiftieth instrument of ratification, acceptance, approval, or accession in October 2014), *available at* http://www.cbd.int/abs/doc/protocol/nagoya-protocol-en.pdf (last accessed 29 Sept. 2014)).

9 Kamau et al., n. 4, at 259.

10 *Id.*

11 *Id.* If the research is "noncommercial," some provider countries may theoretically waive this requirement. However, the line between commercial and noncommercial research is so blurred that the researcher could later be forced to negotiate ABS terms when the provider state knows, *ex post*, that there are commercially valuable prospects, if no such benefit-sharing clause had been inserted *ex ante*.

12 *Id.* at 256–59.

13 *Id.*

14 *Id.* at 259.

In this state of affairs, every transaction becomes unique in legal terms; aggregate transaction costs are very high; and permissions are often not granted or take a lengthy period of time to obtain. Once granted, research proceeds either under complicated, tailor-made agreements negotiated case by case when no one knows the value of the specimens in question, or under a sword of Damocles because the Designated National Authority in the provider state may later lodge serious claims of misappropriation when patented (or otherwise protected) end products appear on the market. In the event that such claims are lodged, evidence of the chain of custody could be difficult to obtain, and meeting the burden of proof could be costly to both sides, with an uncertain outcome.

The net effect of these conditions is to impede would-be users' ability to obtain microbes for public research purposes, sometimes even when the research transpires within the boundaries of a single country.[15] Transborder exchanges of materials for research purposes thus risk breaking down altogether, absent some market-making mechanism, such as a clearinghouse, which would likely raise both transaction costs and impediments of its own.[16] For all these reasons, Dr. E. C. Kamau stresses that, seventeen years after the CBD had entered into force, "barely are there any effectively and efficiently functional measures/regimes for access and benefit sharing" in actual practice."[17]

2. The Flawed Premise of the Proprietary Ethos

The burdensome restrictions on use of microbial genetic resources for research purposes emanating from developing countries under the CBD's bilateral approach – like comparable restrictions imposed under contracts and expanding intellectual property rights in developed countries[18] – are often rooted in a false premise. Administrators in both geopolitical blocs tend to view every *ex situ, and especially in situ,* specimen as if it were of potentially great commercial value, while insufficiently recognizing their functions as basic research tools and as important inputs into the research infrastructure as a whole. There is reason to believe that, as access to high-quality materials increases, so does the number of scientific publications in the scholarly peer-reviewed literature, particularly when

[15] *See, e.g.*, Ninth Meeting of the Ad Hoc Open-Ended Working Group on Access and Benefit-Sharing in the Convention on Biological Diversity, Cali, Colombia, 22–28 March 2010, Side Conference Presentations [hereinafter Cali Presentations], *available at* http://www.cbd.int/wgabs9/events/se-abs9.shtml#tab=0.

[16] Nevertheless, the Nagoya Protocol, n. 8, art. 14, does envision a clearinghouse.

[17] Kamau et al., n. 4, at 259.

[18] Scott Stern, Biological Resource Centers: Knowledge Hubs for the Life Sciences 89 (Brookings Inst. Press 2004); *see* Chapter 2, Section II.

the cost of access remains relatively low. These and other socially valuable uses are unnecessarily compromised by overly restrictive provisions for access and use.

Public research uses of microbial materials are also indispensable to generating optimal for-profit R&D, as is clearly seen with regard to type-strains, which are both reference strains in industrial processes and the basis for taxonomy. Overly restrictive licensing conditions for research purposes based on the hopes of short-term commercial gains could thus end by compromising long-term commercial prospects that would result from a properly constructed research commons.[19] By the same token, a comprehensive access and use policy would not only safeguard the interests of public researchers, it could take into account the needs of the global microbiological research infrastructure to receive a small share of the financial returns from commercial applications of their own research inputs.[20]

The flawed premise underlying the proprietary ethos is its failure to acknowledge that the vast bulk of *ex situ* microbial materials held in both publicly accessible culture collections and those of universities all over the world *have no known or likely commercial applications whatsoever*. It is, accordingly, a blunder to subject them to the same kind of proprietarial regime logically reserved for microbes that do posses known or likely commercial value. There is, instead, a need to devise a more research friendly approach that could stimulate – rather than impede – the kind of basic research that, over time, would produce the socially valuable innovations that all the stakeholders aim for in the end.

To attain that goal, the proper baseline premise is to characterize the bulk of all microbes available from the public culture collections all over the world (or, for that matter, from the informal system of exchanges described in the previous chapter) as "precompetitive" research assets. The meaning is that, although these materials may have elicited some basic scientific interest, they had no known or likely commercial applications at the time of deposit.[21] At later stages of scientific inquiry, when efforts are made to know more about new or previously identified microbial specimens, more focused and refined studies may lead to a deeper appreciation of specific materials, to the discovery of new properties associated with derivatives, and perhaps to the prospect of some commercial applications.

[19] *See, e.g.*, Peter Lee, *Contracting to Preserve Open Science: Consideration-Based Regulation in Patent Law*, 58 *Emory L.J.* 889, 895 (2009); *see also* Jerome H. Reichman & Paul F. Uhlir, *A Contractually Reconstructed Research Commons for Scientific Data in a Highly Protectionist Intellectual Property Environment*, 66 *Law & Contemp. Probs.* 315–462 (2003), *available at* http://scholarship.law.duke .edu/lcp/vol66/iss1/12 [hereinafter Reichman & Uhlir (2003)].

[20] *See, e.g.*, Anthony So et al., *Is Bayh-Dole Good for Developing Countries? Lessons from the U.S. Experience*, 6 *PLoS Biology* 2078–84 (2008 For details, see Section III.

[21] For the potential importance of the "precompetitive" status in forming cooperative research ventures, with potentially big payoffs, see Arti K. Rai et al., *Pathways Across the Valley of Death: Novel Intellectual Property Strategies for Accelerated Drug Discovery*, 8 *Yale J. Health L. Pol'y & Ethics* 1 (2008).

On this view, the bulk of the microbial materials currently and prospectively to be held in public collections may properly be characterized as inputs for basic scientific research; that is, as building blocks of future knowledge, and in some instances, for industrial applications as well. Yet, the current MTAs applicable to these materials in both developing and developed economies increasingly tend to impose restrictive conditions on access, use, and reuse that make scientific research costly and difficult to conduct. This is especially the case in collaborative research projects involving large and diverse microbial populations that may be subject to high-throughput screening or other advanced research methods.[22]

Academics depend on their ability to screen large collections of raw materials against leads developed in their laboratories either by phenotypical observations or by genetic analysis, or by some combination of the two. When proprietary restrictions pervade the upstream research dimension, they undermine and risk defeating the research potential of university scientists everywhere, and thereby adversely affect the interests of all stakeholders. [23]

3. Lessons from the Informal Exchange Practices

As we saw in Chapter 4, microbiologists have often escaped from the restrictive coils of the proprietary ethos by engaging in an informal system of exchanging genetic materials solely on the basis of mutual trust and reciprocity. In effect, this informal system of exchange operates under the premise we pinpointed earlier, namely, that the primary value of most precompetitive microbial genetic resources is to serve as inputs to basic or upstream scientific research.

However, "value" in the eyes of participating scientists is necessarily measured in both reputational and commercial terms. In other words, researchers in the informal system would not readily exchange either microbes relevant to pending, unpublished research, or materials having some known or likely commercial value of direct interest to university technology transfer offices.

In the past, the principal advantage of these informal exchange networks was to lower transaction costs while allowing the re-use and further distribution of the research materials with few, if any, of the strings attached to them as if there were real concerns about future commercial applications. At the same time, the unstated but tacitly recognized quality management standards observed by trusted members of the "club" served to guarantee the authenticity and integrity of the materials exchanged.[24]

[22] *See* Chapter 1, Section II.B. & C.

[23] *Cf.* Rai et al., n. 21.

[24] These guarantees may extend beyond immediate research projects, by means of an informal registration process, to subsequent cumulative projects that build on the initial results.

Despite their practical efficacy, however, these informal networks exhibit a number of serious disadvantages,[25] even disregarding the questions of legality. For example, such exchanges are qualitatively limited because, absent a personal relationship built on trust, the participants would not themselves willingly sustain the case-by-case costs of verifying compliance with acceptable quality standards. They would also expose themselves to the risk that unknown third parties could free-ride on the underlying tacit norms that support the informal system, without affording reciprocal access to collections of equal quality on equivalent terms.[26] If third parties were freely allowed to extract materials from the club's genetic resources, the original providers would lose control over them and thereby forfeit the ability to claim either reputational or commercial benefits from ensuing research uses and industrial applications.

As discussed in Chapter 4, moreover, given the commoditizing pressures on microbiological research, the stability of the club system over time would likely diminish as more academic contributors succumbed to high-protectionist MTAs even in the absence of the CBD.[27] More to the point, the CBD and the resulting international regime of misappropriation perfected by the Nagoya Protocol have deliberately jeopardized continued reliance on any system of informal exchanges in the future. Because the Nagoya Protocol requires international certificates of compliance to accompany all transfers between signatory countries,[28] informal exchanges beyond national boundaries now risk violating international law, and they become subject to punitive action under the enforcement provisions of that Protocol.

Even within national boundaries, the domestic laws needed to fully implement the CBD will increasingly treat the informal system as an outlaw regime that threatens universities and governments with sanctions and liabilities, as well as with the embarrassment of appearing to endorse biopiracy.[29] The future capacity of the informal system of exchanges to liberalize and alleviate the constraints of the formally regulated microbial exchange systems thus seems fatally compromised, just at the moment when pressures to further rigidify the formal system in conformity with the

[25] *See* STERN, n. 18.

[26] For the theoretical and practical importance of reciprocity gains in such exchanges, see Minna Allarakhia et al., *Modeling the Incentive to Participate in Open Source Biopharmaceutical Innovation*, 40 *Research & Dev. Mgmt.* 50 (2009); Bernt Hugenholtz, Owning Science: Intellectual Property Rights as Impediments to Knowledge Sharing, paper presented at "Global Science and the Economics of Knowledge-Sharing Institutions," 2d. Communia Int'l Conference, Turin, Italy, 29–30 June 2009.

[27] *See* Chapter 4, Section II.

[28] *See* Nagoya Protocol, n. 8, art. 17.3 (internationally recognized certificates of compliance). *Cf.* World Intellectual Property Organization, Survey on Existing Forms of Intellectual Property Protection for Traditional Knowledge 123, WIPO/GRTKF/IC/2/5 (2001), *available at* http://www.wipo.int/tk/en/questionnaires/ic-2-5/usa.pdf (last accessed 3 July 2014) [hereinafter WIPO Survey].

[29] *See* Nagoya Protocol, n. 8, arts. 15–16.

CBD are reaching maximum intensity, with serious risks for basic microbiological research on the threshold of a "New Biology" paradigm as discussed in Chapter 1.

The existing system of material exchanges consequently affords microbiology only two unsatisfactory options, namely, either to encumber most genetic resources with the same highly restrictive conditions that are only relevant to a handful of deposits with known or likely commercial opportunities; or to allow informal, relatively unrestricted exchanges among a handful of club members, while limiting the amount of material that is effectively available to, and used by, the global research community. The lesson we draw from this analysis is that managers of the public culture collections operating in a networked environment should consider developing a third option that formalizes the basic norms and benefits of the informal club system, along with the obligations and responsibilities that support them, in compliance with the CBD. This standard package deal could then become uniformly available to all communities and researchers willing to abide by an agreed set of quality standards, consistent with policy guidelines that specify the conditions of access, use, and reuse of the distributed components of a legally established multilateral regime.

B. Formalizing the Informal Sector: Premises for a Multilateral Regime of Facilitated Access to Microbial Genetic Resources

A properly designed research infrastructure along these lines would not treat each specimen within its jurisdictional reach as if it were potentially as valuable as gold. Only when the provider of a specific microbe knows that it has actual or likely commercial value does it make sense to restrict access, use and reuse even for research purposes.[30]

A more enlightened approach would treat microbial materials having no known or likely commercial value as a global public good for research purposes,[31] while securing equitable compensation for providers of genetic resources whose microbes ultimately figured in downstream commercial applications that were unknown or

[30] Many of such specimens are held by private industry in collections whose contents have not been publicly certified. Some of the special collections deposited at ATCC afford good examples of cases in which contractually imposed "absolute permission" rules are needed. *See* Chapter 4, Section II.A. In most cases, however, this hoarding approach is both irrational and self-defeating. When such rules are mandated by statutes, whether those of intellectual property laws or so-called misappropriation laws, they become *de facto* exclusive rights valid against the world ("*erga omnes*").

[31] *See, e.g.*, Peter Drahos, *The Regulation of Public Goods*, *in* International Public Goods and Transfer of Technology under a Globalized Intellectual Property Regime 46–64 (K.E. Maskus & J.H. Reichman eds., Cambridge Univ. Press, 2005); *see generally* Joseph E. Stiglitz, *Knowledge as a Global Public Good*, *in* Global Public Goods: International Cooperation in the 21st Century 308, 308–26 (Inge Kaul et al. eds., Oxford Univ. Press 1999).

unlikely at the time of deposit. On this approach, the public culture collections that pooled genetic resources having no known or likely commercial value in a Microbial Research Commons would make their resources readily available to qualifying users for unrestricted scientific research, whether basic or applied.

A first premise of the proposed multilateral regime is that its access policy would thus not distinguish between "for-profit" and "not-for-profit" research activities.[32] Given that materials with known or likely commercial value were to have been excluded *a priori*, all scientific research on the pooled materials should prove valuable in its own right – whatever the end result may be – while a number of judicial decisions teach us that, today, few research institutions are so pure that they are above commercial concerns, especially universities.[33] The proper approach is not to discourage the formation of a broad, precompetitive, easy access research venue out of fears some unforeseen commercial benefits might later emerge. Rather, the goal is to secure equitable compensation for the entities responsible for maintaining and providing the upstream materials that support the research that leads to those downstream commercial products.[34]

A second fundamental premise is that the participating microbiological communities must not become legally obligated to trade the scientists' right to pursue any legitimate research interests in genetic resources accessed from the multilateral regime in return for the providers' right to receive equitable compensation under the applicable rules.[35] The better solution is to decouple the research right from compensatory obligations and to treat the broadly open research possibilities made available by the advent of a global semicommons as the *sine qua non* of its formation.[36] Together with high quality controls and the preservation of reputational benefits,

[32] Contrast the Belgian common MTA of the BCCM, and the model MTA of the European Cultural Collections, Chapter 4, Section III.A.2, which both distinguish between "for profit" and "not-for-profit" research activities.

[33] *Cf.* Madey v. Duke Univ., 307 F.3d 1351 (2002); Patent and Trademark Law Amendments Act (Bayh-Dole Act), 35 U.S.C. § 200 (1980); Karen E. Nelson et al., A *Catalog of Reference Genomes from the Human Microbiome – The Human Microbiome Jumpstart Reference Strains Consortium*, 328 *Science* 994 (2010). *See also* So et al., above n. 20.

[34] *See further* Section II.C. & III.

[35] As seen in Chapter 3, Section II.C.2, this was the mistake made by the drafters of the ITPGRFA. *See* Laurence A. Helfer, *Using Intellectual Property Rights to Preserve the Global Genetic Commons: The International Treaty on Plant Genetic Resources for Food & Agriculture*, in INTERNATIONAL PUBLIC GOODS AND TRANSFER OF TECHNOLOGY, n. 31, 217–224.

[36] While recipients of genetic resources under a multilateral regime should benefit from nondiscriminatory treatment from all culture collections and research entities, public *ex situ* collections that distribute such resources must meet certain quality and tracking requirements (as well as safety and security standards), which results in a semicommons, rather than a fully open knowledge commons. *See further*, Section II.C.1; *see generally* Chapter 10, Section III.A.

open research opportunities constitute both an inherent right of all participants and the *raison d'être* behind this entire initiative. Equitable compensation for providers of genetic resources from downstream commercial applications serves to reinforce the aggregate reciprocity gains of the research commons as a whole by rewarding those who helped to generate unknown or unlikely commercial benefits arising from participation in that same research commons.

A final basic premise, implicit in the entire arrangement, is that the commercial sector cannot write, rewrite or undermine the "open access" norms governing the semicommons as a whole. Although qualifying firms may profit from finding new and unexpected commercial uses from the pooled resources, their duty to equitably compensate the relevant depositors gives them no corresponding power to negotiate limits on access, use, or reuse of the deposited materials, other than those expressly adopted by the global scientific community. This measure avoids the race to the bottom that occurs when each scientific entity is tempted to trade restrictions on research for better deals with the private sector.[37]

To attain these goals, all the microbial culture collections participating in the multilateral system would have to meet agreed quality and safety standards[38] and to respect measures safeguarding the reputational benefits of both depositors and researchers. Moreover, the system must be designed so as to encourage the participation of collections held by universities and other research institutes that meet these same quality standards, with a view to absorbing parts of the existing network of informal exchanges of microbial materials into the legal architecture of a global Microbial Research Commons.

In short, a primary objective of this initiative is to avoid latent tendencies to hoard or restrict the building blocks of knowledge, in the vague hopes of future discoveries, and instead to pool them in a conscious and collective effort to accelerate future discovery, with a built-in incentive scheme that deliberately avoids restrictions on research uses. The payoff, of course, is that any scientist authorized by virtue of his or her connection to a particular institution becomes instantly enabled to roam and explore the full expanse of microbiological research space,[39] with a view to maximizing his or her future abilities to add to, identify and develop value-adding contributions.

[37] Reichman & Uhlir (2003), n. 19.

[38] Hence, technically speaking, the materials component of the proposed Microbial Research Commons is necessarily a semicommons that evolves outwards as more collections meet minimum quality standards necessary for validated scientific research.

[39] Compare the explanation of "chemical space" that becomes possible by pooling small molecule libraries held under trade secrecy in Rai et al., n. 21.

II. DESIGNING A THIRD OPTION: *EX ANTE* "TAKE AND PAY" RULES FOR STIMULATING RESEARCH AND APPLICATIONS

Based on the premises outlined above, we expect that the participating culture collections and their affiliated scientists and laboratories would not normally contribute microbial genetic resources having known or likely commercial applications to the federated, digitally accessible pool we propose. In principle, once the possibility of commercial gain is realistically envisioned, the scientist has reached the outer edge of the proposed multilateral system of facilitated exchange in the sense that he or she is no longer operating behind a "veil of ignorance."[40] The valuable microbe in question may then be protected by trade secrecy, patents, or other proprietary means, including deposits in special collections to which access is contractually restricted. The values to be gained are negotiable on the basis of actual information, and not merely speculative aspirations.[41]

Rather than distorting the open-access values of the entire research semicommons to accommodate this limited set of materials having downstream commercial opportunities, we presume that holders of microbes with known or likely commercial applications will simply migrate to one or the other of those existing regimes that already deal with restrictions on access and use for this purpose. While we may retain a preference for certain approaches over others (e.g., a standard noncommercial use license with the least restrictive research conditions), this topic lies beyond the scope of the semicommons we are attempting to build.[42]

Access to the common pool resources, which are made available to all qualified scientists under a regime of minimum research restrictions, could then lead to the discovery of later commercial applications of a given material that was neither likely nor foreseeable in advance. Such discoveries are welcome contributions to social welfare, a product of skilled efforts and the investment of time and labor, which are not to be confused with parasitic or free-riding uses that undermine incentives to innovate. They may also entail large investments of additional capital and other resources to bring the research product out of the laboratory and

[40] John Rawls, A Theory of Justice (Harvard Univ. Press, 1971). For the significance of operating behind a "veil of ignorance" to promote cooperative research efforts on precompetitive resources, see Rai et al., n. 21.

[41] *See* Jerome H. Reichman & Tracy Lewis, *Using Liability Rules to Stimulate Local Innovation in Developing Countries: Application to Traditional Knowledge*, *in* International Public Goods and Transfer of Technology, n. 31, 337–67 [hereinafter Reichman & Lewis].

[42] For example, in the case of known or likely commercial value, we would still opt for research-enhancing conditions and, where necessary, the use of nonexclusive, rather than exclusive licenses for research purposes. *See, e.g.*, So et al., n. 20. *But see* Nagoya Protocol, n. 8, art. 8(a) (favoring access for noncommercial research with a duty to notify change of intent).

into the stream of commerce after satisfying burdensome public health or safety requirements.[43]

At the same time, ensuring appropriate protection and reward for downstream commercial investors should not obscure the contributions and efforts made by the providers of the original microbial materials and of the culture collections that maintained them and thereby defrayed the resulting preservation and administrative costs. On the contrary, we envision the emergence of a de facto public-private partnership, in which the downstream investors remain freely entitled to make commercial applications under "take and pay" rules (technically known as "liability rules"),[44] which would, however, oblige them to provide equitable compensation to the upstream entities that enabled these same downstream applications.[45]

A. *Legal and Economic Foundations of a Compensatory Liability Regime*

When one speaks of "intellectual property rights," the term usually refers to exclusive rights that restrict specified uses of certain property without the consent of the owner. By definition, such uses depend on the absolute permission of the rights holder. There is, however, a lesser known, second type of intellectual property that allows third parties to make specified uses of a given property on condition that they also make a specified monetary payment to the rights holder for such uses.[46] These so-called "liability rules" may also operate as true intellectual property rights, in the sense that they confer an *ex ante* entitlement on the rights holder who makes the property available on certain conditions. At the same time, they operate as "take and pay" rules, in the sense that the rights holders cannot exclude qualifying users from making the specified uses, on condition that they pay the compensation required for those uses.[47]

[43] These are, indeed, the typical justification for patents and other exclusive intellectual property rights at the end of the commercialization process. *See, e.g.*, Robert Merges, *Contracting into Liability Rules: Intellectual Property Rights and Collective Rights Organizations*, 84 *Cal. L. Rev.* 1293 (1996).

[44] *See, e.g.*, Guido Calabresi & Douglas Melamed, Property *Rules, Liability Rules, and Inalienability: One View of the Cathedral*, 85 *Harv. L. Rev.* 1089 (1972); *see also* Merges, n. 43; Jerome H. Reichman, *Of Green Tulips and Legal Kudzu: Repackaging Rights in Subpatentable Innovation*, 53 *Vand. L. R.* 1743 (2000) [hereinafter Reichman, *Green Tulips*], *available at* http://scholarship.law.duke.edu/faculty_scholarship/456 (last accessed 3 July 2014); Mark Lemley, *Ex Ante versus Ex Post Justifications for Intellectual Property*, 71 *U. Chi. L. Rev.* 129–49 (2004).

[45] *Cf.* Rai et al., n. 21.

[46] As an historical matter, liability rules have always modulated between exclusive property rights and the public domain. *See, e.g.*, Jerome H. Reichman, *Saving the Patent Law from Itself*, *in* PERSPECTIVES ON PROPERTIES OF THE HUMAN GENOME PROJECT 289 (F. Scott Kieff, ed., Elsevier Press 2003).

[47] *See, e.g.*, Reichman, *Green Tulips*, n. 44; *see also* Reichman & Lewis, n. 41.

While liability rules are most familiar from tort law, especially the sector dealing with nuisance law,[48] Professor Reichman and, lately, numerous others have argued that greater use of such rules in intellectual property law would help to solve many of the thorny problems that arise from allowing patents or other hybrid intellectual property regimes to govern small-scale or "cumulative and sequential" innovation[49] and also from overextending exclusive rights into the upstream research space.[50] At the same time, liability rules frequently operate inside existing intellectual property regimes, often in a subterranean role that passes unnoticed. For example, compulsory licenses frequently used in both patent and copyright laws may convert an "absolute permission" rule into a "take and pay rule,"[51] as will also occur when a patent or a copyright court refuses to issue an injunction to halt infringing activity and instead holds that the payment of damages suffices.[52]

In these examples, however, one must distinguish an *ex ante* automatic license, which is certain to kick in, from compulsory licenses that arise *ex post*, in the sense that they may or may not override the rights holders' otherwise justified expectations of exclusive use.[53] In the former case, rights holders expect that third parties may make certain uses of their property in return for a given payment, in which case the users' business strategy relies *ex ante* on this assumption. For example, record companies in the United States know that, once a musical work has been recorded, any other company can make their own recording by paying a statutory royalty without the consent of the owner of copyright in the musical work.[54] When, instead, a judicial or administrative authority subjects a patented invention to a compulsory license or to the denial of injunctive relief, the payment of damages *ex post* may to some extent skew the investor's *ex ante* business calculus and deprive him or her of expected gains.

[48] *See esp.* Calabresi & Melamed, n. 44.

[49] *See, e.g.*, Jerome H. Reichman, *Legal Hybrids between the Patent and Copyright Paradigms*, 94 *Colum. L. Rev.* 2432 (1994); Jerome H. Reichman, *Charting the Collapse of the Patent-Copyright Dichotomy*, 13 *Cardozo Arts & Ent. L. J.* 475 (1995).

[50] *See, e.g.*, Reichman & Uhlir (2003), n. 19; Jerome H. Reichman, *How Trade Secrecy Law Generates a Natural Semicommons of Innovative Know-How, in* LAW AND THEORY OF TRADE SECRECY: A HANDBOOK OF CONTEMPORARY RESEARCH Ch. 8 (Rochelle Dreyfuss & Kathy Strandberg eds., Edward Elgar Publishing 2011); *see also* Jerome H. Reichman & Ruth L. Okediji, *When Copyright Law and Science Collide: Empowering Digitally Integrated Research Methods on a Global Scale*, 96 *Minn. L. Rev.* 1362 (2012) [hereinafter Reichman & Okediji]; So et al., n. 20.

[51] Jerome H. Reichman, *A Compensatory Liability Regime to Promote the Exchange of Microbial Genetic Resources for Research and Benefit Sharing* [hereinafter Reichman (2011)] in DESIGNING THE MICROBIAL RESEARCH COMMONS: PROCEEDINGS OF AN INTERNATIONAL SYMPOSIUM 48 (P.F. Uhlir ed., Nat'l Acads. Press 2011) [hereinafter DESIGNING THE MICROBIAL RESEARCH COMMONS], at 48.

[52] *See, e.g.*, eBay Inc. v. MercExchange, 547 U.S. 388 (2006).

[53] *See, e.g.*, Lemley, n. 44.

[54] *See generally* Merges, n. 43.

By the same token, even an *ex post* liability rule can sometimes provide more compensation to the rights holder than he or she might have obtained via the exercise of an exclusive property right. This will occur when, in response to market conditions, many different users exercise their "take and pay" rights and thus confer a larger aggregate revenue stream on the rights holder than he or she would have anticipated by licensing the innovation exclusively to one or more users. Liability rules can thus produce a "lottery effect,"[55] as for example occurred when Stanford and the University of California were persuaded to license their Cohen-Boyer gene-sequencing patents nonexclusively to all comers.[56]

This last example also illustrates the fact that intellectual property owners may unilaterally convert an exclusive right into a liability rule, as occurred with the Cohen- Boyer patents, or they may enter voluntary agreements establishing a liability regime in place of even an exclusive rights regime otherwise adopted by applicable statutes. When, for example, many patent owners agree to pool their inventions for a common purpose, in order to avoid patent thickets and other blocking effects,[57] they will have contractually created a pool governed by mutually inclusive liability rules that override the effects of the exclusive rights that constituted the baseline default rules under which each player individually operated at the outset.[58]

In this connection, recent literature has focused attention on the potential advantages that may accrue from voluntarily establishing common pool resources, or semicommons, in which all of the players may benefit from the aggregate resources by paying a specified set of tithes regulating the uses in question.[59] The proposed third option for the large-scale microbial research semicommons under consideration here seems an ideal setting for some variant of "a Compensatory Liability Regime."[60] In fact, such a collocation becomes all the more logical in that

[55] *See, e.g.*, Reichman, *Green Tulips*, n. 44; Reichman & Lewis, n. 41; *see also* Rai et al., n. 21.

[56] *See, e.g.*, So et al., n. 20 (In this case, the NIH did the persuading). See further Chapter 2, Section II.B.2.

[57] Robert P. Merges, *Institutions for Intellectual Property Exchange: The Case of Patent Pools, in* INTELLECTUAL PRODUCTS: NOVEL CLAIMS TO PROTECTION AND THEIR BOUNDARIES (Rochelle Dreyfuss ed., Oxford Univ. Press, 2001).

[58] Rai et al., n. 21; Merges, n. 43; Reichman & Uhlir (2003), n. 19; *see most recently* Jerome H. Reichman, Richard Newell, Arti K. Rai, & Jonathan B. Wiener, *Intellectual Property and Alternatives: Strategies for Green Innovation, in* INTELLECTUAL PROPERTY RIGHTS: LEGAL AND ECONOMIC CHALLENGES FOR DEVELOPMENT 356, 377–80 (M. Cimoli et al. eds., Oxford U. Press 2014).

[59] *See, e.g.*, Lee, n. 19; Robert A. Heverly, *The Information Semicommons*, 18 *Berkeley Tech. L.J.* 1127 (2003); Joan W. Bennett, *Microbiology in the 21st Century, in* DESIGNING THE MICROBIAL RESEARCH COMMONS, n. 11, at 12. *See also* Brett M. Frischmann, *An Economic Theory of Infrastructure and Commons Management*, 89 *Minn. L. Rev.* 917 (2005). For common pooled resources in general see ELINOR OSTROM ET AL., RULES, GAMES, AND COMMON-POOL RESOURCES (Univ. Mich. Press 1994) and the discussion of Knowledge Commons in Chapter 9, Section I.A.

[60] *See* Reichman, *Green Tulips*, n. 44.

the United Nations Food and Agriculture Organization (FAO) already pioneered the application of a rudimentary compensatory liability regime to plant genetic resources in the International Treaty governing plant genetic resources, concluded in 2001,[61] as discussed in Chapter 3. Our proposed third option may, indeed, be understood as an effort to apply such a regime to microbial genetic resources, but without the design flaws outlined in Chapter 3, that have weakened that Treaty.[62]

The economic logic underlying this model is that the providers of microbial materials would presumably obtain more potential reciprocity benefits from the substantial upstream research opportunities generated by the semicommons than would accrue from operating in isolation. Fears of losing unknown future commercial opportunities could undermine the prospect of these potential research gains, however, so we address this concern directly with a built-in provision for benefit-sharing from unknown future downstream commercial applications. That is, we would build a liability rule for downstream commercial applications into the system, yielding equitable compensation for providers (and perhaps for the multilateral system itself), while fulfilling international obligations under the CBD.[63]

A liability rule in this context means that one may freely take the materials for any research purpose, without need of any permission to use, on condition that a duty to pay equitable compensation arises if and when some future commercial application generates financial gains. A liability rule is indicated here precisely because exclusive property rights and equivalent contractual licensing schemes do not work well when the values of potential uses are not known and each party over-values his or her property – or the uses to be made thereof – because nobody knows its true worth.[64]

With a liability rule, the message is not "You cannot use my microbial materials for commercial purposes without permission." It is, instead, the opposite: "Please find commercial uses for my research materials, and, when you do, please pay me a reasonable royalty from your gross sales." Notice that this is not a compulsory license *ex post*. It is a built in automatic *ex ante* license to use and pay – a preexisting obligation to share a small percentage of any eventual economic returns with providers and possibly others who maintain and regulate genetic resources, both of which contributed to the downstream commercial payoffs.[65] Notice, too, that

[61] International Treaty on Plant Genetic Resources for Food and Agriculture, *opened for signature* 3 Nov. 2001, 2400 U.N.T.S. 303 (entered into force 29 June 2004) [hereinafter ITPGRFA]; *see* Chapter 3, Section III.B.

[62] *See id.*, Section III.C.2.

[63] *See* Nagoya Protocol n. 28, art. 4; Chapter 4, Section IV.C; *see also* Chapter 10, Sections II & III.C.2 (on governance and distribution of royalties).

[64] *See* Reichman & Lewis, n. 41.

[65] *Cf.* Reichman, *Green Tulips*, n. 44. For theoretical considerations inherent in the concept of a knowledge commons, *see* Chapter 9, Section I.A.

there is also a built-in possibility of lottery effects if many downstream commercial applications spin off from any given microbial genetic resource.

At the outer edge of our proposed third option, in other words, we protect microbial materials that had no known or likely commercial applications when deposited, but which subsequently turn out to lend themselves to such downstream applications. Here, in addition to reputational benefits, we attempt to build in an equitable compensation model that will ensure that a fair share of the benefits will go to the provider – or to that provider's legal proxies – that enabled the downstream discovery, but without access obligations that would encumber the progress of scientific discovery or create barriers to entry. In short, we envision a Compensatory Liability Regime[66] that would kick in whenever unforeseen commercial applications emerge, in the interests of equity, and to heighten the potential reciprocity gains from contributing to the pooled resources by directly addressing fears of lost commercial opportunities.[67] By building in these reciprocity benefits from the beginning we can reconcile the positive opportunities of commercially exploiting microbial materials with the advantages of open public science.

To achieve this goal, however, the proposed multilateral system must be put on a solid legal and administrative foundation that would immunize it from attacks, especially those sounding in either intellectual property law or the CBD's ABS provisions, without unduly sacrificing its effectiveness.[68] Collective efforts are needed to reduce the risks of all contributors, while enhancing overall research efficacy. Ideally, this approach should promote widespread use of *ex situ* microbial materials, together with related data and information for scientific purposes,[69] with the fewest possible restrictions, and it would make extensive use of science-friendly, standard-form contracts for this purpose.

B. *Operational Logic of a Multilateral Common Pool Resource*

When we posit that a multilateral regime of facilitated access should encompass only deposits of microbial materials that lacked any known or likely commercial value at the time of deposit,[70] we recognize that this criterion ignores the value-adding

[66] See nn. 60–65 and accompanying text.

[67] *See* Allarakhia et al., n. 26 (on the importance of reciprocity benefits in this connection); *see also* Paul A. David, *The Historical Origins of "Open Science": An Essay on Patronage, Reputation and Common Agency Contracting in the Scientific Revolution*, 3(2) *Capitalism & Soc'y art.* 5 (2008), *available at* http://capitalism.columbia.edu/files/ccs/Paul%20A.%20David.pdf [hereinafter David (2008)].

[68] *See generally* Section III and Chapter 10 (governance).

[69] *See generally* Part Three.

[70] With respect to microbial material already having manifested known or likely commercial value, the providers stand to gain more from holding out than from the research opportunities flowing from participation in a multilateral regime of facilitated access. *See* Minna Allarakhia, *Microbial*

contributions of early stage microbiological researchers who discovered *in situ* specimens of interest as well as the preservation and validation efforts of public culture collections in which *ex situ* specimens are eventually deposited.[71] Partial compensation for these efforts arises indirectly from the built-in availability of genetic resources from the multilateral system. As will be seen, the "take and pay" rules can also be adjusted so as to include a revenue stream for the public culture collections. Moreover, it is well to recall that only a small percentage of the world's existing microbial population has actually been identified, and that most of those identified microbes possess largely unidentified properties and characteristics of no known commercial interest.

Once available from the multilateral system, further research on these microbial genetic resources may lead to valuable commercial applications later on, such as biofuels or pollution mitigating agents.[72] One expects that the very capacity for greater numbers of qualified scientists to explore a vast microbiological research space made available by the pooled resources would magnify the prospects for commercial opportunities.[73]

In that event, however, the downstream commercial developer would be bound by a standard MTA that obligated it to pay reasonable royalties from the proceeds of successful commercial applications to the provider country's Designated National Authority. This built-in reach through transaction would not interfere with the proprietary protection of the end product, which results from a negotiated deal between the specific inventor and the relevant commercial investors.[74] The scheme thereby promotes exploratory research on materials made available by providers who are secure in the knowledge that, if commercial applications emerge, they will obtain a fair share of the returns under a built-in reward mechanism.

A primary goal of a multilateral system is thus to avoid unnecessary restrictions on research uses with respect to the bulk of all the materials deposited in the network of the participating collections. It should not be necessary to encumber either the materials exchanged or the research process generally with detailed conditions and requirements that are only appropriate when the parties deal with materials having known or likely commercial value. The scheme should, instead, promote all research uses and reuses of the deposited materials, with the fewest possible

Commons: Governing Complex Knowledge Assets, in DESIGNING THE MICROBIAL RESEARCH COMMONS, n. 51, at 145. However, the research funders themselves might have required that materials resulting from a given project had to be made available for specified public research purposes, as occurs with increasing frequency. See Chapter 8, Section I.

71 See nn. 28–29 & accompanying text. For controversial and still unresolved questions about retroactive application of the CBD, see Chapter 3, Section I.C.

72 *See, e.g.*, Reichman, Rai, Newell & Weiner, n. 58; *see further* Chapter 2, Section II.B.

73 *See, e.g.*, Rai et al., n. 21.

74 *Cf. id.*

restrictions bearing primarily on the need to preserve reputational benefits, plus biosafety and security issues where appropriate.[75]

The standard MTA to be devised for the multilateral system should therefore require little or no *ex ante* negotiations or permissions concerning any aspect of the research process as such, although some *ex post* negotiations may be necessary with regard to the benefit-sharing royalty, as explained later.[76] Permission to make copies and even derivatives as needed, plus the duty to pay a reasonable royalty from commercial applications, should be built into the model MTA. These rights and duties should arise automatically from membership or participation in the multilateral regime and from certified compliance with its quality standards and other legal terms and conditions. Promoting research should thus be the primary goal of the undertaking, and all research uses by nationals of member states should be presumed legitimate unless specifically excluded by the standard MTA or the framework agreement.[77] At the same time, the Compensatory Liability Regime built into the model serves to overcome risk aversion on the part of would-be providers of microbial genetic resources by imposing a duty to share revenues from downstream commercial applications with those same providers or their legal proxies.[78]

This embodiment of a broad, nonnegotiable research exemption in the core provisions of the model thus rejects the approach taken by the International Treaty on Plant Genetic Resources for Food and Agriculture and, in our view, corrects one of that treaty's biggest design flaws. There, it will be recalled, commercial entities accessing plant genetic resources from the Crop Commons can opt out of the duty to pay royalties on commercial applications if they allow use of their end products for research and breeding purposes.[79] Under the proposed Microbial Research Commons, instead, allowing unrestricted research uses of pooled genetic resources is a *sine qua non* of membership itself, and all commercial payoffs are automatically subject to benefit-sharing under the Compensatory Liability Regime.

By the same token, the foundational agreements regulating the Microbial Research Commons, discussed in Chapter 10, must subject the making of copies, modifications, and derivatives from pooled resources to rules that regulate both scientific competition and benefit sharing from possible commercial applications.[80]

75 *See* Section II.C.4.

76 *See* Section II.C.3 and illustrative scenarios in Section III.

77 For example, restrictions for safety and security may be necessary.

78 For details, see Section III.

79 *See* Chapter 3, Section III.C; FAO Conference, Comm'n on Genetic Resources for Food and Agriculture, Standard Material Transfer Agreement [hereinafter SMTA], *available at* http://www.planttreaty.org/content/drafting-standard-material-transfer-agreement. *See also* S. Carrizosa et al., Accessing Biodiversity and Sharing the Benefits: Lessons from Implementing the Convention on Biological Diversity (IUCN Envtl Pol'y & L. Paper Series No. 054, 2004).

80 *See* below Section II.C and the illustrative scenarios in Section III.

These rules are further contingent upon the need to devise an appropriate tracking and registration system, which is intrinsically relevant to the maintenance of quality controls.[81]

Before we proceed to a more detailed illustration of our model, however, it is worth pausing to focus on some important differences between the multilateral regime of facilitated exchanges we envision and the existing options under other forms of standardized or semi-standardized licensing agreements or under informal arrangements outside of any regulatory framework. As demonstrated in Chapter 4, virtually all the MTAs currently used in the formal exchanges of microbial genetic resources risk slowing scientific progress because they impede commercial research generally as well as the distribution of duplicates of the materials exchanged (although many duplicators violate such constraints in practice).[82] Even one version of a proposed Science Commons model seemed not to have allowed any distribution of duplicate materials, and it required explicit permission for a recipient to keep the original material after the specified research project had been accomplished.[83]

[81] *See* Section II.C.1–3.

[82] Facilitating the distribution of duplicates or derivatives is one of the main reasons behind our proposition for a reformed semicommons. Nevertheless, there are various legitimate reasons from a public science perspective to limit some duplication. The most important is the reliability of cumulative follow-on research based on certified biomaterials. Several examples are discussed in the literature where contamination or mutation of biological material after exchange can lead to invalidation of all the results based on the duplicates or derivatives (STERN, n. 18). According to recent estimates, perhaps more than 20 percent of all cell lines remain misidentified, and thousands of articles based on misidentified cell lines are published every year. *See* Amanda Capes-Davis & R. Ian Freshney, *Database of Cross-Contaminated or Misidentified Cell Lines*, ATCC, Sept. 3, 2012, http://standards.atcc.org/kwspub/home/the_international_cell_line_authentication_committee-iclac_/Database_of_Cross_Contaminated_or_Misidentified_Cell_Lines.pdf. Other reasons are related to biosecurity issues and biosafety. Our model of an integrated semicommons aims to address these issues, without creating the unnecessarily burdensome restrictions imposed by the current MTAs.

[83] The basic Science Commons Material Transfer Agreement (now reintegrated with Creative Commons) specified that the recipient may not transfer or distribute the materials, or use the materials for clinical purposes, and may not use the material in connection with the sale of a product or a service. Under these conditions, the recipient was free to use the material for research under his supervision, by himself or by others, and to publish the results of this research. Sci. Commons, Science Commons Material Transfer Agreement, *available at* http://mta.sciencecommons.org/agreements/sc/1.0/legalcode (last accessed 3 July 2014). Some major research institutions have already agreed to use this model contract as the core component of their own MTAs, notably, the Cornell Institute for Medical Research and a set of partner laboratories involved in research into Huntington's disease. These organizations use the Science Commons model contract and a limited set of possible variants, obtained by adding a set of restrictions, which are proxies for degrees of commercial use (the reason for introducing these proxies is the difficulty of defining commercial use). Restrictions that can be added upon the default license include: (1) restrictions that forbid the duplicating of the materials in large quantities; (2) restrictions that forbid retaining the materials after the research is finished; (3) restrictions which only allow use of the material for research on certain specific diseases. Science Commons contracts are available at http://mta.sciencecommons.org/chooser (last accessed

The existing system thus assumes case-by-case negotiations for every commercial contingency with the risk of each party overvaluing what it has to offer when the actual value remains unknown, and is usually unknowable, at the time of the desired research uses. Such negotiations augment the likelihood of many default research constraints, although some transaction costs could otherwise be lowered under the online MTA models that Science Commons originally conceived.[84]

One notable exception to these research restrictions appeared in the standard MTA of the European Union Culture Collection Organization (ECCO) that was approved by the organizations' board in February 2009.[85] Under the rules of this MTA, a participating culture collection could further distribute biomaterials within that semicommons as long as it used the same standard contract under which it received the material in the first place and also complied with an agreed set of quality management and biosafety requirements. However, this standardized agreement to enable "legitimate exchange" primarily allows redistribution of received material to the public service culture collections. It lacked any standard provisions for dealing with commercial applications, and did not expressly provide a broad research exemption for all microbes having no known or likely commercial value.[86]

Under our proposed third option, in contrast, any deposit of an *ex situ* microbe in a participating collection (assuming a qualified provider) should entitle all qualified research users to reproduce the material deposited for scientific purposes, subject to rules concerning attribution, publication of research results, and eventual payoffs from any commercial applications.[87] This standardized approach becomes especially important now that the Nagoya Protocol has entered into force. Without

3 July 2014). These MTAs are directly web-based (the provider can use a simple graphical interface to choose amongst the options), and are disclosed to other parties involved in transactions with the same biological materials (cell lines), and they are machine readable.

84 See n. 83. Recent submissions to the Convention on Biological Diversity build further on the Science Commons model MTAs for designing a so-called ABS commons. *See* Paul D. Oldham, *An Access and Benefit-Sharing Commons? The Role of Commons/Open Source Licenses in the International Regime on Access to Genetic Resources and Benefit-Sharing*, Initiative for the Prevention of Biopiracy (Research Doc., Year IV, No. 11) (2009). As far as it is also based on case – by-case negotiations, it will likely suffer from the risk of a race to the bottom on research restrictions, although, as explained here, it may facilitate licensing of the small subset of materials with known high commercial value.

Based on the principles of open e-Science, the expectation is that providers will gradually opt for the less restrictive licenses, as they will be able to attract more users. The reasoning behind this expectation is that, through the disclosure of all the contracts on the digital interface, users working in a networked environment will be able to compare the different contracts and choose the most science friendly ones.

85 *Cf.* European Union Culture Collections' Organization (ECCO), http://www.eccosite.org/ (last accessed 3 July 2014). *See further* Chapter 4, Section III.A.2. ECCO's standard MTA thus aims to build a regional semi-commons, while preserving the main characteristics that define its microbial materials as authenticated knowledge resources available for further follow-on research and uses.

86 *See further id.*

87 For details, see the illustrative scenarios in Section III.

such an approach, each national provider of genetic resources would tend to impose its own conditions on research uses. The prospect of overlapping and conflicting restrictions on use could then pose heavy burdens on the research community, with fewer downstream applications and fewer payoffs for all the relevant stakeholders.

Conversely, the existence of a third option along the lines we propose would not affect holders of microbial genetic resources having known or likely commercial value, who would normally opt for the more restrictive MTAs currently in use at the time of deposit. But the creation of a third option would give all the stakeholders a legally viable choice. The heightened research benefits available from the resources pooled in the semicommons should then attract more contributors over time, who might otherwise have drifted into a restrictive research regime by inertia. In short, our third option, if successfully implemented, would gradually unite ever larger segments of the scientific community and – if properly designed – could help to build an integrated, one-stop-shop research infrastructure.

The success of this cooperative venture depends, in the first instance, on maintaining the kind of high quality standards that already prevail in both the existing formal and informal sectors. Otherwise, the potential benefits accruing from an open access model would be constantly undermined by the contrasting appeal of higher quality standards imposed by the proprietary models. High quality standards, in turn, necessitate recourse to a semicommons open to qualified participants rather than a full-fledged commons open to all. This and other key components of the proposed Compensatory Liability Regime are discussed in the next section and then illustrated in six hypothetical transactions.[88]

C. *Key Components of the Proposed Multilateral Regime for Facilitated Exchanges of Microbial Genetic Resources*

In order to build a multilateral regime operating on a solid legal foundation, there must be some intergovernmental framework agreement that contractually regulates the relations between all the participating microbial research communities and their member governments based on the foregoing principles, as discussed in Part Four. The contracting parties must also devise a Standard Material Transfer Agreement (SMTA) that specifically ensures conformity with the Access and Benefit Sharing norms of the CBD, which typically operate in favor of developing countries. Enforcement of such a standard-form agreement would be the province of a Governing Body that would become generally responsible for oversight and management of the projected microbial research infrastructure.[89] For present

[88] *See* Section III.

[89] *See further* Chapter 10 (governance). For theoretical and empirical considerations see *generally* Elinor Ostrom, Governing the Commons: The Evolution of Institutions for Collective Action (Cambridge U. Press 1990) and Chapter 9 *passim*.

purposes, it suffices to emphasize that a key condition of the proposed regime is that participating governments must discourage the informal arrangements of the past in favor of a clearly specified transnational agreement that provides the maximum degree of research freedom, consistent with the need for legal and methodological stability.

In what follows, we first identify the key issues that a standard-form MTA devised for a redesigned Microbial Research Commons would need to address, while otherwise avoiding all unnecessary restrictions on research, whether basic or applied, with respect to materials having no known or likely commercial value when deposited in the semicommons. These issues include:

- The requisite quality standards that all participating collections (and users) would have to meet, as a limiting condition on the scope of the commons in the interest of validated scientific research results;
- The need to preserve the reputational benefits of depositors to the fullest extent possible;
- The royalties required to fairly compensate providers of microbial materials for posterior commercial uses under an *ex ante* revenue sharing mechanism;
- The need for measures to track the transfers and uses of deposited microbial materials and to avoid leakage from the system;
- The need for an appropriate governance model and dispute resolution mechanism to ensure that both contractual and treaty obligations are observed.

Needless to say, a standard MTA embodying our proposed liability regime would also need to address restrictions for biosafety and national security, which we do not discuss in detail.[90] We end this chapter with illustrative scenarios based on a sequence of hypothetical transactions. Governance issues are more fully addressed in Part Four.

1. Quality Standards as a Threshold Requirement

A central challenge for life sciences research is how to maintain the integrity of shared biomaterials.[91] To the extent that the public culture collections already operating within the larger context of the WFCC make strains available at the marginal cost of distribution, this practice is subject to stringent requirements of

[90] *See generally* Forum on Microbial Threats, Board on Global Health, Institute of Medicine, Washington, D.C., http://www.iom.edu/Activities/PublicHealth/MicrobialThreats.asp (last accessed April 9, 2015) *Cf.* Instit. Medicine, The Threat of Pandemic Influenza: Are We Ready? (Stacey L. Knobler et al. eds., Nat'l Acad. Press 2005); Institute of Medicine, The Domestic and International Impacts of the 2009 H1N1 Influenza Pandemic – Global Challenges, Global Solutions (David A. Relman et al. Rapporteurs, Nat'l Acads. Press 2010).

[91] Stern, n. 18.

quality management and control. In principle, only entities having the capacity to maintain certain levels of purification, identification and preservation are allowed to further distribute duplicates of any given material.[92]

Considerable attention has focused lately on elevating and harmonizing the quality standards of affiliated culture collections, especially in light of the OECD Best Practices Guidelines for Biological Resource Centers, which also cover biosecurity, capacity building, preservation of biological resources and data management.[93] The WFCC's own revised Guidelines, issued in 2010, aim for an intermediate level of quality controls that would not entail the costly investments needed to become a full-fledged BRC as defined by the OECD.[94]

The WFCC's Guidelines on access, curation, preservation, and validation were explained earlier in Chapter 4.[95] How successfully the applicable quality standards have been implemented in the past remains an open question, however. Such standards have reportedly varied considerably in practice, serious problems have been encountered at various times, and both contamination and misidentification remain problems.[96] Scientists themselves often lack incentives and means to undertake validation of others' research results and may resist pressures to further validate their own previous research.[97] Perhaps more thought could be given to establishing a certification system for ensuring that national culture collections meet internationally agreed quality standards,[98] as well as Access and Benefit-Sharing standards under the CBD.[99]

Meanwhile, some leading collections, such as the ATCC, have earned a reputation for maintaining the highest quality standards, which enhances the appeal of their proprietary models. Moreover, as noted earlier, a distinguishing feature of the informal exchange system still responsible for at least 40 percent of current material exchanges is reportedly the high quality standards practiced by

92 *See* Chapter 4, Section I.A.2.

93 World Fed. Culture Collections (WFCC), *Guidelines for the Establishment and Operation of Collections of Cultures of Microorganisms* 2 (3d. ed., Feb. 2010), *available at* http://www.wfcc.info/guidelines/ [hereinafter WFCC, *Guidelines*] (discussed Chapter 4, Section I.A.2). *See also* OECD, OECD Best Practice Guidelines for Biological Resource Centers (2007) [hereinafter OECD Best Practices], *available at* http://www.oecd.org/sti/biotech/38777417.pdf, discussed Chapter 4, Section I.C.2.

94 For proposals to form a network of only collections meeting the higher BRC standards, *see* the discussion of the proposed Global Biological Research Center Network (GBRCN) in Chapter 9, Section II.C.

95 *See* Chapter 4, Section I.A.2.

96 Stern, n. 18, at 39–40.

97 *Id.*

98 Interview with Micah Krischevsky, April 14, 2009, Washington D.C., citing WHO certification of production standards for pharmaceuticals.

99 *See, e.g.*, Nagoya Protocol, n. 8, art. 17.3 (evidentiary role of internationally recognized certificate of compliance).

those allowed to participate.[100] These standards reinforce the trust indispensable for informal collaboration and sometimes also make it possible to identify and trace the materials in other follow-on activities, including derivatives emerging from the informal system.

In devising a third option between the existing informal, non-MTA model and the more proprietary MTA-based models, it is essential to establish agreed, verifiable, and enforceable quality standards as a precondition for admittance to the system of facilitated exchanges envisioned by the proposed multilateral semicommons. Otherwise, one would forfeit a major benefit that makes the latter proposal more advantageous than the basic noncommercial use license that is otherwise widely available.[101]

Phrased differently, both the WFCC's standards and the reliable quality standards said to prevail in the informal sector of material exchanges must be extended to the emerging global microbial materials pool, which would reach beyond the limited participants in a system of club goods while avoiding the restrictive use and distribution conditions of the proprietary models. Pooling these resources under a set of standardized contracts and procedures should thus enable a rationalization of the quality management requirements prevalent in the informal system. Such pooling should ideally also strengthen validation and enforcement mechanisms under the aegis of a governing authority.[102]

From an economic standpoint, transnational certified conditions of quality management for microbial materials under the proposed sharing scheme would greatly enhance its public good features. Common quality management controls, that provide guarantees of noncontamination for all strains, would have positive network externalities. The more players that use strains managed under common quality criteria, the larger the amount of reliable and standardized scientific research on which further cumulative research can be built. Some microbial materials, such as type strains, also function as basic research and regulatory tools, which exposes them to classic collective action problems of undersupply and free riding by individual players in the commercial sector.

On a more practical level, there are many advantages of such a standardized quality management scheme. It would enhance the trust of the researchers in contributing to and using materials from the pool; it would limit rent-seeking based on informational asymmetries resulting from competition between, instead of coordination among, differing quality standards; and it could limit free riding by players otherwise tempted to use high quality materials from the pool without

[100] *See* Section I.A.3.
[101] *See* Chapter 4, Section III.
[102] *See* Section II.C.3 below and Chapter 10, Section III.D.1 (Mandate of the Governing Body).

due attribution or reciprocal contributions of their own. The end result should be a globally distributed infrastructure for transactions concerning specified research tools, with collective management of the means of ensuring quality control.

To succeed in this endeavor, the governing body of the multilateral system would have to build upon the already existing initiatives and infrastructure developed under the auspices of the World Federation of Culture Collections, on the work of the OECD task force, on a bacterial code developed by a committee of the International Union of Microbiological Societies, and on a botanical code under the auspices of a committee of the International Union of Biological Sciences.[103] In particular, quality management criteria for defining the conditions of entry into the pool should reflect the OECD guidelines developed for accreditation of any given culture collection as a Biological Resource Centre, to the extent feasible. The advantage of this solution would be to extend an already well-developed set of criteria, while contributing to the efforts to implement them in the WFCC community.

However, insistence on verifiable and certifiable quality controls as a condition of admittance to the multilateral system will necessarily slow its growth in the short run. Over time, the global research pool, once established among a sufficiently large number of initial players, could constitute an incentive for new players that want to enter the pool to gradually enhance their own quality management capacities. By the same token, building the globally distributed research commons would further the objective of implementing at least the WFCC guidelines among an ever wider number of players.[104]

An overall package of standardized terms allowing unrestricted research uses plus certifiable quality standards should serve to attract and hold an ever larger pool of qualified collections and scientific entities. This pool would become even more attractive if it also afforded a standardized mechanism for allocating fair and equitable compensation from industrial applications of materials that had no known or likely commercial value at the time they were deposited in the semicommons.

2. Duty to Respect Reputational Benefits

There is a universal understanding in science that the first discoverer must have either a right to first publication or at least an embargoed period of exclusivity

[103] *Cf.* Univ. Catholique de Louvain, Democratic Governance and Theory of Collective Action IAP VI/06 – DEMOGOV (2008) (Annual scientific activity report); Henry E. Smith, *Semi-commons Property Rights and Scattering in the Open Fields*, 29(1) *J. Legal Stud.* 131 (2000).

[104] The WFCC guidelines already embody much of the OECD Guidelines. Attaining even high standards would depend in part on the availability of funds for capacity building, especially in developing countries. *See further* Chapter 10, Sections III.A & E, and Section IV.

during which that discoverer cannot be preempted by second comers having access to the same research materials.[105] These and related norms of science tend to preserve the reputational benefits that are known as the primary motivator of not-for-profit scientific activity. In constructing a third option between informal sharing arrangements and the proprietary MTA-based models, care must be taken to ensure that the proposed microbial research semicommons provides no less support for these norms than other available alternatives.

This premise does not oblige us to recommend maximalist open-access procedures that could unduly discourage scientists from participating in the proposed multilateral regime of facilitated access. For example, a Bermuda-like rule mandating early deposit of newly discovered microbial materials seems inadvisable in a context that seeks to appeal to the broadest number of qualified players.[106] Similarly, no regime of mandatory deposits prior to actual publication seems feasible in microbiology generally, given the distributed and heterogeneous nature of its members and the lack of any universal funding agency, unless a given subcommunity chose to adopt such a rule and to make it binding on all its members, which seems unlikely.[107]

If the principle of voluntary deposits thus seems most consonant with our understanding of what most WFCC members would desire – a premise that needs to be verified – there are nonetheless ancillary procedures that merit consideration, with a view to enhancing the security and confidence of those who agree to make such deposits. For example, under the governance framework we envision – with a single entry portal to all the materials, literature, and databases to be made available[108] – it becomes possible to establish a registration system that could be administered by a governing body, or a trusted intermediary, and regulated by an international database cooperation agreement.[109]

Such a system would provide notice to the world about the nature of specified deposits, and it would yield advantages for preserving both reputational benefits and

[105] *See* NAT'L RESEARCH COUNCIL, BITS OF POWER: ISSUES IN GLOBAL ACCESS TO SCIENTIFIC DATA (Nat'l Acad. Press 1997); NAT'L RESEARCH COUNCIL, SHARING PUBLICATION RELATED DATA AND MATERIALS (Nat'l Acad. Press 2003).

[106] Summary of Principles Agreed at the First International Strategy Meeting on the Human Genome Sequencing, Bermuda, Feb. 25–28, 1996 [hereinafter Bermuda Principles (1996)], *available at* http://web.ornl.gov/sci/techresources/Human_Genome/research/bermuda.shtml#1 (as reported by HUGO). *See* Summary of the Report of the Second International Strategy Meeting on Human Genome Sequencing, Bermuda, 27 Feb.-2 March 1977 (as reported by HUGO) [hereinafter Bermuda Principles (1997)]. *See also* Jorge L. Contreras, *Bermuda's Legacy: Policy, Patents and the Design of the Genome Commons*, 12 *Minn. J. L. Sci. & Tech.* 61 (2011) [hereinafter Contreras, *Bermuda's Legacy*].

[107] However, we agree with Scott Stern that mandating deposit of microbial materials after publication is a norm that research funders should seriously consider. *See* STERN, n. 18, 89–92. *See further* text accompanying n. 114.

[108] *See* Chapter 10, Section III.D.

[109] Interview with Micah Krischevsky, n. 98.

the sharing of commercial benefits to be discussed later in keeping with the Nagoya Protocol.[110] It would inherently furnish some incentives for making a voluntary early deposit prior to publication, but there would be no penalties for later deposit other than the normal risks of anticipation by others who deposit or publish first. The newest edition of the World Data Center for Microorganisms' online portal has already taken major steps in this direction.[111]

A different situation arises after publication of research results based on specified microbial genetic resources. Here the argument for insisting upon a mandatory deposit of the relevant materials becomes compelling for a number of reasons. The norms of science favor disclosure of both underlying materials and relevant data to permit independent verification of the published results. Recent studies have also shown that depositing materials in a Biological Resource Center leads to a significant increase in the citation of articles associated with the deposit.[112] Moreover, disclosure through publication signifies that any relevant patent applications must already have been made (subject to novelty grace periods, where applicable) while some trade secret protection will have been waived.

Once publication has occurred, a deposit of the relevant materials could – under our proposal – nonetheless attract a revenue stream, rooted in the compensatory liability rules discussed later, which would automatically apply to any potential downstream commercial outcomes.[113] Given these strong reasons for pressing scientists to make deposits after publication, both public funding agencies supporting the research project and the journals themselves should consider mandating post-publication deposits in BRCs affiliated with the microbial semicommons we envision.[114]

Even assuming that a voluntary, prepublication deposit would entitle the researcher to a priority right to publish for a specified period, if so required in the standard MTA, there remains some risk that third-parties not bound by WFCC rules could still ignore the depositor's priority claim. This risk could be attenuated

[110] *See* Nagoya Protocol, n. 8, art. 17.3 (International Certificates of Compliance).

[111] *History*, WDCM, http://www.wdcm.org/history.html (last accessed 3 July 2014); *see* Chapter 8, Section II.B.1 (discussing WDCM).

[112] Stern, n. 18, at 29; *see also* Jeffrey L. Furman & Scott Stern, *Climbing Atop the Shoulders of Giants* 8–9 (Nat'l Bureau Econ. Research Working Paper No. 12523, 2006), *available at* http://www.nber.org/papers/w12523.pdf (last accessed 17 Sept. 2012), stating that "[T]he divergence in citations resulting from a BRC deposit grew over time after the deposit occurred." *See also* Nat'l Research Council (NRC), The role and Value of Scientific Data in the Public Domain (Julie M. Esanu & Paul Uhlir eds., Nat'l Acads. Press 2003); Nat'l Research Council (NRC), Toward Precision Medicine: Building a Knowledge Network for Biomedical Research and a New Taxonomy of Disease (Nat'l Acads. Press 2011).

[113] Whether this package also suffices to induce *voluntary* deposits by private sector authors of scientific publications remains to be verified.

[114] *Cf.* Stern, n. 18, at 89–93.

by means of a registration system, as discussed later, that would identify and track all qualified visitors allowed entry to a master portal, and by the willingness of relevant journals to respect priorities that the proposed regime had established.[115]

Nevertheless, there remains some possibility that Scientist A, having gained qualified access to the semicommons, would share the accessed materials with Scientist B, who was not bound by the rules of the game. If Scientist B then sought either a publication or a commercial application based on an unauthorized duplicated version of the same microbial material, that scientist would incur only the risk of peer pressure ostracism or perhaps the possibility that reputable journals might not accept the work. But Scientist B would not otherwise become subject to contractual sanctions or to the compensatory liability rules operating under the proposed regime, unless Scientist A was bound to transfer the materials to Scientist B under a viral copy of the SMTA from the outset.[116] Even thornier questions of this nature would arise if Scientist B's work were based on an unauthorized derivative of Scientist A's deposit, rather than a duplication of it.[117]

There is no perfect solution to this problem under any purely contractual regime, since third parties lacking privity of contract are by definition not bound by its provisions,[118] unless a viral, standard-form license accompanied the transfer of materials.[119] In any event, unique strain identifiers are assigned to each sample deposited in a public culture collection. The record of how the material was originally collected is kept, and each transfer to another collection is added to this record. These records are maintained and made available as catalogues by the culture collections and, increasingly, the information is available online.[120]

Registration and tracking capabilities under the proposed regime should thus serve to expose both those who violated the express provision of a standardized MTA and those who disregarded the norms of science and thereby became exposed to peer pressure, plus some risk of legal sanctions. To the extent that the relevant scientists work within the same and relatively small research communities, peer

[115] This approach does raise questions of overhead and administrative burdens, which are addressed further in Chapter 10, Sections III & IV.

[116] *See* below Section III.B.2 (discussing derivatives).

[117] Tracking mechanisms thus help to ensure that the cheating scientist who makes the derivative would not escape the reach of the duty to pay royalties on commercial applications under the standard MTA. *See* Section II.C.3.

[118] *Cf.* ORG. ECON. CO-OPERATION & DEV. (OECD), BIOLOGICAL RESOURCE CENTERS – UNDERPINNING THE FUTURE OF THE LIFE SCIENCES AND BIOTECHNOLOGY 8 (Mar. 2001) [hereinafter OECD REPORT ON BRCs] (describing the Science Commons noncommercial use license; *see also* Belgian Coordinated Collections of Microorganisms (BCCM), General Conditions of Material Transfer, Jan. 2007 [hereinafter BCCM MTA] and the ECCO MTA, Chapter 4, Section III.A.2.

[119] For a model transaction, *see* below Section III. For principles of the Standard MTA, *see* Chapter 10, Section III.C.

[120] *See* Chapter 4, Section I.A.2.

pressure and a corresponding risk of ostracism can function as a powerful deterrent to deviant behavior.[121] Under the Nagoya Protocol, the scientist's institution can ostensibly be reached through local courts in all CBD member countries.[122] Other possible sanctions available in cases of leakage under the governance provisions of the semicommons (as reinforced by international law) are dealt with in the six scenarios set out later.[123]

Moreover, if scientist B delays a deposit in order to achieve some significant improvement over scientist A's initial deposit, scientist B – by properly crediting scientist A (as identified under the registration system) – will have fulfilled his or her reputational obligations and incurred no risk of opprobrium. In this connection, one should recall that, by definition, materials deposited under our proposed regime lacked any known or likely commercial applications or value at the time of deposit, which further attenuates these risks.[124]

At bottom, our proposed third option affords all participants unrestricted access to a large pool of research resources, subject to contractually imposed obligations to respect reputational norms and, as will be seen later, to share a portion of actual downstream commercial gains that were unlikely or unknown at the outset. We believe that the benefits accruing from this package deal would exceed any small risk of scientific abuse, which is always present to some extent anyway, and is generally addressed as a normative rather than a legal matter. Hence, the proposed scheme represents a logical alternative in virtually all cases except those in which the materials in question possess known or likely high commercial potential, in which case the existing regimes would continue to remain available.[125]

3. Tracking Mechanisms to Maintain the Chain of Custody

A key issue is the need for a reliable means of identifying specific microbes to be made available from the multilateral system and of tracking their further

[121] NAT'L RESEARCH COUNCIL, THE ROLE AND VALUE OF SCIENTIFIC DATA IN THE PUBLIC DOMAIN (Nat'l Acads. Press 2003); NAT'L RESEARCH COUNCIL, BITS OF POWER: ISSUES IN GLOBAL ACCESS TO SCIENTIFIC DATA 17–19, 21–22 (Nat'l Acads. Press 2007). *See also* Wesley M. Cohen & John P. Walsh, *The Real Impediments to Biomedical Research*, 8 *Innovation Pol'y & Econ.* 1–30 (2008).

[122] *See* Nagoya Protocol, n. 8, art. 18.2.

[123] *See* Section III.A; *see further* Chapter 10 on governance.

[124] It can be argued that, in principle, all materials funded by government should eventually be deposited for research purposes after a reasonable period of time, *see, e.g.*, STERN, n. 18 at 89, to be calculated with regard to either publication (or the lack thereof) or commercial development benchmarks, such as release of a final product or the filing of a patent application. In practice, however, there are many mitigating factors undermining this principle – such as quality standards and capacity of collections, while attaining this goal would depend on the funders' willingness to defray the costs. *See further* Chapter 10, Section III.F.1.

[125] Questions of biosecurity are separately discussed. *See* Chapter 10, Section III.F.3.

distribution and use up to, and including, downstream commercial applications. The World Federation of Culture Collection Guidelines for the Establishment and Operation of Collections of Cultures and Microorganisms already recommends that its members record at least the following information for each strain they hold:

- Place
- Substrate or host
- Date of isolation
- Name of the person isolating the strain
- Depositor (or other source of strain, such as from another culture collection)
- Name of the person identifying the strain
- Preservation procedures used
- Optimal growth media and temperatures
- Data on biochemical or other characteristics
- Regulatory conditions applying (relating, for example, to quarantine containment levels and patent status).[126]

As noted in Chapter 4, the most recent proposals for standardized MTAs in the EU would also require public culture collections to record and make available information pertinent to compliance with the Nagoya Protocol.[127] In that context, an internationally recognized certificate of compliance would at least require information about the following rubrics:

- The name of the provider;
- Unique identifier of the certificate;
- The person or entity to whom prior informed consent was granted;
- Subject-matter or genetic resources covered by the certificate;
- Confirmation that mutually agreed terms were established;
- Commercial or noncommercial use.[128]

The costs of obtaining, recording, and communicating such information could add to the financial burdens of the culture collections and further enlarge the growing divide between technically advanced Biological Resource Centers and the much larger number of less advanced WFCC members.[129] Rising administrative and transaction costs encourage many researchers and research institutions to stay outside the ambit of the more formal WFCC system, or to conduct business

[126] WFCC, *Guidelines*, n. 93.
[127] *See* Chapter 4, Sections III.A.2, III.A.3.
[128] Nagoya Protocol, n. 8, art. 18.2.
[129] *See* Chapter 4, Section I.B.

informally, sometimes even within the established culture collections, but only with a small group of trusted colleagues.[130]

However, there is reason to believe that the very establishment of a multilateral system to which both provider and user governments had adhered might lead to more streamlined recording requirements, with some corresponding reduction of administrative costs over time.[131] With specific regard to tracking, a multilateral regime for facilitated access to microbial genetic resources would already benefit from microbiology's existing tracking procedures, unlike the Crop Commons, which dispensed with any obligations to track plant genetic resources from the outset.[132] At present, the WFCC Guidelines not only require careful documentation for each strain held by member collections, they also provide an efficient system of coding, that enables each culture collection to assign Globally Unique Identifiers (GUID) to each strain of it holds and distributes.[133]

More specifically, the WFCC Guidelines state that:

> WDCM [World Data Center for Microbiology] provides for an efficient coding of the strains by defining a collection acronym and WFCC number, which allows each culture collection to give a Globally Unique Identifier (GUID) to each strain of its holdings, combining their acronym with their own internal numbering. The pioneering work of WCDM enables an appropriate recording and management of the document related to the strains. Collections should use this system to be part of the WDCM network and be connected to the international scientific community.[134]

However, there have been problems implementing this initiative. Microbes by their very nature are impossible to see with the naked eye, much less to detect and track with certainty. They also mutate quickly, making their original manifestations difficult to identify over time. The infrastructure needed to maintain them indefinitely in state-of-the-art culture collections (or BRCs), and subsequently to exploit them for various research and applications objectives, is costly. Proving misappropriation in courts or other dispute settlement forums can thus become expensive and uncertain for purely technical reasons.[135]

[130] *See* Per M. Stromberg, Tom Dedeurwaerdere & Unai Pascual, *The Heterogeneity of Public* Ex Situ *Collections of Microorganisms: Empirical Evidence about Conservation Practices, Industry Spillovers, and Public Goods*, 33 *Envtl. Sci. & Pol'y* 3 (Nov. 2013).

[131] *See further* Chapter 10, Section III.

[132] For the Crop Commons, *see* Chapter 3, Section III.B & C.

[133] WFCC, *Guidelines*, n. 93, at 10, art. 11.1.

[134] *Id.*

[135] There have been many high-visibility lawsuits filed by authorities in developing countries or emerging economies that have successfully challenged patents on genetic resources or the overt misappropriation of such resources by firms in OECD countries and elsewhere. *See* Chapter 3, Section I.A.

Disregarding such technical factors, many material transfers are still informally arranged, at least within any given country's borders, not only with the help of thousands of research collections that do not claim a "public collection" or "patent collection" status, but even under the auspices of many officially designated public culture collections.[136] The tracking system is not airtight, even when care is taken to use it and there remain some opportunities for leakage based on unscrupulous parties' theft of commercially valuable strains.

An imperfect system for tracking the flow of microbial materials from *ex situ* collections to downstream users poses significant problems from a number of perspectives. Academic or not-for-profit researchers want assurance that they will be cited or given credit for any research results or breakthroughs that use their microbes. Proprietors in industry and, increasingly, at universities want assurances that their materials and the commercial products and services built around them will not be misappropriated. Estimated losses due to industrial espionage and commercial misappropriation for firms within the United States alone amount to a few billion dollars annually.[137] Governments everywhere are concerned about the tracking of pathogenic microbes, inimical to public health or suitable for terrorist attacks, and they too are potential targets of industrial espionage. As we have already noted, all users want and need assurances of quality, and that they are in fact obtaining the materials they requested.[138]

Most important for present purposes, governments in developing countries would not willingly transfer microbes outside their borders and allow them to be commercially developed without appropriate benefit-sharing guarantees under the Nagoya Protocol.[139] The Compensatory Liability Regime we are proposing must, therefore, seek to track and capture all future downstream uses of microbial materials that had no known or likely commercial value at the time of deposit. Leakage within

[136] Stromberg, Dedeurwaerdere & Pascual, n. 130. However, the Nagoya Protoco will now impede cross-border informal exchanges of genetic resources. *See* Chapter 3, Section IV.C.

[137] Statement made at the meeting organized by U.S. Defense Advanced Research Projects Agency (DARPA) in Arlington, VA on 28 March 2011 [hereinafter DARPA Meeting (2011)]. One consequence of a lack of confidence in the tracking system and related enforcement methods is that it encourages industry to protect economically valuable materials by both actual secrecy and measures needed to trigger domestic and international trade secrecy laws. *See esp.* Agreement on Trade-Related Aspects of Intellectual Property Rights, Apr. 15, 1994, Marrakesh Agreement Establishing the World Trade Organization, Annex 1C, art. 39 The Legal Texts: The Results of the Uruguay Round of Multilateral Trade Negotiations 320 (1999), 1869 U.N.T.S. 299, 33 I.L.M. 1197 (1994) [hereinafter TRIPS Agreement] (imposing global obligation to protect trade secrets under international unfair competition law for the first time). These practices may reduce public disclosures otherwise obtained from patent filings and deposits in patent repositories, and may further discourage broader publication of research results that lead to patent applications.

[138] *See* above Section II.C.1.

[139] See Chapter 3, Section IV.A ("Clarifying the Broad Economic Scope of the CBD").

the system would otherwise operate as a tax on all stakeholders, with a concomitant increase in the costs of maintaining and exchanging materials by those least able to afford them – the developing countries and not-for-profit sectors. The risk of lost opportunity costs could grow for commercial and noncommercial research alike. The potential effects of an unreliable or unaffordable system of tracking are thus far-reaching and could have a, negative impact on the formation and functioning of the Microbial Research Commons generally, and on the successful implementation of the Compensatory Liability Regime, specifically.

On a more positive note, promising research in the United States and elsewhere is now focused on more advanced methods for the tracking and control of microbes that are used in downstream research applications. These efforts may be summarized as follows:[140]

1. **Microbial Steganography**. Unlike cryptography, which is meant to be undecipherable but obvious, steganography is the art of putting secret messages in another object, traditionally in another piece of information, but for present purposes in a genetically engineered microorganism. In 2011, Prof. David Walt and his colleagues at Tufts University reported on the results of an experiment in which they had encoded secret messages in genetically engineered bacteria using fluorescent proteins with three levels of security that could be easily read under a black light. Although no practical applications had yet been developed, the idea proposed was to artificially introduce a watermark through genetic engineering that would remain undetectable by others and could be made irreversible.[141]
2. **Clustered Regularly Interspersed Short Palindromic Repeats**. This novel methodology introduces a memory system to track events that any given microorganism has experienced, such as some manipulation or change in a host. It may become possible to engineer a stable, repeatable process that would be activated after any particular event, such as use of the microbe by an unauthorized lab, to make the microbe self-terminate. A proprietary strain that is genetically tagged for identification purposes or to "impede the transfer of particular nucleic acid sequences (such as phage or plasmid DNA) into a host might be exploited via genetic engineering to specifically preclude the[undesirable] dissemination of ... genetic elements."[142]

[140] DARPA Meeting (2011), n. 137.

[141] David R. Walt, et al, *InfoBiology by printed arrays of microorganism colonies for timed and on-demand release of messages*. 108 *Proceedings of the National Academy of Sciences*, No. 40, at 16510 (Sept. 29, 2011) available at http://www.pnas.org/content/108/40/6510/full.pdf. The article whimsically referred to the encoded messages as SPAM – Steganography by Printed Arrays of Microbes.

[142] Phillippe Hovath & Rodolphe Barrangou, *CRISPR/Cas, the Immune System of Bacteria and Archaea*, 327 *Science* 167–70 (2010).

3. **Gene Guards – Programmable Kill Switches for Microbes**. Another technology similar (at least in application) to the previous one is the use of synthetic riboregulators for tracking and terminating microbes in the event of unauthorized use or misappropriation.[143]
4. **Artificially Expanded Genetic Letters for Intellectual Property Protection**. This technique consists of the use of an unnatural base pair of amino acids in the creation of a semi-synthetic organism that has an increased potential for information storage and retrieval. If an unauthorized party attempts to propagate the organism without a key, this process aims to ensure that it would not work.[144]

Whether these (and perhaps other) advanced tracking or terminating mechanisms for microbial materials can be broadly employed in the future will depend not only on their successful development, but on the extent to which they can be produced and applied inexpensively. Export controls placed on some or all of these technologies could also preclude their global adoption, in the short run. If too complex or expensive to use, only large companies able to pass on the costs to their customers could adopt them.[145]

Meanwhile, the managers of the StrainInfo bioportal in Belgium had begun using semantic web technology to improve the tracking of microbial strains by means of the WFCC's GUID numbering system.[146] By mapping the GUID numbers available from the electronic catalogs of the public culture collections onto the same numbers available from GenBank for any sequenced microbial strain, the culture collections themselves could verify, integrate, and complement the digital tracking of resources made available for research purposes.[147] Leading experts on the Nagoya Protocol have already recognized the importance of this technique for implementing the access and benefit-sharing provisions of that agreement,[148] and the WDCM plans to further perfect this approach.[149]

143 Jarred M. Callura et al., *Tracking, tuning and terminating microbial physiology using synthetic riboregulators*, 107 *Proc. Natl. Acad. Sci.* 15898–903 (2010), *available at* http://www.pnas.org/content/107/36/15898.

144 *See* work done by Floyd E. Romesbery at his laboratory in the Scripps Oceanography Institute, in La Jolla, CA. Denis A. Malyshev et al., *Solution Structure, Mechanism of Replication, and Optimization of an Unnatural Base Pair*, 16 *Chem. Eur. J.* 12650–59 (2010).

145 One possible solution to reduce costs might be the use of third-party trusted intermediaries, where possible. *Cf.* Rai et al., n. 21.

146 For a description of the StrainInfo bioportal, *see* Chapter 8, Section II.B.2.

147 Although not all culture collections have digital catalogs, the WFCC and the WDCM are encouraging all their members to adopt them. *See* WFCC, *Guidelines* above n. 93.

148 *See* Matthias Buck & Clare Hamilton, *Access to Genetic Resources and the Fair and Equitable Sharing of Benefits Arising from their Utilization to the Convention on Biological Diversity*, 20 *Rev. Eur. Community Int'l Env't Law* 47 (2011).

149 *See further* Chapter 8, Section II.B.1

Use of the WFCC's GUID numbering system is thus expected to become an efficient means of qualifying for international certificates of compliance with the ABS/PIC requirements under Article 17(1) of the Nagoya Protocol,[150] at least until the more advanced techniques under study are perfected. What matters for present purposes is that the problem of leakage is potentially manageable by means of the improved techniques already available, as further reinforced by the new compliance obligations impose on CBD members by the Nagoya Protocol itself. How this combination of tracking and related compliance measures might work in practice is illustrated in Section III.

4. The Calculus of Royalties from Commercial Applications

At the outset, we emphasize that, while the microbial genetic resources in question may constitute essential research inputs, they typically lie – by definition – relatively far upstream in the research process when first deposited, and thereby remain relatively distant from the commercial applications phase. In other words, the bulk of the *ex situ* microbes in question, although identified and validated for quality control purposes, still lack value-adding research elements that might merit a larger royalty percentage *ab initio*. This postulate follows "by definition" because the presence of other value-adding research contributions could lead would-be providers to view the microbes in question as having some known or likely commercial value. In that case, the microbes would likely have been deposited in special collections not open to the public, rather than in culture collections that constitute a de facto research commons.

A corollary principle is that, once deposited in the proposed multilateral system, providers cannot withdraw materials that had no known or likely commercial value at the time of deposit. That principle has already been formulated and tested in the FAO's multilateral regime for plant genetic resources, which, as previously indicated, implements one version of a Compensatory Liability Regime.[151] A second corollary principle is that access to the multilateral system necessarily exposes would-be users to a fixed, nonnegotiable, *ex ante* royalty on downstream commercial applications in exchange for the freedom to conduct any form of research on the microbial materials in question, whether commercial or noncommercial in nature.

These principles are important precisely because, at the time of deposit, future downstream applications of the microbial materials in question are still shrouded behind the "veil of ignorance."[152] Would-be researchers must accordingly know

[150] *See* Nagoya Protocol, n. 8, art. 17(1)(iii), 17(3).
[151] *See* Chapter 3, Section III.B.
[152] *See further* Rai et al., n. 21.

in advance exactly how much of a royalty would be owed to the providers' agents if and when some future commercial opportunity arises. The same researchers must then factor this "reach through" obligation into their calculus of expected profits when they subsequently negotiate downstream commercial licenses with firms seeking to develop marketable products and processes from their research results.

One must, therefore, carefully distinguish between the providers' automatic, *ex ante* entitlement to reasonable royalties from eventual downstream commercial applications that never varies and the royalties that scientists or universities *qua* inventors might obtain when transferring research results to industry, with expectations that patented products will likely ensue. Those nonstandard royalties will vary from case to case and must necessarily take account of any reach-through royalty obligations applicable to use of microbial materials taken from the research commons under its standard MTA. These practical considerations reinforce our supposition that the baseline royalty rates under a Compensatory Liability Regime should be relatively modest, lest they unduly discourage downstream investors from the outset.

As any given microbe increases in potential value over time owing to previously successful research endeavors or applications, the value of *ex post* downstream commercial licenses will likely increase. The duty to pay a reasonable royalty to the initial providers, however, will remain unchanged. At the same time, the very success of any downstream commercial applications will likely generate more interest in still other researchers, who will be tempted to conduct further research on potential applications of the microbe in question. Because this specimen cannot be removed from the multilateral system once deposited, it remains available to all would-be future researchers and commercial users under the standard, nonexclusive license, even after that microbe had acquired potential commercial value through prior use. In that event, the initial provider (or its agent) would remain entitled to the baseline royalties under the standard MTA for all future commercial applications of the material in question, with the prospect of multiple revenue streams or lottery effects as multiple downstream investors entered the market with plans to develop new commercial products or to improve on existing products resulting from prior uses of that same material.[153]

Here, however, one must pay attention to the role of intellectual property rights that may attach to the resulting downstream applications. A salutary rule for the

[153] *See* below Section III.A.6; *see generally* Reichman, *Green Tulips*, n. 44. *See also* Rai et al., n. 21. For background, see further Jerome H. Reichman, *Legal Hybrids Between the Patent and Copyright Paradigm*, 94 *Colum. L. Rev.* 2432, 2246–72 (1994) and Jerome H. Reichman, *Charting the Collapse of the Patent-Copyright Dichotomy: Premises for a Restructured International Intellectual Property System*, 13 *Cardozo Arts & Ent. L.J.* 475 (1995).

proposed Microbial Research Commons would require that users benefitting from facilitated access to its genetic resources cannot apply intellectual property rights to those resources in the form in which they were received. This same rule was adopted and implemented in the FAO's Crop Commons.[154] It means that downstream commercial investors can obtain patents on specific functions of a given specimen, or on derivatives and modifications; but they cannot patent or otherwise protect the specimen in the form in which it was accessed from the multilateral system.

It follows that second comers interested in finding still other commercial applications based on a microbial genetic resource that has already elicited one or more successful downstream products must respect prior users' patent rights. Those rights may, in turn, limit the space for future innovations, at least to the extent of the claims recognized in such patents. But these same patentees can never, at least in principle, prevent other researchers – public or private – from returning to the specimens that remain available from the Commons for purposes of conducting further research and seeking additional applications that do not infringe on prior users' patent rights.[155]

Given these premises, a primary task for the Governing Body of a redesigned Microbial Research Commons would be to establish the quantum of nonnegotiable, *ex ante* royalties that all would-be users must pay on future commercial applications, in return for facilitated access to genetic resources made available from the multilateral system. What, in short, would constitute "reasonable" standard royalties under these circumstances?

To answer the question, we first emphasize that the reasonable royalties accruing from commercial uses of microbial materials under a Compensatory Liability Regime must be potentially big enough to motivate providers to continue to deposit microbes having no known or likely commercial value into the commons infrastructure, even though the Nagoya Protocol has significantly strengthened the bilateral approach under the CBD. At the same time, the quantum of reasonable royalties meant to fulfill the benefit-sharing obligations of the CBD must not be set so high as to discourage investors in prospective downstream commercial applications, which would defeat the whole purpose of the liability rule from the start.

A *priori*, we consider that the 0.5–1.1 percent nominal royalty adopted for the liability rule embodied in the FAO's International Treaty on Plant Genetic Resources for Food and Agriculture would remain unacceptably low, even if that Treaty did not allow would-be users to escape this obligation altogether if they

[154] *See* Chapter 3, Section III.B (discussing the International Treaty on Plant Genetic Resources for Food and Agriculture).

[155] We qualify this principle with an "at least" clause because, in biotechnology, patents may otherwise impede the use of research tools. *See* Rebecca S. Eisenberg, *Proprietary Rights and the Norms of Science in Biotechnology Research*, 97(2) *Yale L.J.* 177 (1987).

allow second comers a research exemption under downstream intellectual property rights in appropriate cases.[156] By the same token, a 4 percent reach-through royalty on downstream applications of microbial genetic resources that had no known or likely commercial value at the time of deposit could especially discourage investments by small and medium-sized firms in sectors where the prospects for commercial returns were limited by regulatory considerations, such as those defending the public interest in food security, climate change innovations, and public health.

Between these two extremes, we believe that a standard 2 percent royalty on gross sales of commercial products derived from microbial genetic resources subject to the Compensatory Liability Regime constitutes a viable floor below which developing-country governments adhering to the multilateral system may not be willing to venture.[157] Arguably, a 2 percent royalty should not undermine either the provider's incentive to deposit or the investor's incentive to conduct basic or applied research. Moreover, a flat rate of 2 percent would, in turn, be tied to a fully open-access research semicommons,[158] unlike the FAO's Crop Commons, as noted earlier.

Here, instead, the right to make virtually any basic or applied research uses of deposited microbial materials subject to a duty to pay reasonable compensation for eventual commercial applications is a fundamental tenet of the commons infrastructure that cannot be rewritten or bargained away by either providers or would-be industrial investors.[159] As a result, the actual value of a 2 percent standard royalty on gross commercial sales is magnified by the unfettered rights of all users to conduct future research on the same microbial materials, without negotiation, and by the additional royalties that such research may ultimately generate from still other applications.

That said, one may legitimately ask whether culture collections capable of providing more valuable data and information pertaining to their *ex situ* holdings than most others should be allowed to charge a slightly higher royalty than the standard rate, which we postulated as 2 percent. For example, collections that

[156] *See* Chapter 3, Section III.B & C. While such an exemption is not worthless, we contend that a broad research exemption for uses of the primary genetic resource should be built into the Compensatory Liability rule, from the start, as we have conceived it, and that the quantum of *ex ante* automatic royalties should not depend on posterior agreements affecting research opportunities by private-sector negotiations.

[157] For governance issues, see Part Four, and especially Chapter 10.

[158] Because only qualified recipients can access microbial genetic resources owing to quality, safety, and security standards, it is a "semicommons" by regulatory fiat.

[159] In this respect, the liability rule proposed for the Microbial Research Commons represents a major improvement over the version of that rule embodied in the FAO's International Treaty or Plant Genetic Resources for Food and Agriculture. *See* Chapter 3, Section III.C.2.

had attained the status of BRCs, and thereby guaranteed higher standards in their operations at considerable expense, would support this proposal.[160]

As the impact of molecular biology on microbial research grows, moreover, it seems increasingly likely that both providers and culture collections may progressively add more value to their *ex situ* holdings in the form of genomic data and cross-referencing than in the past, as expressly advocated by the OECD's Task Force in 2005.[161] The Nagoya Protocol itself lays claim to rights in genomic data pertaining to *ex situ* and *in situ* genetic resources subject to the CBD's Access and Benefit Sharing provisions.[162] While such data would hypothetically tend to focus on identification, categorization, and cross-referencing more than utility, they nonetheless represent a form of added research value, comparable in some ways to a research database or even a published article. We believe such added value, if publicly made available either by the provider at the time of deposit or by the culture collection itself, could merit an extra carrot of compensatory royalties that the founding participants in a redesigned Microbial Research Commons would need to negotiate and establish.

Another relevant question is whether the participating culture collections should be allowed to charge a user fee when accessing microbial materials from the multilateral system for research purposes. Such a user access fee, if adopted, would be added to the marginal cost of distribution, which is the usual charge in current practice, and it could help to defray some of the multilateral systems' operating expenses. Answering this question also depends on whether the provider countries insist on obtaining some guaranteed, up-front monetary benefits under the multilateral system, over and above the less certain monetary benefits flowing from the liability rule and the abundant nonmonetary benefits that a knowledge commons generates for the world at large. Similar questions have arisen lately in the context of the Crop Commons,[163] and at least one distinguished economist has publicly endorsed it.[164]

In making these proposals, we stress that the ultimate decisions rest with the Governing Body – and the participating entities – that would constitute the

[160] *See* Chapter 4, Section I.B.

[161] OECD REPORT ON BRCS, n. 118. *See also* Chapter 8, Section II.B.1 (explaining the work of the WDCM); NAT'L RESEARCH COUNCIL, A NEW BIOLOGY FOR THE 21ST CENTURY 49–52 (Nat'l Acad. Press 2009) [hereinafter BIOLOGY FOR THE 21ST CENTURY].

[162] Convention on Biological Diversity arts. 5–8, *opened for signature* June 5, 1992, 1760 U.N.T.S. 79 [hereinafter CBD]; *see further* Chapter 3, Section IV.A (Nagoya Protocol clarifies the broad scope of the CBD).

[163] *See, e.g.*, JULIANNA SANTILLI, AGROBIODIVERSITY AND THE LAW: REGULATING GENETIC RESOURCES, FOOD SECURITY AND CULTURAL DIVERSITY 134–35 (Earthscan 2012) [hereinafter SANTILLI (2012)].

[164] *See* Paul A. David, *Breaking Anti-Commons Constraints on Global Scientific Research: Some New Moves in "Legal Jujitsu," in* DESIGNING THE MICROBIAL RESEARCH COMMONS, n. 51, at 18–34. *See further* Chapter 10, Section II.C.1.

governance model we outline in Part Four of this volume. In the next section, we sketch some preliminary considerations on governance, so that readers may more easily follow the hypothetical scenarios at the end of this chapter. These scenarios illustrate more specifically how we think the proposed Compensatory Liability Regime should operate to the benefit of all stakeholders under the aegis of the Nagoya Protocol.

5. An Enabling Governance Structure

As will be seen from later chapters dealing more specifically with governance in Part Four, a redesigned Microbial Research Commons would necessarily require the support and participation of member governments, as is the case with other existing research commons in various fields, some of whose governance structures are examined in Chapter 9. The precise nature of this governance machinery will depend on a number of variables, which are discussed in Chapters 9 and 10. For purposes of this preliminary discussion, we assume that the participating government entities will at least have signed a foundational agreement with some of the following features.

First, a Governing Body would obviously need to be established. That body in turn, would formulate a standard contractual relationship with each member government, with a view to becoming their agent for implementing a multilateral regime of facilitated access to microbial genetic resources and benefit-sharing within the purview of Article 4.2 of the Nagoya Protocol to the CBD.[165] A primary purpose of this multilateral system would be to "[c]reate conditions to promote and encourage research which contributes to the conservation and sustainable use of biological diversity, particularly in developing countries."[166] Second, pursuant to these agreements, the Governing Body would develop and negotiate a standard MTA with all the participating culture collections under the auspices of the member governments. That SMTA would set out the terms and conditions of the Compensatory Liability Regime, which would henceforth apply to the bulk of *ex situ* holdings subject to the jurisdiction of the Commons, as explained in Chapter 10.

Participating culture collections would designate all *ex situ* holdings having no known or likely commercial value at the time of affiliation with the multilateral system, and they would continue to make similarly situated materials that were subsequently acquired available to the multilateral system. Qualified users of the materials collectively governed by the Commons for research purposes could either approach the member collections individually or enter the system through a master

[165] Nagoya Protocol, n. 8, art. 4.2; *see further* Chapter 3, Section IV.B.
[166] Nagoya Protocol, n. 8, art. 8(a).

portal, which, after registration, would direct them to the appropriate collections. Researchers and other users who approached the member collections directly would nonetheless have to register with the system (and be identified) as if they had entered through the master portal. Either way, registration would entitle the user to make unlimited research uses – whether commercial or noncommercial – of all the holdings in all the collections made available under the Compensatory Liability Regime, subject to the SMTA that would accompany all exchanges of microbial genetic resources drawn from the pool and to the rules entitling users to access the system.

Registration in the system would then further commit the user to the following additional benefits and obligations. First, it would obviate the need for users to negotiate permissions and research agreements with Designated National Authorities, as would be the case under the bilateral approach of the CBD. Instead, the Designated National Authorities in provider countries will already have consented to the terms of a blanket license, as set out in the SMTA, and the culture collections will notify these same authorities (and an International Clearing House if established) of any SMTAs entered into with would-be users. The foundational agreement establishing the mandate of the Governing Body would further ensure that all microbes used for research purposes were properly identified, authenticated, and subject to the tracking mechanisms established by that body and implemented by the collections.[167]

By registering with the Commons, users would empower the Designated National Authority indicated in the SMTA to collect and distribute royalties due from commercial applications in accordance with the relevant Authority's domestic laws. By the same token, the SMTA would require users to report all ensuing commercial transactions to the Designated National Authority, including the quantities and prices of products sold, as well as other relevant information, and it would empower the Designated National Authority to collect and distribute the compensatory royalties arising under the SMTA.

Finally, the SMTA would require all users of microbial materials made available from the Commons to replicate the relevant terms and conditions of the SMTA in posterior agreements undertaken with downstream commercial ventures, who must agree to report to and inform the Designated National Authority about all sales covered by the SMTA. In case of dispute, the SMTA would establish binding mediation and arbitration procedures, failing which it would oblige users' governments to cooperate with the Designated National Authority in pursuing the latter's entitlements under the SMTA in the relevant domestic courts.[168]

Participating members and the Governing Body would eventually have to decide whether single culture collections, meeting the necessary quality standards,

[167] For details, see further Chapter 10, Section III.

[168] *See* Nagoya Protocol, n. 8, art. 18.2.

could join this scheme, even if their governments had not formally adhered to the foundational or other collective action agreements. Admitting such collections to the Commons would greatly expand the microbial materials available for global research, especially if, for example, it enabled collections in the United States (including government-held collections) to participate, whether or not the United States had ratified the CBD or joined the Commons as a contracting party. To this end, the Governing Body would ultimately have to ensure that all such participating collections had contractually accepted the same terms and conditions, whether or not their respective governments were signatories of any framework agreement.

However, in case of disputes, and failing arbitration, an aggrieved Designated National Authority would have no clear legal power to enforce the SMTA in the courts of a country that had not formally adhered to the multilateral system. Even so, the Designated National Authority could seek to impound the goods on which royalties had not been paid in other countries whose governments had adhered to the multilateral system, either under private international law or under express terms to this effect set out in the relevant SMTA, and backed up by the Nagoya Protocol's global regime of misappropriation.[169]

Parenthetically, the benefits of the regulatory scheme we are exploring here are not necessarily confined to transnational exchanges of microbial materials, even though this is the topic of primary focus in this book. Rather, it bears reiterating that the bulk of all materials exchanged probably occurs on an intra-territorial basis, where restrictive MTAs may inflict severe impediments to upstream research and on domestic spillover effects of value to innovators. If properly implemented in domestic law, the regulatory scheme outlined here should accordingly produce equal or greater pro-research benefits for domestic participants as it would at the international level, and this should be regarded as another nonmonetary component of a win-win arrangement.

III. MODELING A SEQUENCE OF HYPOTHETICAL TRANSACTIONS

To better understand how a properly designed Compensatory Liability Regime might actually work in practice, we think it useful to walk the reader through a set of hypothetical transactions. We first describe a relatively simple transaction in a series of discrete steps, starting from the discovery of the microbe in question and ending with the distribution of royalties from successful commercialization of research results based on studies of that microbe. We accompany each step with a

[169] *See* Chapter 3, Section IV.C.

brief discussion of how we think the system should work at that point in the process and why.

We then introduce a number of complications that might arise, in order to flesh out the practical aspects of the scheme, with particular regard to overriding international legal considerations. We end this section with some general observations about the proposed multilateral system as a whole, and we attempt to characterize the perspectives of the various stakeholders whose interests it would aim to promote.

A. *The Standard Deal in Six Scenarios*

1. Identifying and Depositing the Microbe

In 1997, a Ruritanian microbiologist discovered a new species in a local river, which he was prospecting for this purpose under a government grant to his department at the Ruritanian Technological Institute (RTI). He then isolated the type strain and phenotypically characterized it while writing up his findings for publication. In 1999, he deposited the microbe in two different culture collections, as required by WFCC Guidelines pertaining to type strains.[170] The first, Collection A, was the National Culture Collection of Ruritania, and the second, Collection B, was a major public collection in Occitania, well-known for the quality of its bacterial holdings.

Both collections were WFCC members; each of them briefly described the deposit in their catalogues; and each assigned Global Unique Identifying Numbers to the microbe in question, viz. RURI 500 and OCCI 800 respectively. A fuller description of the microbe and certain attributes appeared in the year 2000, when the discoverer published an article about his work in the Ruritanian Journal of Microbiology.

Both Ruritania and Occitania had ratified the Convention on Biological Diversity in 1993. To keep matters simple, we further assume that no traditional knowledge is involved.

COMMENT. Given that the microbe in question was discovered *in situ*, within Ruritania, and deposited *ex situ* in that country, its legal status is determined by the national laws of Ruritania, as the country of origin. Ruritanian sovereignty over the microbial resource is also assured by the CBD. Its initial availability for study or commercial exploitation within or outside the territory will depend on rules that may or may not have been promulgated by the national government.[171] For purposes

[170] *See* WFCC, *Guidelines*, n. 93. Type strains themselves are not patentable, however. See Chapter 4, Section I.A.

[171] CBD websites concerning national ABS legislation and country profiles are the primary source of such rules. *See* Kamau et al., n. 4, at 258 n. 79. Eventually, a clearinghouse mechanism under the

of any eventual requests for use outside the country, we assume that the Ruritanian government will have established a Designated National Authority to administer such transactions, under a bilateral approach, within the purview of the Nagoya Protocol to the CBD.[172]

Collection A, as a member of the WFCC, will have met the minimum quality standards necessary for affiliation with that entity. To meet the latest WFCC standards, the collection would have adopted measures that differentiate its storage capacity from thousands of other collections without necessarily requiring a major capital investment.[173]

Because it meets these quality standards, Collection A was able to verify and authenticate all its holdings, and to assign Global Unique Identifiers (GUID) for tracking each specimen made available to other member collections and approved users. Its taxonomic descriptions are also made available to all potential users, and it may have provided the World Data Center for Microorganisms (WDCM) with more ancillary data and information.

As regards its status in international law, the deposited microbe is now protected under the CBD. Hence, Prior Informed Consent, Mutually Agreed Terms, and Access and Benefit Sharing are required preconditions of use, and unauthorized uses are subject, in principle, to claims of misappropriation under international law. Apart from the Nagoya Protocol, however, there is no additional treaty and few legal precedents to spell out the precise nature of this protection in either national or international laws, despite ongoing efforts to this end at the World Intellectual Property Organization.[174]

2. Collections A and B Join the Proposed Microbial Research Commons

We assume that Collection B joined the proposed multilateral system in 2018, after its government had signed a Memorandum of Understanding that established that system's legal and institutional framework.[175] We further assume that Ruritania

Nagoya Protocol should constitute the primary source of information in this regard. *Id.* at 258. *See* Nagoya Protocol, n. 8, art. 17.2.

[172] *See* Nagoya Protocol, n. 8, arts. 17.2, 17.3.

[173] *See* Chapter 4, Section I.A.

[174] Discussions aiming to convene a Diplomatic Conference to adopt such a treaty have allegedly reached an advanced state at WIPO, but the relevant proposals are still not ripe for action. *See, e.g.*, Catherine Saez, *New WIPO Text on Genetic Resources Misappropriation: Disclosure Still Uncertain*, *IP Watch* (Feb. 6, 2014), http://www.ip-watch.org/?p=34091&utm_source=post&utm=email/&utm_campaign=alerts; Matters Concerning Intellectual Property and Genetic Resources, Traditional Knowledge and Folklore, Establishing the WIPO Intergovernmental Committee (IGC) on Intellectual Property and Genetic Resources, Traditional Knowledge and Folklore, 25 Aug. 2000, WIPO Doc. WO/GA/26/6, *available at* http://www.wipo.int/meetings/en/doc_details.jsp?doc_id=1460.

[175] For the structure and governance of our proposed Microbial Research Commons, *see* Chapters 9 and 10.

had also joined the proposed Commons the following year, and that under the Commons' capacity building program, Collection A has been obtaining some technical assistance to upgrade the capacity of its national collection.[176]

In consultation with scientific experts representing the Commons,[177] Collection A in Ruritania will have designated the bulk of its microbial materials as having no known or likely commercial value, and it accordingly makes these materials publicly available for facilitated access from the multilateral system within the framework of the Compensatory Liability Regime.[178] Presumably these criteria would encompass all specimens in the collection not otherwise covered by patents or expressly reserved in privately held, special collections, or otherwise segregated under a reasonable opt out criterion. As part of this process, Microbe RURI 500 (OCCI 8000) was appropriately tagged (including its status) at the time of deposit in both Collections A and B, and it became available for unrestricted use under the Standard Material Transfer Agreement (SMTA) voluntarily accepted by all members of the multilateral regime.

Comment. Because Ruritania's Collection A has become a member of the redesigned Microbial Research Commons, it must have met the minimum quality standards of that entity. Moreover, we assume that the governing authority of the Commons will want its members gradually to implement the best practices associated with the concept of Biological Resources Centers (BRCs).[179] More substantial investment may be needed to achieve these goals than is required by WFCC minimum standards, although the Commons itself may have established differentiated quality standards to be met over time, according to different levels of affiliation.[180] To this end, Collection A in Ruritania, may seek technical and financial assistance provided under the auspices of the governing authority of the Commons in an effort to improve the functional capacities of culture collections in all member developing countries.[181]

As members of the Microbial Research Commons, both collections A and B must implement the Compensatory Liability Regime in good faith, in conformity with

[176] For the importance of a capacity building program to maximize opportunities for research under the Nagoya Protocol, *see* Chapter 3, Section IV.B. and Chapter 10 (on governance).

[177] *See further* Chapter 10, Section II.D.2 (discussing a Scientific Coordination Council).

[178] Culture collections that operate as Patent Deposit Authorities under the Budapest Treaty on the International Recognition of the Deposit of Microorganisms for Purposes of Patent Procedure of 1977, *as amended on* 26 Sept. 1980, 32 U.S.T. 1241, 1861 U.N.T.S. 361 [hereinafter Budapest Treaty], *available at* http://www.wipo.int/treaties/en/registration/budapest/trtdocs_wo002.html (last accessed 3 July 2014), will of course keep these proprietary holdings separate from the bulk of holdings made publicly available under the liability rule.

[179] *See* Chapter 4, Section I.B.

[180] *Cf.* Chapter 9, Section II.C.1 (discussing the Global Biological Resource Centers Network (GBRCN)).

[181] *See further* Chapter 10, Section III.E.

the foundational agreement establishing a multilateral system and with the Standard Material Transfer Agreement to be negotiated under its auspices.[182] Collection A's management must accordingly possess the authority to contractually agree to abide by all of the rules that the Microbial Research Commons will have established to ensure compliance of its members with the CBD.

With regard to determining how much of Collection A's preexisting holdings will be made available under the liability rule after joining the Commons, there is a presumption that, under a good faith standard, the management will dedicate the bulk of its holding for this purpose, especially if the collection had been publicly funded. It may nonetheless exercise it's opt out right, at the time of joining, with respect to microbes that had already attracted some commercial interest. While Collection A thus decides how much of its holdings to make available from the multilateral system, in consultation with representatives of the Commons administration, that administration will ultimately decide if these criteria have been properly applied in the context of membership.[183]

Once a member of the Commons, Collection A had access to all of the *ex situ* material resources held by all the other members' collections under the Commons SMTA.[184] Its catalog and any literature concerning its holdings will also have been made more visible via the digital services to be provided by the Commons.[185] However, after joining, Collection A cannot change the status assigned any microbe at the time of deposit in the light of posterior events, nor can any deposit made available through the Commons be subsequently withdrawn.[186]

3. Microbe RURI 500/OCCI 8000 Elicits Research Interest

Sometime after 2018, microbiologists at several universities began to investigate the properties of this microbe in greater detail, with a view to constructing an evolutionary model of the species. This research intensified after 2019 as scientists readily obtained specimens from Collections A and B, both of which employed the Standard Material Transfer Agreement of the new Microbial Research Commons. Pursuant to that SMTA, Collections A and B duly notified the Designated National

[182] In point of fact, Collection A, as a member, must comply with these instruments whether or not the Ruritanian government itself had become an official member of the Commons. In most developing countries, a culture collection would not be allowed to join without the government itself having adhered to the foundational agreements. In developed countries, instead, and especially the U.S., collections might join for the research benefits even though the U.S. government – as a nonsignatory of the CBD – might not have joined the Commons.

[183] *See* discussion of the Scientific Coordination Council in Chapter 10, Section II.D.2.

[184] Questions of reciprocity requirements, if any, are also discussed in Chapter 10, Section III.C.6.

[185] *See* discussion of Integrated Digital Services in Chapter 10, Section III.D.

[186] Subject, of course, to the continued operations of the collection.

Authority in Ruritania each time a specimen was shipped, with the supposition that the Designated National Authority would have forwarded a copy of that notice to the Clearing House Mechanism (CHM) in Montreal, established under the Nagoya Protocol.[187]

In the same period, researchers at pharmaceutical supply Company Alpha in Occitania were investigating groups of microbes known to secrete a chemical substance used as an ingredient in the production of broad spectrum antibiotics. Since 2010, a number of supply companies, including Company Alpha, had been obtaining the chemical in question from one particular family of microbes, F_1, under a process well known in the industry and fully approved by the regulatory authorities. By 2019, however, researchers at Company Alpha had begun to suspect that other families of microbes might also produce the same chemical substance after text mining the literature, which covered both phenotypical and genetically sequenced samples of relevant microbes.[188]

Company Alpha's researchers further identified OCCI 8000 as one of a target set of microbes they wanted to test. Later that year, they found that OCCI 8000 was available from Collection B under the SMTA of the Commons, which makes no distinction between commercial and noncommercial research.

Company Alpha sent Collection B a purchase order for OCCI 8000 in August 2019. On dispatching the sample, Collection B duly notified the Designated National Authority in Ruritania, the country of origin, of this transfer, with the understanding that the Designated National Authority would copy the Clearing House Mechanism in Montreal.

Comment. Collections A and B are both bound by the foundational agreement establishing the Commons after 2018, the year in which Collection A, in Ruritania, had also joined the Commons, to which Collection B had adhered in 2017 as well. These dates of adhesion determine the period after which reciprocity under the Commons rules is required by the foundational agreement.[189]

The SMTA of the multilateral system, as we envision it, deliberately makes no distinction between for-profit and not-for-profit research because it wishes to stimulate maximum research use for all purposes, and because there is no clear

187 For the role of the Designated National Authority and Clearinghouse, *see* the Nagoya Protocol, n. 8, arts. 13–14. Because that Protocol establishes the CHM, we think it may become less necessary – although desirable – to maintain a parallel registration system within the administrative framework of the Commons.

188 On the importance of removing intellectual property impediments to the use of automated knowledge discovery tools for such purposes, see Chapters 6–8.

189 The degree of reciprocity to be embodied in the agreement establishing the multilateral system, if any, remains to be determined by the Governing Body. *See further* Chapter 10, Section III.C.6.

boundary between the two. For example, the study of evolutionary microbiology by university students in Step 3 might ultimately produce a major commercial payoff, while the industrial researchers in the same example may ultimately contribute only a basic research paper, depending on their respective findings. What matters is that research should be promoted and facilitated, and that whoever attains a commercial payoff must share a fixed percentage of the gross proceeds with the Designated National Authority in the provider country. In short, the duty to share benefits is detached from the right to undertake research, which is a right built into the Commons system; while the duty to pay is conditioned on the demonstrable existence of financial gain from the relevant research outputs.

The example in Step 3 calls attention to the fact that access to microbes for research purposes under the bilateral system of the CBD has become complicated and extremely onerous in some developing countries, while even in developed countries MTAs have become increasingly complex and restrictive on the whole.[190] The example illustrates the kind of relatively frictionless transfer for all research purposes that can occur under a standard form regulation of microbes having no known or likely commercial value at the time of deposit.

If Ruritania had not joined the Commons, or if it had otherwise not made RURI 500 available in the hopes of holding out for a tailor-made deal at a higher rate of return, it would have no assurance that some future, commercially relevant discovery would be made at all. Indeed, its very hold-out position – or even just the complexities of the bilateral approach – might discourage the very research that could put its microbes to some unknown commercial use. As will be seen in Step 4, finding viable commercial uses for most microbes is a risky and uncertain investment. Adding high transaction costs to the search for possible target microbes discourages the potential innovator from acquiring and testing microbes with no known or likely commercial value. The frictionless access provided under the Commons' SMTA would, instead, facilitate both transfers and research, and thereby augment the chances that there will be benefits to share under the Compensatory Liability Regime in the end.[191]

4. Development of a Commercial Product

Researchers at Company Alpha, using various advanced techniques, such as high-throughput screening and genetic sequencing, determined that microbe OCCI 8000 not only produced the chemical substance of interest, but that it also produced

[190] *See* Chapter 4, Sections II & III.

[191] *See further* steps 4 and 5. *Cf. generally* Rai et al, n. 21.

that substance in surprisingly more abundant quantities than other similar microbes used by other companies as an ingredient in certain antibiotics.[192] The company's marketing department then evaluated the R&D costs of using OCCI 8000, rather than the F_1 family of microbes, to obtain the same product at a more competitive price, taking into account the 2% royalty that must be paid to the provider country for industrial applications of microbes covered by the Commons' SMTA. Company Alpha's researchers also conducted further tests to establish quality and consistency of output and the ability to produce the secreted chemical on an industrial scale. The engineering department, in turn, designed and developed a production process capable of satisfying the business plan approved by the company's directors.[193]

After two years and a considerable investment, the final product was ready for marketing, and Company Alpha's legal office evaluated its intellectual property strategy. Given that no product patent is likely available for the use of a known substance for this purpose in many countries and that it might be obvious in others (and given that type strains cannot be patented), the Company opted to file for a process patent on its improved method of producing the ingredient in question by employing OCCI 8000. The marketing department then began soliciting orders from pharmaceutical companies that currently produce the relevant antibiotics using a more costly essential ingredient.

Comment. Given that microbe OCCI 8000 was obtained under the Commons' SMTA, no upfront licensing negotiations for the R&D uses of the microbe were necessary, and there were no *a priori* contractual impediments to constrain Company Alpha's research options. The absence of such disincentives to invest is important because applied research aiming to translate laboratory results into a commercially viable product or process is often costly, time-consuming, and risky, with no certainty of a successful outcome. The Compensatory Liability Regime thus reduces barriers to entry and has pro-competitive effects because any firm able to innovate and turn a profit after paying the default 2% royalty to the proper beneficiaries, as explained in Step 5, can freely use microbes available from the Commons for any commercial or noncommercial research purpose.

[192] The use of a different microbe to produce a substance also produced by another type strain is reportedly a common occurrence in industrial microbiology. *See* Dagmar Fritze, *A Common Basis for Facilitated, Legitimate Exchange of Biological Materials, Proposed by the European Culture Collections' Organization* (ECCO), 4 *Int'l J. Commons* 507, 521 (2010) [hereinafter Fritze (2010)].

[193] For the sake of simplicity, we assume that the product in question is a bioequivalent of a substance already approved for marketing by the regulatory authorities, and that no further clinical or regulatory trials were necessary. For biologics, this assumption is less tenable at the moment. *See* G.W. Nicholson Price II and Arti K. Rai, *Are Trade Secrets Delaying Biosimilars Science?*, 348 *Science* 188–89 (2015).

In the example as given, Company Alpha's own researchers were able to obtain the innovative product without the aid of university scientists. Had the university developed the key technology, instead, Company Alpha would have been obliged to negotiate a technology transfer agreement with the relevant university. That agreement would have been subjected to the 2% standard reach-through royalty under the SMTA between the university and Culture Collection A, which made Ruritania's Designated National Authority a third-party beneficiary.

The fact that pharmaceutical companies dislike such reach-through obligations is well known in intellectual property circles. Yet, this bias did not stop them from using the Cohen-Boyer patents under nonexclusive licenses from the relevant universities that owned the patents. Nor would it constitute a real – as opposed to hypothetical – barrier to other endeavors, so long as the ultimate commercial outcome were sufficiently profitable, taking into account the need to defray the 2% reach-through royalty from the outset.

Under real-world conditions, the intellectual property strategies available to an innovative company could be more or less complicated than those shown in the example used in Step 4. If no process or product patents were available, for example, a company might have to rely more heavily on trade secrecy law and the lead time it provides. Conversely, product patents on pharmaceuticals, when available, might require massive investments to cover the costs of clinical trials, in addition to the costs specifically attributed to translational medical research.[194] The point is that the Compensatory Liability Regime in no way interferes with the company's evaluation of such options. At the same time it ensures that, under any options chosen the provider country will receive its share of the resulting financial returns, by virtue of the standard reach-through royalty of 2%, and that the company is also contractually obliged to disclose the country of origin when filing patents, whether or not local patent law so requires.

5. Sales of the Product Trigger the Liability Rule and Distribution of Royalties

Sometime in 2021, Company Alpha listed a new product in its catalog and solicited sales from the pharmaceutical industry's antibiotics producers. Company Alpha also notified the Designated National Authority in Ruritania, the country from which Culture Collection B had procured OCCI 8000, that it would provide a statement of sales, gross revenues, and the amount due under a 2% royalty. At the end of its fiscal year, Company Alpha's accountants prepared the relevant statement and sent

[194] *See, e.g.*, Jerome H. Reichman, ***Rethinking the Role of Clinical Trial Data in International Intellectual Property Law: The Case for a Public Goods Approach***, 13 *Marq. Intell. Prop. L. Rev.* 1 (2009), *available at* http://scholarship.law.marquette.edu/iplr/vol13/iss1/1 (last accessed 3 July 2014).

a check for the total royalties owed to the Designated National Authority (or its agents).

Company Alpha's payments of the 2 percent royalty on gross sales to the Designated National Authority in Ruritania, as stipulated in the SMTA, discharged the Commons and its members from any further obligations with regard to the case under consideration.[195] The Designated National Authority in Ruritania will then distribute its share of the royalties to parties entitled to benefits under Ruritania's own national legislation implementing the CBD. Depending on how the Governing Body of the Commons structures the funding obligations of member countries, the Designated National Authority in Ruritania could be obliged to share a small percentage of its royalties on sales of commercial products with the Common's administration, as a component of annual dues, if any.[196]

Comment. Some economists might prefer a flat up-front fee for access and use of the microbes available from the multilateral system instead of the royalty on sales of end products.[197] However, most of the microbes in question will have no value other than as an input to basic research. In that case, any flat fee for use set in advance might exceed the elasticity of demand and only deter the very research we wish to encourage. By the same token, if use of the microbe leads to a major technological advance with a large commercial payoff, any small user access fee set in advance so as to facilitate research could yield the country of origin a grossly inadequate share of the benefits actually obtained by the innovator if it were the only source of benefits to be shared. The Governing Body could nonetheless decide to combine a small user access fee on all transactions to cover operational expenses with the Compensatory Liability Regime that applied the 2% royalty on sales of downstream end products.[198]

The 2% royalty we have suggested costs basic researchers nothing (except for existing costs of procuring specimens from the collections) so long as no commercial gains accrue from their use. Conversely, the 2% rate is low enough so as not to deter innovators from undertaking the investment in advance, when the results are still unknown. Yet, it is also high enough to ensure that provider countries receive a suitable rent from commercial uses of a genetic resource arising from their surrender of territorial jurisdiction in favor of a multilateral regime.

Two further observations follow from this premise. First, the provider country is not typically transferring know-how or other intangible intellectual assets along

[195] For questions about disputes, *see further* Chapter 10, Section III.C.7.

[196] On these and other questions of funding the multilateral system, *see* Chapter 10, Section IV.

[197] *See, e.g.*, David, n. 164.

[198] *See further* Chapter 10, Sections III.C.2 and IV.A.

with the physical transfer of microbial material (unless, for example, traditional knowledge is involved, which is discussed later). That is another reason for keeping the royalty low in the first place, along with ease of administration, but nonetheless high enough to yield significant returns when major downstream applications result.

Viewed as a rent-for-use of a physical asset, rather than as a reward for intellectual creation, there is no compelling reason for the right to that royalty to expire at the end of a fixed period of time. Rather, so long as the microbe is made available *ex situ* through public culture collections, the 2% royalty could apply indefinitely, on analogy to a paying public domain construct.

As noted earlier in this chapter, certain technically advanced culture collections – especially the BRCs – might have added valuable data and information to the microbe originally contributed by Ruritania. Whether this could justify an extra carrot of royalty in favor of, say, Culture Collection B, remains to be determined by the Governing Body and the SMTA it develops.[199]

As matters stand, the culture collections would act as trusted intermediaries on behalf of the Governing Body to be established by the governments participating in the multilateral system (which is discussed in Chapters 9 and 10). As already occurs, each collection charges a fee for its services, in order to recoup the marginal costs of distribution plus some of its overhead, and some price discrimination typically occurs with respect to academic and commercial users.[200]

With regard to revenues generated under the liability rule, we think it proper for the Commons to receive a small percentage of the royalties flowing to provider countries in order to support its own operations over time. From this perspective, the Commons is an agent of the beneficiaries under the CBD (of particular concern to developing countries) and also of the scientific community, both of whose interests it will be designed to protect. It is also an agent of the participating governments, which will be concerned about the long-term sustainability of the organization.[201]

For all these reasons, it seems logical to devolve a small share of the benefits from commercial applications of the protected genetic resources to the management of the Commons, who are contractually responsible for monitoring and compliance with the bylaws and decisions of the Governing Body. The participating governments would, in turn, have to determine the percentage to be subtracted for the functions of the Commons. For purposes of illustration, we suggest no more than a 10 percent share, which could be credited against annual dues.

[199] *See further* Chapter 10, Section II.D (Core Institutional Components). The bylaws of the Commons must in any event provide for cases in which culture collections dissolve.

[200] *See* Chapter 4, Section I.A.3.

[201] *See further* Chapter 10 (which also discusses trust funds for capacity building).

One possibility is that such funds could also be used to enable the Commons' management to provide policy and legal advice in these matters, to all the collections, and to support the provision of essential ancillary services, especially dispute resolution.[202] However, the Governing Body of the Commons could, instead, decide to dedicate its share of royalties accruing from the Compensatory Liability Regime, if any, to the general upkeep and maintenance of the network of participating culture collections as a whole, not to mention capacity building and the provision of digitally integrated data and literature that we explore in Part Three.

As for the royalties flowing to the Designated National Authority in the country of origin, which are to be distributed according to national law, the logical beneficiaries could include Culture Collection A, which preserved and maintained the type strain in question; the actual discoverer of that type strain; and the university or scholar with which that discovery was associated, among other potential claimants worthy of consideration. Indigenous peoples must also be recognized when traditional knowledge is involved.[203] However, we do not think it wise or feasible for our redesigned Microbial Research Commons to adopt a standard distribution protocol, as this will necessarily vary from country to country.

Finally, one must ask what would happen if Company Alpha refused to pay the royalties due under the SMTA, or if disputes arose about the amounts due or other related matters. In such cases, the SMTA will provide for mediation and arbitration,[204] while the Commons will itself make a dispute resolution modality available to its members, as discussed in Chapter 10. As a practical matter, nonetheless, we stress that, under the Nagoya Protocol, claims against a defaulting company would automatically elicit charges of misappropriation that parties to the Protocol have agreed to hear in their domestic courts.[205]

6. Lottery Effects and the Possibility of Leakage

Let us assume that four other pharmaceutical supply companies decided to enter the market for chemicals derived from the use of type strain RURI 500/OCCI 8000 after Company Alpha had begun to sell its product at time 1, i.e., in 2022.

[202] *See* Chapter 10, Section III.C.7.

[203] *See, e.g.*, Chapter 3, Section III.B.1.

[204] *Cf.*, *e.g.*, the SMTA adopted for the WHO's PIP Framework, as discussed above Chapter 4, Sections IV.A. &B; World Health Org. (WHO), *Pandemic Influenza Preparedness Framework for the Sharing of Influenza Viruses and Access to Vaccines and Other Benefits*, World Health Assembly Res. WHA645 (May 24, 2011) [hereinafter *PIP Framework*], *available at* http://www.who.int/influenza/resources/pip_framework/en/index.html (last accessed 23 Feb. 2014).

[205] *See* Nagoya Protocol, n. 8, art. 18.2; *see further* Chapter 3, Section IV.C.

A. MULTIPLE INDUSTRIAL USERS OF THE SAME MICROBE PRODUCE MULTIPLE ROYALTY STREAMS. At time 2, Company Beta licenses both Company Alpha's process patent and the right to sell its product in other countries beyond Occitania.[206] Company Beta subsequently pays 2% of its gross sales to the Designated National Authority in Ruritania as before.

Moreover, at time 3, Company Gamma, with headquarters in Franconia, reverse engineers Company Alpha's product (on which there was no product patent) and conducts R&D of its own, enabling it to develop a different, but equally efficient process for using OCCI 8000 that, in the view of its attorneys, invents around Company Alpha's, process patent. Company Ghamma's researchers, in cooperation with scientists at UniFranc University, also determined that the end product of this process could be employed in the manufacture of certain anti-cancer drugs as well as in antibiotics.

Company Gamma markets its initial product everywhere, in competition with Companies Alpha and Beta. Once marketing approval was finally obtained from the relevant regulatory authorities, Company Gamma, which had filed for a patent on its new use of a known substance,[207] also begins to market its second product for use in the manufacture of certain anti-cancer drugs.

In both cases, Company Gamma had lawfully obtained the needed type strain OCCI 8000 from Culture Collection B. It later sent accounting statements of gross sales and the requisite royalty payments to the Designated National Authority in Ruritania (which would deduct and transmit any small tangent owed to the Commons management, if any).[208]

However, the Designated National Authority in Ruritania has some reasonable grounds for mistrusting the statements supplied by Company Gamma. It demands that Collection B obtain an audit of Company Gamma's accounts.

Comment. Both Companies Beta and Gamma have purchased microbe OCCI 8000 for research and production purposes from Culture Collection B in order to ensure the quality of the end results. Obtaining microbial materials from reputable collections for such purposes is a standard practice in industry.

[206] The product license in these cases saves Company Beta the time and costs of reverse-engineering the product and devising an unpatented process somewhat different product. *See* TRIPS Agreement, n. 157, arts. 28.1(b) and 34.

[207] Whether such patents are available, and on what conditions, varies considerably from one jurisdiction to another. *See* CHRISTOPHE SPENNEMAN & JEROME H. REICHMAN, USING INTELLECTUAL PROPERTY RIGHTS TO STIMULATE PHARMACEUTICAL PRODUCTION IN DEVELOPING COUNTRIES – A REFERENCE GUIDE 49–63 (U.N. Conference on Trade & Dev. 2011).

[208] The Governing Body of the Commons could act as an intermediary between Company Gamma and the Designated National Authority for such purposes. See Chapter 10, Sections III.B & IV.

The two scenarios set out earlier illustrate the cumulative "lottery effects" that may occur under the Compensatory Liability Regime when follow-on firms make further use of a microbe for which an earlier commercial application had been found. No front-end negotiations are needed for this purpose, and no similar regime yet exists for microbes under the CBD.[209]

These examples also illustrate the lottery effects that may occur if future uses for the same microbe are discovered later in time. Under the multilateral framework, all commercial sales from different applications of the same microbe must be compensated via the liability rule.

Both examples illustrate the importance of tracking by means of the Global Unique Identifiers (GUID) that WFCC culture collections assign to each specimen. The examples also show that innovators remain free to choose their own R&D approaches as well as their intellectual property strategies, without negotiations with provider countries, as could not occur under the bilateral approach of the CBD.[210]

With regard to the disputed statement of accounts in the second example, Company Gamma could be asked to submit to an audit under the bylaws of the redesigned Microbial Research Commons, as discussed in Chapter 10, although mediation and arbitration may be the more likely avenues to pursue. The availability of the Commons' own dispute resolution procedures could depend on whether the government of Franconia was also a member of the Commons. Private international law might apply Occitanian law to the issues if no choice of law clause accompanied the purchase and sale of microbe OCCI 8000.

If Company Gamma refused to comply with the request for an audit and Franconia was not a member of the Commons, the Ruritanian government might intervene to press the Franconian government to ensure compliance with the CBD, which always applies by default.[211] If Company C turns out to have misappropriated the microbe in question by violating the default rules of the multilateral system, its products could conceivably be seized in any country adhering to the Nagoya Protocol.[212]

B. ADDRESSING THE POSSIBILITY OF LEAKAGE. At Time 3, pharmaceutical supply Companies Delta and Epsilon both obtain microbes for industrial use from a private broker in Ruritania, who in turn had procured RURI 500 from Collection

[209] For a similar regime covering plant genetic resources, *see* ITPGRFA. n. 61, as discussed in Chapter 3, Section III.B.

[210] This proposition would, of course, not apply in cases where the innovator uses *in situ* genetic resources, which would still require upfront agreements with provider countries. *Cf.* Kamau et al., n. 4.

[211] *Cf.* Brazil's offer of amnesty to companies that take steps to register and negotiate uses of genetic resources previously taken without permission. World Intell. Prop. Rep. (BNA) June 26, 2011, available at http:www:bna.com/international-trade-reporter-pG101.

[212] *Cf.* Nagoya Protocol, n. 8, art. 18.2.

A in Ruritania under false pretenses. Both companies proceed to produce and sell ingredients for manufacturing broad-spectrum antibiotics and for certain anti-cancer drugs in various countries. However, they do not disclose their gross sales to any authority recognized by the CBD, and they do not pay any royalties to the Designated National Authority in Ruritania, the country of origin.

The microbes used by these deviant companies in their respective processes can be identified as RURI 500 or as clones thereof. Moreover, the GUID tracking number, which adheres to all microbes legitimately obtained from the WFCC culture collections at any point in the chain of supply, has identified the broker in question as a purchaser of RURI 500. The Designated National Authority in Ruritania has been asked by the Ministry of Justice to take steps to force the broker to disclose subsequent purchasers of the specimens it bought from the National Culture Collection.

Further investigation reveals that one such purchaser, Company Delta, has its headquarters in Latinia, a CBD member country, whose government has joined the Microbial Research Commons. Another purchaser, Company Epsilon, operates from Nordistan, a country whose government has not joined either the Microbial Research Commons or the CBD.

Comment. As matters stand, the potential risk of leakage under either the CBD or the multilateral system is partly attenuated by the tracking that GUID numbers make possible and by the ability to identify clones through genetic sequencing.[213] If one of the DARPA-funded projects currently under way happened to succeed in the near future,[214] genetically tagged biomarkers might provide compelling evidence of use, legal or illegal, anywhere in the chain of supply. In the meantime, members of the CBD are no less susceptible to leakage than members of the Microbial Research Commons would otherwise be, although the Commons will make dispute resolution procedures available to its members in order to address this problem.

In the case of Companies Delta and Epsilon, their competitive practices clearly violate the foundational agreement of the multilateral system as managed by the Commons, which operates in place of the bilateral regime of the CBD by the consent of the participating governments. The foundational agreement should, in turn, give the Designated National Authority in Ruritania standing to complain to the Governing Body of the Commons, while at the same time triggering the responsibility of the Latinian government to investigate the illicit activities of Company Delta. Resolution of the case could also depend to some extent on any

[213] The question of derivatives is addressed later. *See* Section III.B.2.

[214] *See* nn. 140–145 and accompanying text.

mediation and dispute resolution machinery that the participating governments will have established within the Commons institutional architecture.[215]

If all the relevant countries are members of the CBD, there is also a parallel violation of that treaty. Hence, the Designated National Authority in Ruritania would be entitled to pursue an action for misappropriation of genetic resources against Company Delta by any means eventually provided by member governments under the Nagoya Protocol to the CBD and the various national and regional laws implementing it.[216]

In these circumstances, Ruritania is no worse off, and may be considerably better off, as a member of the Commons than would be the case if it were a CBD member that had declined to join the Commons. This conclusion follows especially if the Commons' own governance machinery provides a more direct and expeditious claim to a clear entitlement to royalties, defined *ex ante*, by the Compensatory Liability Regime, rather than having to rely on a still to be defined action for misappropriation under the CBD. In other words, the worst case scenario of leakage under the proposed Commons infrastructure is tantamount to the best case scenario available from the CBD.

As regards the conduct of Company Epsilon, in contrast, the Designated National Authority in Ruritania will have no direct legal claim against either the company itself or the government of Nordistan under the Commons infrastructure, because Nordistan did not join the Commons, nor under the CBD, because Nordistan has also declined to adhere to that international convention. Given that virtually all other countries have joined the CBD, Ruritania might try to level a claim sounding in a breach of customary international law, although Nordistan might rebut with a claim that it had consistently opted out of that prospect.[217]

A more promising line of attack might entail a claim by Ruritania that Company Epsilon's conduct failed to meet international standards of fair competition as now defined by Article 39 of the TRIPS Agreement (incorporating Article 10*bis* of the Paris Convention) when these provisions are read in the light of the CBD.[218] Ruritania might also decide to take unilateral retaliatory action against Nordistan for unfair trade practices, following the example of the United States' continuing reliance on Section 301 of the Trade Act of 1974 for similar purposes.[219] In any event,

[215] *See* Chapter 10, Section III.B.7.

[216] *See, e.g.*, Nagoya Protocol, n. 8, art. 18.2, 18.3.

[217] *See most recently* Curtis Bradley & Mitu Gulati, *Mandatory Versus Default Rules: How Can Customary International Law Be Improved?*, 120 *Yale L. J.* 421–54 (2011).

[218] *Cf., e.g.*, Jerome Reichman, *The International Legal Status of Undisclosed Clinical Trial Data: From Private to Public* Goods, *in* NEGOTIATING HEALTH: INTELLECTUAL PROPERTY AND ACCESS TO MEDICINES 133–150 (Pedro Roffe et al. eds., Earthscan 2006).

[219] For doubts about the legitimacy of this approach, see SPENNEMAN & REICHMAN, n. 207.

Company Epsilon's products, once exported beyond Nordistan's territorial borders, could remain liable to seizure for violating the global anti-misappropriation regime of the CBD in virtually every other country in the world.[220]

Apart from these international legal considerations about unfair competitive practices, there are also leakage questions concerning the conduct of scientists who might evade the SMTA of the multilateral system once the relevant countries had become members. In principle, risky informal exchanges should be discouraged under the Nagoya Protocol, whereas the Commons' SMTA will legitimize and facilitate research of every kind. If abuses having financial implications nonetheless occurred, there would be considerable peer pressure brought to bear against the offending scientists or scientific entities, with a corresponding risk of disentitlement to future access to the system as well as grants, in addition to any legal liability incurred by members of the Commons (under the CBD) who failed to repress such abusive behavior. In these cases, the potential loss of reputational benefits might impose greater restraint than the fear of legal consequences.

B. *Accommodating More Complicated Transactions*

Not directly covered in the scenarios set out earlier are certain more complicated fact patterns that could occasionally arise. These transactions are briefly discussed in the next sections.

1. Multiple Owners and Possible Royalty Stacking

In the examples so far discussed, the tacit assumption has been that one particular microbe was the source for both research and commercial applications. More complex cases could arise if more than one deposited microbe were involved in any given research project and its subsequent commercial applications. A somewhat similar problem could arise if the commercial applications were derived from an ecology of microbes, or a group of interacting microbes, in which cases there might be multiple owners claiming royalties if the materials in question were drawn from the Microbial Research Commons and subject to its Standard Material Transfer Agreement (SMTA).

In the realm of exclusive intellectual property rights, notably patented inventions, such analogous situations may present a hard problem known as royalty stacking, with multiple occasions for some of the owners to exert "hold out" or blocking

[220] *Cf.* Nagoya Protocol, n. 8, arts. 18.2, 18.3.

effects on the transfer of technology to industry.[221] In the context of liability rules, however, such problems are relatively more manageable in general because no hold outs are permitted by definition. With specific regard to the proposed Compensatory Liability Regime for microbial genetic resources, the problem is further attenuated by a number of ancillary principles.

First, the proposed liability rule rests on a foundationally broad research exemption, unlike statutory patent and most copyright laws, which means that few blocking effects should occur at either the research or application phases, because no licenses are required for use of the relevant microbe beyond the de facto, open-access license set out in the SMTA. As to claims for royalties from multiple owners of the same microbe, or of diverse microbes that all play a part in the successful project, these of course could literally "stack up" if substantiated by evidence that satisfied the standards of the dispute resolution machinery discussed in Chapter 10.

However, the number of claims cannot increase the upper limit of the calculus of royalties wherever the Governing Body of the Commons may decide to set it (e.g., 2 percent in our hypothetical SMTA, possibly 3 percent if BRC data were also supplied). In short, multiple claimants, if successful, would have to divide the maximum royalty rate allowed, and would not obtain any aggregate amount in excess of that limit.[222] Downstream innovators could thus never be burdened by more than, say, an aggregate 2 or 3 percent reach-through claim for benefit sharing purposes under the SMTA.

This approach also helps to resolve the problem that could arise if providers in more than one country claim royalties from commercial use of microbial materials held by two or more collections in different countries. This situation presents, first of all, an evidentiary problem in that the fact that materials were held in two collections does not in and of itself mean that either the researchers or the downstream users actually took materials from both collections. What matters, for our purposes, is who actually used what for which purposes, and not what might possibly have been done with the same microbial materials had they been taken from different collections.

Admittedly, this situation could sometimes lead to challenging evidentiary disputes, as also happens in the realm of patents. But the consequences of any resolution of these evidentiary issues are less dire than in the case of exclusive property rights, for the reasons stated earlier. If more than one claim survives evidentiary scrutiny, each will get a smaller share of the relevant royalties, but the total amount paid by users will not exceed the upper limit of the agreed royalty calculus.

[221] *See, e.g.*, Paul A. David, *Mitigating 'Anticommons' Harms to Research in Science and Technology* (UNU-MERIT Working Paper No. 2011-001) (2010).

[222] We ignore the possible need to determine who gets exactly what if the parties are dissatisfied with equal shares.

2. Derivatives or Modifications that Incorporate Materials Accessed from the Multilateral System

Our survey of existing material transfer agreements adopted by major culture collections identified at least three types of derivatives from the use of microbial genetic resources.[223] One is the case of "unmodified derivatives," such as a component of the original material or a product expressed by the original material. Another is the case of so-called "progeny," which refers to an unmodified descendant of the original material, such as a subculture or a "replicate." The third example is that of "modifications of the original material" that nevertheless continue to embody some component of that material.[224] While these common legal distinctions can become blurry in practice, genetic sequencing, which supplements tracking and chain of custody considerations, helps to verify such claims on the basis of objective evidence.

For present purposes, the first two cases present no particular problem for the Compensatory Liability Regime. Both unmodified derivatives and so-called progeny will continue to attract the established royalty, notwithstanding the added value by the second comer. The third case, in which a modification has been produced and it contains a component of the original material, presents a theoretically more complex situation open to various solutions. However, our survey of MTAs in the previous chapter suggests that a number of the collections have settled on a customary practice that has been deemed acceptable by both public and private sector users. On this approach, the contract could stipulate that the originator of the modification (or whoever the owner may be) must enter into a joint ownership agreement with either the collection or the Designated Authority in the country of origin, or with both.

A parallel survey of stem cell agreements conducted by Carol George at the University of Edinburgh Law School suggests that providers and users of stem cell lines in the United Kingdom have often reached a similar resolution on the same issue.[225] This contractual approach to joint ownership of modifications also bears some resemblance to the treatment of derivatives works in copyright law, in the sense that the author of the derivative work cannot proceed without the consent of the author of the underlying work.[226]

[223] *See* Chapter 4, Sections II and III.

[224] *See, e.g.*, The ECCO Core Material Transfer Agreement for the Supply of Samples of Biological Material from the Public Collection, Feb. 2009 [hereinafter ECCO MTA], *available at* http://www.eccosite.org/wp-content/uploads/2014/07/ECCO_core-MTA_V1_Febr09.pdf]. See also Chapter 4, Section III.A.2. For the ATCC MTA, *see* Chapter 4, Section II.A.

[225] *See* Carol George, Openness and the Governance of Human Stem Cell Lines: A Conceptual Approach (2013) (unpublished Ph.D. thesis, Univ. Edinburgh) (on file with the Edinburgh Research Archive).

[226] *Compare e.g.*, the rules of joint tenancy in U.S. Copyright law, which allow either party to use the jointly owned work subject to an accounting, 17 U.S.C. §201(a) (2010) *with* the rules governing

While, in principle, it seems advisable to accept this increasingly customary arrangement with regard to modifications for purposes of the Compensatory Liability Regime, we emphasize the different results this gives in the context of "take and pay" rules as distinct from any regime – whether contractual or statutory – of exclusive property rights. In the latter case, one joint owner might be able to obstruct either research uses of a given microbe or downstream applications without express permission at the other. Under the proposed liability rule, instead, neither owner can obstruct these uses because a research exemption is built into the multilateral system. However, both would be entitled to share in the royalties accruing from downstream commercial applications of the modifications in questions. The aggregate royalties to be shared cannot exceed the standard royalty of 2 percent or in the case of value-adding data 3 percent, if this calculus were to be adopted by the Governing Body.

3. Modifications Based on Data Pertaining to Microbial Materials Accessed from the Multilateral System

As previously noted, the Nagoya Protocol expressly seeks to implement one tacit aspiration of the CBD, namely to extend the provider countries' assertion of sovereignty over genetic resources to data subsequently extracted from research on such resources, especially when used to produce downstream commercial applications.[227] The extent to which even the Nagoya Protocol succeeds in operationalizing claims of this sort remains open to discussion[228] and could depend in part on evolving methods for evaluating the role of data in patent applications on microbial related inventions generally. Nevertheless, the absence of any agreed approach to the issues this topic raises makes it advisable to consider its implications for the Compensatory Liability Regime that we have so far applied only to biological materials.

To the extent that the microbial data in question pertain purely to basic research, the situation does not differ materially from that of physical specimens available from the proposed Microbial Research Commons. A major purpose of the multilateral regime is to establish a carve out for basic research in return for the benefits accruing from a liability rule on commercial applications; and this principle should suffice to insulate upstream research on both materials and related data from interference by providers. By the same token, when data derived from materials having no known

derivative works, which requires the consent of the owner of the underlying work, 17 U.S.C. §§ 103, 106(2) (2010).

[227] *See* Chapter 3, Section IV.A.

[228] *See, e.g.*, Gerd Winter, *Knowledge Commons, Intellectual Property and the ABS Regime*, *in* COMMON POOLS OF GENETIC RESOURCES, pp. 285–301.

or likely commercial value that were accessed from the Common's multilateral system figure in the development of downstream commercial applications, the Compensatory Liability Regime applicable to materials should apply with equal force to use of the related data.

Although this scenario thus presents no fundamentally different concept of entitlements, it does present tracking and authentication problems that are different from – and more serious than – those familiar from our discussion of materials. For this and other reasons, the burden of proving noncompliance with the Commons' SMTA could become correspondingly more difficult.[229] However, the risk of leakage here seems no greater than in current case-by-case negotiated transactions, where an express legal liability for the use of data remains subject to the comparable difficulties of tracking data actually used and of meeting the relevant burden of proof.

In this connection, it might work to all the stakeholders' advantage if the Governing Body of the Microbial Research Commons could elaborate consensus principles – in agreement with the Designated National Authorities – on a duty to notify the relevant Designated National Authority and the Commons' own data managers when materials provided by member collections had been genetically sequenced.[230] Such a principle could yield at least two important payoffs. First, to the extent that genomic sequence research data were made available on an open-access basis through the Commons, or under its auspices, such data would constitute "a nonmonetary benefit" under the Nagoya Protocol and could be viewed as a constructive repatriation of value to provider countries.[231] Second, early and public notice to both the Designated National Authority and the Commons that a covered specimen had been sequenced could constitute constructive notice to the relevant scientific community that use in future commercial applications had become more feasible for purposes of future policing. That notice policy, in turn, could be strengthened by use of unique identifiers attached to the relevant genetic sequence data, if and when such techniques become more feasible.[232]

This and related measures could become particularly important if, in the future, genomic blueprints began to displace the use of cultured materials in developing downstream commercial applications.[233] Ideally, genomic blueprints having no known or likely commercial value would increasingly be made available to the Commons and could then become subject to the Standard MTA, with perhaps some

[229] *Cf.* problems regarding data obtained by the Global Biodiversity Information Facility (GBIF), discussed in Chapter 9, Section II.B.2, which have discouraged participation by developing countries.

[230] *Cf.* MICRO B3 Consortium, http://www.microb3.eu/ for one example, For data management proposals under the Commons see further Chapter 10, Section III.D.

[231] *See* Chapter 3, Section IV.B.2 ("Recognizing the Importance of NonMonetary Benefits").

[232] *See* Chapter 8, Sections II.B & C.

[233] *See, e.g.*, Daniel Drell, *Research and Applications in Energy and the Environment* (2009), *in* DESIGNING THE MICROBIAL RESEARCH COMMONS, 51, at 121–123.

recalculation of royalties to reflect the value added by the sequencing entity.[234] The more that such blueprints were actually brought within the regulatory framework of the multilateral system, the more secure all the relevant stakeholders – providers, researchers, and commercial users – would be *ex ante* as to their respective freedoms to operate and their duties to compensate *ex post*.

Finally, the CBD itself has fostered a movement to require patentees to disclose the origin of relevant genetic materials when filing for gene-related patents.[235] A universal norm to this effect would discourage biopiracy[236] and would generally facilitate enforcement of the global misappropriation norms solidified in the Nagoya Protocol. Once such a disclosure requirement had become embedded in international industrial property law, as seems increasingly likely to happen, the same principle could logically be extended to identifying the sources of data components used in patented products and processes. Such *ex post* disclosure norms could then reinforce any *ex ante* notification rules devised for genetic sequencing and perhaps significantly reduce or even eliminate the burden of proof problems affecting current uses of microbial data in industrial development processes.

We recognize that this set of issues remains somewhat futuristic. Nevertheless, they require thought when constructing the Commons architecture from the outset, and they raise questions of considerable importance to scientists and universities, as well as to national providers of genetic resources.

C. Advantages of the Scheme

The Compensatory Liability Regime developed for the proposed multilateral system of facilitated access for microbial genetic resources avoids the high transaction costs attendant upon case-by-case negotiations to clear exclusive rights of uncertain value that might otherwise attach very far upstream and thus unduly encumber scientific research. As illustrated in the scenarios set out earlier, this model allows unforeseen commercial applications to be made without protracted negotiation, under a standardized, built-in revenue-sharing arrangement with the providers of genetic resources that intrinsically satisfies the demands of the CBD. A pro rata component of this royalty stream could also help defray the costs of administering the system.

One of the biggest advantages of this approach is that it could halt the spread of ever more restrictive MTAs likely to be adopted as defensive measures by single culture collections and Biological Resource Centers. What instead emerges is an

[234] *See* above Section II.C.4 (proposing extra quantum of royalties for value-adding data).

[235] For WIPO's efforts to establish such a norm, *see* n. 174.

[236] Charles R. McManis, *Fitting Traditional Knowledge Protection and Biopiracy Claims into the Existing Intellectual Property and Unfair Competition Framework*, *in* INTELLECTUAL PROPERTY AND BIOLOGICAL RESOURCES 425–510 (B. Ong ed., Michael Cavendish Int'l Press 2004).

open invitation to the global research community to discover unknown commercial applications from the pooled microbial deposits, in exchange for payment of equitable compensation, without costly case-by-case clearing transactions under the bilateral approach in which each side is likely to overvalue what it has to offer.[237] The primary goal is to empower and enhance public sector research at a time when pressures are mounting to privatize, and thereby unduly restrict, that sector's *ex situ* microbial holdings.

The very availability and efficacy of these common pooled resources may draw private sector researchers and venture capital into the system. This may be done without compromising the public research function and without depriving countries that preserve and contribute genetic resources of the possibility of financial gain under what amounts to a de facto public-private partnership.[238] The more downstream commercial uses that arise from any given set of deposited microbial materials, the more financial benefits would accrue to the relevant contributors, not to mention the benefits to the public at large.

At the same time, we would avoid creating any upstream exclusive rights to impede either the research function of the multilateral system as a whole or the downward and outward flow of innovation, where patents may ultimately be secured.[239] On this approach, both upstream research and future commercial applications are allowed to flow at their own pace, so long as quality standards are maintained, reputational benefits are preserved, and providers of microbial genetic resources stand to receive equitable compensation from downstream applications. This solution also permits both large and small culture collections to interact with each other in a federated system that greatly expands the research capacities of all the players.

The proposed model as outlined earlier thus facilitates transborder exchanges of microbial genetic resources that might otherwise be subject to diverse national rules and regulations establishing different stop signs at every national frontier.

[237] Reichman & Uhlir (2003), n. 19; Reichman & Lewis (2005), n. 41.

[238] Rai et al., n. 21. For a more detailed discussion about the theory and practice of common pool resources, see Chapter 9, Section I.A.

[239] For the importance of this point, *see* MARK HARVEY & ANDREW MCMEEKIN, PUBLIC OR PRIVATE ECONOMIES OF THE KNOWLEDGE TURBULENCE IN THE BIOLOGICAL SCIENCES (Edward Elgar Pub. 2007); *see also* David Smith, *Culture Collections, in* 79 ADVANCES IN APPLIED MICROBIOLOGY, Ch. 4 (2012); Paul A. David, *Mitigating 'Anticommons' Harms to Research in Science and Technology* (UNU-MERIT Working Paper No. 2011-001) (2010). Paul A. David, *The Economic Logic of "Open Science" and the Balance Between Private Property Rights and the Public Domain in Scientific Data and Information: A Primer, in* THE ROLE OF SCIENTIFIC AND TECHNICAL DATA AND INFORMATION IN THE PUBLIC DOMAIN 19, 19–34 (Julie M. Esanu & Paul F. Uhlir eds., Nat'l Acad. Press 2003). Paul A. David, *The Historical Origins of "Open Science:" An Essay on Patronage, Reputation and Common Agency Contracting in the Scientific Revolution*, 3(2) *Capitalism & Soc'y* Chap. 5 (2008), *available at* http://capitalism.columbia.edu/files/ccs/Paul%20A.%20David.pdf.

It deliberately exploits the opening in the Nagoya Protocol that treats research opportunities as a nonmonetary benefit,[240] in order to bring developing-country providers within a global system of exchanges that maximizes their prospects for benefit sharing under the CBD. This scheme, operating within an international framework as envisioned in Part Four, aims to put all the stakeholders in a win-win position.

From a more theoretical perspective, the Compensatory Liability Regime directly addresses the problems of hyper-ownership that have arisen since the enclosure of genetic resources formerly available as international public goods.[241] Technically, this regime intentionally substitutes a variant of the paying public domain concept[242] for both the proliferation of exclusive intellectual property rights since the 1990s that the developed countries have promoted and the countervailing expansion of proprietary claims based on sovereign territorial rights that the developing countries have asserted in response.[243] In so doing, it recognizes the unique importance of genetic resources *qua* carriers of information essential to understanding all living matter;[244] and it thereby contributes to the evolution of a *sui generis* intellectual property model better suited to the age of information technologies than any existing regimes dominated by the patent and copyright paradigms.[245]

Ultimately, the proposed multilateral framework is only as good as the resource contributions it attracts and the cumulative research powers that a federated, digitally interactive semicommons engenders. So long as the research gains from transparency and facilitated access to the common pooled resources, coupled with the cumulative returns from the Compensatory Liability Regime, outweigh losses due to the lack of secrecy and excessive research restrictions rooted in upstream exclusive rights, our third option should appeal to most WFCC culture collections. It should also appeal to growing numbers of individual laboratories in both developed and developing countries, and eventually to most private sector stakeholders as well.

The point is that any properly constructed multilateral system of facilitated access cannot be geared primarily to accommodate the interests of either the private sector

[240] *See* Chapter 3, Section IV.B.2.

[241] *See, e.g.*, Sabrina Safrin, *Hyperownership in a Time of Biotechnological Promise: The International Conflict to Control the Building Blocks of Life*, 98 *Amer. J. Int'l. L.* 641, 644–45 (2004); Kal Raustiala & David G. Victor, *The Regime Complex for Plant Genetic Resources*, 58 *Int. L. Org.* 277 (2004); *see further* Chapter 2, Section II.

[242] *See, e.g.*, Reichman & Lewis, n. 41.

[243] *See, e.g.*, Safrin, n. 241; *see generally* Chapter 2, Sections II.A–B.

[244] Safrin, n. 241.

[245] Reichman, *Green Tulips*, n. 44. *See generally* Reichman et al., *Legal Hybrids*, n. 153, and Reichman, *Charting the Collapse of the Patent-Copyright Dichotomy*, n. 153.

or even those public sector culture collections holding *ex situ* microbial materials having known or likely commercial value.[246] Rather, it must be devised to provide maximum benefits to the research community as a whole, with an open door to the private sector that enables the latter to participate on favorable terms and conditions.[247]

[246] *Cf.* Rai et al., n. 21.

[247] A community buy-in of the scheme is essential to avoid single scientists routinely opting for the high protectionist approach. *See further* Chapter 10 (on governance).

PART THREE

A Digitally Integrated Infrastructure for Microbial Data and Information

6

Legal and Institutional Obstacles Impeding Access to and Use of Scientific Literature and Data

I. POTENTIALLY BOUNDLESS SCIENTIFIC OPPORTUNITIES IN THE DIGITAL ENVIRONMENT

The rapid development and diffusion of digital technologies and global networks has resulted in profound social and economic transformations in practically all countries and sectors of human endeavor. The magnitude of the changes made possible by the shift from print to digital technologies and networks cannot be overstated either quantitatively or qualitatively as illustrated by Box 6.1.

In quantitative terms, digital networks vastly outperform print media. In recent years, the ever expanding production of bits has given rise to the overused, but accurate, label of "Big Data."[1] This deluge of data of all kinds has been accompanied

[1] The explosion in the production of digital bits is now well known as a function of Moore's law, which posits that the computing power of microprocessors doubles every eighteen months. *See, e.g., Moore's law*, WIKIPEDIA, http://en.wikipedia.org/wiki/Moores_law (last accessed 9 Apr. 2014). There are numerous definitions of "Big Data." One well-known consulting firm describes it as "high-volume, high-velocity and high-variety information assets that demand cost-effective, innovative forms of information processing for enhanced insight and decision making." Gartner, *IT Glossary*, http://www.gartner.com/it-glossary/big-data/ (last accessed 9 Apr. 2014). Several scientific programs focus on Big Data initiatives. In the United States, for example, The White House Office of Science and Technology Policy announced a new "Big Data Research and Development Initiative" in 2012, *see* Tom Kalil, "Big Data is a Big Deal," U.S. OFFICE SCI. & TECH. POL'Y (OSTP) (Mar. 29, 2012), http://www.whitehouse.gov/blog/2012/03/29/big-data-big-deal. In biomedical research, the U.S. National Institutes of Health launched an initiative called "Big Data to Knowledge" (BD2K), *see* Office of Strategic Coordination, "Big Data to Knowledge," NAT'L INST. HEALTH (NIH) (July 22, 2013), https://commonfund.nih.gov/bd2k/index.aspx. The NIH defines "big data" as follows:

> The term 'Big Data' is meant to capture the opportunities and challenges facing all biomedical researchers in accessing, managing, analyzing, and integrating datasets of diverse data types [e.g., imaging, phenotypic, molecular (including various '-omics'), exposure, health, behavioral, and the many other types of biological and biomedical and behavioral data] that are increasingly larger, more diverse, and more complex, and that exceed the abilities of currently used approaches to manage and analyze effectively. Big data emanate from three sources: (1) a small number of groups that produce very large amounts of data, usually as part of projects

Box 6.1 Comparison of Some Print and Digitally Networked Paradigm Characteristics

Print Paradigm	Global Digital Networks
• Fixed, static	• Transformative, interactive
• Rigid	• Flexible, extensible
• Physical	• "Virtual"
• Local	• Linear
• Limited content types	• Unlimited contents and multimedia
• Distribution difficult, slow	• Easy and immediate dissemination
• Copying cumbersome, not perfect	• Copying simple and identical
• Significant marginal distribution cost	• Zero marginal distribution cost
• Single user (or small group)	• Multiple, concurrent users/producers
• Centralized production	• Distributed and integrated production
• Slow-knowledge diffusion	• Accelerated knowledge diffusion
• Quasi-private good	• Quasi-public good

This table is adapted from Paul F. Uhlir, *The Emerging Role of Open Repositories as a Fundamental Component of the Public Research Infrastructure*, *in* OPEN ACCESS: OPEN PROBLEMS 50, 52 (G. Sica ed., Polimetrica, 2006), *available at* http://eprints.rclis.org/9656/1/OpenAccess.pdf.

by the instantaneous, concurrent, and global availability of access for each additional user at near zero marginal cost.[2] These quantitative improvements in the amount of time, geographical extent, and cost make it possible to achieve universal availability of data and information.

Just as important, however, are the qualitative advantages of digital technologies and networks in accelerating the dissemination of information and the diffusion of

specifically funded to produce important resources for use by the entire research community; (2) individual investigators who produce large datasets, often empowered by the use of readily available new technologies; and (3) an even greater number of sources that each produce small datasets (e.g. research data or clinical data in electronic health records) whose value can be amplified by aggregating or integrating them with other data.

See further Chapter 8 ("Fully Exploiting Data-Intensive Research Opportunities in the Networked Environment").

2 *See, e.g.*, Michael Mattioli, *Disclosing Big Data*, 99 U. MINN. L. REV. 535, 539–44 (2014) (stressing potential of big data to foster innovation). *See generally* VICTOR MAYER-SCHÖNBERGER & KENNETH CUKIER, BIG DATA: A REVOLUTION THAT WILL TRANSFORM HOW WE LIVE, WORK, AND THINK (2013); BILL FRANKS, TAMING THE BIG DATA TIDAL WAVE (2012).

knowledge. Networks provide opportunities for nonlinear, interactive, and asynchronous communication with multimedia capabilities. In digital formats, information becomes imbued with flexible transformative properties, which make it easy to manipulate and integrate with other types of information in order to create new knowledge that was either not possible or much more difficult to generate in the print context. Moreover, networks enable entirely new forms of collaborative knowledge production on a broadly distributed, and potentially interactive basis, changing or disintermediating the hierarchical and centralized organizational models governing information production and knowledge diffusion in previous eras.[3]

As both the principal inventors and pervasive users of the internet, scientists have a great deal at stake in efforts to fully exploit the potential advantages of this new medium for purposes of both research and applications. Information technology and digital information are transforming all fields of science, with the life sciences perhaps more affected than most, especially in areas of molecular biology, such as genomics and proteomics.[4] Biology has also spawned new subfields, such as metagenomics[5] and metabolomics.[6]

3 Paul F. Uhlir, *The Emerging Role of Open Repositories*, *in* OPEN ACCESS: OPEN PROBLEMS 50–52 (G. Sica, ed., Polimetrica 2006), *available at* http://eprints.relis.org/9656/1/Open-Access.pdf. For a seminal article on the institutional, economic, and legal aspects of the evolving volunteer, distributed, peer-production models online, see Yochai Benkler, *Coase's Penguin, or, Linux and the Nature of the Firm*, 112 *Yale L. J.* 369 (2002).

4 *See, e.g.*, COMM. ON INTELLECTUAL PROP. RIGHTS IN GENOMIC AND PROTEIN RESEARCH & NAT'L RESEARCH COUNCIL, REAPING THE BENEFITS OF GENOMICS AND PROTEOMIC RESEARCH: INTELLECTUAL PROPERTY RIGHTS, INNOVATION AND PUBLIC HEALTH 1 (2006) [hereinafter REAPING THE BENEFITS OF GENOMICS]; COMM. ON A NEW BIOLOGY FOR THE 21ST CENTURY & NAT'L RESEARCH COUNCIL, A NEW BIOLOGY FOR THE 21ST CENTURY 49–52 (Nat'l Acad. Press 2009) [hereinafter NEW BIOLOGY FOR THE 21ST CENTURY].

5 Metagenomics has been defined as "the application of modern genomics techniques to the study of communities of microbial organisms directly in their natural environments, bypassing the need for isolation and lab cultivation of individual species." Kevin Chen & Lior Pachter, *Bioinformatics for Whole-Genome Shotgun Sequencing of Microbial Communities*, 1 *PLoS Computational Biology* 0106 (2005), *available at* http://www.ploscompbiol.org/article/info%3Adoi%2F10.1371%2Fjournal.pcbi.0010024. Advances in bioinformatics, refinements of DNA amplification, and the expansion of computational power have greatly facilitated analysis of DNA sequences recovered from environmental samples. These advances have enabled the adaptation of shotgun sequencing to metagenomics samples, for example, in global ocean sampling expeditions. *See generally* Mya Breitbart et al., *Genomic Analysis of Uncultured Marine Viral Communities*, 99 *Proc. Nat'l Acad. Sci.* 14250 *passim* (2002); J. Craig Venter et al., *Environmental Genome Shotgun Sequencing of the Sargasso Sea*, 304 *Science* 66, 66–67 (2004), *available at* http://www.sciencemag.org/content/304/5667/66.full.pdf.

6 "Metabolomics is the systematic study of the unique chemical fingerprints that specific cellular processes leave behind," i.e., the study of their small-molecule metabolite profiles. Bennett Davis, *Growing Pains for Metabolomics*, SCIENTIST, Apr. 25, 2005, at 25–28. A closely related field is "metabonomics," which extends metabolic profiling at the cellular or any level to include information about perturbations of metabolism caused by environmental factors and other extragenomic influences, such as gut microflora. *See generally* D.G. Robertson, *Metabonomics in*

The combination of massive storage capacity, powerful data manipulation techniques, and graphical capabilities has revolutionized both how research is conducted and how the resulting knowledge is preserved and disseminated in nearly all fields of science.[7] These methodologies have also helped to generate networked communities of users and collaborators, often working in dynamic knowledge hubs,[8] whose interactive communications steer computational applications in potentially more fruitful directions[9] and fill open repositories with new data and information.[10]

In this new research environment, scientists increasingly rely on automated knowledge discovery tools to mine and recombine vast amounts of data and literature that are flowing at rates that exceed the capacity of a single investigator to comprehend and manage.[11] Storing, curating, maintaining, and making this

Toxicology: A Review, 85 *Toxicological Scis.* 809, 809–10, 815–18 (2005) (comparing metabonomics with metabolomics and discussing the latter's impact on toxicology). These disciplines rely heavily on mass spectrometry and nuclear magnetic resonance spectroscopy, among other detection methods, and on complex statistical software programs that analyze the data resulting from the use of these tools. *See, e.g.*, METABOLOMICS: METHODS AND PROTOCOLS vii–viii, 142, 229–46 (W. Weckwerth ed., Humana Press 2007); METABOLOMICS: THE FRONTIER OF SYSTEMS BIOLOGY 2–5, 8, 26–32 (M. Tomita & T. Nishioka eds., Springer 2005). For the aspirations of systems biology and functional genomics to integrate proteomic, transcriptomic, and metabolomic information into a more complete picture of living organisms, see NEW BIOLOGY FOR THE 21ST CENTURY, n. 4, at 21–38.

7 Scholars are still discussing and attempting to understand the impact of this newer ability to share large amounts of scientific research. *See generally* BD. ON RES. DATA & INFO., THE FUTURE OF SCIENTIFIC KNOWLEDGE DISCOVERY IN AN OPEN NETWORKED ENVIRONMENTS: A NATIONAL WORKSHOP, Washington D.C., Mar. 10–11, 2011 (Paul F. Uhlir ed. Nat'l Acads. Press 2011), *available at* http://www.nap.edu/catalog.php?record_id=18258. *See also* Agency for Sci., Tech. & Research, *The Digital Side of Biology*, A* STAR RESEARCH (Mar. 2, 2011), http://www.research.a-star.edu.sg/feature-and-innovation/6291 (describing huge changes to biological research stemming from digital technology).

8 *See, e.g.*, YOCHAI BENKLER, THE WEALTH OF NETWORKS: HOW SOCIAL PRODUCTION TRANSFORMS MARKETS AND FREEDOM 68–90 (Yale Univ. Press 2006); SCOTT STERN, BIOLOGICAL RESOURCE CENTERS: KNOWLEDGE HUBS FOR THE LIFE SCIENCES 36–55 (Brookings Instit. Press 2004); Brett M. Frischmann, *An Economic Theory of Infrastructure and Commons Management*, 89 *Minn. L. Rev.* 917, 1017–20 (2005).

9 *See, e.g.*, James Boyle, *Mertonianism Unbound? Imagining Free, Decentralized Access to Most Cultural and Scientific Material*, *in* UNDERSTANDING KNOWLEDGE AS A COMMONS: FROM THEORY TO PRACTICE 123, 123–40 (C. Hess & E. Ostrom eds., MIT Press 2007) [hereinafter KNOWLEDGE AS A COMMONS]; Paul W. Jeffreys, *The Developing Concept of e-Research*, *in* WORLD WIDE RESEARCH: RESHAPING THE SCIENCES AND HUMANITIES 51, 51–52 (W.H. Dutton & P.W. Jeffreys eds., MIT Press 2010) (noting that cooperation between research groups is necessary to perform complex research and analysis, and describing the pooling of "computational resources and research skills"); *See also* The Metagenomics Rast Server (MG-RAST), *Homepage*, http://metagenomics.anl.gov (last accessed 9 Apr. 2014) (community resource for metagenome data set analysis).

10 *See* Jeffreys, n. 9, at 51 (noting the "data deluge"). THE FOURTH PARADIGM: DATA INTENSIVE SCIENTIFIC DISCOVERY (T. Hay, S. Tansley & K. Tolla eds., Microsoft Research 2001), *available at* http://research.microsoft.com/en-us/collaboration/fourthparadigm/4th_paradigm_book_complete_lr.pdf. *See further* Chapter 8, Section III ("Building Transnational Open Knowledge Environments").

11 *See, e.g.*, Mark Segal, *Accessing Microbiological Data: A User's Perspective*, *in* DESIGNING THE MICROBIAL RESEARCH COMMONS: PROCEEDINGS OF AN INTERNATIONAL WORKSHOP 161, 161–63

huge accumulation of genomic data of interest to microbiology presents unique problems as well as unique opportunities. Once available, there is a pressing need to develop general data and text-mining tools for automated knowledge discovery in the chosen environment, and to establish dynamically updated and flexible portals for disseminating research results.[12]

One of the most promising user-added resources resulting from data accumulation and integration is the establishment of incremental machine learning and automatic interpretation capabilities based on the application of semantic web and integrative techniques to large amounts of data. Such techniques include globally unique identifiers, ontologies, annotation, error correction, and workflow management systems.[13]

Publicly certified, all-inclusive collections of data and information in a given domain are particularly advantageous because of their scope and the fact that they operate under the rules of public science, that is, under testable quality procedures open to scrutiny by the global research community.[14] All-inclusive public or semi-public repositories then extend the range of possibilities for comparing large amounts of information and data by virtue of being open to all available comparable sources. They establish the preconditions for global collaboration in the further development of relevant information infrastructures by adopting rules for, say, data quality, access, and use. They also support cumulative scientific research by promoting standardized quality norms for the certification of data.[15]

To the extent that basic research data and information are then made available under fully open access conditions, such repositories also expand the possibilities

(Paul F. Uhlir ed., Nat'l Acad. Press 2011) [hereinafter DESIGNING THE MICROBIAL RESEARCH COMMONS]; Thinh Nguyen, *The Web-Enabled Research Commons: Applications, Goals, and Trends*, *in* DESIGNING THE MICROBIAL RESEARCH COMMONS, at 91, 94.

[12] Peter Dawyndt et al., *Knowledge accumulation and resolution of data inconsistencies during the integration of microbial information sources*, 17 *IEEE Transactions on Knowledge & Data Eng'g* 1111–26 (2005). For background concerning the production and diffusion of data pertaining to the microbial culture collections, *see generally* David Smith, Dagmar Fritze & Erko Stackebrandt, *Public Service Collections and Biological Resource Centers of Microorganisms*, *in* THE PROKARYOTES – PROKARYOTIC BIOLOGY AND SYMBIOTIC ASSOCIATIONS (E. Rosenberg et al. eds, Springer Verlag 2013), Chapter 11, at 295–97.

[13] J. Uhlir & M. Falc, "Annotating Narratives Using Ontologies and Conceptual Graphs," paper presented at the 15th Int'l Workshop on Databases and Expert Systems Applications, 30 Aug.–3 Sept. 2004, *available at* http://ieeexplore.ieee.org/xpl/login.jsp?tp=&arnumber=1333454&url=http%3A%2F%2Fieeexplore.ieee.org%2Fstamp%2Fstamp.jsp%3Ftp%3D%26arnumber%3D1333454.

[14] *See* Mark Harvey & Andrew McMeekin, "Public or Private Economies of Knowledge: The Economics of Diffusion and Appropriation of Bioinformatic Tools," paper presented to the Microbial Commons Conference, Ghent, Belgium, Jun. 12–13, 2008.

[15] *See generally* Jerome H. Reichman & Paul F. Uhlir, *A Contractually Reconstructed Research Commons for Scientific Data in a Highly Protectionist Intellectual Property Environment*, 66 *Law & Contem. Probs.* 315 (2003) [hereinafter Reichman & Uhlir (2003)].

for further extraction and integration of digital matter otherwise only available from disparate sources. As illustrated in Chapter 8, data mining techniques may then be used to extract data from all known existing sources, with the resulting common pool resource digitally available for still more refined mining and combinatorial manipulation later on.[16]

To fully exploit these new opportunities, in turn, scientists must integrate information and data scattered over a broad range of articles and databases that may or may not be available online for extensive computational research.[17] For example, networked computational techniques for linking global collections of articles and data to generate relevant research results enable investigators to build field-specific knowledge repositories that capture reams of relevant scientific data and technical information on a federated basis and to apply data-mining tools in the chosen environment.[18] Users receive more value when such tools can also be readily applied to the full range of scholarly literature.

The digitization of research inputs and outputs has thus engendered opportunities for the enhanced dissemination of publicly funded scientific data, for the development of advanced search engines that diminish the search time for publications, and for automated cross-linking and text-mining based on standardized metadata, among many other emerging capabilities. The goal of this digital infrastructure should be to maximize these opportunities for public research institutes and universities in both developed and developing countries, while maintaining the classical functions of certification, reputational benefits, and diffusion of research results.

II. COPYRIGHT AND RELATED LAWS AS DIGITAL GRIDLOCK

To make full use of the capabilities offered by search engines, data-mining techniques, and other automated knowledge discovery tools, scientists need unrestricted access to a broad range of journals and databases, and unrestricted rights to extract, use, and reuse the published research results they contain for purposes of future research. The

[16] *See, e.g.*, IAN HARGREAVES, DIGITAL OPPORTUNITY: A REVIEW OF INTELLECTUAL PROPERTY AND GROWTH 46–47 (2011); Mattioli (2014), n. 2, at 541–42 (noting "vast tapestry of electronic devices and services that automatically record information about daily life in the developed world"). See *further* Chapter 8, Section III.C.

[17] *See, e.g.*, Minna Allarakhia, *Microbial Commons: Governing Complex Knowledge Assets, in* DESIGNING THE MICROBIAL RESEARCH COMMONS, n. 11, at 145, 148; Nancy L. Maron & K. Kirby Smith, *Current Models of Digital Scholarly Communication, Assn. Res. Libraries* 27 (Nov. 2008), http://www.arl.org/bm~doc/current-models-report.pdf; Victoria Stodden, *Open Science: Policy Implications for the Evolving Phenomenon of User-Led Scientific Innovation, J. Sci. Comm.* 2–6 (Mar. 22, 2010), http://jcom.sissa.it/archive/09/01/Jcom0901%282010%29A05/Jcom0901%282010%29A05.pdf. For similar applications to digital research in the humanities, see, for example, Matthew Sag, *Copyright and Copy-Reliant Technology*, 103 *Nw. U. L. Rev.* 1607, 1607–08, 1611–12 (2009).

[18] *See, e.g.*, Dawyndt et al., n. 12, at 1111, 1111–12, 1124.

convergence of computerized technologies and telecommunications networks has now made this goal theoretically feasible, and the sharing norms of science generally pull in the same direction.[19] Researchers anywhere should, in principle, be able to locate, analyze, and disaggregate collections of scientific information and data once they have been digitally transmitted and made available to the public, subject only to the prevailing community norms of attributions.[20]

In reality, intellectual property laws, as currently configured, stand in the way of attaining these goals. Since the 1990s, in particular, there has been an unprecedented extension of copyright law and related rights protecting both literature and collections of data into the realm of public science, with no adequate exceptions for governmental or academic research as such.[21] These developments tend to subject the growing profusion of scientific data and information to the same unbridled proprietary impulses that have lately dominated the regulation of creative endeavors in the traditional arts.[22]

For example, global copyright laws automatically confer exclusive proprietary rights on authors of scientific literature,[23] who routinely transfer those rights to commercial publishers.[24] Database protection laws, now enacted in fifty-five countries, simultaneously endow compilers and publishers (as assignees) with

19 *See, e.g.*, Reichman & Uhlir (2003), n. 15, at 319–25; *but see* Mattioli (2014) n. 2, at 544–49 (pinpointing obstacles, especially nondisclosure).

20 The scientists' incentives flow primarily from reputational benefits, not pecuniary interests, with regard to actual publication of upstream research results, and the costs of the research itself are normally borne by public funders, foundations, and universities. However, scientists do have an interest in not sharing either results or data until they can obtain these reputational benefits via publication. *See* Karen A. Jordan, *Financial Conflicts of Interest in Human Subjects Research: Proposals for a More Effective Regulatory Scheme*, 60 *Wash. & Lee L. Rev.* 15, 92–94 (2003) (stressing importance of publication and priority for scientists); Philip M. Davis & Matthew J. L. Connolly, *Institutional Repositories: Evaluating the Reasons for Non-Use of Cornell University's Installation of DSpace*, D-Lib Mag. (Mar./Apr. 2007), http://www.dlib.org/dlib/march07/davis/03davis.html (noting researchers' reluctance to release results before publication).

21 *See generally* Jerome H. Reichman & Ruth L. Okediji, *When Copyright Law and Science Collide: Empowering Digitally Integrated Research Methods on a Global Scale*, 96 *U. Minn. L. Rev.* 1362, 1372–1457 (2012). Sections I and II below are based on that article.

22 *See* Boyle, n. 9, at 123–59 (discussing the harmful consequences of over extending music copyrights); James Boyle, *The Second Enclosure Movement and the Construction of the Public Domain*, 66 *Law & Contem. Probs.* 33 (2003).

23 Agreement on Trade-Related Aspects of Intellectual Property Rights, Apr. 15, 1994, Marrakesh Agreement Establishing the World Trade Organization, Annex 1C, art. 9.1 The Legal Texts: The Results of the Uruguay Round of Multilateral Trade Negotiations 320 (1999), 1869 U.N.T.S. 299, 33 I.L.M. 1197 (1994) [hereinafter TRIPS Agreement]; Berne Convention for the Protection of Literary and Artistic Works arts. 2, 5; Sept. 9, 1886, 1161 U.N.T.S. 31 [hereinafter Berne Convention (1971)] (as last revised at Paris on July 24, 1971). The TRIPS Agreement incorporated substantive provisions of the Berne Convention (1971) into the Marrakesh Agreement Establishing the World Trade Organization, Apr. 15, 1994, 1867 U.N.T.S. 154 [hereinafter Agreement Establishing the WTO].

24 *See* below Section IV.

exclusive rights to the very data that copyright laws traditionally left unprotected.[25] Publishers, in turn, surround both scientific databases and literature with a variety of technological protection measures (TPMs) – so-called electronic fences and digital locks – that cannot be penetrated or pried open even for public research purposes without violating norms rooted in an array of multilateral, regional, and bilateral treaties, as well as in a host of national legislative and regulatory instruments.[26]

The end result is a growing conflict between private rights and public goods at the core of today's most promising research techniques.[27] Enlightened policymakers view these upstream data and information resources as public goods that need to be widely shared in order to produce more downstream commercial applications that advance public welfare.[28] In contrast, intellectual property laws now impede access to publicly generated data and literature, just at the time when developments in scientific research methods require the use of automated knowledge discovery tools that depend on unfettered access and re-use conditions for their successful application.[29]

A. *Two Conceptual Approaches in the Application of Copyright Law to Science*

The well-known philosophical differences between Continental "authors' rights" laws, rooted in natural law tradition, including protection of the author's personality interest, and the copyright laws of common law countries, based on utilitarian notions

[25] *See* Directive of the European Parliament and of the Council of 11 Mar. 1996 on the Legal Protection of Databases, Directive 96/9/EC, 1996 O.J. (L 77) 20, 21 (EC) [hereinafter Database Directive].

[26] *See* Sections II.B & C.

[27] *See* Keith E. Maskus & Jerome H. Reichman, *The Globalization of Private Knowledge Goods and the Privatization of Global Public Goods*, *in* International Public Goods and Transfer of Technology under a Globalized Intellectual Property Regime 3, 3–45 (K.E. Maskus & J.H. Reichman eds., Cambridge Univ. Press 2005) [hereinafter International Public Goods] (discussing IP roadblocks to the diffusion of public knowledge). *See generally* Peter Drahos, *The Regulation of Public Goods*, *in* International Public Goods, at 48, 61–64 (commenting on the impact of IP treaties on developing nations' access to public goods); Joseph E. Stiglitz, *Knowledge as a Global Public Good*, *in* Global Public Goods: International Cooperation in the 21st Century 308, 308–20 (I. Kaul et al. eds., Oxford Univ. Press 1999) (explaining the benefits of treating knowledge as a global public good).

[28] *See, e.g.*, Paul David, *The Economic Logic of "Open Science" and the Balance Between Private Property Rights and the Public Domain in Scientific Data and Information: A Primer*, *in* The Role of Scientific and Technical Data and Information in the Public Domain 19, 19–34 (J.M. Esanu & P.F. Uhlir eds., Nat'l Acad. Press 2003); Paul F. Uhlir, *Discussion Framework*, *in* The Role of Scientific and Technical Data and Information in the Public Domain, 3–4 (discussing public welfare advantages of sharing scientific knowledge and data widely).

[29] *See* Hargreaves Review (2011) n. 16, at 46–47; David, n. 28, at 27–28. Except, of course, for the growing number of scientific journals whose publishers have adopted full or partial open access policies. *See generally* Chapter 7, Section II.

of social welfare,[30] led logically to contrasting views of limitations and exceptions to the basic bundle of authors' rights.[31] These different philosophical foundations, in turn, produced two different approaches to limitations and exceptions bearing on the exclusive rights that copyright law confers on authors of literary and artistic works.

In Europe, the standard approach was to establish a list of enumerated exceptions, with the understanding that activities not covered by any of the listed exceptions were usually proscribed, even if they sometimes appeared to be natural extensions of an existing exception.[32] These codified exceptions thus need updating at regular intervals, and courts tend to interpret them narrowly because they undermine the dominant theme of authors' property rights.[33]

In contrast, U.S. legislation combines a list of fairly specific express exceptions to the exclusive rights of authors with a broad fair use provision that carves out additional space for noninfringing activity, usually transpiring within specified normative guidelines.[34] This open-ended carve-out then applies not only to new situations not directly reached by the codified list of exceptions, but it may sometimes retroactively expand even the scope of those exceptions that are codified.

The differences between these two approaches have diminished over time, as policymakers on both sides of the Atlantic rely on incentives to create the natural-property-rights thinking to justify ever higher levels of copyright protection.[35]

[30] *See* F. Willem Grosheide, *Paradigms in Copyright Law*, *in* OF AUTHORS AND ORIGINS: ESSAYS ON COPYRIGHT LAW 203, 203–28 (B. Sherman & A. Strowel eds., Oxford Univ. Press 1994); Edward C. Walterscheid, *The Nature of the Intellectual Property Clause: A Study in Historical Perspective (Part 1)*, 83 *J. Pat. & Trademark Off. Soc'y* 763, 770 (2001) ("Madison's view that copyrights and patents were monopolies that should be tolerated because of the public good they could produce was in essence the common law justification for these limited-term monopolies.").

[31] *See, e.g.*, Martin Senftleben, *Bridging the Differences Between Copyright's Legal Traditions – The Emerging EC Fair Use Doctrine*, 57 *J. Copyright Soc'y U.S.A.* 521, 524–27 (2010).

[32] *See* Annette Kur, *Of Oceans, Islands, and Inland Water – How Much Room for Exceptions and Limitations under the Three-Step Test?*, 8 *Rich. J. Global L. & Bus.* 287, 295–96 (2009) (contrasting civil and common law approaches to copyright exceptions). Hence, some states carved out more expansive exceptions for science. *See*, SAM RICKETSON, INTERNATIONAL CONVENTIONS AND TREATIES, IN THE BOUNDARIES OF COPYRIGHT – ITS PROPER LIMITATIONS AND EXCEPTIONS 3, 5–10 (L. Baulch et al. eds., 1997) (noting recurring exceptions in national copyright laws for, *inter alia*, "general enhancement of scientific and intellectual discourse").

[33] *See* below text and accompanying nn. 37–53; *see also* Christophe Geiger, *Promoting Creativity through Copyright Limitations: Reflections on the Concept of Exclusivity in Copyright Law*, 12(3) *Vand. J. Ent. & Tech. L.* 515, 519–20 (2010), *available at* http://www.jetlaw.org/wp-content/journal-pdfs/Geiger_online.pdf (noting narrow interpretation of copyright limitations and exceptions in civil law countries).

[34] *See* 17 U.S.C. §§ 106–22 (2012); *see also* William W. Fisher, *Reconstructing the Fair Use Doctrine*, 101 *Harv. L. Rev.* 1661, 1659, 1704 (1988); Ruth L. Okediji, *Toward an International Fair Use Doctrine*, 39 *Colum. J. Transnat'l L.* 75, 117–23 (2000); Pamela Samuelson, *Unbundling Fair Uses*, 77 *Fordham L. Rev.* 2537, 2618 (2009).

[35] *See* PAUL GOLDSTEIN, INTERNATIONAL COPYRIGHT: PRINCIPLES, LAW, AND PRACTICE 10 (Oxford Univ. Press 2001) (stating that the traditions differ "more in emphasis than in outcome"); *see also* Jane

Conversely, scholars in Europe increasingly focus attention on the need for an appropriate balance between protection and free uses.[36] As will be seen later, a degree of harmonization has also been superimposed on all the domestic copyright laws of WTO Members by international law. Nonetheless, these historical foundations help to explain the differences that still characterize the distinctive approaches to limitations and exceptions in the European Union and the United States.

1. Harmonizing the Designated Limitations and Exceptions that Weakly Defend Science in the European Union

A major effort to harmonize limitations and exceptions at the regional level occurred in 2001, with the adoption of the Directive of the European Parliament and the Council of Europe on the Harmonization of Certain Aspects of Copyright and Related Rights in the Information Society (InfoSoc Directive).[37] Ostensibly devised to implement the WIPO Copyright Treaty (WCT) of 1996[38] and the TRIPS Agreement of 1994,[39] this Directive sets out a deliberately exhaustive list of permissible exceptions and limitations to the exclusive rights of authors that European Union member states may enact at their discretion.[40] Besides allowing reproductions for photocopying, subject to payment of fair compensation, and for noncommercial reproductions by public libraries under Article 5(2),[41] the Directive expressly mentions scientific research in Article 5(3)(a). Echoing some prior state practice, this provision allows "use for the sole purpose of illustration for teaching or scientific research," so long as the source, including the author's name, is indicated and "to the extent justified by the noncommercial purpose to be achieved."[42]

C. Ginsburg, *A Tale of Two Copyrights: Literary Property in Revolutionary France and America*, 64 *Tul. L. Rev.* 991, 1014 (1990) (noting that a mix of both utilitarian and natural rights reasoning underlie French and United States copyright laws). For an important attempt to reduce these differences by a fuller interpretation of the Lockean justification for property rights, see Wendy J. Gordon, *A Property Right in Self-Expression: Equality and Individualism in the Natural Law of Intellectual Property*, 102 *Yale L.J.* 1533, 1544–45 (1993).

36 *See, e.g.*, Geiger, n. 33, at 517–18 (citing authorities); Senftleben, n. 31 at 525–26.

37 Council Directive 2001/29, 2001 O.J. (L 167) 10, 16 (EC) [hereinafter InfoSoc Directive]. See also Directive 2004/48/EC of the European Parliament and of the Council of 29 April 2004 on the Enforcement of Intellectual Property Rights, O.J. (L.157/16), *corrigendum*, O.J. (L.195/16), June 2, 2004.

38 World Intellectual Property Organization, Copyright Treaty, adopted Dec. 20, 1996, S. Treaty Doc. No. 105–17 (1997 [hereinafter WCT].

39 TRIPS Agreement, n. 23, arts. 9–14.

40 InfoSoc Directive, n. 37, art. 5(2).

41 *Id.*

42 *Id.* art. 5(3)(a). Technically, the Commission has thus taken the "by way of illustration" language out of Berne Convention (1971), n. 23, art. 10(2), which applies to teaching, and ostensibly applied it to excerpts of scientific research, in addition to the three-step test discussed in text and accompanying nn. 48–53.

The meaning of this ambiguous provision is hard to pin down with any degree of certainty. Even if a broader interpretation were to prevail (by limiting the term "illustration" to exceptions for teaching), it must still overcome the Directive's noncommercial purpose qualifier for scientific research.[43] Because universities now routinely engage in commercial exploitation of their scientific research results in both the European Union and the United States, rights holders (typically publishers) can argue that the bulk of such research is commercial in the strict sense of the word. Such an interpretation was recently upheld in a decision concerning university patents by the United States Court of Appeals for the Federal Circuit,[44] although it is not clear that European courts would take a similarly strict line in regard to either patents or copyrights.

In this unfavorable setting, Article 5(3) of the InfoSoc Directive has done little to strengthen or encourage digital scientific research or the rights of scientific investigators. To the contrary, the Directive may have fatally weakened them by subjecting the old private use exception on which scientists traditionally relied in the print media to a "pay equitable compensation" principle in Article 5(2)(b).[45]

Finally, the exhaustive list of permissible exceptions in the EC's Directive contains no fair use provision that might afford a greater degree of flexibility.[46] On the contrary, Article 5(5) of the EC's InfoSoc Directive imposes three additional requirements that negatively circumscribe all the limitations and exceptions it otherwise allows:

> The exceptions and limitations provided for in paragraphs 1, 2, 3 and 4 shall only be applied in certain special cases which do not conflict with a normal exploitation of the work or other subject-matter and do not unreasonably prejudice the legitimate interests of the right holder.[47]

This three-step test – derived from Article 13 of the TRIPS Agreement and Article 9(2) of the Berne Convention (1971 text)[48] – embodies a narrow reading of the international minimum standard applicable to exceptions and limitations in the copyright laws of all WTO member states.[49] It thus appears to ignore more flexible

43 *See* InfoSoc Directive, n. 37, art. 5(3)(a); *see also id.*, art. 5(2)(b) (restricting private use to noncommercial ends).

44 Madey v. Duke Univ., 307 F.3d 1351, 1361–64 (Fed. Cir. 2002).

45 InfoSoc Directive, n. 37, art. 5(2)(b).

46 *Id.* art. 5(2).

47 *Id.* art. 5(5).

48 TRIPS Agreement, n. 23, art. 13; Berne Convention (1971), n. 23, art. 9(2). *See generally* Reichman & Okediji, n. 21, at 1389–90.

49 "The three-step test thus became a universal norm of world intellectual property law, binding on some 153 signatories to the Agreement Establishing the World Trade Organization. Its enforcement also became subject to the WTO tribunals and cross-sectoral remedies governed by the WTO's Understanding on the Settlement of Disputes (DSU). Marrakesh Agreement Establishing the World Trade Organization, Apr. 15, 1994, 1867 U.N.T.S. 154; *see* Understanding the WTO, World Trade

language later embodied in the WCT of 1996 and more clearly amplified in an accompanying Agreed Statement.[50] The rigidity of the three-step test stems in part from the fact that it embodies no normative guidelines comparable to those in U.S. fair use law, which might privilege scientific research.[51] This rigidity is then magnified by the conventional view that, for any given use to qualify as privileged under the three-step test, the decision maker must answer "yes" to all three questions posed by that test.[52] The end result is that the European Commission's InfoSoc Directive, regardless of how it has been implemented, can significantly cut back on the already narrow sphere of exceptions favoring scientific research in the past, whether or not this was its intended purpose.[53]

2. Limits of the Fair Use Approach in the United States

In contrast, the United States, which did not join the Berne Convention until 1989,[54] adopted a different approach to limitations and exceptions in general and to those bearing on research in particular. The designated provisions in the U.S. Copyright Law of 1976 that are most relevant to scientific research include limitations on the reproduction rights for libraries in § 108[55] and, above all, the fair use exception

Organization, http://www.wto.org/english/thewto_e/whatis_e/tif_e/org6_e.htm (last accessed 9 Apr. 2014). *See also* Understanding on Rules and Procedures Governing the Settlement of Disputes, Marrakesh Agreement Establishing the World Trade Organization, Annex 2, 1869 U.N.T.S. 401 [hereinafter DSU]; Panel Report, United States – Section 110(5) of the U.S. Copyright Act, WT/DS160/R (June 15, 2000) [hereinafter US–Section 110(5) Panel Report].

50 WCT, n. 38, art. 10; World Intellectual Property Organization, Agreed Statements concerning the WIPO Copyright Treaty, *adopted on* Dec. 20, 1996 [hereinafter WCT Agreed Statements], *available at* http://www.wipo.int/treaties/en/ip/wct/statements.html; Reichman & Okediji, n. 21, at 1394–97 (noting that this revised version of the three-step test, including an Agreed Statement to art. 10, was obtained by the U.S. science agencies in order to preserve space for fair use in the digital environment). *See further* below n. 79.

51 *Compare* TRIPS Agreement, n. 23, art. 13 *with* 17 U.S.C. § 107, pmbl. *See generally* Reichman & Okediji, n. 21, at 1390–94 ("Normative Blindness at the World Trade Organization").

52 *See, e.g.*, Senftleben, n. 31, at 530–35. Until recently, this orthodox position went largely unquestioned. Fortunately, the Max Planck Institute has launched a head-on challenge to this position, as we shall explain in our discussion of possible reforms below. *See* Reichman & Okediji, n. 21, at 1454–56.

53 Reto Hilty, *Copyright Law and Scientific Research*, *in* COPYRIGHT LAW, A HANDBOOK OF CONTEMPORARY RESEARCH 315, 318–21 (Paul Torremans ed., 2007) [hereinafter Hilty, *Copyright Law and Scientific Research*]; Reto Hilty, *Five Lessons About Copyright in the Information Society: Reaction of the Scientific Community to Over-Protection and What Policy Makers Should Learn*, 53 *J. Copyright Soc'y U.S.A.* 103, 113–18; (2006) [hereinafter Hilty, *Five Lessons About Copyright*]; *see also* Guido Westkamp, *The Limits of Open Source: Consumer Protection, Exhaustion and Co-Existence with Copyright Law*, 1 *Intellectual Prop. Quarterly* 14, 26 (2008) (stressing that art. 5(5) will likely diminish public interest uses in the digital environment).

54 Berne Convention Implementation Act of 1988, Pub. L. No. 100–568, 102 Stat. 2853 (1988) (effective date of entry Mar. 1, 1989).

55 17 U.S.C. § 108 (2006).

codified for the first time in § 107.[56] By setting out the conditions under which library reproductions and interlibrary loans might be made for purposes of private study, scholarship, or research, § 108 operates in effect as a codified specification of fair use as it applies to libraries in general.

There are no other designated exemptions bearing on quotations, excerpts, or scientific research as such in the 1976 Act, like those under the Berne Convention[57] and the European Commission's InfoSoc Directive.[58] Hence, it is the codified fair use doctrine, as judicially interpreted, that effectively governs the rights of researchers in the United States to avoid or mitigate the exclusive rights of authors and publishers. Any so-called private use exceptions, comparable to those traditionally found in European copyright laws,[59] must stand or fall as fair uses in U.S. law.

Section 107 of the 1976 Act expressly recognizes a set of preambular uses or "purposes such as criticism, comment, news reporting, teaching (including multiple copies for classroom use), scholarship, or research" for which an open-ended fair use exception is deemed particularly suitable.[60] These uses promote public goods in ways that courts must reconcile with the private rights of authors in an appropriately balanced copyright system. Much depends, however, on how judges determine whether the harm incurred by the copyright owner is justified by the benefit to the public from allowing the use in question.[61]

To answer that question, the statute requires courts to evaluate four separate criteria that may pull in different directions and evince different weights, viz.: (1) the purpose and character of the use (such as a noncommercial use or a so-called transformative use); (2) the nature of the copyrighted work (for example, is it of a factual or scientific character to begin with); (3) the amount and substantiality of the portion used (in both quantitative and qualitative terms); and (4) the effect of any given use on the potential market for or value of the copyrighted work.[62]

In the past, and for a fairly long period of time, it was the fourth factor – the so-called market harm test – that predominated in the case law.[63] Following the Supreme Court's decision in *Campbell v. Acuff-Rose*, however, all four factors must now be weighed by the courts.[64] In practice, the first factor increasingly predominates as courts focus on

[56] *Id.* § 107.
[57] Berne Convention (1971), n. 23, art. 10.
[58] InfoSoc Directive, n. 37, art. 5(3)(a).
[59] *See* text and accompanying n. 45. *See generally* Glynn S. Lunney, Jr., *The Death of Copyright: Digital Technology, Private Copying, and the Digital Millennium Copyright Act*, 87 *Va. L. Rev.* 813 (2001).
[60] 17 U.S.C. § 107 (Preamble).
[61] *See, e.g.*, 2 Paul Goldstein, Copyright § 10.1.2 to 10.1.4 (3d ed. 2005).
[62] *See* 17 U.S.C. § 107(1)–(4); Campbell v. Acuff-Rose Music, Inc., 510 U.S. 569, 576–94 (1994) (explaining each of the four factors in § 107).
[63] Jane C. Ginsburg, *Conflicts of Copyright Ownership between Authors and Owners of Original Artworks: An Essay in Comparative and International Private Law*, 17 *Colum.-VLA J.L. & Arts* 395, 401 (1993) (noting that "the inquiry into potential market harm remains dominant").
[64] 510 U.S. at 569. *See most recently* Cambridge Univ. Press v. Patton, 769 F.3d 1232 (11th Cir. 2014).

the presence or absence of a so-called transformative use, i.e., a new use not necessarily envisioned by the original author that enriches culture or the pursuit of knowledge.[65]

This transformative use factor has prevailed in the digital arena, particularly with regard to new technological uses, such as in cases involving search engines that access and index massive amounts of data and information on the internet.[66] Courts in the United States have now routinely held that the use of thumbnail images as markers for search engines, for example, is transformative and that the fair use defense can avail notwithstanding some use for commercial gain.[67]

Despite these achievements, the fair use standard probably offers less help to practitioners of the digital research techniques of primary concern in this chapter. The systematic need that researchers, as users of automated knowledge discovery tools, have to survey vast or, indeed, unlimited amounts of literature and data in virtually every contemporary, large-scale scientific investigation, particularly in the life sciences, could overwhelm the boundaries set by the four-step test of § 107 and stretch the very concept of fair use to the breaking point.

Consider, for example, that the implicit purpose of the substantiality test set out in § 107(3) is to ensure that fair use reproductions of a protected text will be quantitatively and qualitatively reasonable in relation to the work as a whole. In no area, not even parody,[68] can this provision be interpreted to permit wholesale reproduction (as technically defined) of every relevant text in every relevant case, which routinely occurs in computational science or in any scientific research project

[65] Campbell v. Acuff-Rose Music Inc., 510 U.S. 569, 579 (1994) (emphasizing uses that "provide social benefit[s] by shedding light on an earlier work, and in the process, creat[e] a new one"); *see also* Bill Graham Archives v. Dorling Kindersley Ltd., 448 F.3d 605, 609–10 (2d Cir. 2006) (promotional posters used in biography about rock music was "a purpose separate and distinct from the original artistic and promotional purpose for which the images were created"); L.A. News Serv. v. CBS Broad., Inc., 305 F.3d 924, 938–39 (9th Cir. 2002). A nontransformative use may still be viable under the first factor, for example, as an educational use; but the market harm test of factor four may acquire correspondingly greater weight. Cambridge Univ. Press v. Patton, 769 F.3d 1232, 1275 (11th Cir. 2014).

[66] *See, e.g.*, Authors Guild, Inc. v. Google Inc., 954 F. Supp. 2d 282 (S.D.N.Y. 2013); Authors Guild, Inc. v. Hathi Trust, 755 F.3d 87 (2d. Cir. 2014). *See also* Perfect 10, Inc. v. Amazon.com, Inc., 508 F.3d 1146, 1163–67 (9th Cir. 2007) (finding the use of thumbnails as a highly transformative use); Kelly v. Arriba Soft Corp., 336 F.3d 811 (9th Cir. 2003) (holding the use of thumbnail images in search engine as fair use); *see also* A.V. v. IParadigms, L.L.C., 562 F.3d 630, 638–40 (4th Cir. 2009) (finding fair use for archival copies of student papers stored in digital form to help detect and prevent plagiarism); Field v. Google Inc., 412 F. Supp. 2d 1106, 1118 (D. Nev. 2006).

[67] *See* n. 66. *See also* Perfect 10, Inc., 508 F.3d at 1163–68 (holding that search engine compilation of thumbnail-sized photographs was fair use); Arriba Soft Corp., 336 F.3d at 822 (same). For the view that these cases really turn on nonexpressive uses that do not substitute for the author's original expression, see Sag, n. 17, at 1636–37.

[68] *But see, e.g.*, Campbell v. Acuff-Rose Music, Inc., 510 U.S. 569, 593–94 (1994) (holding that entire song may be considered a parody under fair use); Elsmere Music, Inc. v. Nat'l Broad. Co., 482 F. Supp. 741, 746–47 (S.D.N.Y. 1980) (holding that television show "Saturday Night Live" parody of the song "I Love Sodom" to the tune of "I Love New York" constituted fair use).

where automated knowledge discovery tools are employed.[69] By the same token, the market-harm test of § 107(4) may become drained of precedential meaning if the scientific texts thus scrutinized were to be viewed as serving both the research needs of the scientific community and the commercial interests of publishers.[70]

Professor Matthew Sag's brilliant article on copy-reliant technologies[71] sheds considerable light on this conflict of interest. His efforts to reconcile the search engine cases with prior decisions concerning transformative uses of copyrighted works under § 107(1) leads him to posit that nonexpressive, nonsubstitutional uses, in conjunction with copy-reliant technologies, should normally qualify as fair uses across the board, especially if the technologies in question were geared to recognize and implement an opt-in clause,[72] and recent cases have borne out this assessment.[73] When this intriguing proposition is applied to digitally integrated scientific research methods, however, it reveals a number of key differentiating factors.

For example, one must immediately confront the possibility that, from a rights holders' perspective at least, massive copying of published research articles to generate further research by means of automated knowledge discovery tools colorably represents both a substitutional and an expressive use of those same articles. Even if that were precisely what scientists *qua* authors most dearly desired in their relentless pursuit of reputational benefits, gratis fair use on this scale is hardly consistent with the aims of commercial STM (Science, Technology and Medicine) publishers.[74]

If only scientific researchers were involved as both creators and users of their own published outputs, then Professor Sag's default formula for fair uses in regard to copy-reliant technologies could significantly improve the research community's technical legal position, especially if it were underpinned with an opt out, rather than an opt in default condition. Scientists inclined to opt out of such a voluntary pool would immediately incur countervailing peer pressure and perhaps risk

69 Robert C. Denicola, *Copyright in Collections of Facts: A Theory for the Protection of Nonfiction Literary Works*, 81 *Colum. L. Rev.* 516, 536 (1981).

70 17 U.S.C. § 107(4) (2006) (stating that a factor to be considered in determining fair use is "the effect of the use upon the potential market for or value of the copyrighted work"). See, e.g., Cambridge Univ. Press v. Patton, 769 F.3d 1232, 1275 (11th Cir. 2014) (stressing possibility of greater emphasis on substitutional effects if no transformative use occurs).

71 *See* Sag, n. 17.

72 *See id.* at 1675–82. *See also* James Grimmelmann, *Google Books Search Status Conference: Opt-in Settlement in the Works?*, Tech. Acads. Pol'y (TAP), July 26, 2011, http://www.techpolicy.com/Blog/July-2011/Google-Books-Search-Status-Settlement-In-The-Works.aspx.

73 See, e.g., Authors Guild, Inc. v. Google Inc. 954 F. Supp. 2d 287 (S.D.N.Y. 2013); Authors Guild, Inc. v. Hathi Trust, 755 F.3d 87 (2d. Cir. 2014).

74 *See, e.g.*, Letter from Michael Mabe, CEO, Int'l Assoc. of Scientific, Technical & Med. Publishers, to Copyright Review, Dep't. of Jobs, Enter. & Innovation, Dublin, Ireland (July 14, 2011) [hereinafter Letter from STM] ("Consultation on the Review of the Copyright and Related Rights Act 2000") (strongly opposing fair use).

jeopardizing future grants to boot. If, instead, scientists constitute the market for published scientific research, and if that published research cannot be freely and digitally perused without impermissible market harm to publishers, then automated research tools risk becoming instruments of massive and systematic infringement, which no transformative use doctrine could excuse if publishers' customary interests were to be preserved.

That, indeed, poses one of the fundamental questions raised by our present enquiry, namely, should scientific publishers' customary interests be preserved at the expense of the research community's need for wholesale access to, and reuse of, the exploding universe of published scientific literature and data? That question, in turn, raises ancillary questions about what added-value the scientific community obtains from its traditional reliance on external, for-profit publishers, and what the opportunity costs would be if the scientific communities were to break that tie to the publishing industry. These and related questions will be more directly addressed in the final section of this chapter.[75] Nonetheless, for present purposes, what seems undeniable is that the case-by-case approach of the fair use doctrine is potentially overwhelmed by the magnitude and scope of copying that today's digitally empowered research techniques necessitate.[76]

B. Digital Locks and Contractual Overrides in the Online Environment

The WIPO Copyright Treaty of 1996 (WCT),[77] which established new rules governing digital transmissions of copyrighted works, reflects a relatively balanced compromise that resulted from the negotiations of stakeholder coalitions with fairly equal bargaining power on both the publishers' and users' sides.[78] The preamble itself thus recognizes "the need to maintain a balance between the rights of authors and the larger public interest, particularly education, research and access to information."[79]

75 *See* below Section IV.

76 *See* Timothy K. Armstrong, *Digital Rights Management and the Process of Fair Use*, 20 *Harv. J.L. & Tech.* 49, 60–62 (2006) (discussing contemporary copyright after the advent of new technologies).

77 WCT, n. 38.

78 The users' coalition was largely organized and managed by Professor Peter Jaszi, American University School of Law, Washington, D.C. *See* Jerome H. Reichman & Paul F. Uhlir, *Database Protection at the Crossroads: Recent Developments and Their Impact on Science and Technology*, 14 *Berkeley Tech. L.J.* 793, 810–28 (1999) [hereinafter Reichman & Uhlir (1999)] (explaining the negotiations and proposals to resolve database protection issues).

79 WCT, n. 38, pmbl. ¶ 5. Similarly, the agreed statement to Article 10 permits contracting parties "to carry forward and appropriately extend into the digital environment" existing limitations and exceptions in their national laws and "to devise new exceptions and limitations that are appropriate in the digital network environment." WCT Agreed Statements, n. 50 (concerning Article 10). Finally, the very Article 11 that imposed "obligations concerning technological [protection] measures" (TPMs),

However, the WCT said nothing about how states should implement the anticircumvention norms that defend electronic fences surrounding works transmitted online so as to preserve public interest privileges and immunities. When the treaty was translated into the domestic laws of the United States and the European Union, powerful publisher interests persuaded the respective legislatures largely to ignore or override the safeguard provisions otherwise available.[80]

In the Digital Millennium Copyright Act of 1998 (DMCA),[81] for example, the U.S. Congress conditioned the ability of third-party users to invoke public interest measures, such as the idea-expression dichotomy or fair use, on their having first gained lawful access to the work being transmitted online.[82] Yet, the moment a would-be user seeks to gain lawful access to the copyrighted work transmitted online, he or she will normally encounter one-sided electronic contracts of adhesion that strip away most or all of the public-interest user rights nominally available from the domestic copyright law.[83] The DMCA thus arguably created a new exclusive "right of access" subject to virtually no preexisting privileges or immunities of interest to scientific users (or other privileged public-interest users as well).[84]

A similar state of affairs (with different nuances in different jurisdictions) arises in the European Union. Article 6 of the InfoSoc Directive of 2001 expressly enables domestic

also expressly declared that such TPMs were not meant to "restrict acts in respect of [authors'] works which are ... permitted by law." WCT, n. 38, art. 11.

80 Jerome H. Reichman, Graeme B. Dinwoodie & Pamela Samuelson, *A Reverse Notice and Takedown Regime to Enable Public Interest Uses of Technically Protected Copyrighted Works*, 22 *Berkeley Tech. L.J.* 981, 983–85, 1059 (Summer 2007) (explaining why efforts to implement a balancing of interests in the United States and European Union copyright laws were unsuccessful).

81 Digital Millennium Copyright Act, Pub. L. No. 105–304, 112 Stat. 2860 (1998) [hereinafter DMCA].

82 17 U.S.C. § 1201 (2012); *see also* Armstrong (2006), n. 76, at 67–74 (discussing fair use and the DMCA). *See generally* Dan L. Burk & Julie E. Cohen, *Fair Use Infrastructure for Rights Management Systems*, 15 *Harv. J.L. & Tech.* 41 (2001).

83 In effect, once the user is forced through an electronic gateway, the contract of adhesion becomes a privately legislated intellectual property right. *See* Jerome H. Reichman & Jonathan A. Franklin, *Privately Legislated Intellectual Property Rights: Reconciling Freedom of Contract with Public Good Uses of Information*, 147 *U. Pa. L. Rev.* 875, 897–914 (1999) (discussing adhesion contracts for digital technologies); *see also* Dan L. Burk, *Anticircumvention Misuse*, 50 *UCLA L. Rev.* 1095, 1099–102 (2003) (explaining how copyright holders can abuse technological control systems to prevent access to digital content); Nima Darouian, *Accessing Truth: Marketplaces of Ideas in the Information Age*, 9 *Cardozo Pub. L. Pol'y & Ethics J.* 1, 26–46 (2010) (discussing adhesion contracts and virtual marketplaces).

84 Jane C. Ginsburg, *From Having Copies to Experiencing Works: The Development of an Access Right in U.S. Copyright Law*, 50 *J. Copyright Soc'y USA* 113, 125 (2003); *see, e.g.*, Pamela Samuelson, *Intellectual Property and the Digital Economy: Why the Anti-Circumvention Regulations Need to be Revised*, 14 *Berkeley Tech. L.J.* 519, 519–20 (1999) (arguing that the DMCA anti-device provisions are overbroad, unclear, and need to be revised). However, some recent cases have looked askance at this result, and Professors Reichman, Dinwoodie and Samuelson have demonstrated how these precedents could lead courts to a more balanced solution in the future. *See generally* Reichman, Dinwoodie & Samuelson, n. 80 (discussing several cases that have challenged the boundaries of copyright protection for digital works).

legislators to authorize Technical Protection Measures (TPMs) that curtail or override the preexisting limitations and exceptions otherwise available in the hard copy format.[85] Article 6(4) of the same Directive then piously admonishes member states "to ensure that right holders make available to the beneficiary of an exception or limitation provided for in national law … the means of benefiting from that exception or limitation."[86] In practice, however, the Directive provides member states with no solid legal basis for implementing the thrust of Article 6(4), and national legislation concerning TPMs tends to largely ignore Article 6(4) altogether, with a few exceptions.[87]

As a result, technological fencing devices, coupled with electronic contracts, known respectively as TPMs and Digital Rights Management tools (DRMs), enable publishers to automatically protect both data and information delivered through online networks without gaps in enforcement and without any traditional exceptions for science or other public interest purposes.[88] When these technological fences and electronic contracts are further supported by anti-circumvention measures that forbid decryption or other means of cutting through such fences,[89] the publisher's control becomes virtually absolute. Database protection laws enacted in the European Union and elsewhere can then make this absolute control virtually perpetual to boot.

C. *Exclusive Rights in Noncopyrightable Collections of Data*

Compilations of facts and data receive relatively thin protection from the copyright laws of both the United States and the European Union.[90] Under these laws, only a creative selection and arrangement of facts or data qualifies as eligible subject matter, and the disparate facts remain available for use by third-party compilers,[91] at

85 InfoSoc Directive, n. 37, art. 6. *See generally* Guido Westkamp, *Code, Copying, Competition: The Subversive Force of Para-Copyright and the Need for an Unfair Competition Based Reassessment of DRM Laws after INFOPAQ*, 58 *J. Copyright Soc'y USA* 601, 627–43 (2011) (analyzing the aggregate effects of InfoSoc Directive, n. 37, arts. 2, 5(1)–5(5), after the European Court of Justice's decision in Infopaq Int'l A/S v. Danske Dagblades Forening).

86 InfoSoc Directive, n. 37, art. 6(4).

87 For one major exception, see the United Kingdom's new data-mining exception, n. 147 and accompanying text. For a more detailed discussion of other ways to implement art. 6(4), see generally Reichman et al., n. 80.

88 *Id.* at 982–87; Westkamp, n. 85, at 675–77.

89 17 U.S.C. § 1201(b) (2012).

90 Feist Pub's, Inc. v Rural Tel. Serv. Co., 499 U.S. 340, 349–50 (1991) (applying "thin" protection doctrine of functional works cases to factual compilations in general). For statutory support in the United States, see 17 U.S.C. § 101 (defining compilations); *id.* § 103 (defining subject matter of eligible compilations). For the European Union, see Database Directive, n. 25, Part I, § 13 (dealing with harmonization of copyright rules applicable to eligible compilations of data).

91 *See* text and accompanying n. 90; *see also* Key Pub's, Inc. v. Chinatown Today Publ'g Enters., Inc., 945 F.2d 509, 512–14 (2d Cir. 1991) (discussing the test for infringement of original works and compilations).

least in principle, if not always in practice.[92] In a remarkable further development, the U.S. government lobbied successfully to codify both the idea-expression dichotomy and the principle of limited protection for factual compilations, of crucial importance to science, in both the TRIPS Agreement[93] and the WIPO Copyright Treaty.[94] Global copyright law thus, in effect, encourages states to protect so-called factual works against little more than wholesale duplication of an otherwise creatively organized compilation of facts or data, but not the underlying facts or data as such.

In 1996, however, when promulgating its Directive on the Legal Protection of Databases,[95] the European Commission took the unprecedented step of enacting a law that established exclusive rights in the very data that copyright laws had left freely available in the public domain.[96] Ostensibly motivated by the Commission's stated goal of increasing the European Union's share of the global market for directories and compilations in general,[97] which subsequently proved unattainable,[98] this *sui generis* regime introduced radical new restrictions on access to and use of compilations of data that were previously unknown to any intellectual property paradigm.

For example, no element of originality or creativity is required to qualify for this form of protection.[99] Instead, the database laws are triggered by a "substantial

92 *See* e.g., CDN Inc. v. Kapes, 197 F.3d 1256, 1262 (9th Cir. 1999) (allowing copyright protection for estimates of prices for collectible coins); CCC Info. Servs. Inc. v. Maclean Hunter Mkt. Reports Inc., 44 F.3d 61, 67 (2d Cir. 1994) (finding that logically organized price estimates can be original works of authorship).

93 TRIPS Agreement, n. 23, arts. 9.2, 10.2.

94 WCT, n. 38, arts. 3, 5. However, there is remarkably no mention of this same doctrine in the European Union's Infosoc Directive of 2001, notwithstanding the fact that the idea-expression doctrine has now been embodied at the multilateral level in both Article 9.2 of the TRIPS Agreement and in Article 10 of the WCT. For this and other reasons, some commentators express reservations about over-reliance on this doctrine as a buttress to limitations and exceptions under the best of circumstances. *See, e.g.*, ROBERT BURRELL & ALLISON COLEMAN, COPYRIGHT EXCEPTIONS: THE DIGITAL IMPACT (2005), at 20–25.

95 *See* Database Directive, n. 25, arts. 1–11.

96 *See id.*

97 A more realistic motivation arose from the backing of the world's largest publisher of scientific journals, with headquarters in the Netherlands, which spearheaded ultimately unsuccessful efforts to enact a similar law in the United States. Maria Canellopoulou-Bottis, *A Different Kind of War: Internet Databases and Legal Protection or How the Strict Intellectual Property Laws of the West Threaten the Developing Countries' Information Commons*, 2 INT'L J. INFO. ETHICS 1, 10 n. 22 (2004), *available at* http://www.i-r-i-e.net/inhalt/002/ijie_002_07_canellopoulou.pdf (referring to Elsevier's lobbying for database protection).

98 COMM'N OF THE EUROPEAN COMMUNITIES, FIRST EVALUATION OF DIRECTIVE 96/9/EC ON THE LEGAL PROTECTION OF DATABASES (2005) [hereinafter FIRST EVALUATION], *available at* http://ec.europa.eu/internal_market/.copyright/docs/databases/evaluation_report_en.pdf; *see also* Elad Harrison, *Who Owns Enterprise Information? Data Ownership Rights in Europe and the U.S.*, 47 *Info. & Mgmt.* 102, 102 (2010) (stating that the United States continues to dominate the database market).

99 Database Directive, n. 25, art. 7(1).

investment" in obtaining, verifying, or presenting any given collection of facts and data; and unlike copyright or patent laws, the exclusive rights to extract or reuse the data in question protect that investment as such.[100] Despite its anomalously low threshold of eligibility, this regime arises automatically, as if it were part of the copyright infrastructure. It thus poses a direct threat to digitally integrated scientific research by endowing compilers of noncopyrightable collections of data (including, in many countries, even compilers in the government sector using public money) with exclusive rights to extract and reuse the disparate data that their sweat-of-the-brow investment made available to the public.[101]

These exclusive rights to data are potentially stronger and more rigid than those of copyright law.[102] Formally, independent creation remains a perfect defense,[103] as it would under copyright law.[104] In practice, however, independent generation of costly accumulations of scientific data is economically unfeasible, even when conceptually possible, and generally very unproductive.[105] The Directive does allow insubstantial amounts of data to be taken without consequence, but courts have interpreted this exception narrowly, and the Directive also expressly prohibits repeated extractions of even small amounts of data from the same collection.[106]

[100] *Id.*; *see, e.g.*, Daniel J. Gervais, *The Protection of Databases*, 82 *Chi.-Kent L. Rev.* 1109, 1120 (2007) ("The Directive essentially does two things: it confirms the application of copyright to compilations of data and creates a noncopyright, sui generis right in databases to protect the investment of the database maker.").

[101] Database Directive, n. 25. For the rejection of sweat-of-the-brow protection of factual works in U.S. copyright law after a period of experimentation in that regard by some federal appellate courts, especially the Seventh Circuit, see generally Jane C. Ginsburg, *No "Sweat"? Copyright and Other Protection of Works of Information after Feist v. Rural Telephone*, 92 *Colum. L. Rev.* 338 (1992).

[102] *See* Jerome H. Reichman, *Mondialisation et Propriété Intellectuelle: Database Protection in a Global Economy, Revue Internationale de Droit Economique*, 2002 *Int'l Rev. Econ. L.* 455, 455–503 (2002) [hereinafter Reichman, *Database Protection in a Global Economy*] (discussing the evolution of intellectual property legislation and issues relating to database protection). Until these laws were adopted, only the conduct-based liability rules of trade secrecy law were able to protect investment in know-how applied to industry. *See* Jerome H. Reichman, *How Trade Secrecy Law Generates a Natural Semicommons of Innovative Know-How, in* The Law and Theory of Trade Secrecy 185, 186–87 (R. Dreyfuss & K. Strandburg eds., Edward Elgar Pub. 2011) [hereinafter Reichman, *How Trade Secrecy Law Generates a Natural Semicommons*].

[103] Database Directive, n. 25, arts. 7(1), 7(5). *See further* Jerome H. Reichman & Pamela Samuelson, *Intellectual Property Rights in Data?*, 50 Vand. L. Rev. 52 (1997).

[104] Goldstein, n. 61, § 7.2.2 (stating that "conveying evidence" of independent creation constitutes a perfect defense to an action for copyright infringement).

[105] Reichman & Uhlir (1999), n. 78, at 807 n. 80, 814–15.

[106] *See* Database Directive, n. 25, arts. 6, 7(5), 8; British Horseracing Bd. Ltd. v. William Hill Org. Ltd., 2001 E.W.C.A Civ 1268, ¶¶ 29–48, 2001 WL 825162 (July 31, 2001) (finding that copying various pieces of information relating to British horseracing industry constituted extraction of a substantial part of the database, in addition to repeated extraction of insubstantial parts), aff'd Case C-203/02, 2004 E.C.R I-10415, ¶ 87.

Permissible exceptions to exclusive rights of the database regime are paradoxically truncated when compared with those of copyright law.[107] With specific regard to the use of protected data for scientific research, the Directive allows states to adopt an exception couched in the same ambiguous language as that of the InfoSoc Directive of 2001, namely, "for the sole purpose of illustration for teaching or scientific research."[108] As in the InfoSoc Directive, this exception is not mandatory, and major countries, such as France and Italy, have ignored it.[109] Even when governments adopt this exception, it seems to enable only extractions for purposes of illustration, but not for reutilization of scientific data or information in other collections, which is the normal scientific practice.[110]

Once obtained, database protection nominally expires after fifteen years.[111] However, if the compilers make another substantial investment, say, by adding or updating new data to the preexisting collection, their efforts will renew the protection of the entire database for another fifteen-year period.[112] Perpetual protection thus becomes an attainable goal for the first time in the history of intellectual property laws (disregarding, of course, trademark laws, which operate on fundamentally different principles).[113]

In a series of decisions, the European Court of Justice (ECJ) has subsequently introduced an elusive subject-matter distinction between "substantial investment" for purposes of obtaining data that are created (presumably ineligible), and expenditures for purposes of obtaining data that are collected (i.e., developed and maintained in databases as such) and that presumably qualify for protection.[114] In

[107] *See, e.g.*, Miriam Bitton, *A New Outlook on the Economic Dimension of the Database Protection Debate*, 47 *Idea* 93, 141–44, 150–53 (2006).

[108] Database Directive, n. 25, art. 6(2)(b); *see* text and accompanying nn. 42–44.

[109] Reichman & Uhlir (1999), n. 78, at 803–04; Raquel Xalabarder, Copyright Exceptions for Teaching Purposes in Europe (Internet Interdisciplinary Inst., Working Paper WP04-004, 2004), *available at* http://www.uoc.edu/in3/dt/eng/20418/20418.pdf.

[110] *See, e.g.*, Estelle Derclaye, The Legal Protection of Databases: A Comparative Analysis 129–33 (Edward Elgar Pub. 2008) (arguing that the exception is overly narrow and therefore over-protects database makers); *see also* Reichman & Samuelson (1997), n. 103, at 79.

[111] Database Directive, n. 25, art. 10(1).

[112] *Id.*, art. 10(3); *see also* Wesley L. Austin, *A Thoughtful and Practical Analysis of Database Protection under Copyright Law, and a Critique of* Sui Generis *Protection*, 3 *J. Tech. L. & Pol'y* 3,¶ 67 (1997).

[113] *See* Reichman & Samuelson, n. 103, at 86 ("[A]ny publisher who continues to make a substantial investment in updating, improving, or expanding an existing database can look forward to perpetual protection.").

[114] *See* Case C-46/02, Fixtures Mktg. Ltd. v. Oy Veikkaus Ab, 2004 E.C.R. I-10365, ¶ 49 (referred from Finland); Case C-338/02, Fixtures Mktg. Ltd. v. Svenska Spel AB, 2004 E.C.R. I-10497, ¶27 (referred from Sweden); Case C-203/02, British Horseracing Bd. Ltd. v. William Hill Org. Ltd., 2004 E.C.R. I-10415, ¶¶ 50–56 (referred from the United Kingdom); Case C-444/02, Fictures Mktg. Ltd. v. Organismos prognostikon anonon Podosfairou AE, 2004 E.C.R. I-10549, ¶ 27 (referred from Greece).

other words, "only resources used to collect data that [are] already in existence" will qualify for database protection, but not "data compilations that are generated quasi 'automatically' as by-products of other activities."[115] To the extent that scientific databases are characterized as "created" under this slippery distinction, it might conceivably reduce the total number of databases, particularly sole-source databases, eligible for protection.[116] Courts could, for example, exclude some collections of raw scientific data on these grounds.[117]

However, some commentators believe most scientific data are better characterized as "collected" and, therefore, automatically eligible for protection.[118] Even when scientific data are viewed as created, whatever this turns out to mean, entities seeking protection could always spend more money on verification or on improving the conditions of access to, and posterior maintenance of the collection, which might have some scientific value even if undertaken for secondary motives. In other words, there is reason to believe that most collections of scientific data and information could be made to fit within these judicially contrived eligibility requirements by one means or another. If so, any collection of scientific data or information that did qualify would obtain broad and virtually endless protection against value-adding components of a future collection that made unauthorized use of an existing one.[119]

How the Database Directive actually affects science in any given country will then depend on a number of uncertain variables. In the United States, where the scientific community vigorously opposed enactment of database protection bills modeled on the European Union Directive,[120] only copyright law applies to compilations of data,

115 Heather J. Ritch, European Research Infrastructure Consortiums: Privately Ordered and Publicly Funded Research Commons for Data 127 (unpublished S. J. D. dissertation, Duke University (on file with Goodson Library, Duke University) (citing Directmedia Publ'g GmbH v. Albert-Ludwigs-Universitat Freiburg [2009] 1 C.M.L.R. 7 (ECJ 4th Chamber)); *see also* Mark J. Davison & P. Bernt Hugenholtz, *Football Fixtures, Horse Races and Spin-Offs: The ECJ Domesticates the Database Right*, 27 E.I.P.R. 113, 114 (2005) (stating that European Court of Justice discounts investments in collecting data that are indivisibly linked to their creation); Estelle Derclaye, *Databases* Sui Generis *Right: Should We Adopt the Spin Off Theory?*, 26 E.I.P.R. 402, 408–13 (2004) (finding that the database right should only protect investments that are directly attributable to producing a database).

116 For the dangers of protecting sole-source databases under this regime see, for example, Reichman & Samuelson, n. 103, at 113–37.

117 *See* Derclaye, n. 110, at 87–99 (arguing that there is no substantial investment in collecting, verifying or presenting raw scientific data, such as event data, timetables, telephone subscriber data, and the like).

118 *See, e.g.*, Davison & Hugenholtz, n. 115, at 115–18 (arguing that when a large mass of collected data has been created, there are significant costs associated with presentation and verification, which may meet the requirements of the Directive); *see also* Ritch, n. 115, at 127.

119 *Cf.* Derclaye, n. 110, at 255–67 (supporting the database protection regime generally, but strongly criticizing its treatment of science). Preexisting rights in any given component could, of course, be legally waived. *Cf., e.g.*, Creative Commons CCO (waiver of rights).

120 *See* Mark Davison, *Database Protection: Lessons From Europe, Congress, and WIPO*, 57 *Case W. Res. L. Rev.* 829, 853 (2007) ("In the United States, the lack of database protection and, in particular,

although that law, as shown earlier, is much less science friendly today than in the past.[121] In European Union member states and affiliates, however, the *sui generis* database protection laws remain firmly in place despite serious criticism from within the European Union itself.[122] The European Commission has also made strenuous efforts to extend similar database regimes to developing and Least-Developed Countries through a series of regional and bilateral free trade agreements.[123]

Analogies drawn from the historical rhetoric promoting authors' rights, whatever one's view of them, have thus been perversely applied to an investment-based scheme of protection governing the most fundamental building blocks of knowledge.[124] What the *sui generis* database laws actually codified instead was a scheme of powerful exclusive property rights that protect infinitely expandable collections of data from extraction and reuse, with a built-in propensity to favor the emergence of sole-source providers over time.[125] This regime conflicts head on with customary scientific research practices that long antedated the digital universe and the new research opportunities it makes possible.[126]

Nor should one suppose that the social costs of this dismal experiment, which seemingly will not be repealed despite sweeping criticism from the Commission's own officially appointed reviewers,[127] are confined to the some fifty-five countries

its defeat in the Senate in 1998 was the direct product of the input of preexisting, institutionalized, funded, and Congressionally recognized scientific and educational lobby groups, such as the National Research Council.").

121 *See* Section I.A.2, I.B.

122 *See* First Evaluation, n. 98, at 11–27 (listing numerous criticisms of the Directive and proposals for change).

123 *See* Denise Rosemary Nicholson, *Intellectual Property: Benefit or Burden for Africa?*, 32 *Int'l Fed. Libr. J.* 310, 316 (2006), *available at* http://ifl.sagepub.com/content/32/4/310.full.pdf ("[T]he United States and European Union [free trade] Agreements contain a TRIPS-Plus Chapter, which far exceeds all current international obligations for all types of intellectual property.").

124 Disregarding the impact of a powerful lobby, among other factors, see Craig R. Whitney, *European Union's Commission Is Revamped After a Scandal; A 'New Era' Is Promised*, N.Y. Times, July 10, 1999, at A6. The Commission responsible for elaborating the Database Directive completely failed to recognize or observe the systemic limits of the copyright paradigm. *Cf.* Denicola, n. 69, at 518–41 (examining the scope of copyright protection available to writings and exploring the divergent and inconsistently applied rationales used to define property rights in factual works).

125 As correctly predicted by the German government, whose provision to allow compulsory licenses against sole-source providers was deleted, behind closed doors, by the Council of Ministers at the last moment, and without the approval of the European Parliament. *See* Reichman & Samuelson, n. 103, at 86.

126 *See* David, n. 28, at 19–33 (discussing the history and economic logic of "open science"); Reichman & Uhlir (1999), n. 78, at 799–820 (discussing the potential impact of the database protection laws on science and technology); Paul A. David, *The Digital Technology Boomerang: New Intellectual Property Rights Threaten Global "Open Science"* 1–8 (Stanford Dept. of Econ., Working Paper No. 00–006, 2000), *available at* http://ideas.repec.org/p/wpa/wuwpdc/0502012.html.

127 *See* First Evaluation, n. 98, at 11–27 (listing numerous criticisms of the Directive): *see also* Hargreaves, n. 16, at 19 ("The aim was to ensure the EU got a foothold in th[e] growing [database]

that have adopted similar regimes at the behest of the European Communities. Consider, instead, that because science is a global public good,[128] search engines and other digitally empowered research tools must transcend national borders in order to access all publicly available sources of data and information relevant to any given project. Standing in their way are all the formidable legal barriers rooted in the territorial copyright and database protection laws described earlier, which threaten to choke the transnational flow of upstream scientific data and information that would otherwise be capable of digital integration on a global scale.[129]

III. AUTOMATED KNOWLEDGE DISCOVERY TOOLS AS INSTRUMENTS OF MASSIVE INFRINGEMENT

The foregoing analysis of the existing intellectual property framework portrays a set of rules and policies that are diametrically opposed to the needs of scientific researchers in a universe of discourse where automated knowledge discovery tools must freely explore the entire range of thematically relevant, digitally distributed literature and data.[130] Consider, for example, that the Wellcome Trust found that 87 percent of the material housed in the United Kingdom's main medical research database (U.K. PubMedCentral) was unavailable for legal text and data mining as of 2011.[131]

A major independent study undertaken for the British Government reported that existing copyright laws make it virtually impossible to text mine about one thousand journal articles from the first half of the twentieth century that describe malaria in indigenous peoples and soldiers, as well as details of therapeutic measures available at that period. Because of rights clearing requirements that appear out of all proportion to any benefits the rights holders could want, "even if they could be found," researchers could not digitally index or text mine sources that offered potentially significant insights for the development of methods for preventing and treating malaria today.[132] This study actually led the government to adopt the first

sector at an early stage. The European Commission['s] ... evaluation of the Directive in 2006 ... found that EU database creation had declined since introduction of the Directive, whilst it had continued to rise in the US, undermining the rationale for the right in the first place. The EU Database Directive remains unchanged.").

[128] *See generally* Stiglitz, n. 27, at 65–115.

[129] Paul Geller warns that the "interesting choice-of-law issues" are "[i]n practice, a mess – likely to intimidate house counsel for any research institution. Here we approach the bottom line, the chilling effect of the lack of a clear-cut exception with as global an application as possible ..." Letter from Paul Geller to Jerome H. Reichman (Oct. 30, 2011) (on file with authors).

[130] *See* Section I.

[131] HARGREAVES, n. 16, at 47.

[132] *Id.* at 46. *See* Hogarth Chambers, *The Hargreaves Review – Another Mixed Bag*, 33 E.I.P.R. 599, 600 (2011) (criticizing United Kingdom's copyright exceptions).

known data-mining exception to the United Kingdom's copyright law in 2014, as discussed later.[133]

Wittingly or unwittingly, most other copyright laws force scientific researchers to choose between ignoring an unmanageable and unreasonable set of legal constraints, in the interest of pursuing science as a public good, or foregoing research opportunities in order to avoid thickets of rights, burdensome transaction costs, and the fear of stirring up potential law suits down the line. The end result puts both science and the larger public interest in a no-win situation, at a time when the resources available to fund scientific research are shrinking.

If the relevant intellectual property laws were strictly enforced, and the scientific community continued to respect them, scarce public resources earmarked for basic research would be siphoned off to intermediaries from scientists seeking access to and use of their own published research results. In that event, the public pays twice for the same output, plus a surcharge for mushrooming transaction costs, while the "incipient transnational system of innovation" established by the TRIPS Agreement in 1996,[134] is progressively deprived of essential knowledge assets. Less innovation, not more, is the predictable result over time.

Conversely, if restrictive intellectual property laws are ignored by researchers determined to carry on with their work despite unreasonable legal constraints, automated knowledge discovery tools will have become transformed into engines of massive infringement.[135] It is hard to see how systematic disregard of intellectual property laws, coupled with growing contempt for the legislative process that fosters them,[136] will benefit authors, artists, and other creators over time. In this case, the alleged outlaws are not free-riders on costly musical and cinematic productions, but publicly funded scientific researchers in pursuit of knowledge and applications that benefit everyone.

While the pressing need to reform the laws that have produced such anomalous results has not escaped notice,[137] efforts in this regard are confronted with a conflict

[133] The Copyright and Rights in Performances (Research, Education, Libraries and Archives Regulation) 2014, S.I. 2014/1372 (United Kingdom), *available at* http://www.legislation.gov.uk/uksi/2014/1372/contents/made (last accessed Feb. 21, 2015) [hereinafter Copyright and Rights in Performances, U.K. Regulation (2014)].

[134] Maskus & Reichman, n. 27, at 342.

[135] *Cf.* Metro-Goldwyn-Mayer Studios Inc. v. Grokster, Ltd., 545 U.S. 913, 937–38 (2005) (stressing extent to which peer-to-peer music sharing schemes had become instruments of "massive infringement").

[136] "Much of the data needed to develop empirical evidence on copyright . . . is privately held. It enters the public domain chiefly in the form of 'evidence' supporting the arguments of lobbyists ('lobbyonomists') rather than as independently verified research conclusions." Hargreaves, n. 16, at 18.

[137] *See, e.g., id.* at 11–27 (criticizing the InfoSoc Directive); Hilty, *Copyright Law and Scientific Research*, n. 53, at 315–21 (citing problems with European copyright law); Hilty, *Five Lessons About Copyright*, n. 53, at 109–38 (discussing the reaction of the scientific community to copyright over-protection). See further Chapter 7, Sections I & II; Chapter 8 *passim*.

between the interests of scientists and those of publishers. Scientists as authors are primarily interested in the rewards of attribution and integrity – reputation benefits – that the moral rights of copyright laws, together with the norms of science itself, strive to protect.[138] These reputational benefits then serve to attract the kind of financing and status attendant on academic success.[139] Given a conflict between the needs of scientific research and the dictates of copyright and database laws, one can expect scientists normally to opt for the goals of research because their pecuniary interests lie elsewhere, and are, indeed, dependent on the reputational benefits just described.[140]

In contrast, the STM publishers are the main pecuniary beneficiaries of the current state of the law, which they have lobbied hard to obtain, and they would resist any reforms likely to be put on the table.[141] This fact of life makes it logical to ask why the scientific community continues to rely and depend on publishing intermediaries in the first place. Disregarding the historical origins of such reliance, one feels compelled to ask whether the benefits of such reliance still outweigh the costs in today's digitally integrated, totally computerized research environment. No sensible scheme of reform can be devised without addressing these questions, and no specific proposals will make sense unless they are weighed against alternative options that result from such an enquiry.

A. *What Digital Science Would Really Need from Any Serious Legislative Reform*

The extraordinary powers that publishers have obtained under the DMCA in the United States and the InfoSoc Directive in the European Union make an

[138] In the United States, this is true at least in theory, if not in practice. For doubts about the appropriate level of moral rights enforcement in U.S. copyright law, see, for example, Roberta Rosenthal Kwall, *Originality in Context*, 44 *Hous. L. Rev.* 871, 874 (2007) ("Sound reasons may support confining the application of moral rights to a smaller category of works than are covered by copyright law.").

[139] Scientists do have an interest in not sharing either research results or data until they can obtain these reputational benefits via publication. *See* Davis & Connolly, n. 20 (finding that there is some reluctance among researches to use a repository if it could jeopardize one's publication success); Jordan, n. 20, at 82–85 (noting the importance of publication and priority for scientists).

[140] *See* Jordan, n. 20, at 82–85. This is often not the case with patents, where deeper conflicts of interest may arise. *See* Reichman, *How Trade Secrecy Law Generates a Natural Semicommons*, n. 102, at 107–22.

[141] *See* Statement by the Am. Chem. Soc'y, to the Comm. on the Impact of Copyright Policy on Innovation in the Digital Era 5–6 (Oct. 15, 2010), *available at* http://sites.nationalacademies.org/PGA/step/copyrightpolicy/index.htm (opposing sweeping policy changes that undermine peer reviewed publications); Letter from STM, n. 74 (opposing proposals for a fair use exception); *see also* HARGREAVES, n. 16, at 42 ("[C]opyright exceptions for educational purposes and for research are intended to promote knowledge, skills and innovation in the economy, without unduly undermining the incentive for educational and academic publishers to create the works that students, teachers and researchers need.").

industry-wide settlement favorable to science far more difficult now than it might have been prior to the 1990s. In what follows, we discuss possible solutions to the problems that intellectual property laws have created for digitally integrated scientific research from two very different angles. First, we briefly consider the kinds of legal reforms that would be needed if commercial STM publishers continued to act as intermediaries between producers and users of scientific information and data, as they do today, without regard to the likelihood that such reforms would ever be enacted.[142]

We then reconsider the role of publishers as such and ask whether, from a cost-benefit perspective, it should be significantly modified or abandoned altogether. Later, in Chapters 7 and 8, we examine alternative strategies that the scientific community itself could embrace in a concerted effort to manage its own upstream knowledge assets in ways that might avoid, or at least attenuate, the obstacles to digitally empowered scientific research currently flowing from a flawed intellectual property regime.

1. A Tailor-Made Exception for Scientific Research

So long as there is no legislative mandate to deposit publicly funded research results in the public domain,[143] the only workable solution is to adopt a broad and uncompromising exemption for scientific uses that requires no gloss, no fine print, and no elaborately contrived exceptions to a grudgingly acknowledged "exception" for scientific research. To this end, the Max Planck Institute's response to the European Commission's Green Paper in 2008[144] proposed that such a broad and general provision, allowing use and reuse of published research materials for virtually any scientific purpose, should expressly legitimize storage, archiving, data extraction, linking, and the like.[145]

142 For a more detailed analysis of possible incremental reforms and their limits, see Reichman & Okediji, n. 21, at 1430–52.

143 For such a legislative proposal in the United States, *see, e.g.*, Public Access to Science Act, H.R. 2613, 108th Cong. (2003); *cf.* 17 U.S.C. §105 (2012) (mandating public domain status for all copyrightable works generated by government employees in the course of their duties) to the same effect, *see most recently* Memorandum, Office of Sci. & Tech. Pol'y, "Increasing Access to the Results of Federally Funded Scientific Research," Feb. 22, 2013 [hereinafter OSTP Public Access Initiative], *available at* http://www.whitehouse.gov/sites/default/files/microsites/ostp/ostp_public_access_memo_2013.pdf.

144 *See* Reto M. Hilty et al., European Commission – Green Paper: Copyright in the Knowledge Economy – Comments by the Max Planck Institute for Intellectual Property, Competition and Tax Law (2008), *available at* http://www.ip.mpg.de/files/pdf1/comments_on_the_green_paper1.pdf [hereinafter Max Planck Response to EC Green Paper].

145 *See id.*; *see also* Hargreaves, n. 16, at 48 ("The Government should introduce a UK exception in the interim under the non-commercial research heading to allow use of analytics for non-commercial use ... as well as promoting at EU level an exception to support text mining and data analytics for commercial use.").

While endorsing this proposal, which makes a good start, we think even more may be needed. In particular, scientists must be free to subject any published article (and, as we shall see later, any article made publicly available online) to data mining procedures and data manipulation by automated knowledge discovery tools, including virtual scientific experimentation, without any constraint other than attribution under the norms of science.[146] The same exemption must apply to the public release of selectively chosen material in any scientific paper or report.

The United Kingdom's new data-mining exception takes a major step in this direction.[147] This provision allows users to text-mine any literary work available to the public for noncommercial research purposes, notwithstanding the exclusive rights otherwise applicable under the existing copyright and database protection laws. Authors and publishers cannot override this exception by contract. However, the scope of this otherwise enlightened measure is limited by the fact that it does not allow text-mining of works protected by so-called digital locks, as further explained later.

2. Breaking the Digital Locks

No provision exempting scientific research from the exclusive rights of copyright law, as proposed earlier, could fully achieve its purpose unless complementary legislative action were taken to ensure its effectiveness in the online environment. Here we encounter the blocking effects of technical protection measures (TPMs) as implemented in the domestic laws,[148] whose drafters ignored the pro-science mandate expressed in the preamble to the WCT itself, as well as other balancing provisions set out in that treaty.[149]

If rights holders who make scientific works available through digital networks can simply enclose those works behind technological fences and then abolish all user-friendly copyright provisions by contract, little would be gained by clarifying the idea-expression dichotomy or the scope for private and fair uses, or by enacting a broad exception for scientific research and teaching as advocated earlier. The imposition of private intellectual property rights by such technological means also

[146] *See, e.g.*, Stodden, n. 17, *passim*, for a discussion regarding attribution and its problems. *See further* below Section IV and Chapter 7, Section III.

[147] *See* Copyright and Rights in Performances Regulation U.K. (2014), n. 133. *See further* James Boyle, *(When) Is Copyright Reform Possible? Lessons from the Hargreaves Review*, *in* COPYRIGHT LAW IN AN AGE OF LIMITATIONS AND EXCEPTIONS (Ruth L. Okediji, ed. Cambridge Univ. Press, forthcoming 2015), available at http://www.thepublicdomain.org/2015/01/14/15-copyright-reform-possible (last visited Feb. 21, 2015).

[148] *See* text and accompanying nn. 80–89.

[149] See text and accompanying n. 79 (quoting WCT, n. 38, pmbl. ¶ 5); WCT, n. 38, arts. 10–11, and WCT Agreed Statement, n. 50.

raises profound conflicts with constitutional law in the United States and with fundamental rights in Europe. [150]

The text-mining law that the United Kingdom Adopted in 2014 succumbed to this defect. Like the comparable U.S. law on fair use,[151] the U.K. text-mining law only applies online when the would-be user has gained lawful admission to the technical protection measures that surround the relevant work or database. Without lawful entry to the digital locks, in other words, scientists cannot text-mine sources protected by an electronic fence even in the United Kingdom.[152] Unlike the comparable U.S. law, however, once a scientist gains admission to a technically protected source in the U.K., he or she cannot be denied text-mining privileges by contract.[153]

Legislatures enacting appropriate exceptions for scientific research, like the one proposed earlier, should therefore simultaneously implement the proviso set out in Article 11 of the WCT, which expressly exempts "acts … which are … permitted by law" from the obligation of signatories to "provide adequate legal protection and effective legal remedies against the circumvention of effective technological measures."[154] For example, the copyright revision bill languishing at this time in Brazil initially took a major step forward by prohibiting content providers from using TPMs to defeat privileged uses or to impede access to public domain matter.[155] Whether

[150] *See* Reichman & Franklin, n. 83, at 884–914 (discussing the protection of copyright owners' rights through a combination of technological means and adhesion contracts). For fundamental rights in the U.S. *see* David Lange & H. Jefferson Powell, No Law: Intellectual Property in the Image of an Absolute First Amendment 108 (Stanford Univ. P. 2009) (stating that the conflict between intellectual property regimes and constitutional rights is "a conflict in multiple dimensions, in which interests in property are pitted against freedom of expression"); Neil W. Netanel, *Locating Copyright Within the First Amendment Skein*, 54 *Stan. L. Rev.* 1, 30–36 (2001) (discussing developments in First Amendment law as they pertain to copyright law). For the EU, *see, e.g.*, Natali Helberger & P. Bernt Hugenholtz, *No Place Like Home for Making a Copy: Private Copying in European Copyright Law and Consumer Law*, 22 *Berkeley Tech. L.J.* 1061, 1083 (2007) (discussing fundamental rights to be considered in shaping European consumer policy). *See also* Laurence Helfer & Graeme W. Austin, Human Rights and Intellectual Property: Mapping the Global Interface 259–83 (Cambridge Univ. Press 2011) ("Article 10 of the ECHR … provides the principle framework for balancing copyright and the right to freedom of information in European human rights jurisprudence.").

[151] *See* DMCA, n. 81, § 1201.

[152] *See* Copyright and Rights in Performances Regulation (U.K. 2014), n. 133; Boyle (2015), n. 147.

[153] *See* n. 147 and accompanying text.

[154] WCT, n. 38, art. 11.

[155] Law No. 9610 of 19 Feb. 1998, on Copyright and Neighboring Rights, Consolidated with the Bill in Public Consultation since 14 June 2010, *available at* http://www.vgrass.de/wp-content/uploads/2010/07/Brazilian_Copyright_Bill_Consolidated_June_2010.pdf (last accessed 9 Apr. 2014) (English translation); *see also* Pedro Paranaguá, *A Comprehensive Framework for Copyright Protection and Access to Knowledge: From a Brazilian Perspective and Beyond*, *in* How Developing Countries Can Manage Intellectual Property Rights to Maximize Access to Knowledge 103, 106–07 (X. Li & C.M. Correa eds., 2009) (discussing the Brazilian National Copyright Forum).

these and other provisions that seek to expand the copyright misuse doctrine[156] will survive the legislative process in that country remains to be seen, as are the means of implementing them in practice, which future regulations would have to specify.

Meanwhile, one relatively expedient suggestion is the "reverse notice and takedown" regime put forward by Professors Reichman, Dinwoodie, and Samuelson.[157] Under their proposal, bona fide public interest users could avoid passing through a content provider's electronic gateway and, instead, send a request or "flaming arrow" over the electronic fence to catch the copyright proprietors' attention.[158] This notice would signal that the user intended to obtain specified matter held by the proprietor in an online repository for purposes allowed under specified limitations and exceptions.[159] It would give proprietors a period – say fourteen days – in which to accede to the request or deny it on specified grounds that it was willing to defend in court or in an administrative proceeding.[160]

In the latter event, both sides would know that a judicial test of the validity of the request under relevant exceptions would be the likely outcome, and the copyright authorities could establish an expedited judicial or administrative procedure for this purpose.[161] Once the legitimacy of the request was established, the relevant authority or court could enable third parties, if necessary, to disarm or decrypt the TPMs in order to extract the desired scientific material for the specified research purposes.[162] Publishers who needlessly barred the initial request and thereby necessitated a judicial inquiry should bear at least the transaction costs and might be made subject to additional penalties for abuse of TPMs.[163]

However, given the massive amounts of literature and data processed by automated knowledge discovery tools, even the reverse notice and takedown regime – backed by supporting judicial decisions[164] – could break down unless published scientific works in

See generally Pedro Paranagua, Brazil's Copyright Law Revision – Tropicalia 3.0? (2014) (Doctoral thesis on file at the Goodson Library, Duke University School of Law).

156 *See* Burk, n. 83.

157 *See* Reichman, Dinwoodie & Samuelson, n. 80, at 1032–39 (discussing the contours of the proposed "reverse notice and takedown regime").

158 *Id.*

159 *Id.*

160 *Id.*

161 *Cf.* Mark A. Lemley & R. Anthony Reese, *Reducing Digital Copyright Infringement Without Restricting Innovation*, 56 *Stan. L. Rev.* 1345, 1351–52 (2004) (discussing how changing procedures for enforcing copyrights would affect behavior of those infringing them).

162 *See* Reichman et al., n. 80, at 1032–34.

163 *Cf.* Lenz v. Universal Music Corp., 572 F. Supp. 2d 1150, 1154–56 (N.D. Cal. 2008) (requiring publishers who send notice and takedown requests under DMCA § 512 to evaluate fair use considerations in advance); Burk, n. 83, at 1127–32; *see also* Reichman & Franklin, n. 83, at 929–32 (discussing a "public interest unconscionability" doctrine in contract law).

164 For example, two anti-lockout cases have provided various legal bases for overcoming TPMs that deny access to unprotected matter. *See* Storage Tech. Corp. v. Custom Hardware Eng'g & Consulting,

general were governed by some globally effective "digital copyright exchange," like that recommended in the Hargreaves Review.[165] Even then, much would depend on the willingness of funding agencies to insist that science publishers either refrained from surrounding scientific works transmitted online with TPMs and DRMs, or that they made such works automatically accessible to scientists seeking access to them through approved portals for research purposes.

3. Disciplining Contractual Overrides

The foregoing discussion demonstrates that no set of limitations and exceptions enacted by enlightened legislators can achieve the goal of strengthening scientific research so long as the proprietors of scientific publications can contractually override them, whether in print media or increasingly (and often exclusively) in the online environment. For this reason, the Max Planck Institute rightly proposes that both new and existing exceptions favoring scientific research must be made peremptory, mandatory, and nonwaivable, and the United Kingdom's text-mining law of 2014 embodied such a provision.[166]

Short of this logical proposal, other important, if less efficacious measures, remain available. For example, Professor Burk's principle of anticircumvention misuse could be adopted on both sides of the Atlantic to limit private interference with specified public good uses of copyrighted works.[167] To the same end, Professors Reichman

Inc., 421 F.3d 1307 (Fed. Cir. 2005); Chamberlain Grp., Inc. v. Skylink Techs., Inc., 381 F.3d 1178 (Fed. Cir. 2004). *But see* MDY Indus., LLC v. Blizzard Entm't, Inc., 629 F.3d 928, 950 (9th Cir. 2010) (sympathizing with the policy underlying these decisions, but rejecting their legal reasoning), as amended on denial of reh'g, Nos. 09-15832, 09-16044, 2011 U.S. App. LEXIS 3427 (9th Cir. Feb. 17, 2011). One recent district court case has obliged proprietors to take fair use factors into account before sending a request for notice and take down under the existing regime regulating safe harbors and the secondary liability of ISPs. *See* Lenz, 572 F. Supp. 2d at 1154–56.

165 HARGREAVES, n. 16, at 28–35 (proposing a digital copyright exchange); *see also* Joel Smith & Rachel Montagnon, *The Hargreaves Review – A "Digital Opportunity,"* 33 *Eur. Intell. Prop. Rev.* 596, 597 (2011) (stressing need for "digital copyright exchange" to facilitate cross-sectoral and cross-border licensing, plus codes of practice for collection societies). For efforts to implement the Hargreaves Review's recommendation for more science-friendly limitations and exceptions in a reform of the United Kingdom's copyright law, see Intellectual Property Office [U.K.], Hargreaves Implementation: Copyright, Press Release June 2013, *available at* http://www.ipo.gov.uk/types/hargreaves-copyright.htm. See also nn. 151–153 and accompanying text (U.K. text-mining exception of 2014).

166 *Accord.* HARGREAVES, n. 16, at 51 ("Applying contracts in this way means a rights holder can rewrite the limits the law has set on the extent of the right conferred by copyright. It creates the risk that should Government decide that UK law will permit private copying or text mining, these permissions could be denied by contract."); *see also* MAX PLANCK RESPONSE TO EC GREEN PAPER, n. 144, at 11–16 (proposing various exceptions to govern scientific use). For the United Kingdom's text mining law, see nn. 133 & 153, and accompanying text.

167 *See* Burk, n. 83, at 1132–40.

and Franklin's proposals for a "public interest unconscionability" standard for nonnegotiable contracts could be employed to give courts more common law tools for alleviating conflicts between private ordering and the goals of federal copyright and related laws.[168]

Such a response would fit well within certain existing European approaches to consumer protection and contract laws in general.[169] Professor Hilty also stresses the possibility of invoking European competition law, with its concept of abuse of a dominant position, when proprietors leverage their power in the market for scientific articles to inhibit use and reuse of scientific contents by downstream investigators.[170]

What matters is that legislatures concerned with promoting scientific research should take a forthright position against contractual overrides of lawful and permitted uses while also clarifying scientific research as a peremptory example of a lawful and permitted use. In reality, apart from the text-mining exception adopted in the United Kingdom,[171] there is little reason to expect any such enlightened approach in the immediate future. On the contrary, newly proposed measures on enforcement, in their present form, could actually strengthen the proprietors' ability to impose privately legislated intellectual property rights on the scientific research community.[172]

168 *See* Reichman & Franklin, n. 83, at 929–32; *see also* Darouian, n. 83.

169 Mel Kenny, *Globalization, Interlegality and Europeanized Contract Law*, 21 *Penn St. Int'l L. Rev.* 569, 575 (2003) (noting "the trend towards higher standards of EC consumer protection").

170 *See* Hilty, *Copyright Law and Scientific Research*, n. 53, at 315 (calling the European Union Directive "designed one-sidedly to protect the entertainment industry ... thwarting the efforts to make Europe the leading centre for research"). Prospective development of a competition-based limit to the abuse of TPMs and to contractual restraints on use and reuse of noncopyrightable data remains one area where the international regime established by the TRIPS Agreement remains relatively unburdened by the strictures of the three-step test or other legal obstacles to national discretion concerning the design of an appropriate copyright system. TRIPS Agreement, n. 23, art. 40; Estelle Derclaye, *An Economic Approach to What the Conditions of Abuse of a Dominant Position of Copyright Should Be*, Soc'y for Econ. Research on Copyright Issues 6 (2003), *available at* http://www.serci.org/2003/derclaye.pdf (noting "that a dominant position or even a monopoly is (or rather: can be) a natural consequence of the grant of a copyright"); Sara K. Stadler, *Relevant Markets for Copyrighted Works*, 34 J. Corp. L. 1059 *passim* (2009) (arguing that reframing copyright law as a species of competition law would benefit the public interest).

171 *See* nn 133, 151–153 and accompanying text.

172 *See, e.g.*, Directive of the European Parliament and of the Council on the Enforcement of Intellectual Property Rights, Directive 2004/48, 2004 O.J. (L 157) (EC), *available at* http://eur-lex.europa.eu/LexUriServ/LexUriServ.do?uri=CELEX:32004L0048R%2801%29:EN:NOT (attempting to protect member countries from the "growing phenomenon" of counterfeiting and piracy issues); *see also* Anti-Counterfeiting Trade Agreement, Dec. 3, 2010, *opened for signature* Mar. 1, 2011, *available at* http://www.ustr.gov/webfm_send/2417 [hereinafter ACTA]; Charles R. McManis, *The Proposed Anti-Counterfeiting Trade Agreement (ACTA): Two Tales of a Treaty*, 46 *Hous. L. Rev.* 1235, 1235–39 (2009) (discussing the ACTA controversy); The U.S. Trade Representative, The Anti-Counterfeiting Trade Agreement—Summary of Key Elements under Discussion (2009), *available at* http://www.ustr.gov/webfm_send/1479 (summarizing discussions

4. Aligning Database Protection Laws with Tailor-Made Exceptions for Science in Copyright Law

Any legislative reform of domestic copyright laws to facilitate text-mining or other digital research techniques that ignored the database protection laws in the European Union would inadvertently allow the latter to surround the former with a net that would block access to and use of the very facts and data that the copyright paradigm ostensibly left free.[173] It would also impede transnational efforts to pool large collections of scientific data by automatically subjecting contributions from providers in the European Union to a strong regime of exclusive property rights not applicable to other contributors.[174] For these and other reasons, neither science nor culture[175] could fully attain the payoffs that digital technologies make possible without ancillary adjustments of the Database Directive.

When the Max Planck Institute called for a broad exemption from the exclusive rights of the European Union's domestic copyright laws for published scientific information and data, it logically demanded that the Commission should also insert similar language into the Database Directive as well.[176] In effectuating any such alignment, the Institute insists that the exceptions for science in both copyright laws and database protection laws should be preemptory, mandatory, and immune from both contractual overrides and TPMs.[177]

of anti-counterfeiting agreements among different countries). *See* generally Margot Kaminski, *The Capture of International Intellectual Property Law Through the U.S. Trade Regime*, 87 S. CAL. L. REV. 977 (2014). Reichman & Franklin, n. 83, at 913 (writing that "the power to impose privately legislated rights ... becomes a power to determine the competitive boundaries of the underlying intellectual property rights themselves").

173 *See* Database Directive, n. 25, arts. 1 & 3; Reichman & Samuelson, n. 103. The information economy most likely to emerge from an unrestricted exclusive right in data would "resemble models already familiar from the Middle Ages, when goods flowing down the Rhine River or goods moving from Milan to Genoa were subject to dozens, if not hundreds of gatekeepers demanding tribute." Reichman, *Database Protection in a Global Economy*, n. 102, at 484.

174 *See, e.g.*, John Wilbanks, *Public Domain, Copyright Licenses and the Freedom to Integrate Science*, 7 *J. Sci. Comm.* 1, 4 (2008) (discussing legal tools necessary to develop open data sharing). Waivers become necessary to achieve the research goals of the pool, which would otherwise hinge on the lowest common denominator set of default intellectual property rules. *Id.* at 5.

175 For the adverse effects of digital copyright on new forms of cultural expression, see Mira Burri-Nenova, *Trade versus Culture in the Digital Environment: An Old Conflict in Need of a New Definition*, 12 *J. Int'l Econ. L.* 17, 57 (2009) ("Since these [traditional copyright law] models are often too rigid to allow full realization of the possibilities of the digital mode of content production and distribution or render them illegal, obstructing the 'creative play,' some new hybrid models for the protection of authors' rights have emerged."); Senftleben, n. 31, at 521 (arguing that current EC copyright law is likely to frustrate cultural development).

176 *See* MAX PLANCK RESPONSE TO EC GREEN PAPER, n. 144, at 14–15.

177 *Id.* The TPM exclusion was not adopted in the U.K.'s text-mining exception of 2014. See n. 152 and accompanying text.

As was the case with copyright law, a broad exemption that clearly allowed extraction and reutilization of noncopyrightable data for scientific research must expressly empower the use of automated knowledge discovery tools for this same purpose.[178] Such language should ensure the rights of scientists to aggregate data and information in a research commons, to conduct data mining and similar techniques, and to extract data embedded in scientific articles for use in further research.[179]

5. Adjusting the International Legal Framework to Accommodate the Needs of Science

The prevailing international minimum standards of intellectual property protection are not necessarily in conflict with the proposals set out in this chapter. First, the standards themselves are broad and open to interpretation, while both Article 1.1 of the TRIPS Agreement and Article 14(1) of the WCT contain crucial deference provisions that deliberately leave room to maneuver when states make a good faith effort to conform these standards to national needs and policy.[180] Second, the flexibility built into the TRIPS and WCT standards applies in two directions. Although tightening the exclusive rights with more restrictive conditions is still an option,[181] it remains equally possible to flesh out the limitations and exceptions, along with other balancing features, in a manner more favorable to the provision of public goods than has been the case in some OECD countries and in many developing countries as well.[182]

178 *See* text and accompanying nn. 146–47.

179 *See further* Chapter 7, Section III and Chapter 8, Section III.

180 *See* WCT, n. 38, art. 14(1) ("Contracting Parties undertake to adopt, in accordance with their legal systems, the measures necessary to ensure the application of this Treaty."); TRIPS Agreement, n. 23, art. 1.1 ("Members shall be free to determine the appropriate method of implementing the provisions of this Agreement within their own legal systems and practice."). *See generally* Jerome H. Reichman, *Securing Compliance with the TRIPS Agreement After U.S. v. India*, 4 *J. Int'l Econ. L.* 585 (1998) (noting awareness of WTO Appellate Body of this deference provision). The WTO gave significant weight to this deference norm in its TRIPS decision bearing on copyright law in China. *See* Panel Report, China–Measures Affecting the Protection and Enforcement of Intellectual Property Rights, WT/DS362/R (09-0240) (Jan. 26, 2009); *see also* TRIPS Agreement, n. 23, arts. 7 (objectives), 8 (principles); Peter K. Yu, *The Objectives and Principles of the TRIPS Agreement*, 46 HOUS. L. REV. 797, 1008–18 (2009).

181 *See, e.g.*, Bryan Mercurio, *TRIPS-Plus Provisions in FTA's: Recent Trends*, *in* REGIONAL TRADE AGREEMENTS AND THE WTO LEGAL SYSTEM 215, 215–37 (L. Bartels & F. Ortino eds., 2006) (discussing TRIPS-plus provisions affecting many different areas of IP law).

182 *See Development Agenda for WIPO*, WIPO, http://www.wipo.int/ip-development/en/agenda/ (last accessed 9 Apr. 2014) (pledging "to ensure that development considerations form an integral part of WIPO's work"); P. Bernt Hugenholtz & Ruth Okediji, *Contours of an International Instrument on Limitations and Exceptions*, *in* THE DEVELOPMENT AGENDA: GLOBAL INTELLECTUAL PROPERTY AND DEVELOPING COUNTRIES 7 (N. Netanel ed., Oxford Univ. P. 2009) (stressing the need for a reconsideration of balancing principles within the framework of international copyright); Maskus & Reichman, n. 27, at 35 (observing the possibility of governments acting as "defenders and promoters

For these and other reasons, we remain confident that the positive law mandates of the treaties do not necessarily negate the proposals for reform outlined here, so much as a lack of political will and an absence of the kind of collective action needed to stimulate it. However, no serious reform could succeed without some reinterpretation of the three-step test itself, which many consider the biggest obstacle of all.

At least one expert believes that the three-step test already allows more open-ended assessments of both existing and future limitations and exceptions, in the manner of U.S.-style fair use decisions, than many courts and commentators suppose.[183] More promising in this regard are recent proposals from the Max Planck Institute for judges applying the three-step test, which could induce them to undertake a more normative analysis than in the past.[184] That type of analysis is something European positivist courts are unaccustomed to doing,[185] although under a fair use provision, as codified in U.S. copyright law in 1976, for example, courts must routinely perform this very task.[186]

The Max Planck proposals deliberately build on the preamble to the WCT, which recognizes "the need to maintain a balance between the rights of authors and the larger public interest, particularly education, research, and access to information ..."[187] In that vein, the proposal would:

- Mandate that courts applying the three-step test falling under Article 13 of the TRIPS Agreement in copyright cases take into account the interests of third

of a transnational system of innovation in which properly balanced intellectual property rights were not ends in themselves, but rather the means of generating more scientific and technical inputs into a healthy competitive environment").

[183] *See* Senftleben, n. 31, at 543 (observing that "the three-step test sets forth open-ended factors"); *see also* Paul Edward Geller, *A German Approach to Fair Use: Test Cases for TRIPS Criteria for Copyright Limitations*, 57 *J. Copyright Soc'y U.S.A.* 553, 571 (2010) (arguing that neither the idea-expression distinction nor constitutionally rooted exceptions favoring free speech and other uses ought to be subject to the three-step test).

[184] *See* Christophe Geiger et al., *Declaration: A Balanced Interpretation of the "Three-Step Test" in Copyright Law*, 39 *Int'l Rev. Intell. Prop. & Competition L.* 707, 708 (2008) [hereinafter *Max Planck Declaration on the Three-Step Test*].

[185] And should not do, according to some. *See, e.g.*, Mihály Ficsor, The 'Three-Step Test' De Lege Lata – De Lege Ferenda, paper presented at the Fordham Intellectual Property Conference, Cambridge, U.K., Apr. 15, 2009, *available at* http://fordhamipconference.com/wp-content/uploads/2010/08/MihalyFicsor_Three-step_Test.pdf.

[186] One should recall that the relevant WTO Panels do insist that the test has normative content, but without so far specifying its nature, and indirectly limiting its impact. *See, e.g.*, US–Section 110(5) Panel Report, n. 49, ¶ 6.184 (describing the EC's emphasis on potential impact of an exception versus the actual market effects); *cf.* Panel Report, Canada–Patent Protection of Pharmaceutical Products, ¶ 7.54, WT/DS114/R (Mar. 17, 2000) (writing that the panel believes the word normal used in Article 30 "can be understood to refer either to an empirical conclusion about what is common within a relevant community, or to a normative standard of entitlement").

[187] WCT, n. 38, pmbl.

parties, including individual and collective interests of the general public, and not just the interests of rights owners;[188]

- Avoid prioritizing any one step, or requiring an affirmative answer to all steps, but would instead require a judicial balancing of the different prongs, as occurs under U.S. fair use law;[189]
- Give particular weight to unauthorized uses that are underpinned by fundamental rights[190] and other "common interests," notably "in scientific progress and cultural or economic development;"[191]
- Seek to promote competition, especially in secondary markets, by a correct balancing of interests, but without making the three-step test a proxy for competition law;
- Expressly recognize that adequate compensation may be less than market pricing where other public concerns are at stake, including third-party interests or the general public interest.[192]

The Max Planck Institute's carefully considered reforms would introduce a healthy dose of legal realism into the traditional positivism surrounding European copyright jurisprudence. They would counter the prevailing notion in Continental copyright law, which favors narrowly confined exceptions in deference to the interests of authors.[193] They would also curb the European Commission's tendency to fall back on a market

[188] Such a provision was expressly inserted into Article 30 of the TRIPS Agreement with regard to patents. *See* TRIPS Agreement, n. 23, art. 30 (extending the three-step test to patent law for the first time while adding the words "taking account of the legitimate interests of third parties").

[189] *See* 17 U.S.C. § 107 (2012); Section II.A.2 (discussing fair use in the U.S.). *But see* Ficsor, n. 185 (arguing that the legislative history of the Berne Convention prohibits this approach, even though the three-step test itself has now been recodified with significant variations in both art. 30 (patents) and art. 17 (trademarks) of the TRIPS Agreement). It is not clear why the legislative history of the experimental "package deal" that gave us Article 9 of the Berne Convention in 1967 should operate as a deadweight bar to a judicially more enlightened approach to the revised three-step test as now applied, with significant variations, to all of international intellectual property law's major subject-matter categories. Otherwise, we are obliged to assume that only authors' rights remain somehow immune from the need "to take into account the interests of third parties" at the international level.

[190] *Cf.* Hugenholtz & Okediji, n. 182, at 31 (noting fundamental rights must be balanced with other IP rights); Lange & Powell, n. 160, at 171–72 (stressing the First Amendment); *see also* Helfer & Austin, n. 153, at 221–33 (examining interface between fundamental rights and intellectual property rights in both American and international contexts).

[191] *Max Planck Declaration on the Three-Step Test*, n. 184, at 712; *cf.* Margaret Chon, *New Wine Bursting from Old Bottles: Collaborative Internet Art, Joint Works, and Entrepreneurship*, 75 *Or. L. Rev.* 257, 275–76 (1996).

[192] *See Max Planck Declaration on the Three-Step Test*, n. 184, at 712; *cf.* Lea Shaver, *The Right to Science and Culture*, 2010 *Wis. L. Rev.* 121, 169–84 (2009) (calling for a reexamination of the consistency between intellectual property policies and the greater public interest in science).

[193] However, at least one authority questions the ability of courts adjudicating private law disputes to tinker with international public law mandates. Email from Paul Geller to Jerome H. Reichman (Oct. 9, 2011) (on file with the authors); *see also* Ficsor, n. 185.

failure rationale for limitations and exceptions,[194] a tendency from which U.S. courts have increasingly retreated in recent important decisions bearing on fair use.[195]

Nevertheless, even a reinterpreted three-step test along these lines could not securely ensure unfettered use of automated knowledge discovery tools, given the array of legal obstacles outlined in Section II. At the international level, that objective would at least require a well-supported soft law declaration endorsing the broad research exemption for science described earlier,[196] if not an outright amendment to the Berne Convention itself.

B. *The Hard Reality: More, Not Less Protection, Is on the Way*

So far, our focus on measures to make copyright and related laws more science friendly has operated on the premise that publishers would continue to play their traditional role in the process of disseminating research results. This very premise, however, makes it unlikely that the legislative or judicial reforms outlined earlier could be implemented by the OECD countries in the near future, despite growing attention to the conflict between intellectual property laws and the needs of science in a digital age.[197]

To the extent that publishers retain their traditional role as intermediaries, any efforts to reform applicable intellectual property laws must reconcile the needs of science with the needs of commercial publishers to turn a profit.[198] This factor greatly complicates the prospects for reform because the existing copyright and database laws so favor the interests of publishers over those of scientists that merely incremental or piecemeal reforms rooted in traditional exceptions and limitations are unlikely to give the research community what it needs.

194 *See* Burrell & Coleman, n. 94, at 167–87.

195 *See* text and accompanying nn. 63–67. It is worth noting that Dr. Ficsor claims one could interpret the three-step test to yield the flexibility that the Max Planck Declaration on the Three-Step Test seeks to attain, albeit by more traditional means. *See* Ficsor, n. 185.

196 Hugenholtz & Okediji, n. 182, at 49 (discussing the idea "that a joint initiative between the WIPO and WTO could be an ideal and appropriate expression of a soft-law modality with real impact for collective action on an international instrument on L&E's"). In this connection, we would particularly welcome recognition from WIPO that government use of copyrights for, say, science and educational purposes, trumps all other legal or normative considerations. *See* Daniel J. Gervais, *Making Copyright Whole: A Principled Approach to Copyright Exceptions and Limitations*, 5 *U. Ottawa L. & Tech. J.* 1, 22 (2008) (contending that "[c]opyright rights should not prevent governmental use in the public interest").

197 *But see* Hargreaves, n. 16, at 43 (demanding relief for science as a fillip to economic growth) and the U.K.'s text-mining exception it produced, n. 133. *See also* Board on Sci., Tech. & Econ. Pol'y, Copyright in the Digital Era: Building Evidence for Policy (S.A. Merrill & W.J. Raduchel eds., Nat'l Acads. Press 2013) (stating the goal of the Board is "to expand and improve research on the impacts of copyright policy, particularly on innovation in the digital environment").

198 *See, e.g.*, Julie E. Cohen, *Copyright as Property in the Post-Industrial Economy: A Research Agenda*, 2011 *Wis. L. Rev.* 141, 142–44 (2011) (comparing author incentives to capital incentives).

Today, it seems that the lobbying influence of publishers on legislators, especially but not exclusively in OECD countries, has never been greater.[199] Concerns about protecting the interests of the entertainment and cultural industries continue to elicit longer, broader, and stronger intellectual property laws at the national, regional, and international levels, with little or no regard for their potentially deleterious effects on scientific research or the provision of other public goods.[200] Whether reform efforts underway in some emerging economies may create a countervailing trend is hard to predict,[201] but the benefits of such a trend – if it emerges – would likely play out over a lengthy period, and might not extend, at best, beyond certain regional alignments.

Science policy will, accordingly, have to evolve defensive measures of its own in order to neutralize interference from the default rules of copyright, contract, and database protection laws as they stand. Scientists, in short, will increasingly have to manage their own upstream research assets as global public goods, sheltering them within a reinvigorated sharing ethos, in the interests of a more productive downstream innovation system otherwise driven by the incentives of industrial property laws.[202]

As will be seen later, the scientific community, led by many dedicated and visionary individuals and institutions, has already taken steps to widen the choice of open distribution outlets for microbial literature and data.[203] These promising initiatives nonetheless remain hampered by the community's continued reliance on

[199] *See, e.g.*, Kaminski (2014), n. 172. *But see* the recent debacles of SOPA and the failure to approve ACTA in the EU.

[200] *See* nn. 172, 181 (citing EC's Enforcement Directive, ACTA, SOPA, and FTAs). *See also* pending negotiations concerning the Trans-Atlantic Trade and Investment Partnership (TTIP), http://www.ustr.gov.ttip, *and* Trans-Pacific Partnership (TPP), 18th round of negotiations set for July 15–24, 2013, http://www.ustr.gov./tpp. However action on SOPA and ACTA had stalled at the time of writing. In the United States, sponsors of the Sabo Bill would have placed all published articles resulting from publicly funded research results in the public domain, but this proposal has never moved forward. H.R. 2613, 108th Cong. (1st Sess. 2003).

[201] *See, e.g.*, n. 155 (Brazil's copyright reform proposals); *cf.* Amy Kapczynski, *Harmonization and Its Discontents: A Case Study of TRIPS Implementation in India's Pharmaceutical Sector*, 97 *Calif. L. Rev.* 1571 (2009); Rochelle C. Dreyfuss, The Role of India, China, Brazil and Other Emerging Economies in Establishing Access Norms for Intellectual Property and Intellectual Property Lawmaking 1–3 (Inst. for Int'l Law & Justice, Working Paper 2009/5, Pub. L. Research Paper No. 09-53, 2009), *available at* http://ssrn.com/abstract=1442785 (discussing the role of developing nations in the larger context of developing intellectual property law) [hereinafter Dreyfuss Working Paper]; *see also* Graeme B. Dinwoodie & Rochelle C. Dreyfuss, *Designing a Global Intellectual Property System Responsive to Change: The WTO, WIPO, and Beyond*, 46 *Hous. L. Rev.* 1187, 1212 (2009) (remarking on emerging nations discovering different interest-balancing methodologies from those in the developed world); Jerome H. Reichman, *Intellectual Property in the Twenty-First Century: Will the Developing Countries Lead or Follow?*, 46 *Hous. L. Rev.* 1115, 1118–19 (2009).

[202] Reichman & Uhlir (2003), n. 15; Peter Lee, *Contracting to Preserve Open Science: Consideration-Based Regulation in Patent Law*, 58 *Emory L. Rev.* 890, 901 (2009), *available at* http://papers.ssrn.com/sol3/Delivery.cfm/SSRN_ID1698949_code366600.pdf?abstractid=1288183&mirid=1.

[203] *See generally* below Chapter 7.

publishing intermediaries in general. Accordingly, in the next section, we reevaluate the role that these intermediaries should play under existing institutional constraints. We also look at some of the institutional impediments to the pooling of data and digitally networked collaboration in general.

Later, in Chapters 7 and 8, we ask if better solutions are not likely to emerge from a change of paradigm. Such a sea change could result in the outsourced intermediaries being either downgraded or abandoned altogether, with various open-access modes of dissemination taking their place. Knowledge production and scholarly communication functions would thus increasingly be absorbed into digitally integrated, thematic research environments.[204]

IV. INSTITUTIONAL CONSTRAINTS ON DIGITAL KNOWLEDGE RESOURCES

The institutions that govern published scientific literature differ considerably from the institutions responsible for the management of scientific data. In what follows, we first discuss the costs and benefits of the traditional approach to scholarly literature, and we then briefly survey impediments to the interoperability of data and to networking opportunities. In both cases we define institutional constraints broadly to include economic as well as structural impediments.

A. *The Changing Role of Publishing Intermediaries*

Although the bulk of published scientific research is government funded, the customary practice of the scientific community in recent decades has been to rely almost entirely on external publishing intermediaries, who profit from and restrict use of research results.[205] In conformity with this practice, authors of scientific articles normally assigned their copyrights to publishers, who are either commercial entities or learned societies and other not-for-profit scientific organizations.[206] As a

[204] *See, e.g.*, Paul F. Uhlir, *Designing the Digital Commons in Microbiology – Moving from Restrictive Dissemination of Publicly Funded Knowledge to Open Knowledge Environments, in* DESIGNING THE MICROBIAL RESEARCH COMMONS, n. 11, at 77, 83. *See generally* Jorge L. Contreras, *Data Sharing, Latency Variables, and Science Commons*, 25 *Berkeley Tech. L.J.* 1601 (2011) [hereinafter Contreras, *Data Sharing*].

[205] *See, e.g.*, JOHN WILLINSKY, THE ACCESS PRINCIPLE: THE CASE FOR OPEN ACCESS TO RESEARCH AND SCHOLARSHIP 2 (2006) (reporting that NIH itself funded some 60,000 scientific papers per year prior to 2006); Contreras, *Data Sharing*, n. 204, at 1652 (reporting that some "50,000 different scientific journals [were] in print at the end of 2003, many of which are published by commercial entities that charge significant subscription fees").

[206] *See, e.g.*, Contreras, *Data Sharing*, n. 204, at 1652–55 (reporting that the three largest publishers of scientific journals – Elsevier (about 1800 titles), Taylor and Francis (about 1000 titles) and, Springer Verlag (about 500 titles) together control about 60% of scientific research content).

result, it was publishers, rather than authors, that initially determined the conditions for access to these same articles and for reuse of the information and data they contain.[207] In return, authors benefitted from the peer-review mechanisms many publishers manage, which made them reluctant to publish outside traditional, well-established or high-impact outlets, when they had the choice.

Historically, the logic behind this customary arrangement was the need to defray high front-end publishing costs and to perform laborious tasks, such as typesetting and formatting, as well as the physical distribution of printed copies.[208] Another factor was the willingness of many scientific subcommunities to entrust learned societies with the publication task, which in turn became a primary source of revenue for the societies whether they actually performed the publishing service, or, increasingly in recent years, outsourced it to a commercial publisher in return for a share of the proceeds. Over time, the prospects for greater profits enticed commercial publishers to buy out the learned societies, sometimes with continuing royalty payments to the societies.[209]

Lately, scholars have challenged such logic,[210] and many now argue that the value added by intermediaries has reached diminishing returns.[211] The once costly front-end publishing functions have increasingly been reduced by desktop publishing and

[207] *See, e.g.*, Hilty, *Copyright Law and Scientific Research*, n. 53, at 326; Hilty, *Five Lessons About Copyright*, n. 53, at 123–24. Professor Hilty, among others, stresses that for-profit publishers tend to impose greater restrictions on access and use than authors or the scientific community more generally would deem desirable, given that the latter receives motivation through reputation benefits that may accrue from unhindered diffusion.

[208] *See* Hilty, *Five Lessons About Copyright*, n. 53, at 120–21 (discussing the decline of such high-end tasks with the rise of personal computer programs). However, university presses absorbed these or similar functions with respect to specialized books subject to market failure in the normal book trade. Eugene Volokh, *The Future of Books Related to the Law?*, 108 *Mich. L. Rev.* 823, 838–40 (2010) (discussing markets and academic book publishing).

[209] *See* Toby Miller, *Drowning in Information and Starving For Knowledge: 21st Century Scholarly Publishing*, 1 *Int'l J. Comm.* 123, 125 (2007), *available at* http://ijoc.org/ojs/index.php/ijoc/article/viewFile/121/56 ("Since that time, the development of digital technologies has seen for-profit [science] publishers proliferate, as the cost of entering the industry has diminished, and prices have continued to outstrip inflation ...."); Interview by Research Information Staff with Rene Olivieri, CEO, Blackwell Publ'g (Jan./Feb.2005), *available at* http://www.researchinformation.info/features/feature.php? feature_id=92 (stating that "[t]hree quarters of the top 200 and two-thirds of the top 500 ISI-ranked titles are owned by societies or other nonprofit organizations. The majority of these titles are self-published, but between a quarter and a third are contracted out to another publisher").

[210] Among the many excellent analyses, too numerous to cite, see, for example, WILLINSKY, n. 205; Nancy Kranich, *Countering Enclosure: Reclaiming the Knowledge Commons*, *in* KNOWLEDGE AS A COMMONS, n. 9, at 85, 98 (noting the popularity of papers posted on open-access databases versus those not available on such databases); Peter Suber, *Creating an Intellectual Commons through Open Access*, *in* KNOWLEDGE AS A COMMONS, n. 9, at 171, 185–86 (noting the cancellation of expensive databases by libraries at Harvard, Cornell, Duke, and University of California in favor of open access platforms); *see also* Contreras, *Data Sharing*, n. 204, at 1652–55 (citing authorities).

[211] *See e.g.*, Hilty, *Copyright Law and Scientific Research*, n. 53, at 326–27. *See also* Chapter 7, Sections II & III.

automated formatting,[212] while the peer-review process, of great importance to the integrity of science, is performed gratis by scientists who themselves gain power, reputation, and advanced access to new developments from their voluntary labor.[213] Nevertheless, this built-in quid pro quo within the scientific community still tends to perpetuate the dominance of proprietary intermediaries, along with the practice of negotiating the sale of subscriptions (now licensing) directly to libraries on relatively restrictive terms.[214] Meanwhile, the supervisory or editorial role of the learned societies, with some exceptions, has diminished over time, although their dependence on income from publishing seems to have increased.[215]

This web of traditional practices and interests has been carried into the digital age, even though digital networks offer many opportunities to break with the limits of the print model and make whole new dimensions of publishing possible.[216] What really changes in the online environment are not the basic principles of scientific collaboration so much as the burdens and role of publishing intermediaries in the sciences, who increasingly may never publish a physical print copy at all.[217]

The tendency to rely on commercial online distribution in the sciences undermines prior balancing effects of the first sale principle under traditional copyright law.[218] For example, there are fewer printed copies distributed to individuals that may be freely redistributed to others after the initial sale, and the subscription price per journal may rise to levels that academics and even their universities cannot afford.[219] Even

[212] *See, e.g.*, Hilty, *Copyright Law and Scientific Research*, n. 53 at 325–26 (noting that Internet-based web sources reduced the need to produce tangible goods).

[213] Even this traditional form of peer review is now under attack. *Cf.* Linda Hooper-Bui, *A Gulf Science Blackout*, N.Y. TIMES, Aug. 25, 2010, at A21. Note, however, that some journals pay scientists to conduct peer reviews of articles. *The Economic Case for Open Access in Academic Publishing*, FREE ACAD. RES. ASS'N (Feb. 17, 2011), http://www.faraweek.org/?p=6 ("If a journal is highly selective, it must pay for peer review of many articles for each article it accepts.").

[214] Not only are libraries renting the digital journals, often at exorbitant rates, but if subscribers discontinue their license, they may not be able even to retain the electronic copies – frequently the only ones available – of the journals for which they had already paid the costs of subscriptions.

[215] Exceptions occur if the learned society maintains its own editorial subsidiary, as occurs with the Journal of the American Medical Association (JAMA). In Latin America, and probably most other developing countries, scientific journals are still published at universities. Universities in OECD countries have themselves massively entered the book publishing trade to overcome market failure attributable to commercial presses, while remaining aloof from the publication of scientific journals, with rare exceptions.

[216] *See* Section I.

[217] *See, e.g.*, David, n. 28, at 21 (describing the different ethos and norms within various academic fields); Stodden, n. 17, at 33 (stating that "public safeguards should also enable digital telecommunications networks to link the providers of scientific and technical inputs in an endless research commons").

[218] *See, e.g.*, R. Anthony Reese, *The First Sale Doctrine in the Era of Digital Networks*, 44 *B.C. L. Rev.* 577, 577–78 (2003).

[219] *See, e.g.*, Contreras, *Data Sharing*, n. 204, at 1652–53 (discussing the cancellation of subscriptions by academic libraries due to rising costs); *see also* NAT'L ACAD. SCIS. ET AL., ENSURING THE INTEGRITY,

when printed copy distribution continues, the role of publishing intermediaries' in the online environment changes radically, as they add less value to the authors' own research results[220] and become online service providers whose primary contribution to authors, is beyond mere convenience, the quality control aspect inherent in peer review and copy editing.[221]

Notwithstanding these changed conditions, the rules of copyright law have been extended to the digital environment, and the protections available have been greatly strengthened, as demonstrated earlier in this chapter, in order to make the online environment safe for the transmission of printed text. Because STM publishing has drifted along with this tide, the full possibilities of digitally manipulating research results for new scientific discoveries are hamstrung by layers of protection inherited from these legal and institutional developments, and there is a pressing need to avoid the resulting harm to science.[222]

The open-access movement, as evidenced later in Chapter 7, is a major response to this challenge. Today, an ever-growing number of scientific journals, including microbiology journals, are published online, on a fully or partially open-access basis,[223] although these are not yet always the most prestigious journals in their

ACCESSIBILITY, AND STEWARDSHIP OF RESEARCH DATA IN THE DIGITAL AGE 78 (Nat'l. Acad. Press, 2009) (observing rise in subscription prices for scientific, medical, and technical journals).

220 *See, e.g.*, Hilty, *Five Lessons About Copyright*, n. 53, at 123 (questioning the added value of editing electronic-only data compiled and formatted by the researchers themselves); Hilty, *Copyright Law and Scientific Research*, n. 53, at 326–27 (finding a lack of added value within the electronic data management framework); MAX PLANCK RESPONSE TO EC GREEN PAPER, n. 144, at 5–6 (categorizing the divergent roles and interests of intermediaries).

221 This characterization, among others, is of course hotly contested by publishers, who see themselves as indispensable pillars of the scientific endeavor that add considerable value to its research outputs, whereas less rigorous "open access" methods enable less deserving articles to be published. *See* John Ochs, Address Before the Board on Sci., Tech. & Pol'y, "American Chemical Society Submission to the National Academies' Committee on the Impact of Copyright Policy on Innovation in the Digital Era" 2–4 (Oct. 15, 2010), *available at* http://sites.nationalacademies.org/PGA/step/copyrightpolicy/PGA_066845; *see also* Letter from STM, n. 74 (extolling large amounts STM publishers invest in digital technologies to benefit researchers). In reality, not only have publishers sought to configure the online environment on the model of print media, they have also tried to subordinate the new class of intermediaries that digital technology has generated, the Internet System Providers (ISPs), to their own ends, adding yet another layer of potential barriers and transaction costs to the diffusion of research results. *See, e.g.*, Okediji, n. 34, at 116 (calling for more meaningful fair use standards); Ruth Okediji, *The Limits of Development Strategies at the Intersection of Intellectual Property and Human Rights*, *in* INTELLECTUAL PROPERTY TRADE AND DEVELOPMENT: STRATEGIES TO OPTIMIZE ECONOMIC DEVELOPMENT IN A TRIPS-PLUS ERA 349–50 (D. J. Gervais ed., Oxford Univ. Press 2007) (describing the process by which owners used new technological advances to stake claims to previously noncopyrighted material).

222 *See, e.g.*, HARGREAVES, n. 29, at 46–47; Kranich, n. 210.

223 *See, e.g.*, Contreras, *Data Sharing*, n. 204, at 1652–57; Lucie Guibault, *Owning the Right to Open Up Access to Scientific Publications*, *in* OPEN CONTENT LICENSING: FROM THEORY TO PRACTICE 137,

respective fields.[224] To the extent that the learned societies themselves resist the drive for greater use of open-access modalities, their dependence on royalty streams from commercial publishers for scholarly pursuits and other activities may explain their reluctance to change.

While outsourcing the publication of some scientific journals to commercial publishers may still make sense, despite an array of other options, there is a growing trend to subsidize the open-access format, even in an otherwise commercial context, as part of the publicly funded research process. The funding agencies, foundations, and universities that support specific research projects may thus provide supplementary funds to pay commercial publishers a set fee in lieu of royalties or other compensation.[225] In such cases, the funders may – and increasingly will – set open-access terms as the quid pro quo of the subsidy itself.[226] Commercial publishers are increasingly disposed to allow this option, and science funders have begun to insist on it in some disciplines,[227] although the sustainability of this approach obviously depends on the continued availability of financial resources for this purpose.

The point is that desktop-publishing techniques and online transmission have made it technically (if not culturally) feasible to redefine the role of existing intermediaries who benefit from intellectual property rights and practices that impede access to research results. By the same token, once publicly funded research results are made

137–67 (L. Guibault & C. Angelopoulos eds., 2011). For empirical evidence in the field of microbiology journals, see Chapter 7, Sections II.A & B.

224 For statistics on the ten most prestigious microbiology journals, see below Chapter 7, Section II.B.1. Many of these journals are relatively new, while the ISI index (which counts only citations) does not begin tracking impact until a journal has been published at least two years and sometimes for a five-year period. Moreover, some open access journals have achieved high impact in recent years. For pressures by the Harvard faculty advisory council to "move prestige to open access [journals]" in order to offset soaring subscription prices, see Faculty Advisory Council Memorandum on Journal Pricing, Harvard Univ., Apr. 17, 2012, http://isites.harvard.edu/icb/icb.do?keyword=k77982&tabgroupid=icb.tabgroup143448.

225 *See, e.g.*, Contreras, *Data Sharing*, n. 204, at 1655–57.

226 Raym Crow, *Developing an Institutionally-Funded Publishing Channel: Context and Considerations for Key Issues*, ECOMMONS@CORNELL 10–11 (July 1, 2004), http://hdl.handle.net/1813/178; Research Funders' Open Access Policies, SHERPA, http://www.sherpa.ac.uk/juliet/index.php (last accessed 9 Apr. 2014) (showing a number of research funders whose guidelines require open access to funded research).

227 *See* Robert Terry & Robert Kiley, *Open Access to the Research Literature: A Funder's Perspective*, *in* Open Access: Key Strategic, Technical, and Economic Aspects 101, 101–03 (N. Jacobs ed., 2006); *Open Access Policy*, Wellcome Trust, http://www.wellcome.ac.uk/About-us/Policy/Policy-and-position-statements/WTD002766.htm (last accessed 9 Apr. 2014); America Creating Opportunities to Meaningfully Promote Excellence in Technology, Education, and Science Act of 2007 (America COMPETES Act), P.L. 110–69, *available at* http://www.gpo.gov/fdsy/pkg/PLAW-110publ69/content-detail.html (directing the NIH to openly archive its grantees' published research); *see also* Contreras, *Data Sharing*, n. 204, at 1652–55; OSTP Public Access Initiative, n. 143.

available to the scientific community, with due respect for attribution, it becomes logical to ask why scientists as users should ever pay scientists as authors, irrespective of what the default rules of copyright and database laws provide to the contrary.[228]

B. *Impediments to the Pooling of Data and Digitally Networked Collaboration*

In the predigital network era, when scientific databases were comparatively small, it was feasible for individuals and small groups of researchers to compile, annotate, and maintain them by labor-intensive methods that made use of distributed and heterogeneous information sources. Recently, however, the size and complexity of all types of scientific databases have grown enormously, including those used in microbiology, and the potential benefits from exploiting such data have also mounted, although not all aspects of this data deluge are viewed as positive.[229]

Digital, or *e*-science, activities are absorbing ever more resources from publicly funded research programs with growing pressure on researchers to find ways of adding value to, and extracting revenues from, the resulting assets in order to increase the return on public investments.[230] Nevertheless, recent studies show that, with respect to postpublication sharing of data and tools, "good practice is not widespread."[231] Collective action to remedy this problem is needed.

At the same time, with the rise of data science and computational biology, scientific databases have become increasingly valuable both as research tools and as commercially exploitable products. After 1980, when the Bayh-Dole Act and related measures were enacted in the United States, universities and scientists viewed many government funded research outputs as potential sources of revenue and not just contributions to the public domain of science. The new legitimacy of this approach undermined the traditionally espoused sharing ethos of science even with respect to data in many disciplines,[232] especially in the life sciences. It is no accident that

[228] As noted earlier, the scientists' incentives flow almost exclusively from reputational benefits. *See* Davis & Connolly, n. 20 (noting researchers' reluctance to release results before publication); Jordan, n. 20 (noting the importance of publication and priority for scientists).

[229] *See, e.g.*, Mark Sagoff, *Data Deluge and the Human Microbiome Project*, 28 Issues in Sci. & Tech. (Summer 2012); Mattioli (2014), n. 2; for microbiological data, see D. Smith et al. (2013), n. 12. *See further* Chapter 8, Sections I & II.

[230] Reichman & Uhlir (2003), n. 15; *see also* Paul A. David, *Koyaanisqatsi* in Cyberspace, SIEPR Discussion Paper No. 02-29, Stanford Inst. Econ. Pol'y Research (Mar. 2003).

[231] Paul N. Schofield et al., *Post-Publication Sharing of Data and Tools*, 461 *Nature* 171–73 (10 Sept. 2009); *see also* Mattioli (2014), n. 2, at 10–16; Bryn Nelson, *Data Sharing: Empty Archives*, 461 *Nature* 160–63 (2009); Wesley E. Cohen & John P. Walsh, *Real Impediments to Biomedical Research*, 8 *Policy Econ.* 1–30 (2008).

[232] *See* Eric G. Campbell et al., *Data Withholding in Academic Genetics: Evidence from a National Survey*, 287 *J. Am. Med. Assoc.* 473 (2002); Stephen Hilgartner, *Access to Data and Intellectual*

Bayh-Dole emerged about the same time as the United States' Supreme Court's decision allowing patents on living organisms – in this case, genetically modified microbiological organisms – for the first time.[233] As demonstrated in Chapter 2, this Supreme Court decision presaged a stream of patents on biological materials and living matter generally, and in microbiology specifically.[234] It also legitimized and encouraged the growing tendency of universities to patent or otherwise protect and exploit their research results, even when funded by government.

More recently, these commoditizing tendencies have expanded to securing the ownership of both patents and copyrights in university-generated computer programs; to the patenting of molecular biology databases; and to the licensing of scientific databases as research tools on increasingly restrictive terms, which include limits on use and even grant back and reach-through clauses claiming interests in future applications.[235] These pressures to profit from commercial applications of research results have steadily increased as legislation imitating Bayh-Dole proliferates around the world.[236]

The prospect of extracting rents from specialized database products or services has, in turn, stimulated the growth of commercial business practices in the life sciences

Property: Scientific Exchange in Genome Research, in INTELLECTUAL PROPERTY RIGHTS AND THE DISSEMINATION OF RESEARCH TOOLS IN MOLECULAR BIOLOGY: SUMMARY OF A WORKSHOP HELD AT THE NATIONAL ACADEMY OF SCIENCE, Feb. 15–16, 1996 (Nat'l Acads. Press 1996); Stephen Hilgartner & Sherry I. Brandt-Rauf, *Controlling Data and Resources: Access Strategies in Molecular Genetics, in* INFORMATION TECHNOLOGY AND THE PRODUCTIVITY PARADOX (P.A. David & W.E. Steinmueller eds., Harwood Acad. Pub. 1998); *see generally* Reichman & Uhlir (2003), n. 15, at 361–415.

233 Diamond v. Chakrabarty, 447 U.S. 303 (1980). *See generally* DAVID C. MOWERY, RICHARD R. NELSON & BHAVEN SAMPAT, IVORY TOWER AND INDUSTRIAL INNOVATION: UNIVERSITY-INDUSTRY TECHNOLOGY TRANSFER BEFORE AND AFTER THE BAYH DOLE ACT IN THE UNITED STATES (Stanford U. Press 2004).

234 *See* Chapter 2, Section II.B.1. *See generally* Arti K. Rai & Rebecca S. Eisenberg, *Bayh-Dole Reform and the Progress of Biomedicine*, 66 *Law & Contemp. Probs.* 289 (2003); Rebecca S. Eisenberg, *Proprietary Rights and the Norms of Science in Biotechnology Research*, 97 *Yale L.J.* 177 (1987); Rebecca S. Eisenberg, *Noncompliance, Nonenforcement, Nonproblem? Rethinking the Anticommons in Biomedical Research*, 45(4) *Hous. L. Rev.* 1059–99 (2008) (Symposium Issue).

235 *See generally* Paul A. David & Michael Spence, "Designing Institutional Infrastructure for E-Science," SIEPR Discussion Paper No. 07-23, Stanford Inst. Econ. Pol'y Research (Dec. 2007), *available at* http://www-siepr.stanford.edu/papers/pdf/07-23.pdf; Reichman & Uhlir (2003), n. 15, at 366–70, 396–416.

236 *See, e.g.*, India, Utilization of Public Funded Intellectual Property Bill, Bill No. LXVI (2008), *available at* http://www.prsindia.org/uploads/media/1229425658/1229425658_The_Protection_and_Utilisation_of_Public_Funded_Intellectual_Property_Bill__2008.pdf; Republic of South Africa, Intellectual Property Rights from Public Funded Research and Development Bill, Bill No. B46B-2008, *available at* http://www.prng.org.za/bill/20080815-intellectual-property-rights-publicly-financed-research-and-development-10. *See generally* Anthony So et al., *Is Bayh-Dole Good for Developing Countries? Lessons from the US Experience*, 6(10) *PLoS Biology* e262 (2008).

generally.[237] Sometimes these practices are built around direct sales of proprietary databases, with annotated data often protected behind digital fences. More typically, data are bundled with proprietary services, such as software packages or contracts for frequent and early access to updating, editing, and search facilities.[238] These contracts may also provide access to the human know-how needed to use and interpret the results of database searches.

Over time, with the continuing shortfall of public funding in most countries for many types of databases, there has been an increasing perception that commercial opportunities might also help to defray the costs of creating and maintaining public database infrastructures.[239] As a result, even some databases containing raw experimental data, such as certain genetic sequence databases in the early nineties, and some general purpose database services(such as sequence data retrieval systems for use with the public sequence databases developed by European Biotechnology Institute in the mid-nineties), became subject to commercial practices restricting use, reuse, and dissemination of the relevant data and related services.[240]

Reportedly, these and other restrictions on access to genetic sequence data have become a problem for basic research in some areas of microbiology.[241] For example, three biofuels research centers established with public funding later operated as "quasi startup firms" on a proprietary basis.[242] Even with regard to more upstream microbiological research, which benefited from a tradition of open access to some data pertaining to the basic research infrastructure, such as strain names and taxonomic information, these pressures have sometimes fostered the development of commercial business practices and models for heretofore freely available data.[243]

[237] *See* David & Spence, n. 235; Victoria Stodden, *Innovation and Growth through Open Access to Scientific Research: Three Ideas for High-Impact Rule Changes, in* RULES FOR GROWTH THROUGH LEGAL REFORM 416–18 (Ewing Marion Foundation 2011), *available at* http://www.stanford.edu/~vcs/papers/RulesForGrowth-Chapter17-STODDEN.pdf.

[238] David & Spence, n. 235.

[239] MARK HARVEY & ANDREW MCMEEKIN, PUBLIC OR PRIVATE ECONOMIES OF KNOWLEDGE? TURBULENCE IN THE BIOLOGICAL SCIENCES 52, 89 (Edward Elgar Pub. 2007). *See also* Francine Berman & Vint Cerf, *Who Will Pay for Public Access to Research Data?* 341(6146) *Science* 616–17 (2013).

[240] HARVEY & MCMEEKIN, n. 239, at 52, 89. However, there has been a recent countervailing trend to pool genomic and proteomic data as community resources to avoid these problems. *See* Chapter 7, Sections I &II.

[241] Telephone interview with Daniel Drell of the Department of Energy, Washington, D.C. on October 2, 2011. *See also* Ari Patrinos & Daniel Drell, *Strengthening Public-Domain Mechanisms in the Federal Government: A Perspective from Biological and Environmental Research, in* THE ROLE OF SCIENTIFIC AND TECHNICAL DATA AND INFORMATION IN THE PUBLIC DOMAIN 161, 162 (Julie M. Esanu & Paul F. Uhlir eds., National Academies Press, 2003).

[242] Interview with Daniel Drell, n. 241.

[243] *See, e.g., NamesforLife Technology*, NAMES FOR LIFE, http://services.namesforlife.com/about (last accessed April 28, 2015).

An example of this approach was the Names for Life project, developed at Michigan State University. This project aimed to sell services for integrating strain and taxonomic data with relevant literature, based on a set of name disambiguation tools. Names for Life was thus developing a proprietary model of database integration by "bundling" open access components and literature with proprietary databases and data analysis tools,[244] a project that appears to remain subject to many of the resulting restrictions on access and use.[245]

Proprietary restraints affecting basic research tools and upstream data inputs can disrupt public research endeavors. For example, they can magnify the cost of data acquisition and use, just when it has been shown that making digital tools, such as software, algorithms, and formats available at no cost leads to important socioeconomic payoffs for a broad community of users.[246] There is some evidence that restrictive data use and reuse policies may harm even commercial users.[247]

Even when a compiler sees no immediate commercial potential in a given scientific database, he or she may instinctively want to hoard the data beyond the researcher's commonly accepted prepublication period[248] in the hopes of either future economic gain or reputational benefits, rather than making them available for research purposes in conformity with the sharing norms of science.[249] Similarly, many university laboratories and culture collections continue to hold large quantities of data useful for microbiological research under relative degrees of secrecy.[250]

[244] The Names for Life model links strain data (which are openly available) to nomenclature data (not altogether openly available, with the most recent version published by Springer (unavailable in digital format) with the scientific literature (most of which remains available only by subscription). So the component that was openly available (the strain names) could, when combined with proprietary data, end up being locked into a restricted environment that would hinder data mining. Furthermore, there is a 2005 U.S. patent application on the "system and methods" for data processing (which is very similar to the algorithm of the StrainInfo.net Bioportal, already published on open access terms). This result is achieved by replacing the strain names (which might change over time) with DOI numbers (which would remain constant, but which must be purchased from the DOI Foundation, created by the publishing industry for their needs).

The patent in question issued on April 12, 2011, as U.S. Patent No.07,926,444: "Systems and methods for resolving ambiguity between names and entities." The abstract states: "The present invention provides systems and methods that utilize an information architecture for disambiguating scientific names and other classification labels and the entities to which those names are applied, as well as a means of accessing data on those entities in a networked environment using persistent, unique identifiers."

[245] *See NamesforLife Technology*, n. 243.

[246] Improvements in user workflows, for example, have been achieved via the open source policy at the EBI Hinxton research campus in the UK.

[247] Harvey & McMeekin, n. 239, at 64.

[248] *See* Mattioli (2014), n. 2, at 10–16; Cohen & Walsh, n. 231; Schofield et al., n. 231; Campbell et al., n. 232.

[249] *See* Nat'l Research Council, Sharing Publication-Related Data and Materials: Responsibilities of Authorship in the Life Sciences (Nat'l Acads. Press 2003), *available at* http://books.map.edu/catalog/10613.html.

[250] *See* Chapter 3, Sections II.B & IV.C.

Hoarders want to retain control of their data partly to ensure continued reliability, and also to avoid the risks of losing follow on publications or of being scooped. These tendencies to hoard data then lead to the unnecessary duplication of efforts; the use of proprietary and incompatible formats; insufficient attention to the organization and documentation of databases so as to become more useful to others; and technical and semantic obstacles to interoperability.[251]

These well-known obstacles limit the networking capabilities and optimal use of analytical tools needed to manage the deluge of data that has become a cardinal premise of the New Biology paradigm.[252] While hoarding may enable compilers to recoup some of the costs of curation needed for effective disclosure, they deprive the larger research community of access to potentially useful data.

As a result, vast amounts of potentially valuable microbiological data are known to exist, but remain inaccessible for public scientific reuse. Because universities and their scientists increasingly aim to commercialize research products and profit from them, their willingness to exchange data, along with other research tools, has been seriously compromised in some areas. Resistance to sharing is particularly strong in those fields where the boundaries between "basic" and "applied" research have collapsed or become blurred, as in many biomedical and bioengineering disciplines. In the United States, moreover, hoarding tendencies may increase as universities strive to obtain prior user rights under the revised patent law of 2011, which otherwise rewards the first to patent and not the first to invent.[253]

Meanwhile, inter-university exchanges of data are increasingly subject to high transaction costs, delays, and a growing risk of anti-commons effects, that is, too many intellectual property rights and commercial interests making it difficult to build comprehensive or complex databases.[254] This problem is particularly acute in cases of transnational scientific collaboration.[255] As relations between universities and industry become more intense, and as public universities receive a smaller share of their budgets from state legislatures, the universities tend to view each other as

[251] *See generally* Paul F. Uhlir & Peter Schröder, *Open Data for Global Science*, 6 *Data Sci. J.* 17 (2007), *available at* http://www.spatial.maine.edu/icfs/Uhlir-SchroederPaper.pdf. *See further* Chapter 8, Section II.C.

[252] *See* Reaping the Benefits of Genomics, n. 4; Biology for the 21st Century, n. 4.

[253] Leahy-Smith America Invents Act (AIA), P.L. 112–29, §202(a)(2)(2011).

[254] *See* Rai & Eisenberg, n. 234; David (2005) n. 230; and Paul A. David, *A Tragedy of the Public Knowledge 'Commons'? Global Science, Intellectual Property and the Digital Technology Boomerang*, SIEPR Discussion Paper 00-002 (Stanford Univ., Oct. 1, 2000), *available at* https://siepr.stanford.edu/publicationsprofile/667 (last accessed April 28, 2015). *See also* Stephen M. Maurer, Inside the Anticommons: Academic Scientists' Struggle to Commercialize Human Mutations Data, 1999–2001, paper presented at the Franco-American Conference on the Economics, Law, and History of Intellectual Property Rights, Haas School of Business, University of California at Berkeley, Oct. 5–6, 2001.

[255] For the resulting legal complications, see Section II.C in this chapter.

competitors, rather than as partners in a common mission. Their industrial partners are correspondingly more likely to impose their own proprietary terms of exchange on the universities.

We recognize that many data sets may not be of general research interest, or may require considerable effort to make them usable by others. Nevertheless, much of this dark area of data hoarding could be of real use to other scientists, but remains accessible only on a haphazard basis, depending on personal contacts or on the researcher's disposition to share data at any given stage.[256] For example, there is evidence that informal exchanges of data between individual scientists and laboratories in biomedical research have become problematic, with as many as 50 percent denials of requests for information prior to publication, and some 10 percent even after publication.[257]

Science policymakers and leaders of the research community began to respond to these trends in some countries in the 1990s and especially in more recent years as the costs to research became better understood. Various scientific communities have united to resist the proprietary ethos by pooling upstream data resources in a commons open to all, or in semicommons arrangements open to qualifying participants. In addition, a number of public or publicly funded organizations have decided to make their data holdings openly available. These and other recent developments are discussed in Chapter 8. Nevertheless, even these promising initiatives are severely challenged and undermined by national and international intellectual property laws that, as shown earlier in this chapter, remain hostile to the needs of digitally integrated scientific research, and especially to publicly funded and public interest research endeavors.

V. FINAL OBSERVATIONS

Scientific discoveries depend on access to a robust public domain, in which preexisting discoveries become the building blocks of future investigations and existing information and data become inputs to future knowledge assets that cannot be generated nearly as effectively without them.[258] However, the recent tendency to elevate standards of intellectual property protection at both the national and international levels has been motivated largely by interests seeking to protect existing knowledge goods, destined mainly for end users, with insufficient regard to

[256] Mattioli (2014), n. 2, at 544–49; L. R. Hill & Micah Krichevsky, *International Strain Data Networks*, 2 *World J. Microbiology & Biotech.* 341 (1986). *See also* Campbell et al., n. 232; Cohen & Walsh, n. 231.

[257] Campbell et al., n. 232.

[258] *See, e.g.*, James Boyle, The Public Domain: Enclosing the Commons of the Mind 160–78 (Yale Univ. P. 2008); David, n. 28, at 27–28; David Lange, *Recognizing the Public Domain*, 44 Law & Contemp. Probs. 147, 165 (1981). *See generally* Reichman & Uhlir (2003), n. 15, at 332.

the social costs and burdens imposed on future creation and innovation, and with a corresponding bevy of new problems that hinder both objectives.[259] This movement has generated thickets of intellectual property rights, high transaction and litigation costs, receding access to the public domain, growing anticommons effects, and the stifling of privileged uses by means of technological protection measures and digital rights management tools in the online environment.[260]

In this chapter, we have traced the contradictory measures in copyright and related laws that impede upstream scientific investigation and thereby complicate the exploitation of downstream applications of research results. By over-extending the protection of scientific information and data, these laws have made it harder for all investigators to build on, rework, or further elaborate on the contributions of others and to harness the astounding research potential of digital information technologies to their fullest extent.

From this perspective, the worldwide copyright system as it has lately evolved can hardly be said to benefit scientists *qua* authors. On the contrary, authors and compilers of scientific works and databases are still often obliged to surrender their outputs to publishers from whom they must buy back the very information and data they supplied, often at government expense. Rather than opening new vistas for producers of research data and information – as occurred after the printing press was invented and at regular intervals of technological change since then – copyright and database protection laws in the digitally networked environment seem bent on closing off new horizons in order to defend old business models for which many publishers have sought few alternatives.[261]

A. *Bridging the Disconnect Between Private Rights and Public Science*

Given the opportunities that digital networks and automated knowledge discovery tools now make possible, the logical goal for policymakers is to remove obstacles that the existing legal infrastructure poses for twenty-first century scientific endeavor.

[259] *See, e.g.*, David C. Mowery et al., Ivory Tower and Industrial Innovation, n. 233 184–92; David, n. 28, at 27–28; Bhavan H. Sampat, *Patenting and U.S. Academic Research in the 20th Century: The World Before and After Bayh-Dole*, 35 *Res. Pol'y* 772, at 784–86; *see also* Maskus & Reichman, n. 27, at 20–23 (discussing the imbalance in modern intellectual property regimes resulting from a "prolonged effort to strengthen the protection of investors").

[260] *See* James Bessen & Michael Maurer, Patent Failure: How Judges, Bureaucrats, and Lawyers Put Innovation at Risk 1–6 (Princeton Univ. Press 2008) (dealing with patents); Michael Heller, The Gridlock Economy: How Too Much Ownership Wrecks Markets, Stops Innovation and Costs Lives 1–22 (Basic Books 2008); *see also* Paul F. Geller, *Beyond the Copyright Crisis: Principles for Change*, 55 *J. Copyright Soc'y USA* 165, 166 (2008), at 166 ("Copyright law is in crisis ... [I]t has become more and more complicated and less and less reliable, while losing legitimacy."); Lunney, n. 59, at 869–92.

[261] *See* Hargreaves, n. 29, at 41–42.

In this context, copyright law's limitations and exceptions have an important role to play. They are not some nuisance-like sideshow of demands to be appeased as narrowly as possible. Rather, they should at least be viewed as a form of users' rights,[262] which help to supply inputs for scientific discoveries, innovation, and trade that are as indispensable to the dynamic production and dissemination of knowledge goods as suitably crafted incentives for authors and inventors.

A fundamental change of attitudes would be necessary, however. A top priority for policymakers should be to avoid generating legally established fiefdoms, in which a few private rights holders can combine the bulk of all scientific data and literature into monopolized repositories where access and use are restricted and controlled from the top down, and in which the commodified inputs of publicly funded science are distributed on a proprietary basis. Failure to achieve such a shift in priorities places digital and computational science in developed countries at risk of becoming progressively entangled in "copyright thickets"[263] precisely at a time when these countries face stiff challenges from the growing scientific and technological capacities of the emerging economies.[264]

Despite the complexity of these issues, and the countervailing pressures of a powerful publishers' lobby, policymakers should resist the temptation to leave copyright and database protection laws where they stand or to strengthen them further. Few decisions could generate so many unintended harmful consequences. If these laws continue to impede digital science in the ways portrayed earlier in this chapter, the much vaunted comparative advantages that industry and government spokespersons associate with maximalist levels of intellectual property protection could give way to legal strangleholds on the most promising avenues of public digital research, with the predictable result of killing the goose that lays the golden eggs.[265]

[262] *See, e.g.*, Hugenholtz & Okediji, n. 182, at 16–27; *see also* Abraham Drassinower, *Authorship as Public Access: On the Specificity of Copyright vis-à-vis Patent and Trademark*, 2008 *Mich. St. L. Rev.* 199, 199–204 (2008) (arguing that users' rights, instead of simply serving as an exception to copyright, are so integral to the modern copyright system that they entail a redefinition of the wrongs copyright laws were meant to address).

[263] In patent law, such thickets had threatened to undermine information science and such frontier sciences as synthetic biology at least until the U.S. Supreme Court intervened to readjust the most fundamental design principles of preexisting patent law itself. *See, e.g.*, Arti K. Rai & James Boyle, *Synthetic Biology: Caught Between Property Rights, the Public Domain, and the Commons*, 5 *PLoS Biology* e58, 390 (2007); Arti K. Rai & Sapna Kumar, *Synthetic Biology: The Intellectual Property Puzzle*, 85 *Tex. L. Rev.* 1745, 1756–58 (2007).

[264] *See, e.g.*, Dreyfuss, n. 201; Peter Yu, *Sinic Trade Agreements and China's Global Intellectual Property Strategy, in* INTELLECTUAL PROPERTY ASPECTS OF FREE TRADE AGREEMENTS IN THE ASIA PACIFIC REGION 2–4 (C. Antons & R.M. Hilty eds., 2011), *available at* http://ssrn.com/abstract=1333431.

[265] For farsighted comments in this regard, see Tilman Lüder, remarks at the Workshop on Creation and Innovation, Seventeenth Annual Fordham Intellectual Prop. Law Inst. Conference, Cambridge, U.K., Apr. 15–16, 2009 (advocating urgent reforms of copyright law's limitations and exceptions to meet the needs of digital and computational science).

B. *Reconciling the Goals of Innovation Policy with the Needs of Science Policy*

In retrospect, it seems ironic that just as new technologies were producing significant breakthroughs in scientific research, and as digitally networked sites and other information technologies began empowering new models of collaborative investigation, innovation policies that should embrace these developments were instead using intellectual property rights to control or, in many cases, impede them. The use of public and private law to preclude access to basic knowledge resources, as well as knowledge-based goods, has increased the political and social burden of an intellectual property regime that, in theory, remains dedicated to the public interest of society at large.

Meanwhile, within the public-science community, efforts are underway to promote the formation of contractually constructed research commons (or semi-commons, as the case may be), that can flourish in an otherwise highly protectionist intellectual property environment.[266] If successful, the resulting infrastructure could help to maintain a steady flow of downstream research products and socially beneficial commercial applications that do respond positively to the incentives of intellectual property rights.[267] Given this transnational movement, what both the European Union and United States require is a long-term policy perspective that discriminates between the needs of the scientific community, operating within an emerging research commons that is increasingly capable of managing and integrating its own supplies of data and information,[268] and the needs of the downstream technology sectors, which depend on the traditional incentives of intellectual property law to translate scientific discoveries into commercial applications.[269]

The objective is to avoid pushing the exclusive rights that primarily govern those downstream incentives deep into the realm of basic science, where they will fracture and balkanize the research commons.[270] Needed instead are measures that broaden the research commons and enable it to operate both data science and computational

[266] For empirical examples of transnational initiatives, see below Chapter 9 ("Institutional Models for a Transnational Research Commons").

[267] BENKLER, n. 8, at 122–27; BRETT M. FRISCHMANN, INFRASTRUCTURE: THE SOCIAL VALUE OF SHARED RESOURCES (2012).

[268] *See generally* Reichman & Uhlir (2003), n. 15, at 416–56.

[269] *Cf.* Arti K. Rai et al., *Pathways Across the Valley of Death: Novel Intellectual Property Strategies for Accelerated Drug Discovery*, 8 *Yale J. Health Pol'y L. & Ethics* 1 (2008) (articulating a multi-firm, public and private collaboration model for research in small molecule drug discovery).

[270] *See, e.g.*, Charlotte Hess & Elinor Ostrom, *A Framework for Analyzing the Knowledge Commons, in Knowledge as a Commons*, n. 9, at 41, 41–82; Charlotte Hess & Elinor Ostrom, *Introduction: An Overview of the Knowledge Commons, in Knowledge as a Commons*, n. 9, at 3–26; Reichman, *How Trade Secrecy Law Generates a Natural Semicommons*, n. 102, at 185, 185–201. *See further* Chapter 9, Section 1 ("Theoretical Reflections on Designing a Knowledge Commons").

tools in digitally integrated, field-specific communities that span the world, smoothly and without disruption from domestic toll collectors waiving "IP" stop signs.

These projects will require more than tinkering at the edges of copyright law. They will depend on an overall vision, a willingness to remove obstacles to modern research methods, and a determination to fund the necessary operations as a basic component of public research infrastructure. Reforms on this scale will entail more than recognition of "users' rights,"[271] which denote important cultural interests and the public enrichment that ensues from access to literary and artistic works in general. Where science is concerned, information and data function as inputs to the process of discovery and thereby constitute an essential ingredient of future scientific progress.

Exclusive intellectual property rights do not provide the appropriate set of incentives in this upstream research space.[272] Policymakers should accordingly take pains to ensure that domestic and international intellectual property laws no longer undermine or impede the most promising opportunities that databases and automated knowledge discovery tools now make possible. These tools are critical for addressing the major social and environmental challenges of our time.

Making the internet safe for publishers of print media should no longer justify hindering the aggregation of public scientific information and data or the uses of digitally integrated research methods capable of analyzing them on a global scale. Rather, the task is to reconcile the historical values of intellectual property law with the modalities of our digital age, in order to reinforce the needs of scientific investigators operating under twenty-first century conditions, and to stimulate maximum public welfare payoffs from their new technological tools.

C. *Towards a Digitally Integrated Infrastructure for Microbial Literature and Data*

This chapter has briefly surveyed the legal and institutional obstacles impeding access to, and use of, scientific literature and data. The logical question becomes how to remove these obstacles with specific regard to microbiology, which is precisely the task undertaken in the next two chapters.

[271] *See, e.g.*, Rochelle C. Dreyfuss, *TRIPS – Round II: Should Users Strike Back?*, 71 *U. Chi. L. Rev.* 21 (2004) [hereinafter Dreyfuss, *TRIPS*].

[272] However, liability rules (also known as "take and pay" rules) may resolve many conflicts between incentives and user research needs that otherwise seem intractable. *See, e.g.*, Rai et al., n. 269, at 25–27; Jerome H. Reichman, *A Compensatory Liability Regime to Promote the Exchange of Microbial Genetic Resources for Research and Benefit Sharing*, *in Designing the Microbial Research Commons*, n. 11, at 43, 43–53; Reichman, *How Trade Secrecy Law Generates a Natural Semicommons*, n. 102, at 185–200; *see also* Rosa Castro, Ex Post Liability Rules in Modern Patent Law 47–56 (2010) (summarizing arguments for and against liability rules in patent law).

Chapter 7 surveys existing practices of the microbiology journals, and reports on the rising tide of open access publications in that field. We then examine the possibilities for redefining the role of publishing intermediaries altogether, with a view to enabling the scientific community in general, and the microbiology community in particular, to avoid both the legal and institutional constraints identified in this chapter.

Chapter 8 will then consider strategies for more fully exploiting data-intensive research opportunities in the digitally networked environment. After first surveying important public initiatives to promote early release of genomic data, we explore diverse networked sharing options to manage the deluge of microbial and other life science data. We then provide empirical examples of incipient Open Knowledge Environments, whose voluntary strategies to share publicly-funded digital knowledge assets afford a new approach that seems particularly well suited to the goals of the National Research Council's New Biology paradigm.[273]

[273] BIOLOGY FOR THE 21ST CENTURY, n. 4 *passim*.

7

Enabling the Microbiological Research Community to Control Its Own Scholarly Publications

I. RESPONSE OF THE SCIENTIFIC COMMUNITY TO RESTRICTIONS ON PUBLISHED RESEARCH RESULTS

So far, our focus on measures to make copyright and related laws more science friendly has operated on the premise that publishers would continue to play their traditional role in the process of disseminating research results. This very premise, however, makes it unlikely that the legislative or judicial reforms outlined in the previous chapter are attainable in either OECD or developing countries in the near future, despite growing attention to the conflict between intellectual property laws and the needs of science in the digital age.[1]

Notwithstanding the adoption of a text-mining exception to the United Kingdom's copyright law in 2014,[2] the lobbying power of the legacy publishers in this domain, particularly the large commercial STM publishing conglomerates, remains very strong. Concerns about protecting the interests of the entertainment and cultural industries continue to elicit proposal for stronger intellectual property laws at both the national and international levels, with little or no regard for their potentially deleterious effects on scientific research or the provision of other public

[1] *See, e.g.*, IAN HARGREAVES, DIGITAL OPPORTUNITY: A REVIEW OF INTELLECTUAL PROPERTY AND GROWTH 43 (2011) (demanding relief for science from intellectual property constraints as a fillip to economic growth); Press Release, Intellectual Prop. Office [U.K.], Consumers Given More Copyright Freedom (20 Dec. 2012), *available at* http://www.ipo.gov.uk/press-release-20121220 (proposing to allow noncommercial researchers to use computers to study published research results and other data without copyright law interfering); NAT'L ACADS. BOARD ON SCI., TECH. & ECON. POL'Y, COPYRIGHT IN THE DIGITAL ERA: BUILDING EVIDENCE FOR POLICY, at ix (Nat'l Acads. Press 2013), http://www.nap.edu/openbook.php?record_id=14686&page=R9 (last accessed 9 Apr. 2014) (stating the goal of the Board is "to expand and improve research on the impacts of copyright policy, particularly on innovation in the digital environment").

[2] The Copyright and Rights in Performances (Research, Education, Libraries and Archives Regulation) 2014, S.I. 2014/1372 (United Kingdom), available at http://www.legislation.gov.uk/uksi/2014/1372/contents/made (last accessed Feb. 21, 2015).

goods.[3] Whether reform efforts underway in some emerging economies may create a countervailing trend is impossible to predict,[4] but the benefits of such a trend – if it emerges – would likely play out over a lengthy period, and might not extend, at best, beyond certain regional alignments.

Science policymakers will, accordingly, have to evolve defensive measures of their own in order to neutralize interference from the default rules of copyright and database protection laws as they stand. Scientists, in turn, will increasingly need to manage their own upstream research assets as global public goods, sheltering them within a reinvigorated sharing ethos in the interest of a more productive downstream innovation system otherwise driven by the incentives of industrial property laws.[5]

In this chapter, we first survey the existing policies of microbial journal publishers. We then reevaluate the role that these intermediaries can play under existing institutional constraints. We conclude by asking if better solutions are not likely to emerge from a change of paradigm, in which the roles of outsourced intermediaries are either downgraded or abandoned altogether, and new or more pervasive open-access modes of dissemination, to be examined in Chapter 8, were to take their place.

[3] *See* Chapter 6, Sections III.A.3–4. See also Heather J. Ritch, European Research Infrastructure Consortiums: Privately Ordered and Publicly Funded Research Commons for Data 127 (unpublished S. J. D. dissertation, Duke University (on file with Goodson Library, Duke University) (citing Directmedia Publ'g GmbH v. Albert-Ludwigs-Universitat Freiburg [2009], 1 C.M.L.R. 7 (ECJ 4th Chamber)). At the same time, the book and newspaper copyright industries have entered a precipitous decline that is well documented.

[4] *Cf.* Amy Kapczynski, *Harmonization and Its Discontents: A Case Study of TRIPS Implementation in India's Pharmaceutical Sector*, 97 *Calif. L. Rev.* 1571 (2009); Rochelle C. Dreyfuss, The Role of India, China, Brazil and Other Emerging Economies in Establishing Access Norms for Intellectual Property and Intellectual Property Lawmaking 1–3 (Inst. for Int'l Law & Justice, Working Paper 2009/5, Pub. L. Research Paper No. 09-53, 2009), *available at* http://ssrn.com/abstract=1442785 (discussing the role of developing nations in the larger context of improving intellectual property laws) [hereinafter Dreyfuss Working Paper]; *see also* Graeme B. Dinwoodie & Rochelle C. Dreyfuss, *Designing a Global Intellectual Property System Responsive to Change: The WTO, WIPO, and Beyond*, 46 HOUS. L. REV. 1187, 1212 (2009) (remarking on how emerging economies devise different interest balancing methodologies from those in the developed world); Jerome H. Reichman, *Intellectual Property in the Twenty-First Century: Will the Developing Countries Lead or Follow?*, 46 HOUS. L. REV. 1115, 1118–19 (2009) (noting the pressures developing countries face to mimic the intellectual property legislation of OECD countries and the possibilities for exerting new leadership).

[5] *See generally* INTERNATIONAL PUBLIC GOODS AND TRANSFER OF TECHNOLOGY UNDER A GLOBALIZED INTELLECTUAL PROPERTY REGIME (K.E. Maskus & J.H. Reichman eds., Cambridge Univ. Press 2005); Jerome H. Reichman & Paul F. Uhlir, *A Contractually Reconstructed Research Commons for Scientific Data in a Highly Protectionist Intellectual Property Environment*, 66 LAW & CONTEM. PROBS. 315 (2003) [hereinafter Reichman & Uhlir (2003)]. For applications to patented research inputs, see Peter Lee, *Contracting to Preserve Open Science: Consideration-Based Regulation in Patent Law*, 58 EMORY L. REV. 890, 901 (2009), *available at* http://papers.ssrn.com/sol3/Delivery.cfm/SSRN_ID1698949_code366600.pdf?abstractid=1288183&mirid=1 (arguing that upstream patents on research tools in the biomedical arena may adversely affect downstream productivity).

II. SURVEYING THE PRACTICES OF THE MICROBIOLOGICAL JOURNALS

In the not too distant past, the standard publishing deal in the print media was a full subscription model in which the publisher acquired the copyright from the author and then prescribed the terms and conditions under which the author could use his or her own contribution to the journal. With regard to the public at large, the publisher's ability to restrict access, use, and reuse of any given article initially depended on the default rules of copyright law, which are "valid against the world." In other words, would-be users of copyrighted scientific articles would be subject to the publishers' exclusive rights of reproduction, adaptation, and distribution, among others, as narrowed by any applicable designated limitations and exceptions, including the fair use exception where available, with all the consequences discussed in the preceding chapter.

With regard to the authors of scientific articles, however, publishers could further restrict their ability to circulate, use, reuse or authorize reuse of their own articles by means of binding contractual provisions, backed up by ownership of the copyright, that overrode or otherwise reduced the limitations or exceptions generally available to third parties, including fair use. Under this system, carried to the limit, no published article could ever be made freely or openly available for virtually any purpose without the publishers' permission, subject only to nonwaivable exceptions statutorily created for public good purposes, such as libraries, education, and research, among others. Subscription publishers' restrictions were then further reinforced by collection societies that police and license third party uses, with the proviso that some STM publishers may price discriminate in favor of poor developing countries.

Once scientists began to rebel against these restrictions on the availability of their published research in the 1990s, after the World Wide Web opened up unprecedented sharing opportunities along the lines discussed in Chapter 6, they established the rudiments of what has subsequently become known as the "open access" movement.[6] This movement, in turn, spawned two main models for attempting to ease publishers' restrictions on access and use of scientific research, known as the "gold" and the "green" alternatives. Under the former, authors could seek to publish research results in an open access journal from the beginning, potentially subject to the payment of a substantial fee, which may or may not have been covered by the funders who sponsored the research project. Under the latter

[6] The origins of the "open access" movement can be traced back to Stevan Harnad's A Subversive Proposal. Stevan Harnad, A Subversive Proposal, Address at the Network Services Conference (NSC) in London, England (28–30 Nov. 1994). Under the leadership of Harold Varmus, current director of the National Cancer Institute, a series of declarations emerged to be known as the Budapest, Bethseda, and Berlin Declarations in the early 2000s. *See further* Chapter 8, Section II.

or "green" model, authors could bargain for the right to make a copy of their own article – often a preprint that the publisher had not yet edited – available to the public by means of self-archiving on a personal website or by depositing it in a public repository. Such deposits could be subject to embargo or delay periods agreed with the publisher, and over time, the exact conditions of the arrangement could, in part, be determined by agreements between research funders, such as the NIH, and the STM publishing community.[7]

To evaluate the implications of these models, one must first realize that the term "open access" is itself ambiguous and subject to interpretation. At the least it means that an article may be made available to the world (or some part thereof) for reading and private use, but it does not necessarily mean that the article is freed from the exclusive rights of copyright and related laws. Nor does it necessarily mean that the authors who make it available for "open access" are freed from either the default rules of copyright law or specific constraints that the publisher's contract imposed on them that further limit uses otherwise available under copyright law.

More to the point, the green-gold dichotomy arose before the advent of widespread private ordering mechanisms, notably Creative Commons licenses, that enabled copyright owners to waive the default rules of copyright law altogether; to impose standard terms and conditions for access, use, and reuse of their published research that are, in effect good against the world; and above all, to embody these so-called common use licenses in machine readable formats that search can discover automatically. These common use licenses are the real keys to unlocking published scientific literature for digitally integrated research

7 NIH policy preceded U.S. Congressional enactment in 2010 that mandates some versions of the 'green' approach for NIH-funded research, namely, a one-year maximum delay before deposit of a grantee's final article in the NIH's Pub Med Central repository. *See* NIH Public Public Access Policy implementing Division G, Title II, Section 218 of P.L. 110–161 (Consolidated Appropriations Act, 2008); U.S. Dept. of Health & Human Services, *Quick Facts About the NIH Public Access Policy*, Pub. Access NIH (March 2009), http://publicaccess.nih.gov/PublicAccessBrochure.pdf. *See also* Howard Hughes Medical Institute (HHMI), *HHMI Announces New Policy for Publication of Research Articles*, Press Release, June 26, 2007, *available at* www.hhmi.org/about/research/policies/html/#papp (funded research articles to be made freely available in a public repository within six months of publication); Wellcome Trust, *Position Statement for Open and Unrestricted Access to Published Research, available at*: http://www.wellcome.ac.uk/About-us/Policy/Spotlight-issues/Open-access/Policy/index.htm (last accessed May 23, 2015) (requires posting of funded research on public repositories within six months of publication; overrides conflicting journal policies by contract; and guarantees funding for page processing changes under open-access models; Research Councils (U.K.), *RCUK Policy on Open Access and Supporting Guidance* (2012), *available at*: http://www.rcuk.ac.uk/RCUK-prod/assets/documents/documents/RCUKOpenAccessPolicy.pdf (requires funded peer-reviewed papers to be made publicly available within six to twelve months of publication, with statement on how relevant data samples or models can be accessed). Open access mandates have also reportedly been established by the European Research Council, the Institut national de la santé et de la recherche medicale (France), and the Deutsche Forschungsgemeinschaft (DFG), Germany.

purposes, and the availability of such licenses increasingly renders the green-gold dichotomy misleading.

For example, an article may be published in an "open access" journal, but the journal itself may only allow access for purposes of reading the article subject to the default rules of copyright law.[8] If the open access journal does not also carry a common use license, or prevents the author from adopting such a license on his or her own website, then it is not in fact fully open access, even if the author had paid for the privilege of the gold model. More to the point, the article may not even be available for text mining and manipulation by automated knowledge discovery tools.[9]

Conversely, if the author retains the copyright and is not otherwise restricted by conditions imposed in the publisher's contract, then the author can make the article available on a fully open access basis by self-archiving or deposit, so long as the article itself carries a machine-readable common use license, such as a CC-BY license. In that event, however, the so-called "green" approach will actually have attained the result associated with a pure gold model, and it will be more fully open access as a practical matter than an article published under the gold model whose self-styled open access journal does not permit distribution via a machine readable common use license.

The moral of the story is that one must pay more attention to the particular licensing model actually adopted in any given case than to the green-gold distinction that has become increasingly obsolete in the face of private ordering. If a published scientific article is made openly available to scientific readers worldwide, subject to certain restrictions on use imposed by statute or contract, that is already a great improvement over the situation a decade or more ago. But the future of big science – especially the life sciences, including microbiology – depends not just on access, but on use and reuse of published research results by automated machine-readable techniques. Any residual restrictions that block the employment of common use licenses for public research purposes impede the full potential of the digitally integrated research methods discussed at the outset of Chapter 6.

[8] *See, e.g*, JSTOR's "Register & Read" program, which allows registrants to attain read-only access to a limited amount of journal content without payment. JSTOR, *Register & Read*, http://about.jstor.org/rr (last accessed 9 Apr. 2014).

[9] In fact, the major STM firms have surrounded their articles with high-tech electronic fences that can only be unlocked by further payments to use their own materials. When responding to the Hargreaves Review's enquiry about restrictions on access and use of scientific articles, STM publishers resisted a fair use approach in part by arguing that they had spent enormous sums to improve the technical delivery of their articles on line. They failed to add that this delivery was on a for-profit, extra-payment basis, which further charged scientists to use their own intellectual creations. *See* HARGREAVES, n. 1; Jerome H. Reichman & Ruth L. Okediji, *When Copyright Law and Science Collide: Empowering Digitally Integrated Research Methods on a Global Scale*, 96 U. MINN. L. REV. 1362, 1372–1457 (2012).

A. Contractual Provisions of Selected Leading Journals

With these premises in mind, we shall see that, empirically, the licensing conditions of microbiological publications may vary considerably from journal to journal, whether or not they are nominally open access. This situation may afford authors and institutions a choice, as to whether they can publish their findings in open access journals, or, at the very least, pay a fee up front to make an otherwise walled off article free to all.

Among the top-ten microbiology journals by ISI impact factor in 2010,[10] for example, only one, seventh ranked PLoS Pathogens, published by the Public Library of Science,[11] was a fully open access journal from the moment of publication, with authors retaining all rights in the piece. PLoS Pathogens requires that authors publish under a Creative Commons CC-BY license, which allows authors to retain copyright in their works, but permits full free public access subject to reasonable attribution rights.[12] Freedom, however, is not gratis because authors (or their institutions) must pay publication fees of $2,250 up front to ensure open access.[13]

Two other top-ten journals allowed authors to pay a fee at the time of publication to "unlock" their content for immediate "free access," but the meaning of "free access" varied considerably with specific provisions of the contract. For example, Oxford Press, which publishes Clinical Infectious Diseases, allows authors to retain copyright, but it contractually takes the exclusive rights of reproduction and distribution unless authors opt to pay for an open access condition. If authors agree to pay $3,000 in developed countries, or $1,500 in developing countries, Oxford will make the article freely and immediately available under a common use license, namely a Creative Commons Attribution and Non-Commercial Use License (CC-BY-NC). Oxford's clarity in this regard is commendable, as is its policy of allowing users in poor countries free access to all articles. However, a $3,000 fee is steep, and very few authors had actually exercised this option at the time of writing.

In contrast, the FEMS Microbiology Reviews (published jointly by the Federation of European Microbiological Societies and Wiley Blackwell) will also allow authors to purchase an "open access" option for $3,000, but they must transfer the copyright to the publishers. On closer examination, the nominally open access condition

[10] *Journal Impact Factors 2010 – ISI and SJR*, Science Tech Blog (Aug. 1, 2011), http://sciencetechblog.com/2011/08/01/journal-impact-factors-2010-isi-and-sjr.

[11] *About PLoS Pathogens*, PLoS Pathogens, http://www.plospathogens.org/static/information (last accessed 9 Apr. 2014).

[12] *About the Licenses*, Creative Commons, http://creativecommons.org/licenses/ (last accessed 9 Apr. 2014).

[13] *PLoS Publication Fees*, PLoS.org, http://www.plos.org/publish/pricing-policy/publication-fees/ (last accessed 9 Apr. 2014).

turned out to yield only an immediate read only permission with respect to the final published version. All other rights were reserved by the publisher, with some limited allowance for the authors' reuse for educational purposes that are available to all authors. Those authors who do not opt for this limited read-only condition may nonetheless post their submitted non peer reviewed manuscript online – after publication and also on a read-only basis. For NIH grantees, the publisher will post the peer reviewed, accepted (but not final) version on PubMedCentral twelve months after publication. In short, payment of the $3,000 fee only bought the author the right to an immediate posting of the final published version on a read-only basis; but – unlike Oxford – no version could be made available with a common-use license under any circumstances despite payment of the so-called open access license.

The remaining seven top-ten microbiological journals made no provision for the purchase of an open access option at the time of our survey in 2009. Nor would any of them allow authors to make any version available under a common-use license at that time. However, with respect to read only access, these seven journals varied considerably.

For example, the most flexible in this respect was Annual Reviews of Microbiology, which allowed authors to post and share a free link to the final published version on a personal or institutional website immediately after publication. Although this permission enables nothing more than read-only usages, it was the only other top-ten journal whose read-only version was the final, peer-reviewed, error-free incarnation. Given that one does not pay for this option, it is arguable a better deal than any of the others "read only" deals. However, this is an extremely selective journal to begin with, and because the link is only to its own website, would-be users may not necessarily find it through an ordinary online search.

All six of the remaining top-ten journals as of 2009 allowed posting only of accepted or submitted versions for read-only purposes, with varying degrees of embargo periods. They also allowed some uses by the author for personal or educational purposes. Although some access is better than none at all, proponents of open access and open transmission of research results might well argue that access rights only to early versions of articles runs the risk of propagating uncaught inaccuracies that were later removed in peer review, as well as inferior presentations of pertinent data in tables and charts. Furthermore, while free Adobe PDF readers are readily available on the Internet, free readers of documents in Microsoft.DOC or other formats may not be, especially in developing countries.

The upshot here is, first, retention of copyright means little in itself if the publisher takes the author's exclusive rights of reproduction and distribution by contract. Second, without a common-use license in hand, the term "open access" – whether purchased or not – may mean very little in terms of enabling digital manipulation by automated knowledge discovery tools. Third, the complexity and diversity of the contracts adopted by top-ten publishers was so great that even trained lawyers had some difficulty deciphering the precise meaning of their terms. These contracts discourage scientists untrained in the law from making informed decisions concerning their rights or from demanding the rights they might otherwise desire.

Nevertheless, as of 2009, all top-ten microbiological journals allowed at least some read-only access to all their articles in one format or another, before or after publication, which is a step beyond the old, closed subscription models. The extent to which authors who are not so obliged by funders actually make use of even this limited option still remains to be seen. It follows that the one certain way to ensure fully open access treatment with a machine-readable common use license was to publish in a fully open access journal. Happily, the results of our empirical survey showed that a surprisingly large number of fully open access journals already existed at the time of writing, with the movement in that direction appears to have gathered strength in a few short years.

B. Results of Our Broader Survey

In July 2009, we surveyed a total of 303 journals dedicated in whole or in part to microbiology research results.[14] Because we were primarily concerned to discover how many journals in the field of microbiology fully enable the application of digital research tools, our empirical survey declined to accept the labels that the relevant journals themselves adopted. Rather, we attempted to pierce the veil in order to determine the underlying functional realities regardless of the label that was given. Among other goals, this methodology enabled us to better determine the actual extent to which microbial journal publishing remains subject to both contractually imposed limitations on digital research and to science-impeding restrictions flowing from intellectual property laws.

Following this functional approach, we identify three broad categorize in which to classify the published microbial literature. These categories are: (1) fully open access journals; (2) intermediate access journals, and (3) closed subscription journals.

By designating a journal as "Fully Open Access," we mean one that allows any use or reuse of the published materials with the exception that attribution be provided and, with some reluctance, that commercial uses require express consent of the copyright holder.[15] In adopting this notion of "open access," we also recognize that

[14] The authors have a list of microbiological journals on file.

[15] We do not think a commercial/noncommercial science distinction is meaningful in this context, but its adoption does not unduly hinder digital exploitation for research in the public sector, although it can affect private sector research.

some of the hybrid journals (especially those published by Oxford and the Springer Group) allow authors to purchase a fully open access option. In that event, the specific article so optioned in that journal will be made freely available for all scientific uses, as if it had been published in a truly open access journal, even though other articles in that same issue will remain restricted. In a legally appropriate case, in other words, we treat truly open-access publications and hybrid open-access publications as functionally equivalent, as shown in Table 7.1.

The second category of "Intermediate Access Journals" may also be subdivided into two subcategories. First, there was a group of self-styled "open access" journals in our survey that were open only in the sense that scientists might freely read them, but did not entitle these readers to extract, use, or reuse the contents beyond what underlying statutory intellectual property laws (including database protection laws) permit. Also included in this subcategory were journals whose publishers chose to apply a "no derivative work" version of the Creative Commons licenses, for the reason that this license may effectively impede most unauthorized forms of digital manipulation (as well as other follow-on applications).[16]

The second subcategory of "Intermediate Access Journals" consists of those hybrid journals that imposed the same set of limitations as the self-styled open access journals just mentioned. The point is that these two subcategories are functionally equivalent, despite their different labels ("open access" versus "hybrid"). More important, however, the amalgamation of the two functionally equivalent subcategories yielded a more precise estimate of the number of microbiological journals whose research use conditions actually depend on the statutory intellectual property laws applicable in different jurisdictions.

The final category of "Closed Subscription Journals" consists of journals whose publishers normally restricted access, even for reading purposes, to paying customers, with some exceptions as noted later. These journals typically applied other contractual restrictions for research uses of their contents or channeled research uses through proprietary outlets.

In what follows, we present the results of our empirical survey of microbiological journals subdivided according to the three main categories outlined earlier. To this end, we assessed the copyright and access policies of publishers responsible for journals containing primary research articles and reviews in the field of microbiology and related areas, such as immunology. We also selected more general science journals that regularly publish articles in the field of microbiology.[17]

[16] These are CC-BY-ND licenses; it is CC-BY plus no derivatives. Our classification here is partly determined by the still uncertain legal implications of a no derivative work clause in this context. *See About the Licenses*, n. 12.

[17] The authors have the list of microbiological journals surveyed on file.

Most of the open access journals were obtained from the Directory of Open Access Journals (DOAJ) and from individual publisher websites, such as Horizon Press. The hybrid and subscription journals were selected primarily from the publisher websites and a few other web resources. Sixty-four percent of the selected journals included articles about microbiology only, while the remaining journals publish articles from other research areas as well, along with a significant number of microbiology-related articles.[18]

1. A Growing Number of Open Access Microbiology Journals

Table 7.1 shows the distribution of these journals according to the three functional categories identified in the preceding section. Table 7.2 provides the distribution in these categories of journals that publish only microbial results as distinct from more general scientific journals that regularly publish some articles on this topic.

Column 1.0 of Table 7.1 shows that almost a third of these journals were either fully open access or enabled individual authors to purchase an immediate, fully open access option.[19] Considering that the open access movement has only gained traction in the past decade, one can say that a surprisingly large number of existing microbiology journals had already taken serious steps to enable digitally integrated research.

To be sure, 69 of the 97 journals characterized as "Fully Open Access" in Table 7.1 are in fact only hybrid journals that enabled their authors to purchase full open access status for a substantial fee and that otherwise applied restrictive conditions to the rest of the articles for at least a certain period of time.[20] Nevertheless, it remained theoretically possible for all of the authors contributing to all 69 of these journals to opt for full open-access status, assuming they could marshal the necessary funds.[21] In other words, once the funding becomes available, the decision to adopt a fully open access policy in these journals depends on the authors and not on the publishers.

Column 2.0 of Table 7.1 also shows that another 20 percent of existing microbiology journals fell into the category of "Intermediate Access," which means that they provide

[18] We recognize that including these journals somewhat skews the picture of "microbiology" itself; but, as the same time, the importance of some of these journals to microbiology induced us to err on the side of inclusion.

[19] Recall that open access here means attribution only and, in some cases no commercial use, but otherwise allows full access, use, and reuse of the material. Note also that these data were six years old at the time of their publication here.

[20] As of the time of writing, most of these journals were published under Springer's Open Choice Option and some others were published by Taylor & Francis' I Open Access and by Oxford University Press' Oxford Open. For a more comprehensive list of hybrid journals, see *Publishers with Paid Options for Open Access*, SHERPA/RoMEO, Univ. Nottingham, June 19, 2013, http://www.sherpa.ac.uk/romeo/PaidOA.html.

[21] On funding, see Section III.B.

TABLE 7.1. *Classification of journals*

	1.0 Full Open Access		2.0 Intermediate Open Access		3.0 Restrictive Subscription	Total
	31.91%		20.07%		48.03%	
	Open Access	Hybrid-Open Access	Open access-Read only	Hybrid-Read Only	Subscription	
Number of Journals	28	69	22	38	146	303
Percentage	9.21%	22.7%	7.24%	12.83%	48.03%	100

TABLE 7.2. *Classification based on research specialization*

	1.0 Full Open Access		2.0 Intermediate Open Access		3.0 Restrictive Subscription	Total
	Open Access	Hybrid-Open Access	Open access-Read only	Hybrid-Read Only	Subscription	
Microbiology only	24	40	17	32	80	193
Microbiology only (Percentages)	7.89%	13.16%	5.59%	10.53%	26.32%	63.48%
General Biomedical Journals	4	29	5	6	66	110
General Biomedical Journals (Percentages)	1.32%	9.54%	1.64%	2.3%	21.77%	36.51%

open availability of the published results to any readers as well as full text search capacity without any delay periods. This category also revealed noteworthy progress in promoting the dissemination of research results, even if these journals still left further use and reuse of published materials to the disposition of intellectual property laws.

In contrast, column 3.0 of Table 7.1 shows that 146 journals, or almost 50% of the field, were subscription journals that still imposed access restrictions in the online environment as of 2009, in addition to restrictions on use and reuse of contents

TABLE 7.3. *Journal citation and peer review*

	1.0 Full Open Access		2.0 Intermediate Open Access		3.0 Restrictive Subscription
	Open Access	Hybrid-Open Access	Open access-Read only	Hybrid-Read Only	Subscription
Impact factor* (average)	4.02	1.91	2.31	2.99	5.77
Impact factor (range)	2.91 to 9.125	0.443–5.97	1.013 to 2.3	0.7 to 9.38	0.56 to 50.017
Peer review (Showing percentage of journals)	92.86%	100%	68.18%	100%	97.95%
Impact factor official (percentage of journals)	28.57%	76.81%`	22.73%	89.74%	83.56%
Unofficial impact factor (percentage of journals)	14.29%	0%	9.09%	0%	15.06%
No impact factor found (ISI index) (percentage of journals)	57.14%	23.19%	68.18%	10.25%	1.369%

* The impact factor is calculated for journals that have been indexed by the Institute for Scientific Information.

for public research purposes. A small percentage of these journals did allow some form of author self-archiving or the possibility of making their contents available for reading online after some specified period of time.[22]

Table 7.3 also reveals that most of the journals in the "Restrictive Subscription" category had higher Institute for Scientific Information (ISI) impact factors than the journals in columns 1.0 and 2.0. This finding requires some nuancing however. First, and probably most important for present purposes, the ISI calculus does not

[22] *See* Section II.B.2.

TABLE 7.4. *Showing percentage of journals allowing self-archiving on author website and archiving in institutional repositories*

	1.0 Full Open Access		2.0 Intermediate Open Access		3.0 Restrictive Subscription
	Open Access	Hybrid-Open Access	Open access-Read only	Hybrid-Read Only	Subscription
Self-Archiving on personal websites (immediately)	100%	100%	100%	82.05%	67.81%*
Self-Archiving on personal websites (with lag)	0%	0%	0%	2.56%	2.05%**
Self-Archiving on personal websites (not allowed)	0%	0%	0%	0%	19.86%
Self- Archiving on personal websites (not specified)	0%	0%	0%	15.38%	9.59%

* Elsevier allows author self-archiving of preprint version of manuscript.
** Nature Publishing group allows self-archiving of final version six months after publication.

cover journals that are less than five years old, which eliminated a lot of open-access microbiology journals. Likewise, at the time of writing, the ISI Index did not cover any foreign language journals, nor did it track downloads, which are indicative of interest and potential impact. Nevertheless, the subscription journals remain generally better established than the open-access journals, and they still tend to enjoy greater prestige within the immediate research subcommunities.

There are three main types of online locations for authors to self-archive their works – on personal websites, in open institutional repositories (repositories containing an institution's work), and in open thematic repositories (repositories created by research institutions for self-archiving information pertaining to a certain subject, for instance PubMedCentral for biomedical research).[23] Table 7.4 shows that approximately 70% of the journals in the Restrictive Subscription Journals category

[23] *PMC Overview*, U.S. NAT'L INST. HEALTH'S NAT'L LIBRARY OF MED. (NIH/NLM, 14 Nov. 2011), http://www.ncbi.nlm.nih.gov/pmc/about/intro/. Here is a list of open access repositories, listed alphabetically by country, then institution: http://www.opendoar.org/countrylist.php. Here is a list of open-access thematic/disciplinary repositories: http://oad.simmons.edu/oadwiki/Disciplinary_repositories (last modified July 10, 2013).

TABLE 7.4A. *Self archiving by authors publishing in hybrid journals*

	1.0 Hybrid Full Open Access		2.0 Intermediate Open Access	
	Open Access Option	Subscription Option	Open Access Option	Subscription Option
Self-Archiving (immediately)	100%	2.89%	82.05%	7.69%
Self-Archiving (with lag)	0%	8.69%	2.56%	15.38%
Self-Archiving not allowed	0%	88.4%	0%	78.94%
Self-Archiving not specified	0%	0%	15.38%	0%

All values show percentages of journals in the subgroup.

did at least allow authors to self-archive their microbiology research publications on their own personal websites, almost all of them without any time lag.[24] But this also means that almost one third of the subscription journals did not allow even this basic option.

The data presented in Table 7.4a demonstrates that most of the hybrid journals that allowed either a full or intermediate open-access option to be purchased (see Table 7.4, columns 1.0 and 2.0) discouraged any form of self-archiving, except via the purchase of the full open-access option. This practice is consistent with the policy of these journals to encourage author buy-outs, but is inimical to the benefits that an open-access policy otherwise provides.

Almost all the subscription journals prohibited deposits of their articles in authors' institutional repositories, as the results in Table 7.5 show. Interestingly, however, nearly all of the subscription journals in this field did not specify whether authors might or might not deposit articles in external repositories, as appears from Table 7.6, even though these same publishers expressly prohibit such deposits in the authors' own institutional repositories. We surmise that this discrepancy might result from the funding agencies' growing mandates requiring grantees to deposit research results in external repositories, such as PubMed Central or UK PubMed Central.

[24] *See* Table 7.4.

TABLE 7.5. *Percentage of journals that allow archiving in institutional repositories*

	1.0 Full Open Access		2.0 Intermediate Open Access		3.0 Restrictive Subscription
	Open Access	Hybrid-Open Access	Open access-Read only	Hybrid-Read Only	Subscription
Deposit institutional repository (immediately)	100%	100%	100%	82.05%	0.68%
Deposit institutional repository (with lag)	0%	0%	0%	2.56%	2.05%
Deposit institutional repository (not allowed)	0%	0%	0%	0%	87.67%
Deposit institutional repository (not specified)	0%	0%	0%	15.38%	9.59%

TABLE 7.6. *Showing percentage of journals that allow archiving in external repositories*

	1.0 Full Open Access		2.0 Intermediate Open Access		3.0 Restrictive Subscription
	Open Access	Hybrid-Open Access	Open access-Read only	Hybrid-Read Only	Subscription
Deposit external repository (immediately)	100%	100%	100%	74.35%	0%
Deposit external repository (with lag)	0%	0%	0%	0%	10.27%
Deposit external repository (not allowed)	0%	0%	0%	2.56%	1.37%
Deposit external repository (not specified)	0%	0%	0%	23.07%	87.67%

All values show percentages of journals in the subgroup.

2. Self-Archiving by Authors Who Publish in Subscription Journals

Although many journals in this field allow authors to self-archive on their personal websites, this approach entails both advantages and disadvantages from a public interest perspective. Self-archiving on an author's personal website does at least have the advantage of providing immediate access at no charge to the world at large. A prepublication draft archived on a website is also the least objectionable option to the publishers and is one way to ensure additional access for at least read-only purposes.

There are many disadvantages, however, to this approach. Personal websites are not permanent and are typically much more difficult to find by search engines online. This results in the balkanization of the literature as a result of individualized access points, keeping the articles disaggregated, and there is no integration through a searchable, user-friendly, and all-encompassing internet portal (e.g., a single access point). To the extent that the default copyright rules also apply, there is no legal guarantee that users can make one or more copies for educational or research purposes, even in informal exchanges.[25]

An institutional repository is a considerably better option, because it facilitates a larger aggregation of information and is typically more permanent than an individual website. It also helps to standardize the descriptive metadata needed to make the article easier to find and search. As the time of writing, there were 3,924 open institutional repositories registered on the Southampton University Directory of Open Access Repositories,[26] although one can assume there are many more worldwide that are either not self-registered or that are in the process of being created. Many funding organizations are promoting the establishment of open institutional repositories.[27]

[25] *See* Assn. College & Research Libraries (ACRL) Scholarly Commc'n Comm., *Principles and Strategies for the Reform of Scholarly Communication 1*, ACRL ¶ 5, *available at* http://www.ala.org/acrl/publications/whitepapers/principlesstrategies. However, in informal exchanges most users presume that authors will not assert their full copyright, because it is in the authors' interest that their work be used and distributed. In a formal context, such as the use and distribution of self-archived and prepublication works in the classroom, this will probably apply as well. *See* Rollin White, *A Technology Frontier*, OPENDOC 101, http://web.archive.org/web/20071009214704/http://www.sundialsystems.com/articles/opendoc.html (last accessed Feb. 13, 2015).

[26] *See Homepage*, REGISTRY OF OPEN ACCESS REPOSITORIES, http://roar.eprints.org/ (last accessed 19 Feb. 2015). This is an increasingly used mechanism by institutions of higher learning. For example, all Dutch universities and almost all German universities have set up such open repositories.

[27] USpace Inst. Repository, *About Institutional Repositories*, UNIV. OF UTAH, Mar. 7, 2012, http://uspace.utah.edu/aboutIR.php.

TABLE 7.7. *Showing percentage of journals that allow author to keep the copyright*

	1.0 Full Open Access		2.0 Intermediate Open Access		3.0 Restrictive Subscription
	Open Access	Hybrid-Open Access	Open access-Read only	Hybrid-Read Only	Subscription
Copyright with Author	75%	91.3%	18.18%	2.56%	0%
Copyright with Publisher	17.86%	8.7%	68.18%	97.44%	97.95%
Copyright (not specified)	7.14%	0%	13.64%	0%	2.05%

All values show percentages of journals in the subgroup.

While the open institutional repositories are a much better location for authors to self-archive their articles than a personal website, the material on such sites may remain thematically disaggregated and not always easy to find. Subscription journal publishers were also much less likely to allow such deposits, in contrast to personal websites as our study confirmed (see Table 7.5).[28]

A thematic external repository is the preferred option because it is both large-scale and permanent, yet more readily searchable and integrated with other information of the same kind. Some of the microbiological journal literature is deposited in PubMedCentral, especially under grantee mandates where applicable.[29] For these very reasons, however, many subscription STM publishers often discourage or prohibit deposits in such repositories.

3. Disposition of Copyrights

Table 7.7 shows the disposition of copyright ownership across the three main categories set out in Table 7.1. While the fully open-access journals in column 1.0 may logically allow authors to retain copyright in most cases, the read-only access journals in column 2.0 and the subscription journals in column 3.0 almost always require an assignment of copyright. Such assignments enable the publishers to

[28] *See* text and accompanying nn. 17–23.

[29] Unfortunately, there is no integrated thematic repository for microbiological literature for a number of reasons, including the fact that there is no major single funder for microbiology research in each country.

TABLE 7.8. *Open access publishing charges to the author per article*

	1.0 Full Open Access		2.0 Intermediate Open Access	
	Open Access	Hybrid-Open Access	Open Access-Read Only	Hybrid-Read Only
Average (in dollars)	\$1004.48	\$2992.70	–	\$2815
Range (in dollars)	\$0 to \$3250	\$1500 to \$3250	–	\$978 to \$3000

* No open access charge to author is specified.

limit digital use and reuse to the narrow confines of default intellectual property laws, or more typically, to further restrict them by contract, as discussed later in Section C.

Ownership of the copyright also enables the publisher to dictate self-archiving rules, subject perhaps to grantee mandates, and even to restrict the original author's possibilities for use and reuse of their own research. More generally, the publishers' control of the relevant copyrights allows them to resist future progressive open-access developments and related technological improvements, at least to the extent that codified changes in intellectual property laws do not otherwise provide.[30] This said, however, retention of copyrights by the authors means little if the publishers' contract nonetheless requires the same exclusive rights and conditions equivalent to those they would have imposed if they technically owned the copyrights.

4. Costs of the Open-Access Option

Table 7.8 shows that the average costs to authors publishing in fully open access microbial journals vary considerably with the overall policies of the journals in question. To the extent that a journal publishes all its contents in the fully open-access format, average costs in 2009 were about \$1,004 per article and in the range of zero dollars to \$1,470 per piece. In contrast, the hybrid journals, whether fully open access (column 1.0) or read only (column 2.0), tended to charge considerably higher costs (viz., an average price of around \$3,000 per piece). These prices fell within a range of about \$1,000 to \$3,250.

The bulk of these hybrid journals were for-profit in nature (see Table 7.8(a)), which to some extent explains their higher costs. They may also incur higher

[30] For the difficulties in changing the applicable intellectual property laws to support digitally integrated science, see Reichman & Okediji, n. 9.

TABLE 7.8A. *Percentage of journals published by for-profit or nonprofit organizations*

	1.0 Full Open Access		2.0 Intermediate Open Access		3.0 Restrictive Subscription
	Open Access	Hybrid-Open Access	Open access-Read only	Hybrid-Read Only	Subscription
For profit	53.57%	97.1%	13.64%	79.48%	73.97%
Nonprofit	39.29%	2.9%	68.18%	20.51%	19.86%
Not specified	7.14%	0%	9.09%	0%	2.74%

All values show percentages of journals in the subgroup.

editorial and other expenses, although we have gathered no evidence about this question. Nevertheless, one may ask why many of the hybrid journals charge so much more than most of the open-access journals, considering that the former will presumably profit additionally from the subscription charges. Perhaps part of the answer reflects the relatively constrained subscription prices of microbiology related journals offered to institutions generally (which averaged $1,600 to $2,000 annually in 2009, according to Table 7.9).

In any event, Springer at least has reportedly pioneered the practice of reducing subscription rates to libraries as the volume of open-access contents increases over time.[31] We also lack data bearing on the relative editorial and other production and marketing costs in the different formats. Nevertheless, one may question the basis for charging an author some $2,800 when the only purchasable option is a "read only" license (with perhaps downloads) and no license to extract or reuse the contents for follow-on research purposes (Table 7.8).

5. Postscript

As noted at the outset, the foregoing survey, was conducted in 2009. Because we lacked funding to update the survey, it provides a snapshot in time that reflects the status at that time, but does not show more recent developments. Nevertheless, the survey revealed that more microbiology related journals are adopting open-access

[31] *See Springer Open Choice. – Your Research. Your Choice*, SPRINGER, *available at* http://www.springer.com/open+access?SGWID=0-169302-0-0-0 (last accessed 9 Apr. 2014). *See also* Jennifer Howard, *Springer Announces New Open-Access Journals*, THE CHRONICLE OF HIGHER ED.: WIRED CAMPUS (June 28, 2010), http://chronicle.com/blogs/wiredcampus/springer-announces-new-open-access-journals/25156.

TABLE 7.9. *Annual subscription prices of journals*

	1.0 Full Open Access		2.0 Intermediate Open Access		3.0 Restrictive Subscription
	Open Access	Hybrid-Open Access	Open access-Read only	Hybrid-Read Only	Subscription
Issues/year (Average)	6.53	6.55	3.13	11.23	13.23
Issues/year (Range)	1 to 52	1 to 52	1 to 12	1 to 52	1 to 52
Individual subscription (average)	n/a	$971.2	n/a	$258.46	$238.78
Institutional subscription (range)	n/a	$501 to $1730	n/a	$77 to $2254	$30–2,173
Institutional subscription (average)	n/a	$1638.86	n/a	$2038.77	$1710.29
Institutional subscription (range)	n/a	$139 to $6028	n/a	$190 to $7819	$30–7311

policies. However, what journal publishers mean by claiming "open access" varies considerably from case to case.

A large number of *soi disant* "open access" microbiology journals – some 20% in fact – still remained subject only to any limitations and exceptions favoring scientific research that the applicable intellectual property laws in their jurisdictions may have provided.[32] In other words, these journals had not technically waived the application of domestic copyright and database protection laws, which arise automatically where applicable, despite claiming some form of open-access policy.

The end result may thus depend on any contractual riders adopted in connection with a specified open-access regime. In practice, many of these journal publishers contractually override some or most of the science-friendly limitations and exceptions that are provided by statute, thereby tightening the publishers' control over research uses nominally allowed by such laws even for articles published under an open-access arrangement.

[32] Table 7.1, column 2.0.

TABLE 7.10. *Developing country provisions*

	1.0 Full Open Access	2.0 Intermediate Open Access			3.0 Restrictive Subscription
	Open Access	Hybrid-Open Access	Open access-Read only	Hybrid-Read Only	Subscription
Discounts to authors in developing countries	39.29%	2.9%	0%	33.33%	4.11%
No discounts to authors in developing countries	60.71%	97.1%	100%	66.67%	95.89%

All values show percentages of journals in the subgroup.

It also bears emphasizing that, as of 2009, nearly 50 percent of all the relevant journals we surveyed lacked any open-access policy whatsoever.[33] Moreover, these same journals normally protected their contents behind electronic fences,[34] which authorized them to impose the strictest contractual overrides that may impede even fair uses otherwise available from United States copyright law.

III. REDEFINING THE ROLE OF PUBLISHING INTERMEDIARIES UNDER CURRENT INSTITUTIONAL CONSTRAINTS

The foregoing survey showed that there is considerable movement towards more open-access publishing methods and considerable resistance from STM publishers to this movement, despite grudging acquiescence to pressures from funders in the direction of more open-access content. A recent statistical survey of scholarly journals confirmed this trend. It showed that some 660 institutional or funder open-access policies were in effect as of March 2015,[35] and that over half of these policies were mandatory.[36]

[33] Table 7.1, column 3.0.

[34] For the legal significance of electronic fences, see Chapter 6, Section II.B. ("Digital Locks and Contractual Overrides in the Online Environment").

[35] *See* Alma Swan et al., *Working Together to Promote Open Access Policy Alignment in Europe*, Pasteur4OA Work Package 3 Report: Open Access Policies (2015), *available from* http://eprints.sofon.ac.uk/375854/.

[36] *Id.*

The STM publishers' resistance to this trend stems from their efforts to preserve their existing market interests, which depend on their role as intermediaries. This tension in turn, re-proposes the fundamental question raised earlier in this study, namely, should scientific publishers' customary interests be preserved at the expense of scientists' need for wholesale access to, and reuse of the expanding universe of scientific literature and data.

We previously argued that the cost of continuing to rely on commercial publishers now considerably outweighs the value of their contributions. In what follows, we consider different proposals about how to break away from these customary practices and redefine the relationships between scientists as authors and scientists as users of their own research results in a digitally integrated universe of discourse.

Given the diminished costs incurred by today's intermediaries in the online environment, and the shrinking amount of added value they contribute,[37] one questions the entitlements they should be allowed to claim for secondary uses of published scientific research results in either the print media or the online environment, and how such claims should be implemented when recognized. At bottom, science publishers provide measures to maintain quality assurance and control, marketing and distribution, plus certain technical services that the research community could provide for itself, yet typically does not in rich countries, perhaps because of inertia. The reputational benefits of primary importance to authors accrue from the peer-review function that is largely provided gratis by other scientists. Disregarding promotion and publicity, the intermediaries' utility stems from editing, maintaining, and updating electronic collections, possibly also from electronic indexing of these collections, and possibly from the provision of other technical services needed to make embedded data and information available on request.[38]

[37] As providers of digital services, publishing intermediaries increasingly resemble the Red Hat Corporation, which provides services to users of Linux Software but does not control the rights to Linux. Robert Young, *Giving It Away: How Red Hat Software Stumbled Across a New Economic Model and Helped Improve an Industry*, 4(3) *J. Elec. Pub.* (Mar. 1999), *available at* http://quod.lib.umich.edu/cgi/t/text/text-idx?c=jep;view=text;rgn=main;idno= 3336451.0004.304; *see also* NAT'L RESEARCH COUNCIL, BITS OF POWER: ISSUES IN GLOBAL ACCESS TO SCIENTIFIC DATA 111–13 (Nat'l Acad. Press 1997) [hereinafter BITS OF POWER]. However, the science publishers insist that they actually contribute more services than are identified in the text and at considerably greater investment costs than are recognized in the text. *See, e.g.*, Letter from Michael Mabe, CEO, Int'l Assoc. of Scientific, Technical & Med. Publishers, to Copyright Review, Dep't. of Jobs, Enter. & Innovation, Dublin, Ireland (July 14, 2011) [hereinafter Letter from STM Publishers]. The question begged is whether these investments actually benefit research science or merely ensure greater profits to publishers under restrictive copyright laws.

[38] *See* Young, n. 37; Letter from STM Publishers, n. 37. *But cf.* BITS OF POWER, n. 37, at 111–13 (discussing the ways in which the price imposed by private intermediaries for these services is "countercultural" to scientific communities in which "exchange is not monetized but depends on social norms specifying expected and well-understood levels of contribution").

Although science publishers must necessarily charge for these services, funding agencies should, and increasingly do, ensure that government-funded research results remain freely available in public or private repositories, so that to defray these costs, users could perform the needed technical services on their own.[39] Such a policy also serves to attenuate the problems of sole-source providers, who monopolize public science and can pose serious challenges for digitally integrated scientific research.[40]

We recognize that publishers must charge for their technical services and need not extend to endowing them with exclusive rights to downstream uses or reuses of the scientific product they make available. On the contrary, the proprietary restrictions that such rights enable intermediaries to impose in the name of authors' rights, without any palpable authorial contribution, are inconsistent with both the needs of science and the principles of sound exceptions to copyright and database laws, as expounded earlier in Chapter 6.

Because publishers of scientific journals depend, in the first instance, on contractual relationships with the learned societies (or other sponsors of academic journals), these regulatory adjustments can be achieved by contract, without need of legislation.[41] For example, institutional mandates can restrict the transfer of copyrights in publicly funded research results and require that such results be made available in appropriate repositories. Funders of scientific research – whether government agencies, foundations, or academic institutions – should insist on open-access publishing conditions as part of the grant-making process.[42]

39 *See* HARGREAVES, n. 1; Victoria Stodden, *Open Science: Policy Implications for the Evolving Phenomenon of User-Led Scientific Innovation*, J. Sci. Comm. 2–6 (Mar. 22, 2010), *available at* http://jcom.sissa.it/archive/09/01/Jcom0901%282010%29A05/Jcom0901%282010%29A05.pdf (discussing National Science Foundation (NSF)Guidelines, Creative Commons licensing and proposing a new standard contractual template of her own); Michael W. Carroll, *Complying with the National Institutes of Health Public Access Policy: Copyright Considerations and Options*, SPARC/SCIENCE COMMONS/A-RL WHITE PAPER (Feb. 2008), *available at* http://www.arl.org/sparc/bm~doc/NIH_ Copyright_v1.pdf; U.S. Dep't of Health & Human Servs., *Quick Facts About The NIH Public Access Policy* (2009), n. 7

40 *See, e.g.*, HARGREAVES, n. 1; Jerome H. Reichman & Paul F. Uhlir, *Database Protection at the Crossroads: Recent Developments and Their Impact on Science and Technology*, 14 BERKELEY TECH. L.J. 793, 799–812 (1999) [hereinafter Reichman & Uhlir (1999)].

41 For example, universities and publishers have negotiated six to twelve months embargo periods giving the latter a term of exclusivity before articles are deposited in open access repositories. *See, e.g.*, Jorge L. Contreras, *Data Sharing, Latency Variables, and Science Commons*, 25 BERKELEY TECH. L.J. 1601, 1616, 1654 (2011) [hereinafter Contreras, *Data Sharing*] (labeling this practice as "knowledge latency"). For a recent analysis, see Jorge L. Contreras, *Wait for it … Commons, Copyright, and the Private (Re) ordering of Scientific Publishing* 37–38 (Working Paper, Mar. 4, 2012) [hereinafter Contreras Working Paper], *available at* http://papers.ssrn.com/sol3/papers.cfm?abstract_id=2015885 (proposing that scientific authors grant publishers a one year license to recoup costs and make a profit).

42 *See* Section III.B.

Once the current publishing model was thus transformed, only the actual costs of the intermediaries' brokerage services would need to be taken into account, along with a negotiated surcharge for profit.[43] All parties should understand that outer limits on the aggregate online service charges necessarily follow from the fact that taxpayers largely support the entire enterprise; from the need to conserve scarce resources for scientific investigation; and from the implicit threat that, if intermediaries refused to cooperate, the funders themselves could support alternative arrangements, like those discussed later, including some institutionally organized not-for-profit providers.[44]

In fact, the movement to implement open-access scholarly publications has rapidly expanded in the past decade, and now includes over 10,250 journals.[45] Under this approach, authors, research funders, and universities (or some combination thereof) cover the costs of publication. Absent such an approach, care must be taken to avoid fostering sole-source monopolies over unsubstitutable scientific materials that can never realistically be regenerated or otherwise readily obtained from public repositories.[46]

A. *Reflections on the Law Journal Model*

One example worth considering in this regard is the proliferation of law journals at American universities.[47] These student-run journals select and edit the articles

[43] Any such negotiations must take into account the ways in which open access publishing itself is funded, including author pays, research funder pays, or institution pays models. *See, e.g.*, INT'L COUNCIL FOR SCI., COMM. ON DATA FOR SCI. & TECH., http://www.codata.org (last accessed 9 Apr. 2014); *Sponsoring Consortium for Open Access Publishing in Particle Physics*, SCOAP3, http://scoap3.org (last accessed 9 Apr. 2014).

[44] *See* Sections III.B & C. For-profit intermediaries may require some protection from copyright law and unfair competition law in order to prohibit wholesale duplication of an existing proprietary compilation. But such measures should not impede good-faith competitors from accessing public repositories and starting up comparable endeavors of their own, especially if these endeavors add new value to preexisting information. That, indeed, is the true thrust of the "thin copyright" doctrine in U.S. law. Feist Publ'ns, Inc. v. Rural Tel. Serv. Co., 499 U.S. 340 (1991). In that event, negotiations under the contractual setup would presumably determine whose services were of value at what prices to the relevant subcommunities. In our opinion, however, reliance on not-for-profit intermediaries is always the preferable option.

[45] For a browsable directory of such journals, see DIRECTORY OF OPEN ACCESS JOURNALS, 2015, *available at* http://www.doaj.org (last accessed Feb. 19, 2015).

[46] *See* Reto Hilty, *Copyright Law and Scientific Research*, *in* COPYRIGHT LAW, A HANDBOOK OF CONTEMPORARY RESEARCH 315, 353 (Paul Torremans ed., Edward Elgar Pub. 2007); RETO M. HILTY ET AL., EUROPEAN COMMISSION–GREEN PAPER: COPYRIGHT IN THE KNOWLEDGE ECONOMY–COMMENTS BY THE MAX PLANCK INSTITUTE FOR INTELLECTUAL PROPERTY, COMPETITION AND TAX LAW 14–16 (2008), *available at* http://www.ip.mpg.de/files/pdf1/comments_on_the_green_paper1.pdf [hereinafter MAX PLANCK RESPONSE TO EC GREEN PAPER]. *See generally* BITS OF POWER, n. 37 (discussing the adverse impact that strengthened protection of private databases could have on the public-good uses of scientific data).

[47] *See Durham Statement on Open Access to Legal Scholarship*, BERKMAN CTR. FOR INTERNET & SOCIETY, HARVARD UNIV., Feb. 11, 2009, *available at* http://cyber.law.harvard.edu/publications/durhamstatement. "There are now over 570 student-edited journals published at U.S. law schools."

to be published, work directly with the authors during the revision phase, fact check articles carefully and, when necessary, help to amplify the copious citations that accompany each step of legal analysis and argument. Rarely consulting with faculty, student editors shepherd the works through all processing phases, including the generation of camera-ready text, which is then either produced by an external printer or, increasingly, distributed online directly to subscribers. Importantly, the law journals continue to publish student articles on topical developments alongside longer contributions by law professors and other contributors, and all contributions are the result of a competitive selection process.

Costs are borne by the law school's budget, and reputational benefits vary with the citation counts of the articles published, their impact on the respective fields, and the overall reputation of the law school itself. The educational benefits are universally recognized. Participation on a law journal – whether competitively obtained, as is the case with leading journals, or by voluntary membership in other cases – counts significantly in the student's curriculum vitae. Students also have an incentive to meet the quality standards of the journals that enable them to publish in the sections set aside for their work.[48]

These educational benefits have induced many law schools to sponsor a number of different journals, focused on thematic issues of interest to students and on the substantive strengths of the law school, such as the Harvard Journal of Legislation, the Yale Journal of Health Law, Policy and Ethics, and the Law and Contemporary Problems Journal at Duke. Many law schools sponsor at least a second journal devoted to international law, as well as a primary journal devoted to American law.

Increasingly, law journals encourage collaborative undertakings, which take the form of special issues dedicated to important themes or topics, and are often interdisciplinary in character. While this strategy sometimes focuses on a specific area of law, such as developments in administrative law, or on a specific type of practice, such as developments in Supreme Court jurisprudence, the modern trend is to host major symposia, funded by the schools and outside agencies, that gather leading figures in a field to publish on important topics, such as climate change, public health, or new approaches to intellectual property law. These symposia involve extensive interactions between students, faculty, and other legal experts drawn from different universities, government agencies, and private practice, and each symposium may fill an entire volume of the relevant journal.

Other features of the law journal model deserve mention for present purposes, as do certain defects of that model. The selection process for articles is extremely

Legal scholarship: To print or not to print?, DUKE LAW NEWS, Feb. 27, 2009, *available at* http://law.duke.edu/news/legal-scholarship-print-or-not-print/.

48 Such publications can significantly enhance their career opportunities, particularly in academic law.

competitive and, by comparison to the science journals, quite expeditious. Would-be authors can submit their contributions to multiple journals at the same time, and are not confined to soliciting one journal at a time, which shaves months or years off the acceptance process used in the fields of, say, economics, health policy, and science generally. An offer of acceptance at a middle-ranked journal may thus trigger expedited reviews at other, more prestigious journals, which may lead to a bidding process for particularly attractive contributions. By the same token, the fact that all law schools publish law journals means that most serious academic work will eventually find a publisher. In effect, the user community will then determine ultimate reputational benefits through citations and use of the published articles.

Although the most prestigious journals normally attract articles by established academics, other good articles, often by younger scholars, will find suitable outlets for timely publication. Once published, these articles may attract as much attention, on their merits, as those in top journals and thus lead to promotions and recognition. Still another major advantage of this model is that authors usually retain their copyrights and increasingly publish under a Creative Commons license, which further encourages wide distribution and use.[49]

Law schools defray the costs of these journals out of their own funds, which are only partially recouped from low-cost subscriptions.[50] Costs nonetheless remain low because the services of a specialized printing house are very efficient. Law schools also have lots of relatively inexpensive labor that science departments generally lack.

One tradeoff is that law journal articles are often distributed online through commercial outlets, such as Westlaw and Lexis-Nexis (which helps the schools recoup some of their costs). Although these outlets can be costly for users to access, most of the articles are also made available free of charge on the law journals' own websites (after an embargo period, typically six months), on the websites of the authors, and increasingly on the websites of national repositories, such as the Social Science Research Network (SSRN) and Research Gate. Some law journals (such as all those published at Duke Law School) make all their journal articles openly available.

Perhaps the biggest disadvantage of the law journal model is the absence of any systematic peer-reviewed selection process for most publications. This defect skews the selection process in favor of recognized authors (at least at those journals where the process is not subject to anonymized review). It also makes that process vulnerable

[49] *See Durham Statement on Open Access to Legal Scholarship – Frequently Asked Questions*, Berkman Ctr. for Internet & Society, Harvard Univ., April 28, 2009, *available at* http://cyber.law.harvard.edu/publications/durhamstatement/faq.

[50] However, a few journals are supported by the dues paid to a professional group, such as the Law and Society Association or the Law and Economics Society. *See, e.g.*, Journal of Legal Studies, University of Chicago Law School, *available at* http://www.press.uchicago.edu/ucp/journals/journal/jls.html; Journal of International Economic Law, *available at* http://jiel.oxfordjournals.org/.

to student bias favoring one subject-matter area over another and to limitations on students' knowledge, which can have the same effect (e.g., favoring constitutional law issues over, say, commercial law issues). Nevertheless, a growing number of new law journals, often those of an interdisciplinary character, have introduced a peer-reviewed selection process, and some of these journals have quickly moved to the top of their respective fields. Unfortunately, these journals also tend to be more closed-access than the others, but this tendency is mitigated to some extent by self-archiving practices that have sprung up around them. Also unclear is the extent to which the multiple submission policies practiced by the non-peer-reviewed mass of law journals still apply to the peer-reviewed journals.[51]

Today, legal scholarship depends heavily on the distribution of published articles in the SSRN and other public repositories, which has become standard practice. Availability and rapidity of dissemination have thus become greatly enhanced, with sometimes astounding dissemination effects. However, it remains to be seen how the repositories themselves will ultimately be structured and the extent to which they will resist commoditizing pressures that have begun to emerge.[52]

B. *Funders' Ability to Contractually Regulate Access to, Use, and Reuse of Scientific Literature*

Implicit in the foregoing analysis is the premise that governments or nonprofit foundations will have funded most published scientific research results, at least in

[51] It should be noted that there is also a sub-model of the law school model. *See, e.g.*, Journal of Law and Contemporary Problems, Duke University School of Law, *available at* http://lcp.law.duke.edu/. These are very elite, thematic journals that are housed and produced at a law school, but are not part of the law school function as such. They have their own staff, and they are funded by an associated professional society (e.g., the Law and Society group). Another example would be the Journal of the Copyright Society. These entities receive funds, and these funds go to the universities to produce the journals.

There is also a counter example, i.e., some success stories of journals published by member organizations. The Journal of the American Bar Association is one example, *see also* The Journal of Law and Technology published by a section of the ABA; The International Lawyer, published by the International Section of the ABA; and the American Journal of International Law, published by the American Society of International Law which is peer-reviewed and the editorial board consists of law professors. In addition, *see* the Journal of International Economic Law (JIEL), founded by the leading professor in the field, who took it to Georgetown University for funding. This journal is closed access, peer-reviewed, and staffed by a small group of relevant faculty and students with the motive of making Georgetown a major center of international law.

[52] The law journals are linked, but by private intermediaries who charge high prices (Westlaw and LexisNexis). At present, there is a perverse market failure because Westlaw and LexisNexis control access to most statutory materials as well. However, TheLaw.Net Corporation, a new player in the field, could change that situation. *See Virtual Assistant*, THELAW.NET, http://www.thelaw.net/virtual-assistant/?. Another example is LAW360, http://www.law360.com/ (last accessed 9 Apr. 2014).

OECD countries. These entities have the power to impose conditions on the use and reuse of the research results they fund, with respect to literature and data,[53] subject to pushback from publishers and learned societies.

For example, governments can dedicate government-generated work to the public domain, as occurs under Section 105 of the Copyright Act of 1976 in the United States.[54] Funding agencies can mandate the deposit of publications in open-access journals or, at least, in open-access repositories, as is happening ever more frequently in both the United States and the European Union. They can even impose analogs to fair use and to other codified limitations and exceptions by contract, which both publishers and individual scientists, as grantees, have to respect, especially if they wished to qualify for future grants.[55] Funders can also support or reinforce self-archiving practices, and they increasingly provide for the costs of open-access publishing in their grants.[56]

Besides building open-access provisions into their research grants, funding agencies can support the formation of digitally integrated research commons to serve the needs of diverse thematic communities.[57] Universities can lend their own weight to all these initiatives,[58] and many have established open repositories for their employees' scholarly works, as we noted earlier. Individual scientists and research

[53] *See, e.g.*, Contreras, *Data Sharing*, n. 41, at 1641–57 (examining steps taken by the NIH and Department of Energy to ensure that the output of the Human Genome Project was released to the public); Reichman & Uhlir (2003), n. 5, at 331–51 (discussing the formal and informal means by which institutions can shape the use of government-funded data). Patented research results would, of course, be subject to the Bayh-Dole Act, 35 U.S.C. §§ 200–212 (2006 & Supp. III 2009).

[54] 17 U.S.C. § 105 (2012); *see also* HARGREAVES, above n. 1, at 18; Public Access to Science Act, H.R. 2613, 108th Cong. (2003), *available at* http://www.ncbi.nim.nih.gov/pmcl. *See most recently* Memorandum, Office of Sci. & Tech. Pol'y, "Increasing Access to the Results of Federally Funded Scientific Research," Feb. 22, 2013 [hereinafter OSTP Public Access Initiative], *available at* http://www.whitehouse.gov/sites/default/files/microsites/ostp/ostp_public_access_memo_2013.pdf (to the same effect). For similar efforts underway in the European Union, see Miriam Bitton, *Implementing the Public Sector Information Directive*, 34 *E.I.P.R.* 75, 75–86 (2012).

[55] *See, e.g.*, Carroll, above n. 39, at 2–3 (discussing NIH's mandatory policy of public accessibility and Science Commons licenses); Reichman & Uhlir (2003), above n.5, at 331; Stodden, above n. 39, at 9, 20–25 (proposing a Reproducible Research Standard to ensure attribution and facilitate the sharing of scientific works); *see also* Lee, above n. 5, at 963–65 (comparing the freedom of states to regulate the public accessibility of patents as opposed to that of the NIH and the California Institute for Regenerative Medicine).

[56] *See, e.g.*, Contreras, *Data Sharing*, n. 41 at 1653–54, 1656.

[57] *See* Paul F. Uhlir, *Discussion Framework, in* THE ROLE OF SCIENTIFIC AND TECHNICAL DATA AND INFORMATION IN THE PUBLIC DOMAIN (Julie M. Esanu & Paul F. Uhlir, eds., National Academies Press, 2003), at 3–4 (discussing public welfare advantages of sharing scientific knowledge and data widely). *available at* http://www.nap.edu/catalog/10785/The-role-of-science-and-technical-data-and-information-in-the-public-domain. For examples, see Chapter 6, Section IV.A.

[58] *See, e.g.*, Stodden, n. 39, at 48–49; Faculty Advisory Council Memorandum on Journal Pricing, HARVARD UNIV., Apr. 17, 2012, http://isites.harvard.edu/icb/icb.do?keyword=k77982&tabgroupid=icb.tabgroup143448 (describing efforts by Harvard to reduce subscription costs).

institutions can adopt existing Creative Commons licenses when publishing their works.[59] Innovative proposals that go even farther, such as Victoria Stodden's proposed Reproducible Research Standard, should also be tested and perfected.[60]

The common feature of these and other initiatives is that relevant information would become openly and freely available in digital format and online. Often material is made available either under suitably reduced proprietary terms and conditions set out in permissive, common-use licenses[61] (e.g., the GNU licenses for open source software,[62] or Creative Commons licenses[63] for open-access journals or for some works in open repositories), or it will have entered the public domain.[64] Under other mechanisms, such as the delayed open availability option, the works retain full copyright protection, but eventually become freely and openly accessible, at least on a read-only basis.[65]

Taken together, these activities are part of the emerging broader movement in support of both formal and informal peer production and dissemination of publicly funded scientific (and other) information in a globally distributed, voluntary, and open networked environment:[66]

> [They] are based on principles that reflect the cooperative ethos that traditionally has imbued much of [the] academic and government (civilian) research agencies;

59 *See* Mia Garlick, *A Review of Creative Commons and Science Commons*, 40(5) *Educause Rev.* 78–79 (Sept./Oct. 2005); *see also* Niva Elkin-Koren, *Exploring Creative Commons: A Skeptical View of a Worthy Pursuit*, *in* The Future of the Public Domain: Identifying the Commons in Information Law 325, 329–31 (L. Guibault & P.B. Hugenholtz eds., 2006).

60 *See* Stodden, n. 39, at 36–42; Contreras, *Data Sharing*, n. 41 (proposing 1-year license for publishers' subscriptions). *See generally* The Future of the Public Domain, n. 59.

61 For an overview of such permissive licensing approaches spanning all information types, see Lawrence Liang, Guide to Open Content Licenses (2004), http://digital-rights.net/wp-content/uploads/books/ocl_v1.2.pdf.

62 *See Homepage*, GNU (8 July 2013), *available at* http://www.gnu.org/.

63 *See* text and accompanying nn. 12, 39, and 59.

64 *See generally* James Boyle, *Mertonianism Unbound? Imagining Free, Decentralized Access to Most Cultural and Scientific Material*, *in* Understanding Knowledge As a Commons: From Theory to Practice 123, 123–40 (C. Hess & E. Ostrom eds., MIT Press 2007) [hereinafter Knowledge as a Commons] (illustrating several ways in which works enter the public domain). Apart from overt decisions to abandon copy-right protection, information enters the public domain when it meets the following conditions: (1) the information is not copyrightable, such as factual compilations or data sets that lack creativity and originality in their selection and arrangement; (2) the information is produced by a government that does not apply copyright to its own works (e.g., the U.S. federal government); or (3) the statutory period of intellectual property protection has expired, which in many jurisdictions now is the life of the author plus 70 years.

65 *See, e.g.*, Contreras, *Data Sharing*, n. 41, at 1653–54.

66 Yochai Benkler, The Wealth of Networks: How Social Production Transforms Markets and Freedom 2 (Yale Univ. Press 2006) (discussing the role this movement has played in creating "new opportunities for how we make and exchange information, knowledge, and culture"); Elinor Ostrom & Charlotte Hess, *A Framework for Analyzing the Knowledge Commons*, *in* Knowledge as a Commons, n. 64, at 41, 41–82; Michael J. Madison et al., *Constructing Commons in the Cultural Environment*, 95 *Cornell L. Rev.* 657, 669–74 (2010).

> their norms and governance mechanisms may be characterized as those of "public scientific information commons," rather than of a market system based upon proprietary data and information.[67]

How far these open-access initiatives can be carried remains to be seen, however.

For example, the potential unwillingness of intermediaries or grantees to accept permissive contractual templates, in addition to intrinsic constraints on funders' abilities to defray the costs of relevant institutional arrangements over time, effectively limit the regulatory powers of funders to achieve these objectives.[68] With respect to grantees, a requirement to publish only in open-access journals or only under Creative Commons licenses[69] could hinder publication in some high-prestige, peer-reviewed journals and breed resistance from leading members of the relevant scientific communities. Whether funding agencies, and the research community itself, can persuade these journals to become more open remains to be seen, but the evidence suggests that there is considerable momentum in that direction.[70]

C. Integrating Intermediaries' Functions into Transnational Digital Knowledge Environments

Aggressive resort to open-access licensing conditions espoused by funders could, but not necessarily would, persuade some private publishers to abandon the field. This has not happened so far because funders are increasingly willing to enable grantees to purchase open-access conditions from publishers at prices that appear to remain profitable for the latter. Pressure from funders can thus change the commercial

67 Paul A. David & Paul F. Uhlir, *Creating the Information Commons for e-Science: Toward Institutional Policies and Guidelines for Action*, 91 *Codata Newsletter* 1 (Int'l Council for Sci. July 2005), *available at* http://www.codata.info/resources/newsletters/newsltr91A4.pdf; *see also* Benkler, n. 66, at 2 (noting that this broader movement has "increased the role of nonmarket and nonproprietary production").

68 *See* Robert Terry & Robert Kiley, *Open Access to the Research Literature: A Funder's Perspective*, *in* Open Access: Key Strategic, Technical, and Economic Aspects 101, 106–08 (N. Jacobs ed., Chandos Pub. 2006) (arguing that open access initiatives are sustainable). The extent to which funders' actions with regard to copyrighted literature (and data) might or might not be limited by the Bayh-Dole Act depends on how broadly one interprets that Act. *Cf.* Arti K. Rai & Rebecca S. Eisenberg, *Bayh-Dole Reform and the Progress of Biomedicine*, 66 *Law & Contemp. Probs.* 289, 293 (2003) (discussing the limits the Bayh-Dole Act imposes on funders' ability to oversee the use of patents by grantees).

69 *See* nn. 54–56 and accompanying text. See also Peter Dawyndt et al, "StrainInfo.net: Breaking down information barriers into holistic data integration scenarios using globally unique persistent identifiers," paper presented at "Workflows Management: New abilities for the biological information overflow," NETTAB 2005 Conference, Naples, Italy, 5–7 Oct. 2005, at 1–2.

70 David J. Brown, *Repositories and Journals: Are They in Conflict? A Literature Review of Relevant Literature*, ASLIB Proceedings, Mar. 1, 2010, at 116, *available at* 2010 WLNR 25881660 (noting that Springer Science and Business Media acquired BioMed Central, an open-access publisher).

publishers' business model and persuade some to allow scientists to purchase open-access rights and even make a profitable business out of selling such rights at about the same costs as publishing in an open-access journal.[71] Unfortunately, the percentage of grantees that actually opt to exercise this option, when not otherwise mandatory, still remains relatively small.[72]

Although reliance on intermediaries is deeply entrenched in the system, science policymakers might eventually want to reevaluate the costs and benefits of maintaining customary relationships with them and consider alternative strategies for disseminating research results. Such an exercise could, in particular, focus attention on the advantages of absorbing the publishing function, when feasible, into integrated, open-knowledge environments as discussed in Chapter 8.[73]

Once anchored in appropriate institutions and freed from the legal and commercial fetters of both the professional societies and the commercial publishers, the very objective of the publishing exercise could dramatically change. No longer would it need to be bound by obsolete concepts of the print model, which treat each monthly installment as a discrete legal and substantive unit. Rather, every new collection of research results made available to the relevant thematic community could enrich and expand an ever growing, digitally integrated database of aggregate scientific results.

Each of these thematically organized repositories, in turn, could remain fully open to data mining, manipulation, and other automated knowledge discovery tools, with full respect for reputational benefits but without palpable legal or economic

[71] Major funders of research began, from 2005, to introduce policies – some compulsory and some optional – to promote open access to the published findings of the research they fund. For example, the UK Research Councils and other major funders, such as the Wellcome Trust, require any peer-reviewed publications arising from work they fund that are not published in open access or hybrid journals to be made accessible via a repository as soon as possible. REPORT OF THE WORKING GROUP ON EXPANDING ACCESS TO PUBLISHED RESEARCH FINDINGS, ACCESSIBILITY, SUSTAINABILITY, EXCELLENCE: HOW TO EXPAND ACCESS TO RESEARCH PUBLICATIONS 35 (June 2012).

[72] *See* Contreras, *Data Sharing*, n. 41.

[73] *See* Paul F. Uhlir, *Designing the Digital Commons in Microbiology – Moving from Restrictive Dissemination of Publicly Funded Knowledge to Open Knowledge Environments: A Case Study in Microbiology*, *in* DESIGNING THE MICROBIAL RESEARCH COMMONS: PROCEEDINGS OF AN INTERNATIONAL WORKSHOP 83–87 (Paul F. Uhlir ed., Nat'l Acad. Press 2011) [hereinafter DESIGNING THE MICROBIAL RESEARCH COMMONS], at 83–87 (summarizing Open Knowledge Environments (OKEs) thesis, with illustrative examples developed in this volume, Chapter 8, Section III). Obviously, much depends on the availability of funding. For the view that such funding would yield greater benefits per research dollar than the present system, see Paul F. Uhlir et al., *Measuring the Social and Economic Costs and Benefits of Public Sector Information Online: A Review of the Literature and Future Directions*, *in* NAT'L RESEARCH COUNCIL, THE SOCIOECONOMIC EFFECTS OF PUBLIC SECTOR INFORMATION ON DIGITAL NETWORKS: TOWARD A BETTER UNDERSTANDING OF DIFFERENT ACCESS AND REUSE POLICIES 61, 62 (P.F. Uhlir ed., 2009) (listing reports on benefits of open access to government data).

constraints.[74] Moreover, digital portals could link the formally published literature with the so-called grey literature, i.e., conference proceedings and the like, which are not peer-reviewed. This aggregate resource can then be further linked with other data and relevant information bearing on all aspects of the science, including voluntarily contributed data pertaining to research of interest to a given thematic community.[75]

While this is not the place to fully elaborate on this concept, the astounding creative possibilities of unlimited, fully integrated knowledge hubs along these lines can dwarf the gains otherwise to be made from incremental or even structural reforms of the global intellectual property system. We believe that these or similar initiatives, as explained in Chapter 8, are essential for the progress of both science and culture, and would especially be needed to implement the sweeping new research vision that the National Research Council put forward for the life sciences in 2009.[76]

Support for these and other initiatives could further encourage publishing intermediaries either to accommodate the open-access movement or to leave the scientific publishing business as it exists today. By the same token, digitally integrated knowledge hubs could greatly magnify the creative and educational powers of universities and other analogous research institutions.[77]

For all these reasons, we question the customary practices of wholesale reliance on external information brokers. On the contrary, we now live in a scientific world where it has become conceptually and technically feasible to link any given thematic

[74] *See* Designing the Microbial Research Commons, n. 73, at 83–87.

[75] *See id.* at 83–89 (finding that the "logical response is to cut the Gordian knot by retaining ownership and control of all knowledge assets produced by the relevant research community with public funding within the science framework itself, rather than assigning them to external publishing intermediaries").

[76] Comm. on a New Biology for the 21st Century & Nat'l Research Council, A New Biology for the 21st Century 49–52 (Nat'l Acad. Press 2009) [hereinafter Biology for the 21st Century].

[77] In principle, universities themselves could consider reintegrating some academic journals into their publishing operations. Alternatively, one or more universities could jointly produce the journals in question, with direct support of the funding agencies. In so doing, they could integrate the skills and services of different departments, such as the relevant scientific groups, the computer and technical service departments, and especially library services, which could coordinate and manage editorial and publishing functions. Students and postdoctoral candidates could similarly be co-involved at all levels as part of their educational experience, a phenomenon that routinely occurs in U.S. law schools. University librarians so far exposed to these proposals have expressed a positive response. *See, e.g.*, Charlotte Hess, *Institutional Design and Governance in the Microbial Research Commons*, *in* Designing the Microbial Research Commons, n. 73, at 177, 184; Interview with Richard Danner, Duke Univ. Sch. of Law librarian in Durham, N.C. [March 10, 2011]. However, we think more is to be gained from thematically organized digitally integrated knowledge hubs, as indicated in Chapter 8, Section III (discussing the concept of "Open Knowledge Environments"); Uhlir, n. 73, at 83–89 (discussing the "open knowledge environment").

community's essential knowledge resources into a seamless, digitally integrated network of inputs and outputs that remains open to all the contributors to any given research commons or semicommons.[78] The scientific community, now operating within a hostile intellectual property environment, thus faces the challenge of organizing and managing these knowledge assets with a view to establishing a broad upstream research space.[79] In this space, the scientific community's own contractually imposed rules could apply without compromising the possibilities for commercial exploitation of downstream applications of the resulting research results.[80]

Nevertheless, these long-term science policy goals should not obscure nor detract from the pressing short-term need to make the global intellectual property system more science-friendly than at present, along the lines we have explored. Legislatures concerned about the future of scientific research in the digital online environment[81] should take steps now to reconfigure a legal domain that has become increasingly inimical to the needs of the scientific research community. Policymakers in OECD countries should join with key national institutions, such as the U.S. National Institutes of Health, in affirmatively promoting open access to scientific publications.

To this end, the relevant government agencies and private foundations should become funders of first resort for scientific publications and for the institutional repositories and digital commons in which those publications can be collected. Policy-makers should likewise support the process of making government-funded research publications widely available through self-archiving and institutional archiving, with the fewest possible restrictions on use or reuse of published results.[82]

[78] *See, e.g.*, Boyle, n. 64, at 123–44.

[79] *See* BIOLOGY FOR THE 21ST CENTURY, n. 76.

[80] *Cf.* Reichman & Uhlir (2003), n. 5.

[81] *See, e.g.*, European Commission Green Paper, *Copyright in the Knowledge Economy*, 2008 COM 466 (July 2008), *available at* http://eur-lex.europa.eu/LexUriServ/LexUriServ.do?uri=COM:2008:0466:FIN:EN:PDF. *See generally* Ritch, n. 3, at 136–81.

[82] For a positive step in this direction, see the U.K. government's response to the Hargreaves Review's call for a broad research exemption that cannot be overridden by contract, *see* Hogarth Chambers, *The Hargreaves Review – Another Mixed Bag*, 33 E.I.P.R. 599, 600 (2011) (criticizing United Kingdom's copyright exceptions); *see also* HARGREAVES, n. 1.

8

Fully Exploiting Data-Intensive Research Opportunities in the Networked Environment

I. EARLY RELEASE POLICIES TO MANAGE THE DELUGE OF GENOMIC REFERENCE DATA

In Chapter 6, we saw that preexisting legal and institutional constraints on the availability of scientific literature had unforeseen and hard to resolve consequences for research methods needed in the digital era. Meanwhile, the task of managing research driven by exponentially growing quantities of genomic and other reference data had become so important that it constituted a key feature of the New Biology vision in 2009.[1] This comparatively new phenomenon, as discussed in Chapter 1, posed another set of complex issues for science policy that differed from those pertaining to the use of scientific literature. Yet, because so many of the pertinent data sets are controlled by universities and the research community from the start – and are not under the control of proprietary publishers – that community may paradoxically find these problems easier to resolve by contractual and other private ordering solutions. To do so, however, the research community must embrace data sharing as a fundamental policy.

Numerous collections of genomic data have already emerged that may not yet respond to any particular scientific hypothesis, but may nonetheless lend themselves to the formation and pursuit of new hypotheses in the future.[2] The maintenance costs of these collections are sometimes relatively high and, above all, they are constantly expanding on an open-ended basis with no logical termination date. In the past, private institutions were reluctant to host or contribute data to such large-scale repositories, given the costs of the enterprise, their own proprietary interests, and the

[1] National Research Council (NRC), A NEW BIOLOGY FOR THE 21ST CENTURY (Nat'l Acads. Press 2009). See further Chapter 1, Section II.D.

[2] Mark Sagoff, *Data Deluge and the Human Microbiome Project*, 28 ISSUES IN SCI. & TECH., 71 (Summer 2012). *See generally*, VIKTOR MAYER-SCHONBERG, BIG DATA REVOLUTION: A REVOLUTION THAT WILL TRANSFORM HOW WE LIVE, WORK, AND THINK (2014).

risks of free-riding by noncontributors. In response, governments and foundations began to treat fundamental data pertaining to specific areas of genomic and related research as public goods, available to all, including private-sector researchers, on a universal access basis.

Although these initiatives have already made significant contributions, and their impact is growing, as shown later, there are limits to this public-good approach. For example, research subcommunities may only be willing to release some general inputs (or older, thoroughly vetted inputs) to the public at large, while retaining most of the data pertinent to specific areas of ongoing research interest or to possible commercial applications.

Single researchers and laboratories may nonetheless find it unproductive to manage their own data hoards without access to those of similarly focused colleagues who also manage sizeable collections of data that are components of a much larger, but still emerging whole. They are, therefore, increasingly inclined to pool their data resources in order to share the high cost of data maintenance, and to enlarge the research opportunities afforded by their privileged access to aggregated research inputs. These privately ordered and contractually constructed research commons or semicommons, as the case may be, depend on collective reciprocity benefits to overcome the hoarding instincts of individual contributors.[3] Participation in data-sharing initiatives may enhance the ability of researchers to publish, as well as the ability of an entire sub-community to achieve fruitful research results efficiently, and at more sustainable costs.

There is now wide agreement that, in the life sciences generally, what has been called the deluge of genomic data poses an unprecedented challenge to traditional research methods, and microbiology is no exception.[4] For example, as early as 2009, some 4,300 microbiology genome projects were underway around the world, mostly in Archaea and bacterial subjects, with five big sequencing centers performing more than half of the world's sequencing output.[5] Between 2008 and 2012, the cost of producing one megabyte of genomic sequence data reportedly fell from $1,000 to

3 *See, e.g.*, Paul A. David, A Tragedy of the Public Knowledge 'Commons'? Global Science, Intellectual Property and the Digital Technology Boomerang, SIEPR Discussion Paper No. 00-02, Stanford Inst. Econ. Pol'y Research (2000), *available at* http://siepr.stanford.edu/papers/pdf/00-02.html; Minna Allarakhia, *Microbial Commons: Governing Complex Knowledge Assets*, *in* Designing the Microbial Research Commons: Proceedings of an International Workshop 161, 161–63 (Paul F. Uhlir ed., Nat'l Acad. Press 2011) [hereinafter Designing the Microbial Research Commons].

4 *See, e.g.*, Sagoff, n. 2; Nikos Kyrpides, *Digital Research: Microbial Genomics*, *in* Designing the Microbial Research Commons, n. 3; Chris Anderson, *The End of Theory: The Data Deluge Makes the Scientific Method Obsolete*, Wired (23 June 2008), http://www.wired.com/science/discoveries/magazine/16-07/pb_theory.

5 Kyrpides, n. 4, at 121.

about ten cents. Just one of these sequencing machines could now produce in less than one day what it took the Human Genome Project to generate in ten years during the 1990s.[6] As more sequencing takes place in smaller places, and more universities buy their own sequencing equipment and start producing data, questions arise about how to manage, use, and store such enormous quantities of data while optimizing this output for the advancement of science.

Answering these questions depends in part on whether one sees this rapid accumulation of data as a positive or problematic phenomenon in itself. For example, some experts see the universe of proliferating genomic data as divided into two categories. On the one hand, overworked and understaffed investigators instinctively tend to hoard digital resources; on the other hand, huge government-funded (and often government managed) repositories for genomic data continue to expand exponentially.[7]

Within this conceptual framework, scientists are logically inclined to keep their data secret until publication of relevant research findings. They are also tempted to avoid full disclosure even after publication, in order to retain some comparative advantage with regard to follow-up research and potential commercial opportunities, or just for lack of time or resources to properly organize the data for use by others. The costs to the life sciences of these hoarding tendencies can be high, however, and the public research funders, together with leaders from the research and data management communities, have addressed them in several ways.

For example, in 2003, the National Research Council emphasized the need to release data that support published research findings, in order to enable others to verify and replicate those results.[8] That report formulated two basic principles:

Principle 1. Authors should include in their publications the data, algorithms, or other information that [are] central to the publication – that is, whatever is necessary to support the major claims of the paper and would enable one skilled in the art to verify or replicate the claims.

Principle 2. If central or integral information cannot be included in the publication for practical reasons (for example, because a dataset is too large), it should be made freely (without restriction on its use for research purposes and at no cost) and readily accessible through other means (for example, online). Moreover, when necessary to enable further research, integral information should be made

[6] Sagoff, n. 2, at 72; E. Pennesi, *Will Computers Crash Genomics?*, 331 SCIENCE 666–68 (2011).

[7] *See, e.g.*, Sagoff, above n. 2.

[8] Nat'l Research Council, SHARING PUBLICATION-RELATED DATA AND MATERIALS: RESPONSIBILITIES OF AUTHORSHIP IN THE LIFE SCIENCES 5 (Nat'l Acads. Press 2006). *See also id.*, Ch. 3.

available in a form that enables it to be manipulated, analyzed, and combined with other scientific data.

Other studies show that these norms have not been strictly enforced.[9] Nevertheless, some foundations and government research funders have adopted guidelines that require prospective grantees to submit a data management plan as part of their proposals.[10] Equally important for our own study was a major set of data policy initiatives focused on the early release and sharing of data from large-scale biological and genomic research projects, even prior to publication. These policies require relevant data to be deposited in open, public repositories that make the cumulative genomic inputs of a given field available to the relevant scientific community as a whole. The repositories, in turn, constitute basic scientific infrastructures that are costly and sometimes difficult to maintain.[11]

A. *The Bermuda, Fort Lauderdale, and Toronto Data Policy Guidelines*

The first such initiative, known as the Bermuda Principles of 1996, established the proposition that primary genomic sequence data should be deposited in the public domain[12] and rapidly released,[13] preferably within twenty-four hours from generation if the sequence assembly exceeded one kilobase pairs (later two kilobase

9 *See, e.g.*, Eric G. Campbell et al., *Data Withholding in Academic Genetics: Evidence from a National Survey*, 287 J. AM. MED. ASSOC. 473 (2002).

10 *See, e.g.*, NAT'L INST. HEALTH, NIH DATA SHARING POLICY (2006), http://grants.nih.hov/grants/policy/data_sharing, and the NAT'L SCI. FOUNDATION, PROPOSAL AND AWARD POLICIES AND PROCEDURES GUIDE (Oct. 4, 2012), *available at* http://www.nsf.gov/pubs/policydocs/pappguide/nsf13001/gpgprint.pdf. For more comprehensive data access policy initiatives at the U.S. federal agency level, *see* John P. Holdren, Memorandum for the Heads of Executive Departments and Agencies, "Increasing Access to the Results of Federally Funded Scientific Research," Office of Sci. and Tech. Policy (OSTP), February 22, 2013; *see also*, Presidential Executive Order – Making Open and Machine Readable the New Default for Government Information, May 9, 2013 [hereinafter Executive Order 2013].

11 *See, e.g.*, Sagoff, above n. 2; for the importance of infrastructure, *see generally* BRETT M. FRISCHMANN, INFRASTRUCTURE: THE SOCIAL VALUE OF SHARED RESOURCES (2012).

12 Summary of Principles Agreed at the First International Strategy Meeting on the Human Genome Sequencing, Bermuda, Feb. 25–28, 1996 [hereinafter Bermuda Principles (1996)], *available at* http://web.ornl.gov/sci/techresources/Human_Genome/research/bermuda.shtml#1 (as reported by HUGO). It was agreed that all human genomic sequence information to be generated by centers funded for large-scale human sequencing should be freely available and in the public domain, in order to encourage research and development and to maximize benefit to society.

13 Bermuda Principles (1996), n. 12. Sequence assemblies should be released as soon as possible; in some centers, assemblies of greater than 1 Kb would be released automatically on a daily basis. Finished annotated sequence should be submitted immediately to the public databases. A second meeting in 1997 established Sequence Quality Standards, submission and annotation rules, and principles concerning sequence claims and etiquette. *See* Summary of the Report of the Second International Strategy Meeting on Human Genome Sequencing, Bermuda, 27 Feb.–2 Mar. 1997 (as reported by HUGO) [hereinafter Bermuda Principles (1997)].

pairs).[14] These principles applied to "all human genomic sequences generated by large-scale sequencing centres, funded for the public good ... to prevent such centres establishing a privileged position in the exploitation and control of human sequence information."[15]

The Bermuda Principles were widely followed and respected, with benefits that served to "promote the best interests of science and help to maximize the public benefits to be gained from research."[16] Seven years later, at a meeting in Fort Lauderdale, Florida, under the auspices of the Wellcome Trust, major efforts were made to extend the principle of rapid prepublication release "to other types of data from other large-scale production centres [to be] specifically established as 'community resource projects.'"[17] This broadened system of prepublication release was, in turn, expressly subjected to the joint responsibility of sequence producers and sequence users, as well as their funding agencies:

> The contributions and interests of the large-scale data producers should be recognized and respected by the users of the data, and the ability of the production centres to analyze and publish their own data should be supported by their funding agencies.[18]

Specific concessions favoring producers were then recommended in the form of standards for users of the pooled data, including duties of citation, acknowledgement, and to "act responsibly to promote the highest standards of respect for the scientific contributions of others."[19]

The Fort Lauderdale Principles nonetheless insisted that "[t]here should be no restrictions on the use of the data."[20] Moreover, once a project had been designated as a "community resource project," any relevant "set of data, reagents or other materials whose primary utility will be as "a resource to the broad scientific community" became *prima facie* subject to the early release principle.[21] The Fort Lauderdale

[14] Bermuda Principles (1996), n. 12.

[15] *Id. But see* Jorge L. Contreras, *Data Sharing, Latency Variables, and Science Commons*, 25 BERKELEY TECH. L.J. 1601, 1616, 1654 (2011) [hereinafter Contreras, *Data Sharing*], for criticism; *see also* below Section II.C.

[16] *See* The Wellcome Trust, Sharing Data from Large-Scale Biological Research Projects: A System of Tripartite Responsibility, report of a meeting organized by the Wellcome Trust, Ft. Lauderdale. Florida, 14–15 Jan. 2003 [hereinafter Ft. Lauderdale Principles (2003)], *available at* http://www.genome.gov/pages/research/wellcomereport0303.pdf.

[17] *Id.* at 2. The term "community resource projects" implicitly evokes Elinor Ostrom's concept of "common pool resources," often used in the literature concerning knowledge commons. See further Chapter 9, Section I ("Theoretical Reflections on Designing a Knowledge Commons").

[18] Ft. Lauderdale Principles (2003), n. 16, at 4.

[19] *Id.*

[20] *Id.*

[21] *Id.* at 2–3. "The scientific community will best be served if the results of community resource projects are made immediately available for free and unrestricted use by the scientific community to engage in the full range of opportunities for creative science." *Id.* at 3.

Principles further declared that, in "the near future, many other large data sets will be produced as community resources" and, therefore, were likely candidates for application of the same early release requirements. Many other valuable data sets not technically deemed "community resource projects" also would become good candidates for "contribution of the data to the public domain as a resource" on a more voluntary basis.[22]

The strongly worded Fort Lauderdale Principles favored data users over data producer interests, at least in the case of "community resource projects." Data producers became concerned about how to preserve their rights more effectively and especially about their expectations of publication, given the funders' emphasis on prepublication release of reference data. These concerns were addressed and somewhat assuaged at a subsequent Data Release Workshop convened in Toronto, in May, 2009, by Genome Canada and other funding agencies.[23]

The Toronto Statement built on the Fort Lauderdale Principles, but further refined them in at least five important ways. First, it endorsed the principle of extending early data release policies "beyond genomics and proteomics studies to other data sets – including chemical structure, metabolomics and RNA interference data sets," and even to annotated clinical resources in appropriate cases.[24] Here the Statement stressed the importance of "simultaneously releasing metadata (such as environmental or experimental conditions and phenotypes) that will enable users to fully exploit the data."[25]

The Toronto gathering then took pains to flesh out the duties of data users to respect the rights and publishing expectations of producers. It obliged users to respect producers' specified embargo periods on first publication that would "ideally expire within one year."[26] Data producers, in turn, were urged to provide marker papers associated with their database entries that, among other things, would facilitate citations and the tracking of usage of early released data.[27] Given

[22] *Id.* at 5. A similar 2008 Amsterdam meeting extended the principle of rapid data release to proteomic data. *See* Henry Rodriguez et al., *Recommendations from the 2008 International Summit on Proteomics Data Release and Sharing Policy: The Amsterdam Principles*, 209 *J.* PROTEOME RES., 3,689–3,692(2009).

[23] *See* Toronto International Data Release Workshop Authors, *Prepublication Data Sharing*, 461 NATURE 168 (2009) [hereinafter the Toronto Statement (2009)]. This meeting brought together "a diverse multinational group of scientists, ethicists, lawyers, journal editors, and funding representatives." *Id.* at 168.

[24] *Id.*

[25] *Id.* at 169.

[26] *Id.* at 170; *see* Contreras, *Data Sharing*, n. 15. *See also* Jorge L. Contreras, *Bermuda's Legacy: Policy, Patents and the Design of the Genome Commons*, 12 *MINN. J. L. SCI. & TECH.* 61 (2011) [hereinafter Contreras, *Bermuda's Legacy*].

[27] Toronto Statement (2009), n. 23, at 170.

these precautions, the drafters surmised that, if properly handled, the risk of conflict between producers and users was empirically small and worth taking.[28]

However, and for the first time in any major declaration of relevant data-sharing principles, the Toronto Statement expressly distinguished the kind of large-scale reference data sets suitable for the application of early release policies from hypothesis-driven datasets whose release could properly be delayed at least until publication of the relevant findings.[29] Also emphasized were quality problems associated with early release datasets, with an admonition to editors and reviewers to look for possible sources of error.[30] Finally, the Toronto Statement directly addressed issues of enforcement by urging funders to insist on data management plans as part of their grant applications, which should be subject to the peer-review process.[31] The drafters expressly endorsed the funders' willingness to recognize good data release behavior when processing proposals for grant renewals.[32]

The Toronto Statement's attempt to compile and elaborate a set of best practices to better implement early data release policies was a major milestone in this area of science policymaking. While reconfirming the consensus that "rapid prepublication release of sequencing data had served the field of genomics well,"[33] it recognized the outer limits of such policies, the need for them to "evolve with the changing research landscape," and the fact that "actual community behavior (as opposed to intentions)" needed careful and constant scrutiny.[34]

B. *Evaluating the Mandatory Early Release Policies and Their Conceptual Framework*

The Bermuda Principles, as refined at Fort Lauderdale and then Toronto, were widely adopted by the National Human Genome Research Institute(NHGRI) and the major sequencing laboratories worldwide.[35] Initial genomic sequence assemblies started to be routinely deposited into GenBank within twenty-four hours of assembly, and later-stage data were subsequently covered by the now generalized duty to make human genome sequence data publicly available, subject to normative guidelines

[28] *Id.*

[29] *Id.* at 169.

[30] *Id.* at 170 ("Prepublication data are likely to be released before extensive quality control is performed").

[31] *Id.* at 169 (noting that "[s]uch practice is currently the exception rather than the rule").

[32] *Id.*

[33] *Id.* at 170.

[34] *Id.*

[35] Contreras, *Bermuda's Legacy*, above n. 26, at 91. *See, e.g Reaffirmation and Extension of NHGRI Rapid Data Release Policies: Large-Scale Sequencing and Other Community Resource Projects*, Nat'l Human Genome Research Inst. (Feb. 2003), http://www.genome.gov/10506537. The NHGRI is one of the 27 institutes of the U.S. National Institutes of Health (NIH).

TABLE 8.1. *Examples of prepublication data-release guidelines*

Project type	Prepublication data release recommended	Prepublication data release optional
Genome sequencing	Whole-genome or mRNA sequence(s) of a reference organism or tissue	Sequences from a few loci for cross-species comparisons in a limited number of samples
Polymorphism discovery	Catalogue of variants from genomic and/ or transcriptomic samples in one or more populations	Variants in a gene, a gene family or a genomic region in selected pedigrees or populations
Genetic association studies	Genomewide association analysis of thousands of samples	Genotyping of selected gene candidates
Somatic mutation discovery	Catalogue of somatic mutations in exomes or genomes of tumor and non-tumor samples	Somatic mutations of a specific locus or limited set of genomic regions
Microbiome studies	Whole-genome sequence of microbial communities in different environments	Sequencing of target locus in a limited number of microbiome samples
RNA profiling	Whole-genome expression profiles from a large panel of reference samples	Whole-genome expression profiles of a perturbed biological system(s)
Proteomic studies	Mass spectrometry data sets from large panels of normal and disease tissues	Mass spectrometry data sets from a well-defined and limited set of tissues
Metabolomic studies	Catalogue of metabolites in one or more tissues of an organism	Analyses of metabolites induced in a perturbed biological system(s)
RNAi or chemical library screen	Large-scale screen of a cell line or organism analyzed for standard phenotypes	Focused screens used to validate a hypothetical gene network
3-D-structure elucidation	Large-scale cataloguing of 3-D structures of proteins or compounds	3D structure of a synthetic protein or compound elucidated in the context of a focused project

Source: Toronto Statement (2009), n. 23, at 168, as reproduced from 461 NATURE 168–70 (Sept. 10, 2009).

that protected the publication interests of data producers.[36] Follow-up projects with mandatory data-sharing policies multiplied in the United States and abroad, most of them in the field of human genomics or related upstream biomedical research, as appears from Table 8.1.

36 Contreras, *Bermuda's Legacy*, n. 26, at 89.

1. Selected Examples of Compliance in the Field of Microbiology

The microbial genomic research community also adhered to the early release obligations set out in the Fort Lauderdale and Toronto Principles. One example is the Fungal Genome Initiative (FGI), a partnership between the Broad Institute of Harvard and MIT and the larger research community. The FGI aims to provide the sequences of key organisms and their related species across the fungal kingdom and thereby lay "the foundation for work in medicine, agriculture, and industry through comparative studies."[37] Organisms are selected by a steering committee of fungal biologists, and the sequencing is supported by NHGRI, the National Science Foundation (NSF), and the U.S. Department of Agriculture.

FGI's data release policy expressly states that "sequence data for all FGI genomes are made available in advance of assembly, according to the NHGRI policy on rapid data release by regular deposition of traces at the NCBI trace repository."[38] On the websites for each sequenced species in its collection, FGI attaches a notice declaring that its "goal is to make the genome sequence of organisms rapidly and broadly available to the scientific community" in conformance with the genome sequencing community's "recently adopted … statement of principles for the distribution and use of large-scale sequencing data …"[39]

Another example at the foundation level is the Gordon and Betty Moore Foundation's Marine Microbiology Initiative, which aims to "uncover the principles that govern the interactions among microbes as well as those that govern the microbially mediated nutrients flowing in the marine environment."[40] For marine projects specifically, DNA sequencing, assembly data, and annotations, together with associated metadata, should be released within fifteen days to the Community Cyberinstrastrucure for Advanced Marine Microbial Research and Analysis (CAMERA), which manages data released to collaborating laboratories and the general public. Research groups submitting samples to this initiative have exclusive access to the resulting sequence and relevant annotations for an embargo

37 *See* Fungal Genome Initiative (FGI), *Frequently Asked Questions*, BROAD INST., http://broadinstitute.org/science/projects/fungal-genome-initiative/frequently-asked-questions (last accessed 9 Apr. 2014).

38 FGI, A White Paper for Fungal Comparative Genomics 8 (10 June 2003), *available at* http://www.broadinstitute.org/annotation/fungi/fgi/FGI_02_whitepaper_2003.pdf. This statement is apparently consistent with the Broad Institute's overall policy.

39 For FGI website *see* FGI, *Fungal Genome Initiative*, BROAD INST., http://www.broadinstitute.org/scientific-community/science/projects/fungal-genome-initiative/fungal-genome-initiative (last accessed 9 Apr. 2014). For the full data access and use policy, see FGI, *Data Use Policy*, BROAD INST., https://olive.broadinstitute.org/data_policy (last accessed 9 Apr. 2014).

40 Marine Microbiology Initiative (MMI), *Marine Microbiology*, GORDON & BETTY MOORE FOUND., http://www.moore.org//programs/science/marine-microbiology-initiative (last accessed Mar. 26, 2015).

period of six months, following which the data are made publicly available through CAMERA and the National Center for Biotechnology Information (NCBI).[41]

At the government level, the microbiological community's compliance with early data release norms for genomic sequences is very high, as one would expect. For example, the Pathogen Genomics Initiative, under the aegis of the National Institute for Allergy and Infectious Diseases (NIAID), is committed to the rapid release of genomic and other data types because "rapid and unrestricted sharing of data and research resources is essential for advancing research on human health and research on human diseases."[42] Accordingly, NIAID requires all raw genomic data and next generation sequencing data to be submitted as rapidly as possible to either the Trace Archive or, as appropriate, to the Sequence Read Archive at the NCBI.[43] Moreover, "[g]enome and metagenomics full and partial assemblies and their annotations should be deposited in appropriate databases at NCBI after verification by the center or data generator."[44]

Another example is the innovative Genomic Encyclopedia of Bacteria and Archaea (GEBA) at the U.S. Department of Energy (DOE). GEBA is a "large-scale systematic effort to sequence genomes to fill in genomic gaps in the [NSF's] tree of life."[45] It releases all genome sequence data to the community through DOE's Joint Genome Institute (JGI) and GenBank.[46]

2. The International Human Microbiome Consortium

Perhaps the most instructive initiative at the governmental level for present purposes was the International Human Microbiome Consortium (IHMC), which aimed "to coordinate the activities and policies of the international groups studying the human microbiome, with a view to "promot[ing] the generation of a robust ... data

41 For more on CAMERA, see Section III.A.2.

42 *NIAID/DMID Data Sharing and Release Guidelines*, NAT'L INST. ALLERGY & INFECTIOUS DISEASES (NIAID) (11 June 2013), http://www.niaid.nih.gov/LabsAndResources/resources/dmid/Pages/data.aspx.

43 *Id.* These data should also include information on templates, vectors, and quality values for each sequence, as appropriate. This includes RNA seq-transcriptomics data obtained from next generation sequencing.

44 *Id.* Assuming no specific errors are detected during the validation process, final assemblies and final annotations will be submitted to GenBank for individual samples or for defined cohorts of samples as rapidly as possible, and no later than 45 calendar days of being generated, followed by release to other websites, as approved by NIAID.

45 *A Genomic Encyclopedia of Bacteria and Archaea (GEBA)*, U.S. DEPT. OF ENERGY (DOE) JOINT GENOME INST. (JGI) (26 Mar. 2015), http://www.jgi.doe.gov/our-science/science-programs/microbial-genomics/phylogenetic-diversity.

46 *Id.*

resource that is freely available to the scientific community and that can be analyzed across many groups."[47] The IHMC would accordingly formulate a common set of principles and policies supporting the formation of "a shared comprehensive data resource that will enable investigators to characterize the relationship between the composition of the microbiome (or parts thereof) and human health and disease."[48] Here, in other words, the IHMC demonstrated exactly how it intends to apply the Fort Lauderdale Principles to the goal of constructing one major upstream data infrastructure for human microbiology.

One of the first decisions taken by the IHMC Steering Committee was to agree on a set of principles for membership pertaining to data release, intellectual property rules for publications, quality assessment, and privacy standards for consent of participants.[49] As formulated by that Steering Committee, the data release policy required "immediate release of verified (technically accurate) sequence data from isolated microorganisms and/or metagenomic data from samples taken from healthy and/or diseased individuals in an appropriate public database."[50] The IHMC's publication policy further affirmed the principle – later embodied in the Toronto Statement of 2009[51] – that contributors of raw data to public databases will benefit from a "publication moratorium" of up to twelve months, i.e., a period in which only the contributor can publish an analysis of the deposited raw data.[52]

The NIH's description of its own Human Microbiome Project (HMP),[53] a component of the IHMC, was especially instructive on these issues although its policies are not necessarily the same as those of other IHMC members. The NIH statement began by affirming that its national HMP had been designated as a "Community Resource" whose rapid data release policies were based on the Fort

47 Int'l Human Microbiome Consortium (IHMC), "A Description of Its Goals, Operating Structure and Principles," *approved* 20 Jan. 2008 [hereinafter IHMC Description], https://www.human-microbiome.org/fileadmin/Content/Media/Docs/IHMC_Operating__Principles_Doc_FINAL.DOC.

48 *Homepage*, IHMC [hereinafter IHMC Homepage] http://www.human-microbiome.org (last accessed Mar. 26, 2015).

49 *Id.*

50 *Id.* At the Heidelberg Meeting on October 15–18, 2008, the participants agreed that data generated by the IHMC projects would be made available through the public databases at EBI and NCBI, and analyses of those data would be made available through the NIH Human Microbiome Project Data Analysis and Coordination Center, then led by Owen White, University of Maryland School of Medicine, Baltimore, and an equivalent center at the European Molecular Biology Laboratory (EMBL), then led by Peter Bork. *See* IHMC, *Heidelberg 2008*, http://www.human-microbiome.org/index.php?id=58 (last accessed 9 Apr. 2014).

51 *See* Toronto Statement (2009), n. 23.

52 IHMC Description, n. 47.

53 NIH, Human Microbiome Project, https://www.commonfund.nih.gov/hmp/index (last accessed, 27 Mar. 2015).

Lauderdale Agreement on Sharing Data from Large-Scale Biological Research Projects.[54] The HMP's data release policies were then said to apply to two distinct categories, viz, "Consortium Publication and Presentation Policies" and "Resource Sharing Guidelines for Human Microbiome Project Data Production Grants."[55]

As regards what the NIH defined as demonstration projects, all "[s]equence and clinical data must be deposited in the NCBI Sequence Read Archive or Trace Archive and dbGaP within one week of being generated."[56] However, all such data are subject to a publication moratorium (also called a "publication or data embargo"). This embargo confers "a protected period of time during which the data originator has an exclusive right to publish or present on the data, beginning from the date of submission to a public database."[57] The Consortium expected that "no one other than the data originator will submit a paper for publication on the data during the moratorium period."[58] The basic moratorium period set for demonstration projects was up to 12 months from the date of data submission or on publication by the data originator, whichever occurs first.[59] However, it did not apply to reference genome data or to normal human subjects data generated from the HMP Sequencing Center.[60]

While the "publication moratorium" may represent an emerging norm of computational research in the life sciences,[61] the NIH sought to enforce the norm both by moral suasion and by other more direct methods. For example, it would specifically identify databases subject to the moratorium; it would seek to persuade the community to "recognize the moratorium to enhance research and data sharing in the area;" and it would directly press journal editors "to encourage them" to act in accord with its data release guidelines and principles for HMP publications

[54] Ft. Lauderdale Principles (2003), 16; *see* NAT'L INST. HEALTH (NIH), HMP DATA RELEASE AND RESOURCE SHARING GUIDELINES FOR HUMAN MICROBIOME PROJECT DATA PRODUCTION GRANTS (Jan. 1, 2011) [hereinafter NIH HMP Data Release Guidelines], https://www.commonfund.nih.gov/hmp/datareleaseguidelines.aspx (last accessed 27 Mar. 2015).

[55] *See* NIH, HUMAN MICROBIOME PROJECT CONSORTIUM PUBLICATION AND PRESENTATION POLICIES (1 Jan. 2011), https://www.commonfund.nih.gov/hmp/presentationpolicies.aspx [hereinafter NIH HMP Consortium Policies] (last accessed 27 Mar. 2015); NIH HMP Data Release Guidelines, above n. 54.

[56] HMP Consortium Policies, n. 55.

[57] *Id.*

[58] *Id.* For longitudinal studies, with multiple data sets submitted overtime, each of these data sets will have its own moratorium period. Moreover, research participants must have given informed consent to allow the submission of the clinical data to the database. *Id.*

[59] *Id.*

[60] The HMP Sequencing Center was expected to publish global analyses of the reference genomes and the normal human samples from the data that were generated.

[61] *Cf. id.*; Toronto Statement (2009), n. 23; Contreras, *Data Sharing*, n. 15; Contreras, *Bermuda's Legacy*, n. 26.

"by not considering papers submitted for publication before the date on which any applicable moratorium expires."[62]

Implicit in the guidelines is the notion that good relations with the NIH might depend on adherence to these normative requirements. For example, users of any HMP Consortium data, whether or not members of the Consortium, are told to "be aware of the publication status of the data they use and [to] treat the data accordingly."[63] The NIH expressly imposed an obligation to obtain the consent of data originators before using unpublished data in their publications. Investigators outside the Consortium remain free to use its data, but they were expected to observe the 12-month publication moratorium,[64] although no explicit sanctions were imposed.

Finally, the International Human Microbiome Consortium decided to press the scientific journals to respect its data release guidelines and principles. The IHMC further considered drafting a letter of attestation that would accompany journal submissions affirming that the authors had consulted with the data producers, were not violating any community ethic regarding publication of their analysis, and had properly acknowledged the data producers in their manuscript. This letter "would be made available to the journals and microbial research community for their use."[65]

The IHMC Steering Committee also considered whether further steps were needed to protect data producers' interests without undermining the "need to maximize public benefit by minimizing barriers to data access"[66] Meanwhile, in the United States, the NIH's HMP further required attribution by all users of data from Demonstration Projects who "should always cite the source of the data and should acknowledge the data originators from the Consortium."[67] Similarly, all investigators "should enforce a high standard of respect for the scientific contribution of the data originators."[68]

As regards data production grants from the NIH pertaining to the Human Microbiome Project, the guiding principles appeared similar, but their mode of implementation was less clear. Prepublication metagenomic and associated data

[62] HMP Consortium Policies, n. 55.

[63] *Id.* For example, all investigators "should obtain the consent of the data originators before using unpublished data in their individual publications." *Id.*

[64] *Id.*

[65] IHMC, *Publication Policy*, http://www.human-microbiome.org/index.php?id=34 (last accessed 9 Apr. 2014).

[66] *Id.*

[67] NIH HMP Consortium Policies, above n. 55. Outside investigators who analyze consortium data and want to publish before the data producers are encouraged to seek a collaborative agreement, but may not proceed without consent. *Id.*

[68] *Id.* NHGRI issues monthly lists of submitted publications and of presentations of data generated from the Consortium. In this connection, NIH expressly states that data originators who benefit from the publication moratorium for Demonstration Projects will write and submit their papers "as rapidly as possible." *Id.*

were, in principle, to be "released to the scientific community as rapidly as possible via deposition into public databases."[69] To this end, NIH funded a Data Analysis and Coordination Center (DACC) as an HMP informatics resource, as well as a Controlled Access Database – NCBI's dbGaP – for clinical and other data that potentially identify the source.[70] The DACC helped to manage metagenomic data and metadata to facilitate analysis and utilization of tools from HMP-funded projects. It also incorporates information from other relevant sources as needed.

In practice, however, the NIH stipulated that an "appropriate data release plan for each HMP-funded data production cooperative agreement is a condition of the award."[71] This suggests that grantees have some room for negotiating data release policies that was not available to users of the Demonstration Project data.

As noted above in other data contexts all raw genome and metagenome sequence and next generation sequence data that were generated under HMP data cooperative agreements had to be submitted, in a properly formatted fashion, as rapidly as possible (typically on a weekly basis) to the NCBI.[72] Genome and metagenome data and their annotations should be deposited in NCBI after verification by the DACC. Other related data used to characterize the human microbiome should be verified before release "at a broadly accessible site."[73] It appears, however, that analyses performed by grantees were subject to somewhat different rules, as well as to the conditions of the cooperative agreement.[74]

3. Evaluating the Trend

The data release policy established by IHMC represented one of the most advanced efforts to reconcile the needs of public microbiological science for large amounts of community data with the needs of data contributors to maintain reputational benefits, both as contributors and as published analysts of such data. However, the IHMC Description made no mention of any concerted efforts to digitally integrate the human microbiome literature likely to emerge from this initiative into the larger data infrastructure that was envisioned.

[69] NIH HMP Data Release Guidelines, n. 54.

[70] *See*, http://www.hmpdaac.org (last accessed 24 Mar. 2015).

[71] NIH HMP Data Release Guidelines, n. 54.

[72] *Id.* Data that potentially identify donors are deposited in a controlled access database. HMP's data policy here is that "the richest possible set of data should be released to the controlled access database, consistent with the protection of donor privacy." *Id.*

[73] *Id.*

[74] In principle, such analyses should be made available to the public upon acceptance of a manuscript for publication, and the grantee may be told that, in appropriate cases, those data should be housed at the DACC. *See id.* It was not altogether clear whether such published analyses by grantees were normally to be protected by the twelve-month publication moratorium.

The NIH's description of the national HMP does acknowledge the grantees' duty to deposit their articles in publicly available repositories. The DACC also provides some access to the related literature. Nonetheless, more thought might have been given to ensuring full integration of the openly available literature into any resulting digital infrastructure, especially with a view to coordinating the outputs of the different national members.

Looking back at the examples of government-funded mandatory data-sharing ventures treated as "Community Resource Projects" under the Fort Lauderdale Principles, the evidence shows that major digital infrastructure supporting a broad range of research in the life sciences has been created in less than two decades. This achievement rests on a broad consensus that, with regard to certain data-intensive projects, early release of unencumbered upstream data can be "profoundly valuable to the scientific enterprise and lead to public benefits."[75] This proposition applies particularly "when there is a community of scientists that can productively use the data quickly – beyond what the data producers could do themselves in a similar time period, and sometimes for scientific purposes outside the original goals of the project."[76]

Mandatory data-release policies are particularly compelling in the construction of very large-scale reference collections whose validity depends on comprehensiveness and the general absence of holdouts. The resulting scientific benefits are further reinforced by giving data generators priority with respect to publications based on their data and by minimizing patent-related encumbrances on genomic and other upstream datasets.[77]

This broad-gauged approach nonetheless entails costs as well as benefits, and it has lately elicited a growing body of criticism. For example, there remain questions about the quality and reliability of early release data,[78] while tensions between data producers and users, though muted since Toronto, have not altogether disappeared.[79] More recent criticism focuses on the high costs of maintaining and curating endlessly expanding reference sets, now that the costs of generating genomic data in particular have plummeted.[80] The Human Microbiome Project has so far reportedly failed to identify any actual reference sets among the data, which are said to be extremely heterogeneous and change over time in the make-up of individuals.[81]

75 Toronto Statement (2009), n. 23, at 168.

76 *Id.* For additional reasons to support the principle of making data openly available in the life sciences and research generally beyond the mandatory early release framework, see Section III.

77 *See* Contreras, *Data Sharing*, n. 15, at 1658, for measures taken by the U.S. government to minimize tensions with the Bayh-Dole Act.

78 *See, e.g.*, Pennesi, n. 6.

79 For the resolution at Toronto, see Toronto Statement (2009), n. 23.

80 *See, e.g.*, Pennesi, n. 6, at 666–68; Sagoff, n. 2.

81 *See, e.g.*, Sagoff, n. 2.

Above all, there are growing doubts about the scientific payoffs from these initiatives, given that so far at least they have supported mainly nonhypothesis driven research methods whose ultimate fruitfulness remains to be seen.[82] Obtaining additional payoffs could require computational methods well beyond existing capacity, which in turn would require substantial investment again without specific hypothesis-driven goals in mind. The big questions at the moment, as summarized by one critic, are thus how do we deal with the volume of the data that are generated and how do we want the infrastructure that manages those data to be developed?[83]

In our view, some of these critical evaluations may stem from an overly narrow view of the data-sharing landscape, in which the only two possibilities are either privately hoarded data sets or mandated, fully open data sets lodged in massive repositories. In reality, between these two extremes, there is a burgeoning practice of voluntary data sharing and preservation that offers a promising bottom-up alternative. This phenomenon is the subject of the next section.

II. BEYOND EARLY RELEASE: DIVERSE NETWORKED SHARING STRATEGIES TO MANAGE AND EXPLOIT THE DELUGE OF DATA

Outside the field of microbiology, a very promising voluntary initiative is the International Cancer Genome Consortium (ICGC).[84] This body aims to produce a comprehensive description of genomic, transcriptomic, and epigenomic changes in fifty different tumor types and subtypes that are of clinical and societal importance across the globe.[85] Operating under agreed quality standards, members pledge to make the relevant data that targets specific genes and mutations available to the entire research community as rapidly as possible, and with minimum restrictions, in order to accelerate research into the causes and control of cancer.[86] ICGC members also pledge not to make claims to possible intellectual property rights on primary data, and they agree to respect a one-year publication moratorium for initial providers, in conformity with the Fort Lauderdale guidelines.[87]

82 *See id.*; Pennesi, n. 6, at 666–68.

83 Sagoff, n. 2.

84 *Overview*, INT'L CANCER GENOME CONSORTIUM (ICGC), http://www.icgc.org (last accessed 26 Mar. 2015).

85 *Id. See generally* the Int'l Cancer Genome Consortium (ICGC), *International Network of Cancer Genome Projects*, 464/15 NATURE 993 (April 2010) [hereinafter ICGC, *International Network*].

86 *Id.* at 993. Some data is only available from a controlled access database under terms and conditions needed to protect the confidentially of participants. *Id.* at 994.

87 *Id.* at 993–94. "To allow time for a data set to be analyzed and submitted for publication, ICGC members will have at most one year after released data sets reach the specified threshold before third parties are permitted to submit manuscripts describing global analyses." *Id.* at 995.

Users of ICGC data are expected to respect these and other terms and conditions; to cite manuscripts and sources of prepublication data; and to contact member projects directly to discuss publication plans in cases of uncertainty about how the rules apply.[88] To date, funders on four continents have enabled some 39 project teams in 13 jurisdictions to study more than 18,000 tumor genomes under ICGC auspices.[89]

Within the microbial research community itself, there are a growing number of voluntary data-sharing or community resource initiatives under way. Their common denominator is the need to manage and effectively use the exponentially increasing volume and diversity of data resulting from computational and networked research methods.[90] Some of these projects represent a promising institutional response to the twin challenges of data hoarding, on the one hand, and the data deluge on the other.

If, moreover, we look carefully at this still rather amorphous zone of voluntary data pooling initiatives, we see an evolving continuum between projects that operate on a fully open commons basis and a proliferation of others that are better characterized as semicommons open only to those qualified to participate and otherwise committed to given projects. We also see diverse and growing numbers of data-sharing projects that are hypothesis-driven from the outset, and not merely providers of reference data sets or other community resources.[91]

A. *Selected Taxonomic and Related Microbiological Reference Data Collections*

The Encyclopedia of Life (EOL), operating under the auspices of the Smithsonian Institution, is a pooled resource that gathers information about living creatures

[88] *Id.*

[89] ICGC, *Overview*, n. 84.

[90] For data pertaining to microbial materials, *see, e.g.*, David Smith, Dagmar Fritze & Erko Stackebrandt, *Public Service Collections and Biological Resource Centers of Microorganisms*, *in* The Prokaryotes – Prokaryotic and Symbiotic Associations, Ch. 11, at 289–90, 295–97 (Springer-Verlag 2013), discussing the Microbial Information Europe (MINE) project, the Common Access to Biological Resources and Information (CABRI) project, and a related project of the Global Biodiversity Information Facility (GBIF). *See generally* Nat'l Research Council Board on Research Data & Info., The Future of Scientific Knowledge Discovery in Open Networked Environments (Paul F. Uhlir, ed., Nat'l Acads. Press 2012) [hereinafter The Future of Scientific Knowledge Discovery]. For a list of NIH-supported data repositories with links to a list of NIH data sharing policies, some of which are also germane to microbiological research, see http://www.nlm.nih.gov/NIHbmic/. For a self-registered list of all types of data repositories and their terms of access and use, 144 of which are biological, see http://databib.org/ (last accessed 9 Apr. 2014).

[91] Arguably, the hypothesis underlying the International Cancer Genome Consortium (ICGC), n. 84, is that pooled data concerning the relevant genetic mutations will lead to – and has already led to – clinical responses. *See* ICGC, *International Network*, n. 85. A growing number of funders seem to agree. *See* ICGC, *Overview*, n. 84.

stored in partner databases all over the world. EOL organizes this information by the names of the living creatures covered in the pool. It is particularly useful for exploring the plants, animals, and microorganisms found in a particular area or region (i.e., its "biodiversity"). About two million organisms have thus far been named, "with more being identified every day."[92] The ability of EOL to link users to other participating databases makes it a particularly useful taxonomic portal.

In seeking to become "the most comprehensive source for biodiversity information," EOL aims to make the aggregate content of its collection freely available to the public. To this end, its open-access policy urges participating collections to share contents with the rest of the world under at least one of the Creative Commons licenses that allow reuse. Most of its content providers have conformed to this policy.[93]

Two major databases provide "the taxonomic backbone" of the Encyclopedia of Life, namely, the Integrated Taxonomic Information System (ITIS) and the Species 2000 Catalogue of Life. ITIS is the result of a partnership among U.S. federal agencies that was formed to satisfy their mutual needs for scientifically credible taxonomic information.[94] The goal was to create an easily accessible database with reliable information on species names and their hierarchical classification. A Taxonomic Working Group (TWG), which coordinates its efforts with several national and international biodiversity programs, aims to provide a common framework for taxonomic data available to researchers and cooperating government services. The TWG is responsible for ensuring the quality and integrity of the data. Because the ITIS website is operated by the U.S. federal government, the data it produces are placed in the public domain, although some pages may contain material that is subject to copyright or related laws, which may require permission for reuse.

Species 2000 is an autonomous federation of taxonomic database custodians operating throughout the world, whose goal is to collate a uniform and validated index to all known species of organisms on Earth.[95] The Secretariat is located at

[92] *Homepage*, ENCYCLOPEDIA OF LIFE (EOL), http://eol.org (last accessed 9a Apr. 2014). *See generally* Edward O. Wilson, *The Encyclopedia of Life*, 18(2) TRENDS IN ECOLOGY & EVOLUTION 77–80 (2003), *available at* http://www.sciencedirect.com/science/article/B6VJ1-47C8RDN-3/2/befac60e32dd59e55ff8bfc75f9848c6.

[93] All of the EOL's suggested Creative Commons licenses require at least attribution. *See Copyright and Linking Policy*, EOL (Aug. 2011), http://eol.org./info/copyright_and_linking. Users must respect the differentiated restrictions allowed by the different Creative Commons license options, discussed later in Section III.B.2, and any residual restrictions imposed by the few providers that fail to adopt those licenses.

[94] *Background Information*, INTERAGENCY TAXONOMIC INFO. SYS. (ITIS) (Mar. 27, 2015), http://www.itis.gov/info.html. ITIS partners include the Department of Commerce; the Department of Interior's Geological Survey (USGS); the Environmental Protection Agency; the Department of Agriculture's Agricultural Research Service; and the Smithsonian Institution's National Museum of Natural History, among others.

[95] *Homepage*, SPECIES 2000, www.sp2000.org (last accessed 27 Mar. 2015).

the Plant Science Laboratories, University of Reading in the United Kingdom. Its Catalogue of Life (jointly produced with ITIS) is compiled from sectors supported by more than 150 taxonomic databases around the world.[96] Species 2000 and ITIS teams provide peer reviews and databases, select appropriate taxonomic sectors, and integrate these sectors into a coherent catalogue with a single hierarchical classification.[97] They apparently allow noncommercial use of their compilation or any of the species data sets contained therein.[98] However, use of the Catalogue of Life on a public portal or webpage requires correct attribution and credits on three levels, i.e., credit for the complete work, for the contributing database of record, and for the experts who provide the taxonomic information available on the individual record.[99]

In contrast, older and better established taxonomic sources in the field of microbiology are more proprietary than the novel and ambitious initiatives described earlier, but there is some movement toward greater openness even there. For example, J. P. Euzeby publishes his List of Prokaryotic Names with Standing in Nomenclature online, subject to full copyright protection with an "all rights reserved" contractual condition. In practice, the author does make the material freely available on a read-only basis and for personal, noncommercial use.[100] However, no downloading, reuse, or redistribution was allowed without prior written permission.[101] Extraction and reuse of the data would also be subject to the EU's database protection legislation where applicable, as previously described in Chapter 6.[102]

On a more commercial footing, Bergey's Manual of Systematic Bacteriology provides updated classification and descriptive information about the species of bacteria and Archaea and is produced by a nonprofit group of volunteers under the aegis of Bergey's Manual Trust.[103] The editorial office is currently located at the Department of Microbiology, University of Georgia. In recent years, a five-volume

96 The complete work contains contributions from more than 3,000 specialists from an array of taxonomic professions. *About the Catalogue of Life*, Catalogue of Life (25 June 2013), http://www.catalogueoflife.org/col/info/about.

97 *Id.* Alternative taxonomic treatments and classifications are also envisioned, but the primary utility stems from the single, peer-reviewed preferred catalogue approach. *Id.*

98 Inferred from the fact that commercial use requires permission. *See Copyright, reproduction & sale*, Catalogue of Life (27 Mar. 2015), http://www.catalogueoflife.org/col/info/copyright. Guidance on how to incorporate the catalogue's data into one's own system is available from the Secretariat.

99 *Id.* Notice must be given to the Secretariat who will check that appropriate credit is given.

100 List of prokaryotic names with standing in nomenclature (LPSN), *Homepage*, Bacterio, http://www.bacterio.net (last accessed 9 Apr. 2014) (providing a suggested citation form).

101 *Id.*

102 *See* Chapter 6, Section II.C.

103 *See* Fred A. Rainey, *Academic Publications, in* Designing the Microbial Research Commons, n. 3, at 111.

second edition was prepared and released.[104] However, both the first and second editions of Bergey's Manual were published commercially by Springer, with the royalties used to support the activities of the Trust.[105]

Access to and use of Bergey's Manual is thus conditioned on Springer's own copyright and contractual policies, and Springer attempts to obtain assignments of copyright from authors of the underlying sources.[106] Although the first-sale doctrine of copyright law applies in print media, actual use and reuse of the data and information are, in principle, constrained by the narrow limitations and exceptions embodied in such laws, as described in Chapter 6, especially in countries that lack a fair-use provision.[107] Moreover, the Bergey's Trust has only recently begun to consider ways and means of making its data available online, in a database format as well as by chapters on the genera and species.[108] For this and other reasons, the Trust could risk lagging well behind the bolder initiatives described earlier.

B. *Online Aggregators of Data and Information about Microbial Materials Available from Public Culture Collections*

Given the important role that public culture collections affiliated with the WFCC play in microbiology,[109] it seems odd that digitally integrated information about their holdings was not readily available until recently."[110] This contrasts with the policies and practices of the National Library of Medicine (NLM), for example, which has established easy to follow links between genes, the diseases they are likely to correlate with, and the various publications about these genes and diseases.[111] Two initiatives have lately addressed this problem, namely, the World Data Center for Microorganisms and the StrainInfo.net bioportal.

104 *Id.*

105 *Id.*

106 *Cf. id.* at 113.

107 *See* Chapter 6, Section II.

108 *See* Rainey, n. 103, at 116.

109 *See* Chapter 4, Section I.A.

110 Peter Dawyndt, *Straininfo: Reducing Microbial Data Entropy*, *in* DESIGNING THE MICROBIAL RESEARCH COMMONS, n. 3, at 115. For early efforts in this direction, see D. Smith et al. (2003), above n. 90, at 289–290, 295–297 (discussing MINE, CABRI, and related initiatives).

111 *See generally* Jerry Sheehan, *Toward a Biomedical Research Commons: A View from the National Library of Medicine at the National Institutes of Health*, *in* DESIGNING THE MICROBIAL RESEARCH COMMONS, n. 3, at 111–14. In 2011, NLM'S base budget was $56,463,000, including $44,281,000 for research funding and $12,182,000 allocated to contracts for the NN/LM. *See* NIH, NAT'L LIBRARY OF MEDICINE PROGRAMS AND SERVICES: FISCAL YEAR 2011, 64 (2011), *available at* http://www.nlm.nih.gov/ocpl/anreports/fy2011.pdf.

1. The World Data Center for Microorganisms

The World Data Center for Microorganisms (WDCM) was first established in 1966, at the University of Queensland, Australia, as the information component of the World Federation for Culture Collections (WFCC). It was subsequently hosted by the RIKEN Culture Collection and the National Institute of Genetics in Japan. In 2011, after a bidding process, it was moved to its current location at the Institute for Microbiology of the Chinese Academy of Sciences (IMCAS), under the direction of Professor Juncai Ma. The WDCM's goals are to promote standards and principles for data sharing, to develop and maintain appropriate databases, and generally to provide information services to WFCC member collections and research communities.[112]

The WDCM's digital information capacities have grown considerably since its move to IMCAS, and its aspirations have become rather ambitious. Three key existing activities are the Culture Collection Information Worldwide (CCINFO), the Reference Strain Catalogue, and the Analyzer of Bio-Resource Citations, a major new undertaking.[113] A fourth initiative envisions a new and expanded version of the WFCC Global Catalogue of Microorganisms, which was the oldest WFCC project of this kind.[114]

The CCINFO aspires to provide basic information about more than 600 public culture collections in some seventy countries that are affiliated with the WFCC.[115] It is a general world-wide directory of all registered member culture collections that covers such categories as personnel, status of the collection, main subjects of coverage, number of strains preserved, availability of cultures, and available catalogues and services. However, access to this database is limited to WFCC member culture collections.

The WDCM Reference Strain Catalogue enables broader and easier access to the reference stains listed by the ISOTC34SC9 Joint Working Group 5 and by the Working Party on Culture Media for the International Committee on Food Microbiology and Hygiene (ICFMH-WDCM). It provides these bodies with a unique system of identifiers for strains recommended for use in quality assurance.[116]

[112] *See Homepage*, World Data Ctr. for Microorganisms (WDCM), http://www.wdcm.org/ (last accessed 5 July 2014); *see also* Juncai Ma, Recent activities of the World Data Center for Microorganisms (WDCM), presentation at the Second Symposium of WFCC-MIRCEN World Data Center for Microorganisms, Beijing, China, June 7, 2012 [hereinafter Juncai Ma Presentation], slides on file with the authors.

[113] *See* WDCM, *Homepage*, n. 112.

[114] *See* http://gcm.wdcm.org.

[115] *See* Chapter 4, Section I.A.

[116] *See* http://refs.wdcm.org/home.htm. *See also* Juncai Ma Presentation, above n. 112.

As of November 29, 2015, it covered 170 strains, 114 species, 43 culture collections, and 55 ISO and other standards.

The Reference Strain Catalogue primarily serves researchers and industrial users concerned with medical devices and health, foodstuffs, environmental applications, water quality, and cosmetics. This and the other WFCC databases discussed later are openly accessible, but they remain expressly subject to copyright protection (to the extent that copyright applies to the selection and arrangement of data),[117] with all rights reserved.

Of particular interest is a recently created database, the WDCM Analyzer of Bio-resources Citations (ABC), which covers citations to the research conducted on strains held in WFCC Culture Collections and their applications. This database allows users automatically to search and find any relevant publications, patents, gene sequences, and genomic information for each strain. At the time of writing this "data warehouse of microbial resources" covered 120,000 publications from 2,900 journals, in addition to nearly 10,000 patents on some 9,000 strains.[118] With this tool, any interested investigator can obtain statistics on the use of specific strains in both publications and patents, while the culture collections can measure the research and applications pertaining to their holdings.

A new and expanded version of the WFCC Global Catalogue of Microorganisms (GCM) is still in the pilot stage. It aims to provide "a robust, reliable and user-friendly system to help culture collections to manage, disseminate and share the information related to their holdings."[119] It will also provide a uniform platform for the scientific and industrial communities to access the entire range of microbial resource information emanating from affiliated culture collections.[120]

Plans for the GCM call for integrating the three existing WDCM databases into one broader, more comprehensive tool that would offer many new services. In particular, it would provide documents and training materials on standards, catalogues, software and hardware, data quality control, and human resources. It would more fully integrate data and metadata pertaining to strains and species. It would facilitate in-house data management for small collections, promote data sharing, as well as greater industrial utilization of microbial genetic resources. The GCM also aspires to provide digital tools for tracking both patents and benefit-sharing under the Convention on Biological Diversity, as part of a cooperative venture with regional and national nodes.[121]

117 *See* Chapter 6, Section II.C.

118 *See* WDCM, *Homepage*, n. 112.

119 *See id. See also* Juncai Ma Presentation, n. 112.

120 *See generally*, http://gcm.wdcm.org. *See also* Linhuan Wu, et al., *Global Catalogue of Microorganisms (GCM): A Comprehensive Database and Information Retrieval, Analysis, and Visualization System for Microbial Resources*, 14 BMC Genomics 933 (2013).

121 *Id.* For the CBD, see Chapter 3, Sections I & IV.

One of the primary motives behind the Global Catalogue of Microorganisms was to encourage more WFCC member collections to publish online, or at least in printed catalogues, on a regular basis, in order to better disseminate information about strains for science and industry. As matters stand, fewer than one-sixth of the collections registered in CCINFO have their catalogues online, "which greatly hinders the visibility and hence accessibility of strains."[122] At the time of writing, the GCM database covers over 330,000 strains and 43,000 species based on information provided by 67 collections in some 33 countries.[123] It is projected that GCM will cover more than two hundred collections by the end of 2015, which would, in theory, "open up the culture collection community to all microbiologists," and give the collections a higher profile and more visibility.[124]

While the Global Catalogue of Microorganisms thus constitutes a major step forward that is long overdue, and its technical base appears sound, there are nonetheless some problems that bear mentioning. First and foremost, only 200 of over 600 member collections are likely to provide data, while a substantial percentage of even the reporting collections are said to lag well behind either WFCC Global Standards[125] or OECD standards for Biological Resource Centers.[126] The fact that 400 or more culture collections may not participate in GCM poses serious questions about the stability and composition of the world's public collections as basic scientific infrastructure.

As noted in Chapter 3, there are also serious concerns about the future availability of microbial and related genomic data emanating from developing countries under the Nagoya Protocol to the Convention on Biological Diversity, which emphatically assert proprietary claims to such data as well as to biological materials.[127] Finally, the intellectual property policies of the WDCM need reconsideration in view of the growing availability of reference databases and other digital information under a less proprietary and more usable formats.

These queries aside, the GCM as currently being implemented constitutes a promising step toward a digitally integrated system for materials, data, and information in the field of microbiology. As such, it could become a major building block in constructing the redesigned Microbial Research Commons envisioned in Part Four.

[122] *Mission Statement for GCM*, WDMC, www.wdcm.org (last accessed 9 Apr. 2014).
[123] http://gcm.wdcm.org.
[124] *Id.*
[125] *See* further Chapter 9, Section II.B.1 (Governance); *see also* Chapter 4, Sections I.A & B.
[126] *See* Chapter 4, Sections I.B & C.
[127] *See* Chapter 3, Section IV.A.

2. The StrainInfo Bioportal

The StrainInfo.net bioportal (later referred to as "StrainInfo") is based at the University of Ghent in Belgium. It was established to provide a technology platform that would integrate the data from disparate culture collections and attempt to resolve ambiguities surrounding the nomenclature used in the different systems. As the StrainInfo project managers described their own activities:

> ... It brings together the biological material kept at multiple biological resource centers into a single portal interface, with direct pointers to the relevant information at the collections' websites, and provides both historical traces and geographic distribution of the strains they keep in culture. In addition, this information is automatically linked to related sequences in the public domain and refers to all known scientific publications that deal with the organism.[128]

The project thus sought to link and integrate all taxonomic names appearing in the bioportal with key external taxonomic information sources. Additional links would provide direct access to all organisms that were subjects of completed or ongoing whole-genome sequencing projects,[129] with a view to addressing the gap between the availability of genomic information generally and the specific availability of the sequenced organisms in public culture collections.[130]

StrainInfo's funders thus hoped to consolidate strain information with relevant sequences and literature references available from public repositories.[131] In particular, StrainInfo's digital networking system made it possible to incorporate a literature component well beyond the minimum levels elsewhere available. Many scientists could also benefit from the integration of full text search with data mining capabilities, which could further reduce search costs through manual or automated processes. These benefits become increasingly feasible as self-archiving in institutional repositories or on personal websites (with CC licenses) and online open-access journals become more prevalent in microbial science generally.[132]

[128] *See* Bart Van Brabant et al., Navigating microbial space using the StrainInfo.net Bioportal, *available at* http://bioinformatics.cs.vt.edu/~murali/conference-fayfaars/2007-ismb-eccb/ISMBECCB07/Posters/N78Dawyndt.pdf (last accessed 9 Apr. 2014). [hereinafter Van Brabant et al.]. *See also*, *Homepage*, STRAININFO , www.straininfo.net (last accessed 9 Apr. 2014).

[129] Van Brabant et al., n. 128. StrainInfo devoted special attention to detection and correction of errors within equivalence classes due to irregularities in data provided by diverse information sources. It designed automated tools to detect discrepancies in the consistency of the integrated information. Without such quality controls, StrainInfo estimated that about 12 percent of the bacterial type strains would result in erroneous merges into single equivalence classes. *Id.*

[130] Dawyndt, n. 110.

[131] *Id.* at 2.

[132] *See, e.g.*, James Boyle, *Mertonianism Unbound? Imagining Free, Decentralized Access to Most Cultural and Scientific Material*, *in* UNDERSTANDING KNOWLEDGE AS A COMMONS: FROM THEORY TO PRACTICE 123, 123–44 (C. Hess & E. Ostrom eds., MIT Press 2007).

To the extent that StrainInfo would have succeeded in automatically integrating large quantities of knowledge about specific available holdings in response to specific research queries, it could greatly accelerate the research process and enhance the scientific potential of the entire system of culture collections.[133] Of particular importance, StrainInfo digitally tracked and re-encoded the different names assigned to the same materials by different collections and contributors, and it provided links to each of these collections. StrainInfo also added value to the aggregate data by enabling users to map the history of given strains and to develop tools for conducting automatic queries and other research functions pertaining to given strains.[134]

Many of StrainInfo's technical goals depended on its Integrated Strain Database, "a central knowledge base that ... cumulatively learns about the equivalence relation that exists amongst the strain numbers assigned to biological resources in a global research context."[135] At last report, information had been gathered from some sixty cooperating microbial culture collections that covered all the Earth's continents and range from small niche-specific research collections to large general-purpose service collections.[136] Information extracted from two lists of bacterial type strains was also incorporated.[137]

[133] It was one of several major initiatives seeking to implement and exploit emerging semantic web capabilities. *See, e.g.*, John Willbanks, The Digital Commons: Infrastructure for The Data Web, paper presented at "Global Science and the Economics of Knowledge-Sharing Institutions," 2d. Communia Int'l Conference, Turin, Italy, 29–30 June 2009, *available at* http://www.communia-project.eu/node/290.

[134] *See generally* Dawyndt, n. 110. Two technical problems in particular challenged StrainInfo's own efforts to develop an effective digital knowledge hub. One was the persistent problem of ambiguous nomenclature that plagues microbial science more generally. The other was persistent irregularities in the quality of data and the risk of cumulative errors this situation presents. StrainInfo recognized these technical problems and was making efforts to resolve them. For example, they aimed to develop a cross-referencing system to permit integration of different autonomous and heterogeneous data sources, in order to reduce the amount of data duplication among providers and to establish uniform pathways for navigating the cumulative data environment. They also continuously monitored the overall data quality provided by different web services.

Crucial to this operation was the establishment of a durable cross-referencing scheme based on the assignment of globally unique digital object identifiers (DOIs) for the different microbial resources within the relevant digital domain. New unique identifiers are also assigned to strain numbers that were not previously encountered during the integration process, with a view to resolving ambiguities stemming from an initially parochial discovery and identification process. *See* Peter Dawyndt et al., StrainInfo.net: Breaking Down Information Barriers Into Holistic Data Integration Scenarios Using Globally Unique Persistent Identifiers, paper presented at Workflows Management: New Abilities for the Biological Information Overflow, NETTAB 2005 Conference, Naples, Italy, 5–7 Oct. 2005, at 2 [hereinafter Dawyndt et al., StrainInfo.net], *available at* http://www.nettab.org/2005/docs/NETTAB2005_DawyndtAbstract.pdf.

[135] *Id.* at 1–2.

[136] *Id.*

[137] *Id.* at 2.

While StrainInfo aimed to provide a one-stop portal to more than forty digitized catalogues within the WFCC, not all the data on strains kept and managed by the culture collections are made publicly available when reached through this portal. For example, many catalogues in developing countries still may not be digitized or may not be put online even if digitized.[138] Moreover, in-house data collections belonging to research laboratories associated with the culture collections may either keep their data secret for commercial reasons or may lack sufficient incentives to curate and upload the data.

Numerous technical obstacles stand in the way of any integrated data and information portal, such as StrainInfo. The tasks of preserving and curating digital data collections over time present both technical and financial issues that are often hard to resolve. Merging heterogeneous data sources into an interoperable system requires broadly used standards, in addition to the need for technical interoperability of formats, structure, and media. These problems are compounded by the fragmented and distributed character of microbial and other databases, some of which are generated by relatively small laboratories or individuals using different standards, methodologies, and technical solutions. Moreover, all the major databases linked by StrainInfo, and their multiple subsidiary databases, operate under widely different data sharing and access policies.

The refined technical solutions devised by StrainInfo were portable to other initiatives seeking to digitally integrate data, literature, and materials and StrainInfo is accordingly listed as a partner in the WDCM's Global Catalogue of Microorganisms. StrainInfo's technical expertise is thus at the service of the WDCM's much broader, better funded field of operations. Even if StrainInfo fails to survive its reported funding problems, its technical proof of concept and expertise will have provided an essential stepping stone to the WFCC's still more ambitious digital initiatives.

C. Understanding the Data Sharing Movement and Its Future Potential

The examples of microbiology-related databases provided earlier are but some of the burgeoning initiatives in which researchers have been pooling data in order to achieve goals that lie beyond the capacity of single investigators, or even entire laboratories. These initiatives are largely, though not uniformly, voluntary. They may be more or less hypothesis driven, as distinct from purely reference collections. Their data policies may be more or less unrestricted. Their organizational structures may be more or less formal.[139]

[138] *See* Section II.B.1 (discussing WDCM).

[139] *See, e.g.*, Jorge L. Contreras and Jerome H. Reichman, *Sharing by Design: Data and Decentralized Commons–Overcoming Legal and Policy Obstacles*, 350 *Science* 1312 (2015).

While such data sharing initiatives are far too numerous even in the field of microbiology to thoroughly cover here, taken together they represent a movement that defies simple categorization. For example, some operate at the governmental level, where – in the United States and the United Kingdom at least – the funders' concerns to make research data freely available, either before or after publication, are now routinely applied.[140] Others operate at the university level, where the data release policies vary considerably, but funders' pressures for more open availability are palpable.[141] Still others are public-private partnerships, whose pooled data resources may or may not be available to outsiders under a "semicommons" approach.[142]

[140] For example, at the government level in the U.S. is the National Institute of Allergy and Infectious Diseases (NIAID), which maintains the Influenza Research Database. One of five Bioinformatics Resource Centers funded by NIH, it is an iterative and comprehensive publicly available database and analysis resource to search, analyze, visualize, save, and share data for influenza virus research. *See* IRD Mission Statement, INFLUENZA RESEARCH DATABASE (IRD), http://www.fludb.org/brc/staticContent.do?decorator=influenza&type=FluInfo&subtype=Mission (last accessed 9 Apr. 2014). A closely related initiative is the NIAID's BioHealthBase Bioinformatics Resource Center (BRC), which is a public resource for the study of specific biodefense and public health pathogens. *See also*, Burke Squires et al., BioHealthBase: Informatics Support in the Elucidation of Influenza Virus Host-Pathogen Interactions and Virulence, 36 NUCLEIC ACIDS RESEARCH 497 (2008), available at http://nar.oxfordjournals.org/content/36/suppl_1/D497.full.pdf+html.

See generally, OSTP Memorandum 2013 and Executive Order 2013, above n. 10. For the EU's efforts to foster open data policy and a seamless web, *see* EUROPEAN UNION, RIDING THE WAVE: HOW EUROPE CAN GAIN FROM THE RISING TIDE OF SCIENTIFIC DATA (October 2010), available at http://cordis.europa.eu/fp7/int/e-infrastructure/docs/hlg-sdi-report.pdf; and Recommendation of the European Commission of 17 July 2012 on Access to and Preservation of Scientific Information, 2012/417/EU/, L184/19 O.L. E.U (21 July 2012).

[141] At the university level, *see, e.g.*, Ribosomal Database Project (RDP) at the Center for Microbial Ecology, Michigan State University (14 May 2013), http://rdp.cme.msu.edu/ (providing ribosome related data services to the scientific community, including online data analysis, rRNA derived phylogenetic trees, and aligned and annotated rRNA sequences; RDP services' data are freely available under a Creative Commons Attribution – Share Alike License). *See generally*, Royal Society, *Science as an Open Enterprise: Final Report* (June 2012), available at https://royalsociety.org/~/media/policy/projects/sape/2012-06-20-saoe.pdf.

[142] For an example of a public-private partnership that is willing to share data publicly, see the Structural Genomics Consortium, a not-for-profit, public-private partnership with the directive to carry out basic science of relevance to drug discovery; core mandate to determine three-dimensional structures on a large scale and cost-effectively targeting human proteins of biomedical importance on a precompetitive basis; and committed to making research outputs (materials and knowledge) available without restrictions on use. *See further* A.M. Edwards, *Bermuda Principles Meet Structural Biology*, 15 *Nature Structural & Molecular Biology* 116 (2008); A.M. Edwards, *Open-Source Science to Enable Drug Discovery*, 13 *Drug Discovery Today* 731 (2008); J. Weigert, *The Case for Open-Access Chemical Biology: A Strategy for Pre-Competitive Medicinal Chemistry to Promote Drug Discovery*, 10 *EMPO REP.* 941 (2009). *See also Homepage*, Alliance for Cell Signaling http://.atcs.org.aboutus/index/html (last accessed 9 Apr. 2014), a public-private partnership with funding from NISAID, NIH, NGMS; NIH data policy adopted; all AFCS legacy data plus new data developed under Alliance funding to be placed in the public domain via a data download website; available to users for either commercial or noncommercial purposes; and no reach-through rights to future discoveries.

In what follows, we first evaluate the costs and benefits of these diverse data sharing initiatives and then consider where this movement may be headed in the future. In so doing, we will focus particular attention on new path-breaking developments that transcend the confines of existing data sharing initiatives en route to the formation of fully elaborated, thematically driven research commons.

1. Benefits and Drawbacks of the Data Sharing Ethos

At the outset of this chapter, we saw a tendency to view data accessibility in the ambit of molecular biology as trapped between two extremes: the single researcher's proclivity to hoard data, on the one hand, and the mandatory accretion of reference data, especially enormous amounts of publicly funded genomic data on the other. Community pressures, particularly from major funders, have helped to secure and partially enforce a normative duty to release supporting data immediately on or shortly after publication of research findings.[143] We also saw strenuous efforts to ensure early release of basic reference data in genomic (and other) fields, which fuels nonhypothesis driven research while generating formidable problems of curating and maintaining the resulting deluge of data at acceptable costs.[144]

On closer analysis, however, we can discern an evolving continuum of mostly voluntary data sharing practices that fall between these two extremes and defy simple categorization. The efforts of funders pushing top-down directives to pool research data as a community resource proved that data sharing could be made to work and that it paid palpable dividends in the form of research advances. The inability of single researchers to manage the deluge of data often spurred them to look for alternatives and, given the success of the reference data pools, voluntary pooling became an ever more tempting option.

As the data self-generated by single research projects grow larger and less manageable, so do the risks that researchers could not exploit them in sufficiently

For an example of a public-private partnership whose data are available only to qualified participants, see *Homepage*, BIOMARKERS FOR TUBERCULOSIS CONSORTIUM, http://www.biomarkers-for-tb.net (last accessed 9 Apr. 2014). An example of a public-private partnership that evolved away from a partly restricted approach (charging commercial users an annual fee) to freely accessible approach, with a CC-BY-NC Creative Commons license, *see* the Universal Protein Resource (UniProt) databases for metagenomic and environmental data, now funded by EMBL, NIH, and the Swiss government. For the semicommons approach generally, see, e.g., Allarakhia, n. 3, at 145–50; *see also* Minna Allarakhia et al., *Modeling the Incentive to Participate in Open Source Biopharmaceutical Innovation*, 40 *R&D* MGMT. 50–66 (2009); Minna Allarakhia & S. Walsh, *Managing Knowledge Assets under Conditions of Radical Change: The Case of the Pharmaceutical Industry*, 31(2–3) TECHNOVATION 105–117 (2011).

143 *See, e.g.*, Contreras, *Data Sharing*, n. 15 and Contreras, *Bermuda's Legacy*, n. 26.

144 *See* NATIONAL RESEARCH COUNCIL (NRC), ENSURING THE INTEGRITY, ACCESSIBILITY, AND STEWARDSHIP OF RESEARCH DATA IN THE DIGITAL AGE (National Academies Press, 2009).

productive and reliable ways without entering into sharing and pooling arrangements with other like-minded researchers.[145] These arrangements magnify the impact of the whole well beyond any benefits that may be attainable by users of separate data components operating on their own. The combined or pooled data asset that emerges, formally or informally, from such arrangements then stimulates and supports still more joint and single research projects, largely hypothesis driven, that are based on the reciprocity benefits flowing in different directions from the pooled resource.[146] The deluge of data that characterizes the New Biology paradigm thus also generates a still unfolding institutional response that transcends preexisting legal and normative constraints on access to data and whose overall dimensions and future potential still remain largely uncharted.

A. THE PUBLIC-GOODS APPROACH. From a broader perspective, scientific data, when publicly funded acquire public-good characteristics[147] that make them suitable for open availability and reuse on digital networks. Public goods are both nondepleteable and nonrivalrous by definition. Unlike a physical private good that will be appropriated the first time it is used, digital information is by definition nondepleteable, since it cannot be exhausted by use. In fact, networked information gains in value the more that it is used.

Information can become rivalrous, however, when it is legally or technically protected and kept from use by others. In this sense it is not a "pure" public good, although protecting it is often inefficient. Protecting publicly-funded research data and information is especially inefficient, not only because the taxpayer has already

[145] This was a key postulate, for example, in forming the International Cancer Genome Consortium. *See* ICGC, *Overview*, n. 84.

[146] For reciprocity benefits, *see* Paul David, *The Economic Logic of "Open Science" and the Balance Between Private Property Rights and the Public Domain in Scientific Data and Information: A Primer*, *in* THE ROLE OF SCIENTIFIC AND TECHNICAL DATA AND INFORMATION IN THE PUBLIC DOMAIN 19, 19–34 (J.M. Esanu & P.F. Uhlir eds., Nat'l Acad. Press 2003); *see also* Allarakhia, above n. 3. At the investigator level, there are both incentives and disincentives to share research data. Positive incentives and motivations are reported in VEERLE VAN DEN EYNDEN & LIBBY BISHOP, SOWING THE SEED: INCENTIVES AND MOTIVATIONS FOR SHARING RESEARCH DATA, A RESEARCHER'S PERSPECTIVE (Knowledge Exchange, 2014), http://repository.jisc.ac.uk/5662/1/KE_report-incentives-for-sharing-researchdata.pdf (last accessed 27 Mar. 2015) [hereinafter, VAN DEN EYNDEN & BISHOP, SOWING THE SEED]. For disincentives absent a mandate from the top down, *see, e.g.*, CHRISTINE BORGMAN, BIG DATA, SMALL DATA, NO DATA (MIT Press, 2014) [hereinafter, BORGMAN, BIG DATA].

[147] For a detailed definition and discussion of the economic considerations of public goods, see, e.g., HAL R. VARIAN, MICROECONOMIC ANALYSIS (3d. ed., Norton 1992); *see also* Inge Kaul et al., *Defining Global Public Goods*, *in* GLOBAL PUBLIC GOODS: INTERNATIONAL COOPERATION IN THE 21ST CENTURY (I. Kaul et al. eds., United Nations Development Program 1999); Joseph E. Stiglitz, *Knowledge as a Global Public Good*, *in* GLOBAL PUBLIC GOODS, at 308–324.

paid for its production, but also because it gains in value from broad dissemination and it is reusable.[148]

Openness as the default option is especially desirable when the activity that has generated the data – publicly funded research – is itself also a public good, unless there are compelling reasons to keep the resulting data protected, such as national security, privacy, confidentiality, or an economic market that would make it more efficient and productive to invoke intellectual property rights. These premises favor keeping digitally networked publicly, funded research data and information openly available and reusable.

1) *Benefits*. Open availability and unrestricted reuse of research data online affords many advantages.[149] Perhaps the most compelling stem from a purely scientific perspective. It is customary and ethical to make all data that support published research results available for the purpose of independent verification and replicability. Openness overcomes artificial barriers and thereby promotes interdisciplinary, inter-sectoral, and transnational research,[150] which in turn can stimulate new scientific results and new types of research. It facilitates the testing of novel or alternative hypotheses and methods of analysis. These benefits, in turn, empower citizen scientists and help to foster serendipitous results, enabling the exploration and formation of topics not envisioned by initial investigators who produced the data or by the established research community.

The open availability of data is also essential for successfully implementing the integrative digital vision of a "New Biology." So long as data remain in the public domain, or are accompanied by a private-law waiver of rights under a common-use attribution-only license (or by equivalent permissions), automated knowledge discovery becomes legally possible, as explained in Chapter 6. This is an important feature of present-day scientific research as it struggles to address the data deluge, which only automated knowledge discovery tools make manageable.

[148] *See, e.g.*, Joseph Stiglitz et al., The Role of Government in a Digital Age (Computer & Commc'ns Industry. Assn., 2000).

[149] For a description of the advantages of open availability and unrestricted reuse of research data online, *see, e.g.*, Peter Arzberger et al., *An International Framework to Promote Access to Data*, 303(5665) Science 1777–78 (2004). Paul F. Uhlir & Peter Schröder, *Open Data for Global Science*, 6 Data Sci. J. 17 (2007), *available at* http://www.spatial.maine.edu/icfs/Uhlir-SchroederPaper.pdf; Bd. on Research Data & Info. and U.S. Committee for CODATA, The Socioeconomic Effects of Public Sector Information on Digital Networks: Toward a Better Understanding of Different Access and Reuse Policies: Workshop Summary (Paul F. Uhlir, ed., Nat'l Acads. Press, 2009), *available at* http://www.nap.edu/catalog.php?record_id=12687; *see also* Paul F. Uhlir, *UNESCO, Policy Guidelines for the Development and Promotion of Government Public Domain Information* 51 (2004), *available at* http://unesdoc.unesco.org/images/0013/001373/137363eo.pdf.

[150] *See generally*, Van Den Eynden & Bishop, Sowing the Seed, above n. 146.

A nonproprietary status facilitates legal interoperability and the creation of new datasets by combining an existing collection with unrestricted data from other, even multiple sources.

Openness facilitates the education of new researchers who can obtain the broadest possible array of factual information available for instruction and use in experimentation, and it also enables studies on data collection methods and measurement. From an ethical standpoint, it avoids discriminating between those who can and those who cannot afford to pay for data collected by and for the public.

As a socioeconomic matter, open access to publicly funded data delivers still other benefits. It avoids the inefficiencies of erecting artificial barriers to exclude other potential users, including the unnecessary regeneration of data and duplications of research, or the inability of others to reuse unique datasets pertaining to phenomena that cannot be independently observed after they occur. Public data release also helps to curb the space and scope for patents on data-related inventions. More broadly, open data available online supports economic growth and social welfare, as well as democratic governance with meaningful citizen participation. For these and other reasons, the default rule of openness generally provides the greatest returns from public investments in scientific research, except in specific cases where there may be some demonstrated need to capture value through restrictions on access and use.

2) *Disadvantages*. Despite the advantages summarized above, not all publicly funded research data can or should be made openly available, especially online, and there are legitimate competing interests to be taken into account. For example, scientists are notably reluctant to use "other people's data," and often prefer to generate their own data for personal use. From a user perspective, unless the databases are well organized and curated for community use (particularly with regard to the immediate community of researchers contributing data to any given pool), there is a suspicion, often justified, that the data may be of substandard quality, with unknown gaps or defects. These defects may compromise research results or at least impose a steep learning curve for their use, although the participants in a voluntary pool may themselves adopt quality standards to address this risk.[151]

Absent either a mandatory early release policy or the formation of an organized voluntary community resource, there are few incentives – and a number of strong disincentives – to making an individual researcher's data available. For example, one risks forfeiting reputational and reward benefits in releasing data before

[151] *See, e.g.*, ICGC, *Overview*, above n. 84; *see also* Dawyndt et al., StrainInfo.net, above n. 134. For a general review of actual and perceived barriers to sharing research data, *see* Borgman, Big Data, above n. 146.

publication, and perhaps even thereafter, if the same dataset serves as the object of further research. The originating data supplier may be "scooped" by a second comer, which would undermine one of the highest goals of nonprofit, academic, and public-sector research, and would be inconsistent with the ethical norms of the scientific community.

Scientists may also view the efforts needed to provide better data for sharing purposes as an "unfunded mandate." Making the data usable by others can require more organization and documentation of the dataset than the originating scientist has time, money, or human resources to undertake. Although some research funders have begun to recognize this problem in their grant-making policies (over and above the early release mandates and voluntary policy initiatives they have recently supported), most funders still remain reluctant to pay extra for such work. Research institutions are also reluctant to give such efforts much recognition in hiring, promotion, or tenure decisions while publication outlets for pure data-generation work remain limited and lack prestige.

Where the data under consideration are intrinsically of poor quality, the researcher does not want such defects to become broadly known. Conversely, a researcher may be concerned that others will misuse the data and attribute the original source, thus implicating that researcher in erroneous or harmful results. Even when the data are of high quality and the database is properly curated, the downstream users may fail to cite the originators, which robs them of reputational benefits.[152] Still another barrier to data-sharing is the socio-cultural inertia and traditional "small science" mindset associated with some microbiological research projects that we noted in Chapter 1.

B. THE QUASI-PRIVATE GOODS APPROACH. If the success of the large publicly funded data sharing initiatives discussed earlier in this chapter has not only generated an expanding set of parallel initiatives premised on similar open-access policies, it has also stimulated a proliferation of voluntary data-pooling initiatives whose end products are made available only to qualifying contributors or other eligible participants, but not to the research community as a whole. Here, we find another major institutional response to the deluge of data problem, one that often results in the production of so-called "club goods" or semicommons, rather than the public goods associated with the concept of a commons in the full sense of that term. Because

[152] The data citation problem is now being addressed in an organized way. *See, e.g.*, NATIONAL RESEARCH COUNCIL (NRC), FOR ATTRIBUTION – DEVELOPING DATA ATTRIBUTION AND CITATION PRACTICES AND STANDARDS (Paul F. Uhlir ed., Nat'l Acads. Press, 2012), *available at* http://www.nap.edu/catalog.php?record_id=13546. For theoretical aspects of designing a knowledge commons in general, see Chapter 9, Section I.

the data and resulting information still maintain some of their inherent public-good characteristics, despite their more exclusionary aspects, we prefer the "quasi-private goods" designation to that of "club goods."[153]

From an institutional perspective, single investigators who would not willingly contribute data to a fully open repository may nonetheless voluntarily pool their data with others who share common thematic interests. The incentives here are especially strong when such affiliation is driven by the prospect of shared research hypotheses and possible commercial applications that may emerge from a thematically driven pool in the future. Besides the reciprocity benefits flowing from collective management of interrelated and interoperable data resources in these cases, participants share the costs and burdens of curation, management, and maintenance over time[154] and reduce the risks of free-riding by noncontributors – with regard to financial prospects – who do not obtain the same comparative advantages.

To the extent that pooling of data is motivated by a positive cost-benefit analysis of expected gains at both the individual and institutional levels, these gains become potentially greater when contributors to the pool also benefit from a digital infrastructure – geared to the needs of the thematic community – that enables rapid searching, extraction, reuse, and manipulation of all the pooled contents.[155] A common technical infrastructure will likely prove more powerful than any single owner's own software, and it will grow in power when applied to all datasets that are assembled in any given pool. In the long run, technical infrastructure may provide more reciprocity gains to participating researchers than almost any other single component, and it thus becomes crucial to the success of many data pooling initiatives.

Pooling data in a semicommons may seem particularly desirable when the expected research outputs have considerable potential for downstream commercial applications, with the corresponding opportunity to obtain a share of the resulting proceeds by means of reach-through licensing agreements or other legal arrangements. In this and other respects, the semicommons model represents a trade-off between the benefits of open availability summarized earlier and the

[153] *See* Stephen B. Young, Global Exec. Dir., Caux Round Table, *CSR and Public Goods*, CAUX ROUND TABLE (18 May 2010), http://www.cauxroundtable.org/newsmaster.cfm?&menuid=99&action=view&retrieveid=65.

[154] For a game theoretical model concerning reciprocity benefits from sharing data in either commons or semicommons, *see* Allarakhia, n. 3.

[155] *Cf.* FRISCHMANN, n.11. For one promising example, *see* the International Cancer Genome Consortium, ICGC, *Overview*, above n. 84; for another *see* the Micro B3 Consortium (Marine Microbial Biodiversity, Bioinformatics, Biotechnology). http://www.microb3.eu/home (funded by the European Commission) [hereinafter Micro B3]. *See further* the initiatives described later in Section III.A.

risk that major scientific research projects dependent on data sharing might not otherwise become feasible at all.[156]

Although the broader research community arguably stands to benefit less from data-sharing semicommons than from fully open data repositories, at least in the short run,[157] the payoffs to the public are still likely to exceed the benefits to be expected when single researchers, driven by the hoarding instinct, seek to manage the deluge of data on their own. For example, the postulate that information gains in value the more it is used still applies in the semicommons milieu, as it does with even greater force when there are no restrictions on accessing and reusing relevant data at all.

A semicommons arrangement may also help to break down research barriers by promoting interdisciplinary, inter-sectoral, and transnational research strategies. Because the data in question remains "our" data in the eyes of contributors, moreover, there are built-in incentives to keep those data well-organized and curated that fully open repositories may lack or find difficult to sustain. The semicommons approach may also provide – or at least appear to provide – a more direct and certain route to reputational benefits stemming from shared hypotheses and shared research pursuits when members of the subcommunity are obliged not to cheat or scoop one another along the way.

With perhaps fewer prospects of public funding at the outset, semicommons data-sharing initiatives may depend more on private or university funding to stay the course necessary to achieve the subcommunity's predetermined research goals. At the same time, voluntary data-sharing initiatives that operate as semicommons may establish a specified set of attainable goals, beyond which there is no further reason to sustain the costs of maintaining the subcommunities' shared data resources. The contributors might then decide to deposit those shared resources in a public repository, where they may reside long after the semicommons that produced them has wound up its operations.

A particularly promising feature of the semicommons approach to data sharing is the possibility that internal collaboration may generate new data management tools directed to the collective research goals of that same subcommunity.[158] Unexpected commercial opportunities may attend the creation of such tools, and may stimulate the subcommunity to develop its research in new directions that were not previously foreseen. As the sub-community expands its research capabilities and moves in new

[156] For example, data protected by patient confidentiality agreements or regulations could sometimes only be made available to qualified members of a semicommons. *See, e.g.*, ICGC, *Overview*, n. 84.

[157] Recall that the ultimate payoffs from large-scale reference collections have elicited cost-benefit concerns over time, especially when there is a possible shortage of hypothesis-driven research projects based on the collective data tools. *See* Section I.B.3 in this chapter.

[158] Promising examples in this regard are the ICGC, *Overview*, above n. 84, and Micro B3, n. 155.

directions, it may release some preexisting club goods, whose research value it had already exploited, into the public domain, while accumulating newer data assets on a more restricted basis.

The more successful these voluntary initiatives become, the more they may consider the possibility of eventually transforming themselves into some form of hybrid, public-private arrangements, with far greater research capabilities than even their founders initially anticipated. For all these reasons, the formation of voluntary, data-sharing semicommons represents an alternative institutional response to the deluge of data challenge that merits attention and support.

2. Beyond the Public Versus Private Distinction

So far, we have identified several broad types of responses to the deluge of data and the challenges for scientific research that it poses, notably, the large-scale top-down repositories of reference data and the more hypothesis driven, voluntary pooling arrangements that may either make data publicly available or restrict access and use to a specified set of contributors. Looking to the future, one may logically ask how these sharing initiatives may be further stabilized and refined, and where they may ultimately be heading in terms of broader science policy.

In this context, we can envision newer and still more ambitious models of collaborative research that carry the goal of sharing upstream research assets to a higher level in both quantitative and qualitative terms. More specifically, such initiatives would move beyond providing passive, research-supporting infrastructure to spawning dynamic research-generating platforms. This transformation occurs when a given thematic community or subcommunity organizes its upstream research assets, its technical and computational research tools, and its hypothesis-driven research goals in such a way that it successfully generates a set of self-sustaining activities, supported by a constantly evolving infrastructure that transcends the very notion of "repository" or "portal" to become an autonomous knowledge hub.[159]

The empirical evidence suggests that initiatives along these lines are already emerging in the life sciences, including microbiology. A small but growing number of sharing initiatives have moved beyond the familiar data-pooling arrangements. These undertakings combine reference data with hypothesis-driven data; and they can also integrate both materials and even open literature – "gray" or peer reviewed – into their digitized infrastructure. These resources, in turn, serve the entire thematic community or subcommunity as a whole, which necessitates a more formal governance structure to manage both the research assets that constitute a major component of its infrastructure and the research activities that the community

[159] The ICGC may be moving in this direction, especially as it envisions clinical applications. *See* nn. 84–89 & accompanying text. The Micro B3 project, n. 155, has definitely evolved in this direction. For a description of selected other examples, see Section III.A.

supports and generates. We call this digitally networked scientific communications model "Open Knowledge Environments," and we describe and analyze them more fully in the next section.

III. BUILDING TRANSNATIONAL OPEN KNOWLEDGE ENVIRONMENTS

A number of emerging initiatives in the field of microbiology already manifest many of the characteristics we associate with an Open Knowledge Environment (OKE). Most, but not all, of these initiatives operate in a university or public-sector research setting. We have chosen to describe the following: (1) The Genomic Standards Consortium's (GSC) "interactive portal and open access journal" at Michigan State University; (2) the Community Cyberinfrastructure for Advanced Marine Microbial Ecology Research and Analysis (CAMERA) at the University of California at San Diego; (3) the Systems Biology Knowledgebase (KBase) of the U.S. Department of Energy; and (4) the Microbiology of the Built Environment (MoBE), a consortium of four university-based research centers funded by the Sloan Foundation.

All four of these initiatives focus on providing open and interactive information resources of interest to a given thematic community. Each of them possesses different constitutive elements of a fully developed OKE as we envision it. While the GSC initiative confines its attention to standards, it has already developed an online, open-access journal. The other three initiatives do not produce or host a journal, but they do emphasize electronic linkages to open data and literature among a broad set of other services and products provided to the microbiological community.

A. *Examples of Incipient Open Knowledge Environments on the Frontiers of Microbiology*

In what follows, we first discuss the planned contributions that each of these OKEs envisions making to the advancement of microbiology. We then discuss the future of the Open Knowledge Environment model, drawing on the lessons from these empirical examples.

1. The Genomic Standards Consortium (GSC) – Interactive Portal and Open Access Journal

With a tagline of "Innovation through Collaboration,"[160] the three-fold mission of the GSC was initially to (1) implement new genomic standards, (2) pioneer new

[160] Dawn Field et al., *The Genomic Standards Consortium*, 9 PLoS Biology (2011) [hereinafter Field et al. (2011)], *available at* http://www.plosbiology.org/article/info%3Adoi%2F10.1371%2Fjournal.pbio.1001088.

methods of mining and sharing contextual data or metadata pertaining to these standards, and (3) coordinate and harmonize genomic data collection and analysis efforts worldwide.[161] The Genomic Standards Consortium (GSC) thus aspired not only to become an important OKE in its own right, but also to provide vital infrastructure and standardization services that would underpin the efforts of numerous other OKEs.

The GSC's activities initially fell into five categories.[162] First, and perhaps foremost, was the development of so-called "minimum information" checklists intended to facilitate submission of core information along with genomic data when those data were provided in publications and public databases.[163] The MI(x)S standard information template covers such items as habitat, geographic location, data type, environmental information, and how the sequencing was accomplished.[164] Furnishing genomic sequence data with these additional annotations provides considerable value for other researchers working in the field. The primary public genomic sequence databases (for example, the International Nucleotide Sequence Database Collaboration (INSDC)[165]) have relied on data providers to annotate their submitted sequence data.[166] In the absence of guidelines, such as the MI(x)S checklists, most data arrives with little or no added information at all, which greatly reduces the chances that users may efficiently mine or employ the data.[167]

The GSC's second main activity was just as important – fostering so called "implementation projects" that help make compliance with MIGS/MIMS/ MIMARKS standards an actual reality.[168] Some of these implementation projects are designated "GSC Core" projects, initiated by members of the consortium itself; others are "Supporting Projects" in which the GSC reaches out to other research

[161] *See id.* at 2; *see also* Genomic Standards Consortium (GSC) *Mission*, GENOMIC STANDARDS CONSORTIUM (7 Aug. 2010), http://gensc.org; GSC, Bylaws of the Genomic Standards Consortium, art. II, *ratified* Apr. 6, 2011 [hereinafter GSC Bylaws].

[162] *See* GSC, Mission, n. 161; *Genomic Standards Consortium* (GSC) *Projects* (6 Feb. 2012) [hereinafter GSC Projects], http://gensc.org/.

[163] Field et al. (2011), n. 160; *see also* Pelin Yilmaz et al., *Minimum Information about a Marker Gene Sequence (MIMARKS) and Minimum Information about any (x) Sequence (MIxS) Specifications*, 29 *NATURE BIOTECHNOLOGY* 415, 415–419 (2011); *see generally* Dawn Field et al., *The Minimum Information About a Genome Sequence (MIGS) Specification*, 26 *NATURE BIOTECHNOLOGY* 541 (2008) [hereinafter Field et al. (2008)].

[164] Field et al. (2011), n. 160, at 2.

[165] *See* International Nucleotide Sequence Database Collaboration (INSDC), *Homepage*, http://www.insdc.org (last accessed 9 Apr. 2014). Constituent databases of the INSDC currently include the DNA Data Bank of Japan (DDBJ), the European Nucleotide Archive (EBI-ENA) and GenBank. *See also* Yilmaz et al., n. 163, at 415.

[166] Yilmaz et al., n. 163, at 415.

[167] *See id.*

[168] *See* GSC Projects, n. 162.

groups, databases, and OKEs to facilitate MI(x)S compliance.[169] At a more general level, these projects involve what is essentially the creation of a new "language" that genomic researchers can use to make data more readily usable by scientists around the world.[170]

GSC's third level of activity consisted of additional ontological alignment and harmonization efforts intended to create universal formats for exchanges of all types of "omics" data.[171] Fourth, GSC members have participated in the direct generation of their own checklist-compliant genomic sequence data, as well as active crowd-sourcing of annotations for published data. As an example, when genomic sequences for various strains of Rhinovirus (which causes the common cold) were published,[172] the GSC undertook an open community curation effort to gather more information about each strain referenced in the publication.[173]

Finally, a fifth and also very significant undertaking of the GSC was to change the face of genomic data publication. To this end, the GSC created its own journal, an online, fully open-access publication called Standards in Genomic Sciences (SIGS).[174] SIGS is an open, community-based effort that combines biological data, genomic standards, and peer-reviewed content.[175]

SIGS was initially founded with grant support from the U.S. Department of Energy and Michigan State University, but became self-supporting in late 2011.[176] Publishing in SIGS provides researchers with an opportunity to make claims about their submitted datasets, such as significant features, biological niche, or similar unique qualities.[177] The opportunity to expound on the significance of a genomic dataset is often not available in space-limited, general-interest scientific journals, and the fact that SIGS is open access means that the genomics field at large may actively participate in the standardization of an otherwise insurmountable data

169 *Id.*

170 This was also an objective of the StrainInfo project described earlier in Section II.B.2.

171 *See* GSC Projects, http://gensc.org/, above n. 162. Examples include EnvO, an ontology of the environment seeking to create a "controlled, structured vocabulary" for describing environmental data (*Homepage*, EnvO), http://www.environmentontology.org (last accessed 9 Apr. 2014), OBI, a similar "Ontology for Biomedical Investigations," (*Homepage* , OBI (15 Oct. 2012), http://obi-ontology.org), and ISATAB, a universal format for exchanging all types of "omics" data (*Homepage*, ISATAB, http://isatab.sourceforge.net (last accessed 9 Apr. 2014)).

172 Ann C. Palmenberg et al., *Sequencing and Analyses of All Known Human Rhinovirus Genomes Reveal Structure and Evolution*, 324 Sci. Mag. 55 (2009).

173 *See* GSC Projects, n. 162; *see also*, *Community Genome Metadata Curation: Rhinovirus Project*, GSC (6 Aug. 2009), http://gensc.org/.

174 *Homepage*, Standards in Genomics (SIGS), http://www.standardsingenomics.org (last accessed 9 Apr. 2014); *see also* George M. Garrity et al., *Towards a Standards-Compliant Genomic and Metagenomic Publication Record*, 12 *Omics* 157 (2008).

175 *See* SIGS, n. 174.

176 SIGS, *Acknowledgements*, http://standardsingenomics.org/index.php/sigen (last accessed 9 Apr. 2014).

177 *See id.*

deluge problem. As the cost of acquiring a genomic sequence continues to drop, the value of the dataset often resides in its contextual information, which essentially bears on where any given puzzle piece fits into the larger picture.[178]

Membership in the Genomic Standards Consortium is free and open, and defined only by participation.[179] Hundreds of collaborating labs and investigators throughout the world participate in the consortium. As more of a standards-setting, technical service OKE than a full data-generating OKE, the main impact of the GSC's efforts can best be seen in its influence on others. For example, the databases created by the large microbiological OKE, CAMERA,[180] were fully MI(x)S Compliant. The large genomics databases overseen by INSDC, as described earlier, are also MI(x)S-capable, and increasingly datasets submitted there are fully annotated. At least one major research institute, Germany's Max Planck Institute for Marine Microbiology, strongly encourages all of its internal and external collaborators to adopt MI(x)S standards when submitting any genomic or metagenomic sequence to an external database or journal publication.[181]

With such an open membership policy and standardized output, the governance of an OKE such as the GSC becomes particularly important. In April, 2009, the GSC Board was established, consisting of 23 full board members and two institutional liaisons. This Board meets remotely on a frequent basis to formulate strategy and foster additional strategic partnerships.[182] In April 2011, the Board propagated the Bylaws of the GSC, which codified the organization's mission, powers, membership, and governance structure.[183]

Board members were initially further subdivided into three committees made up of Board members only, and three additional committees staffed by a mixture of Board members and general GSC members.[184] Committees of the GSC Board are largely administrative, and respectively provide oversight of incoming funds from the NSF and other funding sources, and of outgoing funds to GSC projects. There is also a general governance and bylaws committee,[185] as well as committees for outreach, technology, and meetings.[186]

[178] This point is stressed in one of the GSC's first journal articles. *See* Field et al. (2011), n. 160, at 2, Box 1. For still other recent activities, *see* n. 160.

[179] *See* Field et al. (2011), n. 160.

[180] *See* Section III.A.2 in this chapter.

[181] *Adopters*, GSC (10 Feb. 2012), http://standardsingenomics.org/index.php/sigen.

[182] GSC, *Governance* (19 Mar. 2012), http://gensc.org/ [hereinafter GSC, *Governance*].

[183] GSC Bylaws, n. 161.

[184] GSC, *Governance*, n. 182.

[185] *Id.*

[186] *Id.*

The OKE as a whole holds plenary meetings twice a year, and some past meetings were held in China, Germany, the United States, and the United Kingdom.[187] At the time of writing, the general GSC membership was divided into seven scientific working groups, which were organized to address specific projects and activities.[188]

Because the GSC remains unattached to any single university or government, funding is particularly important. The GSC was established in 2005 with funding from the UK's National Institute for Environmental *e*Science (NIEeS, defunct as of 2008) and the Natural Environment Research Council (NERC), a governmental agency.[189] NERC later provided further funding for infrastructure, as did the U.S. National Science Foundation.[190] The GSC SIGS journal became self-supporting. As of 2012, the GSC initiated an "Industry Annual Subscription Program," with three levels of private-sector sponsorship.[191] Hence, it can now be said that the GSC is a true public-private partnership.

As the deluge of genomic data continues to grow, one expects the Genomic Standards Consortium will also continue to grow, with its outreach efforts and its fully-integrated journal, and to become even more influential. GSC is potentially a model for other OKEs in its flexible, science-based governance structure, its push for standardization and best practices, and its successful spinoff from university and governmental funding into a full-fledged nongovernmental organization with a mix of public and private institutional members.

2. The Community Cyber-Infrastructure for Advanced Marine Microbial Ecology Research and Analysis (CAMERA)

The CAMERA project had been developing cyber-infrastructure to support data, tools, and resources pertinent to metagenomic analysis from 2001, when it was established, to July 2014, when it ceased to report further operations. In this period, it aspired to become "the definitive repository for metagenomic data and metadata" in the field of marine biology.[192] CAMERA's activities were based on innovative cyber-infrastructure that leveraged "emerging concepts in data storage, access,

[187] *Meetings*, GSC (19 July 2013), http://gensc.org/.

[188] GSC, *Governance*, n. 182.

[189] *Funding*, GSC (11 June 2013), http://gensc.org/.

[190] *Id.*

[191] *Id.*

[192] Community Cyberinfrastructure for Advanced Microbial Ecology Research & Analysis (CAMERA), *Homepage*, http://camera.calit2.net/index.shtm [hereinafter CAMERA, *Homepage*]. However, the project ended in July 2014, and the cited urls in this section are no longer available, although its datasets remain openly available on its residual website. *See* http://camera.calit2.net/. For another ongoing global project in marine microbiology with OKE characteristics, *see* Micro B3, above n. 155.

analysis and synthesis not available in current gene sequence resources." It provided free access to "raw environmental sequence data, associated metadata, precomputed search results, and high-performance computational resources."[193]

This project, based at the University of California at San Diego, avoided the national sovereignty problems under the Convention on Biological Diversity (discussed in Part One)[194] by sampling ocean microbes at 200-mile intervals on the high seas whenever possible.[195] At a lesser range permission would have been needed for exploration of territorial waters. CAMERA's marine metagenomic activities thus operated as a "focus for innovation at the interface of marine environmental science and information technology."[196]

The pace of development and the power of gene sequencing for biological discovery are increasing rapidly with the application of "shotgun" sequencing technology to entire microbial communities. Unlike traditional culture-based sequencing methods, metagenomics arises from a breakthrough sequencing approach that allows scientists to examine the interaction of countless microbial species present at a specific environmental location, in order to better understand the functioning of natural ecosystems. This approach makes it possible to study each gene in the context of its ecology; the composition of the rest of the community; the environmental conditions in which it was discovered; and its relationships to other species with which it is found at other times and places.[197]

Under a seven-year grant of $24.5 million from the Gordon and Betty Moore Foundation in 2006, CAMERA made available "all the metagenomic data being collected by the J. Craig Venter Institute's "Sorcerer 2" Global Ocean Sampling expedition. It sampled diverse microbial communities around the globe, and produced hundreds of full genome maps of marine microbes.[198] The project also developed a suite of tools to enable scientists to analyze resulting data in innovative and more comprehensive ways. As it grew in value to include "new sequences, genes, and gene families, together with their annotations and associated environmental metadata," CAMERA expected its repository and bioinformatics tools to "accelerate understanding of biology and deliver novel biological solutions to important societal challenges in healthcare, energy, and the environment."[199]

[193] *What is CAMERA?*, formerly available at http://camera.calit2.net/about/.

[194] *See* Chapter 3, Section I (CBD) and Section IV (Nagoya Protocol). However, CAMERA expressly agreed to respect CBD requirements when it operated in national waters.

[195] *What is CAMERA?*, n. 193.

[196] *Id.*

[197] *Id. See also* Chapter 1, Section II.B. ("The Revolution in Genetic Science.")

[198] *What is CAMERA?*, n. 193.

[199] *Id.*

Although CAMERA never published a journal, it performed many of the functions that we associate with an Open Knowledge Environment. For example, it provided access to both peer-reviewed and grey literature bearing on its own projects and to the broader literature concerning metagenomics research,[200] in addition to enabling scientists to browse data pertaining to the specific samples it sequenced. With the involvement of scientists from many different countries, it became broadly international and interdisciplinary, in the sense that it expressly depended on continuous inputs from genomics, microbiology, molecular biology, ecology, and related fields.[201] At the same time, CAMERA undertook some training projects, provided interactive online activities that enabled users to report problems with software tools, data, resources, and other services, and to make suggestions for improvements.[202] CAMERA also drew on the expertise and capabilities of various university departments and centers at the host university, including the California Institute for Telecommunications and Information Technology, the San Diego Supercomputer Center, and the Center for Research in Biological Systems.[203] However, CAMERA appears to have lost further funding in 2014.

3. The Systems Biology Knowledgebase (KBase) of the U.S. Department of Energy

The Systems Biology Knowledgebase Project (known colloquially as KBase) is an incipient OKE overseen by the United States Department of Energy under their Genomic Science Program (GSP). KBase was created by the DOE as a means of accessing, sharing, and using the deluge of data that GSP was generating.[204]

KBase is a software and data platform constructed to meet the challenges of systems biology, with regard to predictive behavior and ultimately designing "microbes, plants, and their communities to perform desired functions."[205] By integrating data, tools, associated interfaces, and research results in one unified scalable environment, it enables users to "perform large-scale analyses and combine multiple lines of evidence to model plants and microbial physiology and community dynamics."[206] Moreover, by sharing and publishing both workflows and conclusions,

[200] CAMERA, *Homepage*, n. 192.
[201] *Id.*
[202] *What is CAMERA?*, n. 193.
[203] *Id.*
[204] U.S. Dept. of Energy, *About the DOE Systems Biology Knowledgebase (Kbase)*, http://genomicscience.energy.gov/compbio/kbaseindex.shtml (last accessed 28 March 2015).
[205] U.S. Department of Energy, Kbase. *What Is K Base?*, https:/Kbase.us.whatiskbase, last accessed March 28 2015). See also Folker Meyer et al., "The DOE Systems Biology Knowledgebase: Microbial Communities Science Domain" 1 (2012).
[206] *What Is K Base?*, n. 205.

Kbase aims to provide an integrated environment where knowledge and insights are created and multiplied.

As originally conceived, scientific deliverables for DOE's KBase was closely linked to broader DOE missions, particularly in the area of next-generation biofuels.[207] The DOE estimated that over the next decade, KBase collaborators would work toward greater understanding of microbial metabolic pathways and applications of that knowledge through bioengineering projects.[208] A possible long-term goal was the ability to manipulate expressed genes in engineered bacteria, which would then serve as molecular factories to produce new biofuels and attenuate the harms caused by fossil fuels to human health and the environment.[209] Similar goals were envisioned for plant-related projects, in which plants would be engineered to increase biomass fuel yield and sequester away pollutants in soils and air.[210]

A third focus area in microbial metagenomics would seek to harness enhanced DOE-funded infrastructure in order to more effectively mine the deluge of existing genomic data.[211] Hypothesis-driven projects aim to identify the functions of poorly characterized or unknown genes, as well as to bring all sequenced genes up to GSC MI(x)S annotation standards.[212]

Three "core missions" were initially projected. The first was to create a flexible, high quality framework for inquiry based annotations of genomes,[213] a mission similar to (and operating in conjunction with) that of the Genomic Standards Consortium OKE described earlier.[214] A second core mission was to enable the largely automated creation of metabolic and regulatory computer models that could be compared to existing datasets and then harnessed for specific, applied, hypothesis-driven research.[215] Finally, a third mission was to make all algorithms, software, and data generated by the project completely and openly accessible to the public.[216]

Although not located at a university, KBase is another example of a public OKE that has blossomed into a public-private partnership. Development of KBase began in

[207] *See* U.S. Dept. Energy, Office of Sci., Office of Biological & Envtl. Research, Executive Summary: DOE Systems Biology Knowledgebase Implementation Plan 6 (2010) [hereinafter Kbase Executive Summary], *available at* http://genomicscience.energy.gov/compbio/kbase_plan/KBaseExSum2010.pdf.

[208] *Id.* at 5–7.

[209] *Id.*

[210] *Id.* at 7–8.

[211] *Id.* at 8–9.

[212] *Id.* at 9.

[213] U.S. Dept. Energy, Office of Sci., Office of Biological & Envtl. Research, Overview of the DOE Systems Biology Knowledgebase 3 (2011) [hereinafter Kbase Overview], *available at* http://genomicscience.energy.gov/compbio/ComputingProjects_flyer.pdf.

[214] *See* Section III.A.1.

[215] KBase Overview, n. 213, at 3.

[216] *Id.*

2009 with the help of stimulus funds from the American Recovery and Reinvestment Act (ARRA).[217] Over the next year, five pilot projects were undertaken as proofs of concept, and to illustrate the benefits of integrating and utilizing existing large sets of genomic data.[218] The early Kbase projects (as well as ongoing work) had been led by principal investigators at four DOE National Laboratories: Lawrence Berkeley National Laboratory in California, Argonne National Laboratory in Illinois, Oak Ridge National Laboratory in Tennessee, and Brookhaven National Laboratory in New York.[219] After the success of the pilot projects within these National Laboratories, DOE awarded eleven grants to university and institute-based researchers in 2010 and 2011.[220] The university projects involve a mix of public research universities and both private universities and research institutes.[221]

A major goal of KBase is to enact a "cultural change" in biological research, from cloistered individual laboratories working on duplicative projects to an open community of science where expertise and infrastructure spanning the entire country can be harnessed to tackle bigger problems than any single lab could undertake.[222] This open culture is buttressed by the DOE Office of Science's data-sharing policy, which mandates that, for large-scale genomic sequencing projects, all data must be made fully and publicly available within three months of completion.[223] Notably, the DOE's policy also mentions the efforts of the Genomic Standards Consortium and asks its funded projects to annotate data to GSC standards whenever possible.[224]

For data other than the genomic sequence information generated by the project, temporary embargoes on data release are imposed in order to allow the project participants to publish relevant findings.[225] A default data embargo period for DOE funded projects also appears to be three months after publication for end users.[226] Although there is no specific mention of Creative Commons or similar open-access licensing tools in the DOE project literature, full and open use of KBase data was expected once the project became fully operational.[227]

Governance of the DOE's KBase was still in an incipient stage at the time of writing, when the project had not yet reached full rollout status. As initially envisioned,

[217] *Id.* at 4.
[218] *Id.* at 5–7.
[219] *Id.* at 3.
[220] *Id.* at 8.
[221] *Id.* at 9–11.
[222] KBASE EXECUTIVE SUMMARY, n. 207, at 1.
[223] U.S. DEPT. ENERGY, OFFICE OF SCI., OFFICE OF BIOLOGICAL & ENVTL. RESEARCH, INFORMATION AND DATA SHARING POLICY, GENOMIC SCIENCE PROGRAM 2–3 (9 May 2013) [hereinafter DOE DATA POLICY], *available at* http://genomicscience.energy.gov/datasharing/GTLDataPolicy.pdf.
[224] *Id.*
[225] *See* KBASE EXECUTIVE SUMMARY, n. 207, at 2.
[226] *See* DOE DATA POLICY, n. 223, at 2–3.
[227] *See* KBASE EXECUTIVE SUMMARY, n. 207, at 2–4.

the project was supervised by principal investigators at the aforementioned national laboratories,[228] and, the lead "science team" of the KBase project was based at Lawrence Berkeley National Laboratory.[229] Software and hardware infrastructure was housed at Argonne National Laboratory, while research tool development was headquartered at Brookhaven National Laboratory, with a particular focus on plants.[230] An aggressive scientific outreach program, to be stationed at Oak Ridge National Laboratory, was to train and collaborate with the broader research community at different universities.[231]

The KBase Implementation Plan envisioned a formal Governance Body that would resemble those found at more established OKEs, such as the GSC.[232] The DOE wanted this Governance Body to be drawn from research users, data producers, and research tool developers.[233] Detailed governance responsibilities were set out in the implementation plan, including a timeline for achievement of specified tasks.[234]

Governance Body members were expected to meet twice per year in person and to engage in additional remote teleconferences, at which they would initially tackle issues of organization and definition, and then move on to standardization and compliance among all collaborators, as well as continuing to enhance outreach.[235] In our view, this body would also need to study and devise data and materials licensing strategies for sources outside the United States, especially if microbial and plant genetic resources from developing countries were to be studied.[236] The Governance Body was to be backed by a DOE project management staff, which would more directly oversee the budget and provide hands-on supervision of the distribution of materials and tools among stakeholders in both the public and private sectors.[237]

The DOE Systems Biology Knowledgebase is an ambitious biomedical research project. If successful, KBase will be able to harness public and private resources

[228] KBase Overview, n. 213, at 3.
[229] *Id.*
[230] *Id.*
[231] *Id.*
[232] S. Dept. Energy, Office of Sci., Office of Biological & Envtl. Research, DOE Systems Biology Knowledgebase Implementation Plan 130–32 (2010) [hereinafter KBase Implementation Plan], *available at* http://genomicscience.energy.gov/compbio/kbase_plan/kbaseimplementationplan.pdf; *see also* Section III.A.1.
[233] KBase Implementation Plan, n. 232, at 131.
[234] *Id.* at 131–37.
[235] *Id.*
[236] *See* Chapter 3, Section IV.B; *see also* Section III.B.2 in this chapter.
[237] KBase Implementation Plan, n. 232, at 138–44.

to optimize previously collected genomic data and move the field forward with tactical, hypotheses-driven applications of those data.

4. The Program on Microbiology of the BUILT Environment (MoBe)

The Program on Microbiology of the Built Environment (MoBe) is a large, multi-level OKE funded by the Alfred P. Sloan Foundation to forge a new direction in microbiology. Although today's scientists know a lot more about the microbial ecosystems of the planet than in the past, they know relatively little about microbes that inhabit the indoor world, where we now spend more than 90% of our lives.[238] While indoors, we come into contact with trillions of microbes that mostly differ from those we encounter in nature.

The Sloan Foundation began issuing grants to researchers in late 2005 to participate in the MoBe project,[239] with five preestablished goals, and some of these projects have now ended. The first goal aimed to "push the research frontier" in this field of microbiology by developing standards, tools, and protocols, and by training a small leadership cohort within the community.[240] This component is headquartered at the University of Oregon in a multidisciplinary "Biology and the Built Environment Center."[241]

A second major goal was to establish a national, multidisciplinary network of scientists, engineers, and architects who would all collaborate in studying the microbiomes of buildings. This effort was spearheaded by a laboratory at the University of California, Davis, and has a portal page called microBEnet.[242]

A third MoBe goal focused on improved data analysis and transmission through enhanced storage, visualization, and search capabilities.[243] This goal was pursued by a consortium of four different institutions, known as the Microbiome of the Built Environment Data Analysis Core (MoBeDAC).[244] MoBeDAC was essentially an OKE within an OKE, in the sense that multiple institutions with common goals

[238] Paula Olsiewski, *Sloan Program on Microbiology of the Built Environment*, MICROBENET, http://www.microbe.net/alfred-p-sloan-foundation/ (last accessed 9 Apr. 2014); Paula Olsiewski, *Microbiology of the Built Environment*, ALFRED P. SLOAN FOUND., http://www.sloan.org/major-program-areas/basic-research/mobe/ [hereinafter Sloan] (last accessed 9 Apr. 2014).

[239] *See Sloan Grants*, MICROBENET, http://www.microbe.net/grantees/ (last accessed 9 Apr. 2014).

[240] Olsiewski, Sloan Program, n. 238.

[241] *Id.*

[242] *Id. See Homepage*, MICROBENET, http://www.microbe.net (last accessed 9 Apr. 2014).

[243] Olsiewski, Sloan Program, n. 238.

[244] *Homepage*, Microbiome of the Built Environment Data Analysis Core (MOBEDAC), http://mobedac.org/ (last accessed 9 Apr. 2014). This component of the larger Sloan Program was completed in July 2014, and the website no longer remains active.

pooled resources and tools, and then made their data freely available to the public. Each participating institution brought the functional capabilities of at least one large toolset or database to the consortium, namely:

- University of Chicago: MG-RAST Metagenome Analysis Server[245]
- Marine Biological Laboratory, Woods Hole: Visualization and Analysis of Microbial Population Structures (VAMPS)[246]
- University of Colorado, Boulder: QIIME – Quantitative Insights Into Microbial Ecology[247]
- University of California, Riverside: FungiDB[248]

MoBeDAC's specific goals within the larger MoBe framework were to 1) develop a data archive; 2) establish an intraoperative environment for disparate websites and analytical tools; 3) establish appropriate metadata standards; and 4) develop visualization and new analytical techniques for comparing microbial communities, especially those from indoor environments.[249]

MoBe as a whole had two additional goals. One was to continue funding new, independent researchers, the other was to lobby the federal government to specifically fund more indoor microbiome research.[250] Of note in the Sloan Foundation's Request for Proposals is a specific requirement for grantees to commit to open-access distribution of all "knowledge generated" (including publications) by means of the MoBeDAC databases.[251]

The formal governance structure of the MoBe OKE was less clear at the time of writing than was true of the other examples in this section. Project management staff at the Sloan Foundation (the global funder of this venture) were visible and active,[252] and individual leaders of the various subgroups were in place. In 2011, the Sloan Foundation sponsored a two-day symposium on microbiomes in the built environment, with more than forty attendees and a keynote speech by J. Craig

[245] *Homepage*, METAGENOMICS RAST (MG-RAST), http://metagenomics.anl.gov/ (last accessed 9 Apr. 2014).

[246] *Homepage*, The Visualization and Analysis of Microbial Population Structures (VAMPS), http://vamps.mbl.edu/ (last accessed 9 Apr. 2014).

[247] *Homepage* (2011), Quantitative Insights Into Microbial Ecology (QIIME), http://qiime.org/.

[248] *Homepage*, FUNGIDB (1 June 2013), http://fungidb.org/fungidb/.

[249] Microbiome of the Built Environment Data Analysis Core (MoBeDAC), MICROBENET, http://www.microbe.net/microbiome-of-the-built-environment-data-analysis-core-mobedac/ (last accessed 9 Apr. 2014).

[250] Olsiewski, Sloan Program, n. 238.

[251] *See, e.g.*, Paula J. Olsiewski, Request for Proposal, ALFRED P. SLOAN FOUND. (27 Mar. 2012), *available at* http://www.sloan.org/fileadmin/media/files/olsiewski/mobe_rfp_guidelines_3_27_2012.pdf.

[252] "Microbiomes of Built Environments," Indoor Air 2011 Symposium, Austin, Texas, 8–9 June 2011.

Venter.[253] A workshop at the symposium produced a list of twelve priority research questions that should be addressed.[254]

MoBe is providing another vision of the OKE concept, with its fully open data-sharing infrastructure and related projects spinning off in many directions under a single umbrella entity. It could perhaps benefit from a more formal governance structure and codified bylaws in the manner of the GSC or even KBase, features that could help to promote organizational stability and outreach to future users and collaborators. Overall, MoBe should be applauded for identifying an important, neglected area of research and quickly mobilizing across various disciplines to solve specific problems. Its fully open data-sharing regime and portals that permit interactive communications with users are especially promising.

B. *The Future of Open Knowledge Environments*

Thematic Open Knowledge Environments, like those described earlier, represent some of the most advanced form the diverse institutional responses to the avalanche of data that the genomic revolution has engendered in microbiology. Their research outputs are usually the fruits of common resources that the networked participants have voluntarily pooled from the outset. These outputs are made available for broad use and reuse to an ever-expanding community of interested scientists on terms determined by the thematic community itself, often on a fully open basis. A growing array of digital networks and computational tools and techniques, which are put to a common purpose, then further enhance the productivity of these endeavors.[255]

1. Lessons from the Empirical Models

The OKEs we envision would seek more than reciprocity benefits typically accruing from the joint management of voluntarily pooled collections of data, whether in commons or semicommons arrangements. Rather, the more ambitious OKEs hope to become problem solvers in their own right and, in so doing, to advance substantially the frontiers of the life sciences. For example, the U.S. Department of Energy's Systems Biology Knowledge Database has been assembling a multifaceted

[253] Richard L. Corsi et al., *Microbiomes of Built Environments: 2011 Symposium Highlights and Workgroup Recommendations*, 22 INDOOR AIR 171, 171–72 (2012).

[254] *Id.*

[255] *See most recently* COMM. ON A NEW BIOLOGY FOR THE 21ST CENTURY & NAT'L RESEARCH COUNCIL, A NEW BIOLOGY FOR THE 21ST CENTURY 49–52 (Nat'l Acad. Press 2009) [hereinafter BIOLOGY FOR THE 21ST CENTURY]; *see also*, NRC, FUTURE OF SCIENTIFIC KNOWLEDGE DISCOVERY, above n. 90. *Cf.* YOCHAI BENKLER, THE WEALTH OF NETWORKS: HOW SOCIAL PRODUCTION TRANSFORMS MARKETS AND FREEDOM (Yale Univ. Press 2006).

computational toolkit for collaborative research on systems biology to be undertaken by interdisciplinary teams of scientists working together.[256] It aims to generate research questions that can be addressed by *in silico* experimentation, using the collective resources assembled within the OKE. In this respect, the KBase and other OKEs are implementing the New Biology paradigm discussed in Chapter 1.

While the different OKEs, including others not mentioned in this chapter, marshal different technical and thematic resources according to their respective fields of research interest, they all possess certain common design characteristics worth noting here. One is a need to formulate standards, especially with regard to metadata, curation, and the interoperability of data obtained from diverse sources, both experimental and the product of large-scale sequencing.[257] Their managers thus strive to preserve existing data resources that might otherwise be lost, and to make their cumulative data repositories available for future collaborative research undertakings.

The OKE's infrastructure thus typically enables the massive accumulations of data to become a launch pad for more diverse, hypothesis-driven research initiatives. This goal helps to maximize the returns from public investment in microbiology and related fields, while avoiding duplicative research projects. It also enables research organizers to devote the collective data-management assets available to investigators to the larger research priorities of the relevant thematic community.

A second distinguishing feature is the multidisciplinary role typically played by the thematic community that drives formation of an OKE from the outset. These communities thus tend to transcend preexisting boundaries in the life sciences, as the New Biology paradigm predicates. An underlying assumption is that the inputs and answers to one set of questions posed by one of the stakeholder subcommunities will often prove useful in answering different questions that other constituent groups of investigators may pose. At the same time, conscious effort is made to focus the implications of ostensibly diverse research projects undertaken by different stakeholders on the larger research goals driving the formation of the consortium as a whole.

Although pursing the diverse research activities that OKEs typically support, each subcommunity operates as a component of a larger research scheme. Construction of the broader framework thus aims to elicit collective tools, materials, and other resources that fuel all the research undertaken by all the stakeholders. On the

[256] *See About the DOE Systems Biology Knowledgebase (KBase)*, Genomic Science Program http://www.genomicscience.energy.gov/compbio/ (last accessed 5 July 2014), and Section III.A.3 earlier in this chapter.

[257] *See* Jorge Contreras, *Technical Standards and Bioinformatics*, *in* BIOINFORMATICS LAW: LEGAL ISSUES FOR COMPUTATIONAL BIOLOGY IN THE POST-GENOME ERA (J. L. Contreras & A. J. Cuticchia eds., 2013).

horizon, there is often a possibility of reaching some still unfolding objectives that are perceived to be greater than the sum of any given OKE's existing parts.

Because the OKEs are multidisciplinary in character, they not only enroll subject-matter experts, but also computer engineers, information scientists, librarians, and other potential contributors to help establish and manage both infrastructure and research operations. Once formed, these advanced knowledge hubs could also serve as a vehicle for teaching university students in related departments and for involving students in the development and management of the OKE itself, much as the U.S. law journal model described in the previous chapter.

A third feature of the entire OKE concept is its reliance on specific technical services that support any given multidisciplinary thematic community. Data management tools, graphical capabilities, algorithms, and the like enable the formation of the OKE from the outset. With growing capabilities over time, these tools enhance the OKE's evolutionary drive toward new research directions.

Technical services provide the ability to link all the thematic components and enable participants with diverse backgrounds to develop the pooled data and information into new configurations that elevate the research capacity of the OKE as a whole. This approach was especially evident in the preplanned, technical evolution of the Department of Energy's Systems Biology Knowledge base. But it was also visible in the CAMERA project, which depended on novel shotgun sequencing technologies and digital storage capabilities that generated a virtual materials collection rather than an *ex situ* culture collection.

Ideally, an OKE should involve technical and curatorial services personnel directly in its operations. By the same token, those technical personnel responsible for the service component should themselves qualify as thematic scientists able—to some degree at least—to collaborate at the operational level and to configure the computational instruments to the evolving needs of the thematic community.[258]

The OKEs we envision can readily be hosted at single universities, or their components can be distributed among a consortium of universities having a strong interest in the relevant subject matter. Other not-for-profit research centers or government agencies can (and do) host OKEs, although this may attenuate the educational function that we also seek to promote. As the digital component increases in importance, the research can become more interdisciplinary, dependent on inputs from, say, genomics and proteomics, as well as environmental sciences, among other relevant fields. These interdisciplinary activities, though emanating from a core thematic group at one or more institutional centers, operate across

[258] For a recent OKE that excels in this regard, *see* the Micro B3 project n. 155. *See generally* NATIONAL RESEARCH COUNCIL, PREPARING THE WORKFORCE FOR DIGITAL CURATION (Nat'l Acads Press, 2015).

internal, and even national, boundaries in order to pursue the larger thematic interest on an increasingly global scale.

The payoffs would become even greater if Open Knowledge Environments were deliberately designed to support the educational as well as the research functions of host universities from the start. To achieve these goals would require enhancing the cooperative ethos of the principal players. Specifically:

- The universities or other research institutions acting as hosts would ideally embrace and support an OKE as an integral part of their mission and programs, which, in turn, presupposes a sustainable funding base.
- Universities not managing or hosting an OKE could at least become active participant beneficiaries and contributors, for example, by establishing linked institutional repositories with interoperable data and text mining tools.
- The underlying socio-cultural context should become compatible with this vision of OKEs, in the sense that scientists would view the OKEs as an embodiment of sharing (Mertonian) norms of science.

This approach could also greatly increase the reputational benefits of participating universities, well beyond those of open repositories as such, while fulfilling the goal of rapid dissemination of high-quality, publicly funded research results.

To the extent that a new configuration for both scientific research and education emerges in the form of OKEs, the breakdown of preexisting life science boundaries presupposed by the New Biology paradigm also becomes a prelude to a new form of scientific and educational endeavor, one that requires previously independent departments and services to operate across borders and as an integrated whole.[259] From this perspective, the fact that OKEs are at present likely to be attached to either government ministries (e.g., the Department of Energy) or universities may only be a starting point. If the more ambitious OKEs actually meet their goals and become fruitful and proactive generators of discoveries and applications, they could become a model for the reorganization of universities and even government departments along newer, more cooperative lines. For example, libraries and computer services could become more fully integrated into the actual research delivery process, as could the teaching and educational functions of departments that now rest on separate foundations.

2. Operationalizing the Core Concepts

In what follows, we consider a number of institutional and legal issues that seem common to the OKE concept as a whole. By acknowledging and addressing

[259] See Chapter 1, Section II.D.

these issues early on, the incipient OKE movement might become more solidly established and better able to overcome common obstacles more efficiently. There is also a need to consider how OKEs that succeed as bottom-up initiatives, can become integrated more fully into the larger microbial research infrastructure.

In the thematically integrated undertakings, we characterize as OKEs, data are the glue that holds the parts together and that shape the different components into given research configurations. Data, in this context, are not just a passive resource waiting to be operationalized. They are forged into a cumulative tool or set of tools that help determine the direction of future research and that may be used to generate both new hypotheses and their experimental validations.

Like all knowledge commons in which data are pooled, the organizers of an OKE must, at the outset, determine how to design the infrastructure in which their data and other digital assets will be deposited.[260] Empirical research shows that, when the organizers are fully committed to an agreed open-access data release policy, they may choose from three standard organizational models for this purpose, viz.,

- A single, unified and centralized data repository serving all the contributors;
- A distributed set of repositories that can nonetheless all be accessed from a single portal;
- A distributed set of repositories that can only be accessed one by one.[261]

The second or intermediate model represents a workable compromise in which the costs are subdivided among all the nodes, but there is easy access to them all through the portal, which incentivizes global research endeavors. The existence of a common portal also facilitates opportunities to add value to all the data otherwise available from the distributed repositories.

Once the organizers of any given OKE have determined the appropriate design of their data infrastructure, they must ensure that metadata will adequately support the contributions that the participants will make available. Decisions about digital curation are also necessary at the outset.[262] Like most other data-sharing initiatives, those who manage OKEs must then give considerable thought to both the licensing strategies to be adopted for incoming data and the data release policies applicable to new data generated directly by the OKE and its collaborators as deliverable research outputs.[263] Also needed are licensing strategies applicable to any given data mining

[260] *See* Brett M. Frischmann, Michael J. Madison & Katherine M. Strandberg, Governing the Knowledge Commons (Oxford U. Press 2014).

[261] Contreras & Reichman (2015), n. 139

[262] *See generally*, NRC, Digital Curation, n. 258.

[263] *See, e.g.*, Victoria Tsoukala et al., Policy Guidelines for Open Access and Data Dissemination and Preservation (RECODE, Feb. 15, 2015), http://www.recodeproject.eu/wp-content/uploads/2015/02/RECODE-05.1-POLICY-RECOMMENDATIONS-FINAL.pdf (last accessed 15 April 2015).

or data management tools developed or utilized within the OKE that may turn out to have commercial applications.

A. LICENSING DATA AND TOOLS. A potential limitation on the formation of any OKE is its ability to obtain needed scientific information and data on terms consistent with its own data-release policies. One simplifying model that is common to many different publicly certified, all-inclusive global databases is full open access with the rights to use and reuse data (and information).[264] Such databases are typically either produced from data generated in-house or provided by external contributors. The pooled data may then be made available without any restrictions, or subject to some reserved rights imposed either by disparate contributors or by the organizers of the pooling initiative in question.

Often the implicit assumption is that all or at least most of the contributors to domain-specific global databases are governmental or nonprofit entities, or individual researchers employed by these organizations. These entities or their researchers may either not possess intellectual property rights in the relevant data and databases or, if they do, would not exercise their rights because of the putatively low or non-existent direct commercial value of such databases. However, in light of evidence about increasing pressures to commoditize databases and about the current evolution of database legislation, the validity of this presumption needs to be reexamined when forming complex OKEs.[265]

To the extent that a diversity of data access laws and policies usually govern territorially heterogeneous data contributors, it may hinder not only use and reuse, but also automated knowledge extraction and manipulation of the data. If some providers do retain ownership rights, full automatic integration can be hampered by the need for case-by-case clearance of those rights and successful negotiations of other licensing conditions.[266]

For example, the StrainInfo.net Bioportal attempted to access and integrate externally released microbial data and information in real time. As implemented, however, StrainInfo had to clear these access and redistribution rights with database

[264] The application of copyright and related laws to data and compilations, on the one hand, and to the literature and other more creative works on the other, was discussed in Chapter 6, Section II. Similarly, there are different considerations when applying common-use licenses. We also reviewed the use of Creative Commons licenses to journal articles and other microbial information in Section II of Chapter 7.

[265] *See, e.g.*, Paul N. Schofield et al., *Post-Publication Sharing of Data and Tools*, 461 NATURE 171–73 (10 Sept. 2009); Bryn Nelson, *Data Sharing: Empty Archives*, 461 NATURE 160–63 (2009); Wesley E. Cohen & John P. Walsh, *Real Impediments to Biomedical Research*, 8 POLICY ECON. 1–30 (2008).

[266] *See, e.g.*, Digital Media Project, *The Digital Learning Challenge: Obstacles to Educational Uses of Copyrighted Material in the Digital Age*, BERKMAN CTR. FOR INTERNET & SOC'Y, HARVARD UNIV., http://cyber.law.harvard.edu/media/files/copyrightandeducation.html (last accessed 9 Apr. 2014).

owners or risk infringing national copyright and database protection laws.[267] Moreover, as discussed in Chapter 7, access to published microbial journals that still impose a variety of restrictive conditions on use and reuse of contents is another potentially complicating factor. These burdens require careful internal administration and, if not attended to, could add substantial transaction costs, even when the relevant contributors were publicly funded scientists affiliated with an OKE.[268]

More generally, unless care is taken in advance, legal restrictions applicable to some of the data to be licensed may spread or infect other data in any given database or the repository as a whole. In other words, when "substantial amounts of data are combined from two or more data sources, the resulting data set will incorporate the greatest restrictions from any of the sources used and the accumulated restrictions imposed by each source."[269] This will occur unless the more restricted datasets are completely separable and separated from the evolving database in question.

Such restrictions may arise by operation of law, even when the data provider does not expressly assert them or would not knowingly want to assert them. This premise follows from the fact that a compiler's exclusive rights under national copyright and database protection laws, as well as rights under the Convention on Biological Diversity, where applicable, arise automatically and follow the dataset in question. Unless waived in advance by legislative or regulatory action, or by private ordering, such rights will hinder the OKE's own ability to reuse and redistribute the data without restrictions, assuming that is its policy goal.[270]

Data that are not legally protectable or that have already entered the public domain for one reason or another remain, of course, an exception to the general premise set out earlier. However, the conditions determining the public domain status of any given dataset may vary widely from one jurisdiction to another and may be highly

[267] *See* Chapter 6, Section II.B.2 and Chapter 7.

[268] *See* Chapter 6 *passim*.

[269] GEOSS, Data-Sharing Working Group, *Legal Options for the Exchange of Data Through the GEOSS Data-Core Summary White Paper*, Appendix B (Oct. 30, 2011) [hereinafter GEOSS, *Legal Options*], *available at* http://www.earthobservations.org/documents/dsp/draft_white_paper_geoss_legal_interoperability_30_october_2011.pdf. *See also* John Willbanks, "The Digital Commons: Infrastructure for The Data Web," paper presented at Global Science and the Economics of Knowledge-Sharing Institutions, 2d. Communia Int'l Conference, Turin, Italy, 29–30 June 2009, *available at* http://www.communia-project.eu/node/290.

[270] GEOSS, *Legal Options*, n. 269, §2.3. *See also* UNEP, "A Review of the Barriers to the Sharing of Biodiversity Data and Information with Recommendations for Eliminating Them," paper presented at Conference of the Parties to the CBD, 11th mtg., UNEP/CBD/COP/11/INF/8, Hyderabad, India, 8–19 Oct. 2012 [hereinafter UNEP, *Sharing Biodiversity Data* (2012)], *available at* http://www.cbd.int/doc/meetings/cop/cop-11/information/cop-11-inf-08-en.pdf. For example, the pledge by members of the ICGC not to assert intellectual property rights in genetic mutation data, nn. 84–89, would not in itself affect either data protection rights, copyrights, or rights in genetic resource data that arise automatically under existing legislation and treaties. *See generally* Chapter 6, Sections II & III.

fact-specific in any jurisdiction. Unless the data in question carry a notice in their metadata and on the database owner's server informing potential users of its public domain or common-use status, such as a Creative Commons (CC) Public Domain Mark or a CC0 waiver of rights, this legal uncertainty operates as an inherent source of potential restrictions on use and reuse, including interoperability with other data, regardless of the provider's intentions to the contrary.[271]

To resolve these problems, the OKE's administrative body can devise a legally valid waiver of rights equivalent to a voluntary dedication to the public domain as indicated earlier, or it can formulate a common-use license applicable to all incoming datasets that standardizes the OKE's own use and reuse conditions, as well as its attribution policy.[272] The adoption of standard waivers of intellectual property rights or common-use licenses that meet all the use and reuse needs of any given OKE would help to ensure legal certainty and interoperability of the data, and thus support the OKE's primary research goals.[273]

One tradeoff, however, is that such a standardized legal prerequisite will limit some of the data flow that the OKE might otherwise hope to receive, unless all the key participants have agreed at least to the same prenegotiated licensing conditions. Even then it is now widely recognized that licensing clauses that impose attribution duties may not be fully enforceable for all data used in all jurisdictions. Such clauses may nonetheless retain some force as a standard community practice or norm of proper scientific research conduct.[274]

The OKE must likewise define its own access, use, and reuse policies for the digital tools and research results it generates, and it must implement those policies in appropriate legal instruments. For example, software solutions could presumably be made available under existing open-source public licenses, while the OKE's

[271] *See* GEOSS, *Legal Options*, n. 269. The Creative Commons waiver (CC0) places the data in the public domain and waives all copyright and database rights. *See also* the Open Data Commons Public Domain Dedication and License (PDDL) as created by the Open Knowledge Network Foundation (U.K.). However, neither of these legal instruments would necessarily clear genetic-resource data covered by the CBD. GEOSS, Legal Options, n. 267, §2.5.

[272] GEOSS, *Legal Options*, n. 269, §§ 2.6, 2.7.

[273] *Cf. id.* § 2.7. According to the GEOSS Data Sharing Working Group, the desirable characteristics of either a voluntary dedication to the public domain under a waiver of rights or a common use license would include *inter alia* the following characteristics:

- Clear and simple terms to the data provider and user;
- Easy to recognize and find;
- Embeddable in the data as machine readable metadata;
- Availability in different languages, at a minimum in the language(s) of the country from which the data are made available as well as in English.

GEOSS, *Legal Options*, n. 269, § 2.10.

[274] *See* GEOSS, *Legal Options*, n. 269, §2.8.

information licensing and perhaps own journal(s) could apply the Creative Commons' CC-BY attribution-only license used by many open-access journals.[275] Similarly, that type of license could presumably apply to information in government reports, conference proceedings, and other grey literature produced collaboratively through the interactive functions of the OKE, such as wikis, discussion fora, and postpublication reviews. However, these licenses should be built into the legal infrastructure from the start, because the OKE could only add such a license after the fact to a work that had already qualified for copyright protection with the consent of the copyright owner.

As regards data that the OKE's own research efforts generate, voluntary dedications to the public domain through the waiver of rights, or standardized CC-BY common-use licenses could, of course, greatly facilitate user accessibility, especially if made available from any designated portal that the OKE established for its stakeholders and participants, as well as for the external scientific community at large.[276] Ideally, all such data would in fact be made freely available to all interested parties, as seems to be the case with the DOE's Systems Biology Knowledge Database.[277]

In practice, however, some OKEs may decide to restrict access to and use of some of their research data available to outsiders, at least some if not all of the time,[278] quite apart from an embargo and other delaying conditions that may be imposed to protect the publication interests of participating scientists who contributed data.[279] For the reasons discussed earlier in this chapter, this option, which clearly reduces at least the short-term public benefits of the relevant data-sharing initiative, is best left to the data release policy initially determined by the stakeholders in, and funders of, any given OKE. That said, all OKEs need to formulate a data release policy as early as possible, preferably at the start of the project when the overall data management plan is being formulated. Moreover, funders should in principle encourage organizers and stakeholders to release as much data to the scientific community at large as possible, commensurate with the organizational goals of the OKE itself.

[275] *See* Chapter 7, Section III. Alternatively one could use the "attribution only – no derivative work" clause. *See also* the Open Data Commons Open Database License (ODC-OdbL), which requires attribution of the licensor plus share-alike, so that any adaptations of the database must be licensed under an equivalent license. Memorandum from Andrew Rens, SJD Candidate, Duke Law School, 20 June 2014.

[276] *See* GEOSS, *Legal Options*, n. 269, § 2.7.

[277] *See* Section III.A.3 *Cf.* GEOSS, *Legal Options*, n. 269, § 1.

[278] That seemed to be the case with CAMERA. *See also* ICGC, *Overview*, nn. 84 (restricting access to data that could reveal identity of donor patients).

[279] *See generally* Contreras, *Data Sharing*, n. 15. *See also* NAT'L RESEARCH COUNCIL, BITS OF POWER: ISSUES IN GLOBAL ACCESS TO SCIENTIFIC DATA 111–13 (Nat'l Acad. Press 1997) [hereinafter BITS OF POWER], as well as NAT'L RESEARCH COUNCIL, A QUESTION OF BALANCE (Nat'l Acads. Press 1999).

Finally, it is well to recognize in advance that some datasets produced by the joint operations of an OKE and its stakeholder subcommunities may become research tools of economic value to either the public and private sectors and thereby acquire commercial potential over time. Whether any given OKE is interested in or capable of pursuing such commercial opportunities depends on the nature and goals of the OKE itself, on the rules it has adopted, and the constraints or conditions imposed by its funders. For present purposes, nonetheless, we stress the importance of avoiding patents on databases to be used as research tools whenever possible, in order to avoid thickets of rights, anticommons effects, and burdensome clearance of rights problems that have elsewhere been reported with regard to patents on research tools.[280] Instead, OKE organizers and funders willing and able to commercialize datasets as research tools should normally adopt a nonexclusive licensing strategy on standard conditions that would minimize transaction costs and recognize the funding constraints of the public research sector.[281]

B. BENEFITS OF INTEGRATING THE MICROBIOLOGICAL LITERATURE. In order to take full advantage of its digital infrastructure a fully developed OKE could logically provide links to openly available scientific literature. For example, there is now at least one multidisciplinary "mega-journal," *PLoS ONE*, which operates as an author-pays, fully open-access, peer-reviewed repository, from which anyone can harvest thematically relevant literature.[282] Besides integrating externally published literature from such sources, OKEs can also generate publications of their own research results. Indeed, one can even envision thematically integrated and networked OKEs playing a major role in restructuring the scholarly communication process, as it moves away from the stove-piped journals of the print era.

At the project level, for instance, the CAMERA metagenomic OKE focused more on the task of providing and integrating the sequence data that emerged from its marine bioprospecting endeavor than on producing a journal of its own. Still, CAMERA provided links to openly available literature from other sources that supported its thematic research objectives.[283] Conversely, a given OKE could also incorporate one or

[280] Cf. Chapter 2, Section II.

[281] *Cf.* Jerome H. Reichman & Paul F. Uhlir, *A Contractually Reconstructed Research Commons for Scientific Data in a Highly Protectionist Intellectual Property Environment*, 66 LAW & CONTEM. PROBS. 315, 326–29 (2003) [hereinafter Reichman & Uhlir (2003)]. *See also* National Institutes of Health, *Best Practices for Licensing Genomic Inventions: Final Notice*, 70 Federal Register 18413, 11 April 2005. *Available at*: www.att.nih.gov/sites/default/files/documents/pdfs/70fr18413.pdf (last accessed 28 June 2015).

[282] *See* Public Library of Sci., PLoS ONE, www.plosone.org. In 2012, all PLoS journals together published 23,468 articles, *see* http://blogs.plos.org/everyone/2013/01/03/2012review/. In 2014, that figure rose by about 10,000 articles to over 33,000 with approximately 11.6 million views per month, *see* gtto://blogs.plos.org/plos/2015/02/2014-numbers/ (last accessed 28 June 2015).

[283] *See* Section III.A.2 (describing CAMERA).

more open-access, peer-reviewed scholarly journals, as was the case with the recently launched Journal of the Genomic Standards Consortium.[284]

The open-source software systems tools that support the comprehensive use of both literature and data, as described earlier, could also enable researchers and students to produce new knowledge online by means of networked collaborations. This function could be potentiated by fully exploiting the opportunities that semantic web tools have already begun to provide.[285] Many software tools already exist for such purposes, and these tools can be selected freely and even customized for use on OKE websites.[286]

As we have seen, the core concept of an OKE involves the development of an interactive portal focused on knowledge production and on collaborative research and educational opportunities in specific thematic areas. However, with the exception of the Genomic Standards Consortium, when these joint research activities reach the point of yielding published research results, they are typically outsourced to a journal of a professional society or a commercial publisher that is operating on a closed subscription basis,[287] which then normally triggers all the legal constraints and restrictions we described in Chapters 6 and 7. This customary institutional arrangement limits access to and use of the knowledge assets that the scientific community produces.

Instead, every thematic OKE should consider the possible advantages of publishing its results in fully open journals and other open-access outlets, or even integrating one or more journals into its own operations for the educational and systemic advantages that could ensue. This strategy does entail costs and administrative burdens, which can be alleviated by means of collaborative networks or consortia, but not eliminated altogether. If the relevant journals remain outside the OKEs, then (1) they should operate on an open-access basis and (2) the repository option should be fully developed so that OKEs can harvest all relevant information resources under a contractually constructed research commons.[288] Neither the subscription publishers nor the proprietary professional societies could then obstruct these initiatives, which is why funders should factor in the benefits to be gained from integrating publication of the OKE's research results as part of its general mission.[289]

284 *See* Section III.A.1 (discussing the Journal of the Genomic Standards Consortium).

285 *See* Janet Finch, Report of the Working Group on Expanding Access to Published Research Findings, *Accessibility, sustainability, excellence: how to expand access to research publications*, June 2012 [hereinafter Finch Report], *available at* http://www.researchinfonet.org/wp-content/uploads/2012/06/Finch-Group-report-FINAL-VERSION.pdf.

286 *Id.*

287 *See id.* at 82.

288 *Cf.* Reichman & Uhlir (2003), n. 281.

289 *See* Chapter 7, Section III.B ("Funders Ability to Contractually Regulate Access, Use, and Reuse of Scientific Literature").

An OKE could thus be developed in tandem with one or more thematically linked, open-access journals and augmented by openly available peer-reviewed literature, grey literature, and related data as discussed earlier. In either case, various interactive functions (wikis, discussion forums, blogs, postpublication reviews, perhaps distributed grid computing), all legally supported by appropriate waivers or common-use licenses, could be added to stimulate discussion of specific issues as well as contributions from exernal participants.

Whenever feasible, OKE organizers and their funders should seek to reduce and then eliminate the proprietary subscription publishers' ownership and control of all knowledge assets produced by the relevant thematic community. Researchers affiliated with OKEs would no longer assign their rights to external subscription publishing intermediaries.[290] Once possessed of ownership and control, the participating scientists and their institutions would be in a position to avoid all the technical and legal restrictions described in Chapter 6. They could then organize the use and reuse of these knowledge assets by measures that are specifically designed to promote collaborative research within the fully integrated digital networks of the OKEs.

Such measures, for example, would give organizers of an OKE and their sponsors the power to determine the conditions under which that OKE's research results were created, disseminated, and reused, in a manner consistent with the needs of research and education in microbiology. Under this approach, if external intermediaries were used, they would operate as service providers, on science-friendly terms, with open-access prerequisites, as prescribed by the OKE's governing body.[291] The *quid quo pro* would be the provision of efficient services that the OKE or its host university or other entity did not wish to undertake for various reasons.[292]

The OKEs as described in this chapter build on a number of recent, but already tested, advances in online peer production of knowledge and participative web techniques. The proprietary journal model impedes these capabilities. Under our

[290] *See* Reto Hilty, Copyright Law and Scientific Research, in Copyright Law, A Handbook of Contemporary Research 315, 353 (Paul Torremans ed., Edward Elgar Pub. 2007); Reto Hilty, *Five Lessons About Copyright in the Information Society: Reaction of the Scientific Community to Over-Protection and What Policy Makers Should Learn*, 53 J. Copyright Soc'y U.S.A. 103, 113–18 (2006) [hereinafter Hilty, *Five Lessons About Copyright Law*].

[291] *Cf.* Reto M. Hilty et al., European Commission – Green Paper: Copyright in the Knowledge Economy – Comments by the Max Planck Institute for Intellectual Property, Competition and Tax Law, 14–16 (2008), available at http://www.ip.mpg.de/filespdf1/comments_on_the_green_paper 1.sd [hereinafter Max Planck, Response to EC Green Paper]; Hilty, *Five Lessons About Copyright*, n. 290; Jerome H. Reichman & Ruth L. Okediji, When Copyright Law and Science Collide: Empowering Digitally Integrated Research Methods on a Global Scale, 96 U. Minn. L. Rev. 1362, 1372–1457 (2012).

[292] Stiglitz, n. 147. But the external legal and contractual realities could hinder this outcome.

vision of Open Knowledge Environments, the narrowly stove-piped, print-paradigm journal model could be transformed into a truly interactive and thematically linked networked initiative, while still maintaining high-quality standards of scholarly endeavor and promoting the reputational benefits of the participants and their institutions.

Microbiology journals need not be seen as ends in themselves. Rather, by repositioning many of them within OKEs, the journals could become cogs in, and stepping stones to, the realization of digital knowledge hubs. From this perspective, all the microbiology journals thus repositioned would become open-access by definition, with their contents available for harvesting by others, for thematic reintegration in other collections, and for various forms of digital manipulation.

By thus deconstructing the print publishing model and moving at least some journals and related open-access knowledge resources in-house, one begins to envision the possibility of reconstructing a digitally networked scientific communications model, in which the content providers are the communicators, the intermediaries, and the governors of a dynamically constituted knowledge environment. In the specific context of OKEs, they would also be the primary users of the publishable research outputs, but would still enable the access and use of that knowledge by myriad other external users, thereby increasing the overall value of the project.

If these integrative functions of a fully developed OKE seem futuristic, that is in part because of the status quo of STM publishing in the field of microbiology and elsewhere. As we have shown, the legal terms and conditions in many of the publishers' contracts, buttressed by the larger intellectual property environment, aim tacitly to protect the print model against the challenges – perceived as risks rather than opportunities – of the digital networked environment. The possibility of overcoming this limited vision and obstructive legal culture makes the concept of OKEs a particularly stimulating vision of new scientific opportunities.

3. Funding and Other Governance Considerations

Open Knowledge Environments necessarily require appropriate and sustainable funding. Preferably, this funding should be developed under a "purist" approach, one based on government and foundation or in-kind support, that allows equal access to all users and contributors.[293] The funding of the incipient OKEs discussed in this chapter is instructive in this regard.

[293] *See* the examples of existing OKEs that we described earlier in Section III.A. *Cf. also*. *Homepage*, WIKIMEDIA FOUNDATION (10 July 2013), http://wikimediafoundation.org/wiki/Home.

For example, the CAMERA project was launched with major support from a single foundation, and it obtained some additional funding as well. Another private foundation provided substantial startup funds for the MoBe project. The Genomic Standards Consortium was originally funded by the United Kingdom's National Environmental Research Council, which enabled it to establish an open-access portal. More recently, GSC attracted additional funding to start an open-access journal.[294] The Systems Biology Knowledgebase was set up by the U.S. federal Department of Energy and now operates as part of a major federal government science program. The funding in this case depends on annual appropriations by Congress. To the extent that overall political support for the Department of Energy remains stable, such funding would likely continue so long as its DOE managers and the user community consider the OKE productive.

Once the start-up money runs out, however, as seems to have occurred in the case of CAMERA, the fate of any given OKE could largely depend on its usefulness to the relevant scientific community. In some cases, one would hope that industry might also support or partner with some OKEs if it valued their outputs. While single projects may die, to the extent that the core OKE activities add educational and research value as well as reputational benefits to a host university or other sponsoring entity, they may logically seek funding from that university or entity as well as from the public science funding agencies and foundations. OKEs that outlive their scientific usefulness would fade away or be restructured,[295] but their accumulated data and literature should remain available when financially and institutionally feasible.[296]

More generally, the funding of OKEs cannot and should not be dissociated from governance considerations, which become far more important in this context than in most of the other voluntary data-sharing initiatives surveyed in this chapter. The very complexity of an OKE's thematic mission, coupled with its pressing need for digitally integrated technical support, will likely require some governing body to provide oversight and management well beyond the capacity of traditional lab-to-lab arrangements.

[294] *See* Section III.A.1.

[295] In that case, the fate of the relevant components, including any journal(s) that may be an integrated part of the OKE, would depend on their usefulness plus the willingness of other organizations to support them.

[296] Legacy funding raises many problems. IAN HARGREAVES, DIGITAL OPPORTUNITY: A REVIEW OF INTELLECTUAL PROPERTY AND GROWTH 43 (2011). We note that the CAMERA Project retains open access to many of its data resources through a residual website. See http://??camera.calit2.net/. Providing a repository that could serve this purpose is one goal of the redesigned Microbial Research Commons, as described in Chapter 10.

At different times, for example, OKEs must conduct legal and administrative affairs, including the licensing issues discussed earlier. Management must also deal in some official capacity with their constituent subcommunities, with host or cooperating universities or other sponsoring entities, and with the outside world. They will need to set standards, resolve disputes, and to purchase equipment.

Above all, the participants must periodically reevaluate and validate the OKE's core mission and its deliverables. New research directions may appear unexpectedly, while previously agreed lines of research may prove fruitless. For these and other reasons, governance of an OKE requires scientifically competent and engaged administrators willing and able to work closely with funders. By the same token, success of the venture may depend on the willingness of funders to work closely with these same administrators, with a view to mutually reinforcing decisions prompted by emerging research conditions.[297] In this respect, the governance considerations relevant to OKEs anticipate our discussion of the more complex governance arrangements needed to sustain our proposals for a revised Microbial Research Commons itself, as more fully elaborated in Part Four.

C. Linking the Open Knowledge Environment to the Materials Infrastructure

Up to here, we have envisioned the formation of Open Knowledge Environments as a highly evolved technical and institutional response to the deluge of life sciences data and information, one that enables thematic communities to customize their evolving needs for both internally generated and externally produced knowledge resources bearing on a core research mission or themes. At the same time, if one looks beyond the specific goals of single OKEs, important as these are, one may see them collectively as a microcosm of a larger phenomenon: namely, as artifacts in the construction of the basic scientific infrastructure needed to address the New Biology paradigm, a task specifically recognized by those who conceived the DOE's Systems Biology Knowledge Database.[298]

From this angle, OKEs built around digitally integrated data and information repositories strive primarily for higher levels of virtual *in silico* experimentation. Nevertheless, that goal should not obscure the advantages likely to accrue if researchers involved in these projects were also seamlessly linked to the suppliers of *ex situ* microbial genetic resources held at public culture collections around the world. If, indeed, the existing culture collections were themselves linked within a

[297] *Cf.* the close relationship with funders in the ICGC, n. 84 as described in Sections II.2.A & B. *Cf. also* the International Human Microbiome Consortium, discussed in Section I.B.2.

[298] *See generally* Section III.A.3.

federated and distributed semicommons[299] operating under the aegis of the Nagoya Protocol, as proposed earlier in Chapter 5 and more fully elaborated in Part Four of this volume, the reciprocity benefits from linking those physical assets with the virtual assets of the OKEs and with other open-access resources would flow in all directions. For example, the thematic research-driven OKEs could more readily validate their experimental findings against existing reference strains on a global scale, while the culture collections could themselves benefit from access to the cumulative data repositories and research outputs of single OKEs. One could even foresee the creation of certain OKEs at or affiliated with major culture collections themselves, especially those that had become Biological Resource Centers, as discussed in Chapter 4.

Better integration of the evolving OKEs with the global materials infrastructure would also help to reduce the residual tensions between big and small science mentalities discussed earlier in Chapter 1. It would further provide a logistical foundation to enable university laboratories around the world that maintained informal culture collections to connect with data-intensive research projects on the frontiers of microbiology. Above all, it could stimulate developing-country scientists to participate more directly in microbiological initiatives at a time when their governments are aggressively pursuing national sovereignty claims to both materials and data derived from local genetic resources, as explained in Chapter 3.

Consider, for example, that CAMERA's efforts to genetically sequence previously uncharacterized marine microbes were limited to samples taken from international waters, beyond the territorial boundaries of coastal countries otherwise able to invoke the Convention on Biological Diversity and other relevant laws. Yet, in principle, scientific research in both developed and developing countries would benefit if CAMERA had been able to operate within national boundaries, under sharing arrangements acceptable to local governments.[300] In that event, CAMERA's *in silico* genetic sequencing could have been linked to the full range of both *ex situ* and *in situ* genetic resources in the developing countries, with the results made available to the microbiological research community as a whole under agreed access and benefit-sharing protocols. Such a project has now become feasible within the ambit of the MICROB3 project, as funded by the European Commission, which has promoted the sampling of microbes in territorial waters around the world.[301]

[299] The term semicommons is more accurate owing to the fact that *ex situ* microbial culture collections can provide living specimens only to other collections and laboratories that meet specified quality, security and other restrictions as explained in Chapter 4, Section 1 ("Evolution of Microbial Culture Collections as Basic Scientific Infrastructure").

[300] *See, e.g.*, UNEP, *Sharing Biodiversity Data* (2012), n. 270.

[301] See MicroB3 Project, n. 155.

These insights explain why Open Knowledge Environments provide more than a promising solution to the hard problems of managing vast accumulations of data and avoiding the instinct to hoard. Viewed in isolation, a data pool is only as good as its aggregated single components. But an OKE puts all the strength of the microbiological research community behind a thematically constructed data pool, in the sense that the pool itself becomes only one component of a larger whole that potentially combines data with literature, materials, and technical services in an overarching, community-managed endeavor. In the context of OKEs, the process of exchanging genetic resources could make full use of automated knowledge management tools that are geared to community-determined goals. While these goals evolve and shift over time in keeping with the relevant subcommunities' own research needs, an ever-expanding infrastructure supports and magnifies all the reciprocity gains from formalizing the informal sector with regard to the pooling and exchanging data, information, and materials.[302]

Looking beyond the single OKEs, the move toward a redesigned Microbial Research Commons requires linking the materials and digital data and information resources available from a globally distributed open-access infrastructure, and providing interactive platforms for scientists to build on those resources and contribute to them. In microbiology, as in many other fields, effective links between the different materials and digital components of a research commons are needed to improve the efficacy of cumulative research and increase the speed of the entire research cycle.

In specific cases, the combined use of *in vitro* and *in silico* biology offers new opportunities for research, as we noted in Chapter 1[303] because, in the field of genomics, advances in computing and in molecular analysis go hand in hand to such an extent that "the evolution of one is tied to the existence of the other."[304] From this perspective, the OKEs described in this chapter – not to mention other OKEs within and outside of microbiology that we do not have time to discuss – represent a dynamic, bottom-up process that has only begun to attain its full potential. In our view, that goal cannot be reached without enlightened collective action and coordination across national boundaries, under the aegis of a properly

302 *See* Chapter 5, Section I.B.

303 *See* Chapter 1, Sections I.B & D. For instance, the task of searching for sequence similarities between the results of high-throughput screening and similar sequences with known properties available from public databases has become a key tool of metagenomics research. Without the aid of digital techniques, the full genome sequences, several hundred pages in length when printed, are not interpretable.

304 Adam Bostanci & Jane Calvert, *Invisible genomes: The genomics revolution and patenting practice*, 39 STUDIES IN HISTORY & PHILOSOPHY BIOLOGY & BIOMEDICAL SCI. 109–19 (2008).

redesigned and viable Microbial Research Commons that lays the foundation for digitally integrating materials, data, and literature.

Experience with the nascent OKEs demonstrates that such a structural reorganization would encounter formidable administrative difficulties even if it only entailed issues of science policy. Efforts to link microbial genetic resources, data, and literature in a universally accessible, digitally integrated research commons would also raise difficult legal, political, and institutional issues whose resolution is critical to the success of the New Biology paradigm itself.

In sum, thinking about OKEs in an expanding universe of collaborative research initiatives inevitably raises the central question of how to redesign the existing Microbial Research Commons as it has evolved over time. While the need for such a project was already identified by the OECD's Task Force on Microbiology in 2001,[305] the existing microbial research infrastructure has struggled to keep pace with the bottom-up initiatives described earlier in this chapter. Despite signs of progress, especially the WDCM reorganization in China and such regional projects as ECCO and now MIRRI in the European Union,[306] top-down efforts to redesign basic microbial infrastructure remain in a relatively tentative stage, well behind the major restructuring of the agricultural crop commons that has been underway for more than two decades.[307] Although any project to redesign the Microbial Research Commons would certainly benefit from an understanding of both the successes and failures of the still evolving Crop Commons, it can – and must – also directly address the new challenges and opportunities that the Nagoya Protocol necessarily raises.[308] This and other major governance issues are the topics addressed in the last part of this volume.

[305] *See* OECD initiatives described in Chapter 1, Section III, and Chapter 4, Section I.B.

[306] *See* Chapter 4, Sections I.C, & III.A. For MIRRI, see Chapter 9, Section II.D.

[307] *See* Chapter 4, Sections III & IV.

[308] *See* Chapter 4, Section IV; *see also* UNEP, *Sharing Biodiversity Data* (2012), n. 270.

PART FOUR

Governing Public Knowledge Assets within a Redesigned Microbial Research Commons

9

Institutional Models for a Transnational Research Commons

In the preceding chapters, we argued that the National Research Council's vision of microbiology as an integrated component of a "New Biology"[1] could be frustrated by growing proprietarial restrictions on access to, and use of, materials, data, and literature, even for public research purposes.[2] Such restrictions are rooted in private intellectual property rights reinforced by the WTO's TRIPS Agreement of 1994[3] or in sovereign claims of ownership embodied in the Convention on Biological Diversity of 1992.[4] The resulting proliferation of claims – coupled with pressures on both governments and public research institutions to commercialize their research results – has led to a shrinking public domain for genetic resources as well as related data and literature.[5] Left unaddressed by policymakers, these disaggregating legal, economic, and political trends pose a serious obstacle to the drive for scientific integration in both microbiology and the life sciences as a whole.[6]

1 *See* Chapter 1, Section II.D.

2 *See generally* Parts Two ("Preserving the Public Research Functions of Microbial Genetic Resources After the Nagoya Protocol") and Three ("A Digitally Integrated Infrastructure for Microbial Data and Information").

3 Agreement on Trade Related Aspects of Intellectual Property Rights, 15 April, 15 1994, 1869 U.N.T.S. 299 [hereinafter TRIPS Agreement].

4 Convention on Biological Diversity, opened for signature 5 June 1992, 1760 U.N.T.S. 79 [hereinafter CBD].

5 *See, e.g.*, Sikkina Jinnah & Stefan Jungcurt, *Could Access Requirements Stifle Your Research?*, 323 SCIENCE 464 (2009); Ninth Meeting of the Ad Hoc Open-Ended Working Group on Access and Benefit-Sharing in the Convention on Biological Diversity, Cali, Colombia, 22–28 March 2010, Side Conference Presentations [hereinafter Cali Presentations], *available at* http://www.cbd.int/wgabs9/events/se-abs9.shtml#tab=0; John P. Walsh & Wesley M. Cohen, *Real Impediments to Biomedical Research*, 8 INNOVATION POL'Y & ECON. 1–30 (2008). *See generally* Sabrina Safrin, *Hyperownership in a Time of Biotechnology Promises: The International Conflict to Control the Building Blocks of Life*, 98 Am. J. Int'l L. 641 (2004).

6 *See* Chapter 1, Section III.A ("Recognizing Institutional and Legal Challenges to the Existing Microbial Research Infrastructure"); Chapter 2, Section II ("Impinging Intellectual Property Rights Promoted by the Developed Countries"); and Chapter 3, Section I ("Regulatory Measures to Control Access to Genetic Resources Promoted by Developing Countries").

In recognition of these constraints, science policymakers and the discipline subcommunities have increasingly focused on devising methods of open access to data and literature and on new approaches to organizing and integrating diverse collections of needed material resources. These strategies are intended to better secure and manage upstream research needs without compromising downstream commercial applications. Efforts to expand the cooperative model appear to have gathered considerable intensity in recent years, with both top-down and bottom-up initiatives under way, as amply demonstrated in Chapters 4, 7, and 8.

Nevertheless, we contend that much more needs to be done, given the mounting obstacles to basic microbiological research identified earlier in this book. With regard to materials, for example, efforts to produce more research-friendly material transfer agreements for culture collections at the regional level could ultimately prevail and mature only within an appropriately designed international governance framework that directly meets the objectives of the CBD.[7] Similarly, access to data and information for purposes of public microbiological research, especially with regard to computational science and the use of automated knowledge discovery tools, is only partially addressed by current open-access initiatives in this field. These initiatives, although commendable, insufficiently address the need to make those results readily available within an organized, transnational legal framework.[8]

Moreover, there is still a critical lack of linkage between the culture collections and the emerging digital infrastructure that would be needed to implement the New Biology paradigm discussed in Chapter 1. In effect, there remains a troublesome institutional disconnect between the approaches to physical materials on the one hand and approaches to digital research on the other.[9] What the New Biology paradigm needs is for microbiology to move beyond existing national and regional efforts by embracing a truly international framework, which alone could immunize genetic resources and related data from ownership claims and sovereign interference under both the TRIPS Agreement and the CBD.[10] Such a framework should enable participating research entities to access materials globally under negotiated standard terms consistent with international law and to mine and share

[7] For proposals in this regard, see Chapters 5 and 10.

[8] *See further* Chapters 6–8, and 10.

[9] For a few notable exceptions to this assessment, see Chapter 4, Section IV.A ("Basic Concepts of the WHO's Pandemic Influenza Preparedness Framework (2011)") and Chapter 8, Section III.C ("Building Transnational Open Knowledge Environments").

[10] For important recent steps in this direction by the microbiology community, *see* Chapter 8, Section III.A, and the International Human Microbiome Project described in Section II.B.4 of this chapter. *See most recently* the Micro B3 Project on Marine Microbial Biodiversity, Bioinformatics, and Biotechnology, *Homepage*, http://www.microb3.eu/ [hereinafter Micro B3 Project] (last accessed 2 July 2014).

both data and information wherever situated, empowered by automated knowledge discovery tools operating within the widest possible research space.

To achieve these goals, we contend that the disaggregated knowledge assets of microbiological research should be combined and strengthened within a contractually constructed research commons to be organized and managed by the relevant public science community itself.[11] That community would take charge of its own knowledge assets and agree on operating and governance rules that would reinforce the underlying social norms of science, which have been weakened by the proliferation of strong intellectual property rights and related research policies. While alleviating barriers to accessing and using upstream knowledge inputs, this endeavor would also preserve and defend the opportunities for downstream patents and related rights on commercial applications.

Above all, a properly redesigned Microbial Research Commons could help to resolve current international tensions between the TRIPS Agreement and the CBD, which make research in microbiology (and other life sciences) more difficult. By digitally integrating access to and use of, genetic resources, data and scientific literature, the proposed commons could greatly improve the capacity of research entities in both developed and developing countries. It would thus provide strong incentives for the developing countries to contribute essential microbial genetic resources to the multilateral enterprise, in exchange for both monetary and nonmonetary benefits.[12]

In the rest of this chapter, we address these issues from three angles. First we look at the recent literature devoted to the theoretical underpinnings of a knowledge commons in general. We then survey a number of existing endeavors to form science-related knowledge along the lines we envision. The object here is to identify strengths and weaknesses of these existing models with respect to both institutional design and governance. We conclude the chapter by analysing these empirical findings with a view to identifying the premises for implementing our

[11] *See, e.g.*, BRETT M. FRISCHMANN, INFRASTRUCTURE: THE SOCIAL VALUE OF SHARED RESOURCES 3–60, 61–117 (2012) [hereinafter FRISCHMANN, INFRASTRUCTURE]; Jerome H. Reichman & Paul F. Uhlir, *A Contractually Reconstructed Research Commons for Scientific Data in a Highly Protectionist Intellectual Property Environment*, 66 LAW & CONTEMP. PROBS. 315–462 (2003) [hereinafter Reichman & Uhlir (2003)], *available at* http://scholarship.law.duke.edu/lcp/vol66/iss1/12; *see also* Peter Lee, *Contracting to Preserve Open Science: Consideration-Based Regulation in Patent Law*, 58 *Emory L.J.* 889 (2009) [Lee, *Open Science*]; Peter Lee, *Toward a Distributive Commons in Patent Law*, 2009 *Wisconsin L. Rev.* 917 (2009) [hereinafter Lee, *Distributive Commons*]. For the technical definition of "commons" as an institutional arrangement for managing, or governing specified resources, see below nn. 43–45 and accompanying text.

[12] For elements of a "grand bargain" along these lines, see further Section III of this chapter and Chapter 10, Section I.

proposals for a redesigned Microbial Research Commons, as set out in the last chapter of the book.[13]

I. THEORETICAL REFLECTIONS ON DESIGNING A KNOWLEDGE COMMONS

Much thought has recently been given to the positive role of shared resources in modern economies, in response to what had become conventional wisdom concerning a supposedly ineluctable "tragedy of the commons."[14] The seminal work of Professor Elinor Ostrom and her colleagues focused on commons-based management of natural resources, as regulated by a clearly defined group of local users.[15] Empirically, the formal proprietary schemes underlying the administration of such resources varied in practice from a property-like approach to various forms of collective ownership, including direct state ownership.[16]

Ostrom's work sought to establish the possibility of a sustainable economic alternative for managing common-pool resources, which would be situated midway between market regulated exchanges of private entitlements and pure public goods that typically depend on state-based governance of resources.[17] A key insight here

[13] *See further* Chapter 10 *passim*.

[14] Garrett Hardin, *The Tragedy of the Commons*, 162 (3859) *Science, New Series* 1243–48 (1968).

[15] Elinor Ostrom, Governing the Commons: The Evolution of Institutions for Collective Action (Cambridge Univ. Press 1990) [hereinafter Ostrom (1990)]. *See also* Nat'l Research Council (NRC), The Drama of the Commons (Comm. Human Dimensions of Global Change, Nat'l Acads. Press, E. Ostrom et al. eds. 2002).

[16] Ostrom (1990), n. 15; Jean-Marie Baland & Jean-Philippe Platteau, *Halting Degradation of Natural Resources. Is there a Role for Rural Communities?*, *in* Institutions, Social Norms, and Economic Development (N. Dolšak & E. Ostrom eds. 2000); The Commons in the New Millennium: Challenges and Adaptation (N. Dolšak & E. Ostrom eds. 2003) [hereinafter The Commons in the New Millennium]. *Cf. also* Frischmann, Infrastructure, n. 11 (who develops the concept of "intermediate goods"); Carol M. Rose, *Common Property, Regulatory Property, and Environmental Protection: Comparing Community Based Management to Tradable Environmental Allowances*, *in* The Drama of the Commons 233–58 (Nat'l Acads. Press, E. Ostrom et al. eds., 2002). Typical examples of natural resources subject to commons-based management are long-term irrigation schemes in Spain and Nepal and pastoral grazing schemes in the Swiss Alps. Ostrom (1990), n. 15. For an example of privately-owned natural resources under some form of commons management, see the studies on transferable quotas in fisheries. Tracy Yandle & Christopher M. Dewees, *Privatizing the Commons ... Twelve Years Later: Fishers' Experiences with New Zealand's Market Based Fisheries Management*, *in* The Commons in the New Millennium; Einar Eythorsson, *Stakeholders, Courts, and Communities: Individual Transferable Quotas in Icelandic Fisheries 1991–2001*, *in* The Commons in the New Millennium.

[17] *See, e.g.*, Brett M. Frischmann, *Two Enduring Lessons from Elinor Ostrom*, 387 *J. Inst. Econ.* 1–6 (2013), *available at* http://papers.ssrn.com/sol3/Delivery.cfm/SSRN_ID2252133_code232991.pdf?abstractid=2252133&mirid=1 [hereinafter Frischmann, *Two Lessons*] (citing authorities); Vincent Ostrom & Elinor Ostrom, *Public Goods and Public Choices*, *in* Alternatives for Discovering Public Services: Toward Improved Performance (E.S. Savas, ed. 1977).

was Ostrom's recognition of the need for institutions to solve problems of collective action in managing common-pool resources in ways that neither private enterprise nor state regulatory authorities could typically provide.[18]

As Professor Frischmann pointed out, Ostrom "did not presume that community based institutions were successful or ubiquitous," but stressed instead that such institutions required "systematic study,"[19] Her book, GOVERNING THE COMMONS, analyzed eighty-six case studies of existing commons from different sectors and varying geographical regions.[20] From this analysis, Ostrom formulated an Institutional Analysis and Development (IAD) framework for research on common-pool resources,[21] and she and her collaborators found eight basic design principles that enduring, robust commons initiatives appeared to share, viz.:

- Group boundaries are clearly defined;
- Rules governing the use of collective goods are well matched to local needs and conditions;
- Most individuals affected by these rules can participate in modifying the rules;
- The rights of community members to devise their own rules are respected by external authorities;
- A system for monitoring members' behavior exists and the community members themselves undertake this monitoring;
- A graduated system of sanctions is used;
- Community members have access to low-cost conflict resolution mechanisms;
- For CPRs [Common Pool Resources] that are parts of larger systems, tasks such as appropriation, provisioning, monitoring, enforcement, conflict resolution, and governance activities are organized in multiple layers of nested enterprises.[22]

According to Frischmann, Ostrom's functionalist IAD framework tried to systemize research efforts by facilitating "a more rigorous evaluation by matching and testing of theories and models with observed phenomena, and most generally, enable learning over time."[23] This framework subdivided relevant research inputs into two categories, namely, "exogenous variables" and the "action arena." The former includes

[18] OSTROM (1990), n. 15.

[19] Frischmann, *Two Lessons*, n. 17, at 7.

[20] OSTROM (1990), n. 15.

[21] *See, e.g.*, Elinor Ostrom, *A Diagnostic Approach for Going Beyond Panaceas*, 104 *Proc. Nat'l Acad. Sci.* US 15176, 15181–82 (2007); OSTROM (1990), n. 15, at 58–88 (describing case studies).

[22] OSTROM (1990), n. 15.

[23] Frischmann, *Two Lessons*, n. 17, at 8. It thus avoided theoretical myopia and reductionism á la Hardin's approach and its progeny. *Id. See further* Michael J. Madison, Brett M. Frischmann & Katherine J. Strandburg, *Constructing Commons in the Cultural Environment*, 95 *Cornell L. Rev.* 657, 678 (2010), *available at* http://www.lawschool.cornell.edu/research/cornell-law-review/upload/madison-frischmann-strandburg-final.pdf [hereinafter Madison et al. (2010)].

biophysical characteristics, community attributes, and governance mechanisms; the latter refers to the social space where "participants with diverse preferences interact, exchange goods and services, solve problems, dominate one another, or fight (among the many things that individual(s) do in action arenas)."[24] Understanding how different commons enterprises succeed or fail "may require accounting for all of these factors, even though it may turn out that outcomes are relatively impervious to some of them."[25]

More recently, a group of legal scholars have attempted to apply Ostrom's lessons to "commons in the cultural environment," a term that refers to "the various cultural, intellectual, scientific and social resources/systems that we inherit, use, experience, interact with, change and pass on to future generations."[26] The term "cultural commons," rather than "information environment" or "intellectual commons," aims to capture the contextual, contingent, and social/relational aspects of these resources.[27] The term "commons" in the cultural environment embraces information commons, science commons, knowledge commons, cultural commons, data commons, and still other types of intellectual resource commons.[28]

In their article, Professors Madison, Frischmann, and Strandburg focused on "constructed cultural commons," which refers to environments for developing and distributing cultural and scientific knowledge through institutions that support pooling and sharing that knowledge in a managed construct. This approach tracks the concept of a natural resource commons, which refers to the type of managed sharing environment for natural resources that, for example, a Maine lobster fishery represented. Such environments "are designed and managed with limitations tailored to the character of those resources and to the communities involved rather than left to evolve via market transactions grounded solely in traditional proprietary rights."[29]

In this context, the "commons" concept has been applied to a wide range of tangible research resources, in the life sciences, including pooled genetic resources,[30] and to other intangible information goods that are pooled and distributed

[24] Frischmann, *Two Lessons*, n. 17, at 9 (quoting ELINOR OSTROM, UNDERSTANDING INSTITUTIONAL DIVERSITY 6 (Princeton Univ. Press, 2005)).

[25] *Id.* at 9.

[26] Frischmann, *Two Lessons*, n. 17, at 11. *See generally* FRISCHMANN, INFRASTRUCTURE, n. 11, at 259–314; Madison et al. (2010), n. 23.

[27] FRISCHMANN, INFRASTRUCTURE, n. 11.

[28] Madison et al. (2010), n. 23.

[29] *Id.* at 687.

[30] *See, e.g.*, Derek Byerlee, *Crop Improvements in the CGIAR as a Global Success Story of Open Access and International Collaboration*, 41 *Int'l J. Commons* 452–80 (2010); Jorge L. Contreras, *Bermuda's Legacy: Policy, Patents and the Design of the Genome Commons*, 12 *Minn. J. L. Sci. & Tech.* 61 (2011); Tom Dedeurwaerdere, *Institutionalizing Global Genetic Resource Commons: Towards Alternative Models for Facilitating Access in the Global Biodiversity Regime* (Working Paper, June 2010), *available at* http://ssrn.com/abstract=1611549. *See also* Amy Kapczynski et al., *Addressing Global Health Inequities: An Open Licensing Approach for University Innovations*, 20 *Berkeley Tech. L.J.* 1031–1114 (2005).

through digital networks.[31] Interest in these so called "New Commons" was particularly stimulated once the World Wide Web became a universal tool from the mid-1990s on.[32] Studies have shown that governance by networks or communities of actors can be effective in situations where conventional governance modalities, such as direct government regulation or proprietary market-based incentives, had failed to produce or provide broad access to essential knowledge goods.[33]

Studies in this vein also have attempted to pinpoint some of the ways in which these New Commons have certain characteristics that distinguish them from traditional natural resource commons. For example, many of the former consist of man-made resources, such as open-source software and scientific research inputs and outputs, and they often emerge from the development of new technologies or the growth of new communities.[34] Unlike the traditional natural resource commons, the New Commons "tend to be dynamic ... complex and heterogeneous," often global in scale, with "fuzzy boundaries," and "there is a great deal that we do not know about ... [them], particularly how they work and if they can be sustained."[35]

The growing, if not dominant, importance of diverse "knowledge commons" within the universe of New Commons initiatives has recently spurred searching investigations that seek to determine their unique design features more accurately.[36] Madison, Frischmann, and Strandburg argue that Ostrom's IAD framework must be expanded to account for significant differences between commons in the natural environment and commons in the cultural environment. Most obviously, the pooled resources are of a different nature, and as a result, the obstacles that must be overcome for institutionalized sharing to succeed are different as well.[37]

A key insight is that, unlike commons in the natural resource environment, cultural commons arrangements usually must create a governance structure within which participants not only share existing resources but also engage in producing those same resources.[38] This characteristic of cultural commons yields a more intertwined

[31] *See* Brett M. Frischmann, Michael J. Madison & Katherine J. Strandburg, Governing Knowledge Commons (Oxford U. Press, 2014).

[32] Charlotte Hess, *Institutional Design and Governance in the Microbial Research Commons*, *in* Designing the Microbial Research Commons: Proceedings of an International Symposium 178 (P.F. Uhlir ed., Nat'l Acad. Press 2011) [hereinafter Designing the Microbial Research Commons], *available at* http://www.ncbi.nlm.nih.gov/books/NBK91499/ (last accessed 14 June 2014).

[33] *See generally* Yochai Benkler, The Wealth of Networks: How Social Production Transforms Markets and Freedom (2006) [hereinafter Benkler (2006)]; Frischmann, Infrastructure, n. 11.

[34] Hess (2011), n. 32, at 178.

[35] *Id.* Of particular importance here are Benkler (2006), n. 33 and Madison et al. (2010), n. 23.

[36] Madison et al. (2010), n. 23.

[37] Frischmann, *Two Lessons*, n. 17, at 16.

[38] Madison et al. (2010), n. 23. at 682.

set of exogenous variables, largely because separating the managed resources from the attributes and rules-in-use of the community that produces them is difficult or impossible. Cultural commons are also nested within, and interact with, more complex systems of natural and socially constructed environments, and boundary management becomes more complicated.[39]

These same authors have lately undertaken a series of case studies to further determine the characteristics of cultural commons and to test the validity of their precepts.[40] Preliminary results suggest that, in some of the cases at least, the successful management of cultural commons depends more heavily on the rules of law and the quality of leadership than is the true when natural resource commons are the objects of enquiry.[41]

A. *Applying Commons Theory to the Microbial Research Infrastructure*

Professor Frischmann distinguishes between "infrastructure" as resources and "commons" as a mode of resource management. On this scheme, infrastructural resources within the ambit of a knowledge commons meet the following criteria:

- The resource may be nonrivalrous under some conditions;
- Social demand for the resources is driven primarily by downstream productive activity that requires the resource as an input;
- The resource may be used as an input into a wide range of goods and services that may include private goods, public goods, and social goods.[42]

Commons management, in turn, is one among several strategies for generating value for the public at large from infrastructure resources.[43] While a decision to

[39] *Id.*

[40] *See, e.g.*, Katherine J. Strandburg & Brett Frischmann, "The Rare Diseases Clinical Research Network as a Nested Cultural Commons," paper presented at the Conference on Governing Pooled Knowledge Resources: Building Institutions for Sustainable Scientific, Cultural and Genetic Resources Commons, International Association for the Study of the Commons (IASC), Louvain-la-Neuve, Belgium, Sept. 12–14, 2012.

[41] *See, e.g.*, Michael Madison, Constructing Commons in Intellectual Resources, paper presented at the Conference on Governing Pooled Knowledge Resources: Building Institutions for Sustainable Scientific, Cultural and Genetic Resources Commons, International Association for the Study of the Commons (IASC), Louvain-la-Neuve, Belgium, Sept. 12–14, 2012. See generally FRISCHMANN ET AL., GOVERNING THE KNOWLEDGE COMMONS, n. 31.

[42] FRISCHMANN, INFRASTRUCTURE, n. 11, at xiii; *see generally id.*, Chapter 4. "Social goods" are defined as public goods capable of being delivered as private goods, but that are usually delivered by the government for various reasons, including social policy, and that are publicly funded by taxes. Public good, WIKIPEDIA, http://en.wikipedia.org/wiki/public_good.

[43] In this context, Frischmann defines the term "commons" as "an institutionalized community practice, a form of community management or governance." Frischmann, *Two Lessons*, n. 17, at 141. The term applies to resources, and it involves a group or community of people, "but the commons itself

resort to commons management must be "evaluated carefully and contextually," one particularly relevant criterion for purposes of this and succeeding chapters is that "commons management may maximize the option value of infrastructure when there is high uncertainty regarding sources of future market value."[44] Following in this vein, Frischmann explains why commons management can be a particularly efficient means for supporting the production, use, and distribution of both public and social goods.[45]

First, commons management avoids pressures to pick winners and losers while "leav[ing] it to users to decide what to do with the opportunities (capabilities) provided by infrastructure."[46] Second, given a high degree of uncertainty about which users or uses will generate social value in the future, commons management precludes optimization for an unduly narrow range of activities and "avoids social opportunity costs associated with path dependency."[47] Third, and perhaps most important, commons management

> structures the relationships between infrastructure and infrastructure-dependent systems in a manner that creates a spillover-rich environment, where spillovers flow from the many productive activities of users. These activities yield new and unanticipated innovations, knowledge, social capital, and other public and social goods that leads to economic growth and development, as well as to social welfare improvements not fully reflected in traditional economic measures.[48]

While Frischmann and his colleagues have begun to devote considerable attention to "intellectual infrastructure"[49] and related questions pertaining to cultural commons in general,[50] they recognize that the empirical analysis of upstream

is not the resources, the community, a place or a thing. Commons is the institutional arrangement of these elements." Scholars have adopted a broad definition of the commons to include both pure and quasi-public goods; that is, to include all goods that are not subjected to exclusive and proprietary controls over access and use. BENKLER (2006), n. 33; LAWRENCE LESSIG, THE FUTURE OF IDEAS: THE FATE OF THE COMMONS IN A CONNECTED WORLD (2001); ELINOR OSTROM, UNDERSTANDING INSTITUTIONAL DIVERSITY (Princeton Univ. Press 2005), n. 24; CHARLOTTE HESS & ELINOR OSTROM, UNDERSTANDING KNOWLEDGE AS A COMMONS: FROM THEORY TO PRACTICE (2007). From a technical economic perspective, the designation of commons as nonprivate goods would cover both pure public goods (goods that are both nonexcludable and not-depleteable) and quasi-public goods (goods that have only one of these two characteristics). TODD SANDLER, GLOBAL COLLECTIVE ACTION (Cambridge Univ. Press 2004). Classical examples of pure public goods include certain fully open-access public databases. Examples of quasi public goods are closed knowledge pools or so-called semicommons, which are nondepleteable upon joint consumption, but where it is easy to exclude certain users.

44 FRISCHMANN, INFRASTRUCTURE, n. 11, at xv; *see generally id.*, Chapter 5.

45 FRISCHMANN, INFRASTRUCTURE, n. 11, at xv.

46 *Id.*

47 *Id.*

48 *Id. See generally id.*, Chapter 5. For complications to be considered when evaluating the case for managing infrastructure as commons, *see generally id.*, Part III.

49 *See, e.g.*, FRISCHMANN, INFRASTRUCTURE, n. 11, Part V.

50 *See* text & accompanying n. 26–39.

research assets as global knowledge infrastructures raises a set of issues that have not as yet been explored within the better known literature on the environmental commons.[51] Nevertheless, one can analyze even these knowledge infrastructures in a systematic manner comparing different approaches within the broad category of cultural commons.[52]

In this context, the genetic resources held by the network of public culture collections that the WFCC loosely governs may be conceived as a "common-pool resource" provided by the microbiological research community. Like plant genetic resources, microbial genetic resources may be more or less rivalrous and/or excludable, depending on the conditions in which they are found in nature or artificially preserved *ex situ*.[53] Disregarding microbes held in various degrees of secrecy by industry or academic institutions, we saw, in Parts One and Two of this volume, that the public culture collections represented an important source of basic research inputs operating in an intermediate space between pure public goods (such as the Agricultural Research Service Culture Collection in the United States)[54] and the system of informal microbial exchanges among academic institutions discussed in Chapter 5.[55] Viewed as a common-pool resource, rather than as either a pure public good or club good, the distributed networks of microbial culture collections pose the kind of management and governance issues that the knowledge commons literature has recently been addressing in general terms.[56]

[51] Frischmann, *Two Lessons*, n. 17.

[52] FRISCHMANN, INFRASTRUCTURE, n. 11, Chapter 5; see also Madison et al., n. 23.

[53] Michael Halewood, What Kind of Goods are Plant Genetic Resources for Food and Agriculture? Towards the Identification and Development of New Global Commons, paper presented at the Conference on Governing Pooled Knowledge Resources: Building Institutions for Sustainable Scientific, Cultural and Genetic Resources Commons, International Association for the Study of the Commons (IASC), Louvain-la-Neuve, Belgium, Sept. 12–14, 2012 [hereinafter Halewood (Louvain 2012)].

[54] *See* Cletus P. Kurtzman, *The Agricultural Research Service Culture Collection: Germplasm Accessions and Research Programs*, *in* DESIGNING THE MICROBIAL RESEARCH COMMONS, n. 32, at 55–63.

[55] *See* Chapter 5, Section I.A.3.

[56] *See* Yochai Benkler, *Designing Cooperative Systems for Knowledge Production: An Initial Synthesis from Experimental Economics*, *in* MAKING AND UNMAKING INTELLECTUAL PROPERTY: CREATIVE PRODUCTION IN LEGAL AND CULTURAL PERSPECTIVE (Univ. Chicago Press, Jaszi et al. eds., 2011) [hereinafter Benkler (2011)]; Madison et al. (2010), n. 23. *See generally* LESSIG, n. 43; BENKLER (2006), n. 33; JAMES BOYLE, THE PUBLIC DOMAIN: ENCLOSING THE COMMONS OF THE MIND (2008); HESS & OSTROM, n. 43; FRISCHMANN, INFRASTRUCTURE, n. 11; *Symposium Issue on the Public Domain*, 77 *Law & Contem. Probs.* 1–483 (2003). For particular applications to science, *see also* Paul A. David, *The Economic Logic of "Open Science" and the Balance Between Private Property Rights and the Public Domain in Scientific Data and Information: A Primer*, *in* THE ROLE OF SCIENTIFIC AND TECHNICAL DATA AND INFORMATION IN THE PUBLIC DOMAIN 19, 19–34 (Julie M. Esanu & Paul F. Uhlir eds., 2003); Reichman & Uhlir (2003), n. 11, at 315–462; Peter Dawyndt et al., *Exploring and Exploiting Microbiological Commons: Contributions of Bioinformatics and Intellectual Property Rights in Sharing Biological Information*, 188 *Int'l Social Sci. J.* 249–58 (2006); Lee, *Open Science*, n. 11; Lee, *Distributive Commons*, n. 11.

These issues are made more complex by the fact that microbial genetic resources, like plant genetic resources, are valued both as physical goods and as intangible carriers of information that bear on both scientific research and industrial applications.[57] Viewed as information, genomic expressions of microbial genetic resources are nonrivalrous in ways that are not true of microbial materials that have not been genetically decoded. Yet, when voluntarily shared and made available to the research community as a common pool resource, even these digital resources become subject to the logic of comparable design and governance principles otherwise applicable to other knowledge commons.[58]

Still another complicating factor is that the principles of design, management, and governance of both natural resource commons and cultural commons, which have elicited the most study, may require different nuances and variations when applied to science commons, a topic that has received considerably less study. In 2003, Reichman and Uhlir first proposed a "contractually constructed" research commons for scientific data in a highly protectionist intellectual property environment,[59] a theme that we return to throughout this volume. The concept of a "contractually constructed" research commons was extended to selected areas of patent law by Professor Peter Lee,[60] and then, more generally to commons in the cultural environment by Madison, Frischmann, and Strandburg in 2007:

> The phrase "constructed commons," as we use it, refers to environments for developing and distributing cultural and scientific knowledge through institutions that support pooling and sharing that knowledge in a managed way. [T]hese environments are designed and managed with limitations tailored to the character of these resources and the communities involved rather than left to evolve via market transactions grounded solely in traditional property rights.[61]

These and other scholars have lately begun to focus attention on identifying the governance principles best suited to the formation of sustainable commons in the cultural environment generally, including science commons.[62] They are also studying the complex interrelationship between commons-managed resource infrastructures and resources protected by intellectual property rights or other legal tools that generate semicommons, open to some qualified users as distinct from a "commons" open to all.[63]

57 *See, e.g.*, Halewood (Louvain 2012), n. 53.
58 *See generally* Chapter 8, Section II.C ("Understanding the Data Sharing Movement and Its Future").
59 Reichman & Uhlir (2003), n. 11.
60 Lee, *Open Science*, n. 11; Lee, *Distributive Commons*, n. 11.
61 Madison et al. (2010), n. 23. *See also* BENKLER (2006), n. 33, at 328–55 ("Commons-based Research for Food and Medicines").
62 Madison, n. 41. *See generally* FRISCHMANN ET AL., GOVERNING THE KNOWLEDGE COMMONS, n. 31.
63 *See, e.g.*, Frischmann, *Two Lessons*, n. 17, at 12; FRISCHMANN, INFRASTRUCTURE, n. 11, at 255, 301–05 ("Intellectual Property Laws as Semi-commons Arrangements"). *See also* Paul A. David, *The Historical Origins of "Open Science:" An Essay on Patronage, Reputation and Common Agency Contracting*

As Frischmann observes, "both environmental and intellectual property legal systems construct semicommons arrangements that create and regulate interdependent private rights and public commons. Each does so in very different ways, however."[64]

These studies suggest that, while Ostrom's IAD framework remains a useful analytical tool, the differences between commons managing intellectual resources and those managing natural resources call for empirical analyses specifically tailored to a variety of commons operating in the cultural environment. For example, Charlotte Hess points out that all the sample commons in Ostrom's seminal study "were managed by relatively small, homogenous groups."[65] The extent to which these same principles "scale up" to larger initiatives, such as the one proposed in this volume for the Microbial Research Commons, has yet to be determined.[66] Clearly defined boundaries may be harder to apply in this context, while the importance of monitoring individual behavior may become even greater.[67]

The fact that governance structures in the cultural environment often manage both existing resources and the production and integration of new resources is still another complicating factor. It leads to a set of coordination problems, such as standardization and quality management, that are specific to commons in the cultural environment and that have been discussed at length in this volume. Still another complication arises from the need to regulate use of intellectual resources and related liability issues that are less salient or infrequent in the natural environment, but which pose a key issue for governance of microbial materials, as discussed later in this and succeeding chapters.

1. Distinctive Characteristics of Genetic Materials as a Common-Pool Resource

In the past, and especially before the internet, it was harder to conceive of commons-based management or production of goods on a global scale, due to such factors as the costs of exchange and a lack of global institutional frameworks.[68]

in the Scientific Revolution, 3(2) CAPITALISM & SOC'Y *art.* 5 (2008), *available at* http://capitalism.columbia.edu/files/ccs/Paul%20A.%20David.pdf [hereinafter David (2008)]; Yochai Benkler, *Between Spanish Huertas and the Open Road: A Tale of Two Commons?*, *in* FRISCHMANN ET AL., n. 31; *cf.* the concept of "common pooled resources" that are limited to a specified group of participants. OSTROM (1990), n. 15; Kapczynski et al., n. 30, at 1072 (explaining the related notion of self-binding commons, which "operate by conditioning access to their benefits on reciprocal sharing of appropriately defined improvements. They create a self-binding commons rather than an unrestricted public domain").

64 FRISCHMANN, INFRASTRUCTURE, n. 11, at xvii.

65 Hess (2011), n. 32, at 178.

66 *Id.*

67 *Id. See further* Chapter 10, Section III.

68 ROBERT O. KEOHANE & ELINOR OSTROM, LOCAL COMMONS AND GLOBAL INTERDEPENDENCE (Sage 1995).

Arguably, one of the first major instance of commons-based management on a regional scale was the organization of modern scientific research during the seventeenth century in Europe.[69] In recent decades digital networks have dramatically expanded the opportunities for building and sustaining different kinds of research commons on a global scale, including both networked information in digital environments and networks of *ex situ* genetic resources.[70]

Genetic-resource commons, in particular, have benefited from a combination of technological progress in both the life sciences and informatics.[71] The development of innovative methods for the identification, long-term conservation (e.g., freezing, freeze-drying), and shipping of genetic resources enhanced interest and fostered international cooperation in global life-science research.[72] The information technology revolution in the past two decades dramatically expanded the possibilities of distributed coordination, while diminishing the search costs for locating genetic resources held in collections throughout the world or potentially available *in situ*.[73]

Genetic resources, including microbial materials, are complex goods, with both a physical (the biological entity) and an informational component (the genetic information and information on the biochemical pathways). As biophysical entities, most genetic resources are widely dispersed, whether originally in nature,[74] or as a product of human domestication,[75] and excluding users from accessing these resources *in situ* can become both difficult and costly.

Biological materials are not typically accessed for direct exploitation as such, but for access to the informational components they embody.[76] For example, large quantities of biological samples are collected in order to screen the functions and

[69] David (2008), n. 63.

[70] Benkler (2006), n. 33; Boyle, n. 56; Lessig, n. 43; Hess & Ostrom, n. 44. For genetic resources, *see, e.g.*, David Smith, Dagmar Fritze, and Erko Stackebrandt, Public Service Collections and Biological Resources Centers of Microorganisms, in The Prokaryotes – Prokaryotic Biology and Symbiotic Associations (E. Rosenberg et al, eds., Springer Verlag 2013) [hereinafter D. Smith et al (2013)].

[71] Bronwyn Parry, Trading the Genome (2004).

[72] Scott Stern, Biological Resource Centers: Knowledge Hubs for the Life Sciences 42 (Brookings Inst. Press 2004).

[73] *See, e.g.*, Peter Dawyndt, Tom Dedeurwaerdere & J. Swings, *Exploring and Exploiting Microbiological Commons: Contributions of Bioinformatics and Intellectual Property Rights in Sharing Biological Information*, 188 *Int'l Social Sci. J.* 249–58 (2006) (Introduction to the special issue on the microbiological commons).

[74] Andrew J. Beattie et al., *New products and Industries from Biodiversity, in* 1 Ecosystems and Human Well-Being: Current State and Trends 271–95 (R. Hassan et al. eds., 2005).

[75] Fernand Braudel, Civilization and Capitalism, 15th-18th Century, Vol. I: The Structure of Everyday Life *passim* (2d prtg., 1992).

[76] Tom Dedeurwaerdere, *From Bioprospecting to Reflexive Governance*, 53(4) Ecological Econ. 473–91 (2005); Timo Göeschl & Timothy Swanson, *The Social Value of Biodiversity for Research and Development*, 22 *Envtl. & Resource Econ.* 477–504 (2002).

properties they exhibit against certain targets. Once a new property or function has been discovered, genetic similarity searching can identify the sequences that are involved in the expression of specific properties. Such findings may, in turn, lead to further research on these genes or their properties without having to access any given organism that led to the discovery of the new informational inputs.

Nevertheless, accessing available materials for scrutiny of phenotypical functions often remains important at the end of the research and innovation chain, when biological samples are involved in developing commercial applications. Any regime that regulates access to these resources must necessarily take into account both the precompetitive informational features of pooled resources and the potential commercial uses, if any, of specific biological organisms.[77]

In general, genetic resources provide informational inputs needed for both research and innovation. They serve as stocks of accumulated traits of known utility in the natural environment and as generators of new flows of information based on the discovery of new useful features.[78] However, the regulation of access to, and exchanges of, global genetic resources under international regimes, such as the CBD and the TRIPS Agreement, fail to take these features of global genetic resource networks sufficiently into account.[79]

2. Factoring in the Unprecedented Power of Digital Networks

Studying institutional models developed for digitally networked information commons helps us to better understand the value of the informational component embedded in genetic resources.[80] In this section, we focus on several key design principles of successful digital information commons that scholars have identified, including the role of nonmarket motivations and the modular character of the underlying organizational architecture.[81]

Professor Yochai Benkler has found that, in mixed or complex incentive schemes, such as those at stake in a digital information commons, participants are driven more by social motivations (especially reputational benefits) and intrinsic motivations

77 *Cf.* Arti K. Rai et al., *Pathways Across the Valley of Death: Novel Intellectual Property Strategies for Accelerated Drug Discovery*, 8 *Yale J. Health Pol'y L. & Ethics* 1 (2008), *available at* http://scholarship.law.duke.edu/cgi/viewcontent.cgi?article=2329&context=faculty_scholarship.

78 Timothy M. Swanson & Timo Göeschl, *The Management of Genetic Resources for Agriculture: Ecology and Information, Externalities and Policies* (Ctr. Soc. & Econ. Research on the Global Env't (CSERGE) Working Paper No. GEC 98–12, 1998).

79 *See further* Chapters 2, 3 & 6.

80 Benkler (2006), n. 33. Benkler stresses the voluntary nature of contributions that these initiatives motivate in lieu of monetary incentives.

81 Other design principles include appropriately devised quality controls, widely distributed and/or available physical capital, and investment in social networks. *Cf.* Benkler (2006), n. 33, at 106–226.

(such as ethics, curiosity, and other personal values) than by the prospect of direct monetary rewards alone.[82] In the life sciences, where potential commercial rewards from basic research are usually a factor, especially with regard to university-driven research, Professor Minna Allarakhia found that the reciprocity benefits to be gained from participation in a research commons, or even in a semicommons, are often the key motivational factor.[83] Her findings are particularly relevant to our design principles for a Microbial Research Commons, as will appear in due course.

Mixed motivations are present in a heterogeneous set of initiatives, such as open source software, globally linked genetic sequence databases, and various types of distributed peer-to-peer computational research. Because of the difficulty of putting a precise monetary value on the creative inputs of a large and distributed network of contributors, it is often demonstrably more effective to rely on voluntary, contributions for organizing digital information commons than on proprietary or market-driven incentives.[84] Moreover, empirical research has shown that, when social motivations are involved, such as increasing recognition within a collaborative group or the satisfaction of intrinsic motivation with respect to furthering objectives of general interest, monetary rewards can actually decrease participants' willingness to contribute to the global pool.[85] Moving from social to monetary rewards also entails hidden costs arising from the need to clearly delineate the tasks to be remunerated and to attach monetary value to every contribution to the undertaking.[86]

The true value of the genetic or other resources at issue usually becomes apparent only late in the research and development process, whereas the conceivable monetary value of the same resources, as assessed at the beginning of that process, remains statistically very low.[87] This principle was important in Chapter 5, where we designed a hypothetical Compensatory Liability Regime for the large subset of microbial genetic resources having no known or likely commercial value.

From a broader perspective, another factor in motivating those who organize knowledge commons generally is the fear that heretofore open knowledge assets available as a public good will be enclosed.[88] As we saw in Chapter 6, information

[82] BENKLER (2006), n. 33.

[83] Minna Allarakhia et al., *Modeling the Incentive to Participate in Open Source Biopharmaceutical Innovation*, 40 *R&D MGMT*. 50–66 (2009). *See also* n. 56.

[84] FADI P. DEEK & JAMES A. MCHUGH, OPEN SOURCE TECHNOLOGY AND POLICY (Cambridge Univ. Press 2008); *see also* JANET HOPE, BIOBAZZAR: THE OPEN SOURCE REVOLUTION AND BIOTECHNOLOGY (Harvard Univ. Press 2008).

[85] Bruno S. Frey & Reto Jegen, *Motivation Crowding Theory*, 15 *J. Econ. Surveys* 589–611 (2001).

[86] Edward L. Deci, *The Hidden Costs of Rewards*, 4 *Organizational Dynamics* 61–72 (1976).

[87] R. David Simpson et al., *Valuing Biodiversity for Use in Pharmaceutical Research*, 104 *J. Political Econ.* 163–85 (1996); *see also* Rai et al., n. 77.

[88] Hess (2011), n. 32, at 180. *See generally* James Boyle, *The Second Enclosure Movement and the Construction of the Public Domain*, 66 *Law & Contemp. Probs.* 33–74 (2003).

technologies have already enabled commercial intermediaries to capture and privatize data and information that were relatively unencumbered in the print media. The digitally distributed knowledge commons we envision for microbial genetic resources and related data and literature is thus a response to the threat of enclosure of a public good.[89]

The expanded capacity to adopt modular technical and organizational architectures in the digital environment is another major institutional feature bearing on the success of commons-based knowledge production. Modularity presupposes the possibility of devising a set of independently produced components that can be integrated as a whole. Distributed modular architectures then enable many more participants to effectively pool their efforts and contributions, notwithstanding the fact that these contributions may vary in quality, focus, timing, and geographical location.[90] Such contributions are also often the fruit of nonhierarchical decision-making processes, although the extent to which some managerial hierarchy may in fact be needed varies from case-to-case.[91]

The reciprocity payoffs from distributed modular architecture may vary considerably with the number of potential contributors to the prospective network in question. If there is a large set of relatively small-scale contributors, each of whom only has to invest a moderate amount of additional effort and time into the commons initiative, then the potential benefits of taking a part in global research endeavor is likely to be high.[92] However, if even the smallest contributing entities are relatively big players, and if their participation requires each unit to make a large investment of additional time and effort, the potential reciprocity benefits and cost-effectiveness of joining a collaborative network may diminish, and the universe of potentially willing contributors may shrink, unless careful attention is paid to coordination and governance norms.[93]

Fears of enclosure and the double-edged applications of modularity in the formation of distributed research commons also logically apply to pooled research materials, including microbial genetic resources. These tangible materials can also generate a flow of intangible scientific information.[94] For example, modularity was clearly a factor in the formation of some very successful genetic resource commons,

89 *See generally* Reichman & Uhlir (2003), n. 11; Jerome H. Reichman & Ruth L. Okediji, *When Copyright and Science Collide: Empowering Digitally Integrated Research Methods on a Global Scale*, 96 *Minn. L. Rev.* 1362 (2012) [hereinafter Reichman & Okediji (2012)].

90 Benkler (2006), n. 33, at 100.

91 *Compare* Yochai Benkler, *Coase's Penguin, or, Linux and The Nature of the Firm*, 112 *Yale L.J.* 369 (2002) (stressing nonhierarchical forms of production) *with* Benkler (2006), n. 33 (more nuanced and informed discussion of this variable).

92 For examples, see Benkler (2006), n. 33.

93 *Id.* For an ambitious attempt to devise coordination and governance norms to this end, see the discussion of GEO and the Global Earth Observing System of Systems (GEOSS), Section II.B.3.

94 *See* text accompanying nn. 74–75.

such as the collaborative sequencing of the worm genome by a network of teams distributed around the world in the early days of the genomic revolution[95] or the networks of crop improvers established by the member institutes of the CGIAR.[96]

One should not, however, underestimate the countervailing obstacles to the formation of digital commons. For example, digital information remains surprisingly fragile, and lots of important electronic information is lost every day.[97] Apart from losses due to publishers going out of business or to the patenting of genomic data, digital information is also lost due to inattention, lack of robust preservation strategies, underfunding, and obsolete formats.[98] Universities in OECD countries may not develop storage capacity for the massive datasets produced by their scholarly communities.[99] In developing countries, a general lack of suitable technology, networked infrastructure, and even basic electricity yields unequal access to otherwise readily available digital information.

Other obstacles to the formation of a robust digital infrastructure along the lines envisioned here include the costs and lack of incentives for university scientists to annotate the genomes of organisms in collaborative repositories,[100] not to mention the efforts of university technology transfer offices to commoditize their faculties' research results. While studies by Professor Charles Schweik show that properly elaborated commons norms, rules, and governance structures can overcome these and other obstacles, his work also shows that too much top-down governance may deter, rather than enable the requisite level of collaboration.[101]

3. Potential Payoffs from a Well-Designed Governance Model

With these caveats in mind, the payoffs from specialized investigations conducted within a research commons framework are known to increase with the ability of

[95] John Sulston & G. Perry, The Common Thread: Science, Politics, Ethics and the Human Genome (2003).

[96] Byerlee, n. 30. See Chapter 2, Section I.B ("Early Efforts to Form an Agricultural Research Commons for Plant Genetic Resources").

[97] Hess (2011), n. 32, at 182.

[98] *Id. See generally* Nat'l Research Council (NRC), Preserving Scientific Data on our Physical Universe (1995).

[99] Hess (2011), n. 32, at 182.

[100] *See* Roy Welch & Laura Welch, *If You Build It, They Might Come*, 7 *Nature Revs. Microbiology* 90 (2009).

[101] *See, e.g.*, Charles Schweik, *Schweik Open Source Project*, Nat'l Ctr. for Digital Gov't (Nov. 27, 2009) with links to articles at http://www.umass.edu/digitalcenter/ossuccess/. *See also* Yochai Benkler, *Designing Cooperative Systems for Knowledge Production: An Initial Synthesis from Experimental Economics*, *in* Making and Unmaking Intellectual Property: Creative Production in Legal and Cultural Perspective (Mario Biagioli et al., eds., U. Chicago Press 2011); *see also* Charles Schweik & Robert English, Internet Success: A Study of Open Source Software Commons (2012).

scientists to access large, comprehensive collections of materials, literature, and data.[102] This principle bears directly on the life-science community's prospects for realizing the NRC's vision of a New Biology for the 21st Century,[103] and more specifically, on the potentially important role of microbiology in that vision.[104]

For example, research on many infectious diseases requires access to an entire microbial population to understand the relevant mutational dynamics. Similarly, to exploit one type of microarray technology, the researcher needs access to large amounts of data at the preselection stage for genetic testing.[105] When applying high-throughput screening techniques, the greater the amount of *ex situ* small molecules available for screening, the greater the likelihood of more "hits" from which medicinal compounds can be identified.[106] The same holds true for *in situ* biodiversity resources used when scientists screen for possible medicinal or agricultural applications.[107]

With specific regard to genetic resources as basic inputs into microbiology, the existing modular organization – based on collaboration and specialization across a worldwide network of culture collections – arose mainly in response to the high cost of conserving *ex-situ* microbial genetic resources, and to the fact that ever more infrastructure was needed to hold the *in situ* resources still being collected.[108] This distributed, collaborative infrastructure has recently been digitally empowered by

[102] *See, e.g.*, Mark Harvey & Andrew McMeekin, "Public or Private Economies of Knowledge: The Economics of Diffusion and Appropriation of Bioinformatic Tools," paper presented to the Microbial Commons Conference, Ghent, Belgium, 12–13 June 2008. For data, see Heather J. Ritch, EUROPEAN RESEARCH INFRASTRUCTURE CONSORTIUMS: PRIVATELY ORDERED AND PUBLICLY FUNDED RESEARCH COMMONS FOR DATA (2011) (unpublished SJD Thesis, Duke University) (on file with Goodson Law Library, Duke University); Reichman & Uhlir (2003), n. 11. For small molecule libraries, see Rai et al., n. 77.

[103] *See* Chapter 1, Section II.B.

[104] *See* NATIONAL RESEARCH COUNCIL, A NEW BIOLOGY FOR THE 21ST CENTURY, 41–50 (Nat'l Acads. Press 2009) [hereinafter NRC, A NEW BIOLOGY].

[105] *See e.g.*, Adi Tarca et al., *Analysis of Microarray Experiments of Gene Expression Profiling*, 195(2) *Am. J. Obstetrics & Gynecology* 373–388 (2006).

[106] *See* Rai et al., n. 77.

[107] *See* Ninth Meeting of the Ad Hoc Open-Ended Working Group on Access and Benefit-Sharing in the Convention on Biological Diversity, Cali, Colombia, 22–28 March 2010, Side Conference Presentations [hereinafter Cali Presentations], *available at* http://www.cbd.int/wgabs9/events/se-abs9.shtml#tab=0.

[108] *See, e.g.*, STERN, n. 72. *See also* Dagmar Fritze & André Oumard, "The Pan-European Project, Microbial Resource Research Infrastructure (MIRRI), Has Among Its Goals the Elaboration of Common Policies for BRCs to Comply with the Nagoya Protocol on Access and Benefit Sharing of CBD," paper presented at the Conference on Governing Pooled Knowledge Resources: Building Institutions for Sustainable Scientific, Cultural and Genetic Resources Commons, International Association for the Study of the Commons (IASC), Louvain-la-Neuve, Belgium, Sept. 12–14, 2012, *available at* http://biogov.uclouvain.be/iasc/doc/full%20papers/Fritze.pdf [hereinafter Fritze & Oumard (2012)].

several initiatives, including the World Data Center for Microorganisms and the emerging Open Knowledge Environments discussed in Chapter 8.[109]

Despite the emphasis in the scholarly literature on the importance of nonmarket motivations as a relevant condition for the emergence of effective commons-based production,[110] we believe that is not a sufficient condition in itself. Rather, it is the effectiveness of a modular organizational form, in combination with both market and nonmarket modes of production, that stimulate widely dispersed contributions for integration on a global scale. Research on these general design principles shows that, under conditions of appropriate quality control, and given an initial investment in the creation of social networks,[111] commons based production and management of both tangible and intangible research resources can co-exist with either market or state-based production of knowledge goods. This is especially true in the early stages of research along the innovation and product development chain, when access to multiple upstream inputs, including materials, literature and data, is essential.

In our view, downstream commercial applications of such commons-based outputs remains essential for innovation and the long-term public interest.[112] Even before such applications emerge, it becomes important to institutionalize a link between the upstream contributing entities and the downstream commercial applications for at least two reasons.[113] First, the CBD requires a benefit-sharing option for all genetic resources, especially those originating from the developing countries, that were deposited in public culture collections after 1992.[114] Second, by systematically enabling upstream contributing entities to share some of the financial gains from downstream commercial applications, a contractually constructed research commons can greatly augment the potential reciprocity benefits from participation in

109 Bert Verslyppe et al., *Microbiological Common Language (MCL): A Standard for Electronic Information Exchange in the Microbial Commons*, 161(6) RESEARCH IN MICROBIOLOGY 439–45 (2010). At present, 62 collections (holding more than 300,000 strains) have joined the open-data portal. *See further* Chapter 8, Section II.B.1 ("The World Data Center for Microorganisms"); *id.*, Section III ("Building Transnational Open Knowledge Environments").

110 *See* nn. 84–86 & accompanying text.

111 BENKLER (2006), n. 33.

112 Although the balance between nonmarket and market motivations has shifted in the last decade due to increasing commercial pressures, it is fair to say that the commons-based exchange practices in the microbial field are driven by a mixed set of motivations, such as the scientific-research ethos, biodiversity conservation, animal health and food security, along with monetary recompense. However, this does not imply that the system is only designed for noncommercial uses. As we argued in Chapters 4 and 5, any system for the exchange of microbial resources has to accommodate both potential noncommercial and commercial uses of the same resources, even if the commercial value of the resource is not known at the beginning of the innovation process.

113 *See* Rai et al., n. 77.

114 *See* Chapter 3, Sections I.B & C.

the endeavor.[115] These factors will prove important for our proposed redesign of the Microbial Research Commons, as elaborated later in Chapter 10.

B. *Three Governance Prototypes for Globally Pooled Research Assets*

Economic theory concerning the provision of public goods highlights major collective action challenges for organizing pools of basic research assets on a global scale. Two core ideas bear on potential difficulties for the long-term formation and sustainability of cooperative action on such pools.[116] The first is based on the prisoner's dilemma hypothetical in game theory. It teaches that, without clear guarantees for the other players' cooperative behavior, agents will not cooperate spontaneously even if greater benefits could eventually be achieved from cooperation.[117] The second idea is based on the related free rider hypothesis that attends the provision of public goods. It teaches that some actors will attempt to benefit from the public goods that are collectively produced without contributing in a fair and equitable way to their costs.

One conventional solution to these problems is to introduce an external state authority that could impose collective goals and long-term objectives on individuals who would otherwise only seek to maximize their personal self interest in the short term.[118] For the organization of global pools of basic research assets, this approach could imply the creation of a global authority, through an intergovernmental agreement, which would act as an external rule enforcer (cf. model 1 in Figure 9.1). An important example of such a fully fleshed out intergovernmental solution is the FAO's International Treaty on Plant Genetic Resources for Food and Agriculture, discussed in Chapter 3. When such a transnational authority cannot be established, the alternative solution proposed by the conventional approach is to revert to private appropriation of the research assets under proprietary exclusive rights regimes[119] and to organize collaboration on market-based principles (cf. model 3 in Figure 9.1). An example of such a market-based arrangement in a proprietary framework is the case of global patent pools, in which agreements are made by the patent holders to cross-license the use of the patented technologies to one another.[120]

[115] *See* Reichman & Uhlir (2003), n. 11; Allarakhia et al., n. 83.

[116] Sandler, n. 43.

[117] Elinor Ostrom, *A Behavioral Approach to the Rational Choice Theory of Collective Action, Presidential Address, American Political Science Association*, 92(1) *Am. Political Sci. Rev.* 1–22 (1998).

[118] Garrett Hardin, *The Tragedy of the Commons*, 162(3859) *Science, New Series* 1243–48 (1968).

[119] *See* Chapter 3, Section III.A ("Basic Concepts of the ITPGRFA"). Also relevant here was the WHO's failed treaty governing influenza-related microbial materials, discussed in Chapter 2, Section III.A, which was replaced by a hybrid contractual regime, the Pandemic Influenze Preparedness Framework (PIP) (2011), discussed in Chapter 4, Section IV.A & B.

[120] *See, e.g.*, Gene Patents and Collaborative Licensing Models (G. van Overwalle ed., Cambridge Univ. Press, 2009).

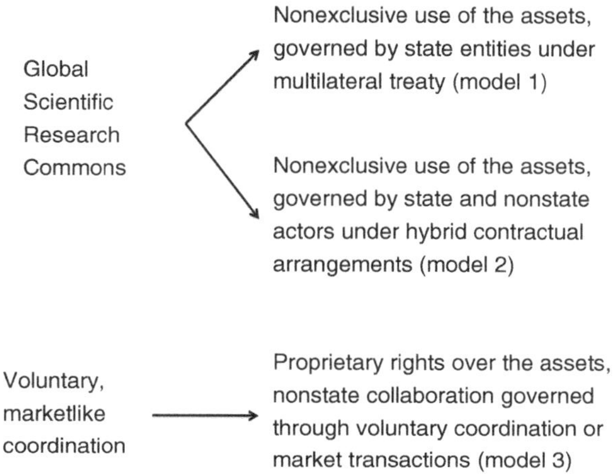

FIGURE 9.1. Theoretical Models of Global Scientific Research Collaboration.

The intergovernmental and market-like solutions for organizing global knowledge pools are, however, not the only possible model. They do not, for example, account for the extensive research collaboration among microbial culture collections discussed in this book, which have proved sustainable even in the absence of proprietary arrangements or of any formally organized intergovernmental authority.[121] In fact, as explained in Chapter 4, many essential knowledge assets for research in microbiology, such as sequence databases and materials, are made available from loosely affiliated transnational pools under nonexclusive use conditions, even though they are governed only by networks of nonstate collective actors. As shown in the literature on the governance of the commons these and other nonstate governance mechanisms are not primarily driven by profit-making incentives or external regulation, but rather by social and reputational motivations.[122]

Despite their demonstrable social benefits, we recognize that commons arrangements are not panaceas that can solve all the problems encountered in efforts to pool global knowledge assets for research purposes.[123] Knowledge commons encounter their own risks of governance failures, such as the need to ensure quality management, sustainable funding, and community involvement. As a consequence, the quality of collective decision-making and coordination in social networks will be important for successfully pooling knowledge goods on a nonexclusive basis.

[121] *See* Chapter 4, Section I ("Evolution of Microbial Culture Collections as Basic Scientific Infrastructure"). Note, however, that most public culture collections remain para-statal at the national level.

[122] BENKLER (2006), n. 33.

[123] Hess (2011), n. 32; OSTROM (2005), n. 43; BENKLER (2006), n. 33.

Two major lessons can thus be drawn from contemporary research on knowledge commons in general for the purposes of this book. First, a functionally efficient commons can be established to deal with the problem that self-interested and opportunistic behavior poses for the provision of public goods, even without state intervention to enforce the rules. Second, from a broader social perspective, the international arrangements that implement a knowledge commons are only a means to realize socially desirable ends, and not ends in themselves.[124] Therefore, as with any institutional tool, actually obtaining desired social benefits through commons-based institutions will largely depend on the organization of effective collective decision-making processes and management procedures, as further discussed later in this and the next chapter.

Given these premises, considerable attention must then be paid to the kind of governance framework in which a redesigned Microbial Research Commons would operate in order to provide the needed incentives through collective action and to achieve the desired goals.[125] Such a governance structure must, from the start, be conceptualized within some transnational legal framework capable of addressing and resolving growing tensions between developed and developing countries. It must reconcile sovereign ownership and control of microbial genetic resources with the needs of scientific research and the equitable management of intellectual property rights flowing from applications of such resources.[126] It must also elaborate internal governance mechanisms that avoid sacrificing the needs of science to the dictates of short-term political and economic expedience.

II. SELECTED EMPIRICALLY RELEVANT GOVERNANCE APPROACHES

To address these issues, we note that recent studies of cultural commons have focused critical attention on governance issues.[127] These studies have identified the following set of variables that should be used to analyze the governance aspects of particular initiatives:

- What are the specific governance mechanisms of the commons in question (e.g., membership rules, resource contribution or extraction standards and requirements, conflict resolution mechanisms, sanctions for rule violation)?
- Who are the decision makers and how are they selected?

[124] HESS & OSTROM, n. 43.
[125] *See, e.g.*, Hess (2011), n. 32.
[126] *See, e.g.*, Chapter 7, Section III ("Redefining the Role of Publishing Intermediaries Under Current Institutional Constraints").
[127] *See, e.g.*, Madison (Louvain 2012), n. 41. Strandburg & Frischmann (Louvain 2012), n. 40. *See generally* FRISCHMANN ET AL., GOVERNING THE KNOWLEDGE COMMONS (2014), n. 31.

- What are the institutions that govern decision making?
- What informal norms govern the commons?
- How do nonmembers interact with the commons? What institutions govern these interactions?
- What legal structures (including intellectual property rules, subsidies, contract and licensing law, antitrust provisions) govern the functioning of the commons?[128]

These questions, provide a useful template for analyzing how intellectual resources can be used as an input into a wide variety of pooling arrangements.[129]

Given this template, it becomes pertinent to ask how these criteria have or have not already been implemented in existing science commons initiatives that may provide viable models and experience for governing a redesigned Microbial Research Commons. To answer this question, we have briefly examined the governance structures of a selected group of recently formed knowledge commons that seem particularly relevant to our own project. Some of these transnational entities primarily govern genetic resources; others primarily govern pooled collections of scientific data; and still others seek to combine data and materials for specific scientific activities.

For purposes of greater analytical clarity, however, we have subdivided the selected entities according to the three basic governance models identified in the preceding section, namely;

- A top-down model, in which pooled assets are governed under a multilateral treaty;
- A hybrid model, in which states and nonstate actors voluntarily pool and govern resources under contractual arrangements; and
- A market-like model, in which all the assets are voluntarily made available under a coordinated scheme of proprietary rights.

In evaluating the results of this empirical review, one must bear in mind that, regardless of which cooperative arrangement is chosen, the organizers must devise a suitable legal and institutional modality for accommodating all the stakeholders' interests within the ambit of the Convention on Biological Diversity.[130] How to navigate this and other pitfalls of international law is more fully analyzed later in this chapter.

[128] Summarized in Frischmann, *Two Lessons*, n. 17.

[129] FRISCHMANN, INFRASTRUCTURE, n. 11, at 280.

[130] *See generally* Chapter 3 ("Tightening the Regulatory Grip: From the Convention on Biological Diversity in 1992 to the Nagoya Protocol in 2010"). In this context, recall that the Nagoya Protocol affords new opportunities to preserve the space for public scientific research with respect to both materials and data. *See* Chapter 3, Section IV.B ("Facilitating Scientific Research").

A. *The Global Crop Commons: A Treaty-Based Intergovernmental Entity*

The Global Crop Commons[131] – now regulated by the FAO's International Treaty on Plant Genetic Resources for Food and Agriculture (ITPGRFA) – grew out of the activities of the Consultative Group on International Agricultural Research (CGIAR), which had become the world's largest provider of *ex situ* plant genetic resources. In focusing on the complex governance structure that this Treaty established, we take note of the often frustrating experiences of the CGIAR, both before and after the Treaty was adopted in 2001.[132] That experience affords a particularly instructive example of the geopolitical tensions with which the governance structure of a redesigned Microbial Research Commons would also have to cope.

1. A Two-Headed Governance Construct

To understand how the Crop Commons operates in practice, one may view it as a kind of two-headed governance construct, which links and coordinates the FAO's International Treaty with the work of the CGIAR's Agricultural Research Institutes,[133] as described in Chapters 2 and 3. Technically speaking, this merger occurred on October 16, 2006, when eleven International Agricultural Research Centers (IARCs) under the aegis of the CGIAR, signed agreements with the Governing Body of the International Treaty on Plant Genetic Resources for Food and Agriculture, placing the *ex situ* collections of IPGRFA they held under the Treaty, in conformity with Article 15.[134] Some 700,000 accessions of the world's most important crops were thus

[131] The term "Global Crop Commons" has recently been adopted in the literature dealing with the FAO's International Treaty, and it is endorsed by major NGOs working in this area, such as Bioversity. *See, e.g.*, Crop Genetic Resources as a Global Commons: Challenges in International Law and Governance (M. Halewood et al. eds. 2013) [hereinafter Crop Genetic Resources]; Emile Frison, "The Role of the Global Crop Commons in Supporting Livelihoods and Food Security in Developing Countries," paper presented at the Conference on Governing Pooled Knowledge Resources: Building Institutions for Sustainable Scientific, Cultural and Genetic Resources Commons, International Association for the Study of the Commons (IASC), Louvain-la-Neuve, Belgium, Sept. 12–14, 2012 [hereinafter Frison (Louvain 2012)]; Halewood (Louvain 2012), n. 53.

[132] *See* CGIAR, The CGIAR at 40 and Beyond (July 2011), *available at* http://www.cgiar.org/www-archive/www.cgiar.org/pdf/cgiar%4040_final_LOWRES.pdf.

[133] For the history of the CGIAR, *see* Chapter 2, Section I.B ("Early Efforts to Form an Agricultural Research Commons for Plant Genetic Resources").

[134] *See* International Treaty on Plant Genetic Resources for Food and Agriculture, *opened for signature* 3 Nov. 2001, 2400 U.N.T.S. 303, art. 15 (entered into force 29 June 2004) [hereinafter ITPGRFA], *available at* http://treaties.un.org/doc/publication/UNTS/Volume%202400/v2400.pdf (last accessed 14 June 2014). This assignment renewed a previous arrangement of 1996, as described in Chapter 2, Section II.B; see also Chapter 3, Section II.B.

legally situated within the multilateral regime of facilitated access that the International Treaty set in place, as previously explained in Chapter 3.

Under these agreements, the Centers recognized the authority of the Governing Body of the Treaty to provide policy guidance pertaining to their *ex situ* collections, while the CGIAR retained autonomous authority over many important areas not covered by the Treaty (for example, soy and bananas).[135] The CGIAR's *ex situ* holdings covered by Annex I of the Treaty nonetheless remain under the regulatory authority of the International Treaty and its intergovernmental organization, which establishes policy for all matters covered by that agreement.[136]

The internal governance structure imposed by the Treaty itself, which mainly consists of a Governing Body and a Secretariat is relatively rigid and cumbersome. The Governing Body is composed of all the member states, and all its decisions must, in principle, be taken by consensus.[137] Each Contracting Party has one vote to be cast by a single delegate (who is allowed to rely on experts and advisers).[138]

The Governing Body meets at least once every two years, with the possibility of holding special sessions as needed.[139] That Body also elects a chairperson and vice-chairpersons (known as the Bureau) in conformity with its own rules of procedure.[140] Observer status at its sessions may be granted to UN Specialized Agencies as well as to intergovernmental organizations (IGOs) or nongovernmental organizations (NGOs) "qualified in fields relating to conservation and sustainable use of ITPGRFA."[141]

The Governing Body is charged with adopting a budget; establishing subsidiary bodies as needed, especially Advisory Committees; organizing a Trust Fund, to be known as the Benefit-Sharing Fund; and establishing relations with other entities, especially the Conference of the Parties (COP) to the CBD.[142] As a practical matter, however, the Governing Body's general duties to oversee implementation of the International Treaty are circumscribed by the "rolling Global Plan of Action for the Conservation and Sustainable Use of Plant Genetic Resources for Food and Agriculture,"[143] which is referenced repeatedly in the Treaty itself.[144]

[135] *See* CGIAR, The CGIAR at 40 and Beyond, n. 132 (noting however, that the coverage of the treaty is "constantly expanding").

[136] CGIAR, *Who We Are – History of the CGIAR*, www.cgiar.org/who/history/index [hereinafter CGIAR, *History*].

[137] ITPGRFA, n. 134, arts. 19.1,19.2.

[138] *Id.* art. 19.4.

[139] *Id.* art. 19.9.

[140] *Id.* art. 19.5.

[141] PGRFA, art. 20.1.

[142] *Id.*, art. 19.3.

[143] *Id.*, art. 14. *See* Danielle Manzella, *The Design and Mechanics of the Multilateral System of Access and Benefit Sharing, in* Crop Genetic Resources as a Global Commons (2013), n. 131, at 150–161.

[144] *See, e.g.*, ITPGRFA, n. 134, art. 14; *see also id.*, art. 13.2 (funds from benefit-sharing to be dispersed taking account of priority activity areas in the ... Global Plan of Action under the guidance of the Governing Body).

The first Global Plan of Action (GPA) was formally adopted in 1996 by representatives of 150 countries during the Fourth International Technical Conference on Plant Genetic Resources in Leipzig, Germany.[145] The FAO facilitated the drafting of the GPA and the monitoring of its implementation under the aegis of the Intergovernmental Commission on Genetic Resources for Food and Agriculture (the Commission) as part of the FAO's Global System for the Conservation and Use of Plant Genetic Resources.[146]

In 2009, the Governing Body of the International Treaty "noted the need to ensure close collaboration between itself and the Commission with regard to the GPA," and it "invited the FAO Commission, in revising the First GPA, to take into account specific issues of relevance to the International Treaty and to reflect adequately the provisions of the International Treaty in the Second GPA."[147] The Second Global Plan of Action (Second GPA), published in 2011, contains 322 Paragraphs dealing with every aspect of PGRFA covered by the International Treaty and beyond.

In implementing this Second GPA, the Governing Body of the Treaty relies heavily on the FAO's Commission on Genetic Resources for Food and Agriculture as well as on its own Secretariat. This de facto integration of treaty implementation into the FAO's own operational framework guarantees a certain degree of stability – given that both the Governing Body and the Commission are intergovernmental entities – and it provides a major source of day-to-day administrative expertise. This nexus, however, also adds to the overall bureaucratic complexity of the Global Crop Commons, and it puts more distance between the decision-making process and the scientific users of that system than may be desirable.

A Secretary to the Governing Body is appointed by the Director General of the FAO, with the approval of that Body. The Secretary, who oversees the implementation of the Treaty and is funded directly by FAO, is also entrusted with administrative support of the Governing Body and any subsidiary bodies it may establish.[148] The Secretary is also expressly mandated to cooperate with the Secretariat of the CBD.[149] In practice, given that the Governing Body is a plenary that meets only every two years, the Secretariat plays a decisive role in everyday governance.

145 United Nations Food and Agricultural Organization (FAO), First Global Plan of Action for the Conservation and Sustainable Use of Plant Genetic Resources for Food and Agriculture (adopted 1996, Endorsed 1997), available at http://ftp.fao.org/docrep/faj63.pdf [hereinafter FAO, First Global Plan of Action (1996)].

146 *See id. See also* United Nations Food and Agricultural Organization (FAO), Second Global Plan of Action for the Conservation and Sustainable Use of Plant Genetic Resources for Food and Agriculture (adopted November 2011), *available at* http://www.fao.org/docrep/fao/meeting/015/i2624a/i2624e00/htm [hereinafter FAO, Second Global Plan of Action (2011)], ¶ 3.

147 FAO, Second Global Plan of Action (2011), n. 146, ¶ 5.

148 ITPGRFA, n. 134, art. 20.2. The Secretary General of the International Treaty is Dr. Shakeel Bhatti.

149 *Id.* art. 20.5.

2. Implementation of the Multilateral Regime

Relations between the members of the International Treaty and the CGIAR are formally regulated by the Treaty itself and by its subsidiary legal instruments. Nevertheless, the CGIAR's *ex situ* collections of the plant genetic resources remain independently funded by their own donors and they are semi-autonomously managed, while conforming to the common policies that the Treaty and its intergovernmental body establishes. Some of the most important policy decisions so far were embodied in the Standard Material Transfer Agreement (SMTA) adopted by the Governing Body at its first meeting in 2006, only two years after the International Treaty took effect. The Treaty's Multilateral System of Access and Benefit-Sharing (ABS) and its SMTA make it the first international instrument that provides a practical method of ABS by facilitating the exchange of genetic resources of 64 essential food crops and forages (Annex I Crops) without the need for complex bilateral negotiations.[150] Despite the Treaty's rigid structure, as described in Chapter 3, the Governing Body's ability to develop and implement this SMTA shows that it can take decisive action to meet its overall goals and specific objectives when needed.

A. THE VIRAL LICENSE. The Standard MTA for transferring plant genetic materials within the multilateral system triggers a viral chain of related MTAs that follow any cultivar taken for research purposes from the Crop Commons to the ultimate recipient who may develop a commercial product from that same genetic resource.[151] The standard viral license, which was analyzed in Chapter 3, sets out the conditions under which commercial users may have to pay preestablished royalties on sales of the resulting products into a Benefit-Sharing Fund established under the ITPGRFA.[152] This embodiment of a "Compensatory Liability Regime"[153] was

[150] *See* Food Agric. Org. (FAO) Conference, Comm'n on Genetic Resources for Food and Agriculture, Standard Material Transfer Agreement (2006), ¶ 6.1 [hereinafter SMTA], *available at* http://www.planttreaty.org/content/drafting-standard-material-transfer-agreement. See generally NINA ISABELLA MOELLER & CLIVE STANNARD, IDENTIFYING BENEFIT FLOWS: STUDIES ON THE POTENTIAL MONETARY AND NON-MONETARY BENEFITS ARISING FROM THE INTERNATIONAL TREATY ON PLANT GENETIC RESOURCES FOR FOOD AND AGRICULTURE (United Nations Food and Agricultural Organization (2013)), Executive Summary at xxi, available at www.plantreaty.org/sites/default/files/Identifying-Benefit-Flows.pdf [hereinafter MOELLER & STANNARD, IDENTIFYING BENEFIT FLOWS (2013)].

[151] *See* SMTA, n. 150, ¶ 6.1, *See generally*, Daniele Manzella, n. 143, at 150, 154–61. *See also* Halewood (Louvain 2012), n. 53; Frison (Louvain 2012), n. 131. *See generally* CROP GENETIC RESOURCES, n. 151.

[152] *See* SMTA, n. 150, ¶6.11; *see also* Chapter 3, Sections III.B.2 and III.C.2. About 800 accessions per day are currently made under the SMTA. Interview with Dr. Shakeel Bhatti, Director General of the ITPGRFA, August 5, 2015 [hereinafter Interview with Dr. Shakeel Bhatti].

[153] The concept of a "Compensatory Liability Regime" was first fully elaborated in Jerome H. Reichman, *Of Green Tulips and Legal Kudzu: Repackaging Rights in Subpatentable Innovation*, 53 Vand. L. Rev. 1743 (2000) [hereinafter Reichman, *Green Tulips*].

explained and critiqued earlier, in connection with our own proposals for the application of a more refined version of that regime to microbial genetic resources under the proposed Microbial Research Commons.[154]

Since January 1, 2007, the CGIAR's research centers have been using the Standard Material Transfer Agreement adopted by the Governing Body in 2006 for transfers of *ex situ* plant genetic resources pertaining to basic crops and forages listed in Annex 1 of the International Treaty.[155] In 2007, the Governing Body of the Treaty decided that the centers should also use the SMTA when transferring non-Annex 1 plant genetic resources for food and agriculture. Consequently, since 2008, the centers have been transferring all the plant genetic resources they hold in trust under the conditions set out in the SMTA, as determined by the Governing Body of the ITPGRFA. Moreover, very substantial nonmonetary benefits have also resulted from fund-raising efforts of the Secretariat under the Treaty.[156]

Given these arrangements, agricultural researchers everywhere have still been able to access the *ex situ* holdings of the CGIAR's centers, despite the threatened instability that had loomed large in the wake of the CBD. However, exchanges under the auspices of the CGIAR have not consistently been immune from complaints about unauthorized commercial applications without benefit sharing, and research on Annex I crops covered by the Treaty is declining vis-à-vis non-Annex I crops.[157] Meanwhile, the CGIAR itself has never succeeded in gaining access to the *in situ* genetic resources in the public domain that member states were supposed to make available for research purposes under the International Treaty.[158] Still another major concern is that, with the growth of the administrative apparatus spawned by the International Treaty and its locus at a United Nations specialized agency, there has reportedly been a growing disconnect between the needs of users – especially scientists and other researchers – and decisions taken by the administration, with diminishing inputs from the research community.[159]

These and other "obstacles and bottlenecks" have elicited mounting criticism from representatives of the CGIAR itself. They have also led to questions about the ability of the rigid intergovernmental apparatus established under the International Treaty to meet the changing needs and conditions of the agricultural research community

154 *See* Chapter 5, Sections II.B and II.C.4.

155 *See, e.g.*, Halewood (Louvain 2012), n. 53.

156 *Id.* About 800 accessions occur per day, and between 800,000 to one million SMTAs have been signed since 2006. Interview with Dr. Shakeel Bhatti. For details, see MOELLER & STANNARD (2013), n. 131, at 57–117.

157 See MOELLER & STANNARD (2013), n. 131, Executive Summary, at xxii–xxv.

158 *See, e.g.*, Chapter 3, Section III.B; Michael Halewood, Isabel López Noriega & Selim Louafi, *The Global Crop Commons and Access and Benefit-Sharing Laws: Examining the Limits of International Policy Support for the Collective Pooling and Management of Plant Genetic Resources*, *in* CROP GENETIC RESOURCES, n. 131, at 1–37 [hereinafter Halewood et al. (2013)].

159 *See, e.g.*, Halewood (Louvain 2012), n. 53.

over time.[160] Because we believe these concerns are of vital importance in redesigning the Microbial Research Commons, we return to them in the next chapter.

B. THE DIGITAL COMPONENT. Having thus put the multilateral system for facilitated access to plant genetic resources into operation, the Secretariat – under the oversight of the Governing Body – has taken steps to implement the global information system envisioned by Article 17 of the International Treaty. This article foresees that the contracting parties will develop a system to facilitate the exchange of information on scientific, technical, and environmental matters related to the PGRFA made available to all the parties. Specifically, this system pursues the following objectives:

1. To create a web-based platform with use-oriented entry points to PGRFA information;
2. To provide a comprehensive overview and facilitate access to sources of PGRFA and associated information;
3. To promote and facilitate interoperability among existing systems by providing clear principles, technical standards, and appropriate tools;
4. To promote transparency on the rights and obligations of users for accessing, sharing, and using PGRFA associated information;
5. To create and enhance opportunities for communication and international and multidisciplinary collaboration to increase knowledge about and add value to PGRFA;
6. To provide capacity development opportunities for conservation, management, and use of PGRFA and associated information and knowledge.[161]

Given that the Contracting Parties are obliged to distribute both *ex situ* and *in situ* plant genetic resources that are under their control and in the public domain, such a system becomes indispensable to enabling would-be users to actually gain access to the contents of the gene pool as a whole.

There were reportedly 2,093,000 accessions of plant genetic resources under the Treaty as of June 2015.[162] Nevertheless, the failure of the Contracting Parties

[160] *See, e.g.*, Halewood et al. (2013), n. 158: Godfrey Mwila, *From Negotiations to Implementation: World Review of Achievements, Bottlenecks and Opportunities for the Treaty in General and for the Multilateral System in Particular*, *in* CROP GENETIC RESOURCES, n. 131, at 226–42.

[161] See ITPGRFA, n. 134, art. 17; *id.*, Sixth Session of the Governing Body, Vision Paper on the Development of the Global Information System, Item 10 of the Provisional Agenda, Rome, Italy, Oct. 5–9, 2015, FAO doc. IT/GB-6/15/7 [hereinafter Vision Paper, Global Information System (2015)].

[162] Interview with Dr. Shakeel Bhatti, n. 152. These accessions break down as follows: 1,347,000 wheat; 549,000 rice; 197,000 maize. *Id.* In 2009, there had been 1.2 million accessions of PGRFA in total, with 440,000 transfers by the CGIAR alone. *See, e.g.*, Halewood (Louvain 2012), n. 53; Frison (Louvain 2012), n. 131.

to effectively contribute new resources to the multilateral system outside the CGIAR's own infrastructure has become a publicly disclosed issue. Whether better information about resources potentially within the reach of the system will prod governments to more fully comply with their treaty obligations remains to be seen.

C. LONG-TERM FUNDING ARRANGEMENTS. The emphasis on conserving biodiversity in the International Treaty, which especially focuses on *in situ* agricultural resources, is supported by the voluntary donations of member governments to a Trust Fund, known as the Benefit-Sharing Fund. Norway, Spain, Italy, and Switzerland, among others, have made substantial contributions to this Fund and about $24,000,000 have so far been contributed to this Fund as of June 2015. The Governing Body has directed some of these contributions to philanthropic foundations and grant-making institutions whose efforts further the conservation and maintenance of plant genetic diversity, especially in developing countries, with a view to enhancing the global gene pool that is the subject of the Treaty. The bulk of these funds have been distributed directly or indirectly to support farmers, especially in developing countries, in addition to nonmonetary benefits provided largely to support capacity building in those countries.[163]

Eventually, the Benefit-Sharing Fund should also receive a revenue stream derived from commercial applications of commonly held seeds under the Compensatory Liability Regime built into the Treaty itself.[164] As of 2015, however, no royalties had yet been paid or collected, and a recent study finds that benefits from commercial applications under present conditions will accrue slowly, and may not reach a target of $23 million annually for at least another 15 years.[165]

Article 18 of the International Treaty sets out very general guidelines for funding, while skirting the extent to which the multilateral system will require mandatory contributions over and above the voluntary contributions to the Benefit-Sharing

[163] Interview with Dr. Shakeel Bhatti, n. 152. *See* ITPGRFA, FIRST MEETING OF THE AD HOC OPEN-ENDED WORKING GROUP TO ENHANCE THE FUNCTIONING OF THE MULTILATERAL SYSTEM, BACKGROUND OF THE WORK UNDERTAKEN BY THE AD HOC ADVISORY COMMITTEE ON THE FUNDING STRATEGY AND ITS FUTURE DEVELOPMENT, Item 4 of the Provisional Agenda, Geneva, Switzerland, May 13–16, 2014 [hereinafter AD HOC FUNDING STRATEGY (2014). *See also* Shakeel Bhatti, *The International Treaty on Plant Genetic Resources*, *in* DESIGNING THE MICROBIAL RESEARCH COMMONS, n. 32, at 139.

[164] *See* ITPGRFA, n. 134, art. 18, 18.4(d) (charging states with a duty to support national activities), (e) and (f) (funds from benefit sharing and voluntary contributions). *See also* Chapter 3, Section III.B.1.

[165] *See, e.g.*, MOELLER & STANNARD, IDENTIFYING BENEFIT FLOWS (2013), n. 150, Executive Summary xxv; see also Frison (Louvain 2012), n. 131. As explained in Chapter 3, Section III.B.2 the International Treaty also allows firms to waive the liability rule in return for a research exception in favor of second comers. For criticism of this provision, *see id*, Section III.C.2.

Fund and any income to be generated by the royalties mentioned earlier. So far, however, all contributions remain on a voluntary basis. The Governing Body is charged with establishing a target for "mobilizing funding for priority activities, plans and program, in particular in developing countries," while taking into account the Global Plan of Action.[166] The Governing Body adopted the objective of raising the very substantial sum of $116 million over a five-year period for the Benefit Sharing fund, with a target of $50 million in the short run, and it also adopted a strategic plan concerning the mobilization of these resources.[167] As noted, however, only about $24 million had actually been collected as of 2015. Strenuous efforts are underway to overcome this shortfall by devising a packet of measures to increase income to the Benefit-Sharing Fund. Proposals under consideration include upfront users' access payments; promoting regular seed sales-based contributions by Contracting Parties; expanding the coverage of the Multilateral System itself; and developing a subscription system for users in place of case-by-case accessions under the SMTA.[168]

Funding of the Secretariat, the Governing Body, and other core administrative operations under the International Treaty are separate from the Benefit-Sharing Fund discussed earlier. These costs amount to $7 million every two years. About 80 percent of these costs are also borne by voluntary contributions of all the Contracting Parties, according to a set annual schedule based on ability to pay, as determined by GDP.[169]

The FAO's Commission on Plant Genetic Resources has, from the outset, also supported implementation of the Treaty, with the approval of the Governing Body, as part of the Commission's own Multi-Year Program of Work. The FAO currently defrays 20 percent of these administrative costs.[170]

Meanwhile, the semiautonomous CGIAR network has taken steps to establish its own funding on a more solid and reliable basis, and it has initiated a coordinating governance body of its own to regulate access to its network of seed banks and to impose standardized practices. This coordinating body will also formulate policy for

166 *See* ITPGRFA, n. 134, art. 18.3.

167 Ad Hoc Funding Strategy (2014), n. 163, at 5, Fig. 2. *Id.*

168 *See id.*, at 79–115. Efforts may also be made to revise the existing SMTA-based approaches. *See id.*, at 84–96. For the proposed subscription system, *see* ITPGRFA, Third Meeting of the Ad Hoc Open-Ended Working Group to Enhance the Functioning of the Multilateral System, Development of a Subscription System for Users of Plant Genetic Resources for Food and Agriculture Under the Treaty (Measure III): Background Information, Brazilia, Brazil, June 2–8, 2015, IT/OWG-FMCS-3/15/Inf.5.

169 Interview with Dr. Shakeel Bhatti, n. 152.

170 *Id. See also*, First Session of the Governing Body of the International Treaty on Plant Genetic Resources for Food and Agriculture, Madrid, Spain, 12–16 June 2008, IT/GB-1/08/Report, ¶42.

major projects affecting the CGIAR Centers' primary constituents, namely farmers, forest and fishing communities, as well as national agricultural research systems.[171]

D. COMPLIANCE AND DISPUTE SETTLEMENT. Formally, Article 21 of the International Treaty charges the Governing Body with approving "cooperative and effective procedures and operational mechanisms to promote compliance with the provisions of the treaty and to address issues of noncompliance."[172] Article 11.4 also authorizes the Governing Body to decide whether facilitated access to the Crop Commons should be denied in the event that countries have not met their obligations to make plant genetic resources available to the Multilateral System.[173]

However, no further action has been reported under these provisions, despite growing complaints that the Contracting Parties have not satisfactorily cooperated in this regard. At the same time, the CGIAR's *ex situ* collections and gene banks operating within the system have continued to make their plant genetic resources available to the rest of the world, without any condition of reciprocity.[174] This state of affairs has elicited growing complaints about free-riding,[175] an issue we address later in this chapter.

With specific regard to the Standard Material Transfer Agreement governing facilitated access to plant genetic resources available from the Multilateral System, Article 8 mandates negotiation, mediation, and arbitration in the case of disputes.[176] However, complaints about the slowness and efficacy of these procedures have also been raised.[177]

B. *Hybrid Pooling Arrangements Among Governments, Para-Statal Entities, and Nongovernmental Stakeholders*

In contrast to the top-down, treaty-based governance structure used to rescue publicly available plant genetic resources from claims of biopiracy under the CBD, the microbiological research community has long depended upon the voluntary collaboration of the networked culture collections discussed in Chapters 2 and 4 of this volume.[178] More recently, several other hybrid intergovernmental entities have

[171] *CGIAR Research Programs*, CGIAR, http://www.cgiar.org/our-research/cgiar-research-programs/ (last visited April 4, 2015); *see generally* D. JOHN SHAW, GLOBAL FOOD AND AGRICULTURAL INSTITUTIONS (Routledge 2008).

[172] ITPGRFA, n. 134, art. 21.

[173] *Id.* at art.11.4.

[174] *See, e.g.*, Halewood (Louvain 2012), n. 53.

[175] *See generally* Halewood et al. (2013) n. 158.

[176] SMTA, n. 150, art. 8.4.

[177] *See, e.g.*, Halewood (Louvain 2012), n. 53.

[178] *See* Chapter 2, Section I.A; Chapter 4, *passim*. *See also* Chapter 3, Section II.A ("The Public Microbial Culture Collections Consider Defensive Options").

been formed by mutual agreement of their state sponsors to promote the voluntary sharing of upstream scientific research assets, with a view to increasing their access and use. These entities are empirically relevant to our thinking about governance strategies for the proposed Microbial Research Commons, as discussed later in this book.

In what follows, we look first at the governance structure of the World Federation for Culture Collections (WFCC). We then turn our attention to three other hybrid entities that have features of particular relevance to this enquiry, namely, the Global Biodiversity Information Facility (GBIF), the Group on Earth Observations (GEO), and the International Human Microbiome Consortium (IHMC).

1. The World Federation for Culture Collections (WFCC)

A. OBJECTIVES AND MEMBERSHIP. As previously noted in Chapters 2 and 4, the WFCC is technically both a Multidisciplinary Commission established by the International Union of Biological Sciences (IUBS) and a Federation within the International Union of Microbiological Societies (IUMS).[179] Its primary objective is to promote and develop culture collections of microorganisms and cells,[180] with a view to facilitating exchanges of microbial genetic resources for purposes of both research and applications.[181] In this section, we focus primarily on aspects of the WFCC's legal structure and its governance apparatus.

Members of the WFCC see themselves as constituting "a unique global network for *ex situ* preservation of microbial diversity, which underpins life on earth."[182] Besides organizing workshops and conferences, they publish newsletters and scientific documents, and generally seek to ensure the long-term perpetuation of important culture collections.[183]

In so doing, the Federation recognizes different membership categories. Any person with an avowed interest in the activities of the microbial culture collections is eligible for "ordinary membership." In addition, individuals or organizations may be invited to join as "sustaining members," usually on the grounds that they have provided extraordinary support for the WFCC goals and activities. Individuals who have demonstrated "long productive service" to the Federation or

[179] WFCC, *About WFCC*, http://www.wfcc.info/about/ [hereinafter WFCC, *About WFCC*] (last accessed 6 July 2012). Both IUBS and IUMS are members of the International Council for Science (ICSU).

[180] WFCC, *About WFCC*, n. 179; *see also* World Fed. Culture Collections (WFCC), *Guidelines for the Establishment and Operation of Collections of Cultures of Microorganisms*, 3d. ed., WFCC (Feb. 2010), http://www.wfcc.info/guidelines/ [hereinafter WFCC, *Guidelines*].

[181] *See further* Chapter 2, Section I.A.1 (discussing the composition and goals of the WFCC).

[182] WFCC, *Guidelines*, n. 180, at 2.

[183] WFCC, *About WFCC*, n. 179.

"outstanding professional contributions" may be appointed as Honorary Lifetime Members.[184]

In determining the eligibility of specific culture collections for membership, the Executive Board now requires that candidate collections implement the WFCC's Guidelines for the Establishment and Operation of Collections.[185] As a result, WFCC member collections fall into two categories, depending on their level of compliance with the quality standards adopted by the organization. "Collaborating Affiliates" will have fully complied with these standards, while "Associated Affiliates" will have conformed to most, but not all of them.[186]

Still another important membership category consists of Affiliated Organizations. Any national or regional federation of culture collections, as well as any national committee or similar organization with interests in culture collection activities, may apply for membership in the WFCC within this category. If approved by the Executive Board, each Affiliated Organization is authorized to appoint one delegate to represent it in the Federation.[187]

When evaluating the central role of the WFCC in microbiology, one should bear in mind that there are thousands of private, institutional, or industrial culture collections that are not members of this organization, many of which are "financially unstable … or lack support in a number of different ways."[188] The WFCC thus seeks to ensure that its members provide access to important collections of microorganisms that meet minimum quality standards. The relevant guidelines were deliberately calibrated to assist in-house and university research collections to become future members, in addition to the public service collections.[189] However, the WFCC standards do not require member culture collections or would-be members to meet the higher OECD standards for Biological Resource Centers,[190] which would entail a more significant investment in equipment and personnel to implement.[191]

In 2010, the WFCC Executive Board promulgated a new set of rules that would further tighten the operational standards that all collections would have to meet in

[184] World Fed. Culture Collections (WFCC), *Statutes* http://www.wfcc.info/index-php./about/statutes .art.VII [hereinafter WFCC, *Statutes*].

[185] WFCC, *Guidelines*, n. 180; WFCC *Newsletter* No. 48, July 2010, at 2 ("It is by the implementation of best practices that WFCC affiliate member collections are distinguished from other culture collections.").

[186] WFCC *Bylaws*, http://wdem.nig.ac.jp./wfcc/bylaws.html, Section E [hereinafter WFCC *Bylaws*].

[187] WFCC, *Statutes* , n. 184, art. VII.

[188] *See* WFCC, *the Endangered Collection Task Group (ECTG)*, http://wdem.nig.ac.jp/wfcc/committee/endangered/home.html, updated June 20, 2001 [hereinafter WFCC *Endangered Collections*].

[189] WFCC, *Guidelines*, n. 180, ¶ 1.4.

[190] OECD, Biological Resource Centers – Underpinning the Future of the Life Sciences and Biotechnology (March 2001) [hereinafter OECD Report on BRCs], *available at* http://www .oecd.org/dataoecd/55/48/2487422pdf.

[191] WFCC, *Guidelines*, n. 180, ¶ 1.4.

the future.[192] These rules require, *inter alia*, that each member collection should implement WFCC Guidelines on Biosecurity; make minimum levels of data about its operations publicly available; implement the Access and Benefit sharing provisions of the Convention on Biological Diversity; keep records of the origin of deposited materials and to whom they are dispatched, preferably under Material Transfer Agreements; respect the intellectual property rights of those who deposit materials, including Prior Informed Consent rules of the CBD; and maintain back-up collections of especially important materials with another collection where feasible.[193]

The WFCC's policies concerning rights and licensing practices applicable to *ex situ* microbial materials were discussed in Chapter 4, Sections II and III. For public accessibility of data from and about the WFCC's member collections, see the discussion of the World Data Center for Microorganisms in Chapter 8.

B. GOVERNANCE. The members of the Federation normally hold a General Assembly, in conjunction with the International Congress of Culture Collections, every three or four years.[194] These assemblies authorize the activities of the Federation. Special meetings of the General Assembly may be called, as needed, by the Executive Board. Each dues paying individual member has one vote, in person or by proxy, concerning the administrative affairs of the Federation.[195] Affiliated members (i.e., some of the culture collections, as described earlier), as well as sustaining members, may appoint one delegate eligible to vote.[196]

The WFCC Executive Board administers decisions taken by the General Assembly between meetings of that Assembly,[197] and generally promotes the objectives of the Federation.[198] The Executive Board consists of eight elected members, plus four additional members appointed by the latter, who should "reflect the interests of the Federation and ... provide a balance of international representation and expertise on the Board."[199] Elected members may serve no more than two consecutive terms.[200] *Ex officio* members of the Executive Board also include the Past-President of the

[192] WFCC *Newsletter* No. 48, n. 185, at 2.
[193] *Id.*
[194] WFCC, *Statutes*, n. 184, art. XVI.
[195] Id., arts. VII, VIII, IX, X ("All members have full participation in the affairs of the Federation, including the right to vote ...").
[196] So-called Adherent Members pay no dues and have no voting rights. WFCC *Bylaws*, n. 186, ¶¶ A, D. We assume that they are given observer status.
[197] WFCC, *Statutes*, n. 184, art. IV.
[198] *Id.* art. XII.
[199] *Id.* art. XIII.
[200] *Id.*

Federation, the Director of the WFCC's World Data Center on Microorganisms, and the editor of the WFCC Newsletter.[201]

The officers of the Board include a President, Vice-President, Secretary, and Treasurer, and they also serve as officers of the organization.[202] The first two are elected, the latter two are appointed by the Board.[203] The Executive Board may establish ad hoc committees as needed, but all such committees are normally dissolved at the time of the Business Meeting of each General Assembly. The new Executive Board then assesses the value and function of each committee and decides which, if any, will continue and what the membership will be.[204]

One committee that seems to have attained a relatively permanent status is the Endangered Collections Task Group. Its purpose is to assist financially unstable collections with advice and expertise. For example, it can help such collections find new funders or a new institutional home, and it can also assist in finding needed personnel. However, the Task Group rarely, if ever, provides direct financial assistance, and it recognizes that "not all collections need to be or indeed can be saved."[205]

Finally, the WFCC often speaks or negotiates on behalf of its member culture collections at international forums, such as the ongoing CBD deliberations. It has also presented its views on packaging and transport regulations concerning the shipment of microbial materials and on biosecurity guidelines at relevant international conferences.[206]

C. FUNDING. The WFCC does not provide funding for member culture collections or for research in microbiology, but leaves these matters to the collections themselves and to their national or regional organizations. As an umbrella organization for qualified, affiliated collections, the WFCC does require these members to seek long-term financial commitments from parent or supporting institutions.[207]

The WFCC itself "is still organized in such a way that most activity is done via the voluntary contributions of its members, including the Executive Board members."[208] Dues in the form of "financial subscriptions" are modest for the collections, which usually assess themselves an annual fee in the range of 150 to 200 euros, at their discretion. Ordinary individual members are assessed dues at the rate of 20 Euros per

[201] *Id.*
[202] *Id.*
[203] *Id.* art. XII.
[204] *Id.* art. XVIII.
[205] WFCC, *Endangered Collection Task Group*, n. 188.
[206] Philippe Desmeth, *News from the Secretary*, WFCC Newsletter No. 47, 2–3 (Jan. 2010).
[207] *See* WFCC, *Guidelines*, n. 180, at 5.
[208] Desmeth, n. 206, at 2.

year.[209] The Federation urges its affiliates to seek sustaining memberships from their respective governing bodies. Donations are also made by participating organizations as well as external bodies.[210]

The important activities of the World Data Center for Microorganisms (WDCM) are partly or largely funded by the host country.[211] With the recent shift of the WDCM from Japan to China, the ambitious program recently undertaken by the WDCM and described earlier in Chapter 8, seems to rest on a solid financial foundation, at least for the immediate future.[212]

D. FUTURE PROSPECTS: THE WFCC AT A TURNING POINT. The leadership of the WFCC has clearly grasped the need to move from a "small science" institutional culture to the "big science" outlook that characterizes the New Biology.[213] They have thus taken steps to better integrate the services of its member collections, to elevate and harmonize quality standards, and to expand the digital infrastructure beyond previous rudimentary levels.[214] Viewing the OECD's Biological Resource Center initiative as "a paradigm shift from traditional culture collections to high quality biological resource centers (BRCs),"[215] the leadership has understood the advantages of forming networks of highly qualified culture collections that could digitally link initiatives in Europe, Asia, Oceania, Africa, and North and South America in a global infrastructure.[216]

The challenge is to keep abreast of developments in taxonomy and systematics (against a background of diminishing expertise in a shrinking workforce) as well as in the identification, authentication, cultivation, and maintenance of existing and new microbial cultures. This challenge will be especially difficult for most of the collections that continue to rely on their own resources. It will require them to cooperate and harness the power of networking on a national, regional, and global scale in order to achieve sustainable levels of needed technical capacity.

Prompted by these considerations (and, in part, by early versions of this book, which were widely circulated at relevant meetings), the previous WFCC leadership pressed for the formation of a rudimentary microbial research commons, to be

209 WFCC *Bylaws*, n. 186, ¶ D.
210 WFCC, *Guidelines*, n. 180.
211 *See News from the WFCC*, WFCC *Newsletter* No. 48, 1 (July 2010) (noting that the host country must have secured long-term funding in order to support the WDCM).
212 *See* Chapter 8, Section II.B.1
213 *See* Chapter 1, Section II.D ("A New Research Paradigm for the Life Sciences").
214 *See, e.g.*, David Smith, Farewell to the Past President, WFCC *Newsletter* No. 49 (Dec. 2010) at 1 ("Consolidating the many initiatives is crucial to establishing a systematic and networked approach.").
215 *Id.*
216 *See* discussion in Chapter 4, Section I.C ("Beyond the WFCC: Regional and Global Networks of BRCs"). See most recently D. Smith et al. (2013), n. 70.

known as the Global Biological Resource Centre Network (GBRCN). This project is described and analyzed later in this chapter.[217]

The WFCC thus needs to decide whether or not it will proceed to build the kind of global infrastructure that appears necessary to achieve the vision of a New Biology, in which microbial science could play a seminal role.[218] The organization's leadership should also decide what kind of commons it actually wants to see emerge, and the extent to which that entity would promote public research on a global scale or become a more proprietary enterprise with less emphasis on the public good mission of its member collections.[219]

2. The Global Biodiversity Information Facility (GBIF)

A. OBJECTIVES AND MEMBERSHIP. Unlike the CGIAR, whose primary concern is the global management of plant genetic resources as common-pool resources for research purposes, GBIF is "a global membership organization, open to all countries and international organizations interested in contributing to and benefiting from more accessible biodiversity data."[220] In this context, "biodiversity data" refers to "scientific data [that are] primarily about biological species and about specimens or observations of individual organisms."[221] By envisioning a "global biodiversity information commons," GBIF's primary task was to create and manage a centralized portal for providing free and open access online to biodiversity data.[222]

GBIF is a multilateral initiative, established in 2001 at the request of the OECD's Megascience Forum,[223] by an intergovernmental agreement that 17 countries initially signed.[224] Its operations are based on a nonbinding Memorandum of Understanding (MoU) renewable every five years.[225] GBIF's objective is "to promote, co-ordinate, design and implement the compilation, linking, standardization, digitization and global dissemination of the world's biodiversity data, within an appropriate

[217] *See* Section II.C. *See also* David Smith, Culture Collection Community Activities, WFCC *Newsletter* No. 49 (Dec. 2010), at 1–3.

[218] *See* Chapter 1, Section II.D.

[219] *See* Section II.C later in this chapter (discussing proprietary tendencies of the GBRCN project).

[220] GBIF – *Outreach* (7–27/2010), *available at* http://www.gbif.org/participation/outreach.

[221] *See, e.g.*, Memorandum of Understanding for the Global Biodiversity Information Facility, Approved at GB12 in Cape Town, South Africa, April 2006; *Id.*, Annex 1, approved at GB12.5, in Madrid, Spain, June 2006 [hereinafter GBIF MoU] *available at* http://www.gbif.org/resource/80661.

[222] GBIF, MoU, n. 221.

[223] *See* WORKING GROUP ON BIOLOGICAL INFORMATICS, FINAL REPORT OF THE OECD MEGASCIENCE FORUM (1999), http://www.oecd.org/sti/sci-tech/2105199.pdf.

[224] Currently there are 38 voting participants. *See* GBIF, *Participation*, *available at* GBIF.org, http://www.gbif.org/participation/participant-list.

[225] *See* GBIF, MoU, n. 221. *See also* Rules of Procedure of the Governing Board of the GBIF (2008), available at http://www.gbif.ort/resources/2713 [hereinafter GBIF, *Rules of Procedure*].

framework for property rights and due attribution."[226] To this end, participants "will establish and support a distributed information system that will enable users to access and utilize quantities of existing and new biodiversity data."[227] As of April 2015, GBIF had 38 voting members, and its portal provided access to almost 530 million indexed records in over fourteen thousand datasets from approximately 660 publishers.[228]

Participants in GBIF are expected to promote standards and software tools; make biodiversity data universally available, while fully acknowledging the contribution made by those gathering and furnishing these data; and to share biodiversity data under a common set of technical standards and within an intellectual property rights framework. They should also contribute to training and capacity development, including implementing specific programs to enhance the biodiversity informatics capacity and technical skills base of developing countries.[229]

B. GOVERNANCE. Perhaps because GBIF does not directly manage physical resources, but only aspires to pool data about such resources, it has avoided entering into any binding international commitments. At the same time, it operates under the authority of the relevant national science and environmental ministries, which have established a relatively formal governance structure.

The key decision-making body is the Governing Board, a forum in which GBIF participants make collective decisions on all relevant matters. This Board consists of one representative from each participating government, but only governments that provide funding according to a predetermined formula qualify as "Voting Participants." While "Associate Participants," including "economies, intergovernmental organizations and international organizations" are encouraged to take part in the deliberations of the Governing Board, only participating countries that make the suggested financial contributions are allowed to vote.[230] Supplementary contributions may also be made by governments, foundations, or other entities.[231]

The Governing Board must seek a consensus for its decisions, and must actually obtain consensus on certain key matters. Where a consensus is not required, a super majority varying from two-thirds of the voting members to a simple majority will suffice.[232] The Governing Board's decisions are implemented by a Secretariat, which is supervised by an Executive Committee that the Governing Board appoints.[233]

[226] GBIF, MoU, n. 221¶ 3.1.
[227] *Id.*
[228] GBIF, www.gbif.org.; see GBIF, *Participation*, n. 224.
[229] *See* GBIF MoU, n. 221, ¶¶ 3.2(d), (f), 3.3(b), (e). Cooperation with the Secretariat of the CBD is expressly envisioned. *Id.* ¶ 3.4.
[230] *See* GBIF, *Rules of Procedure*, n. 225, ¶ 4.
[231] *See* GBIF MoU, n. 221 ¶ 9.
[232] *Id.*, ¶ 4.4.
[233] *Id.* ¶ 4.6.

The GBIF's Secretariat executes its Work Program in accordance with a Strategic Plan, which is updated every five years, as is the annual budget. Led by an Executive Secretary, the Secretariat thus remains accountable "for the execution of all scientific, financial and administrative activities undertaken to implement the GBIF Work Programme."[234]

According to GBIF's Rules of Procedure, the Governing Board may establish a number of Standing Committees, Task Groups, and other advisory committees.[235] At present, there are four Standing Committees, viz., the Science Committee, the Budget Committee, the Rules Committee, and the Participant Node Managers Committee.[236] The Science Committee's key advisory role is further buttressed by three different thematic, time-limited Task Groups and Advisory Committees.[237]

To further its objectives, GBIF has signed partnership agreements, known as Memoranda of Collaboration (MOCs), with diverse organizations operating within the governmental, academic, and private sectors. In the public sphere, for example, it has an MOC with both the Conference of the Parties of the CBD, which promotes collaboration on approaches to accessing, sharing and disseminating biodiversity data via the internet, and with the United Nations' FAO, to support a Global Invasive Alien Species Information Partnership (GIASIP).[238]

C. FUNDING. The cost of becoming a "Voting Participant" at GBIF varies considerably with the Gross Domestic Product (GDP) of member countries.[239] The basic financial contribution of each Voting Participant is determined by a set formula, with a multiplier applied to the country's GDP as established on the World Bank's website. Between the years 2007 and 2011, the multiplier increased every year, from 121.00 in 2007 to 177.36 in 2011.[240] This formula is deliberately skewed to encourage more developing countries to participate in the future.

As of 2015, 38 countries were considered Voting Participants, of which 15 were non-OECD nations.[241] Five members – the United States, Japan, Germany, the U.K. and France – bear the highest contributions, which range from €240,000

234 *Id.* ¶ 6.3. The Secretariat enjoys diplomatic immunity under the laws of the host country, i.e., Denmark.

235 GBIF, www.gbif.org/governance/advisory-committees/.

236 *Id.*

237 GBIF MoU, n. 221, ¶ 8.1.

238 *New Biodiversity Platform Approves Assessments, Data Task Force*, GBIF News (Dec. 19, 2003), http://www.gbif.org/page/3022.

239 See GBIF MoU, n. 221, Annex I, tbl. 1. See also How GBIF Is Funded, GBIF, http://www.gbif.org/governance/finance/#national.

240 *Id.*

241 GBIF, *GBIF Current Participants*, www.gbif.org/governm/governing-board/currentparticipants/ (last accessed Feb. 3, 2015).

to €650,000 per year. Eleven other OECD countries contribute from €79,500 to €120,000 per year. Three countries – Mexico, South Africa, and New Zealand – pledge about €40,000 per year, while the remaining small and mostly poor countries contribute from €500 to €4,100 per year.[242]

D. INTELLECTUAL PROPERTY POLICIES. Despite its commitment to the pooling and sharing of biodiversity data, GBIF's MoU devotes considerable attention to the need to respect intellectual property rights, and the resulting tensions remain visible, much as they do with other international scientific organizations we review in this chapter. Article 8 of the MoU thus proclaims that nothing in its provisions "should be read to alter the scope and application of Intellectual Property Rights and benefit sharing arrangements as determined under relevant laws, regulations and international agreements of the participants."[243] At the same time, this article commits GBIF to become "an open access facility . . . to the greatest extent possible," and it mandates that "[a]ll users ... should have equal access to data in databases affiliated with or developed by GBIF."[244]

In practice, however, while seeking to make its data "freely and openly available with the least possible restrictions on reuse," Article 8 also obliges GBIF to "respect conditions set by data providers that affiliate their databases to GBIF."[245] The possible contradictions in these provisions are tacitly recognized, but only partly reconciled by the following admonition:

> When establishing affiliations or linkages with other databases, GBIF should seek to ensure that the data so made available will not be subject to limitations on the further noncommercial use and dissemination of these data, apart from due attribution of their source.[246]

Even so, the MoU concedes that nothing in its provisions "should be read to restrict the right of owners of databases affiliated with GBIF to block access to any data."[247] Obviously, GBIF has been obliged to tread cautiously in an environment where nations differ considerably in their willingness to share data about biological specimens and biodiversity in general.

Also worth noting is GBIF's own reserved power to assert intellectual property rights in "any tools, such as search engines or other software products" that

[242] *Id.*
[243] GBIF MoU, *n.* 221,¶ 8.3(b), (c).
[244] *Id.* ¶ 8.3.
[245] *Id.*
[246] *Id.*
[247] *Id.* ¶ 8.8.

it develops.[248] GBIF nonetheless commits to the principle of promoting "the nonexclusive transfer on mutually agreed terms, to research institutions, particularly in developing countries, of such informatics technology as it has available."[249]

E. FUTURE PROSPECTS. GBIF views itself as a long-term cooperative endeavor that aspires to "sustain the benefits of access to biodiversity data." However, its survival depends on the willingness of the Voting Participants to renew its MoU every five years, failing which the Secretariat must arrange for liquidation of its assets.[250] GBIF's biodiversity data commons will thus continue to operate only so long as the participating science ministries consider that the benefits outweigh the costs.[251]

3. The Group on Earth Observations (GEO)

A. OBJECTIVES AND MEMBERSHIP. Another recent example of considerable relevance to our survey of selected hybrid commons approaches[252] is that of the Group on Earth Observations (GEO), which is establishing a Global Earth Observation System of Systems (GEOSS).[253] The GEOSS consists of contributed Earth observation systems, ranging from primary data collection systems to systems concerned with the creation and distribution of information products. This organization now comprises the largest consortium of data sources in the world. Although all the contributed satellite and other data collection systems continue to operate within their own national and institutional mandates, they are entitled to leverage each other, so that GEOSS can become a globally effective observational resource.

As a practical matter, GEO's "system of systems" has been designed to enhance the effectiveness of both existing and future Earth observation systems by coordinating and promoting integration of the data outputs contributed by each of the participants. Collectively, GEO aims to:

- address identified common user requirements;
- acquire observational data;
- process data into useful products;

[248] *Id.*
[249] *Id.* ¶ 8.9.
[250] *Id.* ¶ 11.1, 11.2.
[251] *See generally How GBIF Is Funded*, n.239.
[252] *See further* Chapter 10 ("Governing Digitally Integrated Genetic Resources, Data, and Literature").
[253] Group on Earth Observations (GEO), *Strategic Guidance for Current and Potential Contributors to GEOSS* (October 2007), *available at* http://www.earthobservations.org/documents/portal/25_strategic%20Guidance%20Document.pdf [hereinafter GEO, *Strategic Guidance*].

- exchange, disseminate, and archive shared data, metadata, and products; and
- monitor performance against the defined requirements and intended benefits.[254]

GEO's Implementation Boards, Communities of Practice, and Working Groups thus focus on pooling immense amounts of data in a globally distributed networked system, which is becoming accessible by means of a common portal.[255] More precisely, this entity was formed to promote international collaboration "for exploiting the growing potential of Earth observations" to improve human welfare[256] in nine specified "societal benefit areas."[257] Its primary goal in developing the GEOSS was to:

> Ensure comprehensive and sustained Earth observations ... by coordinating ... [existing] efforts, addressing critical gaps, supporting their interoperability, sharing information, reaching a common understanding of user requirements and improving delivery of information to users.[258]

Participants in GEO believe that a synergy will develop as each contributor supports common arrangements designed to make shared observations and products more accessible, understandable and interoperable.[259] As of May 2015, GEO's membership included 96 governments, plus the European Commission.[260] The membership also includes 87 Participating Organizations, i.e., intergovernmental,

[254] Global Earth Observation System of Systems (GEOSS), *10-Year Implementation Plan* (as adopted 16 February 2005), *available at* http://www.earthobservation.org/docs/10-year%20Implemntation%20 Plan.pdf [hereinafter GEOSS, *10-Year Plan*]. *See generally* GEO, at http://www/earthobservations.org (last accessed May 5, 2015).

[255] GEOSS, *10-Year Plan*, n. 254, at 5.

[256] GEO, *About GEO*, http://earthobservations.org/about_geo.shtml.

[257] The current areas of concentration are:

- Reduction and Prevention of Disasters
- Human Health and Epidemiology
- Energy Management
- Climate Change
- Water Management
- Weather Forecasting
- Ecosystems
- Agriculture
- Biodiversity

Group on Earth Observations (GEO), *Strategic Guidance for Current and Potential Contributors to GEOSS*, 1 (October 2007), *available at* http://www.earthobservations.org/documents/portal/25_ strategic%20Guidance%20Document.pdf [hereinafter GEO, *Strategic Guidance*].

[258] GEOSS, *10-Year Plan*, n. 254.

[259] GEO, *Strategic Guidance*, n. 252.

[260] See GEO, http://Earthobservations.org/index.php# (last accessed May 5, 2015).

international and regional organizations with an approved mandate concerning Earth observations or related issues.[261]

Among the methods chosen to achieve its goals, the GEOSS emphasizes a need to engage users in developing countries to improve their opportunities to benefit from the entity's common-pool resources. The members are committed to invest in capacity building to this end.[262]

B. GOVERNANCE. Central to understanding the GEO project as a whole is the fact that, in the system of systems it adopts over time, "virtually all the operational infrastructure is provided through contributions by … Members and Participating Organizations."[263] What emerges is a federated system of linked components whose functions are to acquire observations, to process data into useful information, and to enable the exchange and dissemination of observational data and information. GEO itself does not directly control or manage any of the data and information contributed to the GEOSS, although it provides a portal and certain administrative facilities, which greatly reduces overall operating costs.

In effect, those who contribute data and information to the GEOSS expect to obtain some or all of the following benefits: synergies of discovery from efficient data-sharing mechanisms; collective optimization of observational strategies and cooperative gap filling; worldwide exposure to potential users and collaborators; and enhanced interoperability based on open, international standards. In addition, developing country participants expect to receive both cooperative capacity building with regard to Earth observations and direct societal benefits that might not otherwise be attainable.[264]

GEO is governed by a 10-Year Implementation Plan, which was being updated for another 10-year period at the time of writing.[265] The organization is a voluntary partnership, formed on a legally nonbinding basis, of governments and participating international organizations (both intergovernmental and nongovernmental), which emerged from the 2002 World Summit on Sustainable Development.[266]

Decisions concerning implementation are made by a consensus of member governments that meet at an annual plenary session. Governments are represented by senior officials at the annual meeting and periodically by the relevant ministers.[267] Although all affiliated IGOs and NGOs recognized by the members participate at

[261] *See* GEO, *Participating Organizations*, http://www.earthobservations.org/ag_partorg.shtml (last accessed 16 October 2013) [hereinafter GEO, *Participating Organizations*].

[262] GEOSS, *10-Year Plan*, n. 254, at 5–6, 8–10.

[263] GEO, *Strategic Guidance*, n. 252.

[264] *Id. See also* Paul F. Uhlir, The Value of Open Data Sharing: A White Paper for the Group on Earth Observations (CODATA, 2015).

[265] GEO, *Strategic Guidance*, n. 252.

[266] *Id.*

[267] GEOSS, *10-Year Plan*, n. 254.

these meetings, only the member governments make final decisions. Nevertheless, all decisions implementing the GEOSS 10-Year Implementation Plan, as further elaborated in its Rules of Procedure, are to be "based upon sound scientific and technical advice obtained through appropriate consultation with the research and observation communities."[268]

To support its efforts, the GEO plenary has established an elected Executive Committee (based on a geographically distributed system of representation) and a Secretariat to carry out the work plan between annual plenary meetings. The plenary also appoints subsidiary bodies, as appropriate, "including science and technical advisory mechanisms" and capacity building entities.[269] A Data Sharing Working Group (DSWG) was established in 2006, and initially chaired by members that the international NGO, CODATA, approved. This group was broadened in 2009 to include governmental member representatives from different regions as co-chairs.[270]

C. FUNDING. Funding is on a voluntary basis, with most resources "provided through existing national and international mechanisms."[271] In effect, this means that most of the work is performed by personnel seconded by governments and paid by them as in-kind contributions.

Monetary funding for the work of the organization is placed in a trust fund, which the Secretariat administers. Office space for the Secretariat is provided by the United Nations World Meteorological Organization at its headquarters in Geneva, Switzerland.

D. INTELLECTUAL PROPERTY POLICIES. Among its most important efforts after 2010, the Data Sharing Working Group solicited data contributions from GEO's participants to the GEOSS Data-CORE (Collection of Open Resources for Everyone). The object was to create "a distributed pool of documented datasets with full, open and unrestricted access at no more than the cost of reproduction and distribution."[272]

[268] *Id.* at 11. *See also* GEO, *Rules of Procedure* (updated Nov. 14, 2014), *available at* http://www.earthobservations.org/documents/GEO%20Rules%20of%20Procedure.pdf (last accessed May 8, 2015) [hereinafter GEO, *Rules of Procedure*].

[269] *Id.* For composition and duties of the Executive Committee, *see* GEO, *Rules of Procedure* n.268, ¶3.

[270] *See* GEO, Rules of Procedure, n. 268, ¶5 ("GEO Implementation Boards and Working Groups"). One of the co-authors of this book, Paul Uhlir, was a founding co-chair of this group, as was Robert Chen of Columbia University.

[271] GEOSS, *10-Year Plan*, n. 254. *See also* GEO, *Rules of Procedure* n. 268, ¶6 (which institutes a Trust Fund to support the GEO Secretariat and other activities).

[272] *See* MICHEL SCHOUPPE, GEOSS, DATA SHARING PRINCIPLES: CURRENT AND PROPOSED (2014), slide presentation at the GEO Plenary side event on Open Data and the Developing World (Geneva, Switzerland Nov. 12, 2014), available at http://www.earthobservations.org/documents/dswg/docs.presentations/geo xi se 2 Data Sharing Principles Michel%20Schouppp.pdf (last accessed August 14, 2015).

The Working Group is thus attempting to put this global data commons on a sound foundation and legal footing.

The founders of GEO recognize that its success depends on establishing a common architecture for its data and information providers and on implementing a set of interoperability arrangements, including "technical specifications for collecting, processing, storing and disseminating shared data, metadata, and products" by means of nonproprietary standards.[273] They also promote research and development in key areas to facilitate improvements in Earth observations systems.[274]

To make the scheme work, the GEOSS 10-Year Implementation Plan depends on the effective implementation of its consensus-based Data Sharing Principles (which are currently being updated for the next 10-year period):

- There will be full and open exchange of data, metadata, and products shared within GEOSS, recognizing relevant international instruments and national policies and legislation.
- All shared data, metadata, and products will be made available with minimum time delay and at minimum cost.
- All shared data, metadata, and products for use in education and research will be encouraged to be made available free of charge or at no more than the cost of reproduction.[275]

Also crucial to its data-sharing goals is the willingness of data and information providers to accept and implement a set of data interoperability arrangements. To facilitate these commitments, a GEOSS Clearinghouse has been established to provide registry services that include a description of each of the formally contributed components of the GEOSS; metadata about the datasets and information available from each of the contributed components; technical specifications for using the services provided by the contributed components; and descriptions of key interoperability standards.[276]

E. FUTURE PROSPECTS. The Group on Earth Observation's legal and organizational structure is both innovative and flexible, and its governance arrangements provide clear and effective means of communication between the government agencies involved and the relevant scientific collaborators. However, the voluntary nature

[273] *Id.* at 7–8.

[274] *Id.* at 8.

[275] GEOSS, *10-Year Plan*, n. 254, at 8.

[276] *Id.* The process of contributing a system to become a part of the GEOSS begins by registering it as a "GEOSS Component" through the online registration form. If the system has public service interfaces, these should also be registered. The contributed system and service interfaces are then catalogued by the GEOSS Clearinghouse for discovery and access. *Id.*

of performance and funding commitments built into this structure, which provides maximum flexibility, also leaves the organization with ineffective enforcement machinery and a shortage of funds for key activities.

Despite the carefully elaborated plan of action, and perhaps because of these governance drawbacks, some member countries have so far failed to meet their commitments to supply relevant data in a manner consistent with the principles sketched earlier. Whether the organization's ambitious, but voluntary, data-pooling goals will actually be fulfilled thus remains to be seen.

4. The International Human Microbiome Consortium (IHMC)

A. OBJECTIVES AND MEMBERSHIP. One of the most ambitious attempts to establish a global, digitally integrated data commons to deal with a specific subdiscipline of microbiology was that of the International Human Microbiome Consortium (IHMC).[277] The IHMC was launched at a major scientific congress in Heidelberg, Germany, on October 16, 2008, as "an effort to enable researchers to characterize the relationship of the human microbiome in the maintenance of health and causation of disease." The overall aim was to generate a shared data resource from international projects that would be made freely available to the global scientific community. Research organizations from all nations supporting similar research efforts are invited to become participants. Since its foundation, IHMC has organized periodic international conferences at which research results are shared. The most recent conference was in October 2015.[278]

Membership in the Consortium is open, at any time, to the funders and Principal Investigators (PIs) of human microbiome research programs that have the "capacity to mount a comprehensive analysis of the human microbiome in health and or disease, and that agree to carry out their efforts according to a set of commonly agreed-upon ... IHMC policies."[279] Normally, funders will have identified Consortium members as a research group working on a large scale, doing comprehensive analysis of the microbiome, and as a "Community Resource Project."[280] The latter is defined as a "research project specifically devised and implemented to create a set of data,

[277] Int'l Human Microbiome Consortium, *Homepage*, http://www.human-microbiome.org/ (last accessed May 5, 2015). The human metagenome encompasses the collective genomes of all microorganisms living in the human body.

[278] Interview with Dr. Lita Proctor, Program Director, Human Microbiome Project, NIH, 15 May 2015.

[279] IHMC, *Homepage*, n. 277 ("Membership").

[280] *Id.* (quoting definition of "Community Resource Project," Wellcome Trust, Ft. Lauderdale. Florida, 14–15 Jan. 2003 [hereinafter Ft. Lauderdale Principles (2003)], *available at* http://www.genome.gov/pages/research/wellcomereport0303.pdf).

reagents or other material whose primary utility will be [as] a resource for the broad scientific community."[281]

Affiliation with the IHMC and its goals does not necessarily entail any formal legal obligation, and there is apparently no a priori commitment of funds implicit in the expression of interest to participate.[282] Would-be member research groups not identified by funders must apply to a Steering Committee, which will vet their qualifications and certify that the applicant will abide by IHMC's principles. As of 2015, the following entities had volunteered to participate in the IHMC:

- Australia: Commonwealth Scientific and Industrial Research Organization
- Canada: Canadian Institutes of Health Research
- China: Meta-GUT project (Ministry of Science and Technology (MOST)) Sino-French collaboration; Human Gut Microbiome and Infections Human Gut Microbiome and Infections
- Europe: European Commission
- France: Institut National de la Recherche Agronomique (INRA)
- Gambia: Medical Research Council
- Germany: European Molecular Biology Laboratory (EMBL)
- Japan: Science and Technology Agency
- Republic of Korea: National Research Foundation
- United States: National Institutes of Health.[283]

Because funders and principal investigators write their script as single research proposals evolve, each project is relatively unique. The common denominator for organizational purposes is to enrich the supply of publicly accessible data pertaining to the human microbiome.[284] The Gambian initiative was the only project in a developing country so far covered by the IHMC. However, at its Paris

281 IHMC, *Homepage*, n. 277 ("Membership"). Nonmembers' funders who plan to support relevant initiatives at some future time may become observer members under specified conditions.

282 IHMC, *Homepage*, n. 277 (Membership").

283 *Id.*

284 For example, Gambia's affiliation with the IHMC resulted from a $6,000,000 grant by the Bill and Melinda Gates Foundation to the United Kingdom's Medical Research Council. This grant established the Gates Microbiome Project in Gambia, which looks at the effects of the pneumococcus vaccine on the nasopharyngeal microbiome in Gambian infants. After the Gambia Project proposal was approved, reference genomes from 1,500 nasopharyngea samples were contributed to the IHMC. *From deep end to springboard: Brenda Kwambana's PhD experience*, 10(5) TAMA NEWSLETTER 25 (MRC Unit, Gambia 2011), *available at* http://www.google.com/url?sa=t&rct=j&q=&esrc=s&source=web&cd=1&ved=0CCMQFjAA&url=http%3A%2F%2Fwww.mrc.gm%2Fwp-content%2Fplugins%2Fdownload-monitor%2Fdownload.php%3Fid%3D7&ei=IDp_UM29J-e0oAHI04GQAw&usg=AFQjCNGuah8vUOQGs2K4tHvK1ulPyWaGAQ&sig2=b8WCTwbj1VfMzHNUUqG2hg. *See also*, http://genome.wustl.edu/projects/human_microbiome_project/international_human_microbiome_congress_2011 (brief description and slides).

meeting in 2012, the Steering Committee expressed a strong desire to expand the membership of IHMC to other parts of the world, with a view to enabling analyses of the microbiome in different populations. To this end, the IHMC members were encouraged to develop links with the researchers and funders in these areas.[285]

B. GOVERNANCE. In Chapter 8, we first discussed the IHMC in connection with the NIH's carefully devised data-release policies and their strict enforcement of those policies.[286] Here we return to this entity because of its novel governance approach, which has significantly influenced our own thinking about the proposed Microbial Research Commons. Unlike most of the other entities reviewed in this section, the IHMC initially had a preestablished five-year termination date, scheduled for 2013. Since then, however, the IHMC's management envisions an ongoing mission with willing participants.[287] For this and other reasons, its governance structure seems lighter than most of the others we examine, and it is geared closely to its data pooling objectives, with some novel governance solutions worthy in other contexts of careful consideration in other contexts.[288]

The IHMC's organizational structure consists of a a Steering Committee and a Funders' Committee, with Working Groups to be formed as necessary. The Steering Committee meets annually, when feasible, and at a major conference sponsored by IHMC.[289] These Conferences, which are organized by a Scientific Planning Committee (a subset of the Steering committee), provide a forum in which up to a thousand participants share reports of ongoing research results.[290]

The Steering Committee, which remains the primary governance entity, is comprised of representatives from each of the participating funding agencies or organizations, and of the Principal Investigators leading projects in the IHMC. However, voting rights are vested in one member from each participating country.

[285] IHMC *Newsletter* # 3[del] (Spring, 2012) *available at* http://www.human-microbiome.org [hereinafter IHMC *Newsletter* #3]. Efforts are underway to expand the membership in parts of Africa and Asia. Interview with Dr. Lita Proctor, n.278.

[286] *See* Chapter 8, Section I.B.2.

[287] *See* IHMC, *Homepage* n. 277 (scheduling further scientific conferences); Interview with Dr. Lisa Proctor, n. 278.

[288] *See* Section III.B in this chapter.

[289] *See* IHMC, *Homepage* n. 277 ("Organization").

[290] Interview with Dr. Lita Proctor, n. 278. The First Meeting was held in Heidelberg, Germany, Oct. 2008. The Second Meeting was in Shenzhen, China, March 2010, and the Third Meeting was in Vancouver, Canada, March 2011. The Fourth and most recent meeting and Conference, held in Paris, France on March 19–21, 2012 under UNESCO patronage, gathered more than 620 participants from 36 countries in five continents. A three-day program concluded with a round table on the future of microbiome research. IHMC *Newsletter* # 3, above n. 285, at 1. This was the International Human Microbiome Congress (MetaHIT 2012), hosted by Institut National de la Recherche Agronomique (INRA), France. *See* IHMC *Homepage*, above n. 277.

As thus constituted, the Steering Committee seeks a consensus, in the absence of which issues requiring an organizational decision will be put to a vote. The Funders' Committee may allow more than one PI from large, multi-PI consortia to sit on the Steering Committee.[291]

Because the Steering Committee is made up of both national funding entities and the Principal Investigators of relevant projects, there is no separate scientific advisory body like those that exist in the other selected international research commons we have reviewed. Nor is there a rigid separation of the political and funding dimensions from the formulation of science policy and programs, as often occurs in science driven entities. Rather, decisions appear to be taken jointly by the national funding agencies and the Principal Investigators, who seek to reach a consensus, but will abide by a majority vote, if necessary.

Politically speaking, this governance structure simplifies the transnational administrative complexities of operating a relatively small, highly focused scientific program in which the participating funding entities presumably speak with the approval of their national administrative and regulatory authorities. At the same time, matters bearing on science policy are negotiated directly by the representatives of the relevant research community and the representatives of the funding entities, with a view to ensuring that future operations benefit from the best available scientific and technical expertise.

Once projects and policies are approved by the Steering Committee, funding follows automatically for the research programs that have been approved by the participating national entities. To the extent that additional expertise is needed on either scientific or legal and institutional issues, the Steering Committee can establish Working Groups to report on these matters, as noted earlier. For example, a Working Group on intellectual property was contemplated.[292]

C. FUNDING. As noted earlier, single projects under the aegis of the IHMC are funded on a case specific basis by the participating sponsors, with resulting data normally to become a common-pool resource. For example, the NIH's own Human Microbiome Project (HMP) funded six initiatives under a five-year project ending in 2012 that was renewed for another five years. The NIH's Common Fund Office thus sought new funds to support a second round of microbiome

[291] *See* IHMC, *Homepage* ("Organization") n. 277; Interview with Dr. Lita Proctor, n. 278. Initially, an interim Steering Committee was formed with funders from Australia, Canada, China (MOST), the EU (European Commission), Singapore, the U.S. (NIH) and with investigators nominated by each of those funders. However, China and Singapore may have dropped out.

[292] Working Groups eventually formed to carry out more in-depth analysis of specific issues and policies will be made up of experts in the area under discussion and need not be formal participants in any IHMC project. IHMC, *Homepage* ("Organization"), n. 277.

research (HMP2) that would build on the products and outcomes of the first set of initiatives.[293]

With specific regard to funding the IHMC's internal operations, the general principle is that participants "will obtain their own funding for data generation and specimen collection."[294] The NIH's own contributions were amplified by two EU initiatives, namely, the International Human Microbiome Standards project (IHMS)[295] and the Metagenomics of the Human Intestinal Tract (MetaHIT) project.[296] MetaHIT took a particularly active role in the establishment and functioning of IHMC. The MetaHIT coordinator served as co-chair of the IHMC itself, and several partners participated in working groups pertaining to genome sequencing of gut bacterial strains and the controlled release of clinical data.[297] Both IHMS and MetaHIT were, in turn, funded by the European Commission under its Seventh Framework Program, but MetaHIT's funding ended and it is no longer operational.[298]

D. DATA AND INTELLECTUAL PROPERTY POLICIES. The IHMC has adopted strong and noteworthy measures to ensure the timely public release of all data funded through its operations, and it directly urges publishers of the resulting research results to conform to its data release policies. To this same end, the IHMC facilitated the adoption of informed consent standards, quality standards, and data deposition and release policies that aimed to make high-quality data, tools, and protocols available

293 IHMC *Newsletter* #3, n. 285, at 2.

294 IHMC *Homepage*, n. 277 ("Membership").

295 "To promote the necessary international cooperation and coordination IHMS took an active role in the establishment and functioning of the International Human Microbiome Consortium (IHMC). The IHMS coordinator served as a co-chair of the IHMC and several partners participated in the IHMC working groups, related to genome sequencing of the gut bacterial strains, and the controlled release of the clinical data." *See* Int'l Human Microbiome Standards, *Coordination* (October 21, 2013), http://www.microbiome-standards.org/index.php?id=80. The IHMS is financed by the European Commission under the 7th Framework Programme. The consortium gathers 8 partners from academia and public sector, across 6 different countries. Its total cost has been evaluated at 2,3 million Euros and the funding from the European Commission has been set with an upper limit of almost 2 million Euros. The project was scheduled to last for 4 years, starting from February 1, 2011. *See* http://www.microbiome-standards.org/.

296 MetaHIT is one of the projects financed by the European Commission under the 7th Framework Program. The consortium gathers 13 partners from academia and industry, a total of 8 countries. Its total cost has been evaluated at more than 21,2 million Euros and the funding requested from the European Commission has been set with an upper limit of 11,4 million Euros. The project was funded from January 1, 2008 until June 30, 2012. *See* http://www.metahit.eu/index.php?id=410). *See also* Eur. Comm'n, *International large scale omics research initiatives*, http://ec.europa.eu/research/health/large-scale/omics/international-initiatives-disease-genomics_en.html (last accessed 5 July 2014).

297 MetaHIT, *Homepage*, www.metahit.eu (last accessed 5 July 2014).

298 Interview with Dr. Lita Proctor, n. 278; see also nn. 295–96.

to the entire community as rapidly as possible.[299] In coordinating research efforts to respect the interests and priorities of diverse investigators, funding agencies, foundations, non-profit organizations, and industry; it also seeks to avoid unnecessary redundancy in human microbiome research.[300] These measures were more fully described in Chapter 8.

In principle, data generated by the IHMC will be made available through the University of Maryland School of Medicine's NIH Human Microbiome Project Data Analysis and Coordination Center, in Baltimore, Maryland, and also through the European Molecular Biology Laboratory (EMBL), in Heidelberg, Germany.[301] Data are sometimes also made available through other public databases, such as the National Library of Medicine.[302]

In a brief statement about "intellectual property," the IHMC's management stated that it discourages the filing of intellectual property claims on "pre-competitive basic data of the type produced by a Community Resource Project."[303] As an example, the statement considers sequence or expression data from a bacterial metagenomic study to be "precompetitive," whereas data from follow-up studies of the functional role of the metagenomic bacterial community or single bacteria in that community are potentially available for the filing of intellectual property claims, presumably patents.[304]

However, this intellectual property policy begs some questions. For example, problems could arise from the very success of the IHMC initiative, if and when analysis of its sequence data began to yield target sequences that could lead to pharmaceutical products. Here the IHMC governing body has wisely resisted the temptation to allow patents on these precompetitive datasets.[305] Nonetheless, the very existence of upstream datasets of potentially high commercial value could generate intellectual property tensions later on, precisely because the public-good function of the data infrastructure may also have become a kind of aggregate research tool capable of generating payoffs for the private sector.

Unless some thought is given to mandatory licensing conditions, and perhaps even to structuring a compensatory payoff for the contribution of the Consortium's datasets serving as research tools in such cases, there may be a tendency of the participating members to become less willing to share data over time. They may

[299] *See* IHMC *Homepage*, n. 277 for details; *see also* Chapter 8, Section I.B.2.
[300] *See* IHMC *Homepage*, n. 277; Interview with Dr. Lita Proctor, n. 278.
[301] IHMC website, n. 277.
[302] Metagenomics of the Human Body 80 (K. Nelson ed., Springer, 2011).
[303] IHMC, *Homepage*, n. 277.
[304] *Id.*
[305] *See id.*; *cf.* Rai et al. (2008), n. 77.

fear that their host institutions will lose out on potential payoffs from downstream, perhaps blockbuster medicines.[306]

Moreover, even to achieve such payoffs from a more functional research product, it may be necessary to forge a public-private arrangement that combines the expertise of, say, a pharmaceutical company, with that of the relevant PIs in the consortium. Here, once again, it seems preferable to have predetermined intellectual property arrangements that deal with such prospects *ex ante*, behind "a veil of ignorance," that is, when there is no actual knowledge of commercial applications.[307] Otherwise, attempting ad hoc negotiations *ex post*, when everyone knows that both the data tool and the analysis in question have potentially high-value commercial applications, could become painfully difficult. These tensions could also make it correspondingly harder to preserve the public science interest of the IHMC in specific cases.

As regards materials, such as "reagents and other material whose primary utility will be as a resource for the broad scientific community,"[308] the extent to which members of the Consortium will actually integrate material resources into their sharing arrangements remains unclear. The NIH, in implementing the U.S. component, clearly wants reagents, such as microbial strains, to be sequenced and deposited at the Human Microbiome Project (HMP) Repository prior to actual sequencing."[309] In principle, other material resources and reagents to be shared should be released rapidly, but the Steering Committee and NIH will further elaborate policy in this regard.[310]

E. FUTURE PROSPECTS. Discussions about expanding the scope of IHMC to other regions, especially Africa, South America, and India,[311] confirm that the IHMC is looking beyond the initial five-year period of experimentation. This extension would be consistent with the goal of having some 3,000 reference genomes sequenced, a task that was reportedly going well, and with the goal of increasing the number of reference strains to be generated.[312] The IHMC thus seems to be accomplishing its goals, although the long-term payoffs remain to be evaluated.

306 *Cf.* Reichman & Uhlir (2003), n. 11; Reichman & Okediji (2012), n. 89; *see further* Chapter 8, Section II.C.1.

307 Rai et al. (2008), n. 77.

308 *See* IHMC, *Homepage*, n. 277.

309 See NIH, HMP Data Release and Resource Sharing Guidelines for Human Microbiome Project Data Production Grants, available at https://commonfund.nih.gov/hmp/datareleaseguidelines (last accessed May 7, 2015) [hereinafter HMP Data Release and Resource Sharing Guidelines].

310 At present, the national HMP seems to be depositing relevant materials at the ATCC under a specially administered arrangement, with more research friendly MTA provisions than would normally apply to other ATCC deposits. This program is managed by the ATCC, but remains distinct from their other operations, and it is an NIAID program. *See* Human Microbiome Project: BEI Resources, http://www.beiresources.org/About/Human MicrobiomeProject/tabid/625default.aspx.

311 IHMC *Newsletter* #3, n. 285, at 2.

312 *Id.* at 3.

For present purposes, what seems most striking about this initiative is the funders' ability to avoid top-heavy governance structures to deal with political matters while devising novel, bottom-up procedures for maximizing direct scientific inputs into every decision and initiative. Admittedly, the relatively small size of this project plus the direct participation of funders help to explain this novel approach. Nevertheless, we think the science friendly, anti-bureaucratic, bottom-up approach devised by the organizers of the IHMC needs to be carefully studied, with a view to extrapolating its lessons for a redesigned Microbial Research Commons and for other future knowledge commons dedicated to scientific research. These implications are discussed later in this chapter and implemented in the governance scheme we elaborate in Chapter 10.

C. The Market-Like Nongovernmental Enterprise

As evidenced throughout this volume, microbiology has long depended on a network of public-service culture collections for the conservation, preservation, and exchange of *ex situ* microbial genetic resources, even though only a small percentage of all known microbial resources can actually be cultured.[313] Prominent among the existing networks are the World Federation of Culture Collections (WFCC), and regional networks, such as the European Union Culture Collections Organization (ECCO) and the Asian Consortium for Conservation and Sustainable Utilization of Microbial Resources (ACM).[314] The advent of genomic and metagenomic research techniques has further enlarged the traditional mission of the culture collections. In particular, the living biological materials they preserve are still needed to verify genomic research results generally and also serve as reference specimens against which the results of genomic data analyses can be tested, verified, and improved in the future.[315]

However, many existing culture collections have limited scientific and organizational capacities. They often lack a mandate to implement common policies and practices that would enable the constituent collections to provide high

[313] *See* Chapter 2, Section I.A.1 and Chapter 4, *passim*.

[314] See D. Smith et al. (2013), n. 73.

[315] Global Biological Resource Center Network (GBRCN), A Demonstration Project – An Initial Focus on Microorganisms 10 (Fed. Ministry of Educ. & Research, Germany, 2010) [hereinafter GBRCN, Demonstration Project], *formerly available at* http://www.gbrcn.org. Unfortunately, these and other cited documents have been deleted from the website and no longer appear in the Final Report. See Global Biological Resource Center Network (GBRCN), Final Report on the GBRCN, Demonstration Project (Nov. 20, 2008 – Nov. 30, 2011) (2012), available at http://www.gbrcn.org/fileadmin/gbrcn/media/downloads/GBRCNFinal Report/GBRCN-FinalReport2012/pdf., last accessed May 8, 2015 [hereinafter GBRCN, Final Report (2012)].

quality services in a consistent manner.[316] Their administrators "tend to work in honorary capacities and rely upon individuals to input whatever time they can spare to drive activities," and there is still relatively little coordination among the bulk of them, even with regard to accession policies.[317] Most culture collections, with some exceptions, have not adequately adapted to the challenges of the New Biology, with its dependence on vast amounts of genomic and other data and the corresponding need for collaborative networks dedicated to the analysis of such data.[318]

As reported in Chapter 4, the Organization for Economic Co-operation and Development (OECD) promoted a program, beginning in 1999, designed to upgrade culture collections as traditionally conceived to the status of Biological Resource Centers (BRCs), which could meet uniformly high-quality standards and better address the needs of present-day biotechnology.[319] Subsequent activity culminated in the publication of best practice guidelines for BRCs,[320] which, however, relatively few WFCC collections could realistically aspire to meet in full.[321] The final OECD BRC workshop, accordingly, accepted a proposal to test the possibility of forming a global network of the most highly qualified BRCs, to be known as the Global Biological Resource Center Network (GBRCN) Demonstration Project.

[316] *See* GBRCN, Demonstration Project, n. 315, at 6 (with "the discovery of many new microorganisms ... [it] is essential that representatives of these and other useful organisms be maintained for future use. If a strain is lost, recovery of that strain or even the same species from its natural environment can be difficult, or, for practical purposes, impossible").

[317] *Id.* at 9. For efforts to improve networking, see D. Smith et al. (2013), n. 314, at 284–89.

[318] *See* Chapter 1, Section II.D. According to GBRCN's organizers, the role of biological material in the verification of experimental data is a key issue and a concern in the biological sciences. The literature is full of data that cannot be verified because the material is not available or the material that was used to generate the data has deteriorated. BRCs will encourage timely deposits and thus be able to supply reliable authentic biological material as vouchers for generated data. Additionally, BRCs will ensure that reference strains used to create databases are available for confirmation and further work to protect investments made by funding bodies. GBRCN, Demonstration Project, n. 315, at 14.

[319] *See* Chapter 4, Section I.C.2. Discussions began at a workshop in Tokyo in 1999, under the auspices of the OECD Working Party on Biotechnology and the Task Force on Biological Resources Centers (BRC). *See, e.g.*, David Smith, "Networking Collections to Provide Facilitated and Legislation Compliant Access to Microbial Resources," paper presented at the Conference on Governing Pooled Knowledge Resources: Building Institutions for Sustainable Scientific, Cultural and Genetic Resources Commons for the International Association for the Study of the Commons (IASC), Louvain-la-Neuve, Belgium, Sept. 12–14, 2012 [hereinafter D. Smith (2012)]. OECD, Biological Resource Centers – Underpinning the Future of the Life Sciences and Biotechnology (March 2001) [hereinafter OECD Report on BRCs], *available at* http://www.oecd.org/dataoecd/55/48/2487422pdf. Stern, n. 72.

[320] OECD, OECD Best Practice Guidelines for Biological Resource Centers (2007), *available at* http://www.oecd.org/sti/biotech/38777417.pdf [hereinafter OECD Best Practices].

[321] *See, e.g.*, WFCC, *Guidelines*, n. 180, and discussed in Chapter 4, Section I.A.2.

1. The Global Biological Resource Centers Network (GBRCN) Demonstration Project

This Demonstration Project, which began in 2008 and ended in 2011, was funded by the German Federal Ministry of Education and Research (BMBF).[322] The initial proposal to establish a Demonstration Project envisioned an international network of highly qualified collections[323] to be linked in a distributed global infrastructure that would integrate both microbial materials and related data within a single organizational framework.

A. OBJECTIVES AND MEMBERSHIP. The broad goals set out by the founders in the Demonstration Project Document called for the GBRCN to:

- Help governments to fulfill their commitments arising from international conventions and national legislation, e.g., biosafety and biosecurity challenges and the CBD-ABS development requirements [and provide the science to underpin policy].
- Assist the scientific community in seeking to characterize the range and magnitude of microbial biodiversity, and in implementing legislative requirements, while bridging gaps in knowledge and making better uses of investments in research.
- Help the BRCs keep abreast of modern scientific developments; meet quality standards for research; supply authentic cultures and standardized biological material for testing and quality control; harmonize research methods and procedures on a global basis; and reconcile research and development projects with the need to comply with relevant regulations.
- Generate a critical mass of high-quality data, which in combination with relevant data from other fields, would produce information enabling new scientific findings and innovation when mined by modern interactive tools.[324]

To these ends, according to a Draft Memorandum of Understanding, dated June 7, 2010, the participating governments "working through GBRCN, will establish and support a coordinated and distributed system that will enable users to access and

[322] GBRCN, DEMONSTRATION PROJECT, n. 315. Initially, the GBRCN proposal was to embrace other networks that would eventually focus on human derived materials as well as plants and animals. *See* OECD BEST PRACTICES, n. 320. In reality, however, activities focused only on microbes. D. Smith (2012), n. 319, at 1.

[323] *See* GBRCN, DEMONSTRATION PROJECT, n. 315, at 18–19, for the list of member collections from China, Belgium, the United Kingdom, Canada, Spain, France, Brazil, Germany, Italy, Kenya, The Netherlands, Uganda, Japan, Portugal, and Finland (but not, apparently, U.S. or India).

[324] GBRCN, DEMONSTRATION PROJECT, n. 315, at 8–9.

utilize existing and new *ex-situ* biodiversity and related expertise and data."[325] These governments should further agree that GBRCN's resources would be "shared and distributed, while encouraging cooperation and coherence."[326]

More specifically, the GBRCN would:

- Be global in scale, though implemented nationally and regionally;
- Be accessible by individuals anywhere in the world, offering potential benefits to all, while being funded primarily by those that have the greatest financial capabilities;
- Promote standards and procedures designed to facilitate their adaptation nationally and regionally;
- Serve to disseminate technological capacity by drawing on and making widely available scientific and technical information; and
- Make laboratory-held, living biological material universally available, while fully acknowledging the origin of all value-adding contributions made in the course of gathering and furnishing these materials.[327]

The Draft MoU further envisioned harmonized decisions to implement agreed principles on biosafety, biosecurity, and the "ownership and management of intellectual property in compliance with national and international legislation."[328]

On closer inspection, however, the underlying business model – as revealed in the MoU drafted in 2010 – was, initially at least, to forge the cooperating collections into an integrated set of self-sustaining, quasi-corporate entities.[329] The GBRCN would thus have made both materials and eventually data available under quasi-commercial and potentially research-hostile conditions rooted in

[325] GBRCN, Third Intermediate Report on the Demonstration Project (German Fed. Ministry of Educ. & Research, Nov. 1 – June 30, 2010) [hereinafter GBRCN 3rd Intermediate Report], Annex 8 Memorandum of Understanding for the Global Biological Resource Center Network [hereinafter Draft MOU (2010), no longer available online. For the Final Report (2012), see n. 315.

[326] GBRCN 3rd Intermediate Report, at 325, para. 3. "Facilitated exchange would be promulgated between members of the GBRCN under the GBRCN Cooperation Agreement."

[327] Draft MOU (2010), n. 325, at 40.

[328] *Id.*

[329] *See id.*; *see further* GBRCN, 3rd Intermediate Report, n. 325, Annex 13 – Executive Summary BRC Funding Models – Financial Sustainability of BRCs and Networks of BRCs [hereinafter Annex 13 – Funding Models];, Annex 14 – Executive Summary GBRCN Architecture [hereinafter Annex 14 – GBRCN Architecture]; Annex 15 – Executive Summary Harmonized Approaches [hereinafter Annex 15 – Harmonized Approaches]; Annex 16 – Executive Summary of Secretariat Paper on Sustainability of the Secretariat and the GBRCN [hereinafter Annex 16 – Sustainability]; Annex 17 – Executive Summary BRC General Business Plan [hereinafter Annex 17 – Business Plan]; Annex 18 – Executive Summary – Imbarc and GBRCN Compilation Document on Data Management and Interoperability Standards [hereinafter Annex 18 – Data Management].

national intellectual property laws.[330] In other words, rather than establishing a research or knowledge commons in the sense used earlier in this chapter,[331] careful analysis of subsequently distributed demonstration project documents that are no longer publicly available suggests that the real purpose of the GBRCN initiative was to convert the elite public science collections into a global proprietary club or clearinghouse that would, in effect, emulate – and could eventually compete with – the ATCC model in the United States.[332]

At the same time, the organizers made great efforts to establish high-quality standards for all the accredited culture collections and to push them to implement the OECD Best Practices for Biological Resource Centers.[333] However, because the GBRCN Demonstration Project placed such a strong emphasis on improving quality standards, with a view to converting existing culture collections into full-fledged BRCs, the membership agreement expressly recognized that some or many collections would not qualify for full membership in the proposed organization.[334] Subject to third-party independent inspections, the cooperating collections were accordingly to be classified as entry level or candidate members, Basic Level BRCs, Certified BRCs, Accredited BRCs, or Fully Compliant BRCs.[335]

Once a cooperating collection satisfied the minimum requirements for a BRC, it would not have been further obliged to reach higher levels of certification or accreditation to remain in good standing. Nevertheless, the GBRCN's organizers intended to develop criteria to enable users to distinguish the BRCs at different levels of accreditation.[336] Collections that had not yet attained BRC status but were willing to cooperate, and were demonstrating an intent to implement the agreed best practices, might qualify as associate members, in which capacity they may contribute to the aims of the network.[337]

All cooperating entities at every level were obliged to ensure that "the user receives the high quality materials and a legitimate service" within a common Quality Management System that implemented established guidelines for biosecurity and biosafety.[338] Using a designated point of contact for member BRCs, and subject to

[330] *See, e.g.*, GBRCN, 3RD INTERMEDIATE REPORT, n. 325, Annex 13 Funding Models; Annex 16 – Sustainability; Annex 17 – Business Plan; *see further* Section III.B.1 ("Avoiding the Wrong Incentives") later in this chapter.

[331] *See* Section I of this chapter ("Theoretical Reflections on Designing a Knowledge Commons").

[332] For the ATCC model, *see* Chapter 4, Section II.A.

[333] For OECD Best Practices, *see* n. 320; *see further* Chapter 4, Section 1.B.

[334] "[T]he fundamental principle of the GBRCN is that it gives access to authentic high quality materials in a reproducible manner." GBRCN, 3RD INTERMEDIATE REPORT, n. 325, Annex 9, Draft GBRCN COOPERATION AGREEMENT. *See also* GBRCN, DEMONSTRATION PROJECT, n. 315, at 12.

[335] *Id.*

[336] GBRCN, 3RD INTERMEDIATE REPORT, n. 325, Annex 9.

[337] *Id.*, Annex 9.

[338] *See* GBRCN, DEMONSTRATION PROJECT, n. 315, at 12.

common policies for compliance with national legislation and relevant regulations, the participating BRCs were expected to facilitate the movement of materials and information among themselves while recording both the routes of deposit into each BRC and the users who had received materials from them.[339]

To work efficiently, the founders envisioned that the network would ultimately operate through a series of "clusters." These clusters could be formed either to meet local needs and laws at the national level, or to advance the thematically focused goals of specialist groups or activities, such as "molecular techniques," information technologies, or legal issues and requirements.[340] Existing or future regional groups could also operate as clusters. In that event, each cluster would have to provide its own funding, while the GBRCN Secretariat would provide limited coordination and support for the clusters, in exchange for some income from their funding sources.[341]

Capacity building for developing country participants was built into the program, as a component of "Technology Transfer:"

> The Participants acknowledge that, subject to any relevant Intellectual Property Rights, GBRCN should seek to promote the nonexclusive transfer to research institutions in countries of developing economies of technology ... especially in conjunction with training and capacity development programs.[342]

Capacity building was also listed as a key deliverable of the entire GBRCN project.[343]

The GBRCN's approach to capacity building generally emphasized the "need to manage the transition from culture collection to BRCs."[344] Besides addressing human resources, and the enhancement of tools, facilities, and related policies, the overall goal was to make both the single member BRCs and the networks they sought to form more functional and sustainable over time. This capacity building component extended initially to both emerging economies, such as culture collections in Brazil and China, as well as to collections in other developing countries, such as Kenya and Uganda, which were already partners in the operational network that carried out the Demonstration Project.[345] Efforts were also made to enable the Kenyan and Namibian partners to formulate better biodiversity conservation policies while improving the quality of their culture collections.[346]

339 GBRCN, 3RD INTERMEDIATE REPORT, n. 325, Annex 9.

340 *Id.*, Annex 16.

341 *Id.*

342 *Id.*, ¶ 8.8.

343 GBRCN, 3RD INTERMEDIATE REPORT, n. 325, Annex 15, item 9. *See also id.*, Annex 9, items 12 and 13.

344 *See* GBRCN, DEMONSTRATION PROJECT, n. 315, at 13.

345 GBRCN, *Archive*, http://www.gbrcn.org/news/archive (last accessed May 8, 2015).

346 *Id.*

B. PROPOSED GOVERNANCE STRUCTURE. Because the GBRCN's founders aimed to test an operational concept, rather than to establish a permanent entity, its organizational structure remained at a formative stage. The draft intergovernmental Memorandum of Understanding submitted in June 2010 was explicitly based on the GBIF model.[347] It envisioned a Governing Board, consisting of the government signatories to the Memorandum of Understanding, whose existing BRCs had joined the network during the Demonstration Stage and a Coordinating Secretariat. The membership consisted of a constellation of existing BRCs and other microbial culture collections aspiring to become BRCs, which were called "cooperating entities."[348] These entities were obliged to prove that they had implemented and complied with the OECD Best Practices for BRCs, as well as with the membership requirements of GBRCN. Any qualifying BRC could become a "cooperating entity of GBRCN" even if its host country was not a participating signatory of the MOU.[349]

Signatories to the MOU, which were designated as "Participants," could include "economies" (such as Taiwan), intergovernmental organizations, and unspecified other organizations (presumably NGOs) as well as governments. Technically, the Governing Board would include representatives of all the Participants that had endorsed the GBRCN principles by signing the MoU.[350]

The Governing Board, that consisted of one representative from each "Participant," was to take collective decisions on all matters pertaining to governance and operational scope. Detailed action plans were to be formulated "in agreement with the cooperating entities,"[351] which, at the time, was the only vehicle for direct inputs from representatives of the scientific community. The Governing Board was authorized to appoint an Executive Committee to oversee the Secretariat and to exercise delegated authority between meetings of the Board.[352] The Board could also establish other subsidiary bodies as it deemed necessary, such as a Legal Advisory Board and a Scientific Advisory Group.[353]

The Governing Board was to seek consensus, failing which it could take decisions by a two-thirds super majority of those present and voting. Voting rights were given to participant governments that made preestablished financial contributions to the GBRCN. At the discretion of the Board, voting rights for limited periods could also be given to those economies, IGOs, or other organizations that had negotiated and

[347] *See* Section II.B.2.b. (GBIF governance model).

[348] GBRCN, DEMONSTRATION PROJECT, n. 315, at 12, 15; GBRCN, Draft MoU (2010), n. 325, ¶ 1.9.

[349] GBRCN, Draft MoU (2010) n. 325, ¶¶ 1.3–1.5, ¶ 1.9.

[350] *Id.* ¶¶ 1.3–1, 1.9.

[351] *Id.* ¶ 4.6.

[352] *Id.* ¶¶ 1.4.1.

[353] GBRCN, DEMONSTRATION PROJECT, n. 515, at 12, 15; GBRCN, 3RD INTERMEDIATE REPORT, n. 325, Annex 14.

paid agreed contributions.[354] Participants that failed to make financial contributions could nonetheless take part in deliberations of the Governing Board – as Associate Participants – but could not vote. Other relevant global organizations, such as the Secretariat of the CBD, were invited to designate nonvoting representatives to the Governing Board.[355]

The GBRCN's official Secretariat would eventually consist of a Managing Director and such staff as the Governing Board deemed necessary to implement its Work Program. In effect, the Secretariat would become responsible for carrying out all the approved scientific and administrative activities.[356] A bidding competition would be held to determine the host country in which the GBRCN Secretariat was to be situated.[357]

A small, interim Secretariat was also set up to oversee and coordinate early stage activities of the proposed organization, as an empirical proof of concept.[358] This interim Secretariat worked with candidate microbial BRCs in fifteen countries during the Demonstration Project.[359] Its primary function was to provide for peer-reviewed assessments of the extent to which candidate partners had implemented OECD Best Practices for BRCs.[360] The interim Secretariat was also responsible for further elaborating the infrastructure and governance mechanisms for the GBRCN in light of evidence gained from the Demonstration Project.[361] Since 2011, the interim Secretariat continued its efforts to form and maintain a global umbrella organization that would link the world's most highly qualified BRCs, or candidate BRCs, in a single Collaborative Research Infrastructure.[362]

C. FUNDING AND A BUSINESS MODEL. Ostensibly, the GBRCN's formal governance structure, as described earlier, while perhaps lacking some institutional scientific representatives, could support a major public science infrastructure that would preserve and enhance the public good functions of the microbial culture collections, if that had been the primary objective. In reality, documents made available during the Demonstration Project phase revealed a business model that focused mainly on commercializing materials, data, and related services to be provided by

354 GBRCN, Draft MoU (2010), n. 325, ¶¶ 4.1–4.5.
355 *Id.*, ¶¶ 4.4, 4.5.
356 *Id.* ¶ 6.
357 *Id.* ¶ 5.
358 *See* GBRCN, DEMONSTRATION PROJECT, n. 315.
359 *See* David Smith, Global Networking of Culture Collections: WFCC and GBRCN Perspectives, paper presented at EMbaRC Seminar, Cantacuzino Institute, Bucharest, Romania (Mar. 8–9, 2010), *available at* http://www.embarc.eu/5.DSmith.pdf.
360 *See id.*
361 *See* GBRCN, Final Report (2012), n. 315.
362 D. Smith (2012), n. 319 at 1.

the affiliated BRCs. That model was, at best, indifferent to the legal and institutional arrangements normally required to support digitally integrated public research infrastructure.[363]

In fact, the detailed business plan made available in June 2010 raised serious doubts about the extent to which the GBRCN intended to preserve the public-good functions of the participating BRCs subject to its jurisdiction, notwithstanding the goals originally set out in its Demonstration Project document of February 2009. At the very least, there were recurring contradictions between the formal funding arrangements envisioned in some project documents and the actual business model revealed – perhaps inadvertently – in others.[364]

Formally, the Demonstration Project appeared to endorse specified contributions by participating governments, and it stated that the Secretariat must hold and manage the basic financial contributions of participants, which had initially funded the Work Program. Supplementary financial contributions on a voluntary basis were envisioned to support specific components of the Work Program, or for other purposes to be established by the Governing Board, such as defraying the costs of representatives from developing countries to attend meetings of the Governing Board.[365] The Board could also accept "other income offered for purposes set out in the MOU," while additional income might be generated by GBRCN activities. Otherwise, participants were expected to bear the costs of their own participation in the organization.[366]

Later documents, however, provided a vigorous twist to the notion of "other income" that the GBRCN might raise. For example, the very object of the emphasis on quality standards described earlier was to form an elite, globally linked group of culture collections that could become increasingly self-sustaining over time by distributing both their microbial materials and other value-adding services and products to governments, public science institutions, and private industry on a proprietary and implicitly profit-making basis. The traditional sources of income derived from supply charges (plus the fees for preservation of private collections and patent deposits) were to be greatly augmented in relation to the supply capacity and quality standards of the network as a whole, to the easy access afforded by its single portal and electronic databases, and to the efficiency of its common management protocols, MTAs, and intellectual property policies.[367]

[363] *See generally* FRISCHMANN, INFRASTRUCTURE, n. 11.

[364] *Compare e.g.*, GBRCN, 3RD INTERMEDIATE REPORT, n. 325, at 8–9 *with* GBRCN, 3RD INTERMEDIATE REPORT, n. 325, Annexes 13, 14, 16, and 17.

[365] *See* GBRCN, DEMONSTRATION PROJECT, n. 315, ¶ 9.

[366] *Id.* ¶¶ 9.2–9.5.

[367] *See* GBRCN, 3RD INTERMEDIATE REPORT, n. 325, Annex 9. *See also id.*, Annex 13 (stating that the diversification of activities in moving from the 'Culture Collection' model to the BRC model holds out the expectation of additional sources of revenue, both from existing activities and projects related to new technology-based partnerships).

New sources of income for participating BRCs could then be generated from the following specified initiatives:

- cDNA libraries, genomic libraries, filter sets, clones, plates, PCR products
- Microarrays and reagents
- RNA resources
- Accreditation and standardization practices
- Added value products and services
- Data storage and retrieval
- Software development/collaboration and data mining tools
- Technology development/collaborations – LIMS/robotics
- Sequence database annotation/phenotype analysis
- Linking genomic databases to proteomics
- MLST (multi-locus sequence typing) – population studies.[368]

Consulting services and the "[d]evelopment of spin-off biotechnology companies" could also become potential sources of commercial income, possibly supplemented by research program funding (including payments for deposits in collections), government support (especially for conservation and use of biodiversity commitments), plus some external sponsorships.[369]

In effect, the GBRCN prospectus – as of 2010 – depicted the organization as both a marketing venture for qualified BRCs and a clearinghouse for rights in the quality products and services it aspired to deliver. This goal logically required the GBRCN to charge high fees and to impose restrictive conditions on access, use, and reuse of its materials in the manner of the ATCC.[370] To bolster the network's market power, the GBRCN's leadership must then have logically addressed the risk of leakage that pervades the exchange system, owing to large-scale gratis transfers between academic institutions that we earlier characterized as the "informal exchange system."[371]

The cooperating entities within GBRCN must accordingly have made full use of existing intellectual property rights to reinforce their proprietary claims, whenever possible. They had to satisfy developing country demands for "access and benefit sharing" under the CBD. And, perhaps above all, in order to maximize their medium-term prospects, they needed to extend their reach to genomic microbiology, with a view to modernizing data resources and tools in keeping with the GBRCN's overall commercial goals.

[368] *See* GBRCN, 3RD INTERMEDIATE REPORT, n. 325, Annex 13.
[369] *Id.*
[370] *See* Chapter 4, Section II.A.
[371] *See* Chapter 5, Section I.B ("Formalizing the Informal Sector").

All of these tactics were expressly endorsed in the business plans attached as annexes to the GBRCN's 3rd Intermediate Report of June 2010.[372] For example, the document entitled "Executive Summary BRC Business Plan," recognized that "[c]ulture sales together with preservation contracts income would not be sufficient to cover the BRC running and maintenance costs," in part because of competition from other suppliers (including especially ATCC) and largely "because more microorganisms are exchanged between individual scientists free of charge than are obtained from collections."[373] The economic logic of GBRCN's business model thus leads the founders to conclude that "to support a living collection through culture sales would take a major shift in scientist behavior, which is impossible for one collection to achieve on its own."[374] In other words, the free exchange system of the informal academic research sector needed to be shut down in order to enhance the sustainability of the BRCs aggregated under the umbrella of GBRCN.[375]

By the same token, GBRCN's founders apparently wanted to exploit all the relevant laws, including intellectual property laws, to support both traditional sources of income and new sources of income envisioned in the business plan.[376] Nondiscriminatory MTAs "should regulate supply and exchange of materials, communicating terms and conditions of access and use."[377] No mention of public research needs or exceptions appears in these texts, other than in a passing reference to ECCO's Core MTA and to the obligation to ensure that the source or origin of biological materials and data was properly acknowledged.[378]

Otherwise, intellectual property rights were generally to be respected both for these very purposes and in keeping with the GBRCN's overall policy of deference to national laws and regulations.[379] Looking to the future, a member BRC "could seek exploitable Intellectual Property … through the characterization and screening of its holdings to generate significant income through commercial utilization if compatible with the CBD and the terms under which it receives strains."[380]

[372] *See* n. 325.

[373] GBRCN, 3RD INTERMEDIATE REPORT, n. 325, Annex 17.

[374] *Id.*

[375] It is worth reiterating that, quite apart from these mercenary concerns, the Nagoya Protocol also aimed to shut down the informal system of exchanging biological materials, at least in so far as cross border exchanges were concerned. *See* Chapter 3, Section IV.A ("Clarifying the Broad Economic Scope of the CBD").

[376] GBRCN, Draft MoU (2010), n. 325, ¶ 8.2.

[377] "To the greatest extent possible, GBRCN is fostering facilitated access to holdings at BRCs within an appropriate legal/operational framework." (*Id.*, ¶8.2).

[378] *Id.* ¶¶ 8.2, 8.4.

[379] "The BRC will protect intellectual property rights." GBRCN, 3RD INTERMEDIATE REPORT, n. 325, Annex 9, item 61.

[380] GBRCN, 3RD INTERMEDIATE REPORT, n. 325, Annex 17. *But see* GBRCN, 3RD INTERMEDIATE REPORT, Annex 16 (stating that GBRCN itself would rarely generate products or revenue, and it would depend on contributions from member governments for its financial needs).

To further reconcile these propertizing goals with pressures from developing countries under the CBD, GBRCN's Draft Memorandum of Understanding rightly insisted that cooperating BRCs should ensure the legitimacy of their acquisitions of biological materials.[381] Specifically, access to "new biodiversity resources and associated information" must be consistent with applicable laws, regulations and all relevant requirements pertaining to access and benefit sharing (ABS)."[382] The problems created by these laws and regulations for public research purposes were, however, nowhere addressed.

Finally, all these threads came together in the founders' recognition that data storage and retrieval, sequence database annotation, linking genomic databases to proteomics, and the like were "specific and less common potential income streams for BRCs."[383] Here the propertizing logic of GBRCN's business model, coupled with its unstinting deference to national intellectual property laws, led to visions of commercial exploitation of the very databases and data-mining tools that are indispensable to both the digitally integrated Microbial Research Commons, proposed in this volume, and to the New Biology envisioned by the National Research Council.[384]

Clearly, the GBRCN's business model in the 2010 plan, was at best confusing and at worst troubling. Perhaps one way to resolve the apparent contradictions is to interpret the organization's planned funding as dependent on the member governments' pledged contributions, while the participating BRCs would be allowed – if not encouraged – to pursue the business model identified earlier if they so desired. Such a policy would allow the GBRCN to strike a public-good posture, while actually encouraging a quasi-market-like model at the level of single participating BRCs.[385] That strategy, however, could not disguise the fact that, over time and under its aegis, the ATCC's market-like model would have replaced the public good model at the global level, with serious repercussions for the holistic New Biology paradigm.

381 GBRCN, Draft MoU (2010), n. 325, ¶ 8.7.

382 *Id.*

383 GBRCN, 3RD INTERMEDIATE REPORT, n. 325, Annex 13.

384 For the New Biology paradigm, *see* Chapter 1, Section II.B. See also Chapter 6, Section III ("Automated Knowledge Discovery Tools as Instruments of Massive Infringement") and Section IV.B ("Impediments to the Pooling of Data and Digitally Networked Collaboration"). Contrast the earlier charge to the interim Secretariat to find "a way forward to support research and biotechnology," and to "[d]esign a network to accommodate the future needs of research in life sciences, biotechnology, and biomedicine." *See* GBRCN, DEMONSTRATION PROJECT, n. 315, at 10.

385 *Cf.* D. Smith (2012), n. 319, at 14 (proposing country contributions based upon GDP and/or institutional contributions for GBRCN and stating that there is a need for funding arrangements for the Secretariat "that do not undermine or reduce the already tight funding for BRCs.... History has shown that the collections (BRCs) themselves have no funding for networks," and if it costs too much, they tend not to join.).

2. A Questionable Blueprint for the Future

The GBRCN Demonstration Project, which ended in November 2011, produced a Final Report in 2012 that was prepared by the Interim Secretariat.[386] It also replaced the 3rd Intermediate Report, among other documents, which are no longer available from the GBRCN's website. The Final Report views the GBRCN's "network of networks" concept as the organizing principle around which a future microbial research infrastructure should be built.[387]

The drafters of the Final Report characterized both the GBRCN Demonstration Project itself and the Draft Memorandum of Understanding it elaborates[388] as a "general model for a cooperative infrastructure" linking "the global, regional and national levels."[389] The overriding policy message to be promulgated at the political level "in these economically strained times" is that international collaboration among BRCs "is the only feasible way to implement cost-effective measures to facilitate innovation and discovery to address the bio-economy and the global challenges to society."[390]

As the principal spokesman for the project subsequently explained, rather than acting as a federation of single BRCs, the GBRCN would, instead, act as the overall coordinating body for a growing number of national and regional networks of existing culture collection federations, which already link the most highly qualified BRCs in their respective geographical areas.[391] The overall goal to "provide improved resources for the life sciences to facilitate innovative solutions to global problems"[392] requires

> ... access to high quality biological materials and associated information. It operates on the premise that no one single entity can provide the necessary coverage of organisms and data, therefore, the enormous task of maintaining biodiversity must be shared.[393]

The end vision is thus "a systematic and networked approach" that would not only "bring advantages to both the users and the collections themselves, but importantly provide an infrastructure to underpin research and development."[394]

386 GBRCN, Final Report, n. 315.
387 GBRCN, Final Report, n. 315, at 6 (Executive Summary).
388 *See* GBRCN, Final Report, n. 315, at 63 (Annex: Memorandum of Understanding for the Global Biological Resource Center Network [hereinafter GBRCN Final MoU (2011)].
389 GBRCN, Final Report, n. 315, at 7.
390 *Id.*, at 9.
391 D. Smith (2012), n. 319, at 1–4. The target federations included BCCM (Belgium); SBMC (Brazil); CCCCM (China); FCCM (Federation of Czechoslovak Collections of Microorganisms); CCRB (France); ECCO (EU); SCCCMOMB (Cuba); KFCC (Korea); HPACC (U.K. Health Protection Agency Culture Collection); FORKOMIKRO (Indonesia); ISCC (Japan); PNCC (the Philippines); TNCC (Thailand); UKFCC (U.K.). *Id.*, at 3–4.
392 D. Smith (2012), n. 319, at 1.
393 *Id.*
394 *Id.* at 2.

In putting more emphasis on coordinating regional networks of BRCs than on direct relations with the BRCs themselves, the GBRCN's founders criticized the WFCC's lack of any mandate to effect institutional changes in the policies and practices of its loosely affiliated members. The GBRCN, instead, would introduce "coordinated approaches" at the regional level that would "seek project funding to solve common operational problems or address the common research issues."[395]

With their eyes on the formation of a regional consortium of culture collections in the United States,[396] as well as regional entities that already exist elsewhere, the organizer's plan stressed the "need to harness the properties and products of microorganisms more efficiently ... if we are to tackle the big global challenges of today ... through networked activities and common infrastructure."[397] The aim was "to provide coherence in the application of quality standards, homogeneity in data storage and management and sharing the workload ..." while focusing "expertise and resources to resolve critical problems or towards specific outputs."[398]

From a governance perspective, the Final Report seemed to put more emphasis on a proposed Scientific Advisory Board than appeared in earlier iterations of the project.[399] This Board would meet twice a year and would "support the GBRCN on strategy and operational issues and review proposals for future development" of the entity as a whole.[400] It would also review proposals for new participants and submit recommendations to the Governing Board based on expert analysis.[401] This opening to more direct scientific inputs and bottom-up initiatives is a welcome development, and it tracks some of the governance proposals we envisioned in earlier drafts of this book.[402]

In their post-Demonstration Project articles, the GBRCN's founders also emphasized that the "size of the central Secretariat will be kept small and focused," with much of the work to be performed by members working in "specialized clusters."[403] The Secretariat would manage the day-to-day operations of the research infrastructure, while "reporting to the management board and advised by the

[395] *Id.* (noting in particular the work of the European Union Culture Collections' Organization (ECCO) and the Asian Consortium for Sustainable Use of Microbial Resources).

[396] *See* the United States Culture Collection Network funded by NSF, http://www.usccn.org/, cited by D. Smith (2012), n. 319, at 14.

[397] D. Smith (2012), n. 319, at 2.

[398] *Id.* at 8.

[399] *See, e.g.*, GBRCN, Final Report, n. 315, at 8 ("The remit of a GBRCN should be clearly science based, user community oriented and service driven").

[400] GBRCN, Final Report, n. 315, Annex: GBRCN Architecture, Rationale and Operation, at 4–5.

[401] *Id.*

[402] *See* further, Section III.C in this chapter ("Towards a More Science Driven Organizational Model for the Digital Age") and, Chapter 10, Section II.D.2 ("A Scientific Coordination Council and a Small Secretariat").

[403] *See, e.g.*, D. Smith (2012), n. 319, at 13.

scientific advisory board."[404] In reality, however, some posterior depictions of the Secrtariat's responsibilities seem possibly more ambitious than appeared under the Demonstration Project itself.[405]

The level of prospective Secretariat involvement seems consistent with the GBRCN's increased emphasis on playing a more direct role in research and development, a role that was elsewhere described in rather vague terms.[406] As appears from the last chapter of this book, however, we remain skeptical that the proposed Microbial Research Commons could or should seek to become a research-generating, as well as a research funding institution.[407]

With specific regard to sustainability, the drafters of the Final Report take a more sober and measured view of the problems likely to be encountered than in previous iterations of the GBRCN project, owing perhaps to unforeseen disappointments during the Demonstration stage.[408] The Final Report thus stresses that secured long-term core funding is a prerequisite for the long-term duties of a GBRCN secretariat."[409] To render the GBRCN's operations viable and sustainable, "structured, matching funding should be sought," from "governmental, development and research funding;" and specific budgets should be assigned to "the collaborating entities, the network functioning, [and] the coordinating secretariat."[410] Besides the

404 *Id.*

405 *See* D. Smith (2012), n. 319, at 13–14.
These responsibilities are as follows:

- Managing the technical aspects of BRCs: a quality management system based on international criteria, electronic linkages between BRCs, coordination of catalogues and databases, maintenance and support for BRCs, and development of informatics tools for data analysis, comparison and display;
- Managing the global network of national BRCs: responsible for administration e.g. membership issues, reporting, budget management, inter-laboratory testing and validation of protocols;
- Coordinating the BRC initiative with other international initiatives: coordinating the BRC network with existing international frameworks;
- Providing an intergovernmental forum on BRC issues: facilitating debate, organizing the forum;
- Project development and management; proposal writing, seeking funding, project implementation and management;
- Organization and delivery of capacity building programs: providing programs, tools, resources and activities.

D. Smith (2012), n. 319.

406 *Id.*

407 *See further* Chapter 10, Section II. We suspect that these references to research implicitly presaged reach-through agreements to be imposed on researchers obtaining materials from BRCs, in the manner of ATCC's current MTAs. *See* Chapter 4, Section II.A.

408 *See, e.g.*, GBRCN, Final Report, n. 315, at 14 (noting partners regret that "the project did not provide financial support for the partners" and that "matching national funding was difficult to find." As a result, partners could not always meet their planned commitments and some ambitious aims could not be fully reached. *Id.*).

409 *Id.* at 8.

410 *Id.*

core seed money to trigger activities at participating BRCs, the drafters planned to seek matching short-term funding projects in single countries that would also support a master web portal. To this end, they admonished funding agencies that research infrastructures should be better linked to each other for mutual benefit."[411]

Inexplicably, the propertizing aspirations previously detected in the proposed business plan have disappeared from the GBRCN's Final Report. Instead, while reiterating the public good role that characterized the earlier proposals for the Demonstration Project,[412] the Final Report observed only that the "GBRCN business plan should be modified taking into account deliberations in regional efforts ... and should be adapted to suit national and individual BRC needs and the funding mechanisms available to them."[413]

Given the difficulties that participating BRCs seem to have experienced in obtaining expected funding during the Demonstration Project, this more balanced approach – if implemented – represented a turn in the right direction. Rather than embracing the proprietary trend as previously proposed, the Final Report seemed to suggest that maintaining the public-good approach depends on the availability of public and other supplementary funding sources willing to support it. In light of the financial difficulties from a lack of public funding that led ATCC to embrace the market model in the past, as described in Chapter 4,[414] this more enlightened and balanced approach to sustainability may be the best that one could expect under the circumstances.

3. The Next Step: The Microbial Resource Infrastructure (MIRRI) as a European Stepping Stone to the GBRCN

As indicated earlier, the GBRCN aspired to become a major component of any redesigned Microbial Research Commons by linking key regional associations of Biological Resource Centers into a transnational cooperative network under its guidance. Once the Demonstration phase of that initiative ended, the next step was to construct a pan-European alliance led by major culture collections – that is, collections that already meet, or aspire to meet, the OECD's Best Practices for BRCs – that would embody and implement the principles and goals of the GBRCN on a regional scale.[415]

This new entity, the Microbial Resource Research Infrastructure (MIRRI), was initially funded by the European Commission for three years as part of the European

[411] *Id.*
[412] *See* nn. 329–32 and accompanying text.
[413] GBRCN, Final Report, n. 315, at 8.
[414] *See* Chapter 4, Section II.A.
[415] Fritze & Oumard (2012), n. 108. *See also* D. Smith (2012) n. 319, at 14; D. Smith et al. (2013), n. 70.

Strategy Forum for Research Infrastructures (ESFRI).[416] Once the Preparatory Phase ends (2012–2015), the Construction Phase is expected to run for another three-year period in which MIRRI would coordinate its activities with those of other designated regional entities.

Although initiated by a number of Europe's leading microbial culture collections, the MIRRI aims to become an inclusive network in which both smaller and larger collections become partners on an equal footing.[417] Smaller collections would thus be allowed to enter the network early on, in order to further develop their service capabilities and eventually attain a level of quality they could not otherwise reach. On this view, national culture collections are expected to form and organize networks consisting of all the local culture collections willing and able to join. Each national government is expected to share in funding of the MIRRI during both the construction and operational phases, and these authorities will reportedly insist on a policy of maximum inclusiveness.[418]

If successful, the founders envision that MIRRI would thus become both the model and the organizing vehicle around which the "network of networks" concept, espoused by the GBRCN, would ultimately be built.[419] Although its precise structure and method of operation were still evolving at the time of writing, the MIRRI's general goals and approach will likely emulate that of the GBRCN, viz.[420]

As described by some of its distinguished founders,

> The MIRRI project will provide microbiological services facilitating access to high quality microorganisms, their derivatives and associated data for research, development and application. It will connect resource holders with researchers and policymakers to deliver resources and services more effectively and efficiently to meet the needs of innovation in biotechnology. It will add value to the microbial resources and services needed for research and thus accelerate the discovery

[416] ESFRI emerged from discussions by the Council of Ministers and a study by the EU's Competitiveness Council in 2002. ESFRI's mission is to configure the scientific infrastructure needed by the EU for the next 10 to 20 years, and to identify vital, new European Research Infrastructures of different magnitudes and scope, as needed. *See Support for Policy-Making on Research Infrastructures in the European Research Area*, EUROPEAN STRATEGY FORUM ON RESEARCH INFRASTRUCTURES, http://ec.europa.eu/research/infrastructures/index_en.cfm?pg=esfri-background.

[417] E-mail from Gerard Verklije, 22 Dec. 2014. The MIRRI currently consists of sixteen partners, which include CABI (U.K.), CBS (The Netherlands), CRBIP (France), and DSMZ (Germany) among others. Another 27 institutions are collaborating parties. See *Consortium Partners*, MIRRI, http://www.mirri.org/consortium/partner.html (last accessed April 7, 2015).

[418] E-mail from Gerard Verklije, 22 Dec. 2014.

[419] *See, e.g.*, D. Smith (2012), n. 319, at 14 ("Linking global initiatives to that in Europe with MIRRI the foundation stones for the GBRCN.")

[420] *See* Fritze & Oumard (2012), n. 108, at 12 ("MIRRI is an integrative initiative that brings together a critical mass of loose networks, projects and initiatives to provide a solid structure that can act as a distributed but coordinated service provider").

process. To make this possible in a coordinated and comparable way, the partners will implement the OECD Best Practices for microbial BRCs.[421]

What seems most distinctive about the MIRRI project, is that it deliberately assumes the mantle of leading affiliated culture collections to adopt a standardized mode of implementing the Nagoya Protocol in a research-friendly manner. Fully cognizant of the risks to biological research likely to emerge if the needs of providers and users of genetic resources are not properly aligned with the goals of the CBD, as tightly implemented by the Nagoya Protocol,[422] MIRRI aims to "tackle coordinated approaches to ABS, in microbiology and [to] participate in respective regulatory discussions," with a view to "the harmonization of national legislation under an EU wide framework."[423] It then envisions cooperation to the same end with non-EU regions and partners, in order to elaborate standardized and harmonized regulations and procedures to support research.[424]

The MIRRI would thus build on work already done by the MOSIACC and ECCO projects, as described earlier in Chapter 4,[425] as well as the GBRCN Demonstration Project, and it would also integrate the tracking and cataloguing facilities of the WFCC's WDCM, as described in Chapter 8.[426] By the same token, the MIRRI would directly address some of the most pressing issues raised by this book in order to produce a distributed network of qualified culture collections,[427] operating within a legal framework that met the challenges that the Nagoya Protocol poses for the New Biology paradigm.[428]

What remains to be seen, however is whether the MIRRI, as the European footprint of the GBRCN, will fully embrace the public-good approach of the WFCC[429] or the proprietary, market-like model typified by ATCC.[430] Will the planned "network of networks" create a two-tiered regime, in which a few leading collections will monopolize microbial genetic resources in a way that disadvantages the bulk of existing culture collections and places a new tax on research in addition

[421] *Id.* at 12.
[422] *Id.* at 7–8. *See* Chapter 3, Sections IV.A & B.
[423] Fritze & Oumard (2012), n. 108, at 13.
[424] *Id.*
[425] *See* the MOSAICC Project, Microorganism Sustainable Use and Access Regulation International Code of Conduct, http://www.belspo.be/bccm/mosaicc; ECCO Agreed Core MTA for the Supply of Cultures, www.eccosite.org. *See also* European Consortium of Microbial Resource Centers (EMBARC) and the Common Access to Biological Resources and Information (CABI). *See generally* Chapter 4, Section III.A ("The Research Community Pushes Back").
[426] *See* Chapter 8, Section II.B.1.
[427] Fritze & Oumard (2012), n. 108, at 12–14.
[428] *See* Chapter 3, Section I.V ("New Constraints and Opportunities for Scientific Research Under the Nagoya Protocol").
[429] *See* Chapter 4, Section I, and Section II.B.1 in this chapter.
[430] *See* Chapter 4, Section II.A.

to unnecessary restrictions on access to inputs for basic research? Or will MIRRI and GBRCN reconcile the tensions emanating from the Nagoya Protocol in a way that actually facilitates global biotechnological research, with win-win outcomes for all the stakeholders? How MIRRI – and the founders of the GBRCN – ultimately answer these questions will, in our view, determine the extent to which they could justifiably lead the formation of a redesigned Microbial Research Commons along the lines envisioned in this book, or establish themselves as an obstacle – despite all good intentions – to those goals.

III. IN SEARCH OF A POLITICALLY ACCEPTABLE AND SCIENTIFICALLY PRODUCTIVE OPERATIONAL FRAMEWORK

Ideally, a redesigned Microbial Research Commons would operate on at least two levels in response to the challenges identified earlier in this chapter. From a top-down perspective, such a Commons should reach across national and regional boundaries to establish a multilateral regime that facilitates access to and use of *ex situ* microbial genetic resources, plus available data and literature situated within all participating countries. At the same time, its governance structure should be capable of interfacing with major bottom-up initiatives, especially the thematically organized Open Knowledge Environments like those described in Chapter 8.[431]

To achieve these goals, however, a redesigned Microbial Research Commons must be firmly rooted in a viable international legal and institutional framework under which it could thrive despite the proprietary controversies that currently threaten to undermine the existing microbial research infrastructure.[432] How to configure such a redesigned global infrastructure thus poses fundamental issues of international law and science policy, which are explored here.

A. *Evaluating the Existing Legal and Institutional Landscape*

Earlier in this chapter, we looked at the governance structure of the Crop Commons, which operates under the aegis of the United Nations Food and Agricultural Organization (FAO).[433] As one might expect, we found a relatively

[431] *See* Chapter 8, Section III ("Building Transnational Open Knowledge Environments").

[432] For clear perceptions of this threat, and the need to respond adequately to it, see Christine Godt, *Networks of* Ex Situ *Collections of Genetic Resources, in* COMMON POOLS OF GENETIC RESOURCES: EQUITY AND INNOVATION IN INTERNATIONAL BIODIVERSITY LAW (E.C. Kamau & G. Winter eds. 2013) [hereinafter COMMON POOLS OF GENETIC RESOURCES]; *see also* Fritze & Oumard (2012), n. 108, and D. Smith (2012), n. 319.

[433] *See* ITPGRFA, n. 134, discussed at Section II.A (Governance). *See also* the World Health Organization's Pandemic Influenza Preparedness Framework, discussed at Chapter 4, Section IV.A (nonbinding international agreement).

top-heavy administrative apparatus characteristic of organizations built around an international treaty, whether binding or not binding as the case may be. Nevertheless, we also saw in earlier chapters that this formal, intergovernmental legal architecture conferred a protective mantle over the Crop Commons, which helped it to ensure the participation of developing countries that possess major biodiversity assets.[434]

We then looked at a selection of governance models adopted by a diverse group of research commons initiatives.[435] Despite the absence of a treaty, these contractually constructed pooling arrangements have successfully carried out major transnational undertakings, even when supported only by nonbinding Memorandas of Understanding that provide for easy exit.[436]

All of the hybrid commons initiatives we reviewed espoused public-interest goals, with some open questions about GBRCN's perhaps passing interest in becoming a more commercially oriented network.[437] Disregarding that proprietary option, which we think ought to be avoided, it turned out on closer inspection that these hybrid pooling arrangements had actually adopted rather different governance structures and operational modalities.

For example, both the World Federation of Culture Collections (WFCC) and to some extent, the International Human Microbiome Consortium (IHMC) seemed to operate without top-heavy bureaucratic administrations, and both initiatives are managed largely by scientists. Yet, the IHMC depends directly on the participation of both government agencies and funders, unlike the WFCC.[438] In contrast, both the Global Biodiversity Information Facility (GBIF) and the Group on Earth Observations (GEO) depend heavily on the direct participation of governments in administrative affairs, and both of these entities have erected relatively formal, top-down governance establishments.[439] In the rest of this chapter, we compare and evaluate these similarities and differences, with a view to identifying their implications for a redesigned Microbial Research Commons.

1. Comparing Science-Managed NGOs with a Treaty-Based IGO

From an international legal perspective, the existing microbial culture collections under the umbrella of the WFCC face many of the same problems that confronted

[434] *See* Section II.A.2 ("Implementation of the Multilateral Regime). *See also* Chapter 4, Section IV.B ("Governance of the WHO's PIP Framework").

[435] *See* Section II.B passim.

[436] *See, e.g.*, Section II.B.2.b (discussing GBIF) and Section II.B.3.b (discussing GEO).

[437] *See*, Section II.C.1 & 2.

[438] *Compare* Section II.A ("The Global Crop Commons") *with* Section II.B.4.b ("Governance of IHMC"). *See also* Chapter 3, Section III ("An International Treaty to Rescue and Expand the Global Crop Commons").

[439] *See* Sections II.B.2.b (GBIF) and II.B.3.b (GEO).

the Consultative Group for International Agricultural Research (CGIAR) in the 1990s.[440] Like the plant genetic resources held by the CGIAR, the *ex situ* microbial collections hold essential genetic resources and related data that were traditionally made available as a public good. Both entities systematically engaged, to varying degrees, in providing a public infrastructure that supports basic and applied research. Moreover, strong and conflicting proprietary trends still threaten to disrupt the operations of the *ex situ* microbial collections, as previously occurred with the CGIAR.[441]

As a result, *ex situ* microbial collections – like the CGIAR before them – have experienced a crisis of legitimacy,[442] with corresponding fears that they would become subjected to a patchwork of national, regional and international legal regulations that would impede exchanges for research purposes and gradually impose a cumbersome system of case-by-case negotiations.[443] Challenges to the ownership of existing microbial resources at the international level could also drive the culture collections away from a public goods approach to a more proprietarial approach in their own defense.[444]

This crisis of legitimacy struck the culture collections precisely at a time when advanced microbiology required them to upgrade the quality of their operations, ideally to the level of Biological Resource Centers.[445] In responding to this crisis, a purely science-managed nongovernmental organization (NGO) built around the WFCC might conceivably be better situated to navigate the international legal crosscurrents than was the CGIAR in the 1990s, despite the latter's well-deserved reputation for advancing global agricultural research. The WFCC already benefits from a transnational legal architecture,[446] and its member culture collections

[440] *See* Chapter 3, Section II.B ("The CGIAR's Agricultural Research Infrastructure on the Verge of Collapse").

[441] *See, e.g.*, Graham Dutfield, Intellectual Property Rights and the Life Science Industries (2d ed. 2009). *See generally* Chapters 3 and 4.

[442] For details, *see* Chapter 4 ("The Existing Microbial Research Commons Confronts Proprietary Obstacles").

[443] *See, e.g.*, Dagmar Fritze, *A Common Basis for Facilitated Legitimate Exchange of Biological Materials, Proposed by the European Culture Collections' Organization (ECCO)*, 4 *Int'l J. Commons*, 507 (2010); Fritze & Oumard (2012), n. 108; *cf.* Michael Halewood, *Governing the Management and Use of Pooled Microbial Genetic Resources: Lessons from the Global Crop Commons*, 4 *Int'l J. Commons* 404–36 (2010) [hereinafter Halewood (2010)].

[444] *Cf.* Fritze & Oumard (2012), n. 108 (arguing for the view that microbial culture collections are custodians, and not "owners," of microbial materials). The ATCC went down the proprietary path in the early 1970s, in response to its existential threat. *See* Chapter 4, Section II.A.

[445] *See* Chapter 4, Section I.B & C; *see also* OECD, OECD Best Practice Guidelines for Biological Resource Centers (2007), available at http://www.oecd.org/sti/biotech/38777417.pdf [hereinafter OECD Guidelines].

[446] *See* Chapter 4, Section I, passim; see also Section II.B.1, in this chapter.

usually enjoy the support of local governments. For these and other reasons, the WFCC should logically become a key institutional player in a redesigned Microbial Research Commons under any international legal regime that the participants eventually decide to adopt.

On closer analysis, however, a purely science-managed infrastructure governed by the WFCC without direct buy-ins from governments in both developed and developing countries would likely leave the culture collections in a vulnerable political position. They would be buffeted by international pressures flowing from both the Convention on Biological Diversity (CBD) and the Agreement on Trade Related Aspects of Intellectual Property Rights (TRIPS Agreement), as was the CGIAR before being absorbed into the regulatory framework of the FAO's International Treaty.[447] For example, any negotiations to establish standard licensing terms for the microbial culture collections – with inputs from the relevant scientific communities – would inevitably need the approval of governments, in which the operational role and influence of scientists might be considerably diminished. Meanwhile, external economic, legal, or institutional developments beyond the reach of the scientific community could render some of its decisions irrelevant in practice.

A purely science-driven nongovernmental administrative entity, such as the WFCC or the CGIAR (before the international treaty), inherently lacks the kind of political heft needed to deal with issues such as biosafety, quality, and the tensions between developed and developing countries concerning exchanges of genetic resources for both research and benefit-sharing purposes. Such an NGO could not authoritatively stabilize the legality of the practices adopted by existing culture collections, nor could it ensure future access to essential microbial materials under the Nagoya Protocol to the CBD,[448] while its science policy recommendations might simply fall on deaf ears. Indeed, the position of similar entities is often that of a supplicant, rather than as a partner in the decision-making process. The national science ministries, while logical partners in principle, are unlikely to fund

[447] *See* CBD, n. 4 and TRIPS Agreement, n. 3. *See generally* Chapter 3 ("Tightening the Regulatory Grip: From the Convention on Biological Diversity in 1992 to the Nagoya Protocol in 2010").

[448] *See, e.g.*, Tenth Meeting of the Conference of the Parties to the Convention on Biological Diversity, Nagoya, Japan, 18–29 Oct. 2010, Nagoya Protocol on Access to Genetic Resources and the Fair and Equitable Sharing of Benefits Arising from their Utilization (ABS) to the Convention on Biological Diversity [hereinafter Nagoya Protocol] (entered into force on Oct.12, 2013) *available at* http://www.cbd.int/abs/doc/protocol/nagoya-protocol-en.pdf (last accessed 14 June 2014); *see also* THOMAS GREIBER ET AL., AN EXPLANATORY GUIDE TO THE NAGOYA PROTOCOL ON ACCESS AND BENEFIT SHARING 217, Box 29 (Int'l Union for Conservation of Nature & Natural Res. (IUCN), Envtl. Pol'y & L. Paper No. 13, 2012) [hereinafter IUCN, GUIDE TO THE NAGOYA PROTOCOL (2012)]. See further, Chapter 4, Sections III.B. & IV ("From the Bilateral to the Multilateral Approach").

any initiatives – or support related regulatory or legislative measures – in whose elaboration they themselves had not actively participated as stakeholders.

Given these premises, proposals to establish a redesigned Microbial Research Commons within the governance architecture of a formal international treaty may, at first glance, seem particularly tempting.[449] As previously explained, the CGIAR's genetic materials were rescued by an international treaty, negotiated under the auspices of the UN's FAO.[450] Happily, this multilateral treaty preserved the public domain status of *ex situ* plant genetic resources for food and agriculture held by the CGIAR; it established measures enabling both researchers and plant breeders to access these resources in a relatively standardized manner; and its "multilateral option" was expressly validated by Article 4 of the Nagoya Protocol.[451] The ITPGRFA also subjected these same resources to a rudimentary liability rule that could eventually satisfy the ABS requirements of depositors from developing countries.[452] Some important capacity building initiatives were also undertaken in the developing countries, which, under the Nagoya Protocol, could help to fulfill the technology transfer obligations of the CBD.[453]

While criticism of the FAO's International Treaty has recently grown, as discussed in Chapter 3,[454] the CGIAR's experience at least serves to demonstrate that even the most scientifically productive research centers that depend on access to global genetic materials cannot exist in an international legal vacuum. Rather, they need protection and stability under a viable governance structure that guarantees compliance with international treaties and that anchors scientific research within the resulting legal and institutional framework. It also shows that even a successful international research consortium can succumb to serious financial problems unless its funding is built into the governance structure itself.[455]

The ITPGRFA, rooted in a full-fledged international organization within the framework of the FAO, gave the Crop Commons strong political clout and provided a secure legal foundation for its extensive public domain holdings. Such an IGO obtains all the immunities of an international legal personality, and because one may assume that it enjoys the trust of the participating governments and affiliated

449 For earlier proposals to this effect, *see, e.g.*, Halewood (2010), n. 443. *But see* Halewood (Louvain 2012), n. 53, for later misgivings and concerns.

450 *See* ITPGRFA, n. 448; Chapter 3, Section III.

451 See Nagoya Protocol, n. 3., art. 4: Chapter 3, Sections III & IV.

452 *See* Chapter 3, Section III.B.2 ("Notification, Benefit Sharing, and the Standard Material Transfer Agreement"), see also Section II.A.2 in this chapter ("Implementation of the Multilateral Regime").

453 See Nagoya Protocol, n. 448, art. 2.3

454 *See, e.g.*, Halewood (Louvain 2012), above n. 53; *see generally* Crop Genetic Resources, n. 131. *See further* Chapter 3, Section III.C.2.

455 CGIAR, *Structures and Governance* (discussing CGIAR's organizational structure), *available at* http://www.cgiar.org/who-we-are/ (last accessed Apr. 23, 2015). See also Section II.A.2.c in this chapter ("Long-Term Funding Arrangements" under the Crop Commons).

organizations, it can coordinate responses to international pressures while respecting national sovereignty.[456] A treaty-based entity is thus well-positioned to mediate the conflicts of interest arising under the CBD. Also advantageous are the relative permanence of an international organization once it has been established, and the relative security of its funding, although the ITPGRFA itself did not adequately address funding needs, which remain problematic.[457]

Conversely, the recent CGIAR experience also reveals the drawbacks of an overly formalistic, top-heavy governance structure, which cautions against its unqualified application to microbiological research and resources. First, establishing such an organization is usually a cumbersome and lengthy process (e.g., seven years in the case of FAO's International Treaty),[458] with no guarantee of success. Policy making then depends on a complex and rigid administrative process that may require the unanimous – or virtually unanimous – consent of participating governments for every major decision. Moreover, governments may be less willing to enter a formal treaty organization than to commit to less formal contractual arrangements that can be terminated easily if the payoffs no longer meet their needs or expectations.[459] Whether so-called nonbinding treaties are less daunting or cumbersome in fact remains to be demonstrated.

Finally, the very political nature of a treaty-based organization makes it less amenable to the kind of science-driven initiative we envision for the Microbial Research Commons. For example, complaints surfacing in the follow-up to the FAO's International Treaty focus heavily on the extent to which its deliberations have increasingly sidelined the relevant scientific community.[460] In general, top-heavy political governance schemes tend to alienate and exclude scientists from the process of formulating policies that most affect them as users of the regimes in question.

In this respect, both of the options examined so far – a purely science-managed NGO and a treaty-driven IGO – seem ironically to harbor the risk of a common defect. Both approaches may end up by casting scientists in the role of supplicants seeking government approval and assistance, rather than as autonomous stakeholders authorized to facilitate exchanges of microbial genetic resources on a

[456] Interview with Prof. Laurence Helfer, Co-Director, Duke Law School Center for International and Comparative Law, in Durham, N.C., Sept. 15, 2015. For empirical evidence supporting this premise, *see* Section II.A.2 *passim*.

[457] The CGIAR remains an international multi-donor project with some continued financial problems. *See* Chapter 3, Section III.C.2; see also Section II.A.2 in this chapter.

[458] Halewood (2010), n. 443.

[459] For recent problems with fully implementing the FAO's Crop Commons, and especially with enforcing government commitments to that project, see Section II.A.2 of this chapter.

[460] For these and other complaints, see Halewood (Louvain 2012), n. 53, discussed in Chapter 3, Section III.C.2.

global scale within legal boundaries that both developed and developing countries have endorsed. With that drawback uppermost in our minds, we move on to evaluate another frequently used legal option, namely, that of a hybrid international framework agreement voluntarily undertaken by interested governments.

2. Advantages of a Hybrid International Framework Agreement

Rather than entering into a formal treaty or relying on a purely scientific entity, willing governments can contractually embody a common undertaking at the international level in a framework agreement or memorandum of understanding without necessarily resolving differences in their national laws that might otherwise become problematic. As we have seen, such an arrangement can be made nonbinding,[461] revocable, or otherwise easy to exit, and it can provide many of the advantages of other models while making it possible to avoid or attenuate their disadvantages. Because such an agreement requires the consent and participation of governments, it can also mediate tensions between developed and developing countries, while enabling science ministries and the research community to interact directly.

Our empirical review of selected governance models set out earlier in this chapter shows that OECD governments have themselves been seeking and experimenting with hybrid arrangements of this kind. For example, the GBIF, the GEO, the IHMC and, most recently, the GBRCN Demonstration Project have all adopted this legal device,[462] with widely varying governance characteristics. A third-year review of the GBIF explicitly confirmed that a framework agreement – which the review called a nonbinding Memorandum of Understanding – was a more appropriate legal mechanism than a treaty. Most recently, even the WHO – although itself an IGO – also adopted such an agreement for its Pandemic Influenza Preparedness (PIP) Framework.[463]

[461] Nonbinding arrangements of this kind impose no formal legal consequences if a government decides not to fulfill its commitments. However, once the agreed obligations have been fulfilled, legal consequences may attach to the entity thus established, even though a participating government remains free to withdraw. Interview with Jean Francois Mayence, Head of the Legal Unit, International Relations, Belgian Science Policy Department, Government of Belgium, in Brussels, Belgium (Dec. 6, 2010).

[462] *See* Section II.B.2–4 & II.C.1 in this chapter. For GBRCN, see also D. Smith (2012), n. 319.

[463] No support for elevating the governance arrangement of GBIF to a treaty appeared in Review Committee, *Global Biodiversity – The Global Biodiversity Information Facility Third-Year Review*, 230 KPMG Consulting Co. and the Committee on Data for Science and Technology (CODATA), Feb. 2005 [hereinafter *GBIF Third-Year Review*], *available at* http://www.codata.info/archives/2005/GBIF2005Rept.pdf. (last accessed May 10, 2015). *See also* World Health Org. (WHO), Pandemic Influenza Preparedness Framework for the Sharing of the Influenza Viruses and Access to Vaccines and Other Benefits, World Health Assembly Res. WHA645 (May 24,

A framework agreement can be interest-specific and deliberately attuned to the needs of a particular scientific constituency. This criterion applies to the GBIF, which has established itself as a global portal for open access to biodiversity data. In this capacity, GBIF anticipates one type of digital service provider that we envision for the Microbial Research Commons. It also benefits from close consultations with a scientific advisory board, rather than depending on a bureaucratic arrangement.[464]

However, GBIF's narrow mission focus has proven to be both a strength and a weakness. It provides a technical research infrastructure, rather like the service components of the Open Knowledge Environments discussed in Chapter 8,[465] but it does not support research as part of its core mission, and it does not influence the formation of policy even in the field of biodiversity. As a result, GBIF has so far manifested only a limited policy clout, even with respect to data made available through its own portal. Funding also depends on mandatory contributions by national governments and on delicate, ongoing cost-benefit analyses by those same governments.

For these and other reasons, some governments have reportedly become wary of replicating the GBIF model in part because of a reluctance to divert national research funds to such an intergovernmental organization, and also because of rules requiring consensus on governance issues without a weighted voting provision that would favor the largest dues payers. There has also been some reconsideration of the overall value of the GBIF's output to user communities, at least from the funders' perspective.[466]

A more promising recent example of a narrowly focused research infrastructure that made use of an international framework agreement was the IHMC. This entity relied on specified research projects whose data outputs it aimed to coordinate and magnify. To this end, the members pooled rapidly released sequence data in a research commons designed to serve as basic infrastructure for the field of human microbiome research as a whole.[467]

At the same time, the IHMC manifested a potentially more effective operational configuration than purely science-managed, nongovernmental bodies, such as the WFCC. The IHMC's Steering Committee combines the formation of science policy with inputs from funding agencies that have the approval of national regulatory authorities. The Consortium was thus able to formulate and strictly implement novel procedures governing both the rights of researchers to publish articles based

2011) [hereinafter PIP FRAMEWORK], *available at* http://www.who.int/influenza/resources/pip_framework/en/index.html (last accessed 23 Feb. 2014). *See further* Chapter 4, Section IV.B.

464 *See GBIF Third-Year Review*, n. 463.

465 *See* Chapter 8, Section III ("Building Transnational Open Knowledge Environments").

466 *See GBIF Third-Year Review*, n. 463.

467 *See* Section II.B.4 in this chapter.

on the data they contributed and the duties of these same researchers to make their data available to the community as a whole.[468]

The relative simplicity of its governance structure also enables the IHMC to cut through layers of administrative complexity that characterize most of the other entities whose governance models we have examined. To take but one example, government agencies adhering to the IHMC do not need to pledge large sums of money in advance to support a costly entity the value of whose end products remained uncertain. Funding is instead based on national priorities and deliverables, and the Steering Committee potentiates the value of these deliverables by overseeing the knowledge commons that all the participating entities agreed to establish.[469] The IHMC thus obviated "buyers' remorse" from the outset, and each successfully implemented policy or project seems likely to generate further support for empirically sound, science-driven initiatives.

The strength of the IHMC's approach lies in the efficiency with which it can wholly devote its funds and expertise to the formation of an upstream research infrastructure capable of supporting a whole new field of scientific enquiry and resulting applications. Other strengths lie in the relatively small size of the operation as a whole; in the fact that it would actually sponsor new research projects of its own (unlike the redesigned Microbial Research Commons we envision); and in the fact that it did not, initially at least, project continued operations beyond a specified period of time.[470]

One potential weakness of this streamlined, science-driven governance scheme, however, is that the organizers may give too little thought to the legal and political ramifications likely to arise both from a potentially hostile international legal environment and from successful downstream commercial applications, if and when they actually begin to emerge. For example, once medically relevant genetic sequences are successfully isolated, translational medicine may require recourse to, and further experiments on, the underlying microbial materials from the human gut. However, the availability of these materials for such purposes was not clearly addressed by the publicly available governance norms of the IHMC.[471] The answer might depend on the policies, rules, and MTAs in force at whatever repositories

[468] *See id.*, Section II.B.4.d ("Data and Intellectual Property Policies").

[469] *See* Section II.B.4.b & c in this chapter. *Compare and contrast* the WHO's PIP Framework, Chapter 4, Section IV.B, which, in theory, is largely to be funded by industrial vaccine makers in return for inputs from a knowledge commons that WHO manages, but who have reportedly dragged their feet in meeting these financial obligations.

[470] *See* Section II.B.4 in this chapter.

[471] *See* Section II.B.4.d (discussing the NIH's reliance on materials deposited at BEI within the ATCC). Note that the NIH has now established the National Center for Advancing Translational Sciences (NCATS), https://ncats.nih.gov and disestablished the National Center for Research Resources.

or institutes that happened to hold the relevant materials, not to mention possible restrictions that relevant single donors may also have put on, the items in question, for example, strains, tissues, or other genetic materials. In regard to these and other related concerns, a contractually constructed Microbial Research Commons, with a Standard MTA for exchanges of genetic resources, might have complemented and potentiated the goals of the IHMC, had it been in existence, just as the IHMC would itself complement and further enable the data component of that Commons.[472]

It is useful to compare the relatively light governance approach of a narrowly focused initiative, such as the IHMC, with that of much larger and more ambitious undertakings, such as the GEO.[473] The latter attempts to coordinate voluminous, broad-ranging sets of government-managed geospatial data in order to bring the power of its integrated data resource to bear on nine societal benefit areas over a long period of time. Yet, unlike GBIF, GEO's governance structure remains relatively light, with no mandatory financial contributions to the consortium as such.[474] Both GBIF and GEO benefit from the direct participation of top-level executives in their decision-making processes, and these processes rely heavily on advice from scientists and technical experts. Given a sufficiently enabling political environment, the GEO's leadership could manage its jointly coordinated data resources so as to produce major social payoffs, which is not a comparable goal of the GBIF.

If the GEO's governance model has a structural weakness, it may be the lack of funds to undertake targeted initiatives at its own discretion, unlike the GBIF, or even the IHMC, whose members directly fund their own research projects. Instead, the GEO can only rely on activities undertaken by its participating entities, which it may coordinate more effectively than might otherwise be the case if each operated entirely on its own.

Still another problem with GEO is that many governments have proved reluctant to fully implement their commitments to data sharing in practice.[475] A similar reluctance to make previously committed genetic resources publicly available under the FAO's Crop Commons has also become manifest over time.[476] This lack of follow through thus seems to transcend the specific transnational legal structure adopted by any of the knowledge commons we have examined. It bears, instead, on

472 The IHMC was initially to be a short-term project, in that IHMS, MetaHIT, and the NIH's HMC (IHMC's primary participants) had four-year funding limitations. In fact, the IHMC has continued its operations with renewed funding and changing participants.

473 *See* Section II.B.3.b in this chapter.

474 *Compare* Section II.B.3.b in this chapter (the GEO's governance model) *with* Section II.B.2.b (the GBIF's governance model).

475 Statement of Paul F. Uhlir, Co-Chair, GEO Data Sharing Working Group (May 30, 2014).

476 Frison (Louvain 2012), n. 131; Halewood (2010), n. 443.

enforcement[477] as well as on the incentive structure and value of the core mission as perceived by participating governments throughout the lifetime of the organizations at issue.

The strengths and weaknesses of these different governance models are directly related to the task of redesigning the Microbial Research Commons. How to apply the lessons to be learned from this survey depends, moreover, on the goals that those responsible for establishing this component of the global scientific infrastructure ultimately adopt.[478] We return to this topic in Chapter 10, after first identifying other core premises that must be taken into account if the resulting transnational research commons is to remain legally sustainable within the purview of the Nagoya Protocol to the CBD.

B. *Reconciling National Sovereignty over Microbial Genetic Resources with a Global Public Goods Approach*

The empirical evidence reviewed in the last section suggests that, if the proposed Microbial Research Commons were rooted in an appropriately comprehensive international framework agreement that instrument could – and should – avoid the rigidity of an international treaty, like the ITPGRFA that regulates the Crop Commons. At the same time, the flexibility inherent in such an arrangement should not obscure the need to establish an internal governance apparatus capable of both high-level political engagement and research-promoting operations, backed up by direct inputs from the relevant scientific and technical communities.[479]

In this context, the GBRCN's Demonstration Project arose in part from top-down pressures spawned by the OECD's emphasis on upgrading existing culture collections to BRCs and in part from bottom-up demands by the scientific community for an improved microbial research entity dating back to the 2000s.[480] Promoters of the GBRCN, in turn, laid the foundation for an inter-ministerial framework agreement that could marshal the necessary political and financial resources without the encumbrances of an international treaty organization.[481] The GBRCN model could,

[477] *See, e.g.*, Melanie Dulong de Rosnay & Hervè de Crosner, "An Introduction to the Digital Commons: From Common Pool Resources to Community Governance," paper presented at the Conference on Governing Pooled Knowledge Resources: Building Institutions for Sustainable Scientific, Cultural and Genetic Resources Commons for the International Association for the Study of the Commons (IASC), Louvain-la-Neuve, Belgium, Sept. 12–14, 2012.

[478] *See generally* FRISCHMANN, INFRASTRUCTURE, n. 11 (distinguishing infrastructure from commons governance); see further Chapter 9, Section A ("Applying Commons Theory to the Microbial Research Infrastructure).

[479] *But see* the WHO's PIP Framework, Chapter 4, Section IV.B, for a novel example of a nonbinding international treaty with some of these features.

[480] *See* Chapter 4, Section I.B ("From Culture Collections to Biological Resource Centers").

[481] *See* GBRCN, Draft MoU (2010), n. 325; GBRCN, MoU (2012), n. 388.

in principle, link the national science ministries (including equivalent funding agencies in the United States) directly in a serious effort to stabilize and upgrade the worldwide network of microbial culture collections. Besides facilitating exchanges of genetic resources among cooperating BRCs, it could play a positive role in attenuating the tensions between developed and developing countries with regard to research uses of genetic resources, which became one of the GBRCN's primary objectives in its final iteration.[482]

1. Avoiding the Wrong Incentives

As described earlier in this chapter, the GBRCN's governance structure – at least as outlined in its Demonstration Project – looked top heavy, and relatively bureaucratic in nature, with no formal provision for the participation of the WFCC's representatives or those of other scientific bodies, such as the International Union of Microbiological Societies (IUMS), on the Governing Board. Apart from a brief mention of a possible Scientific Advisory Board, the GBRCN business plan also reflected a commercial appetite for exploiting upstream genetic resources under market-like conditions. For this and other reasons outlined earlier, the GBRCN's internal governance model lacked the kind of dynamic, research-coordinating operational arm, like that which made the IHMC's approach appear so impressive, although the GBRCN Final Report did put more emphasis on research support.[483]

Above all, GBRCN's business plan – at least as initially outlined in the Demonstration Project – suggested that some of its founder collections were willing to abandon the public-good model for a market-like approach. If so, GBRCN's goals would have differed significantly from those of the FAO's International Treaty, which anchored the CGIAR's *ex situ* plant genetic resources in the public domain.[484] They would also differ from the goals of the other resource-pooling initiatives reviewed earlier in this chapter, which seek to establish scientific infrastructures operating on a public good basis. Instead, the GBRCN's business plan, as initially attached to its draft Memorandum of Understanding, would ostensibly have subjected its resources to proprietary controls whose avowed aim was to secure maximum revenues from their commercial exploitation.[485]

[482] *See, e.g.*, Fritze & Oumard (2012), n. 108; D. Smith (2012), n. 319.

[483] *See generally*, Section II.C.1 in this chapter; see also GBRCN Final Report (2012), n. 315.

[484] *Compare* GBRCN's business model, described in Section II.C.1.c *with* the Crop Commons as described in Chapter 3, Section III.A & B.

[485] *See* Section II.C.2 in this chapter. However, the passages supporting these inferences were apparently deleted from the Final Report, n. 315, and from the GBRCN Demonstration Project's website, http://www.gbrcn.org.

For example, while the Demonstration Project recognized that genomic, proteomic, and environmental data had become key components of modern microbiology, it stressed the value of these components as a potential source of income. Otherwise, the Demonstration Project concentrated primarily on materials and some related data directly generated by culture collections.

In this connection, the GBRCN's data policies merit careful scrutiny. Despite one explicit mention of the need to facilitate data sharing among the collections participating in the Demonstration Project (then some fifteen in number),[486] no data access and release policy appeared in its draft Memorandum of Understanding. In contrast, the IHMC, GBIF, and GEO projects all encourage data sharing by one means or another. Perhaps the lack of any similar policy in the GBRCN's framework agreement becomes more understandable in light of its draft business plan. That document expressly proposed to exploit both data and related research tools as cost recovery mechanisms, and the draft Memorandum of Understanding seemed to rely on relevant national intellectual property laws for that purpose.[487]

Given that some 55 countries, including all those countries in or affiliated with the European Union, have enacted strong database protection laws,[488] a plausible inference is that the GBRCN intended to make use of these laws to bolster the commercial opportunities of their cooperating BRCs. Yet, as we showed in Chapter 6, these laws obstruct both traditional research and computational science, and they undermine the drive for digital integration of scientific data in the networked environment as upstream resources for public research purposes.[489] That, indeed, is why the IHMC, GBIF, and GEO projects all make the pooling and sharing of data for research purposes their major objective.

In its defense, the GBRCN's sponsors might argue that its business plan was essential to maintaining the financial stability of cooperating BRCs. The participating governments were obliged to commit funds only to support the Governing Board, a Secretariat, and their activities, but would not have otherwise directly funded the proposed "network of networks."[490] The GBRCN could have further argued that its approach would benefit the developing countries by providing a clearinghouse or quasi-brokerage for exploiting their benefit-sharing

[486] GBRCN, 3RD INTERMEDIATE REPORT, n. 325, at 7.

[487] *Id.*, Annex 13 (as initially drafted); *see* Section II.C.2, in this chapter.

[488] *See* Chapter 6, Section II.C (Exclusive Rights in Noncopyrightable Collections of Data").

[489] *See* Chapter 6, Sections II.C. and III.A.4; *see also* Jerome H. Reichman & Ruth L. Okediji, *When Copyright and Science Collide: Empowering Digitally Integrated Research Methods on a Global Scale*, 96 *Minn. L. Rev.* 1362 (2012) [hereinafter Reichman & Okediji (2012)].

[490] *Cf.* D. Smith (2012), n. 319.

rights in deposits originating from their territories under the provisions of the CBD and its Nagoya Protocol.[491]

Both of these arguments, however, whether or not attributable to the GBRCN's plan, would be shortsighted. Although GBRCN's genetic materials and data would still remain formally available to researchers under that business plan, a market-like approach would complicate the process of acquisition, elevate the costs of research, and strengthen the hoarding mentality of microbial researchers generally.[492] In effect, GBRCN's business plan – as originally formulated – would have treated all microbial deposits and related data as if they possessed known or likely commercial value. Pressures would consequently build to impose further restrictions on access, reuse, and redistribution of resources even among qualifying collections, in the manner of the American Type Culture Collection (ATCC),[493] and transaction costs would mount accordingly, even if a Standard MTA were devised.[494] The existing system of informal exchanges between academic institutions that enables the free exchange of upstream research resources would logically have to be shut down, in order to support the mercenary aims of the cooperating BRCs,[495] without necessarily erecting a global infrastructure that replaced these customary practices with a global public-goods approach.[496]

In reality, as we stressed in Chapter 5, the bulk of the genetic resources thus prospectively subjected to access and use restrictions on research would have no known or likely commercial value at all. Rather, their primary value is to serve as upstream inputs into publicly funded research from which downstream applications and innovation may hopefully follow.[497] The proprietary approach to sustainability

491 *See* Chapter 3, Section IV ("New Constraints and Opportunities for Scientific Research Under the Nagoya Protocol").

492 *Cf.* Chapter 4, Section II ("Contractual Restrictions on Access to and Use of Upstream Microbial Genetic Resources in Both Developed and Developing Countries").

493 *See* Chapter 4, Section II.A ("The Advent of a Proprietary Model in Response to Government Neglect in the United States").

494 *Cf.* Fritze & Oumard (2012), n. 108 (stressing the need for a standard MTA to meet the needs of the Nagoya Protocol). If, for example, ATCC's MTA had become the GBRCN's global model, it would simply have standardized research – complicating terms and conditions. *See* generally Chapter 4, Section II.A.

495 *See* GBRCN, 3RD INTERMEDIATE REPORT n. 325 (where GBRCN expresses this goal).

496 Keith E. Maskus & Jerome H. Reichman, *The Globalization of Private Knowledge Goods and the Privatization of Global Public Goods*, *in* INTERNATIONAL PUBLIC GOODS AND TRANSFER OF TECHNOLOGY UNDER A GLOBALIZED INTELLECTUAL PROPERTY REGIME (K.E. Maskus & J.H. Reichman eds. 2005) [hereinafter INTERNATIONAL PUBLIC GOODS]; Joseph E. Stiglitz, *Knowledge as a Global Public Good*, *in* GLOBAL PUBLIC GOODS: INTERNATIONAL COOPERATION IN THE 21ST CENTURY 308–26 (Inge Kaul et al. eds. 1999).

497 *See generally* Chapter 5, Sections I. & II; see also Chapter 4, Section IV.C.2 ("Opting into a Multilateral Approach in Order to Stimulate More Downstream Benefits from the Bilateral Approach").

initially envisioned in the GBRCN project could discourage the very upstream research that government funders profess to want and could boomerang against both developed and developing countries by diminishing the output of innovation that is the universally desired goal.[498] The GBRCN's proposed Business Plan, as elaborated in the initial Demonstration Project,[499] could thus place leading culture collections – and the microbiological research community itself – at a fork in the road, precisely when international stability can now be achieved by establishing a Microbial Research Commons as a privileged multilateral entity under the Nagoya Protocol to the CBD.[500]

As initially conceived, the GBRCN's mandate extended to a relatively small number of highly qualified BRCs scattered around the world.[501] More recently, however, leading European culture collections have focused attention on a subsequent project, known as the Microbial Research Resource Infrastructure (MIRRI), which grew out of the GBRCN Demonstration Project.[502] As noted earlier in this chapter, the MIRRI aspires to be inclusive rather than exclusive, by gradually upgrading all participating collections. Much could therefore depend on the governance model ultimately adopted and implemented by MIRRI, most of whose European constituents have strongly favored the public-good model in the past.[503]

If, instead, the GBRCN's draft business model as initially elaborated during the Demonstration Stage were implemented with high-level OECD government support, many other microbial culture collections might come under internal and external pressures to emulate the market-like approach, just as some of its elite collections were tempted to imitate the ATCC. These pressures could persuade many culture collections to abandon their open-access norms and practices, especially if GBRCN's draft business model were to win widespread government support and if more research projects became subject to the terms and conditions of a Standard MTA that implemented this approach.[504]

498 *Cf.* Paul A. David, "A Tragedy of the Public Knowledge 'Commons'? Global Science, Intellectual Property and the Digital Technology Boomerang," SIEPR Discussion Paper No. 00–02, Stanford Inst. Econ. Pol'y Research (2000), *available at* http://siepr.stanford.edu/papers/pdf/00-02.html.

499 The initial business plan is no longer available from the GBRCN's website. For the revised plan that remains available, *see* GBRCN Demonstration Project Business Plan, *available at* http://www.cria.org.br/eventos/CRB/presentations/GBRCN.pdf (last accessed 28 Apr. 2015).

500 *See* Chapter 3, Section IV.B ("Facilitating Scientific Research").

501 *See* Section II.C.1 in this chapter. However, interest in the GBRCN project is said to be growing, at least according to one of its principal organizers. *See* D. Smith (2012), n. 319.

502 For MIRRI, *see* n. 417; *see also* Fritze & Oumard, n. 108.

503 *See* Section II.C.3 ("MIRRI as a European Stepping Stone to the GBRCN"); Fritze & Oumard (2012), n. 108 (who, however, do not discuss the business model to be adopted).

504 For the view that the GBRCN should aspire to elaborate a Standard MTA, see, e.g., D. Smith (2012), n. 319.

The microbiological research community must accordingly decide whether to rescue the culture collections and their genetic resources from these privatizing pressures, as the FAO attempted to do with the publicly available contents of the CGIAR's seed banks.[505] Alternatively, the public microbial culture collections could become *soi disant* "nonprofit organizations," like the ATCC, which manages its assets in a self-promoting fashion that seems inconsistent with the goals of the New Biology paradigm. If the GBRCN or its progeny, MIRRI, decide to follow the first path, it could become the foundation for a true Microbial Research Commons, as envisioned in the next chapter. That might conceivably inspire the National Institutes of Health (NIH) to push even ATCC in a direction that could be more favorable to today's global research needs.

If, however, the GBRCN's previously elaborated business model prevails, it could become an obstacle to the formation of such a commons and a problem for microbiology. That model echoes the privatizing trends that have recently expanded exclusive intellectual property rights into the upstream, basic scientific research space.[506] The incentives that private property rights provide make sense when the object is to translate scientific discoveries into the kind of downstream knowledge goods that are best distributed by means of market forces. Inputs into basic scientific research, instead, are best treated as public goods.[507] To the fullest extent possible, such inputs should be openly available to the epistemic communities that respond to a different set of incentives, rooted in the reciprocity gains of sharing and in the reputational benefits accruing from publicly disclosed research results.[508]

The proper approach, as understood by the drafters of all the other governance models surveyed earlier in this chapter, is to reduce or eliminate intellectual property restrictions on pooled upstream research inputs (whether materials or data) having no known or likely commercial value, precisely in order to encourage both public and private researchers to invest time and funds in finding new commercial applications for them. This approach does not mean that the public microbial culture collections can operate without sufficient funds, as we emphasize in the final chapter. We also recognize that government neglect can force national

[505] *See* Chapter 3, Section III ("An International Treaty to Rescue and Expand the Global Crop Commons").

[506] *See* Chapter 3, Section II ("Destabilizing the Exchange of Plant and Microbial Genetic Resources as Global Public Goods"); Chapter 4, Section II ("Contractual Restrictions on Access to and Use of Upstream Microbial Genetic Resources in Both Developed and Developing Countries"); Chapter 6, Section II (Copyright and Database Protection Laws "as Digital Gridlock").

[507] *Cf.* Stiglitz, above n. 496; Maskus & Reichman, n. 496; *see also* Chapter 8, Section II.C ("Understanding the Data Sharing Movement and Its Future Potential").

[508] *See, e.g.*, Paul A. David, n. 56; Allarakhia et al., n. 83. Of course, if governments decline to fund the public good model – as occurred with ATCC in the United States – the GBRCN's initially proposed market-like business model could prevail by default.

collections to adopt a proprietary model in order to survive, as occurred in the case of the ATCC.[509] That is hardly the preferred model, however, for a global knowledge commons to support public scientific research.

2. Facilitated Access to Upstream Research Assets and Benefit-Sharing Under a Multilateral System

By the same token, developing country governments lured into a proprietary approach to *ex situ* genetic resources by expectations of short-term financial returns under the bilateral approach of the CBD could actually frustrate these prospects in the medium or long term. As mounting empirical evidence shows, hoarding genetic and other knowledge resources from public researchers – whether domestic or foreign – diminishes research opportunities, retards innovation, and tends to lower actual returns over time.[510] A more beneficial – and in our view more profitable – solution for the developing countries, with respect to the bulk of their genetic resources that have no known or likely commercial value is to make them readily available to public researchers, in exchange for reasonable royalties on any future downstream applications,[511] as was understood in principle by the drafters of the FAO's International Treaty.[512] Other important benefits for developing countries that participate in a research commons should result from the technical assistance of BRCs in OECD countries, with a view to helping the former upgrade and transform their own research capabilities so as to promote more domestic innovation. While the GBRCN Demonstration Project promised to support such capacity building initiatives, it could have risked making research harder and more costly for the entire scientific community if it had persisted in a strategy of encumbering its essential knowledge assets with proprietary restrictions and conditions.

As we saw in Chapter 3, the Nagoya Protocol itself addressed this imbalance while opening the door to an approach more friendly to scientific research.[513] In

509 For details, see Chapter 3, Section IV.A; see also Chapter 4 Section 4.C. ("Lessons for a Redesigned Microbial Research Commons").

510 *See* Chapter 3, Section IV.B.1 ("Recognizing the Link Between Public Science and Commercial Benefits"). For effects on scientific data access and use in research, see Paul F. Uhlir & Peter Schröder, *Open Data for Global Science*, 6 *Data Sci. J.* 36–53 (2007); Jerome H. Reichman & Paul F. Uhlir, *Database Protection at the Crossroads: Recent Developments and Their Impact on Science and Technology*, 14 *Berkeley Tech. L.J.* 793 (1999) [hereinafter Reichman & Uhlir (1999)]; NAT'L RESEARCH COUNCIL, A QUESTION OF BALANCE (Nat'l Acads. Press 1999).

511 *See* Chapter 5, Section II ("Designing a Third Option: Ex Ante 'Take and Pay' Rules for Stimulating Research and Applications").

512 *See* Chapter 3, Section III ("An International Treaty to Rescue and Expand the Global Crop Commons").

513 *See* Nagoya Protocol, n. 448; Chapter 3, Section IV.B.

Article 4, the Nagoya Protocol explicitly recognized the legality of the multilateral approach to plant genetic resources under the FAO's International Treaty.[514] At the same time, it implicitly invited other research-promoting entities to consider opting out of the restrictive bilateral approach to genetic resources and data by opting into analogously crafted multilateral regimes that promote research while supporting the benefit-sharing requirements of the CBD in all participating countries.[515]

Legal scholars are increasingly convinced that the CBD, as strengthened by the Nagoya Protocol, affords the microbial culture collections only two options, *viz.*, either to fully implement the costly, case-by-case bilateral regime, with its burdensome restrictions on research assets, or to opt into a multilateral regime of access to genetic resources along the lines suggested earlier.[516] On that premise, the culture collections cannot continue to duck their ABS responsibilities, rooted in the CBD, by operating as "trusted intermediaries" under a defective bilateral system. They must instead embrace the concept of a multilateral system of facilitated access and benefit-sharing that becomes fully compliant with Article 4 of the Nagoya Protocol.

That said, there is no reason why the microbial culture collections, united in a multilateral system of access and benefit-sharing, need to repeat possible design defects of the FAO's Crop Commons or even those of the WHO's PIP Framework Agreement.[517] On the contrary, the drafters of the Nagoya Protocol themselves recognized that microbial genetic resources were different from plant genetic resources, although they lacked time and expertise to pursue the implications of such insights.[518] Formation of a redesigned Microbial Research Commons to operate as a multilateral regime of facilitated access under Article 4 of the Nagoya Protocol should, instead, depart from this fundamental premise. The leadership's task is to devise a system that builds on the unique characteristics of the existing microbial research infrastructure while avoiding the design defects that experience with the Crop Commons and with other multilateral approaches to genetic resources may have revealed.

Adopting a multilateral, public-goods approach to the exchange of microbial genetic resources for research purposes does not mean that states opting into a redesigned commons infrastructure must abandon their claims to sovereignty over

[514] *See* Nagoya Protocol, n. 448, art. 4. *See also* IUCN, Guide to the Nagoya Protocol (2012), n. 448.

[515] *See generally* Chapter 3, Section IV.B ("Facilitating Scientific Research").

[516] *See esp.* Christine Godt, n. 432, at 258; *see generally* Chapter 4, Sections III & IV.

[517] *See* Chapter 3, Section III.C ("Strengths and Weaknesses of the International Treaty on Plant Genetic Resources for Food and Agriculture"); Chapter 4, Section IV (Discussing the WHO's Pandemic Influenza Preparedness Framework).

[518] *See* IUCN, Guide to the Nagoya Protocol (2012), n. 448.

the vast quantities of *in situ* resources that originate from their territories. To the extent that the FAO's International Treaty can be understood to impose such a waiver for some *in situ* plant genetic resources held in the Crop Commons,[519] it is not necessarily an appropriate model for the proposed Microbial Research Commons. Given that only about one percent of the aggregate microbial population has so far been isolated and identified,[520] the conditions of release for future research purposes applicable to the undiscovered ninety-nine percent would inherently depend on national laws in the countries of origin unless they were overridden by subsequent regional or international agreements.

These laws, in turn, will presumably have been crafted or modified to accommodate the provisions of the Nagoya Protocol as well as any voluntary agreements governing a microbial research commons that participating governments had adopted.[521] Any standard MTA eventually elaborated under a transnational Framework Agreement by a duly constituted governing body must necessarily reconcile access and benefit-sharing norms of the CBD with the needs of the global scientific research community.[522] Achieving this important goal, however, does not – and should not – require any legal waiver of national sovereignty over microbial genetic resources, other than those *ex situ* resources (and eventually perhaps some *in situ* genetic resources) voluntarily made available through the commons organizational architecture for research purposes. Even then, one may characterize the voluntary deposit of genetic resources into the multilateral system as a waiver of sovereignty rights, conditional on good-faith fulfillment of the built-in benefit-sharing machinery.

Developing countries in particular are unlikely to relinquish control over their *in situ* microbial genetic resources and related data in favor of an international repository in the present economic and legal environment under the Nagoya Protocol. This premise would hold even if interested governments believed that greater long-term benefits might accrue from a public goods approach than from hoarding, as we predict. In the present climate of palpable mistrust, however, governments would likely balk at any such proposal based merely on a promise and

[519] Under the FAO Treaty, the CGIAR's holdings are governed by separate arrangements from those of the treaty itself. See ITPGRFA, n. 131, art. 15. As regards cultivars in the public domain within participating states under the FAO's International Treaty, these are to be made available to the governing body, although actual compliance has reportedly been weak. *See* Halewood (Louvain 2012), n. 53; Frison (Louvain 2012), n. 131. The leading microbial culture collections have tried to finesse the "ownership" issue by self-designating themselves as "custodians." *See, e.g.*, Fritze & Oumard (2012), n. 108; C. Godt (2013), n. 432.

[520] *See* Chapter 1, Section II ("The Changing Nature of Microbial Research").

[521] *See e.g.*, Chapter 4, Section III.A.3 (European Commission's Regulations); *see further* Chapter 10, Section III ("Implementing the Multilateral Regime for Facilitated Access to *Ex Situ* Microbial Genetic Resources").

[522] *Accord.* D. Smith (2012), n. 319; Fritze & Oumard (2012), n. 108.

a hope for better things to come, not to mention national security and public health concerns that already elicit strict export controls over such resources.

For these and other reasons, the governing authority to be established by the stakeholders who participate in the legal and organizational architecture of the proposed Microbial Research Commons must operate along two jurisdictional axes at one and the same time. While primarily concerned with regulating the supply of *ex situ* microbial genetic resources to the global research community, it must also operate as a trustee or agent for the providers of such resources with respect to their benefit-sharing entitlements under the CBD.[523] In this context, rather than a treaty, like that regulating the FAO's Crop Commons, a transnational framework agreement between willing governments that commits the national culture collections to standardized material transfer agreements consistent with both the needs of public research and the benefit-sharing provisions of the Nagoya Protocol remains a far more feasible and promising starting point for a redesigned Microbial Research Commons.[524]

If and when results satisfactory to both developed and developing countries actually emerged from such a legal arrangement, the time might come when a more formal, treaty-based organization would be desirable for long-term stability, as was the case when the Agreement Establishing the World Trade Organization in 1994 replaced the GATT's contractual arrangement of 1947. In that event, states' cooperation in the Commons might become more confident about the governance modalities to be adopted than at present, precisely because of positive scientific and economic outcomes. Until then, we contend that the only workable basis for testing the redesigned Microbial Research Commons envisioned in this book is by adopting such a transnational framework agreement, as the product of direct interaction between the scientific community and the science ministries of the participating governments.

C. *Toward a More Science-Driven Organizational Model for the Digital Age*

Our preference for a negotiated middle ground between a treaty-driven intergovernmental organization and a science-managed nongovernmental organization should not obscure the need to build on the strengths of the FAO's

[523] Bearing in mind that any given providers may also become users of genetic resources provided by others. *See* Fritze & Oumard (2012), n. 108. For considerations about the composition and duties of the Governing Body appropriate for the proposed Commons, see Chapter 10, Section II.D.1.

[524] Whether such a framework agreement is to be technically characterized as "binding" or "nonbinding" depends on decisions by the founding member states. *See further* Chapter 10, Section II.D. ("The Core Institutional Components").

International Treaty, which did, after all, succeed in preserving access to *ex situ* plant genetic resources for food and agriculture under the resulting Crop Commons. One should also consider how the worldwide microbiological community could improve on the FAO's governance model to better address the needs of science while avoiding an overly rigid administrative format in which science was only indirectly represented.[525] This question has become more pressing in light of recent claims that the agricultural research community has become increasingly alienated from the decision-making process established by the International Treaty.[526]

How to organize the Microbial Research Commons so as to best promote these objectives poses a challenge. It will require a critical look at the governance structures that existing commons initiatives have already put in place, as elaborated earlier in this chapter, with a view to avoiding past errors and to devising a less bureaucratic, more science-driven, and financially sustainable administrative model. In so doing, we are reminded that the primary goals of the New Biology paradigm examined at the outset of this study[527] were to break down the barriers between subdisciplines and to integrate different fields of research within a larger scientific whole made more feasible by computational technologies.[528]

1. Avoiding an Unduly Narrow Scientific Mission

In thinking about a suitable governance structure in this light, we are struck by a number of defects that appear in most of the prototypical governance mechanisms described in Section II of this chapter. The negotiated intergovernmental framework agreement we envision for a Redesigned Microbial Research Commons should seek to avoid these design defects.

First and foremost, all the existing infrastructure initiatives described earlier tended to focus on a relatively narrow or circumscribed set of scientific and technical research assets, perhaps largely for historical reasons. For example, even the GBRCN's draft governance proposals during its Demonstration Stage revealed no systematic plan to digitally integrate the full range of microbial materials and related data, as well as "information about functions, structure, localization, and clinical effects of mutation"[529] into the same infrastructure as that which would serve users of its BRCs.[530] Yet, it seems

[525] *See, e.g.*, Halewood (2010), n. 443.

[526] *See especially* Halewood (Louvain 2012), n. 53; *see also* Frison (Louvain 2012), n. 131.

[527] *See* Chapter 1, Section II.D ("A New Research Paradigm for the Life Sciences").

[528] *See* NRC, A New Biology, n. 104.

[529] Christina Chandras et al., *Models for Financial Sustainability of Biological Databases and Resources*, 2009 *J. Biological Databases & Curation*, *available at* http://database.oxfordjournals.org/content/2009/bap017.full.pdf.

[530] At the initial time of writing, the GBRCN's only prominent mention of such data types occurred in Annexes pertaining to its business model, where they were listed among possible new sources of income.

doubtful that microbiology could fulfill the premier role assigned to it by the New Biology paradigm without a research infrastructure that fully integrated materials, data, and literature. Not to do so would also ignore the many spontaneous efforts to form digitally integrated data-sharing platforms that have already begun to emerge, and would afford them no preestablished institutional network or platform through which they could eventually be linked.[531] The longer one postpones the task of more fully integrating materials and related data, the more difficult it will be to overcome barriers rooted in intellectual property rights, and corresponding vested interests later on.[532]

Operationally, we note that the WFCC's World Data Center for Microorganisms (WDCM) has already begun to construct a digitally integrated platform along these lines,[533] and GBRCN's organizers were aware of its potential.[534] The WDCM could logically constitute a core element from which the deep integration of the kind we envision could evolve. To do so, however, WDCM would need to become fully committed to the open access principles that are essential for the public-good functions of a true research commons.[535]

We also note that, with the exception of WDCM, none of the existing organizations surveyed earlier in this chapter had manifested an interest in taking steps to integrate scientific literature into their formally structured service components.[536] This omission is perhaps understandable in view of the power that publishers wield over scientific literature, as described in chapters 6 and 7, and of the efforts needed to change the status quo. Yet, as shown in those same chapters, funding agencies such as the NIH and the Wellcome Trust have adopted ever more stringent measures to encourage Principal Investigators and other researchers to make publicly funded research results widely available.[537]

The European Commission's Directorate for the Information Society recently launched an important initiative favoring greater access to published research results.[538] Moreover, some of the incipient Open Knowledge Environments,

[531] *See* Chapter 8, Section III ("Building Transnational Open Knowledge Environments").

[532] *See* Chapter 2, Section II ("Impinging Intellectual Property Rights Promoted by the Developed Countries"), and Chapter 6, Section II ("Copyright and Related Laws as Digital Gridlock").

[533] *See* Chapter 8, Section II.B.1 ("The World Data Center for Microorganisms").

[534] *See, e.g.*, D. Smith (2012), n. 319.

[535] Interview with Prof. Juncai Ma, Director, World Data Center for Microorganisms (WDCM) in Taipei (Oct. 30, 2012).

[536] Administrators of both GBIF and IHMC mentioned scientific literature in passing, but seemed to have no robust implementation schemes at the time of writing. WDCM, however, does track citations of materials in the literature. See Chapter 8, Section II.B.1.

[537] Efforts are also reportedly under way in the U.S. to ensure that microbes used in government funded research are deposited in public repositories, whether patented or not. Interview with Kevin McCluskey, Curator, Fungal Genetics Stock Center, University of Kansas Medical Center, December 10, 2013.

[538] *See* the OpenAIRE Project, *available at* https://www.openaire.eu/. *See also* Policy Recommendatoins for Open Access to Research Data in Europe (RECODE), Seventh Framework Program for Science

described in Chapter 8, notably the Genome Standards Consortium, MOBEDAC, and KBase, have already begun to integrate open access literature into their thematic hubs.[539] As shown in Chapter 7, a growing number of microbiology journals have already committed to some form of open access publishing.[540] For these and other reasons, we believe that any serious attempt to establish a redesigned Microbial Research Commons should include measures to integrate both scientific data and literature into its core infrastructure.

2. Giving Scientists a Voice in the Decision-Making Process

Another second common defect is that all but one of the organizations surveyed earlier in this chapter established a single governing entity to deal with both political and scientific policy issues, and typically this entity would receive scientific inputs only from advisory boards that lacked any voting rights. The sole exception was the IHMC, which appears to have established a novel and extremely effective scientific governance mechanism with considerable inputs from representatives of both funders and Principal Investigators.[541] However, the flexibility arising from the IHMC's governance model may actually stem from the participating funders' innate ability to support any decisions taken by the Steering Committee.

When the governing bodies of such organizations do not include scientists invested with voting rights, there is a palpable risk that they will perform neither their political nor their scientific responsibilities optimally. Ideally, the governing entities should be animated directly by research scientists operating close to the ground.[542] Otherwise, the distance between administrators and the needs of the relevant scientific community may widen over time, as reportedly has occurred under the Crop Commons,[543] with insufficient scientific inputs at both the top and bottom levels and a concomitant loss of community ownership. There is also a considerable risk that the resulting organizational outputs may appear static, rather than dynamic, to both the funder and user communities over time.

What the empirical review of existing governance models really suggests, in short, is the need for a more innovative governance structure that can operate effectively along two axes at the same time, as postulated earlier. In one dimension, which we can call the "horizontal dimension," administrators must deal with the international

in Society, Final Report Feb. 2, 2015, *available at* www.recodeproject.eu. Both research projects were funded by the European Commission.

539 *See* Chapter 8, Sections II & III.

540 *See* Chapter 7, Section II ("Surveying the Practices of Microbiology Journals").

541 *See* Section II.B.4 in this chapter.

542 *Cf.* IHMC, Section II.B.4 in this chapter.

543 *See* Halewood (Louvain 2012), n. 53; *see also* Halewood et al. (2013), n. 158.

legal and institutional components of a worldwide Microbial Research Commons. In the second dimension, which we call the "vertical dimension" this same entity must be capable of coordinating and supporting an array of interdisciplinary scientific endeavors, which would at least entail the provision of knowledge inputs and technical services to the microbiological community as a whole. That same entity would ideally also become a focal point for coordination with bottom-up initiatives, such as the community-driven Open Knowledge Environments described in Chapter 8.[544]

In practice, the existing governance models that we surveyed often tend to focus on only one of these components, while largely under appreciating the needs of the other. Moreover, although each of the existing models will have generated one or more governance components that, in isolation, seem relatively workable and sometimes ingenious, all of them – taken together – exhibit some shortcomings that one would not wish to emulate when redesigning a large-scale Microbial Research Commons.

The empirical evidence further suggests that these horizontal and vertical dimensions must be suitably combined within a single legal framework. At the transnational political level, for example, the relevant government agencies or ministries must play a crucial role, with strong inputs from the scientific communities. At the operational level, in turn, the tasks of coordinating research and infrastructure activities and of interfacing with bottom-up user communities need to be science-driven, with inputs and support from both funders and governments (to the extent that these are in fact different players).

The importance of this boundary management function[545] has been particularly evidenced at the IHMC, where funders and Principal Investigators operate in tandem on the scientific frontier and remain in contact and open to new developments and proposals.[546] This institutional experiment yields an important lesson for redesigning an up-to-date and effective governance structure for the proposed Microbial Research Commons: namely, that microbiologists with the requisite managerial skills should be placed in both key decision-making and operational positions. How to implement these insights is the task undertaken in the final chapter of this book.

[544] *See* Chapter 8, Section III.A ("Examples of Incipient Open Knowledge Environments on the Frontiers of Microbiology").

[545] *See* Section I.A in this chapter ("Applying Commons Theory to the Microbial Research Infrastructure").

[546] *See* Sections II.B.4, a & b (describing governance and operations of the IHMC).

10

Governing Digitally Integrated Genetic Resources, Data, and Literature

I. PREMISES FOR CONSTRUCTING A COMMON POOL RESOURCE

A. *The Political Economy of a Global Approach*

As indicated in the previous chapter, the main defensive goal of a multilateral approach is to preserve the public-good research functions of the microbial culture collections, which are constrained by the need to navigate between the Convention on Biological Diversity of 1992 and the TRIPS Agreement of 1994.[1] A more positive goal is deliberately to construct the multilateral facilitated-access regime as a "common pool resource"[2] that fully integrates microbial genetic resources, related data, and literature at a time when *in silico* research has become at least as important as traditional *in vitro* methods.[3] Still another positive goal is to ensure that the developing countries that participate in the Commons initiative obtain tangible non-monetary benefits, largely in the form of capacity building, in addition to monetary benefits flowing from the use of their microbial genetic resources in industrial applications.[4]

In the absence of any major new transnational commons initiative, the microbial culture collections remain rooted in their national, or increasingly regional, legal

[1] Convention on Biological Diversity, opened for signature 5 June 1992, 1760 U.N.T.S. 299 [hereinafter CBD]; Tenth Meeting of the Conference of the Parties to the Convention on Biological Diversity, Nagoya, Japan, 18–29 Oct. 2010, Nagoya Protocol on Access to Genetic Resources and the Fair and Equitable Sharing of Benefits Arising from their Utilization (ABS) to the Convention on Biological Diversity [hereinafter Nagoya Protocol] entered into force on 12 Oct. 2014 available at http://www.cbd.int/abs/doc/protocol/nagoya-protocol-en.pdf (last accessed 14 June 2014); Agreement on Trade-Related Aspects of Intellectual Property Rights, Apr. 15, 1994, 2869 U.N.T.S. 299 (1994) [hereinafter TRIPS Agreement]. *See generally* Chapters 2, 3, & 4.

[2] *See* Chapter 9, Section I.A ("Applying Commons Theory to the Microbial Research Infrastructure").

[3] *See* Chapter 1, Section II.B ("The Revolution in Genetic Science").

[4] *See* Chapter 3, Section IV.B.2 ("Recognizing the Importance of Non-Monetary Benefits").

frameworks. They are subject to the vagaries of the diverse Material Transfer Agreements surveyed in Chapter 4, with exchanges of microbial materials among collections dependent on negotiations that risk imposing additional restrictions on access and use for research purposes, with no uniform response to the challenges of the CBD. Even when a more uniform response is attempted at the regional level, as was the case of the European Commission's Regulation No. 511/2014 on Compliance Measures for Users from the Nagoya Protocol,[5] the tendency is to maintain the status quo of the culture collections as trusted intermediaries, rather than to more effectively organize them to deal with the challenges and opportunities posed by the Nagoya Protocol.[6] As indicated in Chapter 4, and recently reconfirmed by spokespersons for leading European culture collections, that state of affairs risks becoming a formula for diminishing returns, that is, fewer exchanges of microbial genetic resources; less research stimulated by such exchanges; fewer deposits of specimens into public collections; and fewer benefits to be shared under the evolving international legal regime.[7]

At the same time, exchanges of both *ex situ* and *in situ* materials are increasingly impeded by patents on microorganisms and trade secret protection, as mandated by Articles 27(3)(b) and 39 of the TRIPS Agreement of 1994[8] and by claims of sovereign rights rooted in the Nagoya Protocol to the CBD.[9] Genomic and other related data are also subject to copyright and database protection laws as amplified by recent international treaties and free trade agreements.[10]

Apart from inappropriate intellectual property restrictions on upstream research inputs, exchanges of microbial materials for research purposes are further hampered by other factors, such as a lack of uniform quality standards applicable to the different culture collections and limits on the capacity of the

5 Council Regulation No. 511/2014 on Compliance Measures for Users From the Nagoya Protocol on Access to Genetic Resources and the Fair and Equitable Sharing of Benefits Arising from Their Utilization in the Union, 2014 O.J.L. 150/59.

6 *Compare* Chapter 3, Section IV ("New Constraints and Opportunities for Scientific Research Under the Nagoya Protocol") *with* Chapter 4, Section III.A.3 ("The European Commission's Regulation on Access to and Use of Genetic Resources").

7 *See most recently* Dagmar Fritze and Andrè Oumard, "The Pan-European Project, Microbial Resource Research Infrastructure/MIRRI, Has Among Its Goals the Elaboration of Common Policies for BRCs to Comply with the Nagoya Protocol on Access and Benefit Sharing of CBD," Paper presented at the Conference on Governing Pooled Knowledge Resources: Building Institutions for Sustainable Scientific, Cultural and Genetic Resources Commons for the International Association for the Study of the Commons (IASC), Louvain-la-Neuve, Belgium, September 12, 2012, available at http://biogov.uclouvain.be/iasc/docfull%twentypapers/Fritze.pdf [hereinafter Fritze & Oumard (2012)].

8 TRIPS Agreement (1994), n. 1, arts 27.3(b) & 39. See Chapter 2, Section II ("Impinging Intellectual Property Rights Promoted by the Developed Countries").

9 *See* Chapter 2, Sections II.B.1 & 2; Chapter 3, Section IV.A.

10 *See generally* Chapter 6, Section II ("Copyright and Related Laws as Digital Gridlock").

public collections to absorb more validated materials, especially from academic institutions. This limited capacity for *ex situ* microbial collections, sometimes referred to as the "Big Refrigerator Problem,"[11] stems in part from the high costs of maintaining and securing physical repositories that meet international quality standards, and from the dependence of most public culture collections on voluntary personnel.[12]

Still another risk is that, under existing arrangements, many culture collections may fail to gear their operations to the needs of data-driven genomic science, at least in the short run. Research would, instead, be better served by encouraging a closer interrelationship between *in vivo*, *in vitro* and *in silico* methods.[13]

A redesigned Microbial Research Commons should address these and other problems surveyed in this book. Through collective action, it would create and maintain a stable upstream research infrastructure organization open to qualified microbiologists. In this space, a multilateral framework agreement, consistent with the Nagoya Protocol and other applicable international legal requirements, would facilitate access to, and use of, microbial genetic resources for research purposes, while encouraging downstream commercial applications and benefit-sharing to the fullest extent possible.

With specific regard to facilitating the exchange of microbial materials, the proposed Research Commons would thus seek to provide the advantages of the model we elaborated in Part Two of this book, namely:

- Establish a multilateral entity that would immunize pooled microbial genetic resources from the bilateral Access and Benefit-Sharing provisions of the CBD;[14]
- Produce a negotiated, standard-form Material Transfer Agreement (SMTA) that would enable scientists to obtain microbial genetic resources from all participating culture collections on research friendly terms;[15]
- Guarantee all governments and entities willing to deposit materials in the federated pool of public research assets a fair and equitable share of any financial benefits accruing from successful downstream commercial applications of

[11] Fiona Murray, Institutional foundations of scientific progress: Implications for collaboration and participation, paper presented at Global Science and the Economics of Knowledge-Sharing Institutions, 2d. Communia Int'l Conference, Turin, Italy, June 29–30, 2009.

[12] *See generally* Scott Stern, Biological Resource Centers: Knowledge Hubs for the Life Sciences 42 (Brookings Inst. Press 2004); Communication from Dagmar Fritze, December 20, 2014.

[13] Interview with Ann Maglia, Nat'l Sci. Foundation, October 12, 2011. *See* Chapter 1, Section II.B.

[14] *See* Chapter 3, Sections I.B & V.

[15] *See* Chapter 5, Section III ("Modelling a Sequence of Hypothetical Transactions"). Presumably, most of these genetic resources would have no known or likely commercial prospects at the time of deposit." If not, they would presumably be already under patent or on the way to patenting or under trade secrecy laws. See Chapter 2, Section II.B.1 ("Increasing Reliance on Patents and Trade Secrecy Laws").

their contributed materials and related data under a built-in Compensatory Liability Regime.[16]

- Encourage all holders of microbial materials of potential research interest, whether located at public collections, universities, government departments or elsewhere, to affiliate with the distributed pool of available research resources under internationally agreed conditions, in an ever-expanding, digitally linked "Big Refrigerator."

In principle, if correctly implemented, holders of *ex situ* microbial resources for research purposes should expect to obtain greater benefits in terms of research capacity and potential returns from commercial applications by participating in the materials semicommons than by operating independently under a hoarding mentality that undermines the sharing norms of science.[17]

The incentives flowing from participation in the redesigned Microbial Research Commons would ideally reinforce efforts already under way to master the enormous quantities of data generated by ever more powerful digital technologies.[18] By establishing a network of networks under the auspices of a suitable governance framework applicable to the *ex situ* collections,[19] the Commons could not only encourage the participating collections themselves to share their data, but could as well provide a central portal through which disparate thematic communities could link data and literature in an ever expanding pool of research resources available everywhere on open-access terms and conditions.[20]

With regard to both microbiological data and literature, the proposed Commons would thus provide the following advantages:

- Encourage the public culture collections and other holders of validated microbial resources to share related data and literature through digitally

[16] See Chapter 5, Section III.

[17] The term "semicommons" is used for the proposed distributed network of pooled microbial genetic resources because, technically speaking, only qualified culture collections can exchange these resources, and only qualified researchers can access them for specified purposes. These restrictions follow from international biosafety and security concerns. Our use of the word "semicommons" thus resembles the "limited common property" concept of Carol Rose, under which members share alike, but outsiders are excluded. Carol Rose, *The Several Futures of Property: Of Cyberspace and Folktales, Emission Trades and Ecosystems*, 83 MINN. L. REV. 129, 132 (1998).

[18] *See further* Chapter 8, Section II ("Beyond Early Release: Diverse Networked Sharing Strategies to Manage and Exploit the Deluge of Data").

[19] For the term "network of networks," *see* the discussion of GBRCN, Chapter 9, Section II.C.1.

[20] Paul Uhlir, *Designing the Digital Commons in Microbiology – Moving from Restrictive Dissemination of Publicly Funded Knowledge to Open Knowledge Environments: A Case Study in Microbiology*, *in* DESIGNING THE MICROBIAL RESEARCH COMMONS: PROCEEDINGS OF AN INTERNATIONAL WORKSHOP, Paul F. Uhlir, ed. (National Academies Press, 2011) available at http://www.nap.edu/catalog/13245/designing-the-microbial-research-commons-proceedings-of-an-international-workshop (last accessed August 18, 2015) [hereinafter DESIGNING THE MICROBIAL RESEARCH COMMONS]. See further Section III, in this chapter.

linked portals on research friendly terms. Such terms would overcome barriers otherwise imposed by the default rules of copyright and database protection laws at the national, regional, and international levels.[21]
- Reinforce the reciprocity benefits that accrue when scientists pool their data in thematically organized public repositories and Open Knowledge Environments by linking their research outputs through a central portal.[22]
- Further reinforce the open-access movement with respect to microbiological literature by creating and maintaining a platform through which such literature is readily made available to the scientific community.[23]

We recognize that with respect to both materials and digital knowledge resources there are often valid limitations on open availability that must be taken into account when implementing this vision. For example, microbial materials from public culture collections that are supplied to researchers may not usually be redistributed unless the researchers in question belong to collaborating laboratories that have met predetermined quality and security standards.[24] This practice serves to ensure the purity of microbial research materials and to enable the tracking and control of specimens needed for verification of research results by other scientists. For this and other reasons, the collections and collaborating laboratories that pool their resources will, technically speaking, have formed an ever-widening semicommons, open only to qualified entrants, rather than a fully open materials commons.[25]

Moreover, lots of materials and data held at universities and other public research institutes are kept secret, either under grants that allow principal investigators some period of exclusive use or until the relevant research results are published.[26] The foreseeable commercial viability of such research assets is typically assessed during this period. As seen in Chapter 8, consortia of likeminded researchers are also increasingly being formed in which the members pool materials and data

[21] *See* above Chapter 6, Section II; *see also* Jerome H. Reichman & Ruth L. Okediji, *When Copyright and Science Collide: Empowering Digitally Integrated Research Methods on a Global Scale*, 96 *Minn. L. Rev.* 1362, 1372–1457 (2012) [hereinafter Reichman & Okediji (2012)].

[22] *See* Chapter 8, Section III.

[23] *See* Chapter 7 ("Enabling the Microbiological Research Community to Control Its Own Scholarly Publications").

[24] *See, e.g.*, European Culture Collections' Org. (ECCO), The ECCO Core Material Transfer Agreement for the supply of samples of biological material from the public collection, Feb. 2009 [hereinafter ECCO MTA], *available at* http://www.eccosite.org/docs/ECCO_core-MTA_V1_Feb09.pdf (discussed in Chapter 4, Section III.A.2).

[25] Materials exchanged among participating collections may be freely redistributed within the commons. Otherwise, materials given to single researchers are limited to use by them in laboratories only.

[26] Wesley E. Cohen & John P. Walsh, *Real Impediments to Biomedical Research*, 8 *Policy Econ.* 1–30 (2008).

in semicommons or clubs open only to members in order to obtain immediate, pre-publication reciprocity benefits through these trading arrangements.[27]

Although early-stage research assets pooled in these semicommons may not always be candidates for inclusion in the Microbial Research Commons, we contend that research outputs from these entities, once voluntarily disclosed to the public, can become highly valuable research resources that should be included in the larger Commons infrastructure whenever feasible. Whether such outputs are new materials, data, or published findings, the principal investigators and their institutions may stand to gain heightened reputational benefits, as well as possible future grants and collaborations, by opening their research outputs for access and use by the global scientific community.

Placement of these new outputs from the various semicommons into the larger pool governed by the Microbial Research Commons could, in turn, enable those same consortia to obtain financial revenues from any unforeseen downstream commercial applications under the Compensatory Liability Regime to be established by the Commons.[28] Participants in the Commons would likewise reciprocally benefit if the members of disparate semicommons were thus incentivized to make their research outputs more generally available throughout the Commons' federated systems.

In sum, we contend that the worldwide microbiological community cannot afford to accept the science-hostile legal and institutional environment as it stands "without organizing a response to the increasing encroachment of a commercial ethos upon its upstream ... [research] resources."[29] Rather, it should endeavor to manage its essential knowledge assets – materials, data, and literature – under a common set of rules, norms, and policies that are deliberately designed by scientists to meet public science goals. The community should accordingly take charge of its own knowledge assets, and not leave them to the vagaries of national, regional, or international intellectual property laws that are typically driven by forces beyond the reach of the public research community.[30] Because science policymakers cannot realistically aspire to change those

[27] *See* Minna Allarakhia et al., *Modelling the Incentive to Participate in Open Source Biopharmaceutical Innovation*, 40 *R&D MGMT.* (2010), 50–66.

[28] *See further* Chapter 5, Section II ("Designing a Third Option: *Ex Ante* 'Take and Pay' Rules for Stimulating Research Applications").

[29] Jerome H. Reichman & Paul F. Uhlir, *A Contractually Reconstructed Research Commons for Scientific Data in a Highly Protectionist Intellectual Property Environment*, 66 *Law & Contemp. Probs.* 317, 417 (2003), available at http://scholarship.law.duke.edu/lcp/vol66/iss1/12 [hereinafter Reichman & Uhlir (2003)].

[30] *Cf. id.* at 416 ("A Contractually Reconstructed Research Commons for Science and Innovation"):

> [T]he other [response] would require science policy to address the challenge by formulating a strategy that would enable the scientific community to take charge of its basic data supply and manage the resulting research commons in ways that preserved its public good functions without impeding socially beneficial commercial opportunities.

laws, the only sure way to manage its public research assets on a global scale is for the microbiological research community to contractually construct its own operational framework so that, by agreement of its constituent members, it overrides or otherwise neutralizes adverse legal, economic, and institutional impediments.

B. *The Critical Role of Effective Leadership*

In addressing these goals, we emphasize that the existing microbial research infrastructure already possesses foundational assets, such as the WFCC's distributed network of culture collections and the expanding capacity of its World Data Center for Microbiology.[31] At the same time, we note that the knowledge commons literature particularly stresses the importance of leadership and governance over and above the role of infrastructure as such.[32]

Efforts to build a redesigned Microbial Research Commons along the lines envisioned in this study would require a coterie of champions to establish the institutional framework, a dedicated and representative group of managers to govern the resulting entity, and a properly designed governance model to achieve the objectives of that entity. Recent empirical studies also demonstrate the importance of leadership in designing and forming sustainable knowledge commons.[33]

At the moment, the most interested candidates for leadership of a redesigned Microbial Research Commons would be active in the World Federation for Culture Collections (WFCC), the Global Biological Resource Centers Network (GBRCN), or its follow on project, the Microbial Resource Research Infrastructure (MIRRI).[34] The WFCC, with some six hundred affiliated culture collections that meet minimum quality standards, already constitutes a de facto commons, albeit one

[31] *See* Chapter 8, Section II.B.1 (WDCM); Chapter 9, Section II.B.2 (WFCC); *see generally* Chapter 4, Section I ("Evolution of Microbial Culture Collections as Basic Scientific Infrastructure").

[32] *See, esp.* Michael J. Madison, Brett M. Frischmann & Katherine J. Strandburg, *Constructing Commons in the Cultural Environment*, 95 *Cornell L. Rev.* 657, 678 (2010), *available at* http://www.lawschool.cornell.edu/research/cornell-law-review/upload/madison-frischmann-strandburg-final.pdf [hereinafter Madison et al. (2010)]. *See also* Brett M. Frischmann, Michael J. Madison & Katherine J. Strandburg, Governing the Knowledge Commons (Oxford U. Press, 2014).

[33] See note 32. *See also* Michael Madison, Constructing Commons in Intellectual Resources, paper presented at the Conference on Governing Pooled Knowledge Resources: Building Institutions for Sustainable Scientific, Cultural and Genetic Resources Commons for the International Association for the Study of the Commons (IASC), Louvain-la-Neuve, Belgium, Sept. 12–14, 2012; Katherine Strandburg, "The Rare Diseases Clinical Research as a Nested Cultural Commons," paper presented at the Conference on Governing Pooled Knowledge Resources, Building Institutions for Sustainable Scientific, Cultural and Genetic Resources Commons, International Association for the Study of the Commons (IASC), Louvain-la-Neuve, Belgium, Sept. 12–13, 2012 [hereinafter Strandberg, 2012].

[34] For a description of GBRCN and MIRRI, see Chapter 9, Sections II.C & D; for the WFCC, see Chapter 9, Section II.B.1.

that operates under a weak institutional architecture that does not include national governments as formal members.[35] However, only a fraction of these collections – perhaps two hundred – are known to meet even the minimum standards to qualify as Biological Resource Centers.[36] It remains to be seen how many WFCC affiliates could muster the technical skills and organizational capacity to accommodate the needs of a transnational governance entity that aimed to standardize, simplify, and expedite exchanges of genetic resources in a manner consistent with the constraints and opportunities afforded by the Nagoya Protocol.[37] Building this sort of capacity is nonetheless a key function of the research commons we envision, and *a priori* leading participants in the WFCC are a logical constituency to undertake these endeavors.

In contrast, the leaders behind the GBRCN project, which is now in a semi-dormant, post-Demonstration Phase, had united some of the most technically advanced culture collections that have already met BRC standards, and they expressed an interest in moving toward the formation of a global research commons. GBRCN's Demonstration Phase, funded by the German government, also provided its leadership with some valuable, real-world experience in this regard.[38] This same group is now seeking to form strategic alliances with regional associations of culture collections in Europe, Asia, Latin America, and the United States,[39] with particular emphasis on the role to be played by MIRRI in the European Union.[40] If successful, these regional alliances could then further coordinate their operations within a formally organized "network of networks."[41]

GBRCN thus had a plan of action that could provide a platform for the formation of a commons infrastructure built around selected, elite culture collections, and it seemed to possess a certain degree of momentum, given that the European Commission has now provided funding for the follow-on MIRRI component. The

35 *See* Chapter 9, Section II.B.1.b ("Governance").

36 For a discussion of BRCs, *see* Chapter 4, Section I.B. The WFCC has devised new technical standards that reflect BRC best practices, but require less investment in heavy-duty equipment than the OECD Standards. See Chapter 5, Section II.C.1.

37 *See* Chapter 3, Section IV ("New Constraints and Opportunities for Scientific Research Under the Nagoya Protocol").

38 *See, e.g.*, Global Biological Resource Ctr. Network (GBRCN), Final Report on the GBRCN Demonstration Project (Nov. 30, 2008–Nov. 30, 2011) (2012) [hereinafter GBRCN, Final Report] (discussed in Chapter 9, Section II.C.1).

39 *See, e.g.*, David Smith, Networking Collections to Provide Facilitated and Legislation Compliant Access to Microbial Resources, paper presented at the Conference on Governing Pooled Knowledge Resources: Building Institutions for Sustainable Scientific, Cultural and Genetic Resources Commons for the International Association for the Study of the Commons (IASC), Louvain-la-Neuve, Belgium, Sept. 12–14, 2012 [hereinafter Smith (2012)]. For regional networks, see Chapter 4, Section I.C.

40 *See* Fritze & Oumard (2012), n. 7.

41 Smith (2012), n. 39.

promoters of the GBRCN project were also committed to addressing the problems of financial sustainability that have become increasingly acute for culture collections everywhere.[42]

However, GBRCN's focus on elite collections tended to undermine the potential universality of the microbial commons project, which could make it more difficult to persuade developing countries to join or otherwise cooperate with a new Microbial Research Commons An alliance of selected, elite collections, even if broadened to include some leading participants from the BRICS countries, could inspire the kind of mistrust among developing country governments that has hindered full implementation of the Crop Commons under the International Treaty for Plant Genetic Resources for Food and Agriculture (ITPGRFA).[43] Above all, the commitment of GBRCN's leadership to maintaining the WFCC's public good approach – rather than the market-like model of ATCC – remains to be demonstrated, as we pointed out in the last chapter.

The leadership of the proposed Microbial Research Commons must also look beyond the activist group previously identified with a view to attracting key microbiologists from different regions in both developed and developing countries, who are not otherwise engaged in WFCC activities.[44] They should also seek champions in government ministeries dealing with science, public health, energy, and the environment, as well as in other microbiological societies and foundations already focused on these issues.[45]

C. *The Need for Political Cover*

In retrospect, one of the fundamental features in thinking about the construction of large-scale science commons projects, especially the Crop Commons discussed in Chapters 3 and 9, has been the emphasis on the need for "political cover" as one of the primary organizational criteria. Because the stakeholders are understandably worried about reconciling governments that have different and conflicting geopolitical interests, they tend to favor top-heavy organizational architectures designed to

[42] *See* Ecological Soc'y Am., Strategies for Developing and Innovating Living Stocks Collections: An ESA Workshop Report, Aug. 2012 (to be published in the ESA Bulletin, Jan. 2013). For more on sustainable funding, see Section IV in this chapter.

[43] *See* Chapter 3, Section III.C ("Strengths and Weaknesses of the ITPGRFA").

[44] For example, at the international level, we would include leaders from the International Union of Microbiological Societies (IUMS), the International Council of Science (ICSU), and relevant specialized United Nations agencies.

[45] In the United States, for example, leaders could be recruited from the Office of Science and Technology Policy (OSTP), the National Science Foundation (NSF), the National Institutes of Health (NIH), and the Department of Agriculture (USDA) (which holds many culture collections), and the Department of Energy (DOE) (which also holds major culture collections).

provide operational security within the confines of a narrow and endlessly negotiated political consensus.[46] In the case of the Crop Commons, the unforeseen result has reportedly been a gradual estrangement of the relevant scientific community from the governance aspects of the Commons.[47] A further risk is that the research community will simply go about its business with a corresponding loss of trust on the part of governments and other stakeholders, and with mounting possibilities for violating international obligations that benefit developing countries.[48]

We do not mean to imply that microbiologists and other scientists needing access to genetic resources in a geopolitically divided world could operate without adequate political cover. The opposite is true. What we mean is that this concern should not become an excuse for constructing a top-heavy, highly politicized entity at the expense of scientific inputs into the design and governance of the Commons enterprise in the first place.

An appropriate governance model for a knowledge commons formed to promote basic scientific research would logically look for political support to the science ministries (or their equivalents) in both developed and developing countries that are willing and able to participate in the venture.[49] The science ministries, in turn, should provide political and financial support because they believe in the mission and expect that the payoffs for both science and industry from the commons project would suffice to justify the costs. Nevertheless, the object of the exercise would be to enable the production of demonstrably valuable scientific payoffs, lacking which the science ministries should retain the option to withdraw their support from the venture.

[46] *See, e.g.*, Clive Stannard, *The Multilateral System of Access and Benefit Sharing: Could It Have Been Constructed Another Way?*, *in* CROP GENETIC RESOURCES AS A GLOBAL COMMONS: CHALLENGES IN INTERNATIONAL LAW AND GOVERNANCE, 243–64 (M. Halewood et al eds., Routledge (2013)) [hereinafter CROP GENETIC RESOURCES].

[47] *See, e.g.*, Michael Halewood, What Kind of Goods are Plant Genetic Resources for Food and Agriculture? Towards the Identification and Development of New Global Commons, paper presented at the Conference on Governing Pooled Knowledge Resources: Building Institutions for Sustainable Scientific, Cultural and Genetic Resources Commons for the International Association for the Study of the Commons (IASC), Louvain-la-Neuve, Belgium, Sept. 12–14, 2012 [hereinafter Halewood (Louvain 2012)].

[48] *See, e.g.*, Michael Halewood, Isabel López Noriega & Sélim Louafi, *The Global Crop Commons and Access and Benefit-Sharing Laws: Examining the Limits of International Policy Support for the Collective Pooling and Management of Plant Genetic Resources*, *in* CROP GENETIC RESOURCES (2013), n. 46 at 99 et seq.; Godfrey Mwila, *From Negotiations to Implementation: World Review of Achievements, Bottlenecks and Opportunities for the Treaty in General and for the Multilateral System in Particular*, *in* CROP GENETIC RESOURCES (2013), n. 46 at 226–42; Isabel López Noriega et al., *Assessment of Progress to Make the Multilateral System Functional: Incentives and Challenges at the Country Level*, *in* CROP GENETIC RESOURCES (2013), n. 46, at 199.

[49] *See* Chapter 9, Section III.C ("Toward a More Science-Driven Organizational Model for the Digital Age").

It follows that the proposed Microbial Research Commons must be legally capable of establishing a governance mechanism with enough authority to defend and promote access to, and use of, essential public knowledge assets that member governments of the Commons make available to the global research community. That mechanism, in turn, would operate as an agent of the multilateral system of facilitated access and benefit-sharing established by the participating governments.[50] A negotiated intergovernmental Framework Agreement should likewise seek to ensure that scientists have a strong, legally protected voice in any governance mechanisms to be established by the Contracting Parties.

While the Framework Agreement should stimulate and support the formation of a network of networks, the single scientific entities and sub-networks would remain independent, despite their reliance for specified purposes on the governance mechanism to be established by the Commons. A major objective is to enable a "Big Science" approach to the fullest extent possible, without erecting a highly centralized public-sector bureaucracy.[51]

In this context, the pivotal role of the national science ministries stems from the fact that they already regulate and fund most of the microbial culture collections as well as the national research agendas. Although the Framework Agreement should provide for transnational coordination, it ought necessarily to rely on these ministries to help formulate and implement the coordination strategies eventually to be adopted by the Commons' management. Any proposed governance structure that circumvented the authority of the national science ministries would surely fail.

The tensions we have focused on throughout this book, which threaten to disrupt public scientific research, arose largely from North-South conflicts in the areas of international trade and intellectual property. The Microbial Research Commons we propose here would rise above these tensions by devising a cooperative approach to basic scientific infrastructure whose resulting research outputs would benefit stakeholders in both developed and developing countries. If successful, the redesigned Microbial Research Commons proposed here might, in the long run, serve to reduce some of the international trade tensions from which it arose and point the way to a more cooperative economic environment, with progressively more favorable outcomes than we see emerging from the drive by OECD countries to foist ever higher intellectual property standards on the developing world.

[50] *See* Thomas Greiber et al., An Explanatory Guide to the Nagoya Protocol on Access and Benefit Sharing, art. 4 (Int'l Union for Conservation of Nature & Natural Res. (IUCN), Envtl. Pol'y & L. Paper No. 13, 2012) [hereinafter IUCN, Guide to the Nagoya Protocol (2012)].

[51] See Chapter 1, Section II ("The Changing Nature of Microbiological Research") and Chapter 8, Section II ("Beyond Early Release: Diverse Networked Sharing Strategies to Manage and Exploit the Deluge of Data").

II. ORGANIZATIONAL AND STRUCTURAL CONSIDERATIONS

When thinking about an appropriate governance structure for the proposed Microbial Research Commons, we note at the outset that the two existing multilateral entities that facilitate access to genetic resources under the CBD, i.e., the Crop Commons[52] and the Pandemic Influenza Preparedness Framework,[53] both rely on the support of United Nations Specialized Agencies, namely, the Food and Agricultural Organization (FAO) and the World Health Organization (WHO), respectively. These agencies serve as implicit or explicit guarantors of the undertakings to which member governments have otherwise subscribed in the relevant legal agreements.[54] Even the Group on Earth Observations (GEO), which has become one of the world's largest data-pooling consortia, receives some modest support from the World Meteorological Organization (WMO), another United Nations Specialized Agency.[55]

By logic alone, one might argue that the Microbial Research Commons could or should also affiliate with an existing United Nations Specialized Agency. Possible affiliations could include the United Nations Educational, Scientific, and Cultural Organization (UNESCO)[56] or the United Nations Environment Programme (UNEP),[57] or alternatively, that it should directly seek an affiliation with the CBD, perhaps under the auspices of the Conference of the Parties.[58]

However, even though UNESCO did support early capacity building efforts in developing countries under its Mircrobial Resource Centers (MIRCEN) project, and generally considers science as within its purview,[59] it has both membership and

[52] International Treaty on Plant Genetic Resources for Food and Agriculture, opened for signature Nov. 3, 2001, 3400 U.N.T.S. 303 (entered into force June 29, 2004) [hereinafter IPGRFA], available at http://treaties.un.org/doc/publication/UNTS/Volume%202400/v.2400.pdf (last accessed Sept. 24, 2014) See further Chapter 9, Section II.A.1 ("A Two-Headed Governance Construct").

[53] World Health Org. (WHO), *Pandemic Influenza Preparedness Framework for the Sharing of the Influenza Viruses and Access to Vaccines and Other Benefits*, World Health Assembly Res. WHA645 (May 24, 2011) [hereinafter PIP Framework], available at http://www.who.int/influenza/resources/pip_framework/en/index.html (last accessed Feb. 23, 2014). *See* Chapter 4, Section IV.A ("Basic Concepts of the WHO's Pandemic Influenza Preparedness Framework (2011)").

[54] *See* nn. 52 & 53.

[55] *See* Chapter 9, Section II.B.3 ("The Group on Earth Observations (GEO)").

[56] *See, e.g.*, Jerome H. Reichman, Paul F. Uhlir, & Heather J. Ritch, *Access to Scientific and Technological Knowledge: UNESCO's Past, Present, and Future Roles, in* STANDARD SETTING IN UNESCO – VOLUME 1, NORMATIVE ACTION IN EDUCATION, SCIENCE AND CULTURE, A. A. Yussuf ed., Martinus Nighoff Pubs. (2007).

[57] *See* United Nations Environment Programmes, *available at* www.unep.org (last accessed Sept. 7, 2015).

[58] *See* Conference of the Parties to the CBD (COP), https://www.cbd.int/COP (last accessed Sept. 11, 2015).

[59] The MIRCEN Project helped to establish a number of microbial culture collections in developing countries in the last quarter of the twentieth century. *See* UNESCO, Science, http://portal.unesco.org/science/en./en.pap-UN-ID=24918 & URL-DD=DO-topic&URL=Section=301.html (last accessed May 7, 2015).

funding problems, as well as programmatic concerns,[60] and could not realistically play as effective a support role as either the FAO or the WHO. UNEP could be a stronger candidate because of its sponsorship of the conferences that led to the CBD.[61] However, microbiology lies beyond UNEP's typical mandate, and short of some pressing emergency,[62] the World Federation of Culture Collections (WFCC), which we view as the logical foundation on which to build, would not likely cede its autonomy to UNEP,[63] even if the idea were otherwise worth considering.

As for affiliation with the CBD, that prospect would likely elicit serious objections from developed country governments who are the primary funders of both culture collection and public microbiological research.[64] It would also elicit even stronger objections from the culture collections themselves, which have been autonomously organized under the umbrella of the WFCC since 1970[65] and would presumably insist on retaining that autonomy while ensuring conformity and compliance with the CBD. Above all, as we explained at length in Chapter 9, there are good reasons to avoid the cumbersome and typically bureaucratic decision-making processes of a treaty-based organization, even if that requires other means of ensuring that the concerns of developing countries are addressed.[66]

Given these premises, the WFCC together with its digital component, the WDCM,[67] would constitute a logical foundation on which to construct a multilateral regime for facilitated access to microbial genetic resources plus related data and literature, even without undertaking either an international treaty or otherwise relying on an existing United Nations agency for support. As an empirical reality, the existing Microbial Research Commons largely depends on the WFCC's own infrastructure, as supplemented by regional networks of culture collections that have become increasingly important in recent years.[68] That is why, throughout this book,

60 See Reichman et al. (2007), n. 56.

61 See Chapter 3, Section I.B ("Foundation of an International Regime of Misappropriation to Govern Genetic Resources).

62 Recall the pressures on the CGIAR, which led to the ITPGRFA under FAO auspices, as described in Chapter 3, Sections II & III.

63 See S.P. Lapage, World Federation for Culture Collections, Xth Congress for Microbiology, Minutes of the Extraordinary Meeting of the Provisional Board, 10 August 1970, in 22 INT'L J. SYSTEMATIC BACTERIOLOGY (1972) [hereinafter Lapage].

64 *See generally* Chapters 3 & 4.

65 *See* D. Smith (2012), n. 39; Lapage, n. 63.

66 *See* Chapter 9, Section III ("In Search of a Politically Acceptable and Scientifically Productive Operational Framework").

67 For the WFCC, see Chapter 9, Section I.A.1 and Section II.B.1; for the WDCM, see further Chapter 8, Section II.B.1.

68 See Chapter 4, Section I.C ("Beyond the WFCC: Regional and Global Networks of BRCs").

we have referred to our proposals for a *redesigned* Microbial Research Commons, rather than suggesting the creation of some completely new entity.

In effect, under this approach, a redesigned Microbial Research Commons would convert the WFCC into an intergovernmental scientific infrastructure[69] overseeing what the leaders of the GBRCN would have characterized as a "network of networks."[70] At the same time, given the international tensions surrounding genetic resources and the role of intellectual property rights generally, the participating culture collections should sign onto a Memorandum of Understanding as members of the Commons, in addition to governments, and apart from the major role we assign to the WFCC, as explained below. Both participating governments and their national culture collections would thus assume obligations and responsibilities by formally adhering to an international framework agreement, to be drafted in the form of a nonbinding Memorandum of Understanding. Regional associations of culture collections would also be invited to join as members.

In this way, the public culture collection would directly undertake compliance with obligations under the CBD in the act of joining the Commons. The participating governments would simultaneously vouch for the compliance of their national collections with the CBD when joining the Microbial Research Commons.

A. *Membership and Decision Making*

The question becomes how to transform the loosely governed WFCC[71] into a more dynamic intergovernmental organization capable of sustaining a multilateral regime of facilitated access to microbial genetic resources under the Nagoya Protocol to the CBD.[72] Here the Group on Earth Observations (GEO) and the Global Biodiversity Information Facility (GBIF) afford instructive legal and institutional models for the project at hand. Both GEO and GBIF were creations of their participating governments, and both operate under nonbinding Memoranda of Understanding (MoU) agreed among those governments.[73] The GBIF, in particular, cooperates with administrators of the CBD, in keeping with Article 17 of that Convention,[74] and it has observer status in deliberations of the members of the CBD. The Secretariat

[69] *See, esp.* Michael J. Madison, *et al*, n. 32; FRISCHMANN *et al*, n. 32.

[70] *See* D. Smith (2012), n. 39. For the Global Biological Resource Centers Network, see Chapter 9, Section II.C.1.

[71] *See* Chapter 9, Section II.B.1.b.

[72] Nagoya Protocol, n. 1, art. 4.

[73] For GEO, *see* Chapter 9, Section II.B.3.b; for GBIF, *see* Chapter 9, Section II.B.2.b.

[74] *See* CBD, n. 1, art. 17.

of the CBD is a nonvoting member of GBIF's Governing Body, and is also a Participating Organization in GEO.[75]

As was the case with both GBIF and GEO, the governments seeking to establish the Microbial Research Commons on a more solid legal footing can negotiate a stand-alone framework agreement that commits members to a common purpose. Unlike GBIF, however, which created a new entity for producing information about biodiversity,[76] the framework agreement for a Microbial Research Commons would seek to preserve and expand the operations of an existing scientific infrastructure. More precisely, it would commit both member governments and participating culture collections to:

- Establishing a multilateral regime of facilitated access to *ex situ* microbial genetic resources and related data for purposes of both basic and applied scientific research within the ambit of the Nagoya Protocol to the Convention on Biological Diversity;
- Ensure the sharing of benefits, both monetary and nonmonetary, ensuing from the research activities to be supported by the multilateral regime;
- Establish a forum for capacity building with the goal of elevating many of the participating culture collections to the status of Biological Resource Centers (BRCs);[77] and
- Further enable microbiology to undertake the crucial role in the life sciences envisioned by the New Biology paradigm.[78]

Such an agreement, or Memorandum of Understanding, should inspire trust and confidence among all the participating members, taking into account the interests of both developed and developing countries. It should strive to establish a more effective, less top heavy, administrative apparatus than exists at GBIF,[79] while promoting the science-driven mission already undertaken by the WFCC and further developed by GBRCN.[80] A Framework Agreement should also allow the participating governments to disengage if they become dissatisfied with the end results, which means that it should take the form of a voluntary, nonbinding legal instrument.[81]

To achieve these goals, the MoU should recognize and protect the interests of three principal sets of stakeholders, namely, developed country governments, developing

[75] *See* n. 73.
[76] *See* Chapter 9, Section II.B.2.a.
[77] For BRCs, *see* Chapter 4, Section I.C.2.
[78] *See* Chapter 1, Section II.D ("A New Research Paradigm for the Life Sciences").
[79] *Cf.* Chapter 9, Section II.B.2.b (Governance of GBIF).
[80] *See* Chapter 9, Section II.C.1.b (Proposed governance structure of GBRCN).
[81] *See* examples of GEO and GBIF cited n. 73.

country governments, and the scientific entities charged with implementing the core mission of the Commons. While this tripartite approach augments the role of scientists in the decision-making process beyond that of other commons organizations that we surveyed in Chapter 9, it deliberately preserves the pivotal role of representatives from the participating government ministries as well.

There are compelling reasons for so doing. Government actors in both developed and developing countries already regulate existing microbial research assets under their territorial jurisdiction, and the approval of the science ministries (or their local equivalents) would encourage other regulators to facilitate transnational exchanges. Governments also remain responsible for observing relevant international laws and norms. Moreover, inputs from the science ministries may be influential or essential in preserving research space for genetic resources in an otherwise contentious atmosphere. Above all, the science ministries largely fund the national public research programs and related infrastructure, and they would logically need to support key components of the Microbial Research Commons' administrative architecture. Hence, it is worth reiterating that any proposal that avoids working through the science ministries is bound to fail.

That said, the interests of developed and developing country governments in the proposed Microbial Research Commons diverge in important respects, and they will want to protect these interests. For example, the developed country governments are the primary funders of major *ex situ* collections operating in their territories.[82] They also fund most of the research undertaken in microbiology, although developing country investments in this field are growing, especially in the BRICS countries.[83] Needless to say, industrial applications of microbiology are mostly carried out by companies in OECD countries, which is a major factor that the Microbial Research Commons must address when supporting compliance with the Access and Benefit Sharing mandate of the CBD.[84]

In contrast, while developing country governments are also funders of important culture collections, they were historically major providers of both *ex situ* and *in situ* microbial genetic resources to scientific researchers in the developed world. In that capacity, they harbor unfulfilled claims against the *ex situ* materials held by the culture collections, especially after 1992, when the CBD was adopted.[85] Moreover, developing countries continue to manage the bulk of still unknown *in situ* genetic resources, and the CBD itself was designed to establish the Access and

[82] D. Smith (2012), n. 39.

[83] *See id. See also* Chapter 4, Section I.C.1 ("Disparities Among WFCC Member Collections").

[84] *See, e.g.*, Chapter 4, Section III.A.3 ("The European Commission's Regulation on Access to and Use of Genetic Resources").

[85] *See* Chapter 3, Section II ("Destabilizing the Exchange of Plant and Microbial Genetic Resources as Global Public Goods").

Benefit-Sharing rights of developing countries as both conservers and providers of genetic resources to the world at large.[86] Ensuring that the developing countries feel confident about obtaining both monetary and nonmonetary benefits from their participation in the multilateral system is essential for its growth over time and for exploiting the favorable opportunities for scientific research that the Nagoya Protocol made possible.[87]

The third constituency consists of all the scientific entities that join or otherwise affiliate with the redesigned Microbial Research Commons, alongside the WFCC, such as the International Union of Microbiological Societies (IUMS), the Committee on Data for Science and Technology (CODATA), and the Open-Access Scholarly Publishers Association (OASPA).[88] Giving established microbiological research organizations a direct legal voice in governance is important for the perceived legitimacy of the project and its normative consequences,[89] lest they be relegated to a lesser position than the political components of the venture, as has reportedly occurred in some other multilateral commons initiatives.[90] It would help to ensure the policies adopted will actually promote the interests of the broader scientific community and thus strengthen the Commons' ability to resist pressures to limit use of research resources to a privileged few.[91]

A scientific delegation empowered with voting rights could also encourage the Common's administrators to take access to both data and literature sufficiently into account when organizing and managing the Microbial Research Commons, and not focus too narrowly on biological materials alone. In any event, scientists with voting rights would necessarily be co-involved in matters pertaining to legal and institutional affairs, just as the political delegates would necessarily be co-involved in matters bearing on the scientific mission.

To achieve its goals while protecting the interests of all these stakeholders, careful attention must be paid to how decisions will be taken by the membership. Voting rights would presumably be given to these three membership categories in conformity with a proportional representation scheme. In principle, decisions by consensus are

86 *See* Chapter 3, Section I.B ("Foundations of an International Regime of Misappropriation to Govern Genetic Resources").

87 *See* Nagoya Protocol, n. 1, art. 4; *see also* Chapter 3, Section IV.B.2 ("Facilitating Scientific Research Under the Nagoya Protocol").

88 For IUMS, see www.iums.org; for CODATA, see www.codata.org; for OASPA, see oaspa.org.

89 *See, e.g.*, Elinor Ostrom, *A Behavioral Approach to the Rational Choice Theory of Collective Action, Presidential Address, American Political Science Association*, 92(1) *Am. Political Sci. Rev.* 1–22 (1998) (stating that the costs of compliance with adopted decisions are likely to decrease with a more inclusive approach to decision-making).

90 *See, e.g.*, Halewood et al., n. 48, with regard to the Crop Commons. Similar concerns have been expressed with regard to both GBIF and GEO.

91 *But see* GBRCN's initial business plan, which contemplated a more commercial approach on the model of the ATCC, as discussed in Chapter 9, Section II.C.1.c.

the most desirable outcome in transnational entities of this kind, and this seems especially apt for an entity responsible for scientific research infrastructure. Failing a consensus, issues of importance should require a supermajority of the three constituencies identified above, perhaps amounting to a total of 75 percent of those entitled to vote.

We further recommend that a proportional voting mechanism be adopted to ensure that the scientific constituency as a whole would be entitled to one-third of the votes needed for any important decision. Member governments would similarly cast two-thirds of the votes on these same issues, always assuming that, in the absence of a consensus, a final decision would require a supermajority of 75 percent of all those entitled to vote.

Whether the government members' two-thirds proportional vote on important issues should be further subdivided by assigning one-third to developing countries (as determined by per capita GDP calculations of the United Nations) or not, is an option that also seems worth considering. As matters stand, the developing country members of the WFCC already outnumber the developed country members, even though most of the important *ex situ* collections still reside in developed countries.[92] A priori, one cannot safely predict how many developing country governments would actually join the redesigned Microbial Research Commons. Nevertheless, assigning one-third of the voting rights to developing countries would provide further incentives to induce them to join the Commons and to open their doors to more scientists conducting research in these countries. At the same time, we recognize that assigning one-third of the decision-making power to the developing countries would constitute a novel experiment whose pros and cons developed country governments would have to evaluate.

Regardless of which of these alternatives are ultimately adopted, the membership of the Commons would have to determine how a proportional voting scheme should be implemented in practice. For example, the scientific constituency would want to decide who speaks and votes for it, as must the political constituencies. There are also some other technical issues concerning membership and decision-making that we discuss separately in the next section.

B. *Ancillary Membership Issues*

Normally, under the foregoing approach, we would expect that governments signing onto the MoU would also be signatories of the CBD,[93] in which case they

[92] *See* Chapter 4, Section I.A.1 ("Aggregate Holdings and Capacity" of the WFCC). *See also id.*, Section I.C.1 ("Disparities among the WFCC Member Collections").

[93] As of June 2015, the CBD had 196 parties, of which 168 were signatories. See CBD List of Parties, https://www.cbd.inf/information/parties/shtml (last accessed June 19, 2015).

would be directly responsible for securing the compliance of their national culture collections with that Convention.[94] For example, the European Commission's Regulation on Access to and Use of Genetic Resources, as discussed in Chapter 4, attempted directly to undertake this responsibility, with debatable results.[95] One major country, the United States, does not formally belong to the CBD,[96] and only some of that country's important culture collections are affiliated with the WFCC.[97] Yet, one hopes that the United States government would adhere to the MoU establishing the Microbial Research Commons, even though it has only signed but not ratified both the International Treaty on Plant Genetic Resources for Food and Agriculture (ITPGRFA) and the CBD.[98]

Organizers of the Commons might also consider allowing public culture collections to join the multilateral regime of facilitated access under the MoU, even if their governments had not signed either the CBD or the MoU establishing the Microbial Research Commons. If governments had already joined the CBD, but had not signed the MoU regulating the Commons, the latter's administrators would be wise to require that any culture collections based in these countries should obtain the consent of their governments as a condition of membership in the Commons. If a government had signed the MoU, but did not adhere to the CBD, as could occur with the United States, the act of signing the MoU would logically constitute both a waiver of objection as to the membership of its national culture collections and a contractual undertaking to respect the rights and duties emanating from the CBD, which the MoU itself explicitly requires. In fact, the United States government has reportedly instructed its numerous culture collections to act as if that country had adhered to the CBD.[99]

Overall, we stress that the Nagoya Protocol, once implemented, will tend to make violations of the CBD directly enforceable in the courts of all the countries adhering to that Protocol.[100] The CBD and the Protocol would then indirectly provide enforcement mechanisms to back the culture collections' commitments under the MoU, even if the state in which the collection operates had not signed that MoU.[101]

94 *See* Chapter 3, Sections I.B. & IV.

95 *See* Chapter 4, Section III.A.3.

96 The U.S. has signed but not ratified the CBD. *See* http://www.org (last accessed Sept. 7, 2015).

97 Twenty-eight culture collections in the U.S. are members of the WFCC. See *Culture Collections Information Worldwide: Statistics*, http://www.wfcc.info/ccinfo/statistical (last accessed June 20, 2005). *See generally* Chapter 4, Sections I.A and I.C.1.a.

98 International Treaty on Plant Genetic Resources for Food and Agriculture, n. 52, *List of Countries*, https://www.planttreaty.org/list-of-countries (last accessed June 20, 2013).

99 Interview with Kevin McCluskey, Curator, Fungal Genetics Stock Center, University of Kansas Medical Center, December 10, 2013.

100 *See* Chapter 3, Section IV.C ("Prescriptions for Strict Enforcement of the Newly Codified Regime of Misappropriation").

101 This state of affairs could leave U.S. culture collections in a legal limbo if the U.S. did not sign the MoU, because it has not ratified the CBD either. We would expect the U.S. to join the Commons

Given the strong enforcement prospects likely to ensue from the Nagoya Protocol, we believe that, in dealing with these marginal cases, the administrators of the Commons should generally admit any public culture collections seeking to join, so long as there are no objections from the relevant national governments.

C. *Observer Status*

Observer status, without voting rights, might be conferred on either a temporary or permanent basis. Observer status would suit government representatives or other nongovernmental scientific entities interested in the work of the project, but not yet ready to commit to either funding or providing genetic resources or fulfilling other membership obligations.[102] Presumably, this category could include even certain entities that do contribute materials, data, literature, or technical support in the form of seconded personnel to the global pool of microbial research resources made available through the Commons. It may also become advisable to establish observer relations with selected governmental and nongovernmental organizations even if they cannot become formal members of the Commons as such. This category would include, for example, the Secretariats of other intergovernmental organizations, such as the CBD, WHO, FAO, UNEP, and UNESCO.[103]

in the interests of its science agencies and the microbiological research community, in keeping with current government policy advising public science agencies to operate in *de facto* compliance with the CBD.

[102] Observer status is conferred by some of the research entities surveyed in Chapter 9. In the case of GBIF, for example, permanent observers are referred to as Associate Country Participants or other Associate Participants (which category comprises intergovernmental organizations, international organizations, and other organizations with an international scope, and economies). In both cases, Associates must sign the MoU, but need not contribute to GBIF's funds and do not vote on the Governing Board. There are 16 Associate Country Participants and 39 other Associate Participants in GBIF. *See GBIF Participant List*, http://www.gbif.org/participation/participant-list (last accessed June 20, 2015).

GEO "may invite other relevant entities to support its work as observers. Observers may be invited to send representatives to GEO Committees and to engage fully in specific GEO activities." Potential observers may contact the GEO Secretariat. At present, one country and six organizations have obtained observer status at GEO. *See What is GEO: Observers*, *available at* https://www.earthobservations.org/observers.php (last accessed June 20, 2015).

[103] The need for close relations with the Secretariat of the CBD is obvious from the text of this book. *See* especially Chapters 2 and 3. Both the FAO and the WHO already support fully operational research commons for genetic resources. *See* Chapter 3, Section III *passim* (ITPGRFA) and Chapter 4, Section IV.A & B (WHO's Pandemic Influenza Preparedness Framework).

The United Nations Environment Programme (UNEP) is the U.N. agency most closely concerned with the CBD's biodiversity mandate. *See* UNEP, http://www.unep.org (last accessed June 20, 2015).

UNESCO has assisted culture collections in developing countries at different times. *See, e.g.*, the MIRCEN Project, above n. 59. UNESCO is also the primary agency for science in general. *See* UNESCO, http://en.unesco.org (last accessed June 20, 2015).

Observer status could also be conferred on relevant industry associations,[104] although their eligibility for full membership is a decision we leave to the Commons itself. In devising this governance framework, we conceive of the Microbial Research Commons as a contractually constructed bargain around the CBD to promote both basic and applied research.[105] The primary parties to this bargain are the WFCC, representing the public microbial culture collections, the WDCM as potential provider of a portal for the Commons' digital assets,[106] and the participating governments, most of which are also members of the CBD. The main goal of the entire exercise is to elaborate a Standard Material Transfer Agreement (SMTA) for microbial genetic resources having no known or likely commercial value at the time of deposit.[107] The multilateral system thus stimulates both basic and applied research without the need to negotiate hundreds of different MTAs that hinder access to such materials for research purposes.

Under the governance structure as further developed below, industry would enter the picture when any given microbial specimen first acquired some known or likely commercial value. In that event, negotiating for commercial applications becomes a private affair, to be conducted on a case-by-case basis between the parties directly involved, subject to the conditions of any applicable commons-based SMTA covering use of genetic resources and related data in return for a standard royalty as prescribed by the SMTA. The administration of the Microbial Research Commons is not directly involved in these case-by-case private negotiations.[108] By the same token, we do not see the need to seat industry representatives at the table of those responsible for governing the Microbial Research Commons, although we do recognize the desirability of hearing interested industry exponents via duly appointed observers and advisory committees, as appropriate.

Industry would benefit from the outcome of these negotiations to the extent that more unfettered upstream research will likely produce more downstream commercial applications. Transfers of upstream research results to industry

[104] For example, in the United States, the Society for Industrial Microbiology and Biotechnology (SIMB), see SIMB, htpp://www.simbhq.org, and the Biotechnology Industry Organization (BIO), see BIO, https://www.bio.org, are logical candidates for observer status. Other geographical areas should, of course, be represented.

[105] *See, e.g.*, J.H. Reichman & Paul F. Uhlir, *A Contractually Reconstructed Research Commons for Scientific Data in a Highly Protectionist Intellectual Property Environment*, 66 LAW & CONTEM. PROBS. 315 (2003).

[106] *See* World Data Center for Microorganisms, https://www.wdsm.org; see further Chapter 8, Section II.B.1.

[107] *See* Chapter 5, Section III ("Modeling a Sequence of Hypothetical Transactions").

[108] However, the administrators of the Commons organization might want to recommend certain standard provisions facilitating academic research on end products, as would be desirable with respect to patented research results generally, especially in the life sciences. *See e.g.*, the proposals discussed in Chapter 4, Section III.A.1.

for commercial applications would be freely negotiated between the relevant academic institutions and their commercial partners as in the past, subject only to a standardized reach-through royalty on use of genetic resources from the Commons that automatically satisfied the benefit-sharing provisions of the CBD.[109]

Just as it would be inappropriate for science agencies or governments to interfere with the transfer of research results from the academy to industry,[110] it would be equally inappropriate for industry to interfere with the facilitated access arrangements devised between the public research community and the CBD's beneficiary governments. A failure to clarify these different spheres of interest and legal responsibility between researchers and industry was, in our view, a defect in constructing the FAO's Crop Commons, one that we wish to avoid here.[111]

D. *The Core Institutional Components*

The description of institutional models in Chapter 9 showed that the typical governance model adopted in recently formed research or infrastructure commons consists of one large, single-entity governing body, supplemented by a robust secretariat, with inputs from a range of specialized committees, including scientific advisory committees, serving at the discretion of the governing body and the secretariat. In contrast, the point of departure for the redesigned Microbial Research Commons, as we envision it, is to avoid that kind of top-heavy governance architecture in favor of a modified version of that model, in which the scientific community would play a more direct and prominent role in both decision making and management.[112]

Specifically, we envision a governance framework in which a Scientific Coordination Council would operate in tandem with a formally constituted Governing Body, with the direct and meaningful participation of scientists in both of these components. If successfully implemented and validated in practice, the governance model sketched out in the rest of this chapter may eventually be of interest to other scientific domains and adapted to the needs of other knowledge commons initiatives.[113]

[109] *See* Chapter 5, Section II *passim*.

[110] *Cf.* Patent and Trademark Law Amendments Act (Bayh-Dole Act), 35 U.S.C. § 200 (1980). See further Chapter 2, Section III.A.

[111] For weaknesses in the Crop Commons' otherwise pioneering adoption of a Compensatory Liability Regime, *see* Chapter 3, Section III.C ("Strengths and Weaknesses of the International Treaty on Plant Genetic Resources for Food and Agriculture").

[112] *Cf.* Halewood (Louvain 2012), above n. 47.

[113] *See, e.g., E-Infrastructures and Data Management Plan Steering Committee, A Place to Stand: eInfrastructure and Data Management for Global Change Research, Belmont Forum eInfrastructures & Data Management Community Strategy and Implementation Plan* (Belmont

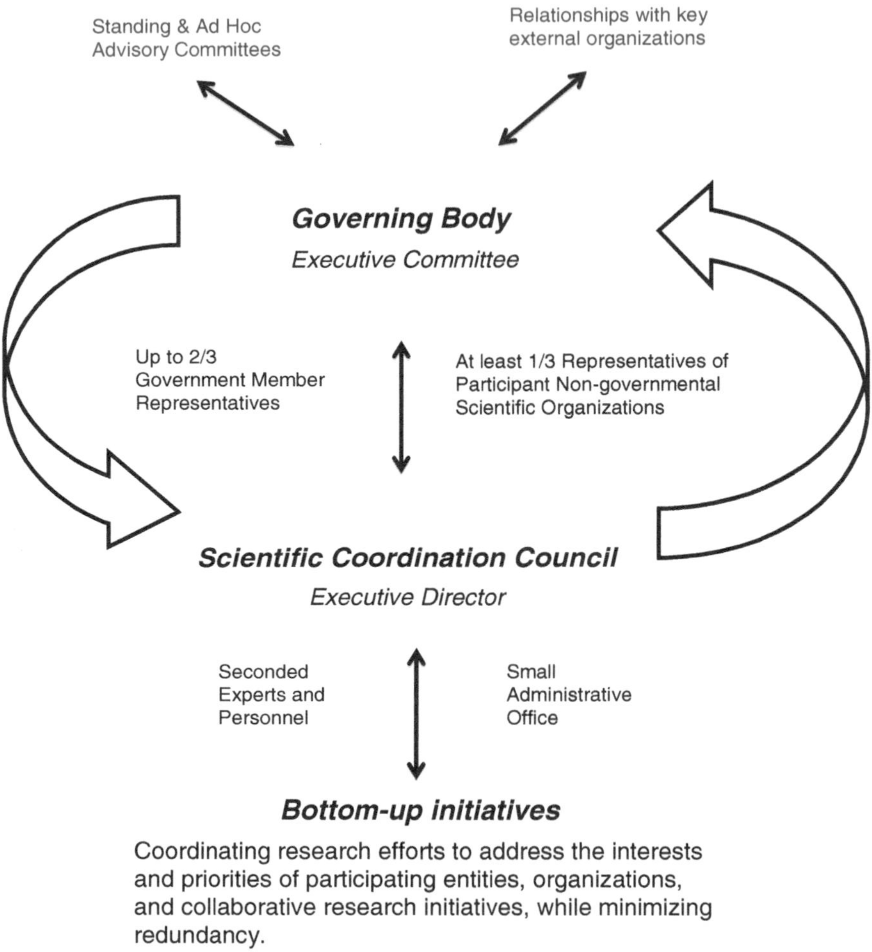

FIGURE 10.1 Proposed Governance Framework.

Figure 10.1 visually represents our proposed governance framework. A Governing Body, made up of both government officials and nongovernmental scientists, would address political, legal, and institutional relations with both

Forum, May 1, 2015), *available at*: http://www.bfe-inf.org/sites/default/files/doc-repository/DRAFT_Belmont%20Forum%20E-Infrastructures%20%26%20Data%20Management%20-%20Community%20Strategy%20%26%20Implementation%20Plan.pdf (last accessed: Sept. 13, 2015). *See also RECODE Project, Policy guidelines for open access and data dissemination and preservation* (European Commission, Feb. 2, 2015), *available at*: http://recodeproject.eu/wp-content/uploads/2015/02/RECODE-D5.1-POLICY-RECOMMENDATIONS-_FINAL.pdf (last accessed: Sept. 13, 2015).

internal and external stakeholders, including the member microbial culture collections. However, given the size of the Governing Body as a committee of the whole, and the complexity of the issues it must address over time, an Executive Committee would be needed to operate on behalf of the Governing Body between its plenary meetings.

The Scientific Coordination Council would implement the policies and programs that the Governing Body had approved. In particular, it would manage the interface between the top-down decisions of the Governing Body and the bottom-up operations of the autonomous scientific entities that depend on, and contribute to, the evolving Microbial Research Commons.

Because the Scientific Coordination Council would thus perform many of the tasks typically assigned to a Secretariat, the figure portrays a relatively small Secretariat that would undertake purely administrative and logistical duties for both of the primary governance entities. Standing, and ad hoc advisory committees of experts would be appointed, as further explained below.

In the rest of this chapter, we briefly describe the configuration of all these administrative components. We then provide a detailed account of the main issues confronting these governing entities, including financial sustainability. The overall objective is to design a more dynamic international governance architecture that remains responsive to the needs of science, one in which representatives of the participating government agencies and of the microbiological research community would deliberate and implement decisions collectively.

1. A Governing Body and an Executive Committee

The Governing Body of the Microbial Research Commons would oversee the creation and overall operations of a vast "paying public domain"[114] comprised of the voluntarily pooled *ex situ* genetic resources held by a federated consortium of participating culture collections. Initially, at least, it should not attempt to regulate access to *in situ* genetic resources, as the Crop Commons has tried to do.[115] In so doing, the Governing Body would remain responsible for decisions

[114] From the French term, *domaine publique payant*, a concept used in some national copyright systems often for eleemosynary purposes that benefit artists and authors. *See, e.g.*, JAMES BOYLE, SHAMANS, SOFTWARE AND SPLEENS: LAW AND THE CONSTRUCTION OF THE INFORMATION SOCIETY (1996).

[115] *See* Chapter 3, Section III.A (Basic concepts of the ITPGRFA). Whether, and to what extent, the Microbial Research Commons would eventually consider regulating the provision of *in situ* genetic resources to the pool would depend on posterior negotiations of the parties. *See* Section III.F.2 in this chapter. In this respect, the proposed Microbial Research Commons is less ambitious *ab initio* than the Crop Commons, but in our view correspondingly more likely to succeed. *Cf.* Chapter 3, Section III.C (Strengths and weaknesses of the ITPGRFA).

concerning membership, policies, programs, and funding. Its primary task would be to promote more unfettered exchanges and research uses of genetic resources under the multilateral system than would otherwise be possible, while safeguarding the providers' benefit-sharing entitlements under the CBD and augmenting the sustainability of the affiliated collections as a whole.[116] However, the Governing Body of the Microbial Research Commons would not have direct operational responsibility for the materials subject to its normative, rule-making authority; and these materials would continue to be housed in and managed by the federated collections operating under their respective domestic laws.[117] Respect for territorial sovereignty over natural resources is thus built into the institutional fabric of the Commons.[118]

In this context, the Governing Body of the microbial research infrastructure should address and help to resolve the common legal and institutional problems of its federated collections, with a view to safeguarding the interests of both its provider and user stakeholders.[119] It should standardize, simplify, and facilitate the general system of access to, and exchange of, its members' *ex situ* genetic resources, with a view to reconciling the demands of the Nagoya Protocol with the needs of the public scientific research community. It should also seek to augment the financial sustainability of the networks of collections as a whole, while elevating quality standards and promoting improvements to their technical capabilities, especially in developing countries. Finally, it should undertake a broader mission that focuses on the sharing of relevant data and literature, including genomic evaluation of the pooled resources.

In carrying out these tasks, moreover, the Governing Body should strive to define the jurisdictional boundaries of the Microbial Research Commons with greater clarity than has reportedly been done under the Crop Commons' administration, particularly with regard to who contributes what, when, and how to the Commons and the permissible uses of the research assets that are made available from the Commons.[120] There could also be some express conditions concerning the minimum contribution of participating entities that entitled them to the benefits they expect to gain from adhering to the Memorandum of Understanding and joining the Commons.[121] Norms applicable to the availability

116 For the importance of innovative research to improve overall sustainability of collections in this regard, see Report of the Ecological Soc'y Am. (ESA), n. 42.

117 In principle, of course, these domestic laws must conform to relevant international laws, including the CBD and the TRIPS Agreement. *See* Chapter 2, Section II.A & B; Chapter 3, Section I.B.

118 *Cf.* Universal Declaration of the Sovereignty of States Over Natural Resources (1964), discussed in Chapter 3, III.B.

119 Bearing in mind that users may become providers and vice versa. *See, e.g.*, Fritze & Oumard (2012), n. 7.

120 *See* Halewood (Louvain 2012), n. 47, at 10.

121 *See id.* at 10. *See further* Section III.C.6 in this chapter.

and use of genetic resources and digital information from the Commons, when adopted and promulgated by the Governing Body, should be monitored for compliance. Graduated sanctions for noncompliance of active members should at least be considered and evaluated.[122] The Governing Body should also consider devising a low-cost and efficient modality for resolving disputes that could help to avoid potentially disruptive activities before too much damage to the system as a whole had occurred.[123] A more detailed, analytical portrait of the Governing Body's responsibilities is set out in Section III.

Because the Governing Body would be, in effect, a committee of the whole, it probably could not afford to meet more than once a year or once every two years. That would necessitate appointing a small Executive Committee – equally representative of the whole in the sense that it mirrors the composition of the Governing Body[124] – that could meet several times per year between plenary sessions.

The Governing Body would also work closely with, and be supported by, a Scientific Coordination Council (SCC), which constitutes a key and integral component of our governance apparatus. The composition and responsibilities of the SCC are discussed in the next section.

Besides appointing the Executive Committee, the Governing Body must also elect the officers of the organization, such as a president, vice-president(s) and a treasurer. The Executive Director of the Scientific Coordinating Council should play a leadership role in the work of the Governing Body.

2. A Scientific Coordination Council (SCC) and a Small Secretariat

As was the case for other knowledge commons surveyed in Chapter 9, the Governing Body should have primary responsibility for formulating the work plan for the Microbial Research Commons and overseeing its implementation, as more fully explained later in Section III. The structure and work plan of the Commons should, in turn, be largely science driven and close to the frontiers of scientific research, as was the case with the International Human Microbiome Consortium,[125] although the Commons would not fund research projects as such.

To this end, the Governing Body would rely primarily on a permanent Scientific Coordination Council (SCC), rather than on the kind of traditional secretariat used

[122] *See* Halewood (Louvain 2012), n. 47, at 11.

[123] *Cf. id.* at 11–12.

[124] The exact composition would depend on the specific proportional voting scheme that the Commons would decide to adopt, i.e., either two-thirds government representatives and one-third scientific representatives or a tripartite regime with one-third of the votes dedicated to developing countries, as discussed earlier in Section II.A.

[125] *See* Chapter 9, Section II.B.4.

in most other commons organizations that we surveyed,[126] except perhaps for GEO. The GEO's Secretariat actually performs a mission somewhat comparable to that we assign to the SCC.[127] However, the SCC would be a continuing, high-level body that would deal directly with the Governing Body and its Executive Committee and with all the scientific entities, culture collections, research institutes, and infrastructure components that potentially interact with the Commons from the bottom up, as indicated in Figure 10.1. The SCC would thus play a still more active role than GEO's Secretariat, and it would constitute an integral governance and implementation unit of the Commons, which may obviate the need for an external Scientific Advisory Committee.

The need for a public-sector, science-driven operational unit to oversee coordination of the activities implementing the mission of the Microbial Research Commons and to take responsibility for technical management of its digital services was partly explored in our discussion of integrating digital knowledge resources in Chapter 8.[128] In keeping the Commons' membership attuned to cutting edge developments in microbiology and genomics in general, the SCC would enable scientists, culture collection managers, technical experts, and funders' representatives to assess and address the proper role of the Commons initiative as they went,[129] rather than entirely entrusting that mission to a more distant Governing Body or to a more bureaucratic Secretariat. The SCC would thus directly manage relations between the Microbial Research Commons and the larger research community, and it would also seek to develop satisfactory relations with private-sector research entities and the industrial microbiology sector. In so doing, the SCC would need to have the requisite expertise to implement the Governing Body's approved plans for technical, legal, and capacity-building programs. The SCC could also be charged with preparing an annual report on implementation of the multilateral system as a whole, with specific regard to its impact on scientific research.[130]

As for microbial materials, the SCC should facilitate and coordinate relations with culture collections in general, whether affiliated with the WFCC or even held externally by government agencies or university laboratories. It would particularly seek to upgrade and certify quality standards and to optimize the exchange of materials for research and other purposes, subject to the benefit-sharing regime

[126] *See* Chapter 9, Section II.A (ITPGRFA); *id.* Section II.B.2 (GBIF), II.B.3 (GEO), and II.C.1 (GBRCN).

[127] *See* Chapter 9, Section II.B.3.b (Governance of GEO).

[128] *See* Chapter 8, Section II.C. ("Linking the Open Knowledge Environments to the Materials Infrastructure"). *See also* the example of the Advisory Group established to assist the WHO with its Pandemic Influenza Framework, discussed in Chapter 4, Section IV.B.

[129] *Cf.* the International Human Genome Project's approach, as described in Chapter 9, Section II.B.4.

[130] *Cf.* the annual reports required by the Advisory Group to the WHO's Pandemic Influenza Framework, Chapter 4, Section IV.B.

adopted by the Governing Body. The SCC should thus promote common technical standards, help to implement the Standard Material Transfer Agreements (SMTAs) that the Governing Body will have to devise,[131] as well as other policies set by the Governing Body, and generally act as an internal clearinghouse for such purposes when necessary.[132]

Here, the MIRRI project recently underway in the European Union could become a valuable pilot model for the role we ascribe to the SCC.[133] It envisions "clusters of expertise" to address priority issues facing culture collections generally; to deliver technical solutions more swiftly and efficiently; to address problems faced by specific Biological Resource Centers and the needs of users; and to liaise with governments. MIRRI also envisions a coordinated response to the different needs of its diverse member collections, and it intends to interact with other research infrastructures servicing the broader scientific user community.[134]

If the SCC were also able to help rescue valuable culture collections in danger of destruction or abandonment, it would make a particularly important contribution. In addition, the SCC should systematically seek to identify especially useful specimens held at universities and other nonmember institutes, which could either be transferred to member collections or kept in suitable quality conditions at their existing institutions and eventually made available through the multilateral system governed by the Microbial Research Commons. In that way, the "Big Refrigerator" problem might be progressively ameliorated through selectively networked affiliations.[135]

Of equal importance, the SCC should take direct responsibility for the online portals that would serve both members and nonmembers of the Commons. The processes of digitally linking the Commons community will take constant oversight and upgrading, although much depends on the successful integration of a workable digital infrastructure, as already partly developed, by the WDCM.[136]

Beyond linking the networked culture collections themselves, the SCC should support the WDCM's plan to continue to forge links to all the microbial data and information websites of all the member entities, both governmental

[131] *See* Section III.C in this chapter.

[132] The issue of who, at the international level, is tasked to certify quality standards, especially those pertaining to BRC standards, remains unsettled and controversial. Interview with Anita Eisenstadt, State Dep't, Washington, D.C. (Nov. 2, 2012).

[133] *See* Chapter 9, Section II.D ("The Next Step: The Microbial Resource Infrastructure (MIRRI) as a European Stepping Stone to the GRBCN").

[134] Dagmar Fritze, *A Common Basis for Facilitated Legitimate Exchange of Biological Materials, Proposed by the European Culture Collections' Organization (ECCO)*, 4 INT'L J. COMMONS 40,7 (2010).

[135] For the limits on the capacity of microbial culture collections to store specimens, see Chapter 4, Section I.A.1.

[136] *See* Chapter 8, Section II.B.1 ("The World Data Center for Microorganisms").

and nongovernmental, that operate on open terms. These links would provide thematically relevant and valuable inputs for both governance deliberations and for attaining the objectives of the research infrastructure itself. Efforts should also be made to link the unified digital portal of the Commons to any other umbrella databases and repositories that emerge over time, as discussed in Chapter 8,[137] and the SCC should encourage other governments and stakeholders to develop more open digital resources.[138]

In this regard, relations with new and existing Open Knowledge Environments (OKEs) of the kind discussed in Chapter 8 could become an important part of the SCC's duties.[139] For example, the SCC could develop standard templates for pooling materials, data, and literature that would make it easier for newly formed OKEs, such as those funded by the private foundations and several U.S. science agencies, to quickly and more efficiently fulfill their aspirations. Once underway, the SCC can help the OKEs to link with each other and to the broader microbial community under the auspices of the Common's own digital portal.[140] In the event that an OKE runs its course and ceases to exist, the SCC should try to ensure that its acquired digital resources remain accessible through the Microbial Research Commons digital architecture, whenever feasible.[141]

The SCC should, in particular, have primary responsibility for implementing capacity-building initiatives designed for developing country members, as discussed later in this chapter.[142] The Governing Body or the SCC could negotiate such projects on a case-by-case basis, with due regard to funding by the member governments, as well as by grants or other sources (such as a Trust Fund from benefit-sharing proceeds).[143] The SCC – with direct inputs from funders and scientific experts drawn from the culture collections – would be the logical entity to coordinate and oversee technical assistance for capacity building initiatives in cooperation with the local beneficiaries. At the same time, the SCC would report to, and be guided by,

137 *See* Chapter 8, Section III.A ("Examples of Incipient Open Knowledge Environments on the Frontiers of Microbiology").

138 *Cf.* Fritze & Oumard (2012), n. 7 ("linking the microbial strain information to other relevant data will allow their full exploitation").

139 See Chapter 8, Section III ("Building Transnational Open Knowledge Environments").

140 For the proposed digital portal, see Section III.D in this chapter.

141 For example, the now defunct CAMERA research project's data could be linked to, and made available through, the Microbial Research Commons. *See* Chapter 8, Section III.A.2. The very existence of the Microbial Research Commons as a potential digital repository of last resort may make funders more willing to establish OKEs designed to meet the research goals of specific thematic communities in the first place. However, the Governing Body of the Commons may not wish to undertake such a commitment. *Cf.* GEO and GBIF, discussed in Chapter 9, Sections II.B. 2 and 3, which will not assert direct control of external digital repositories.

142 *See* Section III.E.

143 *See further* Section III.E ("Relations with Developing Countries").

the Governing Body in all policy matters relating to capacity building, whether in developing or developed countries.

Given its operational importance, the SCC could largely be made up of seconded microbiologists, technical experts, and research administrators drawn from member governments, funding agencies, and public scientific organizations who would serve for perhaps one or two year terms.[144] It should be led and managed by the Executive Director of the Commons mentioned in the previous section.

Finally, we have deliberately downgraded the role that a typical secretariat plays in the organizations described in Chapter 9 (as well as others of which we are aware), in order to upgrade the role of the Scientific Coordination Council as the direct, implementing agent for the policies that the Governing Board will have approved. Nevertheless, a small administrative office would still be needed for routine business matters, such as organizing and servicing meetings, managing any physical facilities and equipment used by the Commons, and managing disbursements and other budgetary matters. In our view, these services, whether in-house or outsourced, should be placed under the responsibility of the Executive Director of the Microbial Research Commons.

3. Advisory Committees

Because the Scientific Coordination Council will have been deliberately invested with broad, ongoing operational authority, the need for permanent advisory committees should be correspondingly reduced. That is also the reason we do not necessarily envision the need to appoint a separate Scientific Advisory Committee. However, some permanent or ad hoc experts may be needed to provide targeted inputs on specific scientific and technical questions.

At least two standing advisory committees appear necessary. A Legal Affairs Committee would be needed to assist the organization with implementing the multilateral regime of facilitated access to microbial genetic resources and related data under Article 1 of the Nagoya Protocol.[145] A standing Committee on Capacity Building may also be advisable, with the understanding that it would take on ad hoc experts to address the needs of each new country seeking admission to or assistance from the organization.[146]

[144] For example, such a secondment arrangement has successfully been used to staff the GEO Secretariat. *See* Chapter 9, Section II.B.3.

[145] *See* Chapter 3, Section IV ("New Constraints and Opportunities for Scientific Research Under the Nagoya Protocol"); Chapter 4, Section IV ("From the Bilateral to the Multilateral Approach"). *See also* Nagoya Protocol, n. 1, art. 4.

[146] *See further* Section III.E in this chapter.

Other ad hoc or permanent advisory committees may also be needed to help with formulating specific organizational policies that the Governing Body would have to approve. Once approved, however, the SCC should retain primary responsibility for implementing such policies and proposals, with the understanding that it could enlist the expertise of these committees or other outside sources as needed. By the same token, the SCC will need to obtain and internalize some legal and capacity-building expertise, in addition to its more scientific and technical expertise.

III. IMPLEMENTING THE MULTILATERAL REGIME FOR FACILITATED ACCESS TO *EX SITU* MICROBIAL GENETIC RESOURCES

The first and most pressing task for a redesigned Microbial Research Commons is to enable its member culture collections to opt out of the CBD's bilateral approach to exchanges of *ex situ* genetic resources by opting into the multilateral approach that Article 4 of the Nagoya Protocol made possible.[147] Once this multilateral regime has been established on a solid legal footing, the Commons membership can address other important issues, some of which were raised by the GBRCN's Demonstration Project discussed earlier in Chapter 9.[148]

Here we emphasize the need to enhance the quality standards of member culture collections, and to facilitate access to genetic sequences and related data, as well as to the relevant literature within a technically advanced digital infrastructure. The Commons would also need to improve relations with provider developing countries through capacity building and the provision of other nonmonetary benefits, in addition to the sharing of monetary benefits from commercial applications of genetic resources made available from the multilateral regime. Maintaining the trust of developing-country participants will also require careful attention to appropriate enforcement and dispute resolution mechanisms. These and other important substantive issues are addressed in this section.

When considering the legal obligations of a redesigned Microbial Research Commons from this perspective, a guiding principle is to avoid all the fragmenting effects of the domestic laws regulating access to, and use of, genetic resources and related information that would otherwise apply under the bilateral approach of the CBD.[149] In so doing, the Governing Body must break with the legal artifice currently adopted by the leading culture collections, which seeks to absolve them of

[147] *See* Nagoya Protocol, n. 1, art. 4. *See generally* Chapter 3, Section IV ("New Constraints and Opportunities for Scientific Research Under the Nagoya Protocol").

[148] For GBRCN, *see* Chapter 9, Section II.C.1. For MIRRI, *see id.*, Section II.C.3.

[149] *See* Chapter 3, Sections I & IV.

responsibility to providers of *ex situ* materials under the bilateral approach by virtue of their status as trusted intermediaries.[150] Because the Nagoya Protocol recognizes no such legal escape hatch,[151] the Governing Body of the proposed Commons must directly assert, enable, and maintain the special status of participating networks of culture collections under the multilateral route that Article 4 of the Nagoya Protocol has expressly carved out. Here, the primary responsibility of the Governing Body is to "support" the Access and Benefit-Sharing (ABS) principles set out in Articles 2 and 4 of the Nagoya Protocol by ensuring that both provider states and user states (including the national and regional networks of culture collections) share responsibility for compliance with the CBD by dint of their membership in the multilateral regime and its Governing Body.[152]

Because the CBD holds user states responsible for their culture collections' compliance with its Mutually Agreed Terms (MAT), Prior Informed Consent (PIC), and Access and Benefit Sharing requirements, user states must themselves shift this responsibility to the Governing Body and the Microbial Research Commons.[153] Otherwise, the culture collections themselves would have to fulfill more onerous obligations to secure the MAT, PIC, and ABS arrangements under the CBD than current practice envisions, with a corresponding increase in transaction costs and restrictions on basic research. Even with the advent of a multilateral approach, the culture collections remain responsible for ensuring transparent dealings between them and their users, which the Governing Body should seek to implement via a Standard Material Transfer Agreement (SMTA).

From this perspective, a primary responsibility of the Governing Body is to serve as the official guarantor of the global Microbial Research Commons compliance with the CBD and its Protocol. The Governing Body must accordingly determine the basic contractual terms and conditions to govern the exchange process as a whole. These provisions would deal chiefly with microbial materials, on the one hand, and the digital access portals for data and literature, on the other.

A. *Promoting and Certifying Quality Standards*

Elevating the quality standards of participating culture collections must constitute a primary goal of a redesigned Microbial Research Commons. For example,

[150] *See* Chapter 4, Section III.A ("Efforts to Negotiate More Research Friendly Material Transfer Agreements").

[151] *See* Christine Godt, *Networks of* Ex Situ *Collections of Genetic Resources, in* COMMON POOLS OF GENETIC RESOURCES: EQUITY AND INNOVATION IN INTERNATIONAL BIODIVERSITY LAW (E.C. Kamau & G. Winter eds., Routledge 2013) [hereinafter COMMON POOLS OF GENETIC RESOURCES (2013)].

[152] *See*, Godt (2013), n. 151, at 258.

[153] *Cf.* Chapter 3, Sections III.B & C.

while efforts to convert elite public "culture collections" into Biological Resource Centers (BRCs) in conformity with the OECD's Best Practices and Guidelines were explicitly promoted by the GBRCN's Demonstration Project,[154] the bulk of the collections in both developed and especially developing countries have not acquired the technical capacities needed to meet even WFCC quality standards, let alone those of the BRCs.[155] A particularly serious problem in this regard is that thousands of non-WFCC collections would fail to provide even the tracking, authentication, and validation services that are the hallmarks of WFCC members.[156]

The WFCC has recently adopted new guidelines for eligible member collections, which aim for a middle ground between the OECD standards for BRCs, requiring heavy capital investment, and the technical capacities of the bulk of its member collections, which cannot afford such investments at all.[157] While the Governing Body must ultimately establish minimum quality standards, a logical policy would be to encourage all participating collections to meet at least the new WFCC guidelines, while welcoming and integrating the full-fledged BRCs that are emerging in different parts of the world.[158]

Once the Governing Body had established minimum quality standards and a process for assisting its network of collections to meet them, it could perform a valuable additional service if it undertook to certify the different levels of quality actually attained by its participating repositories. The Scientific Coordination Council, discussed earlier, could be charged with these responsibilities.[159] The benefits of such a certification process were expressly recognized by GBRCN in its Demonstration Phase submissions following earlier OECD recommendations.[160] However, there is at present no recognized international entity able to manage such a certification process, which is reportedly a political stumbling block.[161] Attention to this issue can thus become a key organizing principle for the redesigned Microbial Research Commons, one that could thus yield political good will for the Governing Body.

[154] *See* Chapter 9, Section II.C.1 & 2; *see also* Chapter 4, Section I.B. ("From Culture Collections to Biological Resource Centers").

[155] World Fed. Culture Collections (WFCC), *Guidelines for the Establishment and Operation of Collections of Cultures of Microorganisms* 11, ¶ 14.3 (3d. ed., Feb. 2010), *available at* http://www.wfcc.info/guidelines/ [hereinafter WFCC, *Guidelines*]. *See further* Chapter 4, Section I.A.2. *See also* id., Section I.C.1 ("Disparities Among the WFCC Member Collections"), and Section I.C.2 ("The Emerging BRC Networks"). *A fortiori*, most of the collections held at research laboratories and academic institutes could not normally meet these standards.

[156] *See* Chapter 4, Section I.A ("The Pivotal Role of the World Federation for Culture Collections").

[157] *See* n.155.

[158] *See* Chapter 4, Section I.C.2 ("The Emerging BRC Networks").

[159] *See* above Section II.D.2 in this chapter.

[160] *See* Chapter 9, Section II.C.1 (describing GBRCN).

[161] *See, e.g.*, Anita Eisenstadt, *International Developments: A Context for the Creation of a Microbiology Commons, in* Designing the Microbial Research Commons, 55–63.

With specific regard to the thousands of important collections held at research centers and academic institutions, the task of the Scientific Coordination Council should be to encourage and assist those having particular scientific value to affiliate with the federated research commons when feasible. By nudging the "informal sector" of material exchanges into conformity with the Nagoya Protocol in this way, the risk of massive violations of the CBD can be attenuated. For this and other reasons, so long as these entities are able to meet minimum quality standards set by the Governing Body, including tracking, authentication and validation requirements, they should be encouraged to attain eligibility for membership in the Commons.

Obviously, the goal of raising the minimum acceptable quality standards for participation in the Microbial Research Commons' network of culture collections would require some funding commitments, as discussed later in this Chapter. With regard to developing countries, it would also necessitate considerable technical assistance and capacity building, which would become the responsibility of the Scientific Coordination Council.[162] The expected payoffs would, however, be correspondingly large, if the *ex situ* holdings of developing countries were more fully integrated into a global system of research promoting exchanges built on the Compensatory Liability Regime, as illustrated in Chapter 5,[163] and discussed in the following sections.

Finally, quality standards will also become important when constructing the digital component of the Microbial Research Commons. These standards must emerge from consultations between the Governing Body and the Scientific Coordinating Council, which, in turn, would collaborate with the relevant data and information providers.[164]

B. *Defining the Conditions of Legitimate Exchange*

Another major preliminary step in organizing cross-border exchanges of microbial materials on research-friendly terms, would require the Governing Body to define what constitutes a "legitimate exchange" under the auspices of the Microbial Research Commons and its institutional affiliates. That definition would then become a fundamental provision of the Standard Material Transfer Agreement that the stakeholders would eventually negotiate.[165]

The concept of legitimate exchange was discussed earlier in Chapter 4, in connection with the SMTA of the European Culture Collections' Organization

[162] *See* Section III.E in this chapter.
[163] *See* Chapter 5, Sections II.A & B.
[164] *See* Section III.D later in this chapter.
[165] *See* Section III.C later in this chapter.

(ECCO).[166] The basic idea is that, once defined, a legitimate exchange policy would apply to all the participating culture collections that met minimum quality standards, with the result that users of any participating collection could access the holdings of all the other member collections on the same facilitated conditions that the Governing Body will have elaborated. In principle, this agreement would authorize any qualified member collection to transfer specified holdings to other member collections for research purposes. Redistribution for similar purposes would also be authorized to and from any qualified member collections under the same contractual conditions embodied in a viral license that would follow any given specimens.

However, single researchers receiving specimens from these participating collections would normally be limited to use in their own labs, or between partners in different institutions collaborating on a defined joint project. They would not have any right to redistribute these specimens,[167] unless the receiving research entity was itself a fully qualified participating member of the Commons. These restrictions are necessary to ensure that the specimens exchanged are effectively tracked, authenticated, and validated for research purposes.

In this and other respects, the materials component of the Microbial Research Commons would operate as an ever-expanding semicommons. It would not become a fully open commons, owing in part to its quality requirements, especially tracking, authentication, and validation, not to mention biosafety and biosecurity obligations, as well as the internal logic of its own "paying public domain" principles. Moreover it could not allow unregulated transfers to the world at large without risk of violating the Nagoya Protocol to the CBD,

Nevertheless, we propose that the legitimate exchange conditions to be devised by the Governing Body would need to deviate from those embodied in the ECCO MTA, despite a common goal of accommodating the CBD. For example, ECCO's license stipulates that the object of the transfer must be for "noncommercial research purposes,"[168] with an express or implied duty to notify of any "change of intent." Instead, we prefer that transfers under the Microbial Research Commons framework would not be subject to any "commercial-noncommercial research" distinction at all, on

[166] *See* Chapter 4, Section III.A.2 ("The Core MTA of the European Union Culture Collections' Organization"). *See also* Dagmar Fritze, *A Common Basis for Facilitated Legitimate Exchange of Biological Materials, Proposed by the European Culture Collections' Organization (ECCO)*, 4 *Int'l J. Commons* 40 (2010).

[167] *See* European Union Culture Collections' Org. (ECCO), The ECCO Core Material Transfer Agreement for the supply of samples of biological material from the public collection, Feb. 2009 [hereinafter ECCO MTA], *available at* http://www.eccosite.org/docs/ECCO_core-MTA_V1_Feb09.pdf, discussed earlier in Chapter 4, Section III.A.2. *See also* Fritze (2010), n. 166. All research collaborators affiliated with the lab would, of course, be allowed to use the specimen in the accessing laboratory.

[168] ECCO MTA, n. 167, "Legitimate Exchange," quoted in Fritze (2010), n. 166, at 11.

the premise that simplified access conditions, with fewer constraints on users, would likely generate more scientific research.[169] To that end, our Commons approach invites research for either commercial or noncommercial purposes because the resulting financial benefits would be shared under the Compensatory Liability Regime implemented in the SMTA as elaborated by the Governing Body.[170] A rudimentary version of this approach was embodied in the SMTA developed for the FAO's Crop Commons, as explained elsewhere.[171]

Alternatively, if the member states of the Commons were nonetheless to insist on retaining a "noncommercial use" clause, with a duty to notify any change of intent[172] the Compensatory Liability Regime would nonetheless authorize any such change of intent, subject to the stipulated reach-through royalty on downstream commercial applications to be devised by the Governing Body. Either way, establishing the Compensatory Liability Regime would be one of the highest priorities of the Governing Body.

C. Drafting an SMTA to Establish the Compensatory Liability Regime: The Critical Issues

One of the Governing Body's most important tasks would be to devise a new SMTA that implemented the Compensatory Liability Regime applicable to exchanges of microbial materials for research purposes. That legal regime should incentivize culture collection managers to make available, for research purposes, the bulk of their materials that had no known or likely commercial applications at the time of deposit.[173] Under the aegis of the Microbial Research Commons, that regime would apply broadly across the whole network of collections that had opted to join the Commons infrastructure and met its minimum quality requirements. At the same time, we assume that the

[169] Gorch Detlef Bevis Fedder, *Biological Databases for Marine Organisms: What They Contain and How They Can Be Used in ABS Contexts*, *in* COMMON POOLS OF GENETIC RESOURCES (2013), n. 151, at 278. *See further* Chapter 5, Sections II.C.4 & III.

[170] *See* Chapter 5, Sections II & III.

[171] *See* Chapter 3, Section III.B.

[172] *See, e.g.*, Nagoya Protocol, n. 1; IUCN, GUIDE TO THE NAGOYA PROTOCOL (2012), n. 50.

[173] *See* Chapter 5, Section I.A.2 ("The Flawed Premise of the Properietary Ethos"). As observed in Chapter 5, this "take and pay" regime invites third parties, whether academics or private firms, to find commercial uses for such materials in return for a duty to share a small percentage of the proceeds with those who initially contributed the material available or with their designated agents. Cf. Jerome H. Reichman, *Of Green Tulips and Legal Kudzu: Repackaging Rights in Subpatentable Innovation*, 53 *Vand. L. Rev.* 1743 (2001). This compensatory liability regime also satisfies the crucial benefit-sharing provisions of the Convention on Biological Diversity with respect to microbial materials originating from developing countries. *See further* Chapter 4, Section IV ("From the Bilateral to the Multilateral Approach").

existing system of informal exchanges of genetic materials would become untenable because of its direct conflict with the Convention on Biological Diversity.[174]

As a preliminary condition, all member governments in the Microbial Research Commons must appoint Designated National Authorities for ABS purposes, and these authorities must be registered with the Governing Body of the Commons. A second preliminary condition is that the Governing Body, in negotiations with prospective members, would have to identify the specific collections that each Party commits to the international system of facilitated access and the nature of their respective holdings of microbial materials. In our view, all the materials held by the collections thus identified should be available to the multilateral system, except only for those special deposit collections having known or likely commercial value or specimens under international patent rules that are expressly excepted at the time each government adheres to the Commons.

As envisioned in Chapter 5, the Governing Body would officially become the agent of the governments participating in the Microbial Research Commons for purposes of implementing the obligations of the Nagoya Protocol to the CBD with respect to facilitated exchanges of the *ex situ* microbial materials held by, and distributed through, its network of affiliated culture collections. The Memorandum of Understanding (MoU) to be signed by both governmental and nongovernmental members should explicitly confer this authority upon the Governing Body. Moreover, as envisioned earlier in this chapter the Governing Body would construct, the legal architecture supporting the Compensatory Liability Regime, but it would not directly manage or engage in its day-to-day operations. The culture collections themselves would thus continue to play the pivotal role in operationalizing cross-border exchanges of microbial materials. However, this role would now be subject to the SMTA to be developed by the Governing Body for the Commons as a whole, and to any oversight and enforcement modalities established by that Body with the consent of its members.

As regards the nature of the Compensatory Liability Regime itself, we provided six detailed scenarios in Chapter 5 with comments to illustrate how it would operate in practice.[175] Many important governance issues were treated there, and they are incorporated here by reference. For example one basic condition was that *ex situ* microbial materials, once contributed to the international system, could not subsequently be withdrawn.[176] Another major premise was that users could not seek

[174] GBRCN's Demonstration Project suggested that the existing system of informal exchanges could largely be replaced by the networked system of Biological Resource Centers it envisioned. *See* Chapter 9, Section II.C.1.c.

[175] *See* Chapter 5, Sections II.C & III.

[176] *See id.*, Section III.A.

to apply intellectual property rights to such materials in the form in which they were received from the multilateral system.[177]

In this connection, the Governing Body must also decide whether all *ex situ* materials held by participating collections that have no known or likely commercial value will be made available for all research purposes, subject to "take and pay" rules as we recommend; or, alternatively, whether a noncommercial research clause, with a duty to inform of any posterior change of intent, also subject to the liability rule, would instead be adopted in the SMTA.[178] If the latter solution were to be adopted – which we disfavor – then the SMTA must address confidentiality issues that could arise with a declaration of changed intent.

In either case, the SMTA must prescribe the use of Global Unique Identifiers for tracking purposes, a feature that was not adopted in the FAO's International Treaty on plant genetic resources.[179] As explained in Chapter 5, tracking has become indispensable for the integrity of microbiology in general, apart from the need to determine commercial uses for purposes of benefit-sharing under the CBD. The Governing Body would, accordingly, need to determine the measures concerning tracking to be inserted in the SMTA. The SMTA must also address the potential need for Certificates of Compliance, as envisioned by the Nagoya Protocol.[180]

As further indicated in Chapter 5, the Governing Body would have to establish these and other rules and procedures for institutionalizing and enforcing the multilateral regime of facilitated access. In particular, that Body must devise the SMTA to implement the "take and pay" liability rule; it must determine the quantum of royalties to be applied, and the modalities for collecting and distributing those assets to the rightful beneficiaries; and it must determine how to oversee the operations of the system, with particular regard to resolving disputes and enforcing compliance with the SMTA and the applicable international legal obligations it implements. These and other related matters are discussed later in this section.

1. The Question of a Users' Surcharge

One of the first decisions the Governing Body would have to take is whether the Microbial Research Commons should, or should not, impose a users' surcharge on all *ex situ* genetic resources made available from the multilateral system, in addition to the culture collections' usual distribution fees.[181] Such a surcharge, if adopted,

[177] *Cf.* Chapter 3, Section III.B (Basic concepts of the ITPGRFA).

[178] *See* Chapter 5. Section II. *See also* Nagoya Protocol n. 1, art. 4.

[179] *See* Chapter 3, Section III.B–C. For tracking mechanisms, see Chapter 4, Section I.A.2 and Chapter 5, Section II.C.3.

[180] Nagoya Protocol, n. 1, art. 17.1(a)(iii).

[181] *See, e.g.*, JULIANNA SANTILLI, AGROBIODIVERSITY AND THE LAW: REGULATING GENETIC RESOURCES, FOOD SECURITY AND CULTURAL DIVERSITY 134–35 (Earthscan 2012) [hereinafter SANTILLI (2012)];

would provide a new source of income for a Global Trust Fund to be managed by the Commons,[182] over and above the royalties from downstream commercial applications that commercial users would devolve to the benefit of provider countries.[183]

Proposals to impose a user access surcharge or tax have arisen under the FAO's Crop Commons, where its liability rule will not begin to elicit significant contributions to the relevant Global Fund from commercial applications of plant genetic resources for several more years.[184] Hence, even a small user surcharge, applied to many thousands of accessions, would generate considerable funds to defray the costs of the multilateral regime, and those proceeds could be placed in a Global Trust Fund established by the Microbial Research Commons. The surcharge could also lessen the need to seek funds from participating governments.

However, an access surcharge could also weaken the public-good approach by imposing additional burdens on academic and other institutional users, especially if it were not kept to a bare minimum. In the microbial context, some users may currently pay nothing (although an absence of fees is rare) to access *ex situ* genetic resources, or more likely, merely the marginal costs of distribution, unless they are purchasing materials from ATCC or certain other collections operating under a more proprietary business model.[185]

User surcharges could particularly trouble researchers in developing countries who, for example, are reportedly the biggest users of plant genetic resources from the Crop Commons, where the line between providers and users may be especially blurred.[186] Even if bulk users of transborder exchanges under the multilateral system might be able to cap their costs by means of an annual subscription fee,[187] that approach must still factor in the vulnerability of research users in developing countries. For these and other reasons, any decision to impose user access surcharges,

Paul A. David, *Breaking Anti-Commons Constraints on Global Scientific Research: Some New Moves in "Legal Jujitsu"*, *in* DESIGNING THE MICROBIAL RESEARCH COMMONS (2011), n. 20; Claudio Chiarolla & Stefan Jungcurt, *Outstanding Issues on Access and Benefit Sharing under the Multilateral System of the ITPGRFA* (Berne Declaration & Dev. Fund Background Study Paper, Mar. 2011), http://www.eub.ch/con_data/ ITPGR_ABS_Study_1.pdf [hereinafter Chiarolla & Jungcurt (2011)].

182 *Cf.* Nagoya Protocol, n. 1, art. 10.

183 *See* Section III.C.2 in this chapter.

184 *See, e.g.*, SANTILLI (2012), n. 181, at 134–35; for the liability rule, *see generally* Chapter 3, Section III.C. For expected returns, see Chapter 9, Section II.A.2.a (The Viral Licenses) and II.A.2.c (Long-Term Funding Arrangements).

185 *See* Chapter 4, Section II.A (ATCC and Progeny), Chapter 9, Section II.C (GBRCN's proposed business model). For the WFCC, *see generally* Chapter 4, Section I.A.

186 Discounts could also be given to those who make their culture collections available to the multilateral system. *See* SANTILLI (2012), n. 181, at 135–38.

187 A proposal that has recently surfaced under the Crop Commons. See Chapter 9, Section II.A.2.c ("Long-Term Funding Arrangements").

over and above a liability rule for commercial applications, depends in part on how the Microbial Research Commons will resolve funding and sustainability issues in general, a topic addressed in Section IV.

In this context, however, one should recognize the potential importance of securing some early monetary benefits for developing-country providers to the Microbial Research Commons. Early returns would avoid some of the mistrust that has been generated by the benefit-sharing provisions of the ITPGRFA[188] and would further encourage developing-country governments to commit their culture collections to the Commons.

2. Quantum and Duration of Royalties

In Chapter 5, we provided detailed illustrations of the proposed Compensatory Liability Regime for research uses of *ex situ* microbial materials having no known or likely commercial value at the time of deposit.[189] In elaborating six implementation scenarios, we evaluated the possibility of adopting a standard royalty rate of 2 percent on gross commercial application of microbial materials accessed from the multilateral system, with the possibility of awarding a slightly higher rate of 3 percent for the provision of value adding data along with relevant materials.[190] These royalties are to be paid by the relevant commercial entities directly to the Designated National Authorities established under Article 10 of the Nagoya Protocol, as discussed later in the next section.[191] We will not amplify our discussion of the pros and cons of that approach here, other than to emphasize that this would remain an important issue for the Governing Body to address when drafting the SMTA.

Questions have also been raised about the duration of the user's obligations to pay royalties under the Compensatory Liability Regime. Specifically, the Governing Body, in drafting its SMTA, would have to determine whether the benefit-sharing provisions under a contractually constructed commons regime should expire after a

[188] *See* Chapter 3, Section III.C ("Strengths and Weaknesses of the International Treaty on Plant Genetic Resources for Food and Agriculture"). Some OECD countries have made substantial voluntary contributions to support farmers under the Crop Commons in response to this issue. *See* Chapter 9, Sections II.B.1.c & d.

[189] *See* Chapter 5, Section III (six scenarios illustrating the application of the Compensatory Liability Regime to microbial genetic resources made available from the proposed multilateral system of facilitated access).

[190] *See* Chapter 5, Section II.C.4 ("The Calculus of Royalties from Commercial Applications"). In the event of multiple claimants for an application of the same microbial specimens, the total royalty to be shared could not exceed four percent, which was presented as an outer limit to royalty stacking. *Cf.* Arti K. Rai et al., *Pathways Across the Valley of Death: Novel Intellectual Property Strategies for Accelerated Drug Discovery*, 8 YALE J. HEALTH POL'Y L. & ETHICS 1 (2008), *available at* http://scholarship.law.duke.edu/cgi/viewcontent.cgi?article=2329&context=faculty_scholarship.

[191] *See* Nagoya Protocol, n. 1, art. 10.

specified period of time, or whether they should continue to apply indefinitely, and at the same rate, so long as the products resulting from upstream research uses of specified microbial genetic resources continued to be sold on relevant markets.[192] This question arises precisely because the Compensatory Liability Regime deliberately substitutes a standard benefit sharing deal for access to and use of *ex situ* microbial genetic resources for research purposes, in lieu of case-by-case negotiations with onerous transaction costs. Answering this question – in the context of a standard deal – could depend in turn on how one conceptualizes the royalty scheme in the first place.

If, for example, one views the *ex-ante* liability rule – a "take and pay" rule – as a contractually agreed analog of, or substitute for, an exclusive intellectual property right,[193] then there is an argument that neither intellectual property rights nor contracts last forever. Even on that view, however, some intellectual property rights last a very long time, such as the life plus fifty-year term of copyright protection under the Berne Convention,[194] now life plus seventy years in the United States and the European Union.[195] To the extent that benefit-sharing under the CBD is intertwined with recognition of the contributions that the traditional knowledge of indigenous people have made to the global store of technical know-how over many generations,[196] a long period of remuneration could be justified even under an intellectual property rationale.

If, in contrast, the benefit-sharing regime embodied in the SMTA is viewed as a form of payment to governments for the preservation of global biodiversity under the CBD,[197] the compensatory royalties accruing from research applications may, in turn, be characterized as a tax that properly follows gainful exploitation of the relevant products, so long as there is any economic demand for them.[198] This characterization is further supported by the concept of a "paying public domain" even under intellectual property laws, which some countries have used as a means to support eleemosynary institutions of interest to authors and artists.[199] In any

[192] *See, e.g.*, SANTILLI (2012), n. 181.

[193] *Cf.* Jerome H. Reichman, *Of Green Tulips and Legal Kudzu: Repackaging Rights in Subpatentable Innovation*, 53 *Vand. L. Rev.* 1743 (2000), *available* at http://scholarship.law.duke.edu/faculty_scholarship/456. *See also*, Chapter 5, Sections II & III.

[194] Berne Convention for the Protection of Literary and Artistic Works, Sept. 9, 1886, *as last revised* July 24, 1971, 828 U.N.T.S. 221, art. 7(1) [hereinafter Berne Convention].

[195] *See* Chapter 6, Section II.A.

[196] Jerome H. Reichman & Tracy Lewis, *Using Liability Rules to Stimulate Innovation in Developing Countries: Application to Traditional Knowledge*, *in* INTERNATIONAL PUBLIC GOODS AND TRANSFER OF TECHNOLOGY UNDER A GLOBALIZED INTELLECTUAL PROPERTY REGIME 337–366 (K.E. Maskus & J.H. Reichman, eds., Cambridge U. Press 2005) [hereinafter Reichman & Lewis].

[197] *See, e.g.*, Nagoya Protocol, n. 1, art. 1.

[198] *See, e.g.*, Reichman & Lewis, n. 196.

[199] *See, id.*

event, the question of the duration of compensatory royalties under the SMTA is one on which the Governing Body would have to reach a negotiated consensus, in the absence of any officially determined benefit-sharing term of duration under the CBD.

3. Protocols for the Distribution of Royalties

Earlier in this book we raised the possibility that the members of the Microbial Research Commons might agree to devolve a percentage of royalties accruing from the Compensatory Liability Regime for the upkeep and maintenance of the Commons' operations.[200] We have also suggested that the Contracting Parties might decide to impose a users' surcharge on accessing microbial materials from the multilateral system, over and above royalties from commercial applications, to help defray the operating costs of the Commons as a whole.[201] It bears emphasizing that, if such a surcharge were to be levied, it would serve a different purpose from that of royalties imposed on commercial applications, the latter representing benefits to be directly shared with provider countries under the multilateral system.

Here, instead, we focus on the need for members, working through the Governing Body, to identify the proper beneficiaries of such commercial royalties and to formulate protocols for their distribution to those beneficiaries. Needless to say, such protocols must be consistent with international law, particularly the Convention on Biological Diversity as implemented in the Nagoya Protocol.

One option available to the Governing Body is to decide that royalties accruing from downstream commercial applications of microbial materials accessed from the Commons should be paid into a trust fund managed by the Commons. Another option is that these proceeds should instead flow directly back to the country that initially provided the materials to the relevant culture collection from which they were accessed. In our view, the better decision would favor paying the royalties directly to provider states, for redistribution locally to those responsible for preserving biodiversity. The absence of any similar provision directing benefits to provider states under the Crop Commons has been identified as a cardinal weakness of the FAO's International Treaty, one that has reportedly disinclined many developing countries from fulfilling their commitments under that Treaty.[202]

Assuming that the Governing Body takes this point, one might argue that, in principle, the Governing Body should consider how the proceeds from commercial applications of microbial research might be distributed in the country of origin.

[200] *See* Chapter 5, Section III.A.5 ("Sales of the Product Trigger the Liability Rule and Distribution of Royalties").

[201] See Section III.C.1 in this chapter.

[202] *See* Chapter 3, Section III.C.2.

For example, rewarding the researchers who first provided genetic resources to the participating culture collections might incentivize other potential providers in the future.[203] In practice, however, attempting to direct revenue streams from commercial applications either to single investigators or even to the culture collections that provided microbial genetic resources for the research in question could pose daunting legal and administrative problems.[204]

Instead, the simplest, most efficient and cost effective solution is the one already suggested by the Nagoya Protocol, which requires all signatory countries to establish a Designated National Authority for purposes of implementing that country's obligations under the CBD.[205] Building on that precedent, the most workable solution available to the Governing Body would be to specify the Designated National Authority under the CBD, whether in developed or developing countries, as the appropriate agent to receive compensatory liability proceeds under the SMTA. Participating governments not members of the CBD – or participating collections whose governments were not formally members of the Commons – would also need to appoint a Designated National Authority for these purposes.

The SMTA should accordingly spell out the information needed by the Designated National Authorities, as well as both the liabilities of commercial users and the enforcement responsibilities of those Designated National Authorities. As indicated in Chapter 5, the SMTA would also take the form of a viral license that imposed its terms of exchange on any posterior users of microbial materials accessed from the multilateral system. By the same token, the SMTA would not empower the culture collections to collect and distribute the resulting royalties. The SMTA should also determine the accounting and reporting obligations imposed on users, and specify binding mediation and arbitration procedures for resolving disputes, as discussed later in this section.

203 Scientists in both developed and developing countries must thus overcome tendencies to hoard genetic materials and related data for further research purposes after publication, or out of fear of forfeiting unexpected commercial prospects later on. These incentives could also help to encourage academic researchers and research laboratories to upgrade the quality of their collections in order to make them available for global exchanges within the Commons infrastructure.

204 In the case of academic providers, for example, their entitlements and shares would likely depend on their employment contracts, or any applicable statutes, and on the specific policies that the relevant laboratories and universities had adopted in this regard. In the case of culture collections, many if not most are funded by governments that might logically see themselves as more entitled to such monetary benefits than the collections they fund. *See, e.g.*, Chapter 4, Section I.A.3 ("The Perennial Problem of Funding"). Moreover, the greater the emphasis that is put on distributing proceeds from the Compensatory Liability Regime directly to either researchers or to relevant culture collections, the higher the tracking and transaction costs are likely to be, with a concomitant reduction of the net amounts actually pocketed by provider states. *See, e.g.*, Chapter 5, Section II.C.3 ("Tracking Mechanisms to Maintain the Chain of Custody").

205 Nagoya Protocol, n. 1, art. 13 (requiring National Focal Points and Designated National Authorities or one entity for both purposes).

In all cases, each of the national authorities so designated would then distribute royalties received according to its own priorities, as determined by national governments. Local governments would thus determine the persons or institutions that most deserved to share in the benefits received, including any indigenous communities whose traditional knowledge may have contributed to the commercial applications in question.[206]

Governments whose Designated National Authorities were thus entitled to receive monetary benefits could conceivably waive their rights in favor of other options. As indicated, for example, under the FAO's International Treaty on Plant Genetic Resources for Food and Agriculture, which has already adopted a variant of the Compensatory Liability Regime, the signatories have agreed to pay any resulting proceeds into a Trust Fund that will support future research grants.[207] By the same token, under the voluntary Memorandum of Understanding to establish the Commons that we propose, participating governments entitled to proceeds under the Compensatory Liability Regime applicable to microbial materials could waive the payment in favor of some other distribution arrangement. Nevertheless, absent such a waiver, the Governing Body of the Commons would remain obligated to direct all proceeds covered by the SMTA to the CBD's Designated National Authorities. The Governing Body would then have no say as to how those royalties were ultimately to be distributed, or who the final beneficiaries were to be, unless it had negotiated an up-front agreement with Designated National Authorities to this effect.

Another corollary of this approach is that the Designated National Authorities in provider countries would logically assume direct and primary responsibility for enforcement of the benefit-sharing provisions of the SMTAs. In so doing, those Authorities would rely on the strong enforcement provisions added to the CBD by the Nagoya Protocol, including the possibility of direct access to national courts for this purpose.[208] This approach would cure serious defects reported in the governance of the FAO's Crop Commons, where national authorities must rely on the FAO as a third-party enforcer, rather than on their own legal initiatives.[209] It would also ensure direct payments of royalties from commercial users to the stakeholders most interested in those receipts, with less pressure on formal enforcement machinery being provided by the Microbial Research Commons itself.[210]

[206] *See, e.g., id.*, art. 12; IUCN, GUIDE TO THE NAGOYA PROTOCOL (2012), n. 50, at 137–41.

[207] *See* Chapter 3, Section III.B (regarding plant genetic resources under the Crop Commons).

[208] *See* Chapter 3, Section IV.C ("Prescriptions for Strict Enforcement of the Newly Codified Regime of Misappropriation").

[209] *See id.*, Section III.C (discussing the role of the FAO as Third Party Enforcer).

[210] *See further* Section III.C.6 & 7 in this chapter.

Besides administrative simplicity, this reliance on Designated National Authorities would lower transaction costs for the Microbial Research Commons, although transaction costs in the receiving countries could rise at the expense of some incentive effects. Another administrative advantage would lie in having essentially parallel distribution mechanisms in all user and provider countries. This solution would also largely insulate the Governing Body from disputes about distribution of proceeds at the local level, which would remain the primary responsibility of the Designated National Authority or its equivalent in developed countries.

Payments to provider governments would have a direct incentive effect on their willingness to build the Commons, whatever the subsequent internal distribution of proceeds and its incentive effects might be. Conversely, evidence that commercial applications were actually yielding financial rewards to some provider countries by such means would encourage other countries to make their microbial materials more available over time.

Finally, in Chapter 5, we suggested that the Designated National Authority might be put under an obligation to share a small portion of the resulting proceeds with the relevant culture collection from which materials were accessed, or with the Commons' own management, or conceivably both. Such a percentage would be particularly appropriate if no up-front users' surcharges were to be collected for the Microbial Research Commons. If provider countries did earmark a small percentage of royalties to cover the costs of the collection plus a share for the Commons itself, a Global Trust Fund could use that income stream for maintaining the Commons' activities and for capacity building purposes in developing countries. Whether the participating governments would approve such a division of compensatory royalties, and in which percentages if they did, would depend on the outcome of negotiations between the Governing Body and its member governments.

In any event, the Governing Body will have to embody its decisions about these matters in the SMTA applicable to all participants. In practice, some variations on the standard deal may necessarily arise in negotiations with new countries acceding to the MoU, or with collections affiliating with the Commons even if their governments had not signed the MoU.[211] The Governing Body would also have to consider how best to oversee implementation of the Compensatory Liability Regime, and it should provide for internal dispute resolution mechanisms, as discussed later in this chapter.[212]

[211] *See* Section II.B ("Ancillary Membership Issues").
[212] *See* Section III.C.7.

4. New Uses of Pre-1992 Microbial Materials

As indicated in Chapter 3, the status of microbial materials collected from provider countries before 1992, when the CBD took effect, remains controversial. Arguments for and against their coverage under the CBD have been advanced with no clear resolution of this issue.[213] However, Article 10 of the Nagoya Protocol itself envisions that new uses of such pre-1992 materials are to be covered by the benefit-sharing provisions of the CBD, with the proceeds to be paid into a Global Fund that the CBD would use "to support the conservation of biological diversity and the sustainable use of its components globally."[214]

Regardless of the uncertainties that remain under international law, the Governing Body can resolve this issue by agreement of its membership in consultation with the Secretariat of the CBD. If they agree that new uses of pre-1992 *ex situ* materials will be subject to the Compensatory Liability Rule, in keeping with Article 10 of the Nagoya Protocol, then the obligation to pay the resulting royalties from commercial applications to a Global Trust Fund should be embodied in the SMTA, along with the standard notification procedures. Such funds should be used "to support the conservation of biological diversity and the sustainable use of its components globally,"[215] which suggests that they could be made available to support both *ex situ* and *in situ* conservation of microbial genetic resources, especially in developing countries. However, this decision might require the approval of the CBD's own administrators, given that Article 10 of the Nagoya Protocol can be read to imply that payments for new uses of pre-1992 materials should be made to the CBD's own Global Trust Fund.

Once the Governing Body of the Microbial Research Commons established its own Global Trust Fund it could be used to accommodate other borderline situations that the Protocol may or may not envision. For example, so-called "transborder situations" may arise where the specific origin of the specimens at issue cannot be determined or is disputed by several countries.[216] There may also be pre-1992 situations in which documentable traditional knowledge was used without prior informed consent.[217] By covering these cases, the SMTA can ensure that a measure of equitable benefit-sharing from downstream applications of such resources is secured.

[213] *See* Chapter 3, Section IV.A ("Clarifying the Broad Economic Scope of the CBD"); *see also* IUCN, Guide to the Nagoya Protocol (2012), n. 50, at 127 (stating that "there is enduring controversy around the issues that Article 10 expressly and implicitly addresses").

[214] Nagoya Protocol, n. 1, art. 10.

[215] *Id.*

[216] *See, e.g.*, IUCN, Guide to the Nagoya Protocol (2012), n. 50, art. 10.

[217] *See* Nagoya Protocol, n. 2, art. 10.

5. Genetic Sequences and Other Related Data

As indicated in Chapter 3, the Nagoya Protocol is understood to clarify the application of the CBD to all relevant data – including especially genetic sequence data and derivatives – related to the microbial materials available for transborder exchanges under existing or future access arrangements of either a bilateral or a multilateral nature.[218] Professor Gerd Winter has recently called attention to the resolve of the CBD's Conference of the Parties to ensure compliance with this interpretation.[219] The growing importance of *in silico* research methods underscores the importance of this view.[220]

The Compensatory Liability Regime applicable to downstream commercial applications of genetic resources taken from the international system could conceivably be drafted to cover such data. The use of persistent identifier tags, if this technology becomes further developed, would make this decision more feasible.[221] If so, the SMTAs would have to be crafted to deal with this potential source of revenue.

In practice, however, as Professor Winter points out, enforcement of such claims would face daunting obstacles, given the current lack of available means to track the results of product development and commercialization based on data derived from pooled microbial materials.[222] Any notification and record-keeping provisions to be adopted for microbial materials, as discussed later,[223] would necessarily need to extend coverage to the relevant data as well. But new questions would also arise; for example, whether the mere genetic sequencing of microbes taken from the multilateral system should automatically be notified to the Designated National Authority.[224] How to craft specific legal terms and conditions covering genetic resource data for these purposes in the SMTA would at best have to be worked out in consultations between the Governing Body and the CBD's authorized representatives.[225]

[218] *See* Chapter 3, Section IV.A ("Clarifying the Broad Economic Scope of the CBD"). Nagoya Protocol, n. 1.

[219] *See, e.g.*, Gerd Winter, *Data Banks of Genetic Information: How They Are Organised and Affected by ABS Issues, in* COMMON POOLS OF GENETIC RESOURCES (2013), n. 151 Ch. 20 [hereinafter Winter, *Data Banks of Genetic Information*] (noting difficulties of enforcement); *see also* Remarks of Prof. Gerd Winter, 2d. Gen. Assembly Meeting, Micro B3 Project, Max Planck Inst. Marine Microbiology, Bremen, Germany, April 23–25, 2014.

[220] *See* Chapter 1, Section II.B ("The Revolution in Genetic Science").

[221] Winter, *Data Banks of Genetic Information*, n. 219; George M. Garrity et al., *Towards a Standards-Compliant Genomic and Metagenomic Publication Record*, 12 OMICS 157 (2008); Fedder, n. 169, at 278. *See also* Chapter 5, Section III.B.3.

[222] *See* Winter, n. 221.

[223] *See* Section III.E.

[224] For notification in general, *see* Chapter 5, Section III.A.3–5.

[225] For the CBD's permanent observer status at the Microbial Research Commons, *see* Section II.C in this chapter.

Even then, the difficulties of authoritatively tracing data usage back to specific materials from specific provider states may be impracticable. According to Professor Winter, still another option to consider is the possible "introduction of a general biodiversity charge to be laid on any product derived from the utilization of G[enetic]R[esources] and paid into a global biodiversity fund."[226] This proposal resonates with the possibility of adopting user surcharges for access to materials from the multilateral system in general, as discussed earlier.[227]

6. Prescribing Minimum Conditions of Reciprocity

When forming any given "knowledge commons," the founders should consider the extent to which those who agree to pool resources should be obliged to meet and maintain minimum threshold requirements concerning their respective contributions as a condition of membership.[228] From a theoretical perspective, such conditions bear on expectations of attaining the collective benefits that motivate single researchers to participate in a knowledge commons from the start.[229] They also satisfy the perceived need for clear boundaries concerning the perimeters of the Commons and access to its resources, as well as on the need to ensure that operations of the Commons conform to the rule of law.[230]

Critics of the FAO's Crop Commons have recently focused on this very issue.[231] They complain that member countries have so far failed to contribute *in situ* plant genetic resources covered by the International Treaty, or even to provide data concerning the existence or availability of such resources in their respective territorial public domains.[232] While the *ex situ* resources held by the Consultative Group on Agricultural Research (CGIAR) centers and other European gene banks remain freely and widely available to agricultural researchers in both member and nonmember countries alike, this discrepancy reportedly encourages free-riding and reduces the incentives to join the International Treaty in the absence of sanctions for noncompliance.[233]

226 *See* Winter, n. 219.

227 See Section III.C.1.

228 *See* Chapter 9, Section I.A.3 ("Potential Payoffs from a Well-Designed Governance Model").

229 Minna Allarakhia et al., n. 27, at 50–66.

230 *See, e.g.*, ELINOR OSTROM, GOVERNING THE COMMONS: THE EVOLUTION OF INSTITUTIONS FOR COLLECTIVE ACTION (Cambridge Univ. Press 1990); FRISCHMANN, ET AL, n. 32. *See also*; Halewood (Louvain 2012), n. 47, at 10.

231 *See, e.g.*, Halewood (Louvain 2012), n. 47, at 10; Chiarolla & Jungcurt (2011), n. 181.

232 *See* Halewood (Louvain 2012), n. 47, at 10 (stating that "[o]f the 127 country members, only approximately 20% have shared information about what materials are available from them, in the system, on a website maintained by the treaty secretariat.").

233 *Id.* at 10 (stating that it also "undermines the sense of cohesiveness and potential shared purpose of the countries that are members.").

In posing this question with specific regard to redesigning the proposed Microbial Research Commons, we begin by recalling the need for the Governing Body to adopt minimum quality standards that would necessarily limit the number of culture collections eligible for membership.[234] Adding a negotiated reciprocity requirement as a minimum threshold obligation affecting the member countries' *ex situ* microbial culture collections' commitments could, in principle, risk further shrinking the number of qualified collections willing to join the semicommons. It could thus limit the total amount of microbial materials made available for research purposes under an SMTA with standard benefit sharing arrangements.

For this and other reasons, the Governing Body should approach this question with caution. For starters, we contend that the redesigned Microbial Research Commons as envisioned here should not initially seek to regulate access to *in situ* resources at all. Efforts in this direction might subsequently become more feasible if the proposed common pool of *ex situ* resources proved successful and fostered a high level of trust among the participating countries. Unlike the Crop Commons, in other words, we recommend that the Microbial Research Commons leave *in situ* genetic resources to the bilateral approach, at least for the short and medium term.[235]

Another consideration is that the existing public culture collections – or at least those affiliated with the WFCC – already make their microbial genetic resources available to researchers everywhere, without regard to reciprocity, as matters stand.[236] Efforts to unite these collections within a federated distributed network of digitally integrated providers should in principle aim to augment, rather than to diminish, the aggregate quantity of microbial materials available to the global research community, absent some overriding considerations to the contrary that we do not foresee at the present time.[237]

If the Governing Body decided not to negotiate any minimum threshold requirements bearing on the quantum of materials that a participating government must pledge to contribute initially and to continue to make available thereafter, certain disincentives to free ride would nonetheless be built into the system, at least as we have described it so far. For example, nonmember culture collections could not participate in the multilateral regime of facilitated access to be implemented by the SMTA. On the contrary, collections that stayed outside the multilateral system would remain subject to the bilateral regime that the CBD imposes by default.[238] These nonmember collections would thus continue to incur the risk of liability for

[234] *See* Section III.A.

[235] For the treatment of *in situ* resources under the Crop Commons, *see* Chapter 3, Section III.A.

[236] *See* Chapter 4, Section I.A.2 ("Servicing the Broad Microbial Research Community").

[237] *But see* GBRCN's Demonstration Project, Chapter 9, Section II.C, which initially at least aimed to form a chain of elite collections to this end.

[238] *See* Chapter 3, Section I.B ("Foundations of an International Regime of Misappropriation to Govern Genetic Resources").

their users' failure to comply with the CBD, despite their attempts to obtain immunity via exculpatory contractual clauses that cast them as trusted intermediaries only.[239]

Users of microbes from these nonmember collections must likewise discharge all the burdens of the bilateral regime via case-by-case negotiations *ex ante* with host country providers of the *ex situ* materials in question.[240] We believe that most researchers would generally prefer easy access under the multilateral system, with no obligation to negotiate case-by-case terms until actual commercial opportunities had emerged. If so, this state of affairs could gradually marginalize nonmember collections due to a loss of comparative advantage.

Meanwhile, those non-member collections would have none of the advantages of membership in the Commons. For example, they would not qualify for any scientific and technical assistance, including certification of quality standards and capacity building;[241] they would not obtain the full array of benefits arising from digital integration of all the Commons assets;[242] and they would have no voice in the governance of the Commons at any level.

If researchers in these non-member countries needed access to the aggregate genetic resources held by the federated network of microbial culture collections, they would nonetheless remain bound by the viral SMTA of the Commons and its benefit sharing provisions. Any commercial applications ensuing from the research in question would thus require outward payments of reasonable royalties to Designated National Authorities under the Compensatory Liability Regime.[243] Moreover, a fundamental working hypothesis underlying the incentives to form such a research commons in the first place was that frictionless access to upstream genetic resources via the SMTA would generate more research, more commercial applications, and correspondingly more royalties from the use of microbes having no known or likely commercial value at the time of deposit than would otherwise accrue from case-by-case negotiations.[244]

[239] *See, e.g.*, Chapter 4, Section III.A.3 ("The European Commission's Regulation on Access to and Use of Genetic Resources") *and id.*, Section III.B ("Opting Out or Opting In? Limits of the Trusted Intermediary Approach").

[240] *See* Chapter 3, Section I.B.2 ("Access and Benefit-Sharing Under the Convention on Biological Diversity").

[241] *See* Section III.A in this chapter.

[242] *See* Section III.D in this chapter.

[243] Presumably the conditions imposed by the SMTA would become justiciable under the Nagoya Protocol at the national level, with or without adherence of either provider or user country to the Commons architecture, because states members of the CBD must enforce its ABS provisions via the duty to collaborate under the Nagoya Protocol. *See* Chapter 3, Section IV.C ("Prescriptions for Strict Enforcement of the Newly Codified Regime of Misappropriation").

[244] *See* Chapter 5, Section III.C ("Modelling a Sequence of Hypothetical Transactions"). Ex-hypothesis, microbes having some known or likely commercial value will have been directed to so-called "special collections," where access and use are logically determined via case-by-case negotiations.

The Governing Body could nonetheless seek to attenuate some residual risks of free-riding by direct negotiations with all collections that intended to join the proposed Microbial Research Commons. For example, the Governing Body could require a census of the holdings of existing culture collections, and the Scientific Coordination Council could periodically update and validate reports on the current contents of participating collections.[245] The Governing Body could also require a commitment to include specific qualifying collections identified by the Commons as a condition of initial membership, subject to the terms and conditions applicable to all participating collections. Further commitments concerning efforts to upgrade the quality standards of certain collections should also figure in negotiations for initial adherence to the organization.[246] In general, the Governing Body should have the authority to conduct such negotiations on a case-by-case basis at the time each state or its culture collections apply for membership in the Commons, subject to conditions deemed mutually satisfactory to the other members.

7. Mediation and Dispute Resolution

As indicated in Chapter 5 and elsewhere, the strong transborder regime of misappropriation established by the Nagoya Protocol makes it logical to entrust enforcement of the SMTA's obligations concerning the payment of royalties under the Compensatory Liability Regime to the member governments' own National Focal Points on Access and Benefit Sharing and their Designated National Authorities.[247] The Protocol itself expressly makes the National Focal Points "responsible for liaison with the Secretariat" of the CBD,[248] and also allows the Parties to designate a single entity to fulfill the functions of both focal point and competent national authority.[249] Governments that had not adhered to the CBD would nonetheless need to appoint a Designated National Authority if some of their culture collections sought membership in the Commons.[250]

Because the Governing Body would not directly engage in or otherwise manage the daily operations of the Microbial Research Commons,[251] the first point of contact between users of materials and the Commons would normally occur at the level of

[245] *Cf.* Chiarolla & Jungcurt (2011), n. 181. The WFCC affiliates might have to make similar reports on *ex situ* holdings to the WDCM as a matter of course. *See* Chapter 8, Section II.B.1 (describing the World Data Centre for Microorganisms).

[246] *Cf.* GBRCN's Demonstration Project to the same effect, Chapter 9, Section II.C.1.a.

[247] *See* Nagoya Protocol, n. 1, art. 13.

[248] *Id.* art. 13.1.

[249] *Id.* art. 13.3.

[250] For this possibility, see Section II.B. ("Ancillary Membership Issues").

[251] *See generally* Chapter 5, Section III.A ("The Standard Deal in Six Scenarios").

the single participating culture collections. Under the MoU, the collections would continue to view themselves not as owners of the genetic resources in question, but as custodians or intermediaries. Now, however, they would also become agents on behalf of the global Microbial Research Commons.[252]

In this capacity, the culture collection would notify the Designated National Authority in the provider's country of origin that an SMTA had been signed by a given user of the microbial material for a specified purpose. If the country of origin were unknown, as could easily occur for deposits made prior to 1992 when the CBD was signed, the culture collection should instead notify the Designated National Authority in its own country, which would operate as a trustee of the Commons in overseeing compliance with the Commons' own legal obligations and those of the Convention on Biological Diversity.

On reflection, it seems logical for the collections to routinely notify the Designated National Authorities (or National Focal Points as the case may be) in both the provider and user countries.[253] This approach follows from the fact that the Nagoya Protocol seems to make cooperation between such countries a cardinal principle for enforcing Mutually Agreed Terms.[254]

Legitimate questions may arise as to whether a separate registration system should be used for accessions from the multilateral system, over and above the single culture collections' duty to notify the Designated National Authority in the country of origin. In Chapter 5, for example, we suggested that the Governing Body might require all accessions from the multilateral system to be initiated via a master portal, with a registration system that would notify either the Commons management or the prospective Global Clearing House, or both, of SMTAs issued for *ex situ* materials.[255] Such a registration system, if adopted, could reinforce the tracking mechanisms also discussed in Chapter 5, for purposes of enforcing the payment of royalties covered by the SMTAs in question.[256] It could also enhance the users' reputation benefits that are of primary importance to scientific researchers. The costs of maintaining a registration system along these lines could be more manageable if it were part of the WDCM's existing data-management activities on behalf of the WFCC.[257]

[252] For the view that this claim cannot legally be sustained under the bilateral approach, *see* Godt (2013), n. 151; *see also* Chapter 4, Section III.B (stressing limits to the trusted intermediary approach).

[253] Nagoya Protocol, n. 1, art.13. If and when an ABS Clearing House were established, *see id.* art. 14, that would also be a logical point for notification.

[254] Nagoya Protocol, n. 1, arts. 18(2), 18(3).

[255] *See* Chapter 5, Sections II.C.2 & 3. *See also* SANTILLI (2012), n. 181, at 157 (discussing SMTAs under the FAO's International Treaty).

[256] For tracking, *see* Chapter 5, Section II.C.3 ("Tracking Mechanisms to Maintain the Chain of Custody").

[257] *See* Chapter 8, Section II.B.1 (describing the WDCM).

Once any given exchange transaction were completed and the research begun, any monitoring and reporting on the progress of commercial applications, up to the point where sales of products or services occur, would remain the responsibilities of commercial users and the Designated National Authorities in the provider and possibly the user countries, as determined by the SMTA.[258] The extent to which the intermediary culture collection remained involved in this process, if at all, could be left either to the collections themselves to decide or to the Governing Body of the Commons. In any event, once sales of products or services had triggered the user's duty to pay reasonable royalties under the SMTA, that user must notify the Designated National Authority in the country of origin, and it must then account for the proceeds from commercial applications and transmit the agreed royalties to that Authority or its agent for this purpose.[259]

Although the primary responsibility for enforcing the SMTAs thus rests with the Designated National Authorities, the Governing Body of the Commons should nonetheless retain a supplementary interest of its own in the fulfillment of these obligations. In case of disputes arising about ownership of specific microbes or about entitlements to royalties under the liability regime, for example, it is the rules of the Commons – under an international Framework Agreement – that would be tested and interpreted, as well as the international rules on ABS. All members of the Commons, accordingly, would have a vested interest in the methods by which decisions concerning these rules are to be taken and enforced.

In approaching these issues, the laws applicable to any given dispute require some clarification. As we have already indicated, an SMTA regulating relations between providers and users of microbial genetic resources itself implements an agreement by governments and other participating entities to bargain around the Convention on Biological Diversity. The SMTA may thus logically be construed as creating a "safe harbor"[260] from the rigors of the bilateral approach under the Nagoya Protocol,

[258] This would include any notifications of "changed intent," if the SMTA adopted that approach, i.e. from noncommercial to commercial uses.

[259] The Designated National Authority could appoint the providing culture collection its agent for purposes of monitoring and collecting royalties, if it so desired. If the SMTA provided for a share to be devolved to a Trust Fund on behalf of the Microbial Research Commons, the Designated National Authority would be responsible for this remittance. For statistical purposes, Designated Authorities would notify the Commons and/or the Clearing House of payments made under the Compensatory Liability Regime (unless otherwise subject to notification under the SMTA).

[260] A "safe harbor" is a legal term indicating an agreed means of avoiding certain legal consequences that would otherwise occur. For example, if online service providers in the U.S. comply with the Notice and Takedown Regime of Section 512 of the Digital Millennium Copyright Act of 1998, they obtain a "safe harbor" against complaints by copyright owners sounding in contributory infringement. *See, e.g.*, Jerome H. Reichman, Graeme B. Dinwoodie & Pamela Samuelson, *A Reverse Notice and Takedown Regime to Enable Public Interest Uses of Technically Protected Copyrighted Works*, 22 *Berkeley Tech. L.J.* 981, 981–1060 (Summer 2007).

provided that users strictly comply with the provisions and obligations of that SMTA. Breach of the SMTA would give rise not only to a contractual dispute between the commercial users of the relevant research results and the Designated National Authority in the provider country or its agent; it would also deprive the commercial user of its right to shelter within the safe harbor created by the SMTA. In that event, both the terms of the Framework Agreement supporting the Microbial Research Commons and the provisions of the Nagoya Protocol could apply to the dispute in question, with the possibility of triggering the responsibility of the violator's own state for a breach of international legal obligations under the CBD.[261]

The governments that agree to establish the Microbial Research Commons by adhering to the Framework Agreement have, in effect, undertaken that their nationals will either comply with the rules of the Commons, as embodied in its SMTA, or suffer the consequences under the Nagoya Protocol to the CBD. It is, in fact, this express or implied undertaking that makes the SMTA a safe harbor against the rigors of the bilateral approach under the CBD.

Against this background, it might greatly simplify matters if, in the case of disputes arising under the SMTA, the Designated National Authorities of either the provider or the user state could directly lodge a cause of action, alleging breach of that SMTA by a given commercial entity, in the courts of the state having jurisdiction over that commercial entity. In principle, the Nagoya Protocol would already have obliged that commercial entity's own state to provide a jurisdictional foundation for such actions in its courts anyway, independently of its obligations under the Framework Agreement establishing the Microbial Research Commons.[262] If so, violations of the SMTA, including a failure to pay the promised royalties, could lead to seizure of the relevant goods by order of the courts in member countries.[263] In that event, the culture collections that had provided materials under the SMTA could assist the local authorities to reach a just result, but no dispute resolution machinery of the Commons would be needed.

Now that the Nagoya Protocol has entered into force, and further assuming that the Contracting Parties fully implement it, the role of any dispute resolution mechanism to be established by the Commons would be supplementary in nature. It would primarily serve to bolster the confidence of members that the multilateral

[261] Contracting Parties in the Commons thus accept a Framework Agreement that implements the rules and principles of the CBD via a multilateral system, and in so doing, they waive recourse to claims rooted solely in the bilateral approach of the CBD. If, however, the rules of the Commons were themselves violated, or inadequately enforced, members would, in principle, have the residual right to resurrect claims under the CBD they had otherwise waived.

[262] *See* Nagoya Protocol, n. 1, arts. 18(2), 18(3). For the risk that a defaulting company's products might be seized for violating the SMTA and the CBD, see Chapter 5, Section III.A. & B.

[263] *See, e.g.*, Margo A. Bagley & Arti K. Rai, *The Nagoya Protocol and Synthetic Biology Research: A Look at the Potential Impacts*, (2013), available at http://law.duke.edu/faculty_scholarship/3230.

system of exchange would deliver both the monetary and nonmonetary benefits expected from it. Rules adopted by the Governing Body for this purpose would be backed up by the duties of cooperation in enforcing ABS rules under Article 18 of the Nagoya Protocol.[264]

That said, compliance with other obligations undertaken by the Commons itself, not necessarily required by the Nagoya Protocol, could also necessitate access to a dispute resolution forum to be established by the Governing Body. For example, if that Body had negotiated state-by-state commitments regarding the specific culture collections and microbial holdings to be committed to the Microbial Research Commons as a condition of membership, as discussed earlier,[265] a failure to meet these obligations could trigger recourse to such internal dispute settlement machinery.

8. Recognizing the Importance of Nonmonetary Benefits

Article 5 of the Nagoya Protocol expressly declares that "[b]enefits may include monetary and non-monetary benefits."[266] Because the Microbial Research Commons aims to develop a multilateral access and benefit-sharing regime to promote research under Article 4 of the Nagoya Protocol, nonmonetary benefits flowing from the promotion of research are as or more important than the monetary benefits so far discussed. As reported earlier in Chapter 3, it was the drafters' belated recognition of the crucial role of public research in the overall calculus of benefits to be shared under the CBD that prompted them to facilitate the use of a multilateral approach dedicated to generating such benefits.[267]

To their credit, the drafters of the Nagoya Protocol now specifically spell out the nonmonetary benefits expected to flow from facilitating scientific research under a multilateral approach. Article 22 identifies "Bioprospecting, associated research and taxonomic studies" and "Technology transfer, and infrastructure and technical capacity to make such technology transfer sustainable" as measures consistent with the treatment of capacity-building as nonmonetary benefits.[268] The Annex to the Protocol concerning Monetary and Nonmonetary Benefits further specifies the need for assistance from the scientific community in the following areas:

- Sharing of research and development results;
- Collaboration, cooperation and contribution in scientific research and development programs, particularly biotechnological research activities;

264 *See* Nagoya Protocol, n. 1, art. 18.
265 *See* Sections II.A. and III.B of this chapter.
266 Nagoya Protocol, n. 1, art. 5(4).
267 *See* Chapter 3, Section IV.B.2 ("Recognizing the Importance of Nonmonetary Benefits").
268 *See* Nagoya Protocol, n. 1, art. 22.

- Transfer to the provider "under fair and most favorable terms of knowledge and technology that make use of genetic resources, including biotechnology;"
- Institutional capacity-building;
- Training related to genetic resources;
- Access to scientific information relevant to conservation and sustainable use of biological diversity, including biological inventories and taxonomic studies;
- Research directed toward priority needs, such as health and food security . . .;
- Institutional and professional relationships;
- Joint ownership of intellectual property rights.[269]

The lesson here is that capacity building is not only in everybody's interest, it is also crucial to establishing the legality of the multilateral approach to compliance with the CBD.

In this context, we think it advisable that the Governing Body devote a section of the SMTA to listing nonmonetary benefits that could potentially be available when microbial genetic resources were accessed from the multilateral system. Researchers could then indicate the items applicable to specific transactions, if any, by ticking the applicable boxes, in addition to their obligations to pay a share of royalties from commercial applications to the Designated National Authority.[270] Such a list of expected nonmonetary benefits should facilitate access by academic users and encourage provider states to make materials available for scientific research.

In effect, the existence of this rubric on the SMTA would focus the researchers' attention on specific undertakings regarding the provision of nonmonetary benefits within the scope of their institutional authority. It would also remind provider countries, especially developing countries of the research-promoting benefits they were receiving irrespective of any monetary benefit that may or may not be forthcoming.

That said, the Governing Body must negotiate and specify the nonmonetary benefits otherwise to be provided by the Microbial Research Commons as a whole, notably in the area of capacity building, which remain independent of any specific transactions under the SMTAs. This topic is discussed later in this chapter. Here we stress that the very existence of a multilateral system of facilitated exchanges for research uses of microbial genetic resources, with built-in monetary benefits from commercial applications, would in and of itself constitute a significant nonmonetary benefit to all the participating stakeholders.

[269] Nagoya Protocol, n. 1, Annex – Monetary and Nonmonetary Benefits arts. 2 (a), (b), (f), (h), (j), (k), (l), (n), (o).

[270] *See* Section II.C.2 ("Quantum and Duration of Royalties").

D. Digitally Integrating Knowledge Assets Available from the Multilateral System

As we explained in Chapter 5, the Nagoya Protocol implicitly obliges microbial culture collections to track uses of the knowledge assets to be made available from the multilateral system envisioned in this chapter, in order to ensure fulfillment of the benefit-sharing provisions of the CBD.[271] This undertaking does not mean, however, that the culture collections can freely disclose users and uses specified in the relevant SMTA to the world at large without possibly violating confidentiality obligations, which may be explicitly or implicitly a condition of the underlying exchange transactions.[272] Nevertheless, by digitally integrating publicly available information about both the material held in the collections that participate in the Commons, as well as related data and literature, the Commons could become an increasingly valuable research tool in its own right.

1. The Core Project

Ideally, the redesigned Microbial Research Commons would afford scientific researchers a single portal through which they could access all publicly available information and data pertaining to the genetic resources available from all the collections participating in the multilateral system. This goal would, in fact, take a major step in implementing the National Research Council's vision of a New Biology paradigm, in which microbiology was expected to play a major role.[273]

At present, the WDCM has taken major steps in this direction. Building on previous initiatives at the regional level,[274] including the StrainInfo project,[275] the WDCM portal identifies the materials available online from WFCC member collections, and it also provides information about publications, patents, gene sequences and other genomic information pertaining to these strains.[276] The WFCC's Global Catalogue of Microorganisms is expected to help member culture collections "to manage, disseminate and share the information related to their holdings."[277]

[271] *See* Chapter 5, Section III.B.3 ("Tracking Mechanisms to Maintain the Chain of Custody").

[272] *But see* David Smith, Dagmar Fritze, & Erko Stackegrandt, *Public Service Collections and Biological Resource Centers of Microorganisms*, *in* THE PROKARYOTES – PROKARYOTIC BIOLOGY AND SYMBIOTIC ASSOCIATIONS (E. Rosenberg et al., eds., Springer Verlag 2013) [hereinafter D. Smith et al. (2013), advocating the need to compile information about users in general.

[273] *See* Chapter 1, Section I.D ("A New Research Paradigm for the Life Sciences").

[274] *See* D. Smith et al. (2013), n. 272, at 283–93.

[275] For the now defunct StrainInfo project, *see* Chapter 8, Section II.B.2.

[276] *See* WDCM, Analyses of Bio-Resources Citations (ABC), available at htpp://abc.wfcc.info (last accessed August 3, 2015).

[277] Linhuan Wu et al., *Global Catalogue of Microorganisms (GCM): A comprehensive Database and Information Retrieval, Analysis, and Visualization System for Microbial Resources*, 14 BMC

Currently, however, these promising initiatives are limited by the number of culture collections willing or able to participate in relevant digital projects. Only 71 of the WFCC's member collections currently participate in the GCM.[278] Apart from inertia, many nonparticipants either lack the funds and technical expertise to join the system or they remain reluctant to digitally disclose their holdings to the world at large.

In our view, establishing the redesigned Microbial Research Commons should encourage and enable more culture collections to overcome these obstacles. For example, collections still worried about unauthorized uses of their holdings should welcome the opportunities to participate in a tightly regulated Compensatory Liability Regime, which is designed to promote commercial uses of materials available from the multilateral system.[279] As for technical obstacles, the Microbial Research Commons would directly address them through the capacity building initiatives outlined later in this chapter.[280] As implemented by the Scientific Coordination Council, technical assistance under the Commons would be tailor made to the needs of single collections in different countries. The Commons' own governance arrangements – with the participating science ministries – would provide a more robust foundation for coordinating and developing digital management and related policy initiatives than would otherwise be possible within the confines of the WFCC alone.

We envision the participation of at least those culture collections that had met the minimum quality standards of the Commons and that already had established, or expressed interest in establishing, a digital presence. These collections are the most likely to affiliate with the Commons in the first place. They will be concerned that the Nagoya Protocol is understood to encompass data, especially genomic data, pertaining to microbial genetic resources,[281] and that making their data available to all researchers from the multilateral system falls into the category of "specialized access and benefit sharing agreements … that … are supportive of and do not run counter to the objectives of the Convention [on Biological Diversity] and … [the Nagoya] Protocol."[282]

Culture collections that affiliate with the Commons should expect to derive considerable advantages from the advent of a single portal, especially if it were

GENOMICS 933 (2013); Global Catalogue of Microorganisms (GCM) *available at* htpp://gcm.wfcc.info/overview/ (last accessed August 3, 2015) [hereinafter GCM CC List].

[278] GCM, http://scm.info/eclist (last accessed August 3, 2015) [hereinafter GCM List].

[279] For detailed scenarios implementing this regime, *see* Chapter 5, Section III; for the rationale and basic premises underlying this approach, *see id.*, Section II.

[280] *See* Section III.E in this chapter.

[281] *See* Section III.C.5 in this chapter. *See also* chapter 5, Section III.B.3 ("Modifications Based on Data from Microbial Materials Accessed from the Multilateral System").

[282] Nagoya Protocol, n. 1, art. 4.2.

open to the world at large – as we advise – and not just to researchers in member countries.[283] With all the catalogs of all the member collections freely available from one central place, both the availability of the network's holdings and related data, as well as the terms for commercial applications specified in the SMTAs would become widely known, with added incentives for research use and eventual benefit sharing.

The portal we envision would constitute a logical extension of the WDCM's current initiatives, and the Governing body would presumably appoint WDCM as its agent for this purpose. In that event, WDCM would coordinate its activities with the mandate of the Governing Body and with the assistance of the SCC.

Once such a portal was established, the Commons governance architecture becomes the proper forum for discussing and implementing coordinated policies regarding the availability and reuse of both data and literature pertaining to the members' own holdings. For example, users of such data could be obliged to specify the biological and geographical sources of the underlying materials.[284] This goal would require attention to the requirements for legal interoperability among multiple data sets from different sources.[285] The Governing Body, in tandem with the Scientific Coordination Council discussed earlier in this chapter, would in turn need to devise "public law statutory, regulatory and policy approaches, as well as private law instruments, such as waivers, [standard] licenses and contracts that may be used to place … [such] datasets in the public domain or otherwise make them publicly available for use and re-use without restrictions,"[286] as envisioned by a core data policy to be formulated by the Governing Body.

We suggest that in setting the rules for data to be made available from the multilateral system, the Governing Body should take the following criteria into consideration. The applicable open-access policy should require that access to the data be free of charge and that there should be no restrictions on use or reuse, other than a duty of attribution where the data provider so requires. If the data are not already in the public domain, a Creative Commons Public Domain Waiver (CC0) or a Creative Commons Attribution Only (CC-BY 4.0) license for data should suffice.[287] In no

[283] This was a limitation established by GBRCN's Demonstration Project, which also covered data, but stressed opportunities to charge for accessing data products, rather than making data freely available. See Chapter 9, Section II.C.1. Value added services, of course, could generate revenue for the culture collections without compromising research.

[284] *See, e.g.*, Fedder, n. 169.

[285] Catherine Dolderina et al., *Mechanisms to Share Data as Part of GEOSS Data-Core*, Draft White Paper, Group on Earth Observations, June 18, 2014, Executive Summary, at 3.

[286] *Id.* Here the Governing Body must also consult with both UNEP and the Secretariat of the CBD regarding their recommendations for the sharing of biodiversity data and information.

[287] *Id. See also* the Creative Commons Public Domain Mark and the Open Data Commons Public Domain Dedication and License (PDDL).

case should it be possible to contribute data through the core portal if the applicable license required deference to the local laws of the originators, especially if the default rules of those local laws would conflict with the Governing Body's data policy for the core portal, as for example would occur where the EU Database Directive automatically applied.[288]

That said, it must be borne in mind that data pertaining to *ex situ* microbial genetic resources held by member culture collections are normally subject to the CBD and the Nagoya Protocol, as explained in Chapter 3. It follows that the Compensatory Liability Regime governing materials available from the multilateral system must also apply with equal force to related data.[289] How to implement this mandate was explored at some length in Chapter 5,[290] and the Governing Body would necessarily have to resolve ambiguities identified there.

Although devising and implementing an appropriate data policy constitutes a primary function of the Governing Body, we think that Body should seize the opportunity to develop a complementary policy for open access microbial literature and other forms of scientific information generated by the member collections. To date, none of the existing scientific commons empirically surveyed earlier in Chapter 9 have paid sufficient attention to this topic.[291]

The microbiological literature, like the relevant data, may conveniently be subdivided into two categories. The first constitutes those reports, catalogues and other publications generated by the participating culture collections themselves. The second category is the body of externally generated literature by the microbiologists whose institutions might eventually be directly or indirectly affiliated with the Commons.

As with data, the Governing Body's policies in this regard can only apply to information falling into the first category. This relatively small but key component of the literature should clearly be digitally integrated into, and made available, through the Commons' open access portal. In this way, the catalogues and related reference materials of all the participating collections would become instantly available to all users of the Commons in an ever-expanding and up-to-date inventory of the collective *ex situ* holdings managed by the distributed member collections.

288 *See* Chapter 6, Section II ("Copyright and Related Laws as Digital Gridlock").

289 *See* Chapter 3, Section IV ("New Constraints and Opportunities for Scientific Research under the Nagoya Protocol").

290 *See, e.g.*, Chapter 5, Section III.B.3 ("Modifications Based on Data Pertaining to Microbial Materials Accessed from the Multilateral System").

291 However, the International Human Microbiome Consortium did take some strides in this direction. *See* Chapter 9, Section II.B.4.d. See also the WDCM, discussed in Chapter 8, Section II.B.1.

2. Optional Longer Term Projects

If the Governing Body were sufficiently ambitious, it could eventually consider building a digital directory with links to externally generated microbiological data and literature not related to the genetic resources held by culture collections participating in the Commons. This project could encompass a master directory and a search engine for selected available microbiological data repositories, as well as peer-reviewed and "grey" literature, which are fully open access.[292]

The exact contents to be made available through this optional portal would depend in part on both the financial and human resources available and on the thematic interests of participating members. The Governing Body and the Scientific Coordination Council would accordingly need to devise and implement policies concerning the selection and diffusion of the knowledge assets to be made accessible to the research community as a whole.

In this context, the data and information generated by the kind of Open Knowledge Environments discussed in Chapter 8 would qualify for inclusion in such a project.[293] For example, the Community Cyber-Infrastructure for Advanced Marine Microbial Ecology Research and Analysis Project (CAMERA) in the United States which ended recently, leaves behind a major data set that it makes openly available online.[294] The Marine Microbiological Diversity, Bioinformatics, Biotechnology Project (Micro B3) in the European Union will also be winding up its operations in 2015, with a major output of data concerning *in situ* marine microbes from coastal territorial waters.[295] Linking the open-data repositories from such projects to the digital portal of the Microbial Research Commons could serve a useful purpose and indicate a direction for the future.

E. Relations with Developing Countries

Apart from the mounting international tensions surrounding research uses of genetic resources, described in Part One of this book, special attention to the needs of microbiologists in the developing countries would be in the interest of the global scientific community. Scientists in OECD countries need access to both *ex situ* and *in situ* microbial materials originating from the developing countries. They also want to study the uses that indigenous populations have made of local genetic resources for agricultural and medicinal purposes.[296] Scientists in developing countries need

[292] *See generally* Chapter 7 (describing open-access microbiological literature); *see also* Chapter 8, Section II (dealing with mandatory and voluntary deposits of data in selected research commons).

[293] *See generally* Chapter 8, Section III ("Building Transnational Open Knowledge Environments").

[294] *See* Chapter 8, Section III.A.2.

[295] *See* Micro B3, https://www.microb4.eu/ (last accessed 3 Auguts 2015).

[296] *See, e.g.*, Chapter 3, Section I.A ("Bioprospecting or Biopiracy?").

to improve the quality standards of their public culture collections, as well as access to materials and digital knowledge resources mostly from the OECD countries.[297] These reciprocal interests in scientific cooperation are often obscured by conflicting demands for intellectual property protection of upstream microbial materials and related data.[298]

This said, one must first recognize the ambiguity inherent in the term "developing countries." Economists have found it useful to subdivide these countries into upper, middle, and lower per capita GDP groups.[299] The TRIPS Agreement of 1994, instead, recognized only one large group of developing countries subject to all the minimum intellectual property standards imposed on OECD countries after a short transitional period,[300] plus a very small group of Least-Developed Countries that are still subject to virtually none of these same minimum international standards of intellectual property protection.[301] In practice, the literature concerning implementation of TRIPS obligations acknowledges the particular importance of the BRICS countries in driving the "development agenda" for poorer countries as a whole.[302]

These and other standard categorizations must be used with caution when formulating science policy, especially with regard to the exchange of genetic resources for research purposes. In principle, they remind us that selected microbial culture collections in some BRICS countries are technically superior to those in most other developing and least-developed countries and that the technical capacity of culture collections in developing countries as a group lags far behind that of collections in OECD countries as a group.[303]

Paradoxically, however, the potential ability of poor countries to supply microbes of interest to both science and industry may exceed that of all the OECD countries

[297] *See, e.g.*, Chapter 4, Section I.C.1 ("Disparities Among the WFCC Member Collections").

[298] *See, e.g.*, Chapter 2, Section II. ("Impinging Intellectual Property Rights Promoted by the Developed Countries"); Chapter 3, Section I ("Regulatory Measures to Control Access to Genetic Resources Promoted by the Developing Countries").

[299] *See* The World Bank, Country and Lending Groups, *available at* http://data.worldbank.org/about/country-dash-and-lending groups (last accessed August 4, 2015).

[300] *See* Agreement on Trade-Related Aspects of Intellectual Property Rights, art. 9.1, Apr. 15, 1994, 108 Stat. 4809, 1869 U.N.T.S. 299, art. 65 [hereinafter TRIPS Agreement] (five year transition period for all developing countries, 10 years in the specific case of patentable pharmaceuticals, food, and agriculture for developing countries that did not previously grant product patents on these subject matters.)

[301] *See id.*, art. 66. LDCs determined by per capita GDP (waiver extended to 2023 for TRIPS and 2033 for pharmaceutical products).

[302] The BRICS countries include Brazil, Russia, India, China and South Africa. *See* Rochelle C. Dreyfuss, The Role of India, China, Brazil and Other Emerging Economies in Establishing Access Norms for Intellectual Property and Intellectual Property Lawmaking 1–3 (*Inst. for Int'l Law & Justice*, Working Paper 2009/5, Pub. L. Research Paper No. 09–53, 2009), *available at* http://ssrn.com/abstract=1442785; Jerome H. Reichman, *Intellectual Property in the Twenty-First Century: Will the Developing Countries Lead or Follow?*, 46 Hous. L. Rev. 1115, 1118–19 (2009); WIPO, *Development Agenda for WIPO*, http://www.wipo.int/ip-development/en/agenda/ (last accessed 9 Apr. 2014)

[303] *See* Chapter 4, Section I.C.1 ("Disparities Among the WFCC Member Collections").

combined. The greater the investment made in building microbial research capacity in developing countries, and the more widely dispersed that investment becomes, the larger the potential payoffs are likely to be for global scientific research and commercial applications everywhere.

There is, in short, a global public interest in building up the microbial research capabilities of even the poorest countries (as there was with regard to plant genetic resources from the 1960s on),[304] quite apart from the demands of these countries with respect to sovereign rights to national resources under the CBD. Given these demands, this global scientific interest makes capacity building a valuable bargaining chip – a "nonmonetary benefit" – when reconciling the legal mandates of the CBD with the needs of the research community, as expressly recognized in Article 22 of the Nagoya Protocol and as we discussed in the previous section on nonmonetary benefits.[305] The lesson here is that capacity building is clearly in everybody's interest.

A number of international, regional, and national capacity building programs have already focused on improving microbial research capacity and related infrastructure in developing countries. At the international level, notable examples include the activities of the World Federation of Culture Collections' Education and Capacity Building Committee,[306] and UNESCO's Microbial Resource Centers (MIRCEN) – a network in environmental, applied microbiological and biotechnological research – which UNESCO carried out in collaboration with UNEP, starting in 1975.[307] The MIRCEN program helped to establish over thirty mature collections, mostly in the developing world.

Regional associations of culture collections, such as the European Union Culture Collection Organization (ECCO) and the Asian Biological Resource Centres Network, focus mostly on helping their own members.[308] National funding agencies, foundations, single culture collections, and professional societies also tend to concentrate on assisting microbial research and infrastructure within their respective national borders, although some notable instances of collaboration with and assistance to microbiologists in developing countries have occurred.[309]

304 *See* Chapter 2, Section I.B.1 ("Emergence of an International Consortium for the Preservation and Improvement of Cultivars Essential for Food Security").

305 *See* Nagoya Protocol, n. 1, art. 22 (stating that "The Parties shall cooperate in capacity-building, especially development and strengthening of human resources and institutional capacities to effectively supplement this Protocol in developing country Parties . . ."); *see also* Section IX.C.8 in this chapter.

306 *See* WFCC, *Education and Capacity Building Members*, http://www.wfcc.info/index.php/committees/education (last accessed 7 May 2015).

307 For additional information on the MIRCEN project, *see* UNESCO, *Science*, http://portal.unesco.org/science/en.lev.pap-URL-ID=2491&URL-DO=DO-topic&URL-Section=301.html (last accessed 7 May 2015).

308 *See* Chapter 4, Section I.A & C.

309 For example, in the United States the Fogarty International Center of the National Institutes of Health has projects with different developing countries. *See* Flora Katz, *Proposal for a Microbial*

The assistance so far provided by these and other entities to their counterparts in developing countries spans the gamut of policies and practices in the conduct of microbiology, including some direct funding of infrastructure development.[310] These capacity building activities have been undertaken in a relatively uncoordinated way, however. Even the broader WFCC and UNESCO programs operated within their own organizational purviews and did not seek to integrate these different initiatives.

Looking to the future, the governments and scientific organizations likely to join the Microbial Research Commons would logically seek to further promote capacity building in developing countries. Such activities would encourage developing country governments to join the Commons, together with their national culture collections, and it would provide tangible up-front benefits to them and build trust in the multilateral system as a whole. Funding permitting, such a broadly constituted forum could provide the opportunity to coordinate capacity building objectives and implement them more effectively than in the past, and the Governing Body of the Commons should devote considerable attention to this end.

The Governing Body of the Commons, in consultation with the CBD's administration, should encourage members to fund and become involved in approved capacity-building measures that favor scientific entities in developing countries.[311] In this connection, evidence shows that direct, hands-on contributions of know-how, expertise, and related training to institutes in developing countries from science entities in developed countries is an especially welcome and fruitful way to promote the exchange of genetic resources.[312] Moreover, the open access policies we recommend the Governing Body adopt for both materials and related information assets could particularly benefit scientific researchers in poor countries, and it could likewise encourage research communities in those countries to implement similar policies. Technical cooperation to enhance digital capabilities in developing countries should also prove especially productive.

More generally, all parties should thus bear in mind that by providing both monetary benefits under the Compensatory Liability Regime and nonmonetary benefits via capacity building, the proposed Microbial Research Commons would afford developing country members a better deal than they could otherwise obtain

Semi-Commons: Perspectives from the International Cooperative Biodiversity Groups, *in* DESIGNING THE MICROBIAL RESEARCH COMMONS, n. 20, at 129–37. Culture collections in the more economically developed countries also host microbiologists from developing countries in joint activities that frequently have capacity building aspects.

310 *See, e.g.*, MIRCEN, n. 307.

311 The GBRCN demonstration project also stressed the importance of capacity building initiatives, as does the GBIF. *See, e.g.*, the discussion of GBRCN in Chapter 9, Section II.C.1. As regards GBIF, capacity building initiatives have been augmented by grants from international cooperation agencies and other development aid donors. *See* Capacity Enhancement in GBIF, *available at* http://www.gbif.org/capacity-enhancement/summary.

312 *See, e.g.*, F. Katz (2011), n. 309.

under the CBD's bilateral approach.[313] Instead, the Commons would aim to provide both capacity building and a system for sharing in financial gains from commercial applications. Finally, it is worth reiterating the substantial benefits to science likely to flow from this arrangement, which would obviate a patchwork quilt of national restrictions on access to microbial genetic resources.

F. Other Issues for the Governing Body to Consider

Our list of issues that the Governing Body might need to deal with is not meant to be exhaustive. On the contrary, an important function of the Scientific Coordination Council and any Advisory Committees, as described earlier in this chapter, is to apprise the Governing Body of topics that concern the microbiological community in general and to propose agenda items for that Body's deliberations.

We nonetheless call attention here to certain issues that seem particularly important. One is the possibility of defining a policy for the release of microbial research materials to the public at the time of publication of research results. Another concerns the desirability of considering access to *in situ* microbial genetic resources that initially remain beyond the scope of the proposed Microbial Research Commons. Finally, we note in passing the need to consider aspects of biosafety and security that inevitably affect exchanges of certain microbial materials.

1. Devising Policies for Earlier Release of Materials Used in Basic Research

In Chapter 8, we saw that the dependence of biological science on pooled genomic data had led the funding agencies to devise, and increasingly enforce, standards for the early release of data obtained from publicly funded research projects.[314] Voluntary data pooling initiatives seeking reciprocal research benefits have also proliferated.[315] This same genomic revolution has also focused growing attention on the need to formulate standard policies for the release of microbial materials used in basic research.[316]

A priori, genomic mapping and sequencing of microbial strains might be expected to lessen the dependence of microbiologists on deposits of living materials held by

[313] *See* Chapter 3, Section II ("Destabilizing the Exchange of Plant and Microbial Genetic Resources as Global Public Goods").

[314] *See* Chapter 8, Section I ("Early Release Policies to Manage the Deluge of Genomic Reference Data").

[315] *See id.*, Section II ("Beyond Early Release: Diverse Networked Sharing Strategies to Manage and Exploit the Deluge of Data") and Section III ("Building Transnational Open Knowledge Environments").

[316] *See, e.g.*, Jeffrey L. Furman, Fiona Murray, and Scott Stern, *More for the Research Dollar*, 468 Nature 757 (2010) [hereinafter Furman et al. (2010)].

public culture collections.[317] Although this has been the case in some areas, such as marine exploration of living matter,[318] in other areas a proliferation of experimental derivatives seem to have made the deposit of underlying reference strains and other microbial materials (especially those associated with synthetic biology) in public culture collections more important than ever.[319] Empirical evidence suggests that for these or other reasons, more microbes used in some publicly funded research areas are being deposited in various public or restricted collections than was the case prior to the genomic revolution.[320]

There is accordingly a growing perception in the microbiological community that standard protocols for the release of materials supporting publicly-funded research need to be devised.[321] Formulating such standards, however, poses problems for science policymakers. One is the difficulty of reconciling the single researcher's need to keep using undisclosed materials for follow-up research with the larger research community's needs to verify and build upon published research findings.[322] Another problem is the limited physical capacities of existing public culture collections, with corresponding pressures to accept deposits of microbial materials from researchers on a selective basis.[323]

Nevertheless, past experience with the pooling of genomic data suggests that the managers of funding agencies and policymakers, working together, can find ways to promote the release of microbial research materials without unduly sacrificing reputational benefits contingent upon follow-up research and related publications.[324] As regards the physical capacity of the culture collections, one goal of the proposed Microbial Research Commons is precisely to expand the overall capacity of the system by eventually including more upgraded collections at universities and research institutes within a distributed, digitally integrated, regulatory framework whose cross-border exchanges of materials automatically complied with international law. Greater participation of universities in the Commons initiative could also help to preserve the public research goals of their collections against the privatizing pressures of their technology transfer offices.[325]

317 *See, e.g.*, Daniel Drell, *Research and Applications in Energy and the Environment* (8–9 Oct. 2009), *in* Designing the Microbial Research Commons, n. 20, at 12.

318 *See* Chapter 8, Section III.A.2 (discussing the CAMERA project).

319 *See, e.g.*, Michael Fischbach & Christopher A. Voigt, *Prokaryotic Gene Clusters: A Rich Toolbox for Synthetic Biology*, *in* Inst. Medicine, The Science and Applications of Synthetic and Systems Biology 449 (Nat'l Acads. Press, 2011).

320 Interview with Kevin McCluskey, Curator, Fungal Genetic Stock Center, University of Kansas Medical Center, Jan. 14, 2014.

321 *See, e.g.*, D. Smith (2012), n. 39

322 *See* Chapter 8, Section II.C.1 ("Benefits and Drawbacks of the Data Sharing Ethos").

323 *See* Chapter 4, Section I.B.

324 *See, e.g.*, Furman et al. (2010), n. 316.

325 Interview with Kevin McClusky, n. 320.

It would be entirely appropriate for the Governing Body of the Commons to devote attention to this issue. Obviously, that Body could not formulate binding policies for either the funding agencies or the microbiological community at large. Nevertheless, the very existence of a formally organized Microbial Research Commons under a transnational Framework Agreement would provide a unique opportunity to bring the stakeholders together, including funders and science policymakers, with a view to negotiating a normative platform for early release of microbial materials. The end result could then be applied, tested, and further refined by funding agencies in different countries, with a view to developing common material release policies like those applicable to genomic data.[326]

2. Possible Negotiations Concerning Access to *In Situ* Microbial Genetic Resources

As discussed in Chapter 3, core provisions of the International Treaty on Plant Genetic Resources for Food and Agriculture (ITPGRFA) required substantial contributions of *in situ* plant genetic resources legally residing in the public domain, or otherwise under the control of the Contracting Parties, to the multilateral system of facilitated access that this Treaty established in 2001.[327] These provisions have not been fulfilled, which adds to the mistrust that reportedly hampers implementation of the Crop Commons Treaty.[328]

Because we believe that these commitments concerning access to *in situ* genetic resources under the ITPGRFA were premature and possibly ill-advised, we have abstained from following that model in our proposal for a redesigned Microbial Research Commons. That does not mean that access to *in situ* microbial genetic resources for research purposes should not eventually be addressed within the governance framework of this Commons.

In our view, this topic should not become an action item on the Governing Body's agenda until and unless all the stakeholders have become satisfied with the regime adopted and implemented to facilitate exchanges of *ex situ* microbial materials for research purposes, as set out in this chapter. If this regime failed to satisfy stakeholders, and especially the provider countries, no formal set of rules concerning access to *in situ* resources could likely be implemented. If, instead, our proposals for exchanges of *ex situ* materials with built in benefit-sharing provisions,

[326] *See* Chapter 8, Section I ("Early Release Policies to Manage the Deluge of Genomic Reference Data").

[327] *See* Chapter 3, Section III.A ("Basic Concepts of the ITPGRFA").

[328] *See id.*, III.C. (discussing major weaknesses of the International Treaty).

were to satisfy the Contracting Parties, especially the developing countries, the latter would be better disposed to consider facilitated access to *in situ* materials with a more positive, self-interested outlook.

One transitional undertaking that the Governing Body could consider in this regard is the possibility of establishing a multilateral benefit-sharing arrangement for microbes discovered in areas beyond the reach of claims based on territorial sovereignty. For example, marine microbiology has entered a period of exploratory growth, which is producing correspondingly large amounts of materials and data that are not necessarily subject to the bilateral approach of the CBD.[329] If these resources could be brought within the operations of the Microbial Research Commons with the consent of participating governments, they could become a source of both monetary and nonmonetary benefits for all the stakeholders.[330]

In any event, under the cautious approach we suggest, access to *in situ* microbial genetic resources within territorial boundaries would normally remain subject to the bilateral approach under the CBD, which requires Prior Informed Consent and Mutually Agreed Terms.[331] To facilitate research, the Governing Body could eventually try to develop a recommended SMTA for these *in situ* resources, in cooperation with the CBD. We leave this topic to the prospective discussions of the Governing Body, in the hopes that successful implementation of our proposals for exchanges of *ex situ* materials would have laid the foundation on which such negotiations could profitably build.

3. Biosafety and Security Considerations

Efforts to facilitate cross-border exchanges of microbial materials for research purposes under the proposed Microbial Research Commons must necessarily take account of existing or prospective biosafety and security constraints. For example, pathogens require special treatment, handling, and procedures.[332]

[329] *See, e.g.*, The Micro B3 Project, *Homepage*, http://www.microb3.eu/ [hereinafter Micro B3 Project] (last accessed 2 July 2014,) (with funding provided by the EU); *see also* Chapter 8, Section III.A.2 (discussing the CAMERA project).

[330] *See, e.g.*, Fedder, n. 169. *Cf.* Nagoya Protocol, n. 1, arts. 4.2, 10.

[331] *See* Chapter 3, Section I.B ("Foundations of an International Regime of Misappropriation to Govern Genetic Resources").

[332] *See, e.g.*, United Nations Model Regulations on the Transport of Dangerous Goods (2007), http://www.unece.org/trans/danger/publi/unrec/rev15/15files e html#c190. For European agreements on the international carriage of dangerous goods, see http://www.unece.org/trans/danger/dangerhtml. International treaties already impose restrictions on access to microbial materials for reasons of biosafety and national security. In some countries, notably the United States, national security has become a major regulatory concern for microbiology.

The special arrangements needed to maintain pathogens and other dangerous microbes in restricted areas necessarily affect access and reuse for any research purposes.[333]

We cannot estimate the extent to which the Governing Body of the Commons would become actively engaged in these issues. However, we do expect it and the Scientific Coordination Council to monitor actions being taken in other fora on these issues and to apprise members of their implications for scientific research when relevant.[334] More generally, biosafety and security concerns, together with related import-export controls raise the costs of microbial exchanges for all researchers. They also reinforce the need for member culture collections to track the uses to which specimens are put and the resulting chain of custody, and they further reinforce the drive for higher quality standards and scientific reliability discussed elsewhere in this volume. Such regulations can also make the transfer or rescue of endangered collections very costly or impossible.

How to preserve a proper regulatory balance between restrictions on access to microbial materials for biosafety and security and the need to promote benefit sharing from commercial applications of research results nonetheless constitutes a concern that the Governing Body and the Scientific Coordination Council should monitor on a regular basis. A more detailed discussion of this topic lies beyond the scope of this volume.

[333] *See, e.g.*, Interagency Working Group on Scientific Collections, Scientific Collections: Mission-Critical Infrastructure for Federal Science Agencies (2009). In the United States, the National Select Agent Registry oversees the activities of those who possess biological agents and toxins that could become a threat to public, animal or plant health. It also maintains supervision over the use of dangerous pathogens for research purposes. *See* National Institutes of Health (NIH), Office of Science Policy, National Select Agent Registry (NSAR) website, *available at* http://osp.od.nih.gov/externa/-resource/national-select-agent-registry-nasar-website (last accessed 5 August 2015). *See also* NIH, *Office of Biotechnology Activities*, http://oba.od.nih.gov/oba/rac/guidelines_02/NIH_Guidelines_Apr_02.htm (last accessed 7 May 2015); OHSU, *Institutional Biosafety Committee*, http://www.ohsu.edu/xd/research/about/integrity/ibc/ (last accessed 7 May 2015).

Another layer of oversight and supervision is provided by the institutional biosafety committees that operate under the Office of Biotechnology Activities (OBA) at NIH. These committees provide oversight to research groups working on recombinant DNA, infectious agents or biological toxins, and they help to implement the NIH guidelines for such research. *See* NIH Office of Science Policy, Institutional Biosafety Committees, available at http://osp.od.nih.gov-office-biotechnology-activities/biosafety/institutional-biosafety-committees (last accessed 5 August 2015). Taken together, these layers of supervision act as a filtering process that generally allows scientists to use microbial materials for research while helping to identify experiments of particular concern. *See also* Forum on Microbial Threats, Inst. Medicine, Infectious Diseases in a Borderless World (Nat'l Acads. Press, 2010).

[334] For example, close relations with the WHO's Pandemic Influenza Preparedness Framework would seem desirable. *See* Chapter 4, Section IV.A.

IV. FUNDING AND INSTITUTIONAL STABILITY

It is important to reiterate two foundational premises of the redesigned Microbial Research Commons that bear directly on the funding required for its operations. First, the Commons would not become a research generating institution; it would not conduct, fund, or make grants for specific research projects.[335] By pooling genetic resources and other knowledge assets, it would consolidate and enhance basic scientific infrastructure that would enable and support microbiological research generally.

Second, the Commons would not provide any in-house or centralized physical repository for the microbial genetic resources it made available from the multilateral system. Rather, it would strengthen the existing, federated network of microbial culture collections and bind them collectively within a multilateral contractual framework consistent with Article 4 of the Nagoya Protocol, which would avoid restrictions on both basic and applied research that the Convention on Biological Diversity would otherwise mandate.[336] At the same time, the proposed Commons would make publicly available data and literature pertaining to microbial genetic resources of at least its participating members accessible to the research community at large by means of a central portal.

In this section, we consider the prospects for obtaining the funds needed to establish the multilateral entity on a sound and sustainable financial basis. We also highlight some of the lost opportunity costs likely to result if such a project is not undertaken.

A. *The Need for Adequate and Dependable Funding*

The primary objective of a redesigned Microbial Research Commons is to shelter the existing system of exchanging *ex situ* microbial materials within a multilateral regime, much as the CGIAR's seed banks were sheltered under the Crop Commons in 2001.[337] From this perspective, many of the activities falling within the ambit of the multilateral system are already funded by a variety of different sources.

For example, governments already fund most of the public culture collections.[338] Official government entities, such as the NIH and the NSF in the United States, and

[335] However, as described later, the administrators could make grants pertaining to capacity building and to organizational and implementation aspects of the framework agreement.

[336] *See generally* Chapter 3, Section IV ("New Constraints and Opportunities for Scientific Research under the Nagoya Protocol").

[337] See Chapter 3, Section III ("An International Treaty to Rescue and Expand 'The Global Crop Commons'").

[338] *See* Chapter 4, Section I.A.3 ("The Perennial Problem of Funding").

the European Commission in the European Union, as well as private foundations and trusts, are the major funders of research results in the form of published literature and data.[339] Government aid agencies and private foundations already provide some financial support for capacity-building initiatives in developing countries for biomedical and microbiological research and infrastructure.[340] Universities partially support some culture collections of leading scientists and increasingly fund digital repositories for their own researchers' output. Industry funds its own collections and relevant data, as well as special collections kept at public repositories in different countries.

As matters stand, however, and despite some existing coordination efforts underway, the results of traditional funding methods risk remaining fragmented within circumscribed institutional and legal boundaries under a "small science" approach inherited from the past.[341] For example, the WFCC – a nongovernmental umbrella organization – is underfunded in our view and largely dependent on voluntary expertise.[342] A growing array of regional networks appear to have obtained some promising public funds at least in the short term.[343]

What funders should now consider is the possibility of spending a little more up front in order to support the policies and plans for the existing actors to reorganize themselves. They could thus catalyze the formation of a "Big Science" infrastructure that would pool microbial materials, data, and literature in the federated and digitally integrated research infrastructure envisioned earlier in this Chapter and that would transcend territorial, institutional, and discipline-imposed boundaries. By enabling relatively unfettered access to, and use of, these resources on a global scale, funders should recognize that the added value flowing from this infrastructure would greatly augment the projected returns to all participants from existing levels of investments in microbiology and applications. This calculus of magnifying long-term returns on short-term investments is eminently consistent with the New Biology vision of the National Research Council, which has been one of the key underlying premises of this book.[344]

339 *See generally* Chapter 7 ("Enabling the Microbiological Research Community to Control Its Own Scholarly Publications") and Chapter 8 ("Fully Exploiting Data-Intensive Research Opportunities in the Networked Environment").

340 *See, e.g.*, U.S. Agency for International Development, the U.K. Department for International Development, and the German Development Initiatives in Biomedicine and Agriculture; *see also* Flora Katz, n. 309 on NIH capacity building projects in microbiology.

341 *See* Chapter 1, Section III.A ("Recognizing Institutional and Legal Challenges to the Existing Microbial Research Infrastructure").

342 *See* Chapter 4, Section A.3; *see also* D. Smith et al (2013), n. 272.

343 *See* Chapter 4, Section I.C ("Beyond the WFCC: Regional and Global Networks of BRCs"); D. Smith et al. (2013), n. 272 ("Toward a Global Network").

344 *See* NATIONAL RESEARCH COUNCIL, A NEW BIOLOGY FOR THE 21ST CENTURY 41–50 (Nat'l Acads Press 2009) [hereinafter NRC, A NEW BIOLOGY].

As we have described the organizational functions of the proposed Commons, it need not burden any one funder, or even any group of funders, with major additional expenditures. On the contrary, we believe that funders would only have to add some incremental support to their existing portfolios (including also contributions in kind and seconded services), and that the value of the returns on these contributions would greatly exceed their costs. How much in aggregate additional funding would be needed in the first instance could depend on whether or not the Contracting Parties and the Governing Body decided to impose a user surcharge on all microbial materials available from the multilateral system, as discussed earlier[345] and whether they were also willing to undertake annual contributions to support its governance and operations.[346]

Disregarding for a moment the possibility of a users' surcharge, under our institutional design governments would need to contribute some supplementary funds to cover the costs of the Governing Body and Executive Committee, the Scientific Coordination Council, the Advisory Committees, and other operations of the Commons, as well as any necessary physical facilities. However, these contributions, whether voluntary or mandated by agreement, could to some extent be offset by other means. For example, physical facilities may be donated by a host country or organization, as occurred in the case of GBIF and GEO.[347] Expert and administrative personnel could be seconded from member governments and nongovernmental scientific organizations, as routinely occurs at GEO.[348] Most of the travel costs could be borne by employing organizations of the participants, as is the case in many scientific bodies; and committees should be composed of volunteers whenever feasible. In general, the leadership should seek to control costs and to be as efficient as possible, while fostering new revenue streams, such as grants and donations,[349] as well as the Trust Funds discussed later.

Scientific funding entities, together with governments, would presumably try to defray the costs of major operational functions. Primary cost drivers would include efforts to boost and harmonize quality standards and best practices among

[345] *See* Section III.C.1 ("The Question of a Users' Surcharge").

[346] *Cf.*, *e.g.*, the voluntary annual contributions to both the Crop Commons and the Group on Earth Observations discussed in Chapter 9, Section II.A.2.c and II.B.3.c. For mandatory annual contributions to the GBIF, *see* Chapter 9, Section II.B.2.c.

[347] GBIF is hosted by the University of Copenhagen, Denmark. GEO occupies office space donated by the World Meteorological Organization (WMO) in Geneva, Switzerland.

[348] *See* Chapter 9, Section II.3.c.

[349] *Cf.* Ecological Soc'y Am. (ESA), Strategies for Developing and Innovating Living Stocks Collections 3 (ESA Workshop Report, 2012), *available at* http://www.esajournals.org/doi/abs/10.1890/0012-9623-94.1.118. The Secretariat of the ITPGRFA has been particularly successful in obtaining grants and donations, in addition to voluntary annual contributions. Interview with Dr. Shakeel Bhatti, Director General of the ITPGRFA, Rome, Italy, on Aug. 5, 2015.

participating culture collections. Upgrading existing culture collections to BRC levels, and accrediting those collections that qualify, would require major investment however.[350] Actual costs would depend on whether OECD standards for BRCs or the more flexible standards of the WFCC were applied.[351] In any event, national governments in developed countries should fund the bulk of these costs on the assumption that by lifting all boats, they would obtain better returns from science in the end.

Costly efforts to design, build, and operate the basic portal that would link participating culture collections with other entities contributing open digital resources have already been undertaken by the World Data Center for Microorganisms.[352] Substantial funds for this purpose are provided by the Institute for Microbiology at the Chinese Academy of Sciences (IMCAS). Additional revenue may also be derived from other institutions willing to sponsor the open portal.[353]

Additional costs would accrue mainly from meeting various interoperability standards for the data and literature generated by the member culture collections. More broadly, expanding the central portal to include publicly available data and literature not provided by the members themselves would entail higher costs as well, although this could be purely discretionary.

Beyond these items, capacity building projects that will be of considerable long-term importance for relations between the Commons and developing countries under the CBD[354] would require substantial external funding from government aid agencies that already deal with developing countries, along with continued funding from science agencies and intergovernmental organizations. These financial commitments would be incurred on a voluntary basis, over and above some foundational contributions for this purpose. In practice, as scientific entities in different developing countries indicated their interest in joining the Commons, tailor-made capacity building programs would need to be negotiated with them, their governments, and willing sources of funds. This approach builds on successful initiatives undertaken by the U.S. National Institutes of Health[355] and those of other

[350] *See* Chapter 4, Section I.B ("From Culture Collections to Biological Resource Centers"); *see also* Chapter 9, Section II.C.1 ("The Proposed Global Biological Resource Centers Network (GBRCN)"). *See generally* Scott Stern. Biological Resource Centers, n. 12.

[351] For OECD practices *see* Chapter 4, Sections I.A.2 & 3 and I.B.

[352] *See* Chapter 8, Section II.B.1.

[353] *Cf.* the sponsorship of the GEO portal by the European Space Agency. *See also* Christina Chandras et al., *Models for Financial Sustainability of Biological Databases and Resources*, 2009 Database article id. bap017, *available at* http://www.ncbi.nlm.nih.gov/pmc/articles/PMC2790311/.

[354] *See, e.g.*, IUCN, Guide to the Nagoya Protocol (2012), n. 50, at 72–73, 127.

[355] *See e.g.*, F. Katz, n. 309.

international scientific organizations, such as GBIF, and especially the experience of the ITPGRFA's Secretariat.[356]

Offsetting some of these added expenses, however, are the prospects of future revenues from several other different sources. One is the possibility of user access fees imposed on all exchanges of materials available from the multilateral system, which we discussed earlier in this chapter.[357] Another is the possibility of broadly applying the Compensatory Liability Regime to new downstream uses of all the materials held in participating culture collections. Whether the Compensatory Liability Regime could thus help to support the upkeep of the Commons depends on decisions the Governing Body would have to make in this regard, as discussed earlier in this chapter.[358] In principle, revenues from commercial applications covered by the SMTA to be adopted by the Governing Body would flow to the Designated National Authorities in the countries providing the relevant *ex situ* genetic resources made available from the multilateral system.[359] However, the Contracting Parties could agree that a small share of the royalties paid to Designated National Authorities, under the Compensatory Liability Regime should be reserved for sustaining the costs of the Commons.[360]

Another thorny question that directly affects the long-term sustainability of the Microbial Research Commons is whether the participating governments should establish preset mandatory financial contributions as occurs, for example, at GBIF,[361] or whether such contributions should be placed on a purely voluntary basis, as occurs under the ITPGRFA and GEO.[362] For example, a mandatory dues schedule could be based on relative per capita GDP or on other indices of ability to pay, or a similar scheme can be used to elicit suggested voluntary contributions. Although a fixed schedule of mandatory dues appears to provide greater certainty and stability, governments that have tried this approach in other transnational scientific organizations that pool research assets have expressed reservations about incurring similar obligations in the future.[363] Moreover, current budgetary

356 For GBIF's *see* Chapter 9, Section II.B.2. For the commendable capacity building initiatives of the ITPGRFA, *see* Chapter 9, Section II.A.2.c.

357 See Section III.C.1.

358 See Section III.C.3 ("Protocols for the Distribution of Royalties").

359 For the calculus of royalties under the Compensatory Liability Regime, *see* Sections III.C.2 & 3.

360 *See* Section III.C.2 ("Quantum and Duration of Royalties) and III.C.3 ("Protocols for the Distribution of Royalties"). *See also* Chapter 5, Section III.A.5 ("Sales of the Product Trigger the Liability Rule and Distribution of Royalties"). *See also id.*, Section II.B. ("The Calculus of Royalties from Commercial Applications").

361 *See* Chapter 9, Section II.B.3.c (GBIF Funding).

362 For ITPGRFA, *see* Chapter 9, Section II.A.2.c. For GEO, *see id.*, Section II.B.3 ("Funding").

363 *See, e.g.*, the discussion of GBIF, Chapter 9, Section II.B.2.c.

constraints in most OECD countries could reinforce resistance to such mandatory commitments.

Meanwhile, major pooling initiatives – notably both the ITPGRFA and GEO – have implemented a scheme of voluntary contributions to an operational trust fund with considerable success.[364] Clearly, this approach would enable the members to evaluate the benefits of their contributions in terms of real deliverables and perceived added value, and it would also provide a simple exit option for any dissatisfied member, whether a governmental or nongovernmental entity.

However, total reliance on a voluntary funding option could compromise even the medium-term stability of the Microbial Research Commons. Hence, it might be desirable for the Governing Body to consider devising a mixed funding profile, in which some minimal dues were mandatory, including possible offsets by in-kind contributions, and others remained voluntary, with the understanding that active membership presupposed a moral obligation to contribute to both types of funding streams.

Much depends on how the multilateral regime of facilitated access to microbial genetic resources is perceived by all the relevant stakeholders. The more successful and useful its services become to basic and applied research, the more governments and other funders may become willing to defray its costs. That is to some extent demonstrated by fund-raising efforts under the ITPGRFA, discussed earlier in Chapter 9.[365] A formal assessment of the Commons' operations, both internal and external, would need to be carried out periodically for this purpose. All the same, OECD governments will need to ante up considerable capacity building funds in order to win the trust of the developing countries long before the Compensatory Liability Regime has begun to pay dividends to those same developing countries.[366]

B. Hidden Costs of Not Funding a Redesigned Microbial Research Commons

Science policymakers and research funders should not imagine that scarce resources would actually be saved by not investing in the formation of a redesigned Microbial Research Commons. The opposite is true for several compelling reasons.

First, the transaction costs of doing microbiological research will likely increase substantially in the absence of an organized effort to establish a multilateral system

[364] *See* GEO, Chapter 9, Section II.B.3.c *See also* the FAO's International Treaty, Chapter 9, Section II.A.3.c.

[365] *See* Chapter 9, Section II.A.2.c.

[366] Experience under the Crop Commons teaches this lesson. Interview with Dr. Shakeel Bahti, n. 349.

of facilitated exchange of microbial genetic resources under Article 4 of the Nagoya Protocol. This conclusion follows from the fact that the only alternative for researchers needing *ex situ* microbial materials and related data is to negotiate case-by-case access and benefit-sharing arrangements under the bilateral regime of the CBD.[367] Apart from the costs of negotiating ABS agreements in this manner, the willingness of provider governments to release materials under such agreements is always uncertain and subject to changing political conditions. By focusing only on possible downstream commercial applications, negotiations under the bilateral approach are more likely to ignore the nonmonetary benefits that flow from scientific research.[368] Instead, a contractually constructed research commons facilitates research and development, while ensuring that provider countries obtain both monetary and nonmonetary benefits under a standard Access and Benefit Sharing deal at the multilateral level.[369]

Moreover, *ex situ* microbial materials and related data not made available from the multilateral regime remain subject to growing tendencies to assert intellectual property rights in upstream research assets. The resulting costs are magnified by the disparate intellectual property rules and claims applicable under different national, regional, and international systems that largely ignore the needs of public science.[370]

Unless the microbial culture collections remain relatively insulated from these pressures within a contractually constructed research commons that includes government protection, they may also be pushed or pulled into a more political and proprietary approach. In that event, both developed and developing countries could expect to see the public-goods model that most culture collections have pursued replaced by a market-like approach.[371]

From a science policy perspective, unless the community of microbiologists takes charge of its own basic research assets, proprietarial suppliers could eventually price the exchange of materials and related data well beyond today's rates while adding

[367] *See* Nagoya Protocol, n. 1, art. 4; Chapter 3, Section IV.B ("Facilitating Scientific Research"); Chapter 4, Section IV.C.2 ("Opting Into a Multilateral Approach in Order to Stimulate More Downstream Benefits from the Bilateral Approach").

[368] *See* Chapter 2, Section I.C.2 ("The Threat to Public Scientific Research on Plant and Microbial Genetic Resources").

[369] *See generally* Chapter 5 ("Facilitating Transnational Exchanges of Microbial Genetic Resources under a Redesigned Multilateral Research Infrastructure"). For details, *see id.*, Section III ("Modeling a Sequence of Hypothetical Transactions").

[370] *See* Chapter 2, Section II ("Impinging Intellectual Property Rights Promoted by the Developed Countries"); Chapter 6 ("Legal and Institutional Obstacles Impeding Access to and Use of Scientific Literature and Data").

[371] *See* Chapter 4, Section II. ("Contractual Restrictions on Access to and Use of Upstream Microbial Genetic Resources in Both Developed and Developing Countries"). For the example of the American Type Culture Collection (ATCC)), *see id.*, Section II.A.

burdensome research restrictions in the process.[372] These costs are certain to rise if action is not taken to prevent the privatization of upstream microbial research assets that should be treated as global public goods and as inputs into downstream commercial applications and further research. Spillover effects and positive externalities for research and innovation would also likely decline.[373]

Second, the rising costs of obtaining permissions to access microbial genetic resources, coupled with likely restrictions on use and reuse of data even for basic research purposes under a bilateral approach, will generate hidden lost opportunity costs as scientists steer away from research projects that depend upon access to large quantities and diverse sources of such public knowledge assets.[374] Even with regard to *in silico* research, public investments in data that have a potential for reuse and that are eventually pooled and made interoperable are far more cost efficient and productive than is the case when such data are kept in closely held and uncoordinated databases.[375] The public investment in automated knowledge discovery tools will similarly yield less than optimal results owing to these same transaction costs and to the legal barriers to accessing and especially using large amounts of basic knowledge resources.

These lost opportunity costs would, in turn, become further magnified by tendencies to hoard both microbial materials and related data in both developed and developing countries. Such tensions have already delayed and even shut down important research projects in the field of microbiology.[376] To the extent that both basic and applied microbiological research are hindered by these obstacles, there is a risk that innovation in fields such as agriculture, medicine, energy, and environmental protection would be correspondingly diminished.[377]

Finally, the vision of a New Biology itself – with which we began this volume – largely depends on breaking down interdisciplinary, inter-sectoral, and international boundaries that hinder scientific collaboration.[378] Given all the potential obstacles identified earlier, it is hard to see how this vision for an integrated approach could

372 *See* Chapter 9, Section I ("Theoretical Reflections on Designing a Knowledge Commons").

373 *See, e.g.*, Brett M. Frischmann & Mark A. Lemley, *Spillovers*, 100(2) *Colum. L. Rev.* 101 (2006), *available at* http://ssrn.com/abstract=898881. *See also* Paul A. David, n. 181.

374 Jeffrey L. Furman & Scott Stern, Climbing Atop the Shoulders of Giants 8–9 (Nat'l Bureau Econ. Research Working Paper no. 12523, Sept. 2006), available at http://www.nber.org/papers/w12523.pdf (last accessed 1 Oct. 2014).

375 *See generally*, Chapter 8 Section I.C ("Understanding the Data Sharing Movement and Its Future Potential").

376 *See, e.g.*, Katz (2011), n. 309.

377 *See further* Chapter 1, Section II.C ("Cutting Edge Applications of Microbiology in Response to Major Global Challenges").

378 *See* Chapter 1, Section II.D ("A New Research Paradigm for the Life Sciences").

effectively be realized without the legal, economic, institutional, and socio-cultural benefits flowing from a redesigned Microbial Research Commons.

V. CONCLUDING OBSERVATIONS

In writing this book, we have taken note of several trends that are working in parallel to shape the future of microbiology. These trends will continue, whether the scientific community takes advantage of the opportunities and minimizes the threats that they present, or merely reacts to them in some haphazard or even counter-productive manner. One of the biggest factors has been the rising awareness of leaders in the developing world, mostly in the southern hemisphere, regarding the great potential wealth of their biodiversity and the need to both protect and exploit it in myriad applications.

Although the extent and diversity of microbes remains largely unknown and unknowable, a small percentage has been tamed in a network of *ex situ* culture collections for utilitarian and humanitarian purposes. Microbial materials, like many other life forms, consequently have become increasingly valuable and a focus of strong economic and political forces. The 1992 Convention on Biological Diversity (CBD) was a major milestone and recognition of this fact. More recently, the Nagoya Protocol to the CBD, which was adopted in 2010 and entered into force in 2014, will establish an enforceable international misappropriation regime to protect the interests of providers (especially those in developing countries) of plant, microbial, and animal genetic resources for both research and applications.

The Nagoya bombshell should be evaluated against the changing nature of intellectual property laws generally, which have become broader, longer, and stronger, largely at the instigation of multinational economic interests in the northern hemisphere. In the past thirty years or so, these laws and their increasingly onerous restrictions on users of knowledge goods have invaded the upstream research dimension of public science, negatively affecting the potential benefits for public and private researchers alike that broad and largely unfettered access to, and use of, these inputs can generate. Privatizing interests have also been extended to many of these public-sector resources by means of laws that promote and enforce the patenting of genetic materials, database protection rights in genomic and other data, and digital locks on publicly funded research results in the networked environment. Whatever positive effects this confluence of unchecked proprietary trends in both developed and developing countries might have if managed rationally, it has actually distorted and undermined the increasing potential of public science as a whole, and microbiology specifically.

At the same time, there has been a movement from small towards big science and the greater integration of the life sciences, in what has been called the "New

Biology." This quest for greater research efficiency in the public sphere has been accompanied by plummeting costs in the digital production of data, which has heralded a change in the research paradigm from phenotype to genotype.

The rise of global digital networks has further magnified the growing but still unfulfilled promise of cheap, universal access to research data and information. Tremendous and ubiquitous strides in scientific and technical capabilities have already been achieved, many of them fueled by government investments in academic research. Any failures to convert these advantages into socioeconomic benefits can largely be attributed to shortcomings in social organization and institutional design, rather than to any lack of scientific and technical advances.

Moreover, we appear to have entered into an extended period of austerity in public-sector budgets in most countries that would inexorably elicit greater accountability and demand for results from public expenditures. Taken together, these trends oblige us to ask how we can maximize public investment in science, especially microbiology, to provide more opportunity for research and innovation at a time of intense budgetary constraints. What seems clear is that much more needs to be done with less.

We, therefore, have used both an empirical and analytical approach in developing novel proposals that take all these trends into account and try to arrive at win-win solutions in what we refer to as a redesigned Microbial Research Commons. In the sphere of *ex situ* microbial genetic resources, dominated in large part by a range of formal and informal culture collections, we have sought to establish a multilateral regime of facilitated exchanges, legally situated within the space created by the Nagoya Protocol's explicit legitimization of the Crop Commons established by the United Nations Food and Agricultural Organization in 2001. We then seek to stimulate broad access and use of microbial materials and related data by both public and private research communities, with lower transaction costs, by urging adoption of an *ex ante* Compensatory Liability Regime, which would be embodied in standard material transfer agreements (SMTAs).

Academics and the managers of culture collections are not experienced negotiators, and their bargaining leverage is weak. Under our legal proposal for microbial genetic resources having no known or likely commercial value at the time of deposit, there needs to be one SMTA adopted for a multilateral system of facilitated access. A Standard MTA would preserve the value of public upstream research, instead of a multitude of different licenses with onerous transaction costs under the bilateral approach that the CBD otherwise mandates.

The Compensatory Liability Regime applicable to *ex situ* genetic resources deposited in this globally distributed set of repositories should especially benefit developing countries. The very existence of such an endeavor would reconfirm their sovereign proprietary rights in both *in situ* and *ex situ* materials. Consistent with the

CBD, it would provide full transparency and tracking for purposes of upstream and downstream research transactions. Perhaps most important, it would create a de facto partnership between genetic resource contributors and commercial exploiters that would help to delegitimize "biopiracy" and to institutionalize the Access and Benefit-Sharing provisions of the CBD on a sound and effective legal basis.

Measures to enforce this regime may seem relatively complex in the short term, but only because the Nagoya Protocol is not yet fully operational. Once operational, SMTAs emanating from the multilateral system would become enforceable in the courts of CBD member countries.

Besides seeking to establish a multilateral system for facilitated exchanges of *ex situ* microbial materials and data on a sound legal foundation, we also consider ways to facilitate access to, and use of, the digital databases and published research results generated by the global microbiological community. In so doing, we have built on the emerging institutional and legal approaches of new publishing intermediaries and of the research community itself.

We encourage greater use of early release policies for research data that can be compiled and used as community resources. Other publicly developed databases, not amenable to an early release approach, should be made as freely and openly available as possible using creative financing through consortia and distributed crowd-sourced designs. The private ordering of rights under common-use licenses and waivers that allow maximum freedom and provide legal certainty to users, and especially reusers, of factual information can help to achieve these goals.

As regards microbiology journals, in 2009 we conducted an empirical survey of publishing models and licensing conditions that showed a trend toward much greater openness in a surprisingly short period of time. Open-access publishing models and even read-only open repositories of digital research results, established by science funders, universities, and some publishers, can go a long way toward making the microbiological literature widely available online. Similar to the redesign of research data, open-access approaches can be further facilitated by common-use licenses, and experimentation with new institutional designs that have begun to flourish from the bottom up.

Nonetheless, the digital sphere continues to suffer from hangovers inherited from the print paradigm. Evolutionary social practices have not kept up with revolutionary digital capabilities. A wholesale deconstruction of the scholarly communication process should be followed by a reconstruction of institutions and publishing models that take full advantage of the disruptive computational and network technologies that continue to emerge. This reorganization cannot be accomplished by fiat, overnight, nor in the inflexible manner of a one-size-fits-all solution. The U.S. law school journal model should be explored more fully for useful insights in this regard, while preserving the strengths of the peer-review system in the science publishing model.

We also identify an emerging approach to scholarly communication that builds on all the technological, institutional, and legal capabilities being developed in different research contexts. We characterize this new model as "Open Knowledge Environments" (OKEs) and believe that it could eventually supersede the stove-piped journals and the existing disaggregated data and literature models. The empirical examples of selected OKEs in microbiology reviewed here provide a more holistically integrated and thematically interactive approach that broadly serves the interests of research and applications on digital networks.

In the last part of the book, we examine international governance structures, first looking at the theory of common pool resources and then undertaking an extensive empirical review and analysis of selected international scientific infrastructure organizations. We then draw on their positive institutional features while minimizing their negative aspects in developing a suitable governance model for a redesigned Microbial Research Commons. Our goal has been to construct a science-friendly, legally and politically rational organization, in a fiscally prudent way. In particular, our objective has been to motivate the stakeholders to move away from their increasingly intransigent positions.

Attaining this objective depends on the extent to which leaders of the microbiological research community and science policymakers become persuaded that they will obtain more from a "grand bargain" than from holding out. Developing countries will not waive their rights of sovereign control over *ex situ* genetic resources acquired under the CBD without real and substantial countervailing benefits. OECD governments will not undertake legal and funding obligations without clear and tangible cost-saving benefits from facilitated access to microbial genetic resources and related digital research assets. Scientists will not surrender their autonomy in return for access to any international arrangement that fails to give them a strong participatory voice in governance, while facilitating their access to and use of public research assets. And the private sector will resist any arrangement that raises the costs of doing business or that undermines the perceived incentives of intellectual property rights, unless facilitated access to precompetitive genetic resources stimulates more profit-making commercial applications than would otherwise occur.

The evidence marshaled in this book shows that the formation of a properly managed multilateral system under the aegis of a redesigned Microbial Research Commons could reconcile the interests of all these stakeholders. Without lowering the barriers to global access to *ex situ* microbial materials and related data scattered throughout the world, a disaggregated research community risks duplication of efforts, blockage of many potentially fertile lines of research, and reduced interdisciplinary research opportunities. Public welfare, in turn, will suffer from these lost research opportunities, and scarce public expenditures will yield less innovation.

In contrast, a digitally integrated Microbial Research Commons would build upon the strengths of existing institutional networks, especially the World Federation of Culture Collections and other networks of culture collections emerging at the regional level. It would provide much of the impetus and procedures to help upgrade the quality and usefulness of the *ex situ* collections held in the developing countries and enable their scientists to apply the growing stock of digitally available knowledge to the needs of their own countries. It would stimulate the pace of worldwide innovation needed to address such major social challenges as improving global health, mitigating the effects of environmental degradation, and augmenting food security. And it would lead to more productive industrial applications with a more equitable distribution of the resulting economic benefits than occurs under the balkanized bilateral approach. Besides maintaining scientific autonomy and integrating the developing countries into the larger biological community, a global Commons along these lines should progressively foster trust and reciprocity among the various stakeholders, while reducing the tensions that flow from policy options based on perceived national interests alone.

If, as we contend, the international microbiological research communities were to form such a Commons to avoid the aforementioned threats of disintegration and to maximize the opportunities that digital science now make possible, they should accept the CBD as an integral part of its basic legal platform. This premise applies whether or not a few outlier national governments, notably the United States, have formally adhered to that agreement. We accordingly envision a transaction in which all stakeholders would bargain around the CBD, with a view to contracting a win-win outcome for public science that nonetheless remains consistent with the goals of the CBD, and that promotes – not blocks – economic growth.

A Microbial Research Commons founded on these premises would provide participating governments – in both developed and developing countries – with immediate and tangible research benefits, including support for capacity-building and digital infrastructure. It would also generate a reciprocally beneficial Standard Material Transfer Agreement governing upstream biological resources and downstream commercial applications in the future. The resulting multilateral system of facilitated access and benefit-sharing would guarantee non-OECD countries the possibility of greatly improving their physical and digital scientific infrastructures in place of the speculative benefits of hoarding. All participating countries would emerge with strengthened scientific capacity in microbiology.

Finally, our Microbial Research Commons model could be used by other fields of science, especially those that use collections of materials in the life sciences, such as stem cells or the geosciences, and beyond. The details of such an exploration into analogous domains, however, are properly the topic of another initiative and further analysis.

Index